The Haynes
Diesel Engine Repair Manual

by Ken Freund
and John H Haynes Member of the Guild of Motoring Writers

General Motors and Ford V8 diesel engines:
GM 350 cu in (5.7L), 397 cu in (6.5L) and 379 cu in (6.2L)
Ford 420 cu in (6.9L) and 445 cu in (7.3L)

(2E17 - 10330)
(1736)

ABCDE
FGHIJ
KLM
2

Haynes Publishing Group
Sparkford Nr Yeovil
Somerset BA22 7JJ England

Haynes Publications, Inc.
861 Lawrence Drive
Newbury Park
California 91320 USA

Acknowledgments

We are grateful to the General Motors Corporation and the Ford Motor Company for assistance with technical information and certain illustrations. The Federal-Mogul Corporation supplied the illustrations of various engine bearing wear conditions. Technical writers who contributed to this project include Mike Stubblefield, Robert Maddox and Larry Warren.

A book in the Haynes Automotive Repair Manual Series

Printed in the U.S.A.

ISBN 1 56392 188 X

Library of Congress Control Number 97-70905

00-288

Contents

Chapter 1
Introduction

Chapter 2
GM 5.7L, 6.2L and 6.5L V8 engines

Chapter 3
Ford 6.9L and 7.3L engines

Chapter 4
Engine overhaul procedures

Chapter 1 Introduction

About this manual

The purpose of this manual is to help you service and overhaul the Ford 6.9L and 7.3L light truck diesels and General Motors automobile and light pickup truck 5.7L, 6.2L and 6.5L diesel engines.

The manual is divided into Chapters. Each Chapter is sub-divided into Sections, some of which consist of consecutively numbered Paragraphs (usually referred to as "Steps", since they're normally part of a procedure). If the material is basically informative in nature, rather than a step-by-step procedure, the Paragraphs aren't numbered.

The first three Chapters contain material on servicing and maintaining Ford and General Motors engines, electrical, cooling and emissions systems. Chapter 4 covers the specifics of the overhaul procedure. Comprehensive Sections covering tool selection and usage, safety and general shop practices have been included.

The term **"see illustration"** (in parentheses), is used in the text to indicate that a photo or drawing has been included to make the information easier to understand (the old cliche "a picture is worth a thousand words" is especially true when it comes to how-to procedures). Also, every attempt is made to position illustrations directly opposite the corresponding text to minimize confusion. The two types of illustrations used (photographs and line drawings) are referenced by a number preceding the caption. Illustration numbers denote Chapter and numerical sequence within the Chapter (i.e., 3.4 means Chapter 3, illustration number four in order).

The terms **"Note"**, **"Caution"**, and **"Warning"** are used throughout the text with a specific purpose in mind - to attract the reader's attention. A "Note" simply provides information required to properly complete a procedure or information which will make the procedure easier to understand. A **"Caution"** outlines a special procedure or special steps which must be taken when completing the procedure where the Caution is found. Failure to pay attention to a Caution can result in damage to the component being repaired or the tools being used. A **"Warning"** is included where personal injury can result if the instructions aren't followed exactly as described.

Even though extreme care has been taken during the preparation of this manual, neither the publisher nor the author can accept responsibility for any errors in, or omissions from, the information given.

How diesel engines are different

There are many similarities between diesel and gasoline engines but it's the differences involved with the ignition of the fuel that makes the servicing and repairing of a diesel engine unique. The diesel engines covered by this manual operate on the same four-stroke principle (intake, compression, power and exhaust) as most gasoline powered engines. But, instead of igniting the fuel and air in the cylinder during the third (power) stroke by introducing a spark from the spark plug as on a gasoline engine, the diesel engine uses compression ignition. Compression ignition occurs when the extremely high compression ratio of the diesel engine (18:1 to 24:1 as opposed to gasoline engine ratios of 5:1 to 14:1) compresses the air in the cylinder until it's hot enough (around 1000-degrees F) to ignite when fuel is injected.

A gasoline engine has both an ignition system and air/fuel mixture induction system for controlling engine speed. The diesel is much simpler because it needs only the fuel injection pump and injectors for delivering the fuel at the precise moment when ignition can occur. An unrestricted supply of air is supplied by the intake manifold.

Another unique feature of the diesel engines covered by this manual is that an electric glow plug system operates when starting the engine and until combustion chamber temperatures sufficient to sustain self-ignition are reached. Starting fluid must never be used with the diesel engines covered by this manual, because the glow plugs remain hot when the intake valves are open and can ignite the fluid, causing a flame to travel back through the intake manifold. Personal injury and/or engine damage could occur.

Diesel fuel comes from a lower petroleum distillate than gasoline and it lubricates the fuel pump and injectors, reducing wear in the engine. The cetane number of diesel fuel indicates it's ignition quality; the higher the number, the faster it burns, making for easier starting and smoother running. Diesel fuel contains more energy and burns more air per volume than gasoline, which partially accounts for a diesel's greater efficiency.

There are, however, several disadvantages to diesel fuel. It contains paraffins (waxes), it's volatility is affected by temperature more than gasoline is, it absorbs moisture readily and it contains ten times more sulfur than gasoline. At lower temperatures (10-degrees F and below) number two diesel fuel will become cloudy (this is called the "cloud point") because the parafins crystallize, clogging the fuel filter. If cold enough, the fuel can't be poured or pumped ("pour point"), so different blends of fuel are required for summer and winter driving.

Water in diesel fuel can corrode the injectors and encourage bacteria (which feed off the sulfur) to grow in the fuel, leading to contamination problems. The sulfur can also accelerate wear because it forms corrosive deposits in the engine. Consequently, devices such as water sensors and separators are used in diesel fuel systems and care must be taken to change the fuel filter and engine oil and filter on a regular basis.

Gasoline engines use intake manifold vacuum to operate components such as the brake booster. Since diesel engines have unrestricted air in-take, there is very little vacuum. Vacuum pumps are required to operate accessories.

Diesel engines offer several advantages over gasoline engines, particularly in the area of fuel efficiency, low-speed torque and reliability. One of the reasons diesel engines tend to get better mileage than gasoline engines is because they burn fuel more completely. Diesel fuel exhaust, when the engine is properly tuned, actually have fewer hydrocarbons and carbon monoxide than gasoline exhaust.

Glossary of terms

A

Additive - A material added to fuels and lubricants, designed to improve their properties such as viscosity, cetane, pour point, film strength, etc.

Afterglow - The period during which the glow plugs continue to operate after the engine is started.

Ambient - Surrounding on all sides.

Atmospheric pressure - The weight of the air, usually expressed as pressure; the air pressure at sea level is 14.7 psi.

Atomization - Breaking up of the fuel into fine particles, so it can be mixed with air.

B

BDC - Bottom Dead Center.

Biocide - A bacteria-killing compound used in diesel fuel.

Black smoke - Incompletely burned fuel in the exhaust.

Blow-by - A leakage or loss of pressure past the piston into the crankcase.

Blue smoke - Caused by blowby allowing crankcase oil in the combustion chamber due to bad rings, valve seals, or other faulty components.

BTU - British Thermal Unit.

By-pass oil filter - An oil filter that removes soot and carbon by continually filtering 20 percent of the engine oil.

C

Calibration - Adjusting the rate, speed and timing of fuel delivery in an injection system.

Calibration oil - Oil which is used in a tester for checking injection nozzles, meeting SAE J967D specifications.

Cetane number - The number indicating the ignition quality of diesel fuel, similar to the octane rating of gasoline.

Ceton filter - A sock-type filter in the fuel tank capable of wicking diesel fuel, but not water; keeps water from the rest of the fuel system until the sock is 90 percent submerged in water,

Clearance volume - At Top Dead Center, the volume measurement of a combustion chamber.

Cloud point - The temperature at which paraffin crystals in diesel fuel separate out of solution and start to crystallize.

Coalescing action - The process of smaller water droplets merging together into larger droplets which takes place in a water separator.

Compression ignition - The burning an air/fuel mixture in the combustion chamber by compressing it sufficiently to raise its temperature above the flash point of the fuel and injecting into the combustion chamber so that it begins to burn spontaneously.

Compression ratio - The clearance volume of an engine cylinder divided by its total volume.

Control module - A device used when starting a diesel engine which controls the fast idle, glow plug temperature, preglow time and afterglow time; also known as a controller.

Crankcase Depression Regulator (CDR) - A device which aids in the control of crankcase gases by maintaining a specific amount of vacuum in the crankcase.

D

Delivery lines - Fuel lines used to carry fuel from the injection pump to the injector nozzles.

Delivery valve - An injection pump valve that rapidly decreases injection line pressure to an achieve an abrupt fuel cutoff at the injector.

Diesel fuel - A petroleum-based middle distillate suitable as a fuel for diesel engines.

Diffusion - Mixing the molecules of two gases by thermal agitation.

Diode - An electronic device that permits current to flow through it in one direction only.

Dispersant - Dispersing or scattering in various directions; a state of matter in which finely divided particles of one substance (disperse phase) are suspended in another (dispersion medium) substance.

Dissipated - Scattered in various directions.

Distributor injection pump - An injection pump using pistons which pressurizes fuel for injection in the proper cylinder based on the relative port position of the rotating shaft in the hydraulic head.

Dribble - Insufficiently atomized fuel issuing from the nozzle at or immediately following the end of the main injection phase.

Duration - The period of time during which anything lasts.

Dynamic timing meter - A diesel tool used for measuring timing while the engine is running by using a quartz sensor in the combustion chamber that measures the point of combustion and converts this to timing in degrees of crankshaft rotation through the use of a magnetic crankshaft pickup and microprocessor.

E

Eccentric - One circle within another circle not having the same center.

Exhaust Gas Recirculation (EGR) - Redirecting exhaust gases back into the combustion chamber to reduce peak combustion temperatures.

Exhaust Pressure Regulator (EPR) - A device for increasing exhaust backpressure at specific times to increase exhaust flow to the EGR valve.

F

Flash point - The temperature at which fuel self-ignites.

Fuel advance system - Advances fuel delivery during cold starts on diesel vehicles. Consists of a thermal-sensitive solenoid on the intake manifold which sends a signal to the cold advance solenoid terminal, which opens a ball-check valve on top of the injection pump housing. With pump housing pressure reduced, the timing mechanism has less resistance to overcome and operates earlier, advancing fuel delivery 3 degrees.

Fuel return line - The fuel line that returns excess fuel from the injectors back to the fuel tank or inlet side of the injector pump.

G

Glow plug - An electrical heating device that helps diesel engines start and run smoothly when cold by creating heat in the pre-combustion chamber.

Governor - A device that controls the speed of an engine within specified limits.

I

Injection lag - the time interval (usually expressed in crankshaft degrees) between the nominal start of injection pump delivery and the actual start of injection at the nozzle.

Injection nozzles - See "Injector nozzles."

Injection pump - A pump which delivers fuel to the pre-combustion chambers at a high pressure so the injector can overcome compression and combust during the injection process.

Injection pump governor - A device which controls fuel delivery to limit the minimum and maximum engine speeds, as well as intermediate throttle positions.

Injection timing - The matching of the pump timing mark, or the injector timing mechanism, to some index mark on the engine components, so that injection will occur at the proper time with reference to the engine cycle. Injection advance or retard is respectively an earlier, or later, injection pump delivery cycle in reference to the injection cycle.

Injector nozzles - Small spring-loaded valves with nozzles which inject fuel into the combustion chamber.

Injector opening pressure - The point at which injection pump fuel pressure overcomes nozzle valve-spring resistance, or combustion chamber pressure, so that fuel is injected into the pre-combustion chamber.

In-line fuel heater - A 100-watt heater which is integral to the fuel line. This heat warms the fuel prior to the filter to keep paraffin crystals from stopping fuel flow. The heater warms the fuel by 20 degrees.

L

Leak-off pressure - Manufacturer-specified pressure used to test injector leakage on a pop tester.

M

Manometer - Instrument used for measuring the pressure of liquids and gases.

Micron - One-millionth of a meter.

Min/max governor - Controls the idle speed and prevents overspeed.

Module - Electronic control unit that controls the glow-plug system.

Moisture content - The amount of water contained in diesel fuel.

N

NOx - Oxides of nitrogen, a pollutant; a component of diesel exhaust.

Number one diesel fuel - Diesel fuel used in cold climates; sometimes blended with number two diesel fuel to increase number one's energy and two's cold-weather performance.

Number two diesel fuel - Diesel fuel used in moderate climates.

O

Orifice - A restriction to flow in a line or tube.

P

Paraffin - A semi-transparent, waxy mixture of hydrocarbons, derived principally from the distillation of petroleum; any hydrocarbon of the methane series.

PCV - Positive Crankcase Ventilation.

Peak pressure period - The phase of diesel combustion lasting from about five degrees before top dead center to about 10 degrees after top dead center; the majority of diesel fuel burns during this period.

Pencil-type injector - An early type injection nozzle shaped like a pencil.

Pop tester - An injector testing tool used for measuring opening pressure, leakoff pressure and spray patterns of injectors.

Positive Crankcase Ventilation (PCV) system - A system for crankcase ventilation that returns crankcase blowby gases to the intake manifold where they are sent to the combustion chambers for reburning.

Pour point - The temperature at which diesel fuel can no longer be poured or pumped.

Pre-chamber - A precombustion chamber built into the cylinder head that creates turbulence in the incoming air. Sometimes called a swirl chamber.

Pre-glow - The period of time when the glow plugs are heating to operating temperature.

S

Sock - The fuel pick-up strainer in the fuel tank. The sock is made of saran, so water won't enter until the sock is virtually engulfed in water.

Supply pump - A pump that transfers fuel from the tank and delivers it to the injection pump.

Swirl - Rotation of the mass of air as it enters the cylinder is known as "swirl." This is one form of turbulence.

T

Thermal switch - A bimetal switch that controls glow plug operation.

Transfer pump - Part of the fuel injection pump; boosts fuel pressure from around 20 psi to about 130 psi, depending on the pump and the engine speed.

V

Vacuum pump - A device for creating vacuum to power various systems such as power brakes and other components.

Vacuum regulator valve - A device for controlling vacuum from the vacuum pump to the transmission, cruise control and other components that need a regulated vacuum source based on throttle position.

Viscosity - A liquid's resistance to flow.

W

Water separator - A device for removing water from the fuel, located in the fuel line of diesel engines.

White smoke - Unburned fuel emitted by the exhaust that indicates low combustion chamber temperatures.

Buying parts

Commonly replaced engine parts such as fuel filters, piston rings, pistons, bearings, camshafts, lifters, timing chains, oil pumps and gaskets are produced by aftermarket manufacturers and stocked by retail auto parts stores and mail order houses, usually at a savings over dealer parts department prices. Many auto parts stores and mail order houses offer complete engine kits, often at a considerable savings over individual parts.

Less-commonly replaced items such as injectors and lines may not be available through these same sources and a dealer service department may be your only option. Keep in mind that some parts will probably have to be ordered, and it may take several days to get your parts; order early.

Wrecking yards are a good source for major parts that would otherwise only be available through a dealer service department (where the price would likely be high). Engine blocks, cylinder heads, crankshafts, manifolds, etc. for these engines are commonly available for reasonable prices. The parts people at wrecking yards have parts interchange books they can use to quickly identify parts from other models and years that are the same as the ones on your engine.

A place to work

Establish a place to work. A special work area is essential. It doesn't have to be particularly large, but it should be clean, safe, well-lit, organized and adequately equipped for the job. True, without a good workshop or garage, you can still service and repair engines, even if you have to work outside. But an overhaul or major repairs should be carried out in a sheltered area with a roof. The procedures in this book require an environment totally free of dirt, which will cause wear if it finds its way into an engine.

The workshop

The size, shape and location of a shop building is usually dictated by circumstance rather than personal choice. Every do-it-yourselfer dreams of having a spacious, clean, well-lit building specially designed and equipped for working on everything from small engines on lawn and garden equipment to cars and other vehicles. In reality, however, most of us must content ourselves with a garage, basement or shed in the backyard.

Spend some time considering the potential - and drawbacks - of your current facility. Even a well-established workshop can benefit from intelligent design. Lack of space is the most common problem, but you can significantly increase usable space by carefully planning the locations of work and storage areas. One strategy is to look at how others do it. Ask local repair shop owners if you can see their shops. Note how they've arranged their work areas, storage and lighting, then try to scale down their solutions to fit your own shop space, finances and needs.

General workshop requirements

A solid concrete floor is the best surface for a shop area. The floor should be even, smooth and dry. A coat of paint or sealant formulated for concrete surfaces will make oil spills and dirt easier to remove and help cut down on dust - always a problem with concrete.

Paint the walls and ceiling white for maximum reflection. Use gloss or semi-gloss enamel. It's washable and reflective. If your shop has windows, situate workbenches to take advantage of them. Skylights are even better. You can't have too much natural light. Artificial light is also good, but you'll need a lot of it to equal ordinary daylight.

Make sure the building is adequately ventilated. This is critical during the winter months, to prevent condensation problems. It's also a vital safety consideration where solvents, gasoline and other volatile liquids are being used. You should be able to open one or more windows for ventilation. In addition, opening vents in the walls are desirable.

Electricity and lights

Electricity is essential in a shop. It's relatively easy to install if the workshop is part of the house, but it can be difficult and expensive to install if it isn't. Safety should be your primary consideration when dealing with electricity; unless you have a very good working knowledge of electrical installations, have an electrician do any work required to provide power and lights in the shop.

Consider the total electrical requirements of the shop, making allowances for possible later additions of lights and equipment. Don't substitute extension cords for legal and safe permanent wiring. If the wiring isn't adequate, or is substandard, have it upgraded.

Give careful consideration to lights for the workshop. A pair of 150-watt incandescent bulbs, or two 48-inch long, 40-watt fluorescent tubes, suspended approximately 48-inches above the workbench, are the minimum you can get by with. As a general rule, fluorescent lights are probably the best choice. Their light is bright, even, shadow-free and fairly economical, although some people don't care for the bluish tinge they cast on everything. The usual compromise is a good mix of fluorescent and incandescent fixtures.

The position of the lights is important. Don't place a fixture directly above the area where the engine - or the stand it's mounted on - is located. It will cause shadows, even with fluorescent lights. Attach the light(s) slightly to the rear - or to each side - of the workbench or engine stand to provide shadow-free lighting. A portable "trouble-light" is very helpful for use when overhead lights are inadequate. If gasoline, solvents or other flammable liquids are present - not an unusual situation in a shop - use special fittings to minimize the risk of fire. And

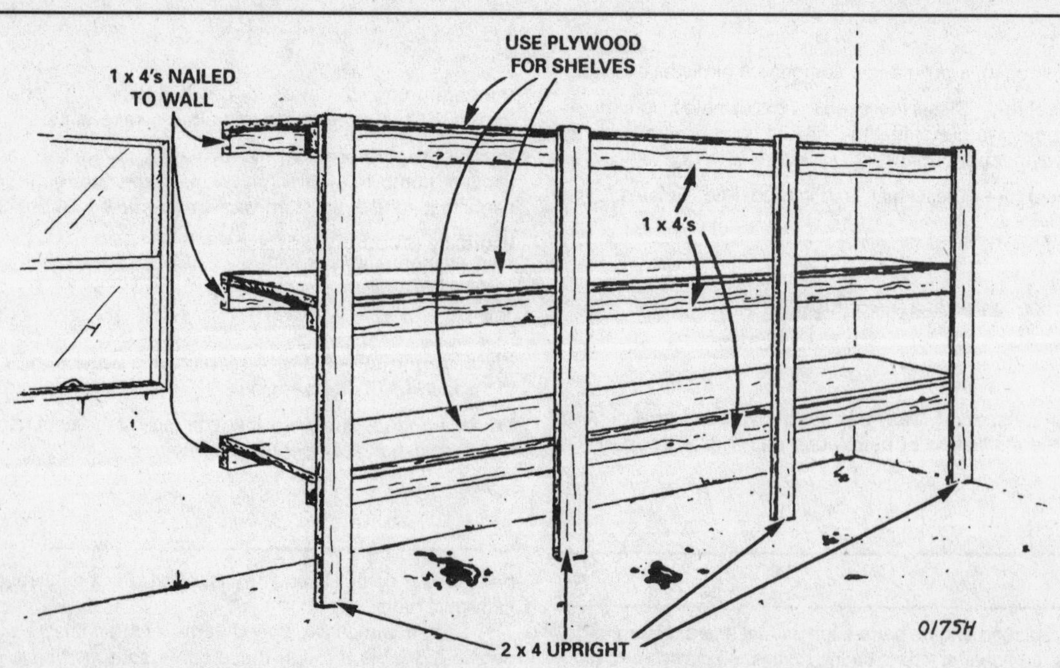

1.1 Homemade wood shelves are relatively inexpensive to build and you can design them to fit the available space, but all that wood can be a fire hazard

don't use fluorescent lights above machine tools (like a drill press). The flicker produced by alternating current is especially pronounced with this type of light and can make a rotating chuck appear stationary at certain speeds - a very dangerous situation.

Storage and shelves

Once disassembled, an engine occupies more space than you might think. Set up an organized storage area to avoid losing parts. You'll also need storage space for hardware, lubricants, solvent, rags, tools and equipment.

If space and finances allow, install metal shelves along the walls. Arrange the shelves so they're widely spaced near the bottom to take large or heavy items. Metal shelf units are pricey, but they make the best use of available space. And the shelf height is adjustable on most units.

Wood shelves **(see illustration)** are sometimes a cheaper storage solution. But they must be built - not just assembled. They must be much heftier than metal shelves to carry the same weight, the shelves can't be adjusted vertically and you can't just disassemble them and take them with you if you move. Wood also absorbs oil and other liquids and is obviously a much greater fire hazard.

Store small parts in plastic drawers or bins mounted on metal racks attached to the wall. They're available from most hardware, home and lumber stores. Bins come in various sizes and usually have slots for labels.

All kinds of containers are useful in a shop. Glass jars are handy for storing fasteners, but they're easily broken. Cardboard boxes are adequate for temporary use, but if they become damp, the bottoms eventually weaken and fall apart if you store oily or heavy parts in them. Plastic containers come in a variety of sizes and colors for easy identification. Egg cartons are excellent organizers for tiny parts like valve springs, retainers and keepers. Large ice cream tubs are suitable for keeping small parts together. Get the type with a snap cover. Old metal cake pans, bread pans and muffin tins also make good storage containers for small parts.

Workbenches

A workbench is essential - it provides a place to lay out parts and tools during repair procedures, and it's a lot more comfortable than working on a floor or the driveway. The workbench should be as large and sturdy as space and finances allow. If cost is no object, buy industrial steel benches. They're more expensive than home-built benches, but they're very strong, they're easy to assemble, and - if you move - they can be disassembled quickly and you can take them with you. They're also available in various lengths, so you can buy the exact size to fill the space along a wall.

If steel benches aren't in the budget, fabricate a bench frame from slotted angle-iron or Douglas fir (use 2 x 6's rather than 2 x 4's) **(see illustration)**. Cut the pieces of the frame to the required size and bolt them together with carriage bolts. A 30 or 36 by 80-inch, solid-core door with hardboard surfaces makes a good bench top. And you can flip it over when one side is worn out.

An even cheaper - and quicker - solution? Assemble a bench by attaching the bench top frame pieces to the wall with angled braces and use the wall studs as part of the framework.

Regardless of the type of frame you decide to use for the workbench, be sure to position the bench top at a comfortable working height and make sure everything is level. Shelves installed below the bench will make it more rigid and provide useful storage space.

Tools and equipment

For some home mechanics, the idea of using the correct tool is completely foreign. They'll cheerfully tackle the most complex overhaul procedures with only a set of cheap open-end wrenches of the wrong type, a single screwdriver with a worn tip, a large hammer and an adjustable wrench. Though they often get away with it, this cavalier approach is stupid and dangerous. It can result in relatively minor annoyances like stripped fasteners, or cause catastrophic consequences like blown engines. It can also result in serious injury.

A complete assortment of good tools is a given for anyone who plans to overhaul engines. If you don't already have most of the tools listed below, the initial investment may seem high, but compared to the spiraling costs of routine maintenance and repairs, it's a deal. Besides, you can use a lot of the tools around the house for other types of mechanical repairs. We've included a list of the tools you'll need and a detailed description of what to look for when shopping for tools and how to use them correctly. We've also included a list of the special factory tools you'll need for engine rebuilding.

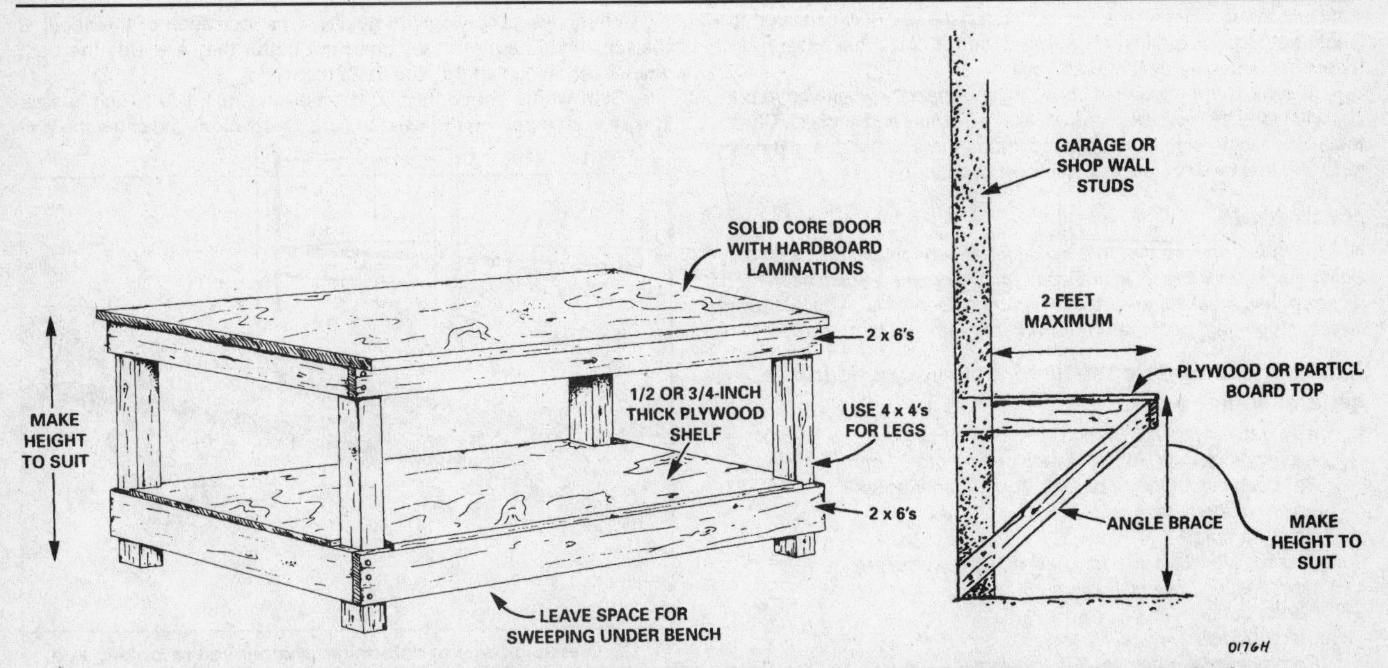

1.2 You can build a sturdy, inexpensive workbench with 4 X 4s, 2 X 6s and a solid core door with hardboard laminations - or build a bench using the wall as an integral member as shown

Buying tools

There are two ways to buy tools. The easiest and quickest way is to simply buy an entire set. Tool sets are often priced substantially below the cost of the same individually priced tools - and sometimes they even come with a tool box. When purchasing such sets, you often wind up with some tools you don't need or want. But if low price and convenience are your concerns, this might be the way to go. Keep in mind that you're going to keep a quality set of tools a long time (maybe the rest of your life), so check the tools carefully; don't skimp too much on price, either. Buying tools individually is usually a more expensive and time-consuming way to go, but you're more likely to wind up with the tools you need and want. You can also select each tool on its relative merits for the way you use it.

You can get most of the hand tools on our list from the tool department of any large department store or hardware store chain that sells hand tools. Blackhawk, Cornwell, Craftsman, KD, Proto and SK are fairly inexpensive, good-quality choices. Specialty tools are available from mechanics' tool companies such as Snap-on, Mac, Matco, Kent-Moore, Lisle, OTC, Owatonna, etc. These companies also supply the other tools you need, but they'll probably be more expensive.

Also consider buying second-hand tools from garage sales or used tool outlets. You may have limited choice in sizes, but you can usually determine from the condition of the tools if they're worth buying. You can end up with a number of unwanted or duplicate tools, but it's a cheap way of putting a basic tool kit together, and you can always sell off any surplus tools later.

Until you're a good judge of the quality levels of tools, avoid mail order firms (excepting Sears and other name-brand suppliers), flea markets and swap meets. Some of them offer good value for the money, but many sell cheap, imported tools of dubious quality. Like other consumer products counterfeited in the Far East, these tools run the gamut from acceptable to unusable.

If you're unsure about how much use a tool will get, the following approach may help. For example, if you need a set of combination wrenches but aren't sure which sizes you'll end up using most, buy a cheap or medium-priced set (make sure the jaws fit the fastener sizes marked on them). After some use over a period of time, carefully examine each tool in the set to assess its condition. If all the tools fit well and are undamaged, don't bother buying a better set. If one or two are worn, replace them with high-quality items - this way you'll end up with top-quality tools where they're needed most and the cheaper ones are sufficient for occasional use. On rare occasions you may conclude the whole set is poor quality. If so, buy a better set, if necessary, and remember never to buy that brand again.

In summary, try to avoid cheap tools, especially when you're purchasing high-use items like screwdrivers, wrenches and sockets. Cheap tools don't last long. Their initial cost plus the additional expense of replacing them will exceed the initial cost of better-quality tools.

Hand tools

Note: *The information that follows is for early-model engines with only Standard fastener sizes. On some late-model engines, you'll need Metric wrenches, sockets and Allen wrenches. Generally, manufacturers began integrating metric fasteners into their vehicles around 1975.*

A list of general-purpose hand tools you need for general engine work

Adjustable wrench - 10-inch
Allen wrench set (1/8 to 3/8-inch or 4 mm to 10 mm)
Ball peen hammer - 12 oz (any steel hammer will do)
Box-end wrenches
Brass hammer
Brushes (various sizes, for cleaning small passages
Combination (slip-joint) pliers - 6-inch
Center punch
Cold chisels - 1/4 and 1/2-inch
Combination wrench set (1/4 to 1-inch)
Extensions - 1-, 6-, 10- and 12-inch
E-Z out (screw extractor) set
Feeler gauge set
Files (assorted)
Floor jack
Gasket scraper
Hacksaw and assortment of blades
Impact screwdriver and bits
Locking pliers
Micrometer(s) (one-inch)
Phillips screwdriver (no. 2 x 6-inch)
Phillips screwdriver (no. 3 x 8-inch)
Phillips screwdriver (stubby - no. 2)
Pin punches (1/16, 1/8, 3/16-inch)
Piston ring removal and installation tool
Pliers - lineman's
Pliers - needle-nose
Pliers - snap-ring (internal and external)
Pliers - vise-grip
Pliers - diagonal cutters
Ratchet (reversible)
Scraper (made from flattened copper tubing)
Scribe
Socket set (6-point)
Soft-face hammer (plastic/rubber, the biggest you can buy)
Standard screwdriver (1/4-inch x 6-inch)
Standard screwdriver (5/16-inch x 6-inch)
Standard screwdriver (3/8-inch x 10-inch)
Standard screwdriver (5/16-inch - stubby)
Steel ruler - 6-inch
Straightedge - 12-inch
Tap and die set
Thread gauge
Torque wrench (same size drive as sockets)
Torx socket(s)
Universal joint
Wire brush (large)
Wire cutter pliers

What to look for when buying hand tools and general purpose tools

Wrenches and sockets

Wrenches vary widely in quality. One indication of their cost is their quality: The more they cost, the better they are. Buy the best wrenches you can afford. You'll use them a lot.

Start with a set containing wrenches from 1/4 to 1-inch in size. The size, stamped on the wrench **(see illustration)**, indicates the dis-

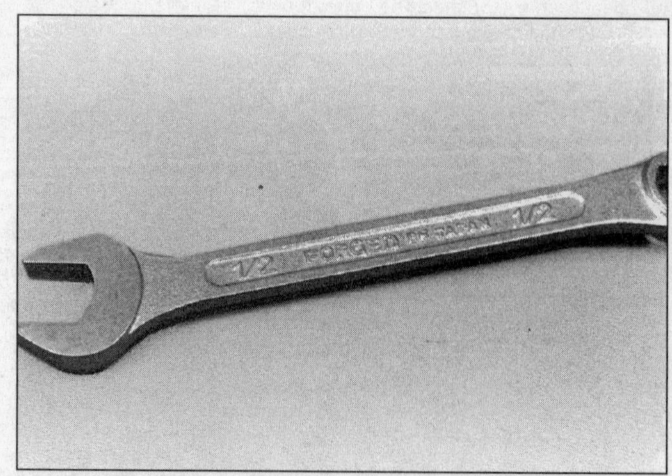

1.3 One quick way to determine whether you're looking at a quality wrench is to read the information printed on the handle - if it says "chrome vanadium" or "forged," it's made out of the right material

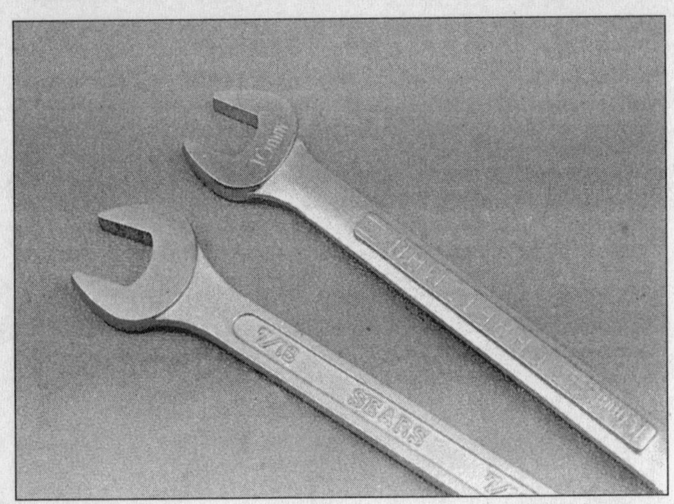

1.4 The size stamped on a wrench indicates the distance across the nut or bolt head (or the distance between the wrench jaws) in inches, not the diameter of the threads on the fastener

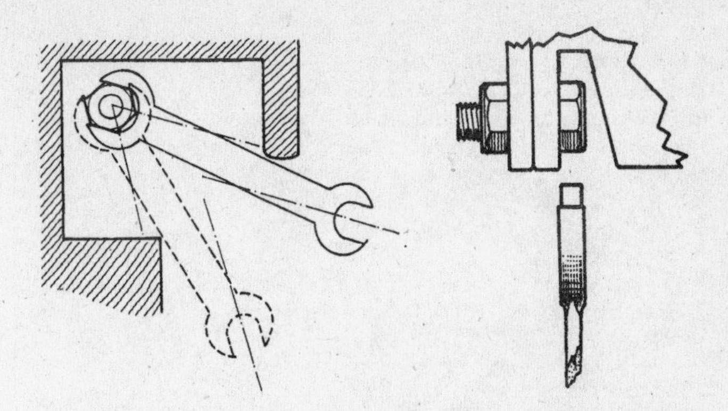

1.5 Open-end wrenches can do several things other wrenches can't - for example, they can be used on bolt heads with limited clearance (left) and they can be used in tight spots where there's little room to turn a wrench by flipping the offset jaw over every few degrees of rotation

tance across the nut or bolt head, or the distance between the wrench jaws - not the diameter of the threads on the fastener - in inches. For example, a 1/4-inch bolt usually has a 7/16-inch hex head - the size of the wrench required to loosen or tighten it. However, the relationship between thread diameter and hex size doesn't always hold true. In some instances, an unusually small hex may be used to discourage over-tightening or because space around the fastener head is limited. Conversely, some fasteners have a disproportionately large hex-head.

Wrenches are similar in appearance, so their quality level can be difficult to judge just by looking at them. There are bargains to be had, just as there are overpriced tools with well-known brand names. On the other hand, you may buy what looks like a reasonable value set of wrenches only to find they fit badly or are made from poor-quality steel.

With a little experience, it's possible to judge the quality of a tool by looking at it. Often, you may have come across the brand name before and have a good idea of the quality. Close examination of the tool can often reveal some hints as to its quality. Prestige tools are usually polished and chrome-plated over their entire surface, with the working faces ground to size. The polished finish is largely cosmetic, but it does make them easy to keep clean. Ground jaws normally indicate the tool will fit well on fasteners.

A side-by-side comparison of a high-quality wrench with a cheap equivalent is an eye opener. The better tool will be made from a good-quality material, often a forged/chrome-vanadium steel alloy **(see illustration)**. This, together with careful design, allows the tool to be kept as small and compact as possible. If, by comparison, the cheap tool is thicker and heavier, especially around the jaws, it's usually because the extra material is needed to compensate for its lower quality. If the tool fits properly, this isn't necessarily bad - it is, after all, cheaper - but in situations where it's necessary to work in a confined area, the cheaper tool may be too bulky to fit.

Open-end wrenches

Because of its versatility, the open-end wrench is the most common type of wrench. It has a jaw on either end, connected by a flat handle section. The jaws either vary by a size, or overlap sizes between consecutive wrenches in a set. This allows one wrench to be used to hold a bolt head while a similar-size nut is removed. A typical fractional size wrench set might have the following jaw sizes: 1/4 x 5/16, 3/8 x 7/16, 1/2 x 9/16, 9/16 x 5/8 and so on.

Typically, the jaw end is set at an angle to the handle, a feature which makes them very useful in confined spaces; by turning the nut or bolt as far as the obstruction allows, then turning the wrench over so the jaw faces in the other direction, it's possible to move the fastener a fraction of a turn at a time **(see illustration)**. The handle length is generally determined by the size of the jaw and is calculated to allow a nut or bolt to be tightened sufficiently by hand with minimal risk of breakage or thread damage (though this doesn't apply to soft materials like brass or aluminum).

Common open-end wrenches are usually sold in sets and it's rarely worth buying them individually unless it's to replace a lost or broken tool from a set. Single tools invariably cost more, so check the sizes you're most likely to need regularly and buy the best set of wrenches you can afford in that range of sizes. If money is limited, remember that you'll use open-end wrenches more than any other type - it's a good idea to buy a good set and cut corners elsewhere.

Box-end wrenches

Box-end wrenches **(see illustration)** have ring-shaped ends with a 6-point (hex) or 12-point (double hex) opening **(see illustration)**. This allows the tool to fit on the fastener hex at 15 (12-point) or 30-degree (6-point) intervals. Normally, each tool has two ends of different sizes, allowing an overlapping range of sizes in a set, as described for open-end wrenches.

Although available as flat tools, the handle is usually offset at each end to allow it to clear obstructions near the fastener, which is normally an advantage. In addition to normal length wrenches, it's also

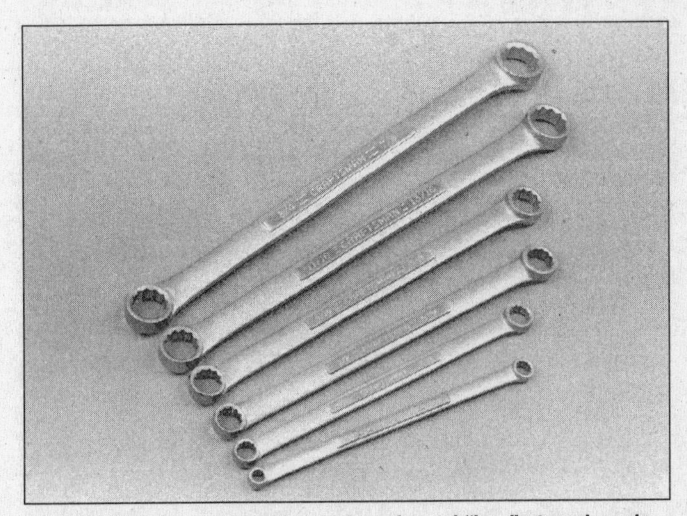

1.6 Box-end wrenches have a ring-shaped "box" at each end - when space permits, they offer the best combination of "grip" and strength

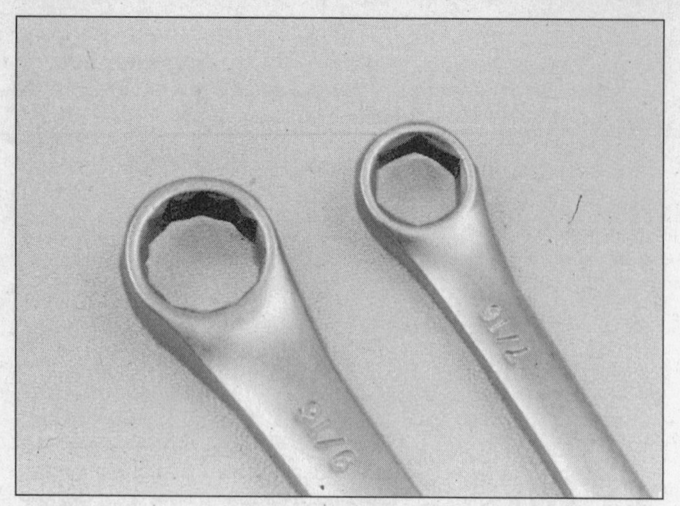

1.7 Box-end wrenches are available in 12 (left) and 6-point (right) openings; even though the 12-point design offers twice as many wrench positions, buy the 6-point first - it's less likely to strip off the corners of a nut or bolt head

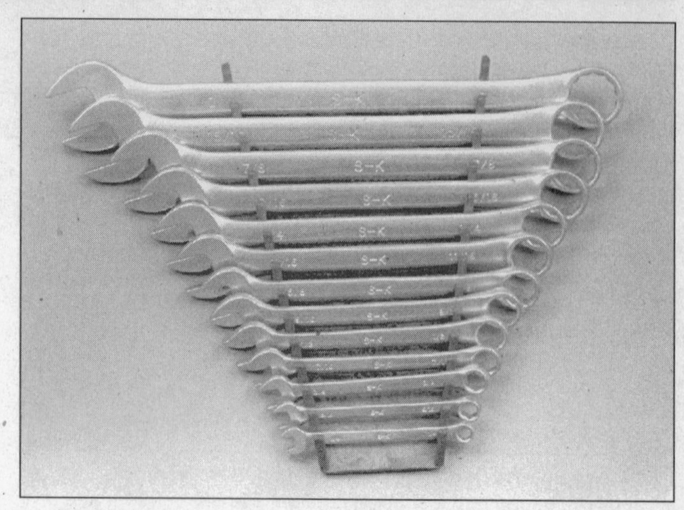

1.8 Buy a set of combination wrenches from 1/4 to 1 -inch

possible to buy long handle types to allow more leverage (very useful when trying to loosen rusted or seized nuts). It is, however, easy to shear off fasteners if not careful, and sometimes the extra length impairs access.

As with open-end wrenches, box-ends are available in varying quality, again often indicated by finish and the amount of metal around the ring ends. While the same criteria should be applied when selecting a set of box-end wrenches, if your budget is limited, go for better-quality open-end wrenches and a slightly cheaper set of box-ends.

Combination wrenches

These wrenches **(see illustration)** combine a box-end and open-end of the same size in one tool and offer many of the advantages of both. Like the others, they're widely available in sets and as such are probably a better choice than box-ends only. They're generally compact, short-handled tools and are well suited for tight spaces where access is limited.

Adjustable wrenches

Adjustable wrenches **(see illustration)** come in several sizes. Each size can handle a range of fastener sizes. Adjustable wrenches aren't as effective as one-size tools and it's easy to damage fasteners with them. However, they can be an invaluable addition to any tool kit - if they're used with discretion. **Note:** *If you attach the wrench to the fastener with the movable jaw pointing in the direction of wrench rotation* **(see illustration)**, *an adjustable wrench will be less likely to slip and damage the fastener head.*

The most common adjustable wrench is the open-end type with a set of parallel jaws that can be set to fit the head of a fastener. Most are controlled by a threaded spindle, though there are various cam and spring-loaded versions available. Don't buy large tools of this type; you'll rarely be able to find enough clearance to use them.

Ratchet and socket sets

Ratcheting socket wrenches **(see illustration)** are highly versatile. Besides the sockets themselves, many other interchangeable accessories - extensions, U-drives, step-down adapters, screwdriver bits, Allen bits, crow's feet, etc. - are available. Buy six-point sockets - they're less likely to slip and strip the corners off bolts and nuts. Don't buy sockets with extra-thick walls - they might be stronger but they can be hard to use on recessed fasteners or fasteners in tight quarters. Buy a 3/8-inch drive for work on the outside of the engine. It's the one

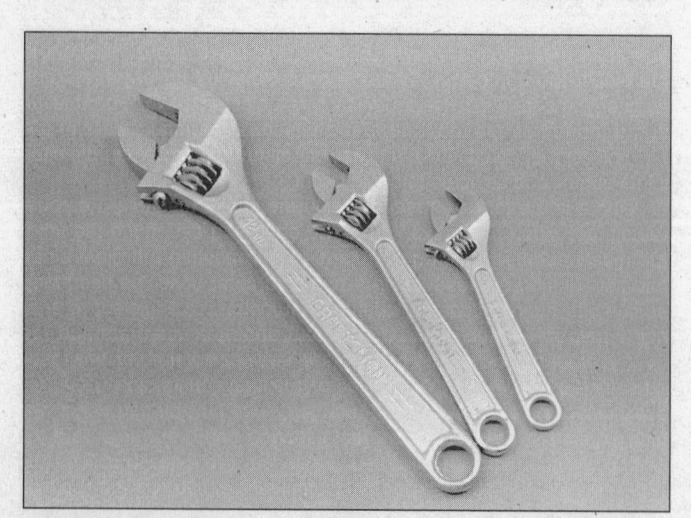

1.9 Adjustable wrenches can handle a range of fastener sizes - they're not as good as single-size wrenches but they're handy for loosening and tightening those odd-sized fasteners for which you haven't yet bought the correct wrench

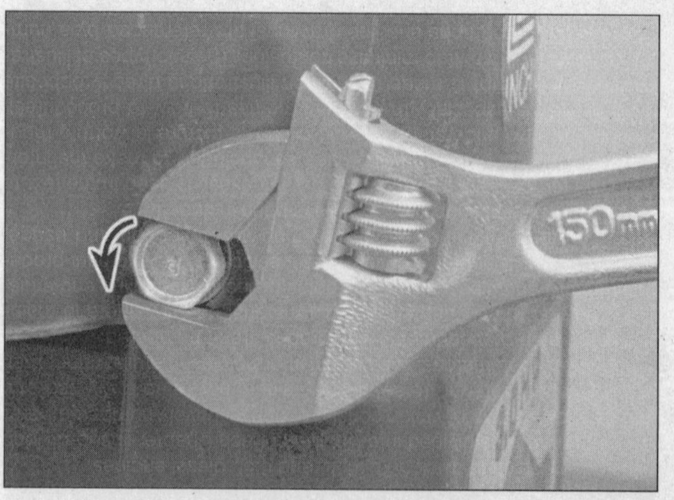

1.10 When you use an adjustable wrench, make sure the movable jaw points in the direction the wrench is being turned (arrow) so the wrench doesn't distort and slip off the fastener head

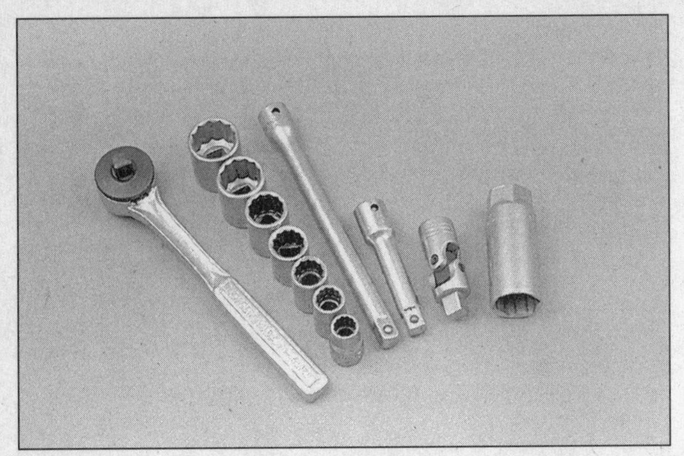

1.11 A typical ratchet and socket set includes a ratchet, a set of sockets, a long and a short extension, a universal joint and a spark plug socket

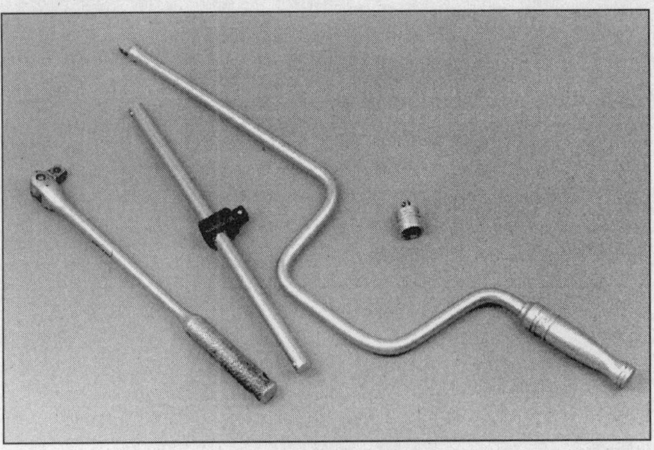

1.12 Lots of other accessories are available for ratchets: From left to right, a breaker bar, a sliding T-handle, a speed handle and a 3/8-to-1/4-inch adapter

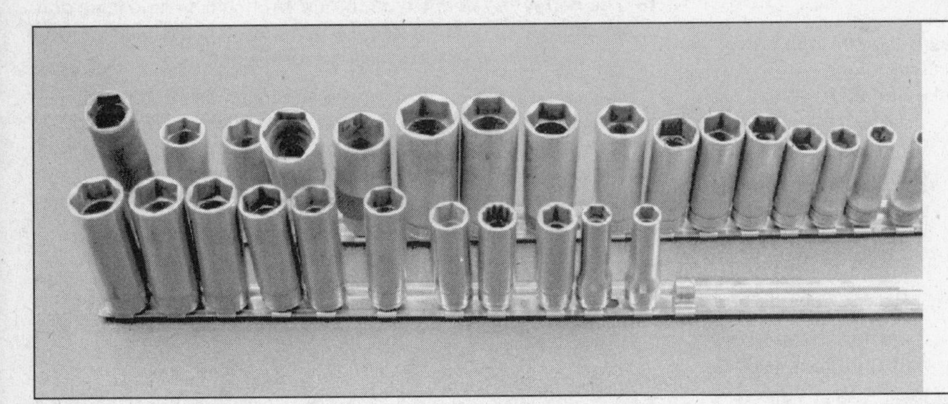

1.13 Deep sockets enable you to loosen or tighten an elongated fastener, or to get at a nut with a long bolt protruding from it

you'll use most of the time. Get a 1/2-inch drive for overhaul work. Although the larger drive is bulky and more expensive, it has the capacity of accepting a very wide range of large sockets. Later, you may want to consider a 1/4-inch drive for little stuff like ignition and carburetor work. Interchangeable sockets consist of a forged-steel alloy cylinder with a hex or double-hex formed inside one end. The other end is formed into the square drive recess that engages over the corresponding square end of various socket drive tools.

Sockets are available in 1/4, 3/8, 1/2 and 3/4-inch drive sizes. A 3/8-inch drive set is most useful for engine repairs, although 1/4-inch drive sockets and accessories may occasionally be needed.

The most economical way to buy sockets is in a set. As always, quality will govern the cost of the tools. Once again, the "buy the best" approach is usually advised when selecting sockets. While this is a good idea, since the end result is a set of quality tools that should last a lifetime, the cost is so high it's difficult to justify the expense for home use. As far as accessories go, you'll need a ratchet, at least one extension (buy a three or six-inch size), a spark plug socket and maybe a T-handle or breaker bar. Other desirable, though less essential items, are a speeder handle, a U-joint, extensions of various other lengths and adapters from one drive size to another **(see illustration)**. Some of the sets you find may combine drive sizes; they're well worth having if you find the right set at a good price, but avoid being dazzled by the number of pieces.

Above all, be sure to completely ignore any label that reads "86-piece Socket Set," which refers to the number of pieces, not to the number of sockets (sometimes even the metal box and plastic insert are counted in the total!).

Apart from well-known and respected brand names, you'll have to take a chance on the quality of the set you buy. If you know someone who has a set that has held up well, try to find the same brand, if possible. Take a pocketful of nuts and bolts with you and check the fit in some of the sockets. Check the operation of the ratchet. Good ones operate smoothly and crisply in small steps; cheap ones are coarse and stiff - a good basis for guessing the quality of the rest of the pieces.

One of the best things about a socket set is the built-in facility for expansion. Once you have a basic set, you can purchase extra sockets when necessary and replace worn or damaged tools. There are special deep sockets for reaching recessed fasteners or to allow the socket to fit over a projecting bolt or stud **(see illustration)**. You can also buy screwdriver, Allen and Torx bits to fit various drive tools (they can be very handy in some applications) **(see illustration)**. Most socket sets include a special deep socket for 14 millimeter spark plugs. They have rubber inserts to protect the spark plug porcelain insulator and hold the plug in the socket to avoid burned fingers.

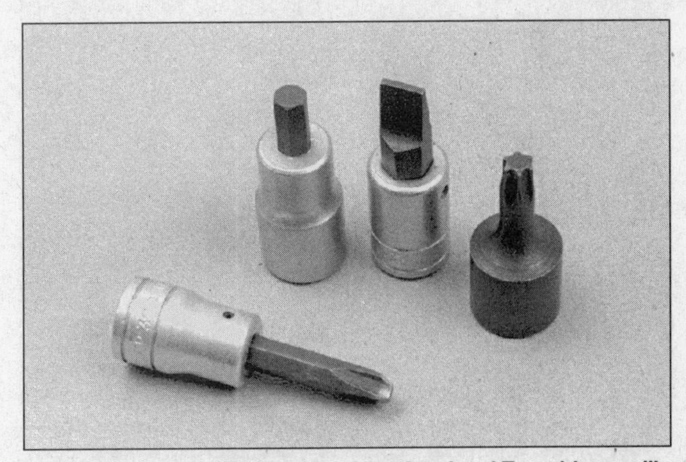

1.14 Standard and Phillips bits, Allen-head and Torx drivers will expand the versatility of your ratchet and extensions even further

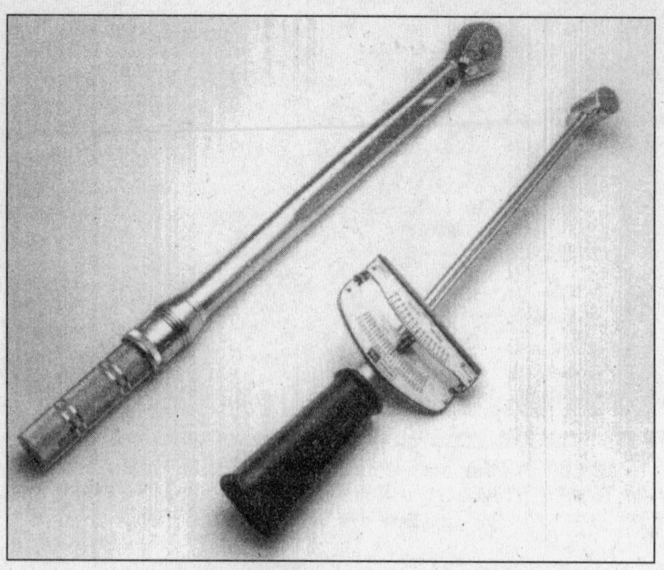

1.15 Torque wrenches (click-type on left, beam-type on right) are the only way to accurately tighten critical fasteners like connecting rod bolts, cylinder head bolts, etc.

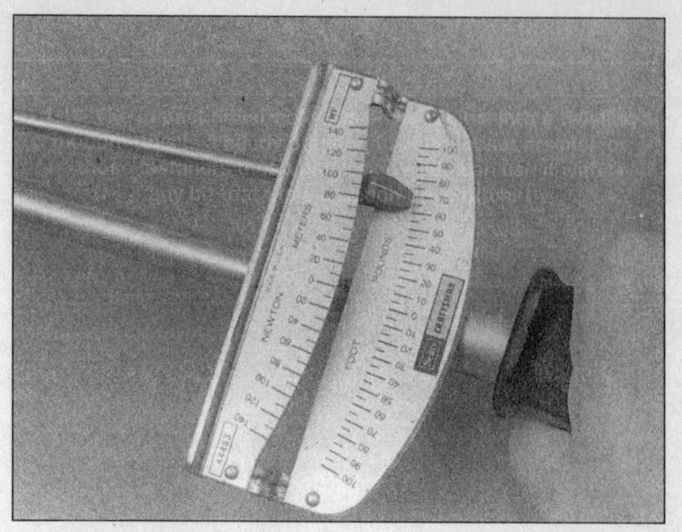

1.16 The deflecting beam-type torque wrench is inexpensive and simple to use -just tighten the fastener until the pointer points to the specified torque setting

Torque wrenches

Torque wrenches **(see illustration)** are essential for tightening critical fasteners like rod bolts, main bearing cap bolts, head bolts, etc. Attempting an engine overhaul without a torque wrench is an invitation to oil leaks, distortion of the cylinder head, damaged or stripped threads or worse.

There are two types of torque wrenches - the "beam" type, which indicates torque loads by deflecting a flexible shaft and the "click" type **(see illustrations)**, which emits an audible click when the torque resistance reaches the specified resistance.

Torque wrenches are available in a variety of drive sizes and torque ranges for particular applications. For engine rebuilding, 0 to 150 ft-lbs should be adequate. Keep in mind that "click" types are usually more accurate (and more expensive).

Impact drivers

The impact driver **(see illustration)** belongs with the screwdrivers, but it's mentioned here since it can also be used with sockets (impact drivers normally are 3/8-inch square drive). As explained later, an impact driver works by converting a hammer blow on the end of its handle into a sharp twisting movement. While this is a great way to jar a seized fastener loose, the loads imposed on the socket are excessive. Use sockets only with discretion and expect to have to replace damaged ones on occasion.

Using wrenches and sockets

Although you may think the proper use of tools is self-evident, it's worth some thought. After all, when did you last see instructions for use supplied with a set of wrenches?

Which wrench?

Before you start tearing an engine apart, figure out the best tool for the job; in this instance the best wrench for a hex-head fastener. Sit down with a few nuts and bolts and look at how various tools fit the bolt heads.

A golden rule is to choose a tool that contacts the largest area of the hex-head. This distributes the load as evenly as possible and lessens the risk of damage. The shape most closely resembling the bolt head or nut is another hex, so a 6-point socket or box-end wrench is usually the best choice **(see illustration)**. Many sockets and box-

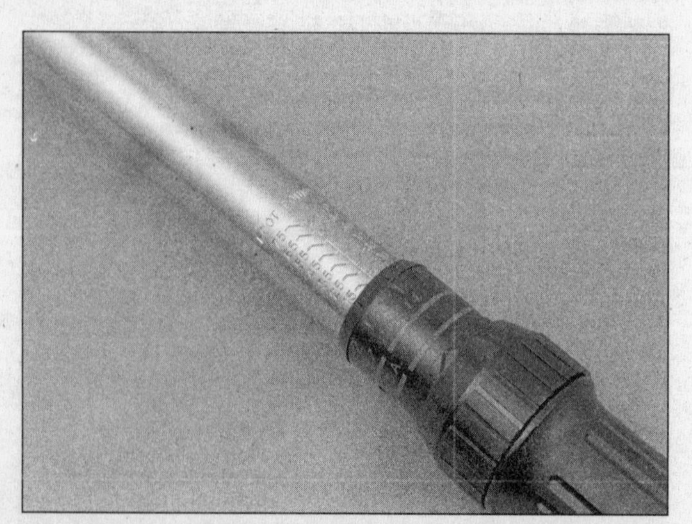

1.17 "Click" type torque wrenches can be set to "give" at a preset torque, which makes them very accurate and easy to use

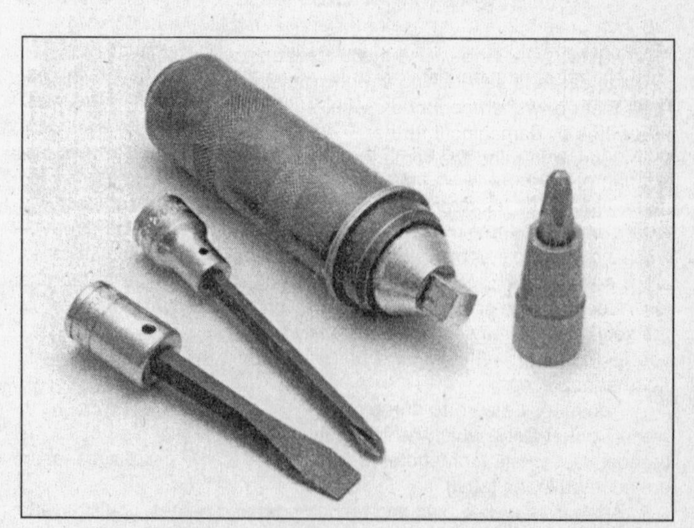

1.18 The impact driver converts a sharp blow into a twisting motion - this is a handy addition to your socket arsenal for those fasteners that won't let go - you can use it with any bit that fits a 3/8-inch drive ratchet

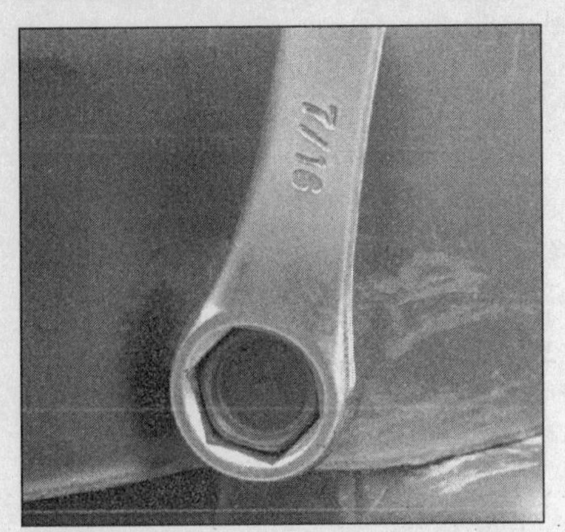

1.19 Try to use a six-point box wrench (or socket) whenever possible - its shape matches that of the fastener, which means maximum grip and minimum slip

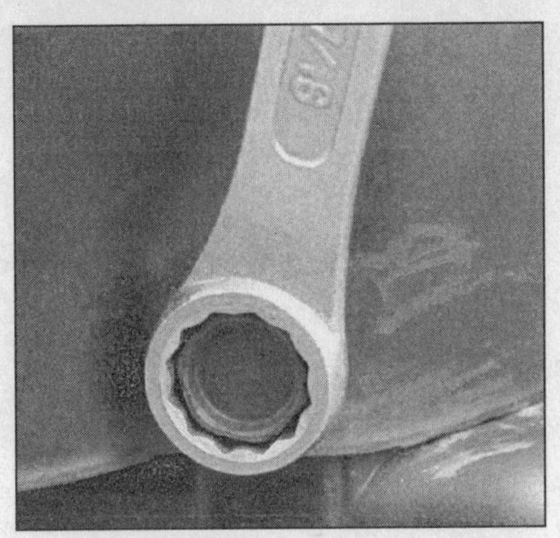

1.20 Sometimes a six-point tool just doesn't offer you any grip when you get the wrench at the angle it needs to be in to loosen or tighten a fastener - when this happens, pull out the 12-point sockets or wrenches - but remember: they're much more likely to strip the corners off a fastener

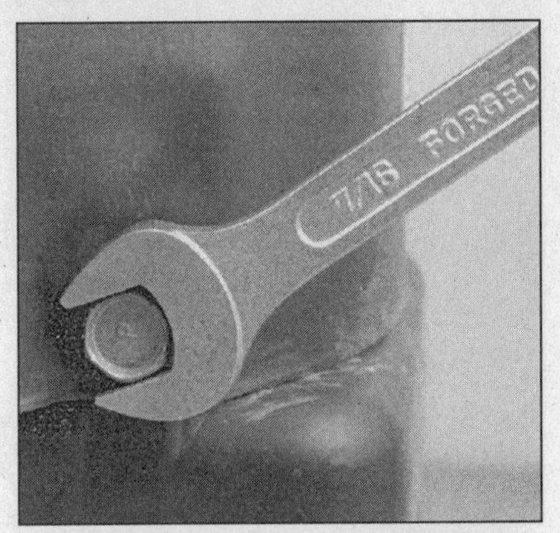

1.21 Open-end wrenches contact only two sides of the fastener and the jaws tend to open up when you put some muscle on the wrench handle - that's why they should only be used as a last resort

end wrenches have double hex (12-point) openings. If you slip a 12-point box-end wrench over a nut, look at how and where the two are in contact. The corners of the nut engage in every other point of the wrench. When the wrench is turned, pressure is applied evenly on each of the six corners **(see illustration)**. This is fine unless the fastener head was previously rounded off. If so, the corners will be damaged and the wrench will slip. If you encounter a damaged bolt head or nut, always use a 6-point wrench or socket if possible. If you don't have one of the right size, choose a wrench that fits securely and proceed with care.

If you slip an open-end wrench over a hex-head fastener, you'll see the tool is in contact on two faces only **(see illustration)**. This is acceptable provided the tool and fastener are both in good condition. The need for a snug fit between the wrench and nut or bolt explains the recommendation to buy good-quality open-end wrenches. If the wrench jaws, the bolt head or both are damaged, the wrench will probably slip, rounding off and distorting the head. In some applications, an open-end wrench is the only possible choice due to limited access, but always check the fit of the wrench on the fastener before attempting to loosen it; if it's hard to get at with a wrench, think how hard it will be to remove after the head is damaged.

The last choice is an adjustable wrench or self-locking plier/wrench (Vise-Grips). Use these tools only when all else has failed. In some cases, a self-locking wrench may be able to grip a damaged head that no wrench could deal with, but be careful not to make matters worse by damaging it further.

Bearing in mind the remarks about the correct choice of tool in the first place, there are several things worth noting about the actual use of the tool. First, make sure the wrench head is clean and undamaged. If the fastener is rusted or coated with paint, the wrench won't fit correctly. Clean off the head and, if it's rusted, apply some penetrating oil. Leave it to soak in for a while before attempting removal.

It may seem obvious, but take a close look at the fastener to be removed before using a wrench. On many mass-produced machines, one end of a fastener may be fixed or captive, which speeds up initial assembly and usually makes removal easier. If a nut is installed on a stud or a bolt threads into a captive nut or tapped hole, you may have only one fastener to deal with. If, on the other hand, you have a separate nut and bolt, you must hold the bolt head while the nut is removed. In some areas this can be difficult, particularly where engine mounts are involved. In this type of situation you may need an assistant to hold the bolt head with a wrench while you remove the nut from the other side. If this isn't possible, you'll have to try to position a box-end wrench so it wedges against some other component to prevent it from turning.

Be on the lookout for left-hand threads. They aren't common, but are sometimes used on the ends of rotating shafts to make sure the nut doesn't come loose during engine operation (most engines covered by this book don't have these types of fasteners). If you can see the shaft end, the thread type can be checked visually. If you're unsure, place your thumbnail in the threads and see which way you have to turn your hand so your nail "unscrews" from the shaft. If you have to turn your hand counterclockwise, it's a conventional right-hand thread.

Beware of the upside-down fastener syndrome. If you're loosening a fastener from the under side of a something, it's easy to get confused about which way to turn it. What seems like counterclockwise to you can easily be clockwise (from the fastener's point of view). Even after years of experience, this can still catch you once in a while.

In most cases, a fastener can be removed simply by placing the wrench on the nut or bolt head and turning it. Occasionally, though, the condition or location of the fastener may make things more difficult. Make sure the wrench is square on the head. You may need to reposition the tool or try another type to obtain a snug fit. Make sure the engine you're working on is secure and can't move when you turn the wrench. If necessary, get someone to help steady it for you. Position yourself so you can get maximum leverage on the wrench.

If possible, locate the wrench so you can pull the end towards you. If you have to push on the tool, remember that it may slip, or the fastener may move suddenly. For this reason, don't curl your fingers around the handle or you may crush or bruise them when the fastener moves; keep your hand flat, pushing on the wrench with the heel of your thumb. If the tool digs into your hand, place a rag between it and your hand or wear a heavy glove.

If the fastener doesn't move with normal hand pressure, stop and try to figure out why before the fastener or wrench is damaged or you hurt yourself. Stuck fasteners may require penetrating oil, heat or an impact driver or air tool.

Using sockets to remove hex-head fasteners is less likely to result in damage than if a wrench is used. Make sure the socket fits snugly over the fastener head, then attach an extension, if needed, and the ratchet or breaker bar. Theoretically, a ratchet shouldn't be used for loosening a fastener or for final tightening because the ratchet mechanism may be overloaded and could slip. In some instances, the location of the fastener may mean you have no choice but to use a ratchet, in which case you'll have to be extra careful.

Never use extensions where they aren't needed. Whether or not an extension is used, always support the drive end of the breaker bar with one hand while turning it with the other. Once the fastener is loose, the ratchet can be used to speed up removal.

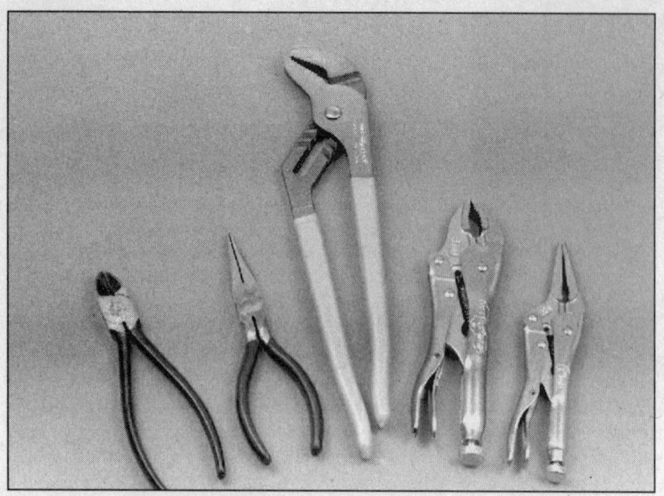

1.22 A typical assortment of the types of pliers you need to have in your box - from the left: diagonal cutters (dikes), needle-nose pliers, Channel-lock pliers, Vise-Grip pliers, needle-nose Vise Grip pliers

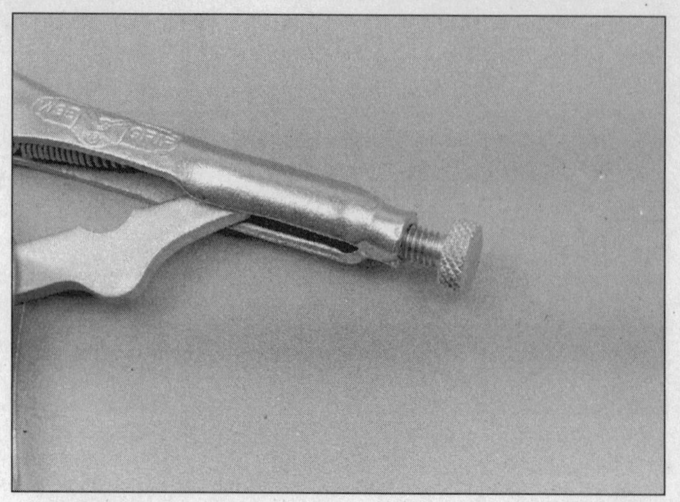

1.23 To adjust the jaws on a pair of Vise-Grips, grasp the part you want to hold with the jaws, tighten them down by turning the knurled knob on the end of one handle and snap the handles together - if you tightened the knob all the way down, you'll probably have to open it up (back it off) a little before you can close the handles

Pliers

Some tool manufacturers make 25 or 30 different types of pliers. You only need a fraction of this selection **(see illustration)**. Get a good pair of slip-joint pliers for general use. A pair of needle-nose models is handy for reaching into hard-to-get-at places. A set of diagonal wire cutters (dikes) is essential for electrical work and pulling out cotter pins. Vise-Grips are adjustable, locking pliers that grip a fastener firmly - and won't let go - when locked into place. Parallel-jaw, adjustable pliers have angled jaws that remain parallel at any degree of opening. They're also referred to as Channel-lock (the original manufacturer) pliers, arc-joint pliers and water pump pliers. Whatever you call them, they're terrific for gripping a big fastener with a lot of force.

Slip-joint pliers have two open positions; a figure eight-shaped, elongated slot in one handle slips back-and-forth on a pivot pin on the other handle to change them. Good-quality pliers have jaws made of tempered steel and there's usually a wire-cutter at the base of the jaws. The primary uses of slip-joint pliers are for holding objects, bending and cutting throttle wires and crimping and bending metal parts, not loosening nuts and bolts.

Arc-joint or "Channel-lock" pliers have parallel jaws you can open to various widths by engaging different tongues and grooves, or channels, near the pivot pin. Since the tool expands to fit many size objects, it has countless uses for engine and equipment maintenance. Channel-lock pliers come in various sizes. The medium size is adequate for general work; small and large sizes are nice to have as your budget permits. You'll use all three sizes frequently.

Vise-Grips (a brand name) come in various sizes; the medium size with curved jaws is best for all-around work. However, buy a large and small one if possible, since they're often used in pairs. Although this tool falls somewhere between an adjustable wrench, a pair of pliers and a portable vise, it can be invaluable for loosening and tightening fasteners - it's the only pliers that should be used for this purpose.

The jaw opening is set by turning a knurled knob at the end of one handle. The jaws are placed over the head of the fastener and the handles are squeezed together, locking the tool onto the fastener **(see illustration)**. The design of the tool allows extreme pressure to be applied at the jaws and a variety of jaw designs enable the tool to grip firmly even on damaged heads **(see illustration)**. Vise-Grips are great for removing fasteners that've been rounded off by badly-fitting wrenches.

As the name suggests, needle-nose pliers have long, thin jaws designed for reaching into holes and other restricted areas. Most needle-nose, or long-nose, pliers also have wire cutters at the base of the jaws.

Look for these qualities when buying pliers: Smooth operating handles and jaws, jaws that match up and grip evenly when the handles are closed, a nice finish and the word "forged" somewhere on the tool.

Screwdrivers

Screwdrivers **(see illustration)** come in a wide variety of sizes and price ranges. Anything from Craftsman on up is fine. But don't buy screwdriver sets for ten bucks at discount tool stores. Even if they look exactly like more expensive brands, the metal tips and shafts are made with inferior alloys and aren't properly heat treated. They usually bend the first time you apply some serious torque.

A screwdriver consists of a steel blade or shank with a drive tip formed at one end. The most common tips are standard (also called straight slot and flat-blade) and Phillips. The other end has a handle attached to it. Traditionally, handles were made from wood and secured to the shank, which had raised tangs to prevent it from turning in the handle. Most screwdrivers now come with plastic handles, which are generally more durable than wood.

The design and size of handles and blades vary considerably. Some handles are specially shaped to fit the human hand and provide

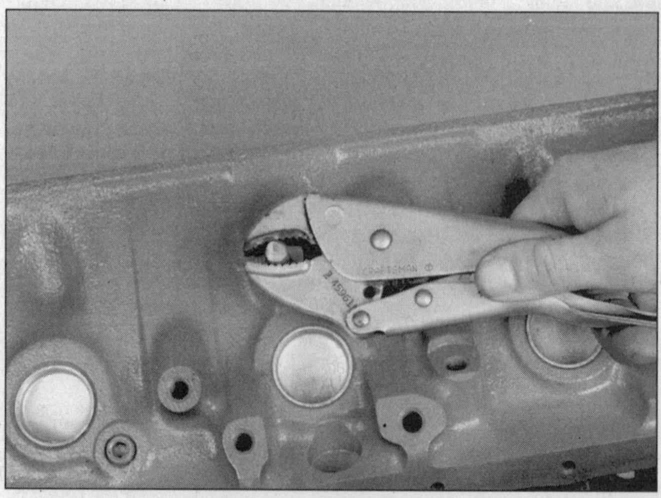

1.24 It you're persistent and careful, most fasteners can be removed with Vise-Grips

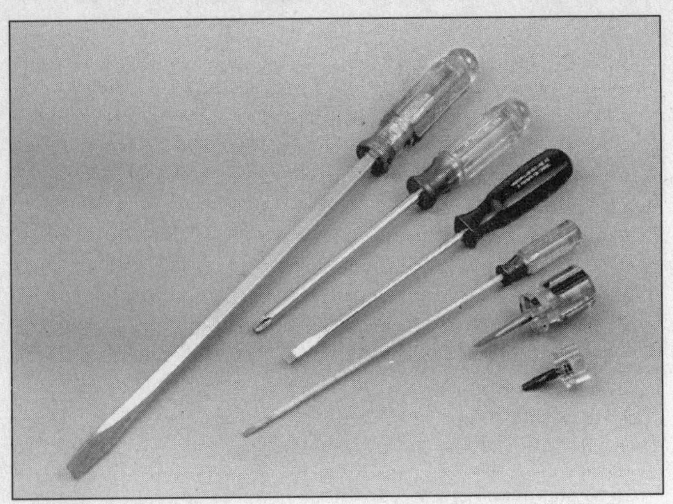

1.25 Screwdrivers come in myriad lengths, sizes and styles

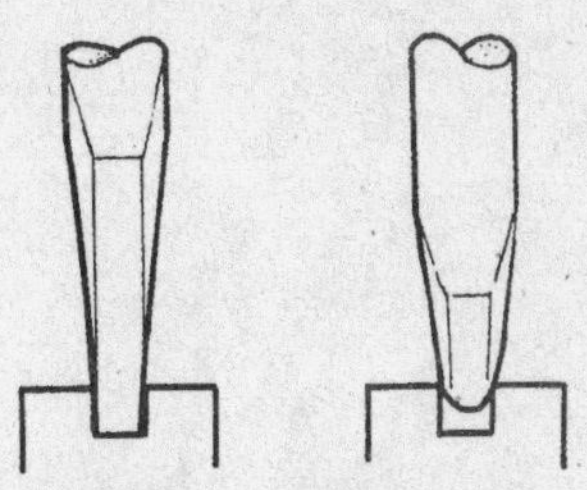

Misuse of a screwdriver – the blade shown is both too narrow and too thin and will probably slip or break off

The left-hand example shows a snug-fitting tip. The right-hand drawing shows a damaged tip which will twist out of the slot when pressure is applied

1.26 Standard screwdrivers - wrong size (left), correct fit in screw slot (center) and worn tip (right)

a better grip. The shank may be either round or square and some have a hex-shaped bolster under the handle to accept a wrench to provide more leverage when trying to turn a stubborn screw. The shank diameter, tip size and overall length vary too. As a general rule, it's a good idea to use the longest screwdriver possible, which allows the greatest possible leverage.

If access is restricted, a number of special screwdrivers are designed to fit into confined spaces. The "stubby" screwdriver has a specially shortened handle and blade. There are also offset screwdrivers and special screwdriver bits that attach to a ratchet or extension.

The important thing to remember when buying screwdrivers is that they really do come in sizes designed to fit different size fasteners. The slot in any screw has definite dimensions - length, width and depth. Like a bolt head or a nut, the screw slot must be driven by a tool that uses all of the available bearing surface and doesn't slip. Don't use a big wide blade on a small screw and don't try to turn a large screw slot with a tiny, narrow blade. The same principles apply to Allen heads, Phillips heads, Torx heads, etc. Don't even think of using a slotted screwdriver on one of these heads! And don't use your screwdrivers as levers, chisels or punches! This kind of abuse turns them into very bad screwdrivers.

Standard screwdrivers

These are used to remove and install conventional slotted screws and are available in a wide range of sizes denoting the width of the tip and the length of the shank (for example: a 3/8 x 10-inch screwdriver is 3/8-inch wide at the tip and the shank is 10-inches long). You should have a variety of screwdrivers so screws of various sizes can be dealt with without damaging them. The blade end must be the same width and thickness as the screw slot to work properly, without slipping. When selecting standard screwdrivers, choose good-quality tools, preferably with chrome moly, forged steel shanks. The tip of the shank should be ground to a parallel, flat profile (hollow ground) and not to a taper or wedge shape, which will tend to twist out of the slot when pressure is applied **(see illustration)**.

All screwdrivers wear in use, but standard types can be reground to shape a number of times. When reshaping a tip, start by grinding the very end flat at right angles to the shank. Make sure the tip fits snugly in the slot of a screw of the appropriate size and keep the sides of the tip parallel. Remove only a small amount of metal at a time to avoid overheating the tip and destroying the temper of the steel.

Phillips screwdrivers

Phillips screws are sometimes installed during initial assembly with air tools and are next to impossible to remove later without ruining the heads, particularly if the wrong size screwdriver is used. And don't use other types of cross-head screwdrivers (Torx, Posi-drive, etc.) on Phillips screws - they won't work.

The only way to ensure the screwdrivers you buy will fit properly, is to take a couple of screws with you to make sure the fit between the screwdriver and fastener is snug. If the fit is good, you should be able to angle the blade down almost vertically without the screw slipping off the tip. Use only screwdrivers that fit exactly - anything else is guaranteed to chew out the screw head instantly.

The idea behind all cross-head screw designs is to make the screw and screwdriver blade self-aligning. Provided you aim the blade at the center of the screw head, it'll engage correctly, unlike conventional slotted screws, which need careful alignment. This makes the screws suitable for machine installation on an assembly line (which explains why they're sometimes so tight and difficult to remove). The drawback with these screws is the driving tangs on the screwdriver tip are very small and must fit very precisely in the screw head. If this isn't the case, the huge loads imposed on small flats of the screw slot simply tear the metal away, at which point the screw ceases to be removable by normal methods. The problem is made worse by the normally soft material chosen for screws.

To deal with these screws on a regular basis, you'll need high-quality screwdrivers with various size tips so you'll be sure to have the right one when you need it. Phillips screwdrivers are sized by the tip number and length of the shank (for example: a number 2 x 6-inch Phillips screwdriver has a number 2 tip - to fit screws of only that size recess - and the shank is 6-inches long). Tip sizes 1, 2 and 3 should be adequate for engine repair work **(see illustration)**. If the tips get worn or damaged, buy new screwdrivers so the tools don't destroy the screws they're used on **(see illustration)**.

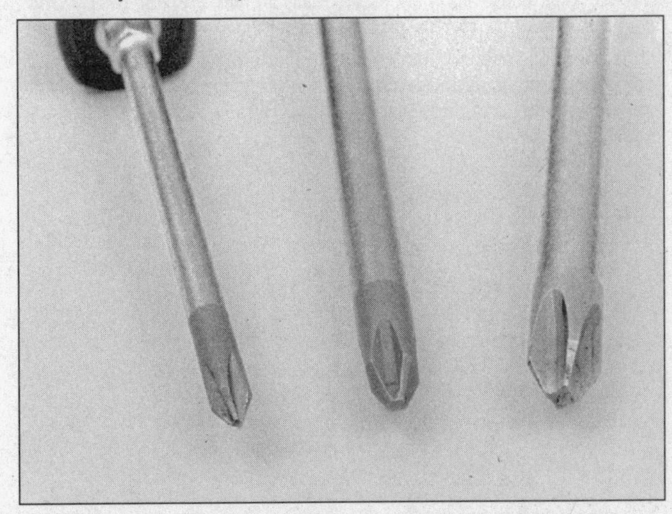

1.27 The tip size on a Phillips screwdriver is indicated by a number from 1 to 4, with 1 the smallest (left - No. 1; center - No. 2; right - No. 3)

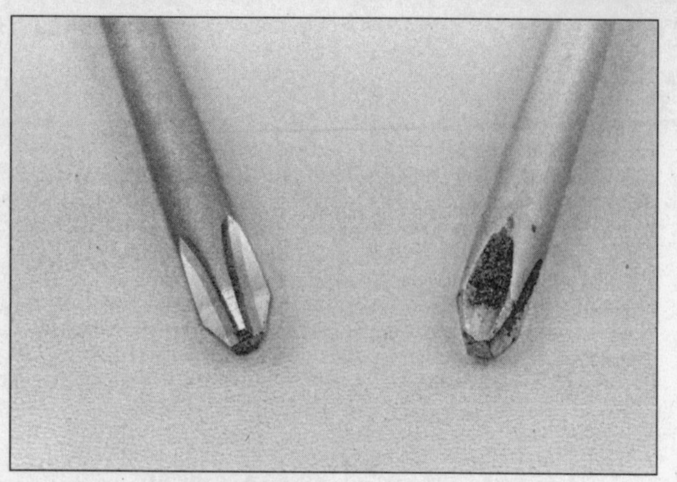

1.28 New (left) and worn (right) Phillips screwdriver tips

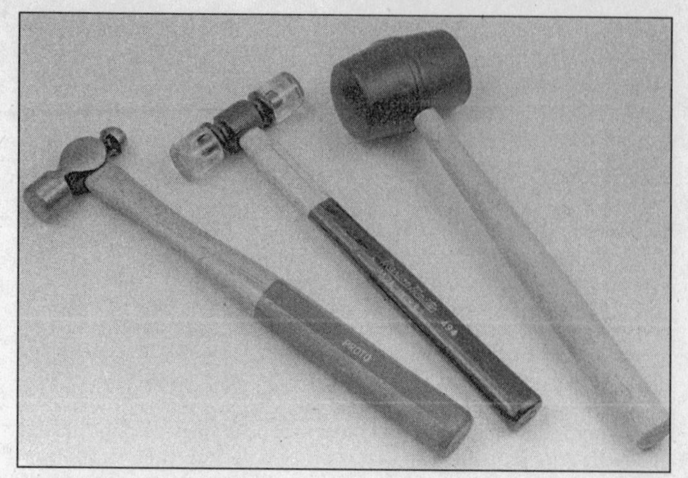

1.29 A ball-peen hammer, soft-face hammer and rubber mallet (left-to-right) will be needed for various tasks (any steel hammer can be used in place of the ball peen hammer)

Here's a tip that may come in handy when using Phillips screwdrivers - if the screw is extremely tight and the tip tends to back out of the recess rather than turn the screw, apply a small amount of valve lapping compound to the screwdriver tip so it will grip the screw better.

Hammers

Resorting to a hammer should always be the last resort. When nothing else will do the job, a medium-size ball peen hammer, a heavy rubber mallet and a heavy soft-brass hammer **(see illustration)** are often the only way to loosen or install a part.

A ball-peen hammer has a head with a conventional cylindrical face at one end and a rounded ball end at the other and is a general-purpose tool found in almost any type of shop. It has a shorter neck than a claw hammer and the face is tempered for striking punches and chisels. A fairly large hammer is preferable to a small one. Although it's possible to find small ones, you won't need them very often and it's much easier to control the blows from a heavier head. As a general rule, a single 12 or 16-ounce hammer will work for most jobs, though occasionally larger or smaller ones may be useful.

A soft-face hammer is used where a steel hammer could cause damage to the component or other tools being used. A steel hammer head might crack an aluminum part, but a rubber or plastic hammer can be used with more confidence. Soft-face hammers are available with interchangeable heads (usually one made of rubber and another made of relatively hard plastic). When the heads are worn out, new ones can be installed. If finances are really limited, you can get by without a soft-face hammer by placing a small hardwood block between the component and a steel hammer head to prevent damage.

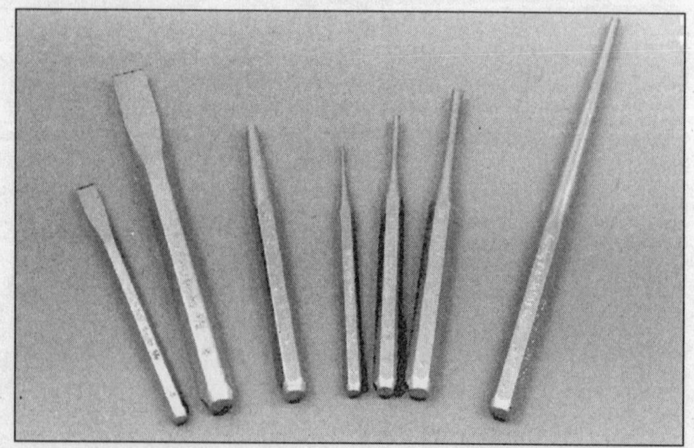

1.30 Cold chisels, center-punches, pin punches and line-up punches (left-to-right) will be needed sooner or later for many jobs

Hammers should be used with common sense; the head should strike the desired object squarely and with the right amount of force. For many jobs, little effort is needed - simply allow the weight of the head to do the work, using the length of the swing to control the amount of force applied. With practice, a hammer can be used with surprising finesse, but it'll take a while to achieve. Initial mistakes include striking the object at an angle, in which case the hammer head may glance off to one side, or hitting the edge of the object. Either one can result in damage to the part or to your thumb, if it gets in the way, so be careful. Hold the hammer handle near the end, not near the head, and grip it firmly but not too tightly.

Check the condition of your hammers on a regular basis. The danger of a loose head coming off is self-evident, but check the head for chips and cracks too. If damage is noted, buy a new hammer - the head may chip in use and the resulting fragments can be extremely dangerous. It goes without saying that eye protection is essential whenever a hammer is used.

Punches and chisels

Punches and chisels **(see illustration)** are used along with a hammer for various purposes in the shop. Drift punches are often simply a length of round steel bar used to drive a component out of a bore in the engine or equipment it's mounted on. A typical use would be for removing or installing a bearing or bushing. A drift of the same diameter as the bearing outer race is placed against the bearing and tapped with a hammer to knock it in or out of the bore. Most manufacturers offer special drifts for the various bearings in a particular engine. While they're useful to a busy dealer service department, they are prohibitively expensive for the do-it-yourselfer who may only need to use them once. In such cases, it's better to improvise. For bearing removal and installation, it's usually possible to use a socket of the appropriate diameter to tap the bearing in or out; an unorthodox use for a socket, but it works.

Smaller diameter drift punches can be purchased or fabricated from steel bar stock. In some cases, you'll need to drive out items like corroded engine mounting bolts. Here, it's essential to avoid damaging the threaded end of the bolt, so the drift must be a softer material than the bolt. Brass or copper is the usual choice for such jobs; the drift may be damaged in use, but the thread will be protected.

Punches are available in various shapes and sizes and a set of assorted types will be very useful. One of the most basic is the center punch, a small cylindrical punch with the end ground to a point. It'll be needed whenever a hole is drilled. The center of the hole is located first and the punch is used to make a small indentation at the intended point. The indentation acts as a guide for the drill bit so the hole ends up in the right place. Without a punch mark the drill bit will wander and you'll find it impossible to drill with any real accuracy. You can also buy automatic center punches. They're spring loaded and are pressed against the surface to be marked, without the need to use a hammer.

Pin punches are intended for removing items like roll pins (semi-hard, hollow pins that fit tightly in their holes). Pin punches have other uses, however. You may occasionally have to remove rivets or bolts by cutting off the heads and driving out the shanks with a pin punch. They're also very handy for aligning holes in components while bolts or screws are inserted.

Of the various sizes and types of metal-cutting chisels available, a simple cold chisel is essential in any mechanic's workshop. One about 6-inches long with a 1/2-inch wide blade should be adequate. The cutting edge is ground to about 80-degrees **(see illustration)**, while the rest of the tip is ground to a shallower angle away from the edge. The primary use of the cold chisel is rough metal cutting - this can be anything from sheet metal work (uncommon on engines) to cutting off the heads of seized or rusted bolts or splitting nuts. A cold chisel is also useful for turning out screws or bolts with messed-up heads.

All of the tools described in this section should be good quality items. They're not particularly expensive, so it's not really worth trying to save money on them. More significantly, there's a risk that with cheap tools, fragments may break off in use - a potentially dangerous situation.

Even with good-quality tools, the heads and working ends will inevitably get worn or damaged, so it's a good idea to maintain all such tools on a regular basis. Using a file or bench grinder, remove all burrs and mushroomed edges from around the head. This is an important task because the build-up of material around the head can fly off when it's struck with a hammer and is potentially dangerous. Make sure the tool retains its original profile at the working end, again, filing or grinding off all burrs. In the case of cold chisels, the cutting edge will usually have to be reground quite often because the material in the tool isn't usually much harder than materials typically being cut. Make sure the edge is reasonably sharp, but don't make the tip angle greater than it was originally; it'll just wear down faster if you do.

The techniques for using these tools vary according to the job to be done and are best learned by experience. The one common denominator is the fact they're all normally struck with a hammer. It follows that eye protection should be worn. Always make sure the working end of the tool is in contact with the part being punched or cut. If it isn't, the tool will bounce off the surface and damage may result.

Hacksaws

A hacksaw **(see illustration)** consists of a handle and frame supporting a flexible steel blade under tension. Blades are available in various lengths and most hacksaws can be adjusted to accommodate the different sizes. The most common blade length is 10-inches.

Most hacksaw frames are adequate. There's little difference between brands. Pick one that's rigid and allows easy blade changing and repositioning.

The type of blade to use, indicated by the number of teeth per inch, (TPI) **(see illustration)**, is determined by the material being cut. The rule of thumb is to make sure at least three teeth are in contact with the metal being cut at any one time **(see illustration)**. In practice,

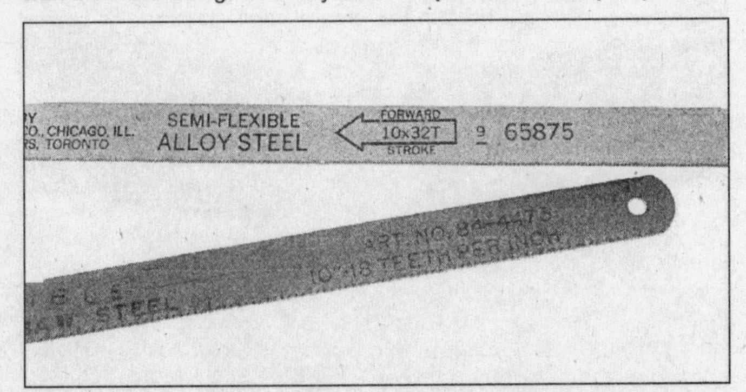

1.33 Hacksaw blades are marked with the number of teeth per inch (TPQ - use a relatively coarse blade for aluminum and thicker items such as bolts or bar stock; use a finer blade for materials like thin sheet steel

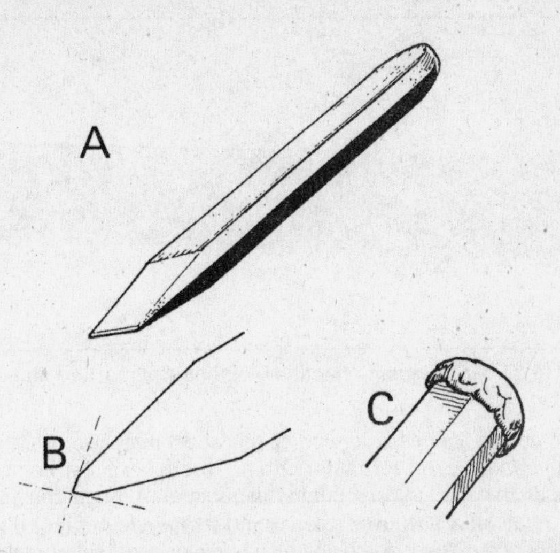

1.31 A typical general purpose cold chisel (A) - note the angle of the cutting edge (B), which should be checked and resharpened on a regular basis; the mushroomed head (B) is dangerous and should be filed to restore it to its original shape

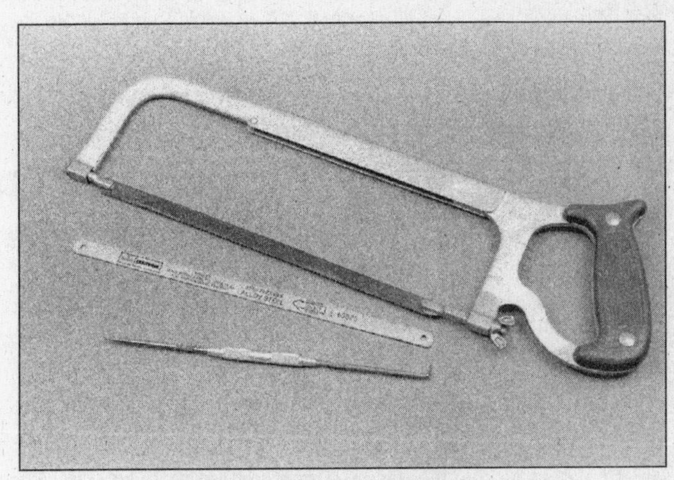

1.32 Hacksaws are handy for little cutting jobs like sheet metal and rusted fasteners

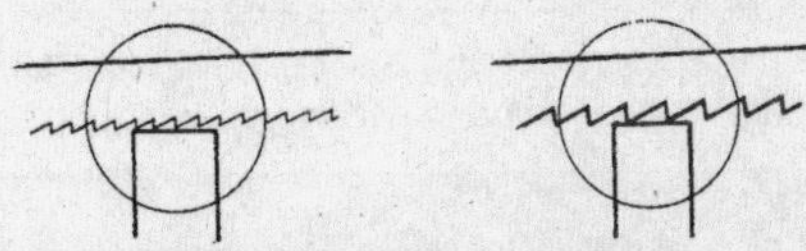

When cutting thin materials, check that at least three teeth are in contact with the workpiece at any time. Too coarse a blade will result in a poor cut and may break the blade. If you do not have the correct blade, cut at a shallow angle to the material

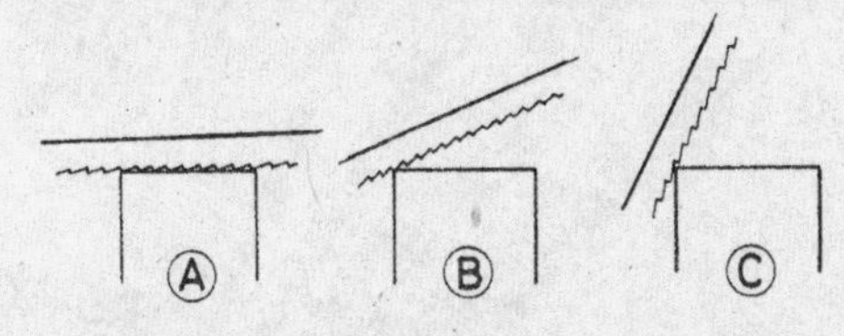

The correct cutting angle is important. If it is too shallow (A) the blade will wander. The angle shown at (B) is correct when starting the cut, and may be reduced slightly once under way. In (C) the angle is too steep and the blade will be inclined to jump out of the cut

1.34 Correct procedure for use of a hacksaw

1.35 Good quality hacksaw blades are marked like this

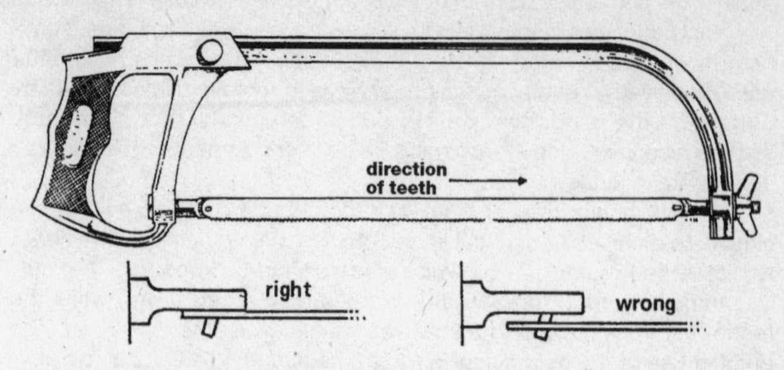

1.36 Correct installation of a hacksaw blade - the tooth must point away from the handle and butt against the locating lugs

this means a fine blade for cutting thin sheet materials, while a coarser blade can be used for faster cutting through thicker items such as bolts or bar stock. When cutting thin materials, angle the saw so the blade cuts at a shallow angle. More teeth are in contact and there's less chance of the blade binding and breaking, or teeth breaking.

When you buy blades, choose a reputable brand. Cheap, unbranded blades may be perfectly acceptable, but you can't tell by looking at them. Poor quality blades will be insufficiently hardened on the teeth edge and will dull quickly. Most reputable brands will be marked "Flexible High Speed Steel" or a similar term, to indicate the type of material used **(see illustration)**. It is possible to buy "unbreakable" blades (only the teeth are hardened, leaving the rest of the blade less brittle).

Sometimes, a full-size hacksaw is too big to allow access to a frozen nut or bolt. On most saws, you can overcome this problem by turning the blade 90-degrees. Occasionally you may have to position the saw around an obstacle and then install the blade on the other side of it. Where space is really restricted, you may have to use a handle that clamps onto a saw blade at one end. This allows access when a hacksaw frame would not work at all and has another advantage in that you can make use of broken off hacksaw blades instead of throwing them away. Note that because only one end of the blade is supported, and it's not held under tension, it's difficult to control and less efficient when cutting.

Before using a hacksaw, make sure the blade is suitable for the material being cut and installed correctly in the frame **(see illustration)**. Whatever it is you're cutting must be securely supported so it can't move around. The saw cuts on the forward stroke, so the teeth must point away from the handle. This might seem obvious, but it's easy to install the blade backwards by mistake and ruin the teeth on the first few strokes. Make sure the blade is tensioned adequately or it'll distort and chatter in the cut and may break. Wear safety glasses and be careful not to cut yourself on the saw blade or the sharp edge of the cut.

Files

Files **(see illustration)** come in a wide variety of sizes and types for specific jobs, but all of them are used for the same basic function of removing small amounts of metal in a controlled fashion. Files are used by mechanics mainly for deburring, marking parts, removing rust, filing the heads off rivets, restoring threads and fabricating small parts.

File shapes commonly available include flat, half-round, round, square and triangular. Each shape comes in a range of sizes (lengths) and cuts ranging from rough to smooth. The file face is covered with rows of diagonal ridges which form the cutting teeth. They may be aligned in one direction only (single cut) or in two directions to form a diamond-shaped pattern (double-cut) **(see illustration)**. The spacing of the teeth determines the file coarseness, again, ranging from rough to smooth in five basic grades: Rough, coarse, bastard, second-cut and smooth.

You'll want to build up a set of files by purchasing tools of the required shape and cut as they're needed. A good starting point would be flat, half-round, round and triangular files (at least one each - bastard or second-cut types). In addition, you'll have to buy one or more file handles (files are usually sold without handles, which are purchased separately and pushed over the tapered tang of the file when in use) **(see illustration)**. You may need to buy more than one size handle to fit the various files in your tool box, but don't attempt to get by without them. A file tang is fairly sharp and you almost certainly will end up stabbing yourself in the palm of the hand if you use a file with-

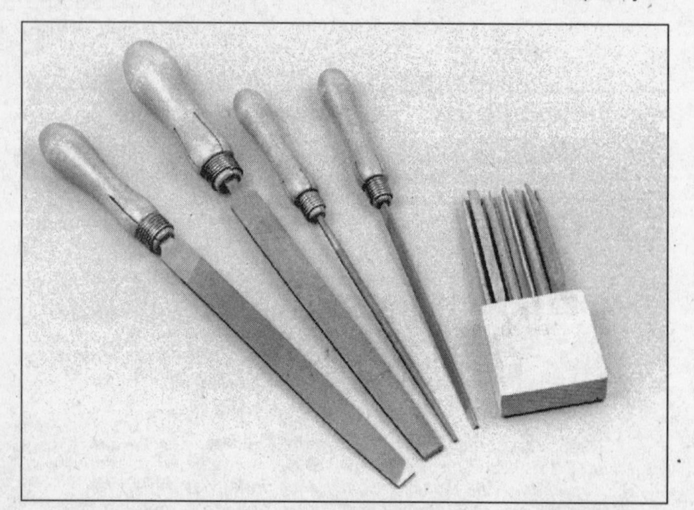

1.37 Get a good assortment of files - they're handy for deburring, marking parts, removing rust, filing the heads off rivets, restoring threads and fabricating small parts

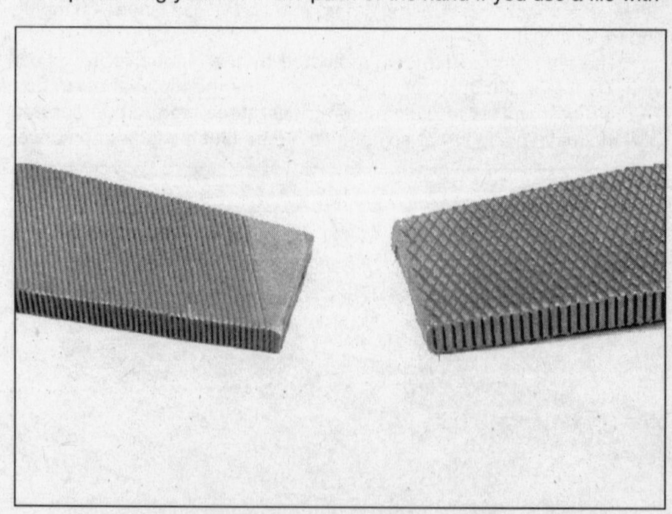

1.38 Files are either single-cut (left) or double-cut (right) - generally speaking, use a single-cut file to produce a very smooth surface; use a double-cut file to remove large amounts of material quickly

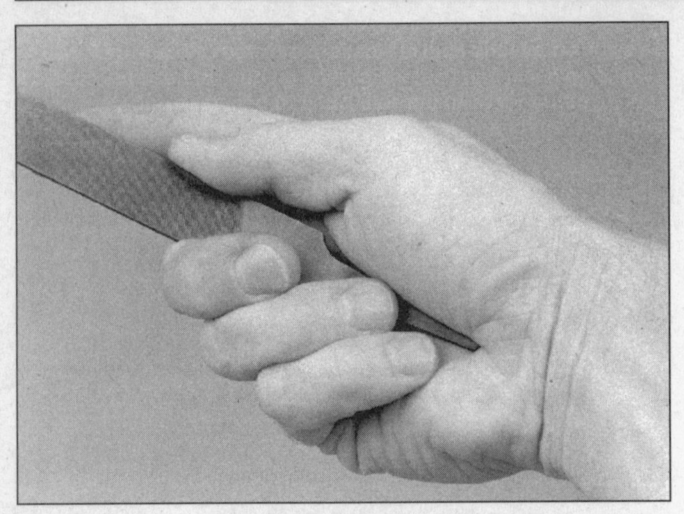
1.39 Never use a file without a handle - the tang is sharp and could puncture your hand

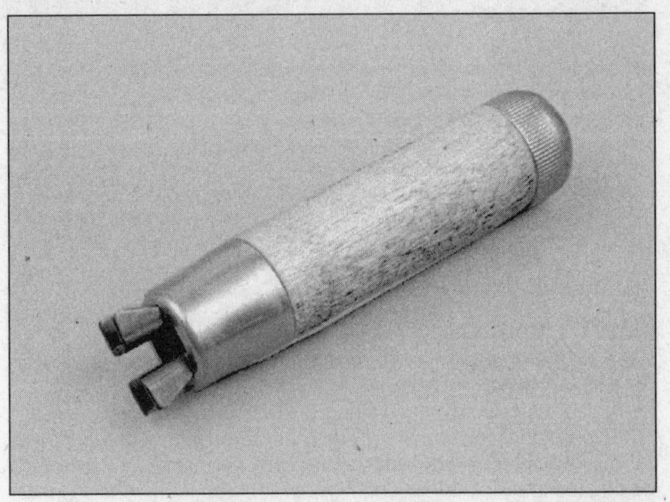
1.40 Adjustable handles that will work with many different size files are also available

out a handle and it catches in the workpiece during use. Adjustable handles are also available for use with files of various sizes, eliminating the need for several handles **(see illustration)**.

Exceptions to the need for a handle are fine swiss pattern files, which have a rounded handle instead of a tang. These small files are usually sold in sets with a number of different shapes. Originally intended for very fine work, they can be very useful for use in inaccessible areas. Swiss files are normally the best choice if piston ring ends require filing to obtain the correct end gap.

The correct procedure for using files is fairly easy to master. As with a hacksaw, the work should be clamped securely in a vise, if needed, to prevent it from moving around while being worked on. Hold the file by the handle, using your free hand at the file end to guide it and keep it flat in relation to the surface being filed. Use smooth cutting strokes and be careful not to rock the file as it passes over the surface. Also, don't slide it diagonally across the surface or the teeth will make grooves in the workpiece. Don't drag a file back across the workpiece at the end of the stroke - lift it slightly and pull it back to prevent damage to the teeth.

Files don't require maintenance in the usual sense, but they should be kept clean and free of metal filings. Steel is a reasonably easy material to work with, but softer metals like aluminum tend to clog the file teeth very quickly, which will result in scratches in the workpiece. This can be avoided by rubbing the file face with chalk before using it. General cleaning is carried out with a file card or a fine wire brush. If kept clean, files will last a long time - when they do eventually dull, they must be replaced; there is no satisfactory way of sharpening a worn file.

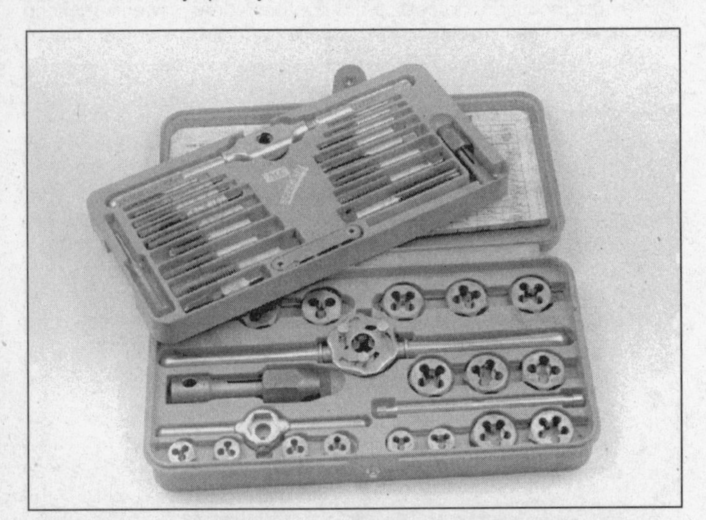
1.41 Tap and dies sets are available in inch and metric sizes - taps are used for cutting internal threads and cleaning and restoring damaged threads; dies are used for cutting, cleaning and restoring external threads

Taps and dies

Taps

Tap and die sets **(see illustration)** are available in inch and metric sizes. Taps are used to cut internal threads and clean or restore damaged threads. A tap consists of a fluted shank with a drive square at one end. It's threaded along part of its length - the cutting edges are formed where the flutes intersect the threads **(see illustration)**. Taps are made from hardened steel so they will cut threads in materials softer than what they're made of.

Taps come in three different types: Taper, plug and bottoming. The only real difference is the length of the chamfer on the cutting end of the tap. Taper taps are chamfered for the first 6 or 8 threads, which makes them easy to start but prevents them from cutting threads close to the bottom of a hole. Plug taps are chamfered up about 3 to 5 threads, which makes them a good all around tap because they're relatively easy to start and will cut nearly to the bottom of a hole. Bottoming taps, as the name implies, have a very short chamfer (1-1/2 to 3 threads) and will cut as close to the bottom of a blind hole as practical. However, to do this, the threads should be started with a plug or taper tap.

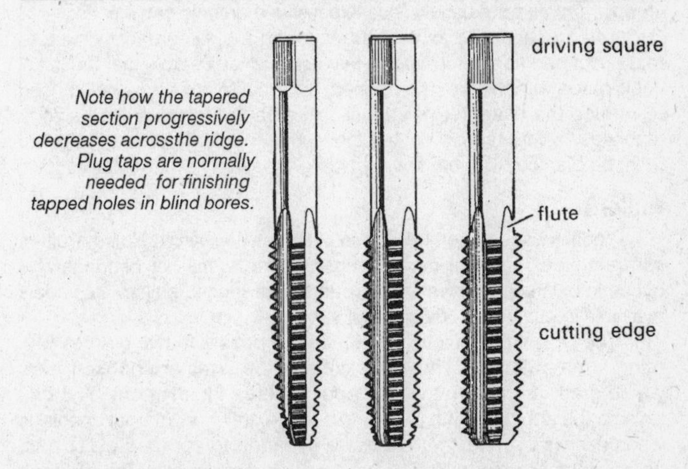

1.42 Taper, plug and bottoming taps (left-to-right)

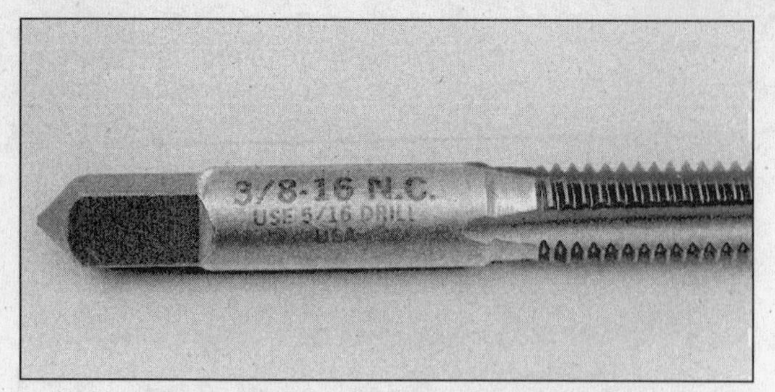

1.43 If you need to drill and tap a hole, the drill bit size to use for a given bolt (top) size is marked on the tap

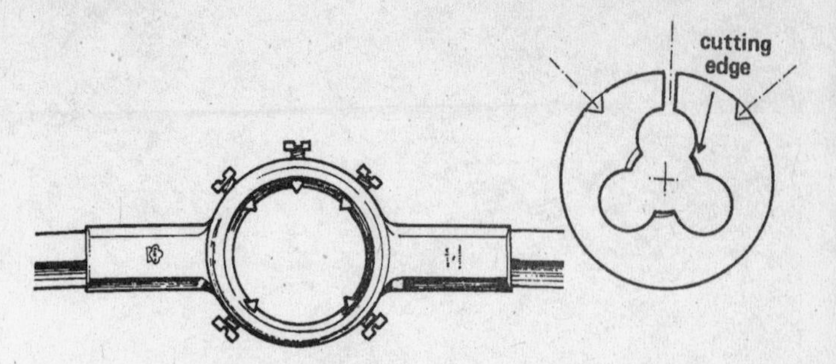

1.44 A die (right) is used for cutting external threads (this one is a split-type/adjustable die) and is held in a tool called a die stock (left)

Although cheap tap and die sets are available, the quality is usually very low and they can actually do more harm than good when used on threaded holes in aluminum engines. The alternative is to buy high-quality taps if and when you need them, even though they aren't cheap, especially if you need to buy two or more thread pitches in a given size. Despite this, it's the best option - you'll probably only need taps on rare occasions, so a full set isn't absolutely necessary.

Taps are normally used by hand (they can be used in machine tools, but not when doing engine repairs). The square drive end of the tap is held in a tap wrench (an adjustable T-handle). For smaller sizes, a T-handled chuck can be used. The tapping process starts by drilling a hole of the correct diameter. For each tap size, there's a corresponding twist drill that will produce a hole of the correct size. This is important; too large a hole will leave the finished thread with the tops missing, producing a weak and unreliable grip. Conversely, too small a hole will place excessive loads on the hard and brittle shank of the tap, which can break it off in the hole. Removing a broken off tap from a hole is no fun! The correct tap drill size is normally marked on the tap itself or the container it comes in **(see illustration)**.

Dies

Dies are used to cut, clean or restore external threads. Most dies are made from a hex-shaped or cylindrical piece of hardened steel with a threaded hole in the center. The threaded hole is overlapped by three or four cutouts, which equate to the flutes on taps and allow metal waste to escape during the threading process. Dies are held in a T-handled holder (called a die stock) **(see illustration)**. Some dies are split at one point, allowing them to be adjusted slightly (opened and closed) for fine control of thread clearances.

Dies aren't needed as often as taps, for the simple reason it's normally easier to install a new bolt than to salvage one. However, it's often helpful to be able to extend the threads of a bolt or clean up damaged threads with a die. Hex-shaped dies are particularly useful for mechanic's work, since they can be turned with a wrench **(see illustration)** and are usually less expensive than adjustable ones.

The procedure for cutting threads with a die is broadly similar to that described above for taps. When using an adjustable die, the initial cut is made with the die fully opened, the adjustment screw being used to reduce the diameter of successive cuts until the finished size is reached. As with taps, a cutting lubricant should be used, and the die must be backed off every few turns to clear swarf from the cutouts.

Pullers

You'll need a general-purpose puller for engine rebuilding. Pullers can removed seized or corroded parts, bad bushes or bearings and dynamic balancers. Universal two- and three-legged pullers are widely available in numerous designs and sizes.

The typical puller consists of a central boss with two or three pivoting arms attached. The outer ends of the arms are hooked jaws which grab the part you want to pull off **(see illustration)**. You can reverse the arms on most pullers to use the puller on internal openings when necessary. The central boss is threaded to accept a puller bolt, which does the work. You can also get hydraulic versions of these tools which are capable of more pressure, but they're expensive.

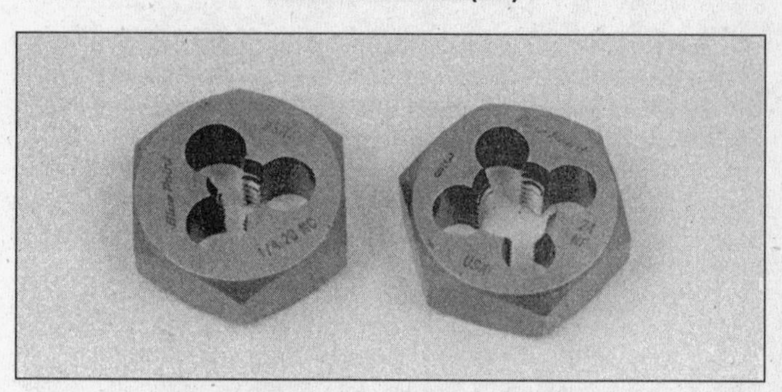

1.45 Hex-shaped dies are especially handy for mechanic's work because they can be turned with a wrench

You can adapt pullers by purchasing, or fabricating, special jaws for specific jobs. If you decide to make your own jaws, keep in mind that the pulling force should be concentrated as close to the center of the component as possible to avoid damaging it.

Before you use a puller, assemble it and check it to make sure it doesn't snag on anything and the loads on the part to be removed are distributed evenly. If you're dealing with a part held on the shaft by a nut, loosen the nut but don't remove it. Leaving the nut on helps prevent distortion of the shaft end under pressure from the puller bolt and stops the part from flying off the shaft when it comes loose.

Tighten a puller gradually until the assembly is under moderate pressure, then try to jar the component loose by striking the puller bolt a few times with a hammer. If this doesn't work, tighten the bolt a little further and repeat the process. If this approach doesn't work, stop and reconsider. At some point you must make a decision whether to con-

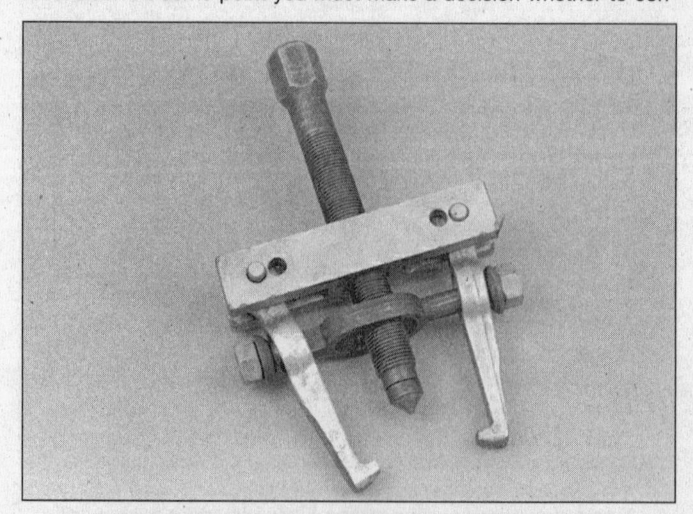

1.46 A two or three-jaw puller will come in handy for many tasks in the shop and can also be used for working on other types of equipment

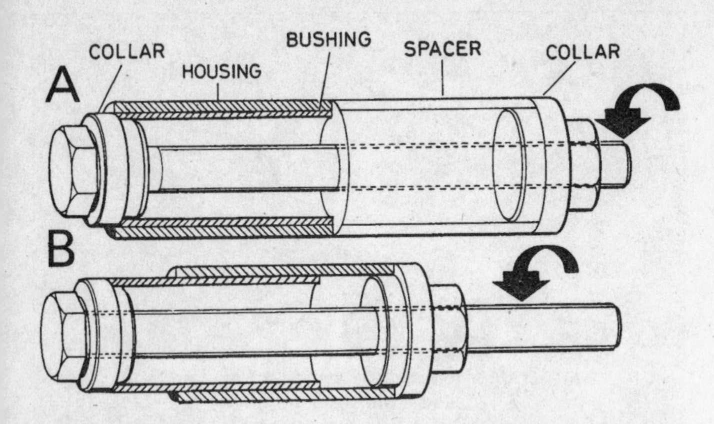

1.47 **Typical drawbolt uses - in (A), the nut is tightened to pull the collar and bushing into the large spacer; in (B), the spacer is left out and the drawbolt is repositioned to install the now bushing**

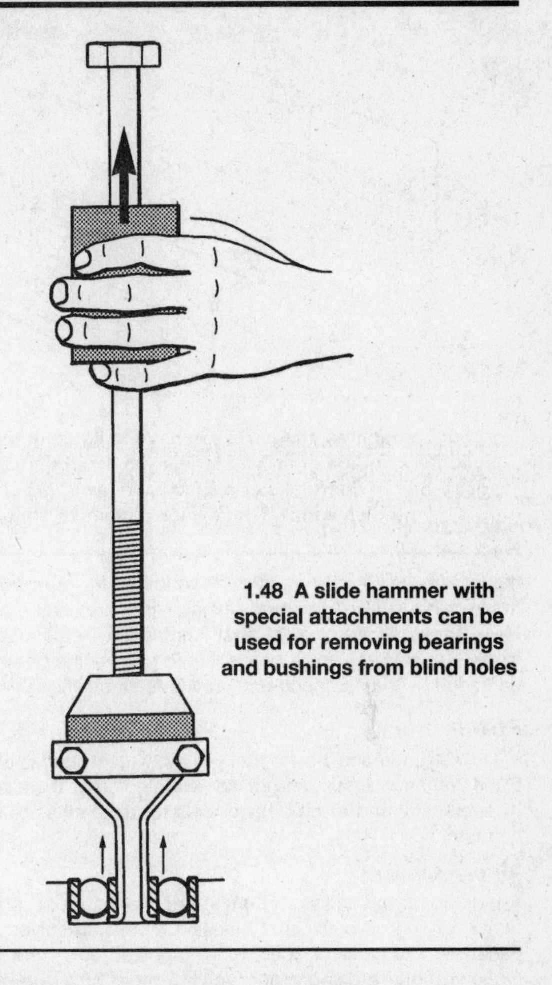

1.48 A slide hammer with special attachments can be used for removing bearings and bushings from blind holes

tinue applying pressure in this manner. Sometimes, you can apply penetrating oil around the joint and leave it overnight, with the puller in place and tightened securely. By the next day, the taper has separated and the problem has resolved itself.

If nothing else works, try heating the area surrounding the troublesome part with a propane or gas welding torch (We don't, however, recommend messing around with welding equipment if you're not already experienced in its use). Apply the heat to the hub area of the component you wish to remove. Keep the flame moving to avoid uneven heating and the risk of distortion. Keep pressure applied with the puller and make sure that you're able to deal with the resulting hot component and the puller jaws if it does come free. Be very careful to keep the flame away from aluminum parts.

If all reasonable attempts to remove a part fail, don't be afraid to give up. It's cheaper to quit now than to repair a badly damaged engine. Either buy or borrow the correct tool, or take the engine to a dealer and ask him to remove the part for you.

Drawbolt extractors

The simple drawbolt extractor is easy to make up and invaluable in every workshop. There are no commercially available tools of this type; you simply make a tool to suit a particular application. You can use a drawbolt extractor to pull out stubborn piston pins and to remove bearings and bushings.

To make a drawbolt extractor, you'll need an assortment of threaded rods in various sizes (available at hardware stores), and nuts to fit them. You'll also need assorted washers, spacers and tubing. For things like piston pins, you'll usually need a longer piece of tube.

Some typical drawbolt uses are shown in the accompanying line drawings **(see illustration)**. They also reveal the order of assembly of the various pieces. The same arrangement, minus the tubular spacer section, can usually be used to install a new bushing or piston pin. Using the tool is quite simple. Just make sure you get the bush or pin square to the bore when you install it. Lubricate the part being pressed into place, where appropriate.

Pullers for use in blind bores

Bushings or bearings installed in "blind holes" often require special pullers. Some bearings can be removed without a puller if you heat the engine or component evenly in an oven and tap it face down on a clean wood- den surface to dislodge the bearing. Wear heavy gloves to protect yourself when handling the heated components. If you need a puller to do the job, get a slide-hammer with interchangeable tips. Slide hammers range from universal two or three-jaw puller arrangements to special bearing pullers. Bearing pullers are hardened steel tubes with a flange around the bottom edge. The tube is split at several places, which allows a wedge to expand the tool once it's in place. The tool fits inside the bearing inner race and is tightened so the flange or lip is locked under the edge of the race.

The slide-hammer consists of a steel shaft with a stop at its upper end. The shaft carries a sliding weight which slides along the shaft until it strikes the stop. This allows the tool holding the bearing to drive it out of the bore **(see illustration)**. A bearing puller set is an expensive and infrequently-used piece of equipment, so take the engine to a dealer and have the bearings/bushings replaced.

Bench vise

The bench vise **(see illustration)** is an essential tool in a shop. Buy the best quality vise you can afford. A good vise is expensive, but

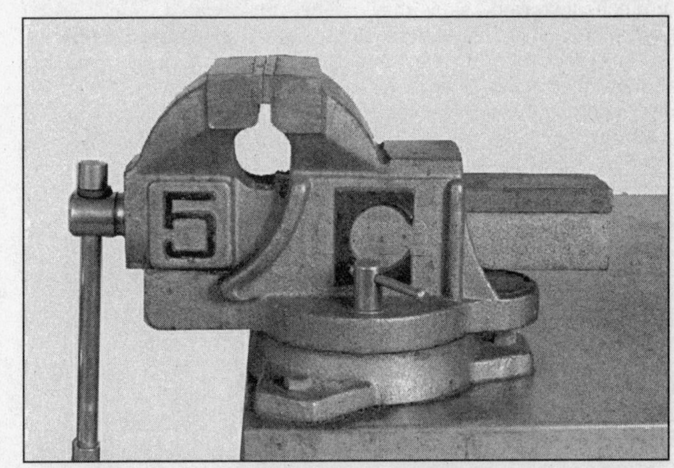

1.49 A bench vise is one of the most useful pieces of equipment you can have in the shop - bigger is usually better with vises, so get a vise with jaws that open at least four inches

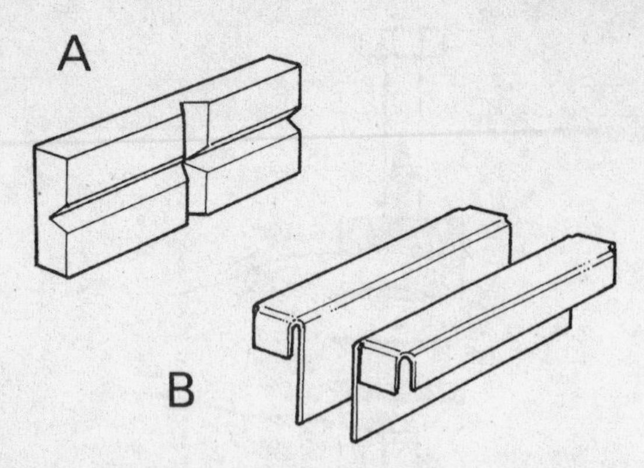

1.50 Sometimes, the parts you have to jig up in the vise are delicate, or made of soft materials - to avoid damaging them, got a pair of fiberglass or plastic "soft laws" (A) or fabricate your own with 1/8-inch thick aluminum sheet (B)

the quality of its materials and workmanship are worth the extra money. Size is also important - bigger vises are usually more versatile. Make sure the jaws open at least four inches. Get a set of soft jaws to fit the vise as well - you'll need them to grip engine parts that could be damaged by the hardened vise jaws **(see illustration).**

Power tools

Really, the only power tool you absolutely need is an electric drill. But if you have an air compressor and electricity, there's a wide range of pneumatic and electric hand tools to make all sorts of jobs easier and faster.

Air compressor

An air compressor **(see illustration)** makes most jobs easier and faster. Drying off parts after cleaning them with solvent, blowing out passages in a block or head, running power tools - the list is endless. Once you buy a compressor, you'll wonder how you ever got along without it. Air tools really speed up tedious procedures like removing and installing cylinder head bolts, crankshaft main bearing bolts or vibration damper (crankshaft pulley) bolts.

Bench-mounted grinder

A bench grinder **(see illustration)** is also handy. With a wire wheel on one end and a grinding wheel on the other, it's great for cleaning up fasteners, sharpening tools and removing rust. Make sure the grinder is fastened securely to the bench or stand, always wear eye protection when operating it and never grind aluminum parts on the grinding wheel.

1.52 Another indispensable piece of equipment is the bench grinder (with a wire wheel mounted on one arbor) - make sure it's securely bolted down and never use it with the rests or eye shields removed

1.51 Although it's not absolutely necessary, an air compressor can make many jobs easier and produce better results, especially when air powered tools are available to use with it

Electric drills

Countersinking bolt holes, enlarging oil passages, honing cylinder bores, removing rusted or broken off fasteners, enlarging holes and fabricating small parts - electric drills **(see illustration)** are indispensable for engine work. A 3/8-inch chuck (drill bit holder) will handle most jobs. Collect several different wire brushes to use in the drill and make sure you have a complete set of sharp *metal* drill bits **(see illustration).** Cordless drills are extremely versatile because they don't force you to work near an outlet. They're also handy to have around for a variety of non-mechanical jobs.

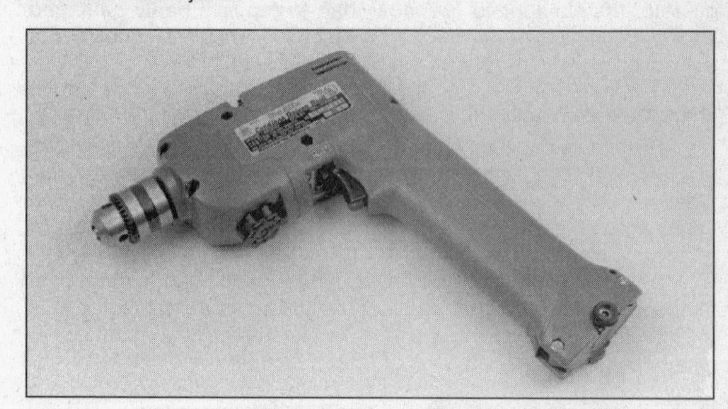

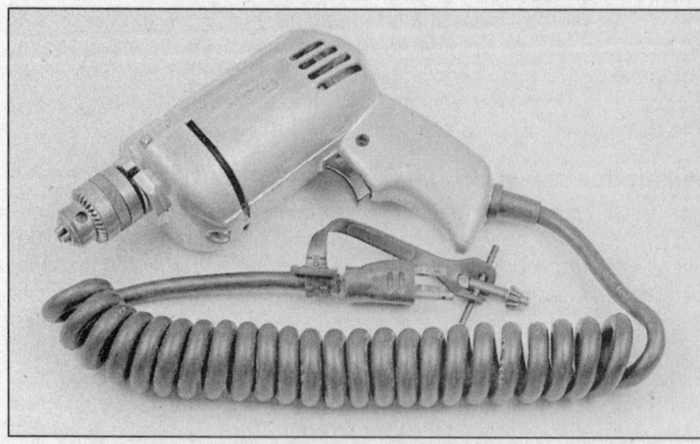

1.53 Electric drills can be cordless (above) or 115-volt, AC-powered (below)

1.54 Get a set of good quality drill bits for drilling' holes and wire brushes of various sizes for cleaning up metal parts - make sure the bits are designed for drilling in metal

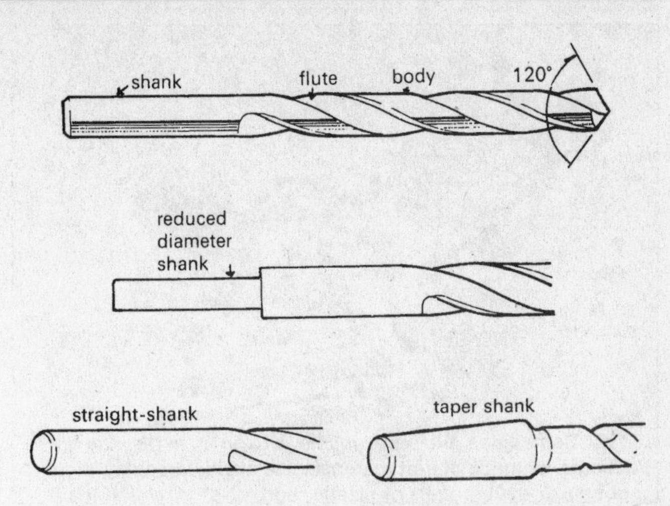

1.55 A typical drill bit (top), a reduced shank bit (center), and a tapered shank bit (bottom right)

Twist drills and drilling equipment

Drilling operations are done with twist drills, either in a hand drill or a drill press. Twist drills (or drill bits, as they're often called) consist of a round shank with spiral flutes formed into the upper two-thirds to clear the waste produced while drilling, keep the drill centered in the hole and finish the sides of the hole.

The lower portion of the shank is left plain and used to hold the drill in the chuck. In this section, we will discuss only normal parallel shank drills **(see illustration)**. There is another type of bit with the plain end formed into a special size taper designed to fit directly into a corresponding socket in a heavy-duty drill press. These drills are known as Morse Taper drills and are used primarily in machine shops.

At the cutting end of the drill, two edges are ground to form a conical point. They're generally angled at about 60-degrees from the drill axis, but they can be reground to other angles for specific applications. For general use the standard angle is correct - this is how the drills are supplied.

When buying drills, purchase a good-quality set (sizes 1/16 to 3/8-inch). Make sure the drills are marked "High Speed Steel" or "HSS". This indicates they're hard enough to withstand continual use in metal; many cheaper, unmarked drills are suitable only for use in wood or other soft materials. Buying a set ensures the right size bit will be available when it's needed.

Twist drill sizes

Twist drills are available in a vast array of sizes, most of which you'll never need. There are three basic drill sizing systems: Fractional, number and letter **(see illustration)** (we won't get involved with the fourth system, which is metric sizes).

Fractional sizes start at 1/64-inch and increase in increments of 1/64-inch. Number drills range in descending order from 80 (0.0135-inch), the smallest, to 1 (0.2280-inch), the largest. Letter sizes start with A (0.234-inch), the smallest, and go through Z (0.413-inch), the largest.

This bewildering range of sizes means it's possible to drill an accurate hole of almost any size within reason. In practice, you'll be limited by the size of chuck on your drill (normally 3/8 or 1/2-inch). In addition, very few stores stock the entire range of possible sizes, so you'll have to shop around for the nearest available size to the one you require.

Sharpening twist drills

Like any tool with a cutting edge, twist drills will eventually get dull **(see illustration)**. How often they'll need sharpening depends to some extent on whether they're used correctly. A dull twist drill will soon make itself known. A good indication of the condition of the cutting edges is to watch the waste emerging from the hole being drilled. If the

1.56 Drill bits in the range most commonly used are available in fractional sizes (left) and number sizes (right) so almost any size hole can be drilled

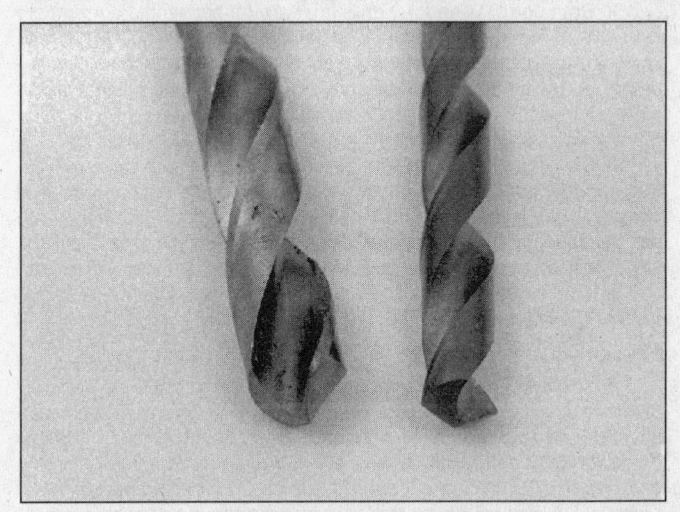

1.57 If a bit gets dull (left), discard it or resharpen it so it looks like the bit on the right

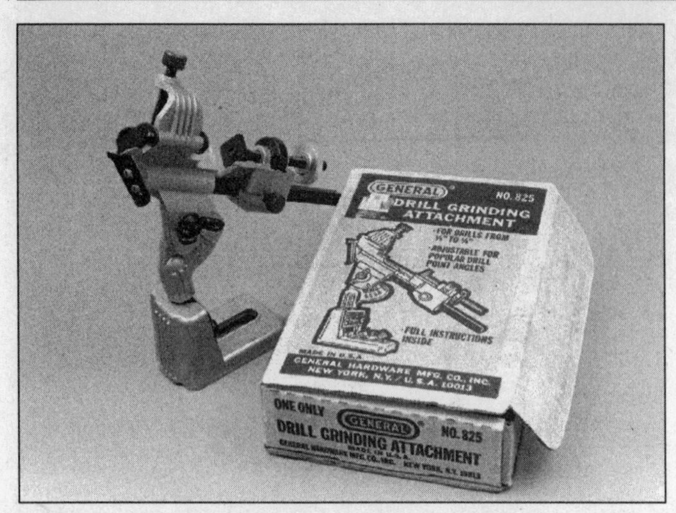

1.58 Inexpensive drill bit sharpening jigs designed to be used with a bench grinder are widely available - even If you only use it to resharpen drill bits, it'll pay for itself quickly

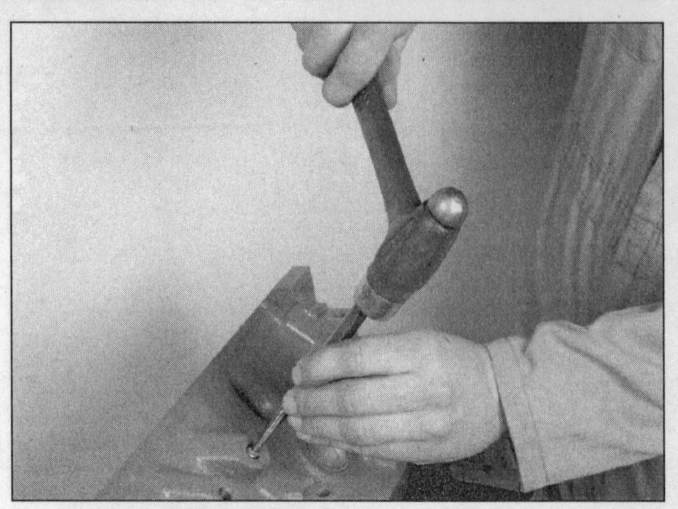

1.59 Before you drill a hole, use a centerpunch to make an indentation for the drill bit so it won't wander

tip is in good condition, two even spirals of waste metal will be produced; if this fails to happen or the tip gets hot, it's safe to assume that sharpening is required.

With smaller size drills - under about 1/8-inch - it's easier and more economical to throw the worn drill away and buy another one. With larger (more expensive) sizes, sharpening is a better bet. When sharpening twist drills, the included angle of the cutting edge must be maintained at the original 120-degrees and the small chisel edge at the tip must be retained. With some practice, sharpening can be done freehand on a bench grinder, but it should be noted that it's very easy to make mistakes. For most home mechanics, a sharpening jig that mounts next to the grinding wheel should be used so the drill is clamped at the correct angle **(see illustration)**.

Drilling equipment

Tools to hold and turn drill bits range from simple, inexpensive hand-operated or electric drills to sophisticated and expensive drill presses. Ideally, all drilling should be done on a drill press with the workpiece clamped solidly in a vise. These machines are expensive and take up a lot of bench or floor space, so they're out of the question for many do-it-yourselfers. An additional problem is the fact that many of the drilling jobs you end up doing will be on the engine itself or the equipment it's mounted on, in which case the tool has to be taken to the work.

The best tool for the home shop is an electric drill with a 3/8-inch chuck. Both cordless and AC drills (that run off household current) are available. If you're purchasing one for the first time, look for a well-known, reputable brand name and variable speed as minimum requirements. A 1/4-inch chuck, single-speed drill will work, but it's worth paying a little more for the larger, variable speed type.

All drills require a key to lock the bit in the chuck. When removing or installing a bit, make sure the cord is unplugged to avoid accidents. Initially, tighten the chuck by hand, checking to see if the bit is centered correctly. This is especially important when using small drill bits which can get caught between the jaws. Once the chuck is hand tight, use the key to tighten it securely - remember to remove the key afterwards!

Drilling and finishing holes

Preparation for drilling

If possible, make sure the part you intend to drill in is securely clamped in a vise. If it's impossible to get the work to a vise, make sure it's stable and secure. Twist drills often dig in during drilling - this can be dangerous, particularly if the work suddenly starts spinning on the end of the drill. Obviously, there's little chance of a complete engine or piece of equipment doing this, but you should make sure it's supported securely.

Start by locating the center of the hole you're drilling. Use a center punch to make an indentation for the drill bit so it won't wander. If you're drilling out a broken-off bolt, be sure to position the punch in the exact center of the bolt **(see illustration)**.

If you're drilling a large hole (above 1/4-inch), you may want to make a pilot hole first. As the name suggests, it will guide the larger drill bit and minimize drill bit wandering. Before actually drilling a hole, make sure the area immediately behind the bit is clear of anything you don't want drilled.

Drilling

When drilling steel, especially with smaller bits, no lubrication is needed. If a large bit is involved, oil can be used to ensure a clean cut and prevent overheating of the drill tip. When drilling aluminum, which tends to smear around the cutting edges and clog the drill bit flutes, use kerosene as a lubricant.

Wear safety goggles or a face shield and assume a comfortable, stable stance so you can control the pressure on the drill easily. Position the drill tip in the punch mark and make sure, if you're drilling by hand, the bit is perpendicular to the surface of the workpiece. Start drilling without applying much pressure until you're sure the hole is positioned correctly. If the hole starts off center, it can be very difficult to correct. You can try angling the bit slightly so the hole center moves in the opposite direction, but this must be done before the flutes of the bit have entered the hole. It's at the starting point that a variable-speed drill is invaluable; the low speed allows fine adjustments to be made before it's too late. Continue drilling until the desired hole depth is reached or until the drill tip emerges at the other side of the workpiece.

Cutting speed and pressure are important - as a general rule, the larger the diameter of the drill bit, the slower the drilling speed should be. With a single-speed drill, there's little that can be done to control it, but two-speed or variable speed drills can be controlled. If the drilling speed is too high, the cutting edges of the bit will tend to overheat and dull. Pressure should be varied during drilling. Start with light pressure until the drill tip has located properly in the work. Gradually increase pressure so the bit cuts evenly. If the tip is sharp and the pressure correct, two distinct spirals of metal will emerge from the bit flutes. If the pressure is too light, the bit won't cut properly, while excessive pressure will overheat the tip.

Decrease pressure as the bit breaks through the workpiece. If this isn't done, the bit may jam in the hole; if you're using a hand-held drill, it could be jerked out of your hands, especially when using larger size bits.

Once a pilot hole has been made, install the larger bit in the chuck and enlarge the hole. The second bit will follow the pilot hole - there's no need to attempt to guide it (if you do, the bit may break off). It is important, however, to hold the drill at the correct angle.

After the hole has been drilled to the correct size, remove the burrs left around the edges of the hole. This can be done with a small

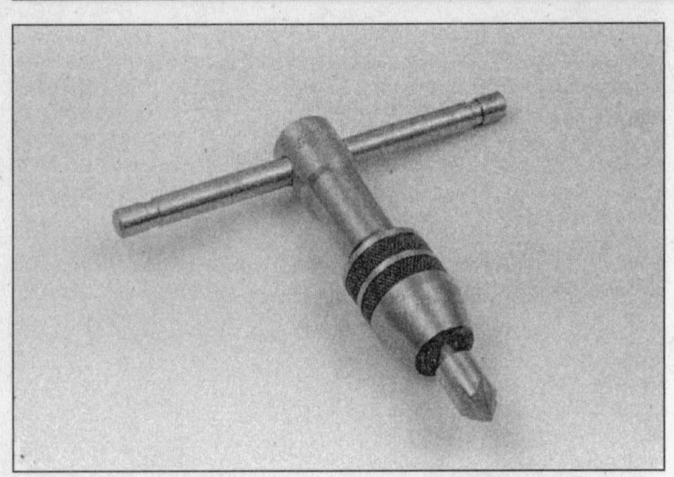

1.60 Use a large drill bit or a countersink mounted in a tap wrench to remove burrs from a hole after drilling or enlarging it

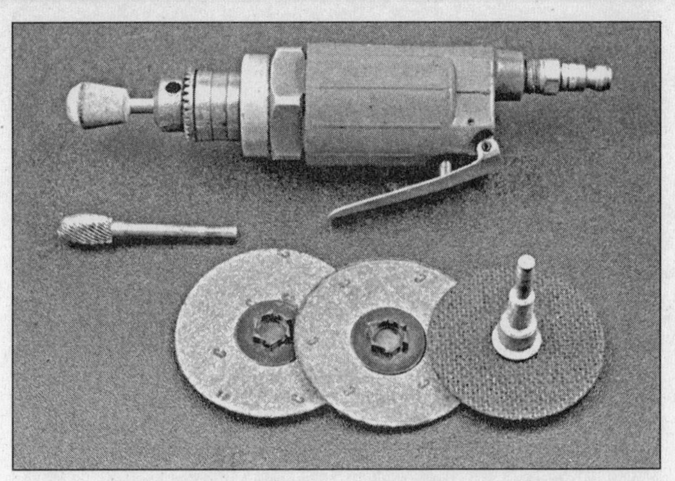

1.61 A good die grinder will deburr blocks, radius piston domes, chamfer oil holes and do a lot of other little jobs that would be tedious if done manually

round file, or by chamfering the opening with a larger bit or a countersink **(see illustration)**. Use a drill bit that's several sizes larger than the hole and simply twist it around each opening by hand until any rough edges are removed.

Enlarging and reshaping holes

The biggest practical size for bits used in a hand drill is about 1/2-inch. This is partly determined by the capacity of the chuck (although it's possible to buy larger drills with stepped shanks). The real limit is the difficulty of controlling large bits by hand; drills over 1/2-inch tend to be too much to handle in anything other than a drill press. If you have to make a larger hole, or if a shape other than round is involved, different techniques are required.

If a hole simply must be enlarged slightly, a round file is probably the best tool to use. If the hole must be very large, a hole saw will be needed, but they can only be used in sheet metal.

Large or irregular-shaped holes can also be made in sheet metal and other thin materials by drilling a series of small holes very close together. In this case the desired hole size and shape must be marked with a scribe. The next step depends on the size bit to be used; the idea is to drill a series of almost touching holes just inside the outline of the large hole. Center punch each location, then drill the small holes. A cold chisel can then be used to knock out the waste material at the center of the hole, which can then be filed to size. This is a time consuming process, but it's the only practical approach for the home shop. Success is dependent on accuracy when marking the hole shape and using the center punch.

1.62 Buy at least one fire extinguisher before you open shop - make sure it's rated for flammable liquid fires and KNOW HOW TO USE IT!

High-speed grinders

A good die grinder **(see illustration)** will deburr blocks, radius piston domes and chamfer oil holes ten times as fast as you can do any of these jobs by hand.

Safety items that should be in every shop

Fire extinguishers

Buy at least one fire extinguisher **(see illustration)** before doing any maintenance or repair procedures. Make sure it's rated for flammable liquid fires. Familiarize yourself with its use as soon as you buy it - don't wait until you need it to figure out how to use it. And be sure to have it checked and recharged at regular intervals. Refer to the safety tips at the end of this chapter for more information about the hazards of gasoline and other flammable liquids.

Gloves

If you're handling hot parts or metal parts with sharp edges, wear a pair of industrial work gloves to protect yourself from burns, cuts and splinters **(see illustration)**. Wear a pair of heavy duty rubber gloves (to protect your hands when you wash parts in solvent).

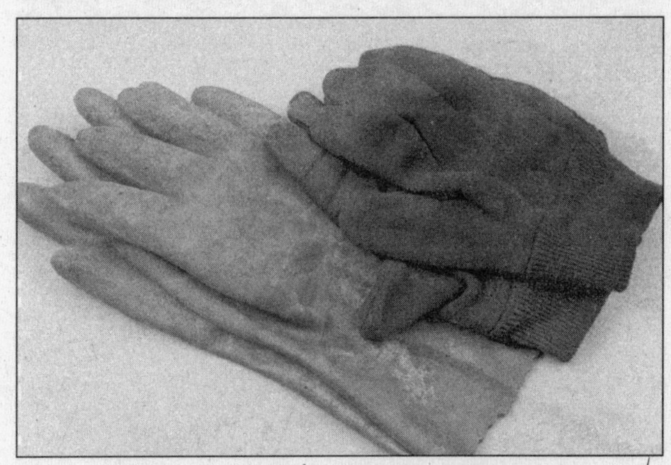

1.63 Get a pair of heavy work gloves for handling hot or sharp-edged objects and a pair of rubber gloves for washing parts with solvent

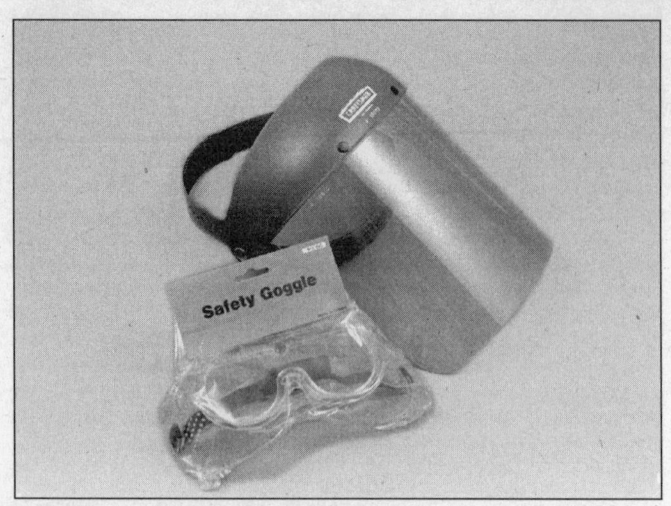

1.64 One of the most important items you'll need in the shop is a face shield or safety goggles, especially when you're hitting metal parts with a hammer, washing parts in solvent or grinding something on the bench grinder

1.65 The combustion leak block tester tests for cracked blocks, leaky gaskets, cracked heads and warped heads by detecting combustion gases in the coolant

Safety glasses or goggles

Never work on a bench or high-speed grinder without safety glasses **(see illustration)**. Don't take a chance on getting a metal sliver in your eye. It's also a good idea to wear safety glasses when you're washing parts in solvent.

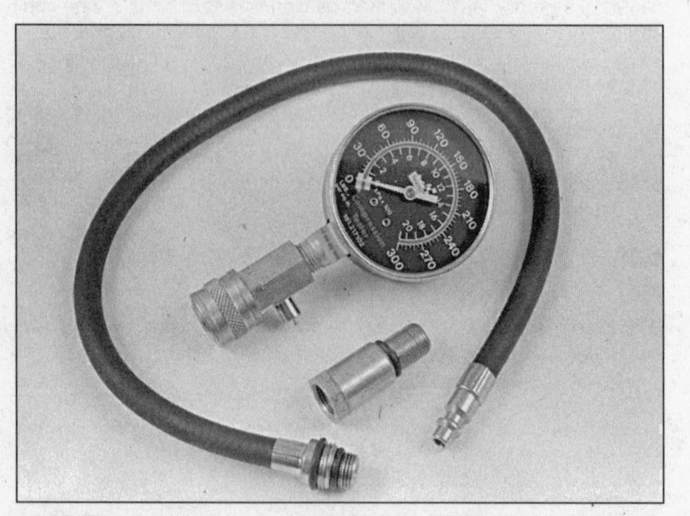

1.66 The compression gauge indicates cylinder pressure in the combustion chamber - you'll need to get one made for diesels (because of their higher compression)

Special diagnostic tools

These tools do special diagnostic tasks. They're indispensable for determining the condition of your engine. If you don't think you'll use them frequently enough to justify the investment, try to borrow or rent them. Or split the cost with a friend who also wants to get into engine rebuilding. We've listed only those tools and instruments generally available to the public, not the special tools manufactured by Ford for its dealer service departments. Occasional references to Ford special tools may be included in the text of this manual. But we'll try to provide you with another way of doing the job without the special tool, if possible. However, when there's no alternative, you'll have to borrow or buy the tool or have the job done by a professional.

Combustion leak block tester

The combustion leak block tester **(see illustration)** tests for cracked blocks, leaky gaskets, cracked heads and warped heads by detecting combustion gasses in the coolant. For information on using this tool, see Chapter 3.

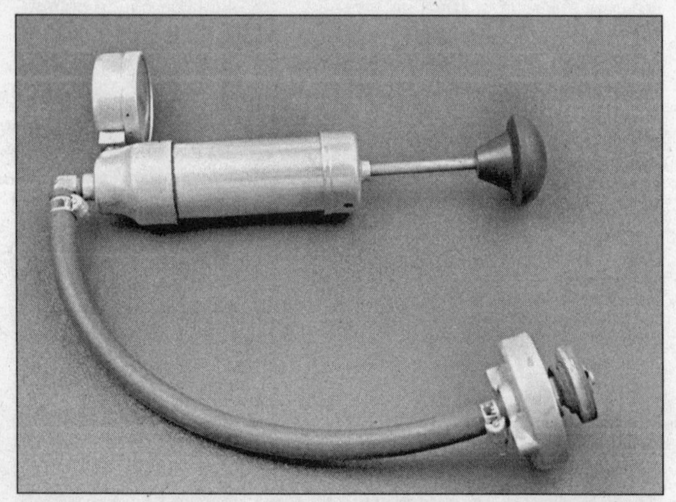

1.67 The cooling system pressure tester checks for leaks in the cooling system by pressurizing the system and measuring the rate at which it leaks down

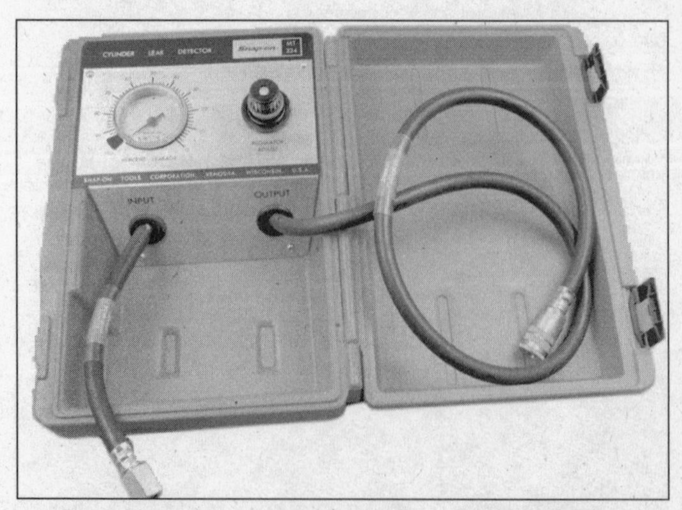

1.68 The leakdown tester indicates the rate at which pressure leaks past the piston rings, the valves and/or the head gasket

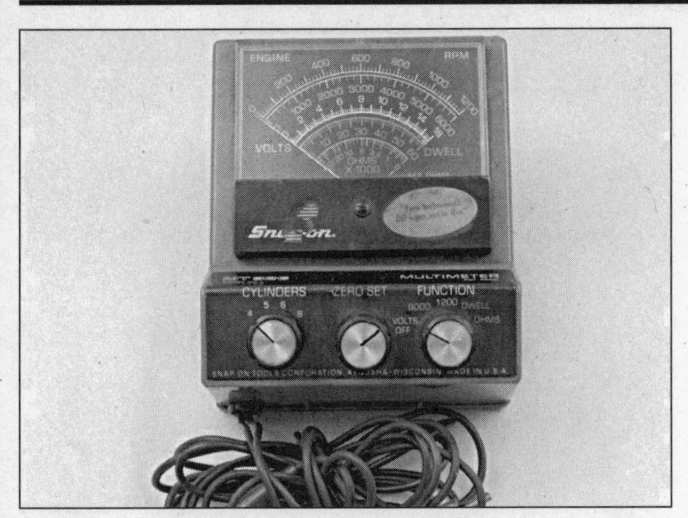

1.69 The multimeter combines the functions of a volt-meter, ammeter and ohmmeter into one unit - it can measure voltage, amperage or resistance in an electrical circuit

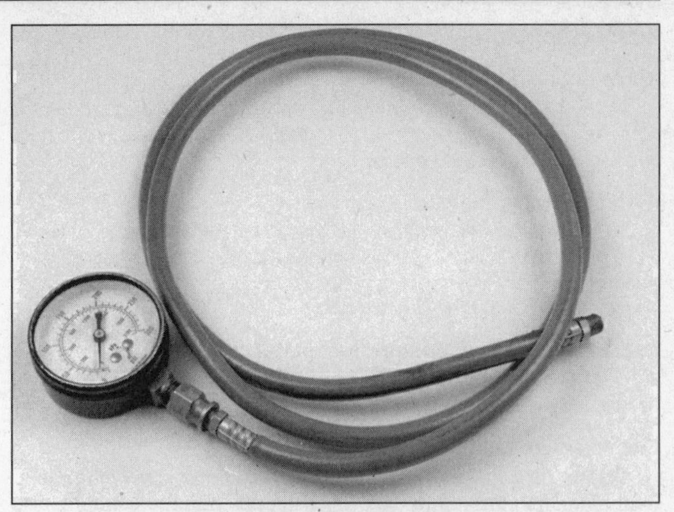

1.70 The oil pressure gauge, which screws into an oil pressure sending unit hole in the block or head, measures the oil pressure in the lubrication system

Compression gauge

The compression gauge **(see illustration)** indicates cylinder pressure in the combustion chamber. The adapter (depending on engine) screws into the glow plug hole. Because diesel engines have very high compression, a special diesel compression gauge is required.

Cooling system pressure tester

The cooling system pressure tester **(see illustration)** checks for leaks in the cooling system.

Leakdown tester

The leakdown tester **(see illustration)** indicates the rate at which pressure leaks past the piston rings, the valves and/or the head gasket, in a combustion chamber.

Multimeter

The multimeter **(see illustration)** combines the functions of a voltmeter, ammeter and ohmmeter into one unit. It can measure voltage, amperage or resistance in an electrical circuit.

Oil pressure gauge

The oil pressure gauge **(see illustration)** indicates the oil pressure in the lubrication system.

Stethoscope

The stethoscope **(see illustration)** amplifies engine sounds, allowing you to pinpoint possible sources of pending trouble such as bad bearings, excessive play in the crank, rod knock, etc.

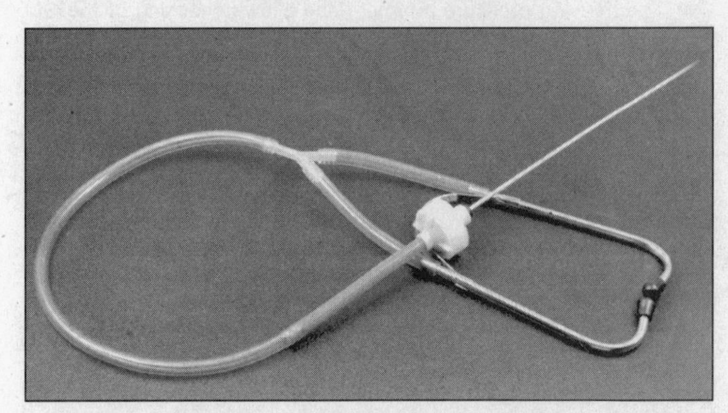

1.71 The stethoscope amplifies engine sounds, allowing you to pinpoint possible sources of trouble such as bad bearings, excessive play in the crank, rod knocks, etc.

Vacuum gauge

The vacuum gauge **(see illustration)** indicates intake manifold vacuum, in inches of mercury (in-Hg).

Vacuum/pressure pump

The hand-operated vacuum/pressure pump **(see illustration)** can create a vacuum, or build up pressure, in a circuit to check components that are vacuum or pressure operated.

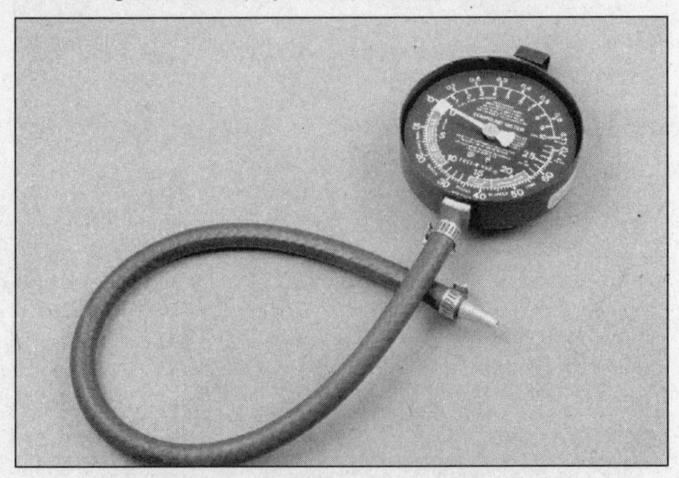

1.72 The vacuum gauge indicates intake manifold vacuum, in inches of mercury (in-Hg)

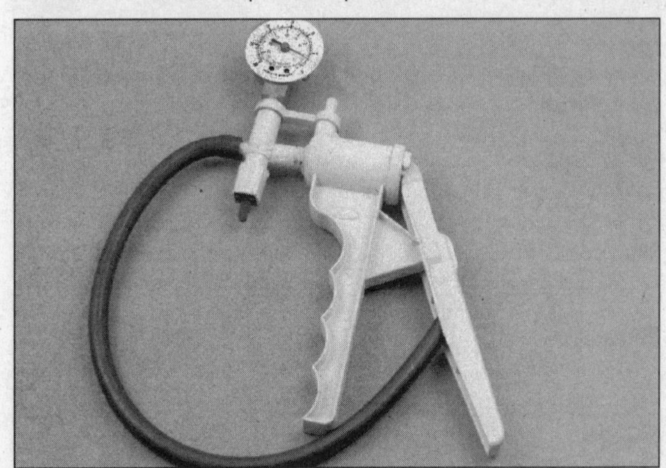

1.73 The vacuum/pressure pump can create a vacuum in a circuit, or pressurize it, to simulate the actual operating conditions

Fuel system tools

Angle gauge and adapter

When checking and adjusting the GM 5.7L vacuum regulator valve, an angle gauge and mounting adapter are required in conjunction with a hand vacuum pump.

Dynamic timing meter

The dynamic timing meter is used to adjust the timing on a diesel while the engine is running. It determines the start of the injection/combustion and crankshaft location signals and displays the engine rpm and injection pump timing in degrees.

Fuel injection pump adapter seal installer

On GM engines, a tool is required to properly install the seal evenly into the fuel injection pump adapter.

Fuel injection pump adjusting tool

Both the Ford and GM diesel engines require special tools to rotate the often difficult to reach injection pump to the correct timing position **(see illustration).**

Fuel injection pump housing pressure adapter

The housing pressure adapter hose allows checking of the GM 5.7L engine fuel injection housing pressure without removing the metering valve sensor. It also can be used to measure the housing pressure on both GM 5.7L and 6.2L engines with the air crossover in place.

Fuel injection nozzle tester

This is used to check the opening pressure of the fuel injector(s). Other pressure gauges may be required to check the high and low injection system pressures **(see illustration).**

Fuel injection pump wrench

Offset thinwall wrenches are required when installing, removing or adjusting the fuel injection pump nuts/bolts **(see illustration).**

Fuel line nut wrench

Because of the high fuel pressures used in diesel fuel injection systems, special wrenches designed for use with a torque wrench, socket or breaker bar are needed when loosening or tightening the line connections **(see illustration).**

Tachometers

There is no electrical ignition system on diesel engines, so a special tachometer using a magnetic pickup on the crankshaft is required, or a photo-electric tach, which detects a piece of reflecting tape on the crank pulley, is needed to check engine speed **(see illustration).**

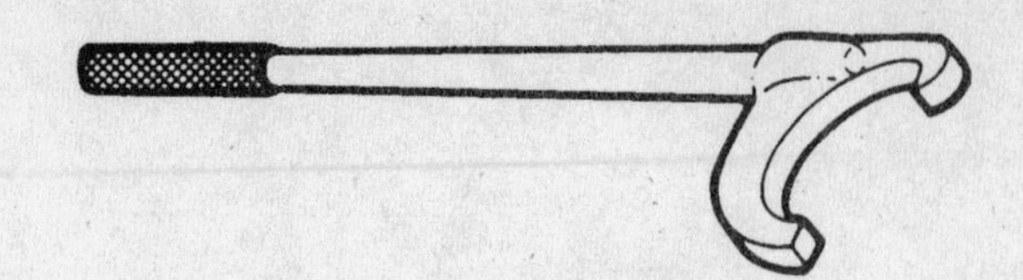

1.74 Adjusting the fuel injector pump requires a special tool, such as this one

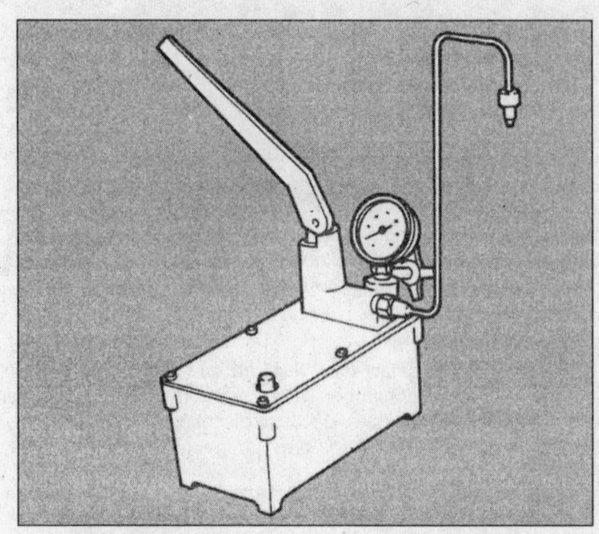

1.75 The nozzle tester is used for-checking the opening pressure of the injector nozzles

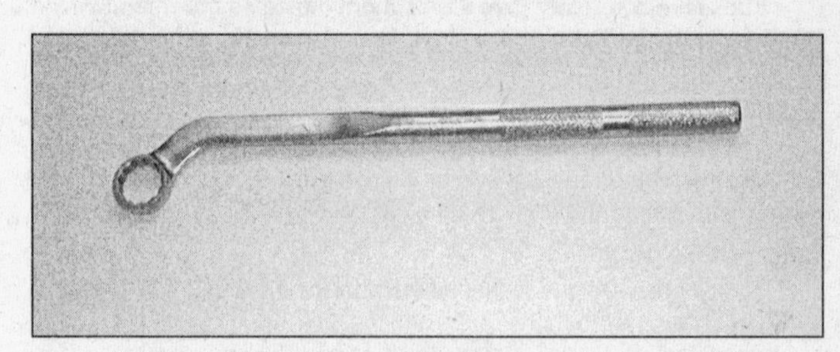

1.76 The fuel injection pump bolts and nuts are invariably hard to got at, so special wrenches, such as this one for GM engines, are necessary

1.77 Fuel lines are hard to reach using regular wrenches, so special "crow's foot" ones like this, used with a 3/8-inch socket drive and extension, are required

1.78 Diesel engines have no electrical Ignition system, so a photo electric (shown) or magnetic pickup tachometer is necessary to check the engine speed

1.79 Get an engine hoist that's strong enough to easily lift your engine in and out of the engine compartment - an adapter, like the one shown here (arrow), can be used to change the angle of the engine as it's being removed or installed

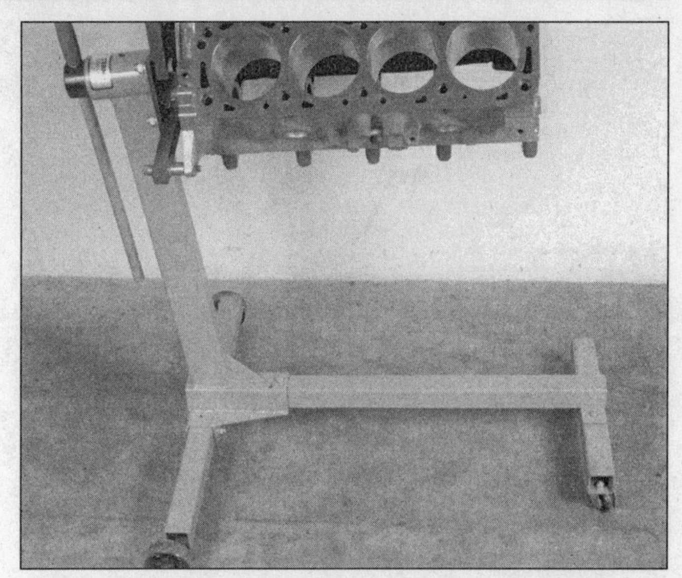

1.80 Get an engine stand sturdy enough to firmly support the engine while you're working on it - stay away from three-wheeled models - they have a tendency to tip over more easily - get a four-wheeled unit

Static timing gauge

Static timing on GM engines is set by installing a timing gauge tool in place of the fuel pump, rotating engine until the number one piston is at Top Dead Center (TDC), then marking the pump position with the punch scribe built into the tool.

Throttle shaft cam timing adapter

On GM engines, a throttle shaft cam timing adapter is required to measure the throttle shaft cam timing position before replacing the throttle shaft seals. The injection pump can then be disassembled and the seals replaced without changing the cam timing position.

Engine rebuilding tools

Engine hoist

Get an engine hoist **(see illustration)** that's strong enough to easily lift your engine in and out of the engine compartment. A V8 diesel is far too heavy to remove and install any other way. And you don't even want to think about the possibility of dropping an engine!

Engine stand

A V8 diesel is too heavy and bulky to wrestle around on the floor or the workbench while you're disassembling and reassembling it. Get an engine stand **(see illustration)** sturdy enough to firmly support the engine. Even if you plan to work on small blocks, it's a good idea to buy a stand beefy enough to handle big blocks as well. Try to buy a stand with four wheels, not three. The center of gravity of a stand is high, so it's easier to topple the stand and engine when you're cinching down head bolts with a two-foot long torque wrench. Get a stand with big casters. The larger the wheels, the easier it is to roll the stand around the shop. Wheel locks are also a good feature to have. If you want the stand out of the way between jobs, look for one that can be knocked down and slid under a workbench.

Engine stands are available at rental yards; however, the cost of rental over the long period of time you'll need the stand (often, weeks) is frequently as high as buying a stand.

Clutch alignment tool

The clutch alignment tool **(see illustration)** is used to center the clutch disc on the flywheel.

Cylinder surfacing hone

After boring the cylinders, you need to put a cross-hatch pattern on the cylinder walls to help the new rings seat properly. A flex hone with silicone carbide balls laminated onto the end of wire bristles **(see illustration)** will give you that pattern. Even if you don't bore the cylinders, you must hone the cylinders, since it breaks the glaze that coats the cylinder walls.

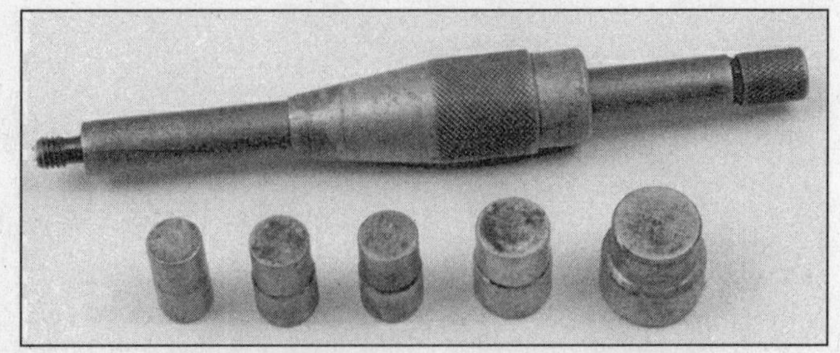

1.81 A clutch alignment tool is necessary if you plan to install a rebuilt engine mated to a manual transmission

1.82 A cylinder hone like the one shown here is easier to use, but not as versatile as the type that has three spring-loaded stones

1.83 If there are ridges at the top of the cylinder wall created by the thrust side of the piston at the top of its travel, use a ridge reamer to remove them

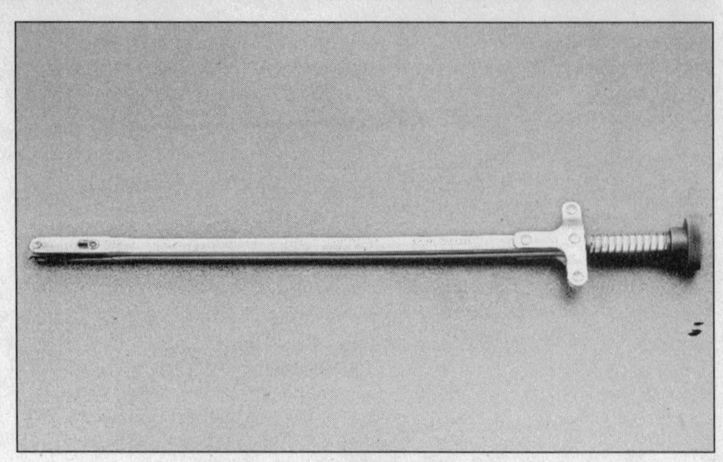

1.84 Here's a typical hydraulic lifter removal tool

1.85 The best universal ring compressor is the plier-type - to fit different bore sizes, simply insert different size compressor bands in the plier handles

1.86 The band-type ring compressor is as easy to use as the plier type but is more likely to snag a ring

Cylinder ridge reamer

If the engine has a lot of miles, the top compression ring will likely wear the cylinder wall and create a ridge at the top of each cylinder (the unworn portion of the bore forms the ridge). Carbon deposits make the ridge even more pronounced. The ridge reamer **(see illustration)** cuts away the ridge so you can remove the piston from the top of the cylinder without damaging the ring lands.

Hydraulic lifter removal tool

Sometimes the lifters are gummed up with varnish and become stuck in their lifter bores. This tool **(see illustration)** will get them out.

Piston ring compressor

Trying to install the pistons without a ring compressor is almost impossible.

The best "universal" ring compressor is the plier-type, like the one manufactured by K-D Tools **(see illustration)**. To accommodate different bore sizes, simply insert different size compressor bands in the plier handles. Plier-type spring compressors allow you to turn the piston with one hand while tapping it into the cylinder bore with a hammer handle.

The band type ring compressor **(see illustration)** is the cheapest of the two, and will work on a range of piston sizes, but it's more likely to snag a ring.

Piston ring groove cleaner

If you're reusing old pistons, you'll want to clean the carbon out of the ring grooves. This odd-looking tool **(see illustration)** has a cutting bit that digs the stuff out.

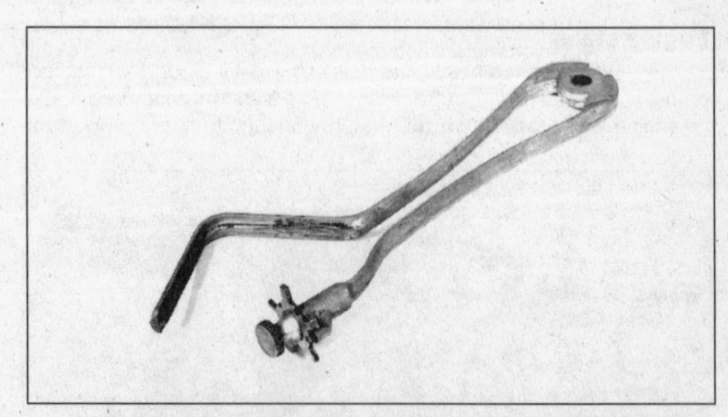

1.87 Sometimes you find that you can re-use the same pistons even when the rings need to be replaced - but the piston ring grooves in the old pistons are usually filthy - to clean them properly, you may need a piston ring groove cleaning tool

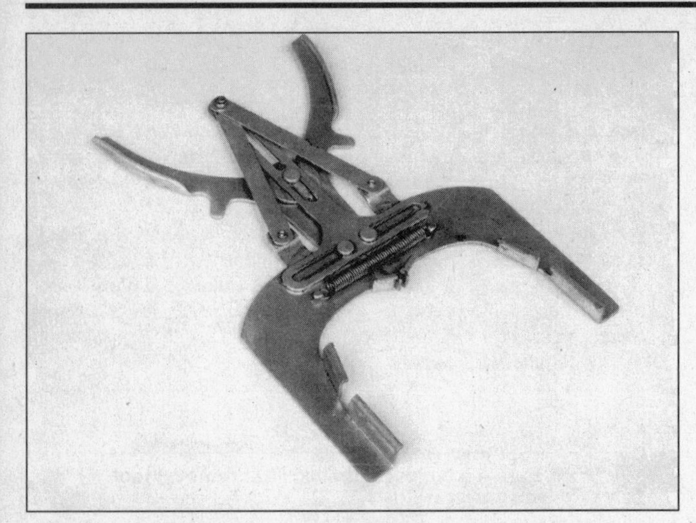

1.88 The piston ring expander pushes the ends of each ring apart so you can slip it over the piston crown and into its groove without scratching the piston or damaging the ring

Piston ring expander

This plier-like tool **(see illustration)** pushes the ends of the ring apart so you can slip it over the piston crown and into its groove.

Valve spring compressor

The valve spring compressor **(see illustration)** compresses the valve springs so you can remove the keepers and the retainer. For engine overhaul, if you can afford it, get a C-clamp type that's designed to de-spring the head when it's off the engine. Cheaper types also work, but they're more time consuming to use.

Precision measuring tools

Think of the tools in the following list as the final stage of your tool collection. If you're planning to rebuild an engine, you've probably already accumulated all the screwdrivers, wrenches, sockets, pliers and other everyday hand tools that you need. You've also probably collected all the special-purpose tools necessary to tune and service your specific engine. Now it's time to round up the stuff you'll need to do your own measurements when you rebuild that engine.

The tool pool strategy

If you're reading this book, you may be a motorhead, but engine rebuilding isn't your life - it's an avocation. You may just want to save some money, have a little fun and learn something about engine building. If that description fits your level of involvement, think about forming a "tool pool" with a friend or neighbor who wants to get into engine rebuilding, but doesn't want to spend a lot of money. For example, you can buy a set of micrometers and the other guy can buy a dial indicator and a set of small hole gauges.

Start with the basics

It would be great to own every precision measuring tool listed here, but you don't really need a machinist's chest crammed with exotic calipers and micrometers. You can often get by just fine with nothing more than a feeler gauge, modeling clay and Plastigage. Even most professional engine builders use only three tools 95-percent of the time: a one-inch outside micrometer, a dial indicator and a six-inch dial caliper. So start your collection with these three items.

Micrometers

When you're rebuilding an engine, you need to know the exact thickness of a sizeable number of pieces. Whether you're measuring the diameter of a wrist pin or the thickness of a valve spring shim or a thrust washer, your tool of choice should be the trusty one-inch outside micrometer **(see illustration)**.

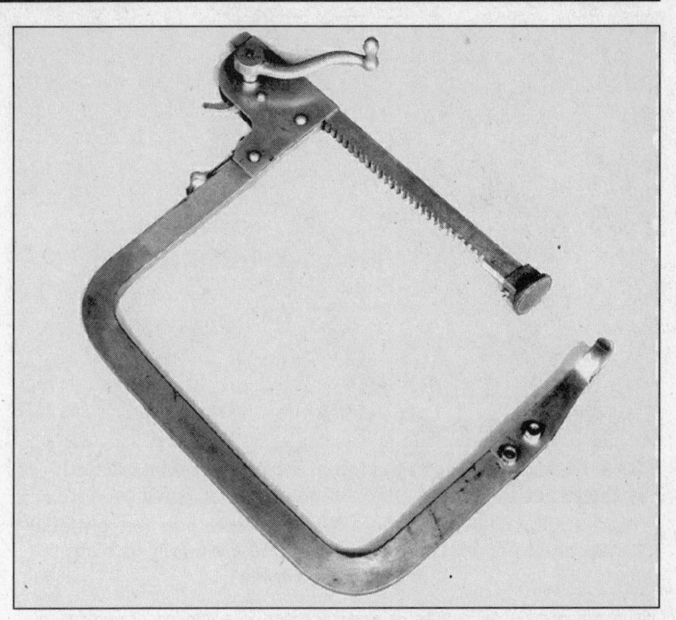

1.89 The valve spring compressor compresses the valve springs so you can remove the keepers and the retainer - the C-type (shown) reaches around to the underside of the head and pushes against the valve as it compresses the spring

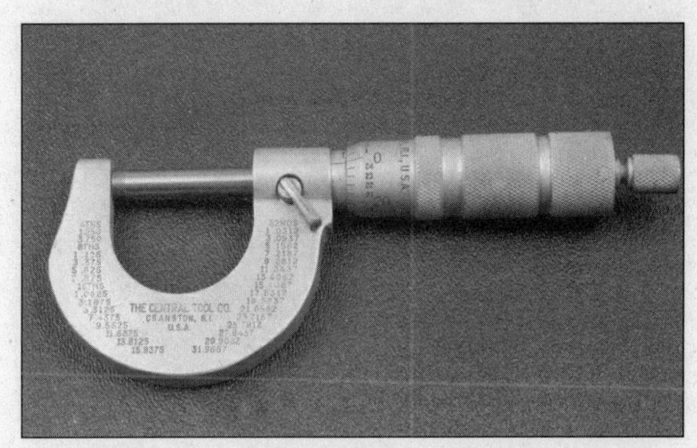

1.90 The one-inch micrometer is an essential precision measuring device for determining the dimensions of a wrist pin, valve spring shim, thrust washer, etc.

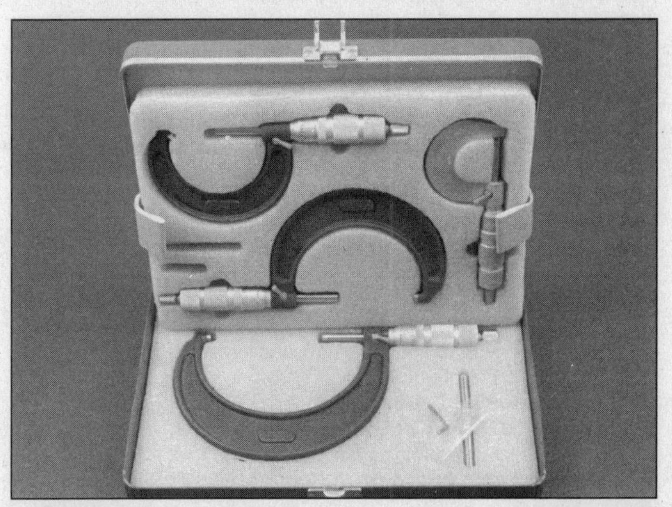

1.91 Get a good-quality micrometer set If you can afford it - this set has four micrometers ranging in size from one to four inches

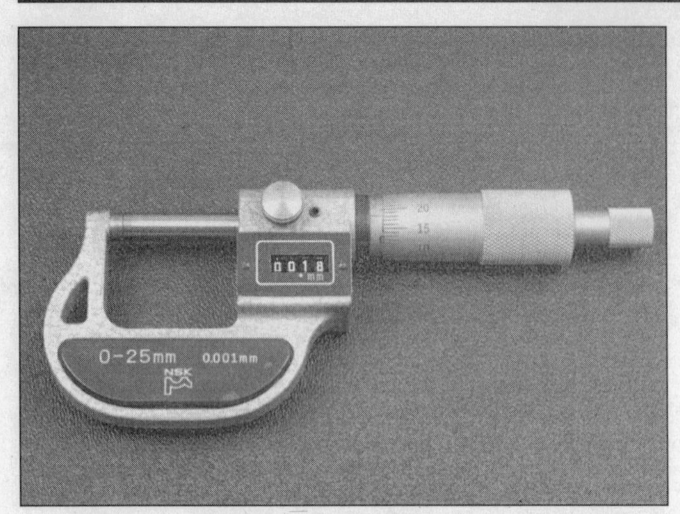

1.92 Digital micrometers are easier to read than conventional micrometers, are just as accurate and are finally starting to become affordable

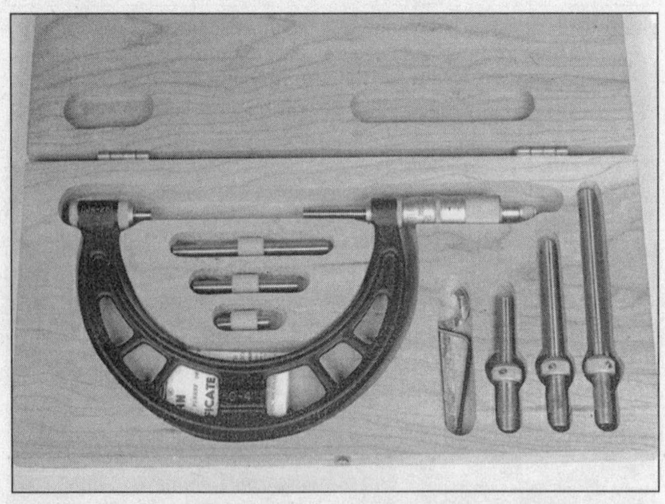

1.93 Avoid micrometer "sets" with interchangeable anvils -they're awkward to use when measuring little parts and changing the anvils is a hassle

Insist on accuracy to within one ten-thousandths of an inch (0.0001-inch) when you shop for a micrometer. You'll probably never need that kind of precision, but the extra decimal place will help you decide which way to round off a close measurement.

High-quality micrometers have a range of one inch. Eventually, you'll want a set **(see illustration)** that spans four, or even five, ranges: 0 to 1-inch, 1 to 2-inch, 2 to 3-inch and 3-to-4-inch. On engines bigger than about 350 cu. in., you'll also probably need a 4-to-5-inch. These five micrometers will measure the thickness of any part that needs to be measured for an engine rebuild. You don't have to run out and buy all five of these babies at once. Start with the one-inch model, then, when you have the money, get the next size you need (the 3 to 4-inch size or 4 to 5-inch is a good second choice - they measure piston diameters).

Digital micrometers **(see illustration)** are easier to read than conventional micrometers, are just as accurate and are finally starting to become affordable. If you're uncomfortable reading a conventional micrometer **(see sidebar)**, then get a digital.

Unless you're not going to use them very often, stay away from micrometers with interchangeable anvils **(see illustration)**. In theory, one of these beauties can do the work of five or six single-range micrometers. The trouble is, they're awkward to use when measuring little parts, and changing the anvils is a hassle.

How to read a micrometer

The outside micrometer is without a doubt the most widely used precision measuring tool. It can be used to make a variety of highly accurate measurements without much possibility of error through misreading, a problem associated with other measuring instruments, such as vernier calipers.

Like any slide caliper, the outside micrometer uses the "double contact" of its spindle and anvil **(see illustration)** touching the object to be measured to determine that object's dimensions. Unlike a caliper, however, the micrometer also features a unique precision screw adjustment which can be read with a great deal more accuracy than calipers.

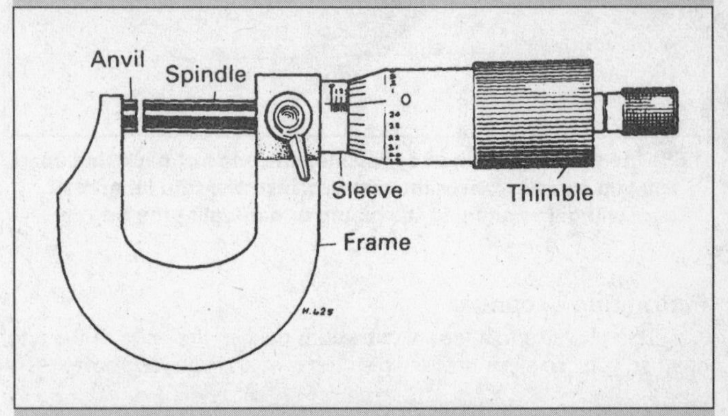

1.94 This diagram of a typical one-inch micrometer shows its major components

Why is this screw adjustment so accurate? Because years ago toolmakers discovered that a screw with 40 precision machined threads to the inch will advance one-fortieth (0.025) of an inch with each complete turn. The screw threads on the spindle revolve inside a fixed nut concealed by a sleeve.

On a one-inch micrometer, this sleeve is engraved longitudinally with exactly 40 lines to the inch, to correspond with the number of threads on the spindle. Every fourth line is made longer and is numbered one-tenth inch, two-tenths, etc. The other lines are often staggered to make them easier to read.

The thimble (the barrel which moves up and down the sleeve as it rotates) is divided into 25 divisions around the circumference of its beveled edge and is numbered from zero to 25. Close the micrometer spindle till it touches the anvil: You should see nothing but the zero line on the sleeve next to the beveled edge of the thimble. And the zero line of the thimble should be aligned with the horizontal (or axial) line on the sleeve. Remember: Each full revolution of the spindle from zero to zero advances or retracts the spindle one-fortieth or 0.025-inch. Therefore, if you rotate the thimble from zero on the beveled edge to the first graduation, you will move the spindle 1/25th of 1/40th, or 1/25th of 25/1000ths, which equals 1/1000th, or 0.001-inch.

Remember: Each numbered graduation on the sleeve represents 0.1-inch, each of the other sleeve graduations represents 0.025-inch and each graduation on the thimble represents 0.001-inch. Remember those three and you're halfway there.

For example: Suppose the 4 line is visible on the sleeve. This represents 0.400-inch. Then suppose there are an additional three lines (the short ones without numbers) showing. These marks are worth 0.025-inch each, or 0.075-inch. Finally, there are also two marks on the beveled edge of the thimble beyond the zero mark, each good for 0.001-inch, or a total of 0.002-inch. Add it all up and you get 0.400 plus 0.075 plus 0.002, which equals 0.477-inch.

Some beginners use a "dollars, quarters and cents" analogy to simplify reading a micrometer. Add up the bucks and change, then put a decimal point instead of a dollar sign in front of the sum!

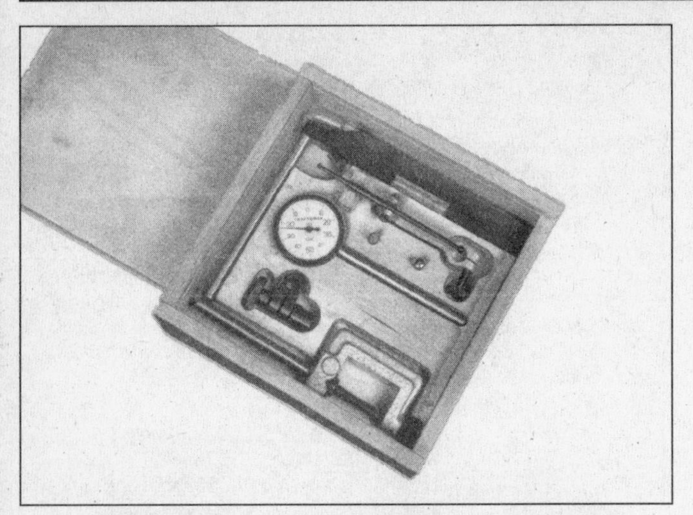

1.95 The dial indicator is indispensable for degreeing crankshafts, measuring valve lift, piston deck clearance, crankshaft endplay and a host of other critical measurements

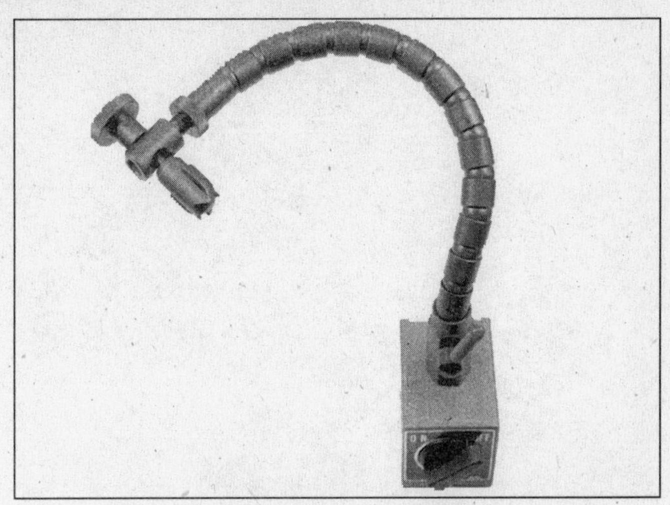

1.96 Get an adjustable, flexible fixture like this one, and a magnetic base, to ensure maximum versatility from your dial indicator

Dial indicators

The dial indicator **(see illustration)** is another measuring mainstay. It's indispensable for degreeing camshafts, measuring valve lift, piston deck clearances, crankshaft endplay and all kinds of other little measurements. Make sure the dial indicator you buy has a probe with at least one inch of travel, graduated in 0.001-inch increments. And get a good assortment of probe extensions up to about six inches long. Sometimes, you need to screw a bunch of these extensions together to reach into tight areas like pushrod holes.

Buy a dial indicator set that includes a flexible fixture and a magnetic stand **(see illustration)**. If the model you buy doesn't have a magnetic base, buy one separately. Make sure the magnet is plenty strong. If a weak magnet comes loose and the dial indicator takes a tumble on a concrete floor, you can kiss it good-bye. Make sure the arm that attach the dial indicator to the flexible fixture is sturdy and the locking clamps are easy to operate.

Some dial indicators are designed to measure depth **(see illustration)**. They have a removable base that straddles a hole. This setup is indispensable for measuring deck height when the piston is below the block surface. To measure the deck height of pistons that protrude above the deck, you'll also need a U-shaped bridge for your dial indicator. The bridge is also useful for checking the flatness of a block or a cylinder head.

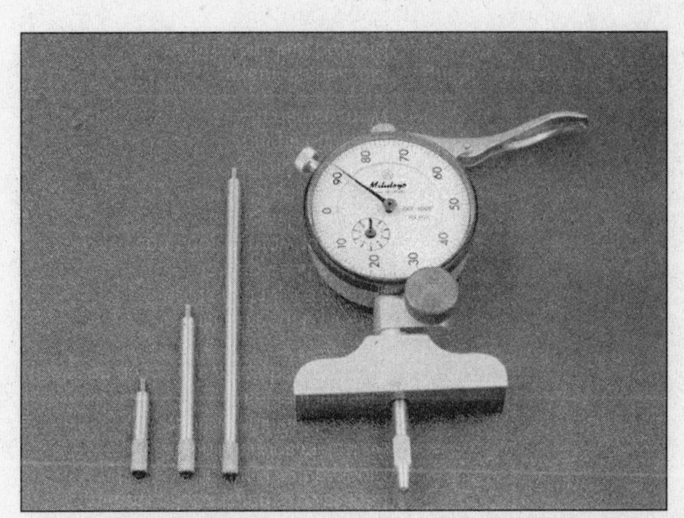

1.97 This dial indicator is designed to measure depth, such as deck height when the piston is below the block surface - with a U-shaped bridge (the base seen here is removable), you can measure the deck height of pistons that protrude above the deck' (U-shaped bridges are also useful for checking the flatness of a block or cylinder head)

Calipers

Vernier calipers **(see illustration)** aren't quite as accurate as a micrometer, but they're handy for quick measurements and they're relatively inexpensive. Most calipers have inside and outside jaws, so you can measure the inside diameter of a hole, or the outside diameter of a part.

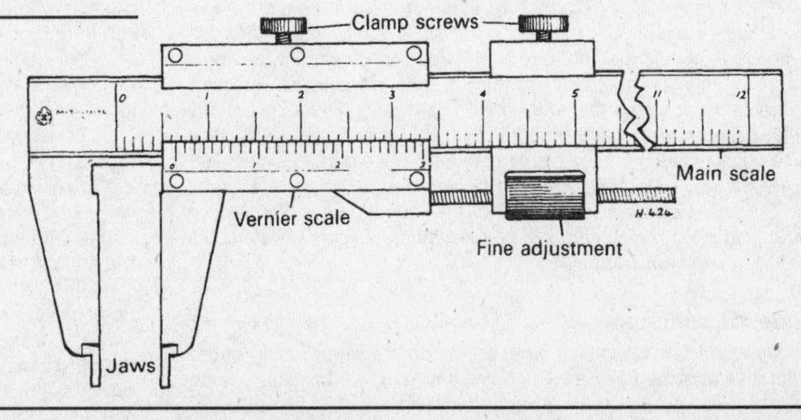

1.98 Vernier calipers aren't quite as accurate as micrometers, but they're handy for quick measurements and relatively inexpensive, and because they've got jaws that can measure internal and external dimensions, they're versatile

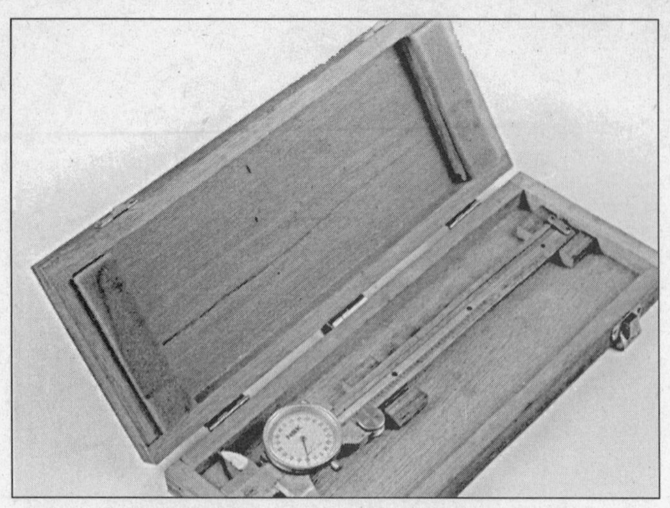

1.99 Dial calipers area lot easier to read than conventional vernier calipers, particularly If your eyesight isn't as good as it used to be!

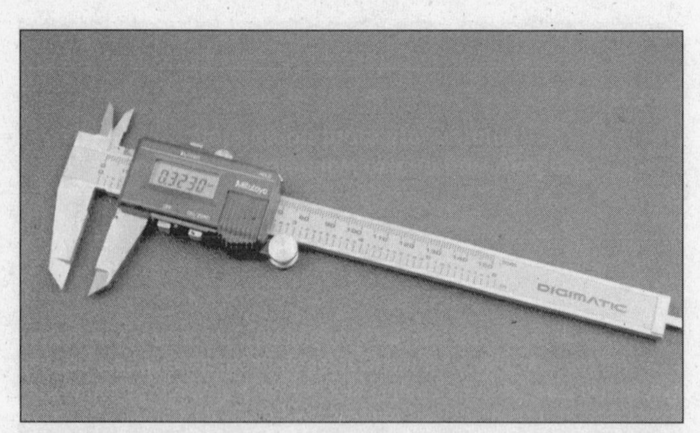

1.100 The latest calipers have a digital readout that is even easier to read than a dial caliper - another advantage of digital calipers is that they have a small microchip that allows them to convert instantaneously from inch to metric dimensions

How to read a vernier caliper

On the lower half of the main beam, each inch is divided into ten numbered increments, or tenths (0.100-inch, 0.200-inch, etc.). Each tenth is divided into four increments of 0.025-inch each. The vernier scale has 25 increments, each representing a thousandth (0.001) of an inch.

First read the number of inches, then read the number of tenths. Add to this 0.025-inch for each additional graduation. Using the English vernier scale, determine which graduation of the vernier lines up exactly with a graduation on the main beam. This vernier graduation is the number of thousandths which are to be added to the previous readings.

For example, let's say:

1) The number of inches is zero, or 0.000-inch;
2) The number of tenths is 4, or 0.400-inch;
3) The number of 0.025's is 2, or 0.050-inch; and
4) The vernier graduation which lines up with a graduation on the main beam is 15, or 0.015-inch.
5) Add them up: 0.000
0.400
0.050
0.015
6) And you get:0.46-inch.

That's all there is to it!

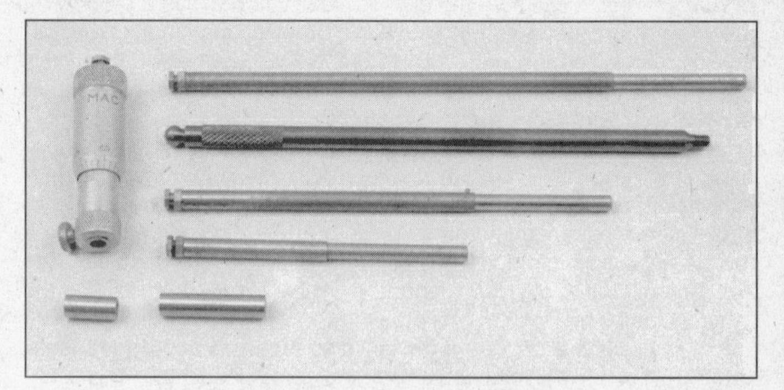

1.101 Inside micrometers are handy for measuring holes with thousandth-of-an-inch accuracy

Better-quality calipers have a dust shield over the geared rack that turns the dial to prevent small metal particles from jamming the mechanism. Make sure there's no play in the moveable jaw. To check, put a thin piece of metal between the jaws and measure its thickness with the metal close to the rack, then out near the tips of the jaws. Compare your two measurements. If they vary by more than 0.001-inch, look at another caliper - the jaw mechanism is deflecting.

If your eyes are going bad, or already are bad, vernier calipers can be difficult to read. Dial calipers **(see illustration)** are a better choice. Dial calipers combine the measuring capabilities of micrometers with the convenience of dial indicators. Because they're much easier to read quickly than vernier calipers, they're ideal for taking quick measurements when absolute accuracy isn't necessary, Like conventional vernier calipers, they have both inside and outside jaws which allow you to quickly determine the diameter of a hole or a part. Get a six-inch dial caliper, graduated in 0.001-inch increments.

The latest calipers **(see illustration)** have a digital LCD display that indicates both inch and metric dimensions. If you can afford one of these, it's the hot setup.

Inside micrometers

Cylinder bores, main bearing bores, connecting rod big ends - automotive engines have a lot of holes that must be measured accurately within a thousandth of an inch. Inside micrometers **(see illustration)** are used for these jobs. You read an inside micrometer the same way you read an outside micrometer. But it takes more skill to get an accurate reading.

To measure the diameter of a hole accurately, you must find the widest part of the hole. This involves expanding the micrometer while rocking it from side to side and moving it up and down. Once the micrometer is adjusted properly, you should be able to pull it through the hole with a slight drag. If the micrometer feels loose or binds as you pull it through, you're not getting an accurate reading.

Fully collapsed, inside micrometers can measure holes as small as one inch in diameter. Extensions or spacers are added for measuring larger holes.

Telescoping snap gauges **(see illustration)** are used to measure smaller holes. Simply insert them into a hole and turn the knurled handle to release their spring-loaded probes, which expand out to the walls of the hole, turn the handle the other way and lock the probes into position, then pull the gauge out. After the gauge is removed from the hole, measure its width with an outside micrometer.

For measuring really small holes, such as valve guides, you'll need a set of small hole gauges **(see illustration)**. They work the same way as telescoping snap gauges, but instead of spring-loaded probes, they have expanding flanges on the end that can be screwed in and out by a threaded handle.

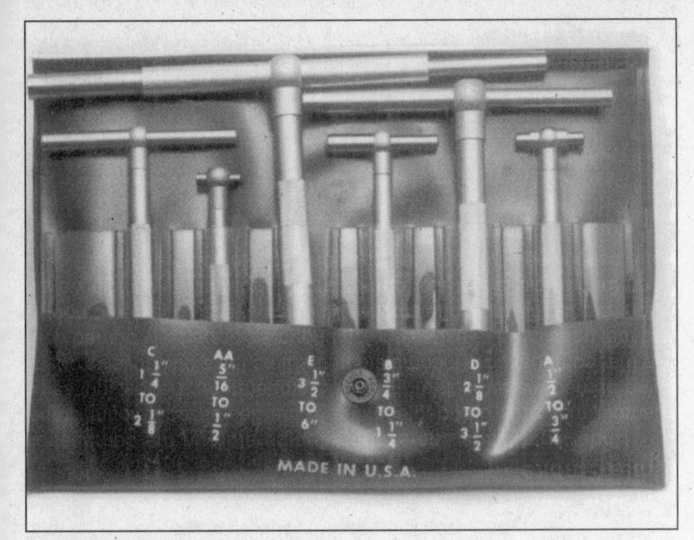

1.102 Telescoping snap gauges are used to measure smaller holes simply insert them into a hole, turn the knurled handle to release their spring-loaded probes out to the wall, turn the handle to lock the probes into position, pull out the gauge and measure the length from the tip of one probe to the tip of the other probe with a micrometer

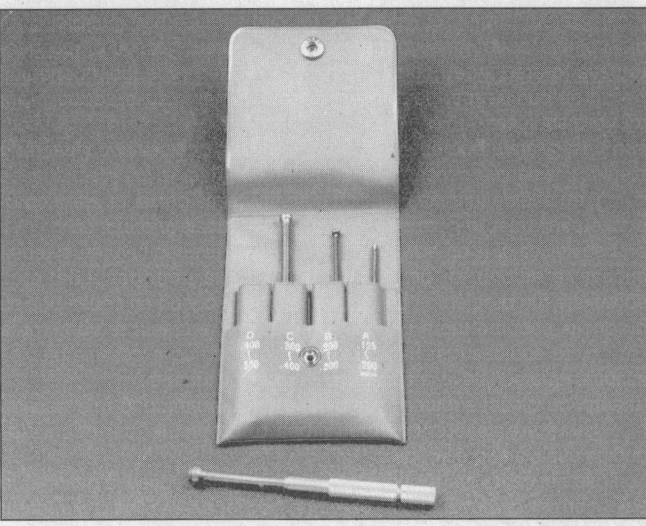

1.103 To measure really small holes, such as valve guides, you need a set of small hole gauges - to use them, simply stick them into the hole, turn the knurled handle until the expanding flanges are contacting the walls of the hole, pull out the gauge and measure the width of the gauge at the flanges with a micrometer

Dial bore gauge

The dial bore gauge **(see illustration)** is more accurate and easier to use - but more expensive - than an inside micrometer for checking the roundness of the cylinders, and the bearing bores in main bearing saddles and connecting rods. Using various extensions, most dial bore gauges have a range of just over 1-inch in diameter to 6 inches or more. Unlike outside micrometers with interchangeable anvils, the accuracy of bore gauges with interchangeable extensions is unaffected. Bore gauges accurate to 0.0001-inch are available, but they're very expensive and hard to find. Most bore gauges are graduated in 0.0005-inch increments. If you use them properly, this accuracy level is more than adequate.

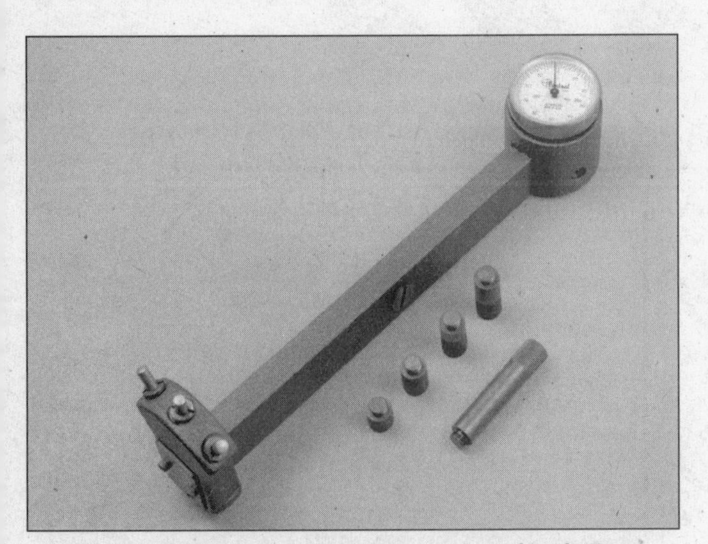

1.104 The dial bore gauge is more accurate and easier to use then an inside micrometer or telescoping snap gauges, but it's expensive - using various extensions, most dial gauges have a range of measurement from just over one inch to six inches or more

Storage and care of tools

Good tools are expensive, so treat them well. After you're through with your tools, wipe off any dirt, grease or metal chips and put them away. Don't leave tools lying around in the work area. General purpose hand tools - screwdrivers, pliers, wrenches and sockets - can be hung on a wall panel or stored in a tool box. Store precision measuring instruments, gauges, meters, etc. in a tool box to protect them from dust, dirt, metal chips and humidity.

Fasteners

Fasteners - nuts, bolts, studs and screws - hold parts together. Keep the following things in mind when working with fasteners: All threaded fasteners should be clean and straight, with good threads and unrounded corners on the hex head (where the wrench fits). Make it a habit to replace all damaged nuts and bolts with new ones. Almost all fasteners have a locking device of some type, either a lockwasher, locknut, locking tab or thread adhesive. Don't reuse special locknuts with nylon or fiber inserts. Once they're removed, they lose their locking ability. Install new locknuts.

Flat washers and lockwashers, when removed from an assembly, should always be replaced exactly as removed. Replace any damaged washers with new ones. Never use a lockwasher on any soft metal surface (such as aluminum), thin sheet metal or plastic.

Apply penetrant to rusted nuts and bolts to loosen them up and prevent breakage. Some mechanics use turpentine in a spout-type oil can, which works quite well. After applying the rust penetrant, let it work for a few minutes before trying to loosen the nut or bolt. Badly rusted fasteners may have to be chiseled or sawed off or removed with a special nut breaker, available at tool stores.

If a bolt or stud breaks off in an assembly, it can be drilled and removed with a special tool commonly available for this purpose. Most automotive machine shops can perform this task, as well as other repair procedures, such as the repair of threaded holes that have been stripped out.

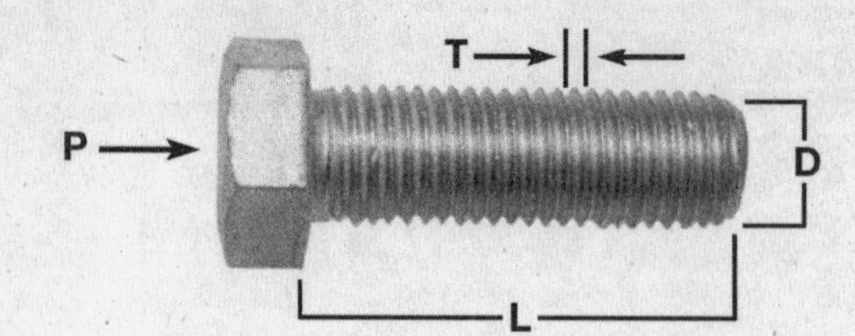

1.105 Standard (SAE and USS) bolt dimensions/grade marks

G *Grade marks (bolt strength)*
L *Length (in inches)*
T *Thread pitch (number of threads per inch)*
D *Nominal diameter (in inches)*

Metric bolt dimensions/grade marks

P *Property class (bolt strength)*
L *Length (in millimeters)*
T *Thread pitch (distance between threads in millimeters)*
D *Diameter*

Bolt strength marking (standard/SAE/USS; bottom - metric)

Grade	Identification
Hex Nut Grade 5	 3 Dots
Hex Nut Grade 8	 6 Dots

Standard hex nut strength markings

Grade	Identification
Hex Nut Property Class 9	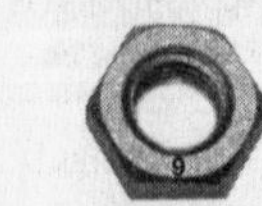 Arabic 9
Hex Nut Property Class 10	 Arabic 10

Metric hex nut strength markings

Metric and strength markings

1.106 Bolt strength markings

Fastener sizes

For a number of reasons, automobile manufacturers are making wider and wider use of metric fasteners. Therefore, it's important to be able to tell the difference between standard (sometimes called U.S. or SAE) and metric hardware, since they cannot be interchanged.

All bolts, whether standard or metric, are sized in accordance with their diameter, thread pitch and length **(see illustration)**. For example, a standard 1/2 - 13 x 1 bolt is 1/2 inch in diameter, has 13 threads per inch and is 1 inch long. An M12 - 1.75 x 25 metric bolt is 12 mm in diameter, has a thread pitch of 1.75 mm (the distance between threads) and is 25 mm long. The two bolts are nearly identical, and easily confused, but they are not interchangeable.

In addition to the differences in diameter, thread pitch and length, metric and standard bolts can also be distinguished by examining the bolt heads. The distance across the flats on a standard bolt head is measured in inches; the same dimension on a metric bolt or nut is sized in millimeters. So don't use a standard wrench on a metric bolt, or vice versa.

Most standard bolts also have slashes radiating out from the center of the head **(see illustration)** to denote the grade or strength of the bolt, which is an indication of the amount of torque that can be applied to it. The greater the number of slashes, the greater the strength of the bolt. Grades 0 through 5 are commonly used on automobiles. Metric bolts have a property class (grade) number, rather than a slash, molded into their heads to indicate bolt strength. In this case, the higher the number, the stronger the bolt. Property class numbers 8.8, 9.8 and 10.9 are commonly used on automobiles.

Strength markings can also be used to distinguish standard hex nuts from metric hex nuts. Many standard nuts have dots stamped into one side, while metric nuts are marked with a number **(see illustrations)**. The greater the number of dots, or the higher the number, the greater the strength of the nut.

Metric studs are also marked on their ends **(see illustration)** according to property class (grade). Larger studs are numbered (the same as metric bolts), while smaller studs carry a geometric code to denote grade. It should be noted that many fasteners, especially Grades 0 through 2, have no distinguishing marks on them. When such is the case, the only way to determine whether it's standard or metric is to measure the thread pitch or compare it to a known fastener of the same size.

Standard fasteners are often referred to as SAE, as opposed to metric. However, it should be noted that SAE technically refers to a non-metric fine thread fastener only. Coarse thread non-metric fasteners are referred to as USS sizes.

Since fasteners of the same size (both standard and metric) may have different strength ratings, be sure to reinstall any bolts, studs or nuts removed from your vehicle in their original locations. Also, when replacing a fastener with a new one, make sure that the new one has a strength rating equal to or greater than the original.

Tightening sequences and procedures

Most threaded fasteners should be tightened to a specific torque value **(see accompanying charts)**. Torque is the twisting force applied to a threaded component such as a nut or bolt. Overtightening the fastener can weaken it and cause it to break, while undertightening can cause it to eventually come loose. Bolts, screws and studs, depending on the material they are made of and their thread diameters, have specific torque values, many of which are noted in the Specifications at the beginning of each Chapter. Be sure to follow the torque recommendations closely. For fasteners not assigned a specific torque, a general torque value chart is presented here as a guide. These torque values are for dry (unlubricated) fasteners threaded into steel or cast iron (not aluminum). As was previously mentioned, the size and grade of a fastener determine the amount of torque that can safely be applied to it. The figures listed here are approximate for Grade 2 and Grade 3 fasteners. Higher grades can tolerate higher torque values.

Metric thread sizes	Ft-lbs	Nm
M-6	6 to 9	9 to 12
M-8	14 to 21	19 to 28
M-10	28 to 40	38 to 54
M-12	50 to 71	68 to 96
M-14	80 to 140	109 to 154

Pipe thread sizes		
1/8	5 to 8	7 to 10
1/4	12 to 18	17 to 24
3/8	22 to 33	30 to 44
1/2	25 to 35	34 to 47

U.S. thread sizes		
1/4 – 20	6 to 9	9 to 12
5/16 – 18	12 to 18	17 to 24
5/16 – 24	14 to 20	19 to 27
3/8 – 16	22 to 32	30 to 43
3/8 – 24	27 to 38	37 to 51
7/16 – 14	40 to 55	55 to 74
7/16 – 20	40 to 60	55 to 81
1/2 – 13	55 to 80	75 to 108

If fasteners are laid out in a pattern - such as cylinder head bolts, oil pan bolts, differential cover bolts, etc. - loosen and tighten them in sequence to avoid warping the component. Where it matters, we'll show you this sequence. If a specific pattern isn't that important, the following rule-of-thumb guide will prevent warping.

First, install the bolts or nuts finger-tight. Then tighten them one full turn each, in a criss-cross or diagonal pattern. Then return to the first one and, following the same pattern, tighten them all one-half turn. Finally, tighten each of them one-quarter turn at a time until each fastener has been tightened to the proper torque. To loosen and remove the fasteners, reverse this procedure.

How to remove broken fasteners

Sooner or later, you're going to break off a bolt inside its threaded hole. There are several ways to remove it. Before you buy an expensive extractor set, try some of the following cheaper methods first.

First, regardless of which of the following methods you use, be sure to use penetrating oil. Penetrating oil is a special light oil with excellent penetrating power for freeing dirty and rusty fasteners. But it also works well on tightly torqued broken fasteners.

If enough of the fastener protrudes from its hole - and if it isn't torqued down too tightly - you can often remove it with vise-grips or a small pipe wrench. If that doesn't work, or if the fastener doesn't provide sufficient purchase for pliers or a wrench, try filing it down to take a wrench, or cut a slot in it to accept a screwdriver **(see illustration)**. If you still can't get it off - and you know how to weld - try welding a flat piece of steel, or a nut, to the top of the broken fastener. If the fastener is broken off flush with - or below - the top of its hole, try tapping it out with a small, sharp punch. If that doesn't work, try drilling out the broken fastener with a bit only slightly smaller than the inside diameter of the hole. For example, if the hole is 1/2-inch in diameter, use a 15/32-inch drill bit. This leaves a shell which you can pick out with a sharp chisel.

If THAT doesn't work, you'll have to resort to some form of screw extractor, such as E-Z-Out **(see illustration)**. Screw extractors are sold in sets which can remove anything from 1/4-inch to 1-inch bolts or studs. Most extractors are fluted and tapered high-grade steel. To use a screw extractor, drill a hole slightly smaller than the O.D. of the extractor you're going to use (Extractor sets include the manufacturer's recommendations for what size drill bit to use with each extractor size). Then screw in the extractor **(see illustration)** and back it - and the broken fastener - out. Extractors are reverse-threaded, so they won't unscrew when you back them out.

A word to the wise: Even though an E-Z-Out will usually save your bacon, it can cause even more grief if you're careless or sloppy. Drilling the hole for the extractor off-center, or using too small, or too big, a bit for the size of the fastener you're removing will only make things worse. So be careful!

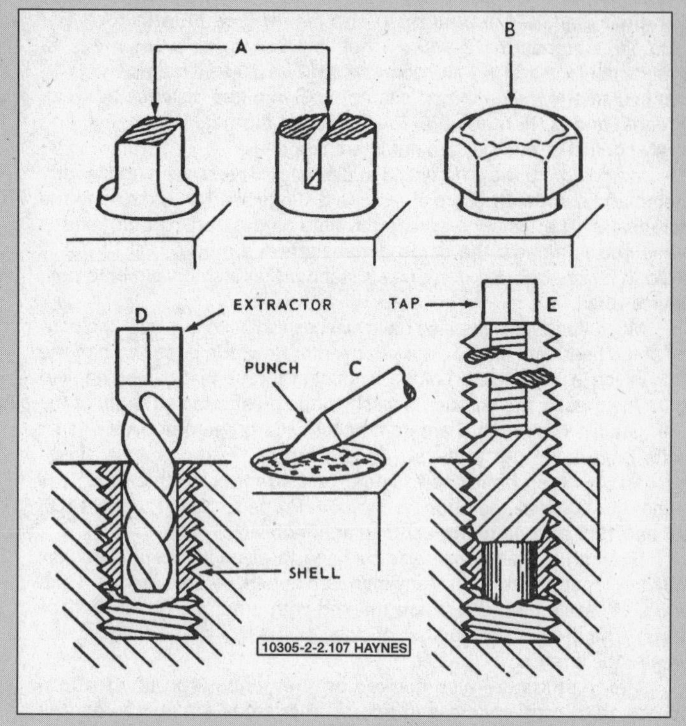

1.107 There are several ways to remove a broken fastener

A File it flat or slot it
B Weld on a nut
C Use a punch to unscrew it
D Use a screw extractor (like an E-Z-Out)
E Use a tap to remove the shell

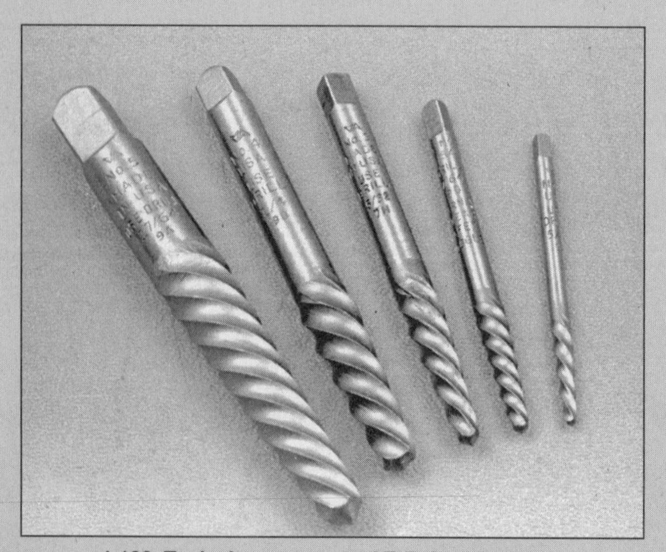

1.108 Typical assortment of E-Z-Out extractors

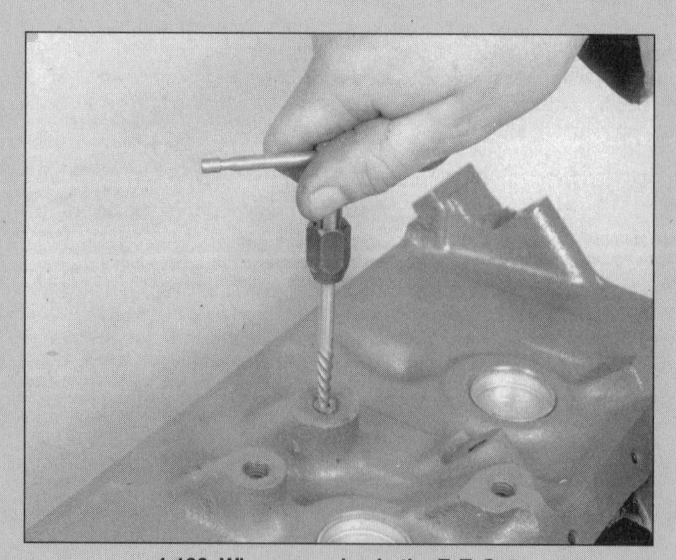

1.109 When screwing in the E-Z-Out, make sure it's centered properly

How to repair broken threads

Sometimes, the internal threads of a nut or bolt hole can become stripped, usually from overtightening. Stripping threads is an all-too-common occurrence, especially when working with aluminum parts, because aluminum is so soft that it easily strips out. Overtightened spark plugs are another common cause of stripped threads.

Usually, external or internal threads are only partially stripped. After they've been cleaned up with a tap or die, they'll still work. Sometimes, however, threads are badly damaged. When this happens, you've got three choices:

1) *Drill and tap the hole to the next suitable oversize and install a larger diameter bolt, screw or stud.*
2) *Drill and tap the hole to accept a threaded plug, then drill and tap the plug to the original screw size. You can also buy a plug already threaded to the original size. Then you simply drill a hole to the specified size, then run the threaded plug into the hole with a bolt and jam nut. Once the plug is fully seated, remove the jam nut and bolt.*
3) *The third method uses a patented thread repair kit like Heli-Coil or Slimsert. These easy-to-use kits are designed to repair damaged threads in spark plug holes, straight-through holes and blind holes. Both are available as kits which can handle a variety of sizes and thread patterns. Drill the hole, then tap it with the special included tap. Install the Heli-Coil* **(see illustration)** *and the hole is back to its original diameter and thread pitch.*

Regardless of which method you use, be sure to proceed calmly and carefully. A little impatience or carelessness during one of these relatively simple procedures can ruin your whole day's work and cost you a bundle if you wreck an expensive head or block.

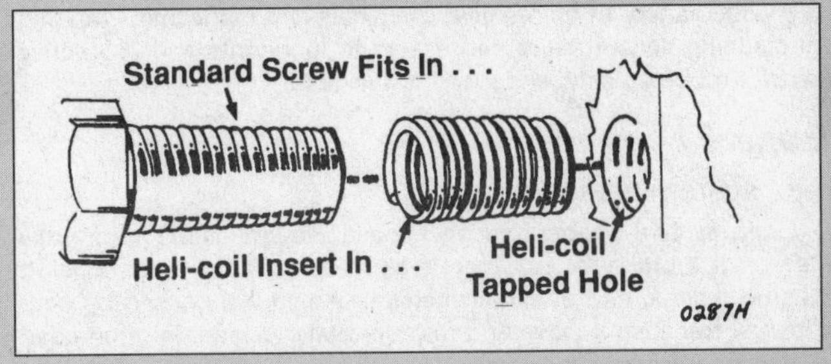

1.110 To install a Heli-Coil, drill out the hole, tap it with the special included tap and screw in the Heli-Coil.

Component disassembly

Disassemble components carefully to help ensure that the parts go back together properly. Note the sequence in which parts are removed. Make note of special characteristics or marks on parts that can be installed more than one way, such as a grooved thrust washer on a shaft. It's a good idea to lay the disassembled parts out on a clean surface in the order in which you removed them. It may also be helpful to make sketches or take instant photos of components before removal.

When you remove fasteners from a component, keep track of their locations. Thread a bolt back into a part, or put the washers and nut back on a stud, to prevent mix-ups later. If that isn't practical, put fasteners in a fishing tackle box or a series of small boxes. A cupcake or muffin tin, or an egg crate, is ideal for this purpose - each cavity can hold the bolts and nuts from a particular area (i.e. oil pan bolts, valve cover bolts, engine mount bolts, etc.). A pan of this type is helpful when working on assemblies with very small parts, such as the carburetor or valve train. Mark each cavity with paint or tape to identify the contents.

When you unplug the connector(s) between two wire harnesses, or even two wires, it's a good idea to identify the two halves with numbered pieces of masking tape - or a pair of matching pieces of colored electrical tape - so they can be easily reconnected.

Gasket sealing surfaces

Gaskets seal the mating surfaces between two parts to prevent lubricants, fluids, vacuum or pressure from leaking out between them. Gaskets are often coated with a liquid or paste-type gasket sealing compound before assembly. Age, heat and pressure can cause the two parts to stick together so tightly that they're difficult to separate. Often, you can loosen the assembly by striking it with a soft-face hammer near the mating surfaces. You can use a regular hammer if you place a block of wood between the hammer and the part, but don't hammer on cast or delicate parts that can be easily damaged. When a part refuses to come off, look for a fastener that you forgot to remove.

Don't use a screwdriver or pry bar to pry apart an assembly. It can easily damage the gasket sealing surfaces of the parts, which must be smooth to seal properly. If prying is absolutely necessary, use an old broom handle or a section of hard wood dowel.

Once the parts are separated, carefully scrape off the old gasket and clean the gasket surface. You can also remove some gaskets with a wire brush. If some gasket material refused to come off, soak it with rust penetrant or treat it with a special chemical to soften it, then scrape it off. You can fashion a scraper from a piece of copper tubing by flattening and sharpening one end. Copper is usually softer than the surface being scraped, which reduces the likelihood of gouging the part. The mating surfaces must be clean and smooth when you're done. If the gasket surface is gouged, use a gasket sealer thick enough to fill the scratches when you reassemble the components. For most applications, use a non-drying (or semi-drying) gasket sealer.

Hose removal tips

Warning: *If the vehicle is equipped with air conditioning, do not disconnect any of the A/C hoses without first having the system depressurized by a dealer service department or a service station (see the Haynes Automotive Heating and Air Conditioning Manual).*

The same precautions that apply to gasket removal also apply to hoses. Avoid scratching or gouging the surface against which the hose mates, or the connection may leak. Take, for example, radiator hoses. Because of various chemical reactions, the rubber in radiator hoses can bond itself to the metal spigot over which the hose fits. To remove a hose, first loosen the hose clamps that secure it to the spigot. Then, with slip-joint pliers, grab the hose at the clamp and rotate it around the spigot. Work it back and forth until it is completely free, then pull it off. Silicone or other lubricants will ease removal if they can be applied between the hose and the outside of the spigot. Apply the same lubricant to the inside of the hose and the outside of the spigot to simplify installation. Snap-On and Mac Tools sell hose removal tools - they look like bent ice picks - which can be inserted between the spigot and the radiator hose to break the seal between rubber and metal.

As a last resort - or if you're planning to replace the hose anyway - slit the rubber with a knife and peel the hose from the spigot. Make sure you don't damage the metal connection.

If a hose clamp is broken or damaged, don't reuse it. Wire-type clamps usually weaken with age, so it's a good idea to replace them with screw-type clamps whenever a hose is removed.

Automotive chemicals and lubricants

A wide variety of automotive chemicals and lubricants - ranging from cleaning solvents and degreasers to lubricants and protective sprays for rubber, plastic and vinyl - is available.

Cleaners

Brake system cleaner

Brake system cleaner removes grease and brake fluid from brake parts like disc brake rotors, where a spotless surfaces is essential. It leaves no residue and often eliminates brake squeal caused by contaminants. Because it leaves no residue, brake cleaner is often used for cleaning engine parts as well.

Carburetor and choke cleaner

Carburetor and choke cleaner is a strong solvent for gum, varnish and carbon. Most carburetor cleaners leave a dry-type lubricant film which will not harden or gum up. So don't use carb cleaner on electrical components.

Degreasers

Degreasers are heavy-duty solvents used to remove grease from the outside of the engine and from chassis components. They're usually sprayed or brushed on. Depending on the type, they're rinsed off either with water or solvent.

Demoisturants

Demoisturants remove water and moisture from electrical components such as alternators, voltage regulators, electrical connectors and fuse blocks. They are non-conductive, non-corrosive and non-flammable.

Electrical cleaner

Electrical cleaner removes oxidation, corrosion and carbon deposits from electrical contacts, restoring full current flow. It can also be used to clean spark plugs, carburetor jets, voltage regulators and other parts where an oil-free surface is necessary.

Lubricants

Assembly lube

Assembly lube is a special extreme pressure lubricant, usually containing moly, used to lubricate high-load parts (such as main and rod bearings and cam lobes) for initial start-up of a new engine. The assembly lube lubricates the parts without being squeezed out or washed away until the engine oiling system begins to function.

Graphite lubricants

Graphite lubricants are used where oils cannot be used due to contamination problems, such as in locks. The dry graphite will lubricate metal parts while remaining uncontaminated by dirt, water, oil or acids. It is electrically conductive and will not foul electrical contacts in locks such as the ignition switch.

Heat-sink grease

Heat-sink grease is a special electrically non-conductive grease that is used for mounting electronic ignition modules where it is essential that heat is transferred away from the module.

Moly penetrants

Moly penetrants loosen and lubricate frozen, rusted and corroded fasteners and prevent future rusting or freezing.

Motor oil

Motor oil is the lubricant formulated for use in engines. It normally contains a wide variety of additives to prevent corrosion and reduce foaming and wear. Motor oil comes in various weights (viscosity ratings) from 5 to 80. The recommended weight of the oil depends on the season, temperature and the demands on the engine. Light oil is used in cold climates and under light load conditions. Heavy oil is used in hot climates and where high loads are encountered. Multi-viscosity oils are designed to have characteristics of both light and heavy oils and are available in a number of weights from 5W-20 to 20W-50. Some home mechanics use motor oil as an assembly lube, but we don't recommend it, because motor oil has a relatively thin viscosity, which means it will slide off the parts long before the engine is fired up.

Silicone lubricants

Silicone lubricants are used to protect rubber, plastic, vinyl and nylon parts.

Wheel bearing grease

Wheel bearing grease is a heavy grease that can withstand high loads and friction, such as wheel bearings, balljoints, tie-rod ends and universal joints. It's also sticky enough to hold parts like the keepers for the valve spring retainers in place on the valve stem when you're installing the springs.

White grease

White grease is a heavy grease for metal-to-metal applications where water is present. It stays soft under both low and high temperatures (usually from -100 to +190-degrees F), and won't wash off or dilute when exposed to water. Another good "glue" for holding parts in place during assembly.

Sealants

Anaerobic sealant

Anaerobic sealant is much like RTV in that it can be used either to seal gaskets or to form gaskets by itself. It remains flexible, is solvent resistant and fills surface imperfections. The difference between an anaerobic sealant and an RTV-type sealant is in the curing. RTV cures when exposed to air, while an anaerobic sealant cures only in the absence of air. This means that an anaerobic sealant cures only after the assembly of parts, sealing them together.

RTV sealant

RTV sealant is one of the most widely used gasket compounds. Made from silicone, RTV is air curing, it seals, bonds, waterproofs, fills surface irregularities, remains flexible, doesn't shrink, is relatively easy to remove, and is used as a supplementary sealer with almost all low and medium temperature gaskets.

Thread and pipe sealant

Thread and pipe sealant is used for sealing hydraulic and pneumatic fittings and vacuum lines. It is usually made from a teflon compound, and comes in a spray, a paint-on liquid and as a wrap-around tape.

Chemicals

Anaerobic locking compounds

Anaerobic locking compounds are used to keep fasteners from vibrating or working loose and cure only after installation, in the absence of air. Medium strength locking compound is used for small nuts, bolts and screws that may be removed later. High-strength locking compound is for large nuts, bolts and studs which aren't removed on a regular basis.

Anti-seize compound

Anti-seize compound prevents seizing, galling, cold welding, rust and corrosion in fasteners. High-temperature anti-seize, usually made with copper and graphite lubricants, is used for exhaust system and exhaust manifold bolts.

Gas additives

Gas additives perform several functions, depending on their chemical makeup. They usually contain solvents that help dissolve gum and varnish that build up on carburetor, fuel injection and intake parts. They also serve to break down carbon deposits that form on the inside surfaces of the combustion chambers. Some additives contain upper cylinder lubricants for valves and piston rings, and others contain chemicals to remove condensation from the gas tank.

Oil additives

Oil additives range from viscosity index improvers to chemical treatments that claim to reduce internal engine friction. It should be noted that most oil manufacturers caution against using additives with their oils.

Safety first!

Regardless of how enthusiastic you may be about getting on with the job at hand, take the time to ensure that your safety is not jeopardized. A moment's lack of attention can result in an accident, as can failure to observe certain simple safety precautions. The possibility of an accident will always exist, and the following points should not be considered a comprehensive list of all dangers. Rather, they are intended to make you aware of the risks and to encourage a safety conscious approach to all work you carry out on your vehicle.

Essential DOs and DON'Ts

DON'T rely on a jack when working under the vehicle. Always use approved jackstands to support the weight of the vehicle and place them under the recommended lift or support points.

DON'T attempt to loosen extremely tight fasteners (i.e. wheel lug nuts) while the vehicle is on a jack - it may fall.

DON'T start the engine without first making sure that the transmission is in Neutral (or Park where applicable) and the parking brake is set.

DON'T remove the radiator cap from a hot cooling system - let it cool or cover it with a cloth and release the pressure gradually.

DON'T attempt to drain the engine oil until you are sure it has cooled to the point that it will not burn you.

DON'T touch any part of the engine or exhaust system until it has cooled sufficiently to avoid burns.

DON'T siphon toxic liquids such as gasoline, antifreeze and brake fluid by mouth, or allow them to remain on your skin.

DON'T inhale brake lining or clutch disc dust - it is potentially hazardous (see Asbestos below)

DON'T allow spilled oil or grease to remain on the floor - wipe it up before someone slips on it.

DON'T use loose fitting wrenches or other tools which may slip and cause injury.

DON'T push on wrenches when loosening or tightening nuts or bolts. Always try to pull the wrench toward you. If the situation calls for pushing the wrench away, push with an open hand to avoid scraped knuckles if the wrench should slip.

DON'T attempt to lift a heavy component alone - get someone to help you.

DON'T rush or take unsafe shortcuts to finish a job.

DON'T allow children or animals in or around the vehicle while you are working on it.

DO wear eye protection when using power tools such as a drill, sander, bench grinder, etc. and when working under a vehicle.

DO keep loose clothing and long hair well out of the way of moving parts.

DO make sure that any hoist used has a safe working load rating adequate for the job.

DO get someone to check on you periodically when working alone on a vehicle.

DO carry out work in a logical sequence and make sure that everything is correctly assembled and tightened.

DO keep chemicals and fluids tightly capped and out of the reach of children and pets.

DO remember that your vehicle's safety affects that of yourself and others. If in doubt on any point, get professional advice.

Asbestos

Certain friction, insulating, sealing, and other products - such as brake linings, brake bands, clutch linings, torque converters, gaskets, etc. - contain asbestos. Extreme care must be taken to avoid inhalation of dust from such products since it is hazardous to health. If in doubt, assume that they do contain asbestos.

Batteries

Never create a spark or allow a bare light bulb near a battery. They normally give off a certain amount of hydrogen gas, which is highly explosive.

Always disconnect the battery ground (-) cable at the battery before working on the fuel or electrical systems.

If possible, loosen the filler caps or cover when charging the battery from an external source (this does not apply to sealed or maintenance-free batteries). Do not charge at an excessive rate or the battery may burst.

Take care when adding water to a non maintenance-free battery and when carrying a battery. The electrolyte, even when diluted, is very corrosive and should not be allowed to contact clothing or skin.

Always wear eye protection when cleaning the battery to prevent the caustic deposits from entering your eyes.

Fire

We strongly recommend that a fire extinguisher suitable for use on fuel and electrical fires be kept handy in the garage or workshop at all times. Never try to extinguish a fuel or electrical fire with water. Post the phone number for the nearest fire department in a conspicuous location near the phone.

Fumes

Certain fumes are highly toxic and can quickly cause unconsciousness and even death if inhaled to any extent. Gasoline vapor falls into this category, as do the vapors from some cleaning solvents. Any draining or pouring of such volatile fluids should be done in a well ventilated area.

When using cleaning fluids and solvents, read the instructions on the container carefully. Never use materials from unmarked containers.

Never run the engine in an enclosed space, such as a garage. Exhaust fumes contain carbon monoxide, which is extremely poisonous. If you need to run the engine, always do so in the open air, or at least have the rear of the vehicle outside the work area.

Household current

When using an electric power tool, inspection light, etc., which operates on household current, always make sure that the tool is correctly connected to its plug and that, where necessary, it is properly grounded. Do not use such items in damp conditions and, again, do not create a spark or apply excessive heat in the vicinity of fuel or fuel vapor.

Keep it clean

Get in the habit of taking a regular look around the shop to check for potential dangers. Keep the work area clean and neat. Sweep up all debris and dispose of it as soon as possible. Don't leave tools lying around on the floor.

Be very careful with oily rags. Spontaneous combustion can occur if they're left in a pile, so dispose of them properly in a covered metal container.

Check all equipment and tools for security and safety hazards (like frayed cords). Make necessary repairs as soon as a problem is noticed - don't wait for a shelf unit to collapse before fixing it.

Accidents and emergencies

Shop accidents range from minor cuts and skinned knuckles to serious injuries requiring immediate medical attention. The former are inevitable, while the latter are, hopefully, avoidable or at least uncommon. Think about what you would do in the event of an accident. Get some first aid training and have an adequate first aid kit somewhere within easy reach.

Think about what you would do if you were badly hurt and incapacitated. Is there someone nearby who could be summoned quickly? If possible, never work alone just in case something goes wrong.

If you had to cope with someone else's accident, would you know what to do? Dealing with accidents is a large and complex subject, and it's easy to make matters worse if you have no idea how to respond. Rather than attempt to deal with this subject in a superficial manner, buy a good First Aid book and read it carefully. Better yet, take a course in First Aid at a local junior college.

Diesel Engine Troubleshooting

General information

Correct diagnosis is an essential part of every repair; without it you can only cure the problem by accident. This Section is devoted to engine checks and diagnosis.

There are several common types of problems that occur frequently on diesel engines. The diagnosis information in this Section assumes the engine is in satisfactory condition internally.

Worn engines that have low compression and excessive blowby gradually become more difficult to start. Low ambient temperatures aggravate the problem. If the engine you are working on consumes an excessive amount of oil, has internal knocks from bearings, etc., burned valves or is otherwise suspected to be in need of internal repair or overhaul; refer to Chapter 4 for further diagnosis information.

If an engine that was running normally loses power or stalls completely, it is usually caused by fuel starvation. Refer to the Maintenance and Fuel System portions of Chapter 2 for GM engines and Chapter 3 for Ford engines.

If an engine that has been functioning well cranks normally but refuses to start when cold, it is usually caused by a malfunctioning glow plug system. Refer to the Glow Plug System Section of the Electrical System portion of Chapter 2 for GM engines and Chapter 3 for Ford engines.

Problems that occur gradually tend to be caused by mechanical wear, whereas problems that occur suddenly are more likely to be caused by an electrical component failure. Always check the simplest items first. Be sure there is clean, good quality diesel fuel in the tank(s). Make sure the fuel filter(s) are clean and no water has accumulated in the water separators. Ensure that the batteries are fully charged and the engine oil is at the correct level and is of the proper viscosity.

Sometimes, in extremely cold weather, the only practical way to get a diesel engine started is to put the vehicle into a warm garage to thaw out. Thickened fuel will clog the filters and excessive cranking can burn out the starter motor.

Rough Idle Diagnosis

A rough idle is caused by uneven power output between cylinders as they fire. The following items can vary the fuel flow to each cylinder and change its relative power output.

Air in fuel system
Nozzle opening pressure
Nozzle tip leakage
Injection pump output
Injection pump low speed governor sensitivity

Smoke Diagnosis

Three different types of smoke will be reviewed in this section. Black, white and blue.

Black smoke is the most common smoking complaint. Under full load conditions, the exhaust smoke contains a large quantity of unburned carbon (soot) formed by excess fuel in the over-rich mixture in the cylinder.

Any variable that increases fuel or reduces the amount of air taken into the cylinder will increase the tendency to produce black exhaust smoke.

Some causes of black smoke are:
Air into injection pump
Restricted fuel return
Injection pump timing advanced (usually will be accompanied by excess combustion noise)
Wrong fuel
Excess fuel delivery from nozzles due to low opening pressure or stuck nozzle
Low fuel pump pressure
Clogged air inlet
EGR stuck open (at full throttle only) - GM engines
Low compression
Missing prechamber (causes black smoke when hot and white smoke when cold)
Restricted exhaust

White Smoke

Under light loads, the average temperature in the combustion chamber may drop 500 degrees due to the decreased amount of fuel being burned. As a result of the lower temperature, the fuel ignites so late that combustion is incomplete at the time the exhaust valve opens and fuel goes into the exhaust in an unburned or partially burned condition, producing the white smoke. White smoke is considered normal when the vehicle is first started but should stop as the vehicle warms up. A continuing white smoke condition could indicate:

Loss of compression
Retarded injection pump timing
Restricted fuel return
Thermostat stuck open

Blue Smoke

Blue smoke indicates that engine oil is burning in the cylinders and may be accompanied by excessive oil consumption. Some conditions which should be considered are:

Cracked pistons
Worn or damaged cylinder walls
Stuck piston rings
Worn or broken piston rings
Failed valve seals or guides
Faulty crankcase vent
Oil level too high
Fuel in crankcase oil
Inaccurate dipstick (causing the crankcase to be overfilled)

Power balance test

A power balance test may be used to determine which cylinder(s) are not doing their share of the work when there is a rough and uneven idle and/or skipping under load. The injector line for each cylinder is loosened momentarily at the injector and the resultant drop in idle speed is noted. If a cylinder is not producing power, the idle speed won't drop when the fuel to it is shut off.

1.111 Loosen the fuel line very carefully; the fuel is under very high pressure - it's a good Idea to have a rag nearby to catch the leaking diesel fuel

Allow the engine to warm up completely. Set the parking brake and put the transmission in Neutral (or Park on automatics).

Loosen the fuel line fittings at the injectors one at a time **(see illustration)** until fuel leaks from the line and notice the rpm drop for each one, then compare the results. **Warning:** *Wear eye protection and position a cloth to catch the fuel spray.*

On a healthy engine, the rpm drop will be about the same for each cylinder. If the rpm doesn't drop when an injector is shorted out, there is a problem with that cylinder.

Perform a compression test as described in Chapter 4. If compression is low, repair the engine as necessary. If compression is normal, test the fuel injector. If the injector and compression test normal, have the injection pump tested by an authorized Roosa-Master repair facility.

The following charts will help you diagnose your engine more completely.

Condition	Possible causes
Won't start	Out of fuel Air leak in suction line(s) Restricted fuel line or filter Defective fuel supply pump Incorrect fuel Paraffin deposit in filter Defective injection pump Incorrect oil grade for climate Low compression Discharged batteries Inoperative glow plugs
Hard to start	Restricted air intake Air leak in suction line(s) Blocked fuel line or filter External fuel leak Clogged or defective injection nozzle(s)s Defective fuel supply pump Incorrect fuel Paraffin deposit in filter Defective injection pump Incorrect oil grade for climate Low compression Advanced timing Retarded timing Worn camshaft Inoperative glow plugs
Starts, then stops	Restricted air intake Excessive exhaust back pressure Out of fuel Blocked fuel return line Air leak in suction line(s) Restricted fuel line or filter External fuel leaks Inoperative fast idle Defective fuel supply pump Paraffin deposit in filter Low idle speed Defective injection pump

Condition	Possible causes
Rough idling	Blocked fuel return line Air leak in suction line(s) External fuel leak Clogged or defective injection nozzle(s) Inoperative fast idle Defective fuel supply pump Low idle speed Defective injection pump Broken or worn piston rings Leaking valve(s)s Low compression Loose timing chain Advanced timing Retarded timing Worn camshaft
Missing	Blocked fuel return line Air leaks in suction line(s) Blocked fuel line or filter External fuel leak Clogged or defective injection nozzle(s) Defective fuel supply pump Incorrect fuel Defective injection pump Leaking valve(s)s Low compression Loose timing chain Worn camshaft
Oil diluted	Defective fuel supply pump Defective injection pump Broken or worn piston rings
Knocking	Clogged or defective injection nozzle(s) Incorrect fuel Defective injection pump Broken or worn piston rings Incorrect bearing clearance Damaged bearings Advanced timing

Condition	Possible causes
Low power	Blocked air intake Excessive exhaust back pressure Blocked fuel return line Air leak in suction line(s) Blocked fuel line or filter External fuel leak Clogged or defective injection nozzle(s) Incorrect fuel Paraffin deposit in filter Defective injection pump Leaking head gasket Broken or worn piston rings Leaking valve(s)s Loose timing chain Advanced timing Retarded timing Worn camshaft
Black smoke when idling	Blocked air intake Blocked fuel line or filter Clogged or defection injection nozzle(s) Broken or worn piston rings Advanced timing
Black smoke under a load	Blocked air intake Excessive exhaust back pressure Blocked fuel line or filter Clogged or defective injection nozzle(s) Incorrect fuel Defective injection pump Low compression Advanced timing Worn camshaft EGR stuck open
White smoke	Blocked fuel return line Air leak in suction line(s) Blocked fuel line or filter Defective injection pump Incorrect oil grade Loose timing chain Retarded timing Inoperative glow plugs
Excessive fuel consumption	Blocked air intake Excessive exhaust back pressure Blocked fuel return line Blocked fuel line or filter External fuel leak Clogged or defective injection nozzle(s) Defective fuel supply pump Incorrect fuel Defective injection pump Leaking head gasket Broken or worn piston rings Leaking valve(s)s Low compression Loose timing chain Advanced timing Retarded timing Worn camshaft
No heat from heater	Leaking head gasket Thermostat stuck open

CONDITION	POSSIBLE CAUSES	CORRECTIVE ACTION
The engine cranks normally, but won't start	a) Incorrect starting procedure	Use the correct starting procedure
	b) Inoperative glow plugs	Refer to the glow plug section in Chapter 2 or 3
	c) Inoperative glow plug control system	Refer to the glow plug section in Chapter 2 or 3
	d) No fuel to cylinders	Remove a glow plug, depress the throttle slightly and crank the engine for five seconds. If no fuel vapor comes out of the glow plug hole, go to step e. If fuel vapors do come out, remove the rest of the glow plugs and see if fuel vapor comes out of each hole when the engine is cranked. If fuel comes out of just one glow plug hole, clean and test the injection nozzle for that cylinder. Crank the engine and note whether fuel vapor is coming out of every glow plug hole. If fuel vapor is coming from every cylinder, go to step k.
	e) Plugged fuel return system	Disconnect the fuel return line from the injection pump and run it into a metal container. Connect a hose to the injection pump connection and run it to the metal container. Crank the engine. If it starts and runs, there's a restriction in the fuel return line. If it doesn't start, remove the top of the injection pump and make sure that is it not plugged. **Note:** *If the fitting is plugged and/or there are small black particles in the pump, a governor weight retainer flex ring might be needed.*
	f) No fuel to the injection pump	Loosen the line coming out of the filter and crank the engine. Fuel should spray out of the fitting. Make sure the fuel is directed away from any open flame or other source of ignition. If fuel sprays from the fitting, go to step j.

CONDITION	POSSIBLE CAUSES	CORRECTIVE ACTION
	g) Blocked fuel filter	Loosen the line going to the filter. If fuel sprays from the fitting, the filter is blocked and must be replaced. Make sure the fuel is directed away from an open flame or other ignition sources.
	h) Inoperative fuel pump	Remove the inlet hose to the fuel pump. Connect a hose to the pump inlet fitting and put the other end of the hose in a container with fuel in it. Loosen the line going to the filter. If fuel does not spray from the fitting replace the pump. Make sure the fuel is directed away from an open flame or any other source of ignition.
	i) Plugged fuel tank filter	Remove the fuel tank and clean or replace the filter.
	j) No voltage to fuel solenoid	1) Hook up a voltmeter to the wire at the injection pump solenoid and to ground. The voltage should be at least 9 volts. If the voltage is less, look for an open circuit in the ignition switch or the switch circuit. 2) Unplug the pink lead from the terminal on top of the injection pump. Turn the key to the ON position. Touch the pink lead to the terminal momentarily. You should hear an audible clicking sound from inside the pump. If you don't, turn off the key and remove the governor cover. Verify that the solenoid arm and plunger move freely. Repair or replace the solenoid as necessary. **NOTE:** *The plunger solenoid sometimes sticks because of an accumulation of metallic debris in the mechanism. Before replacing an inoperative solenoid, blow off the debris with compressed air and recheck it for correct operation by applying 12 volts to the terminal and grounding the cover.*
	k) Incorrect or contaminated fuel	Flush the fuel system and then add the correct fuel. To verify contaminated or poor quality fuel, connect a hose to the inlet of the fuel supply pump and route it to a container of known good quality fuel. If the engine starts and runs, drain and flush poor fuel from vehicle.
	l) Pump timing incorrect	Make certain that the pump timing mark is aligned with the mark on the adapter or the front cover. Check the timing with a timing meter.
	m) Low compression	Check the compression to determine the cause. Repair as necessary.
	n) Bent upper compression ring	Replace the rings.
	o) Injection pump malfunction	With the pump installed on the engine, check the transfer pressure during cranking. If it's insufficient, remove the pump from the engine and have the calibration checked by an authorized repair facility. Pay particular attention to delivery and transfer pressure at cranking speed.
	p) Injector nozzle malfunction	Remove the nozzles from the engine and have them checked on a nozzle tester in accordance with the manufacturer's instructions.
	q) Air in fuel supply lines	Connect a good hose to a container of known good fuel. If the engine starts, locate the source of the air leak in the supply lines.
The oil warning light comes on at idle	a) Oil cooler or oil cooler line blocked	Remove the restriction from the cooler or the cooler line.
	b) Oil pump pressure low	Refer to the oil pump inspection and repair procedure in Chapter 2 or Chapter 3.
The engine won't shut off when the key is turned to OFF. Note: *To shut off the engine In an emergency, pinch the fuel return line at the flexible hose while the engine is at idle.*	The injection pump fuel solenoid does not return the metering valve to the OFF position	Disconnect the wire at the solenoid. If the engine now shuts off, check for a short circuit in the ignition switch or in the ignition circuit. If the engine still doesn't shut off, remove the injection pump and have it repaired.

CONDITION	POSSIBLE CAUSES	CORRECTIVE ACTION
The engine starts, but won't idle, and stalls	a) The slow idle is incorrectly adjusted	Adjust the idle to specification.
	b) The fast-idle solenoid is inoperative	Start the engine (the engine must be cold). The solenoid should move to hold the injection pump lever in the fast-idle position. If the solenoid doesn't move, check for a defective solenoid or look for an open in the fast-idle circuit.
	c) The HPCA switch or the solenoid is inoperative (GM only)	Start the engine (the engine must be cold). Disconnect and connect the green wire on the injection pump. The engine speed and/or sound should change.
	d) Restricted fuel return system	Disconnect the fuel return line at the injection pump and redirect the hose to a metal container. Connect a hose to the injection pump connection and route it to the metal container. Crank the engine and allow it to idle. If the engine idles normally, there's a restriction in the fuel return line. If the engine doesn't idle normally, remove the return line check valve fitting from the top of the pump and make sure it's not plugged. **Note:** *If the fitting is plugged and/or small black particles are visible in the pump, a governor weight retainer flex ring might be bad.*
	e) The glow plugs turn off too soon	Refer to the glow plug section in Chapter 2 or 3
	f) The pump timing is incorrect	Make sure that the timing mark on the injection pomp is aligned with the mark on the adapter or the front cover.
	g) Insufficient fuel supply to the injection pump	Test the engine fuel pump and check the fuel lines. Replace or repair as necessary.
	h) Low compression	Check the compression to determine the cause.
	i) The fuel solenoid closes in the run position	Check the ignition switch and the switch circuit.
	j) Injection pump malfunction	Remove the injection pump and have it repaired by an authorized repair facility.
	k) Incorrect, contaminated or poor quality fuel	To verify contaminated or poor quality fuel, connect a hose to the inlet of the fuel supply pump and route it to a container filled with known good quality fuel. Start the engine. If engine performance improves, drain and flush the system and refill with the correct fuel.
	l). Air in the fuel	To check for air in the fuel lines, disconnect the fuel return line from the top of the pump and connect a cleat hose to the container. Start the engine and let it idle. Watch the return fuel for air bubbles. If bubbles are present, find the source of the air leak in the fuel supply system and repair as necessary. If the engine only stalls during cold starts, check for fuel leaking backwards or air leaking into the fuel lines (a defective injection pump ball check regulator valve will allow fuel to siphon backwards).
The engine stalls under deceleration or heavy braking	a) The idle speed is too low	Adjust the idle to specification. Also, check and adjust the fast-idle solenoid.
	b) Defective governor weight retainer ring	Remove the governor cover and look for small black particles. If they're present, a governor weight retainer flex ring might be the problem.
	c) Binding condition between the Min-Max block and the throttle shaft	To check for binding between the min-max block and the throttle shaft, remove the governor cover, put the throttle in the low-idle position and slide the min-max governor back and forth on the guide stud. It should move freely without binding.
	d) Sticky metering valve or linkage in the injection pump	Remove the pump from the engine and have the calibration checked at an authorized repair facility. Ask the repair shop to pay particular attention to the low-idle setting and the action of the governor at low-idle speed. The shop will repair or replace the metering valve or other governor components as necessary.
Excessive surge at light throttle, under load	a) Torque converter clutch engages too soon	Remove the transmission and have it repaired.
	b) Retarded timing	Make sure that the timing mark on the injection pump is aligned with the mark on the adapter or the front cover.
	c) Restricted fuel filter	Check fuel pump pressure at the inlet and outlet sides of the filter.
	d) Injection pump housing pressure too high	Look for and repair the obstruction in the return line. Replace the back leak connector.
	e) Insufficient injection line volume	Replace the bad injection line(s)
	f) Insufficient injector nozzle opening pressure	Replace the bad injector nozzle(s)

CONDITION	POSSIBLE CAUSES	CORRECTIVE ACTION
Engine starts, but idles roughly, in Neutral or Drive. There is no abnormal noise or smoke, and the engine is warmed up	a) Incorrectly adjusted slow idle	Adjust the slow idle to specification
	b) Leaking injection line(s) for leaks. Repair as necessary.	Wipe off the injection lines and connections. Run the engine and watch
	c) Restricted fuel return systems	Disconnect the fuel return line at the injection pump and redirect the hose to a metal container. Connect a hose to the injection pump connection and run it into the metal container. Start the engine and allow it to idle. If the engine idles normally, locate and repair the obstruction in the fuel return line. If the engine doesn't idle normally, remove the return line check valve fitting from the top of the pump and make sure it's not blocked.
	d) Air in the system	Install a section of clear plastic tubing on the fuel return fitting from the engine. Start the engine. The presence of bubbles in the fuel during cranking or running indicates the presence of an air leak in the suction fuel line. Locate and repair the leak as follows: 1) Raise the vehicle and disconnect both fuel lines at the tank unit. 2) Plug the smaller disconnected return line. 3) Attach a low pressure (preferably, a hand-operated) pump) to the larger 3/8-inch fuel hose and apply 8 to12 psi. 4) Once pressure has been applied, watch the pump gauge. A decrease in pressure indicates that the pressure is pushing fuel out at the leak point. Look for and find the leak. 5) Repair as necessary. Make sure that all hose clamps are the correct size. A burr on the edge of a pipe could rip the inside of a hose and create a leak. Pay particular attention to incorrectly installed or defective auxiliary filters or water separators.
	e) Incorrect or contaminated fuel	Flush the fuel system and refill with the correct fuel.
	f) Defective injector nozzle(s)	To locate the missing cylinder, perform a glow plug resistance test or crack open the nozzle inlet fittings.
	g) Incorrect timing	1. Check and, if necessary, adjust the timing. 2. If the pump is equipped with a mechanical light load advance, check for a sticky or stuck advance mechanism (internal timing) by depressing the rocker level on the side of the injection pump with the engine at 2000 rpm. If the engine sound doesn't change, remove the pump and send it to an authorized repair facility.
	h) Governor weight retainer flex ring fault	Remove the governor cover and look for small black particles. If they're present, a governor weight retainer flex ring might be at fault.
	i) Low or uneven engine compression	Check the compression (see Chapter 4).
	j) Internal injection pump fault	Remove the pump from the engine and have the calibration checked by an authorized repair facility.
When cold, the engine idles roughly after start-, up but smoothes out as it warms up (often accompanied by white exhaust smoke)	a) Incorrect starting procedure	Refer to your owner's manual for the correct starting procedure.
	b) Fast idle solenoid inoperative or incorrectly adjusted	Test and readjust according to the vehicle emissions sticker.
	c) Air in the fuel	Check for air leaks.
	d) One or more glow plugs are not working	Troubleshoot the glow plug system (see Chapter 2 or 3).
	e) The injection pump isn't timed to the engine	Check the alignment of the timing mark on the pump with the engine front cover.
	f) Insufficient engine break-in time	Break-in the engine for 2,000 or more miles.
	g) The injector nozzle(s) is/are sticking open (usually accompanied by a knocking sound)	Remove the injector nozzles from the engine and repair or replace as necessary.
The engine misfires above idle, or runs roughly while driving, but idles okay	a) Injection pump incorrectly timed	Check and adjust timing to specifications.
	b) Air in the fuel	To check for the presence of air in the fuel, disconnect the fuel return line from the top of the pump, connect a clear hose to the return fitting and route the hose to a metal container. Start the engine, allow it to idle and watch for air bubbles in the return fuel. If bubbles are present, locate the source of the air leak in the fuel supply system and repair as necessary.
	c) Blocked fuel return system	Measure the pump housing pressure at idle. The pressure should be no more than 12 psi. If the pressure is excessive, locate and repair the obstruction in the fuel return system.
	d) Blocked fuel supply	Test the fuel supply pump and look for an obstruction in the fuel filter.
	e) Incorrect or contaminated fuel	Flush the fuel system and refill with the correct fuel.

CONDITION	POSSIBLE CAUSES	CORRECTIVE ACTION
Noticeable lack of power	a) Blocked air intake	Inspect the air cleaner element.
	b) Incorrectly adjusted timing	Make sure the timing mark on the injection pump is aligned with the mark on the adapter or front cover
	c) Defective EGR or EPR (GM only)	Refer to Emissions Diagnosis, Chapter 2.
	d) Blocked or damaged exhaust system	Check the system and replace as necessary.
	e) Obstruction in fuel filter	Replace the filter.
	f) Blocked fuel tank vacuum vent in fuel cap	Remove the fuel tank filler cap. If you hear a loud hissing sound (a *slight* hissing sound is normal), the vacuum vent in the cap is plugged. Replace the cap.
	g) Insufficient fuel supply from fuel tank to injection pump	Inspect the fuel supply system to determine the cause of the restriction. Repair as necessary.
	h) Obstruction in fuel tank filter	Remove the fuel tank and clean or replace the filter.
	i) Pinched or otherwise restricted return system	Inspect the system for restrictions and repair as necessary.
	j) Incorrect or contaminated Fuel	Flush the fuel system and refill with the correct fuel.
	k) External compression leaks	Check for compression leaks at all injector nozzles and glow plugs. If you find a leak, tighten the nozzle or glow plug.
	l) Obstruction in injector nozzle(s)	Remove the injector nozzles. Have them checked for obstructions and then repair, replace or clean (where applicable) as necessary
	m) Low compression	Check the compression to determine the cause.
The engine stalls on deceleration or at idle	Sticking metering valve	Remove the metering valve. Clean with 400 or 500 grit sandpaper. Wet the sandpaper with diesel fuel and turn the metering valve in the wet sandpaper no more than 5 or 6 turns.
The engine won't return to idle	a) The external linkage is binding	Free up the linkage. Adjust or replace as necessary.
	b) Incorrect fast idle	Check the fast-idle adjustment.
	c) Defective injection pump	Remove the injection pump for repair.
Fuel leaks on the ground	a) Loose or broken fuel line or connection	Inspect the entire fuel system, including the tank, hoses, lines and injection lines. Determine the source and cause of the leak, and repair it as necessary.
	b) Injection pump internal seal leak	Remove the injection pump and have it repaired by an authorized repair facility.
Rapping noise from one or more cylinders (sounds like rod bearing knock)	a) Injector nozzle(s) sticking open or with very low nozzle opening pressure	Remove the injector nozzles and have them tested. Clean or replace as necessary
	b) Mechanical problem	Refer to mechanical diagnosis.
	c) Piston hitting cylinder head	Replace the defective parts. Break in the engine for 2000 miles.
Louder than normal operating noise	Incorrectly adjusted timing	Make sure that the timing mark on the injection pump is aligned with the mark on the front housing.
Louder than normal operating noise, accompanied by excessive black smoke	a) Defective EGR (GM)	Refer to the Emission Section in Chapter 2.
	b) Incorrect injection pump housing pressure	Check the housing pressure.
	c) Defective injection pump	Remove the injection pump and have it repaired by an authorized repair facility.
Internal or external engine noise	Engine, fuel pump, alternator, water pump, valve train, vacuum pump, bearing(s), etc.	Repair or replace as necessary. If the noise is internal, see preceding diagnoses for noise, and for rough idle with excessive noise and/or smoke.
Engine Overheats	a) Leaking coolant system or oil cooler system, or defective coolant recovery system	Inspect these systems for leaks and repair as necessary.
	b) Slipping or damaged belt	Adjust or replace the belt as necessary.
	c) Thermostat stuck closed	Check and, if necessary, replace the thermostat.
	d) Leaking head gasket	Check and, if necessary, replace the head gasket.
Poor performance, excessive hot cranking time, no W.O.T. upshift	Kink in the fuel supply hose between the fuel tank and the body	Shorten the fuel supply hose at the kinked area.
Excessive engine blowby	Bent upper compression ring	Check the compression. If it's about 100 psi low, replace the piston rings.

Chapter 2
GM 5.7L, 6.2L and 6.5L V8 engines

Specifications

Recommended lubricants and fluids

Note: *Listed here are manufacturer recommendations at the time this manual was written. Manufacturers occasionally upgrade their fluid and lubricant specifications, so check with your local auto parts store for current recommendations.*

Engine oil	
Type	API grade SF/CD or SF/CE
Viscosity	See accompanying chart
Capacity	7.0 qts
Engine coolant	50/50 mixture of water and ethylene glycol based antifreeze

General

Cylinder numbering (front to rear)	
Left (driver's) side	1-3-5-7
Right side	2-4-6-8
Firing order	
5.7L	1-8-4-3-6-5-7-2
6.2L and 6.5L	1-8-7-2-6-5-4-3

Camshaft

Bearing journal diameter	
5.7L	
Number 1	2.0357 to 2.0365 in
Number 2	2.0157 to 2.0165 in
Number 3	1.9957 to 1.9965 in
Number 4	1.9757 to 1.9765 in
Number 5	1.9557 to 1.9565 in
6.2L and 6.5L	
Number 1, 2, 3 and 4	2.1642 to 2.1663 in
Number 5	2.0067 to 2.0089 in

Torque specifications	Ft-lbs (unless otherwise noted)
Camshaft sprocket bolt	
5.7L	65
6.2L and 6.5L	
1985 through 1993	74
1994 through 1996	126
Camshaft thrust plate bolt (6.2L and 6.5L)	17
Connecting rod cap nuts	
5.7L	42
6.2L and 6.5L	48
Crankshaft damper bolt	200
Crankshaft pulley bolts	30

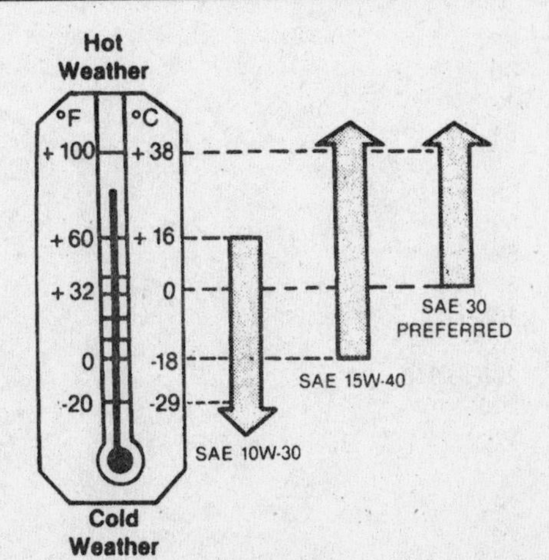

RECOMMENDED SAE VISCOSITY GRADE ENGINE OILS

SELECT THE SAE GRADE OIL BASED ON THE EXPECTED TEMPERATURE RANGE BEFORE NEXT OIL CHANGE

API SERVICE SF CD
SAE 30

SF/CD QUALITY PREFERRED
SF/CC QUALITY ACCEPTABLE
DO NOT USE SAE 10W-40 GRADE OIL, OR ANY OTHER GRADE NOT RECOMMENDED

Engine oil viscosity chart

Torque specifications (continued)

	Ft-lbs (unless otherwise noted)
Cylinder head bolts	
5.7L	
First step	100
Second step	130
6.2L and 6.5L	
First step	20
Second step	50
Third step	Tighten an additional 90-degrees
Engine block to transmission	30
Engine mount to cylinder block bolts	75
Engine mount to frame bolts	50
Exhaust manifold bolts	26
Flywheel/driveplate bolts	65
Fuel injection pump mounting nuts	31
Front cover bolts	33
Fuel injector to cylinder head	
5.7L	25
6.2L and 6.5L	50
Glow plug	
5.7L	124 in lbs
6.2L and 6.5L	156 in lbs
Hydraulic lifter guide plate bolts	18
Injector line fitting	18
Injector nozzles	50
Injector pump gear bolts	17
Injector pump gear nuts	31
Intake manifold bolts	
5.7L	40
6.2L and 6.5L	31
Main bearing cap bolts	
5.7L	120
6.2L and 6.5L	
Inner	110
Outer	100
Oil pump drive - clamp bolt	31
Oil pump cover screws	12
Oil pump to bearing cap bolts	
5.7L	35
6.2L and 6.5L	65
Oil pan to crankcase bolts	
5.7L	10
6.2L and 6.5L	
All except two rear	84 in lbs
Two rear	17
Oil pan to timing cover bolts	
6.2L and 6.5L	84 in lbs
Oil pan drain bolt	20
Oil cooler lines to oil filter base	12
Oil cooler lines to radiator	25
Pulley to vibration damper bolts	20
Rocker arm shaft bolts	
5.7L	28
6.2L and 6.5L	40
Secondary flywheel bolts (dual mass flywheel only)	12
Thermostat housing bolts	31
Timing cover bolts	
5.7L	35
6.2L and 6.5L	33
Turbocharger to exhaust manifold bolts	37
Vacuum pump clamp bolt	31
Valve cover bolts	16
Vibration damper bolt	
5.7L	200 to 310
6.2L and 6.5L	200
Water pump plate to water pump bolts	17
Water pump to front cover bolts	
5.7L	13
6.2L and 6.5L	32
Water pump plate to front cover bolts	17

Introduction

This Chapter contains everything you need to know to service your GM 5.7L or 6.2L/6.5L V8 diesel engine. Included are sections on routine maintenance, on servicing the cooling, fuel, electrical and emission control systems, and even the engine repairs you can perform with the engine still installed in the vehicle (you'll find general engine overhaul procedures - those jobs that require engine removal - in Chapter 4).

Look at the mileage/time master maintenance schedule below. It tells you what to do, when to do it and how to do it. Each recommended service or maintenance procedure - a visual check, and adjustment, replacement of a component, etc. - is explained later in this Chapter.

Servicing your engine in accordance with this sensible maintenance schedule will significantly prolong its service life. Keep in mind that this is a comprehensive plan: servicing selected items - but skipping others - will not produce the same results.

When you service your engine, you'll find that many of the maintenance procedures can be grouped together because they're logically related, or because they're located next to each other.

Before you get started, read through the service items you're planning to do, familiarize yourself with the procedures and gather up all the parts and tools you'll need. If it looks like you might run into problems during a particular job, seek advice from a mechanic or an experienced do-it-yourselfer.

Maintenance schedule

Every 250 miles or weekly, whichever comes first

Check the engine oil level
Check the engine coolant level

Every 5000 miles or 6 months, whichever comes first

All items listed above, plus:
Change the engine oil and filter*
Check and adjust the engine drivebelts

Every 15,000 miles or 12 months, whichever comes first

Check the cooling system
Check and service the batteries
Replace the air filter
Replace the fuel filter
Inspect and replace, if necessary, all underhood hoses

** This item is affected by "severe" operating conditions as described below. If the vehicle is operated under severe conditions, perform all maintenance indicated with an asterisk (*) at 3000 mile/3 month intervals. Severe conditions exist if you mainly operate the vehicle.*

In dusty areas
Towing a trailer
Idling for extended periods and/or driving at low speeds
When outside temperatures remain below freezing and most trips are less than four miles long

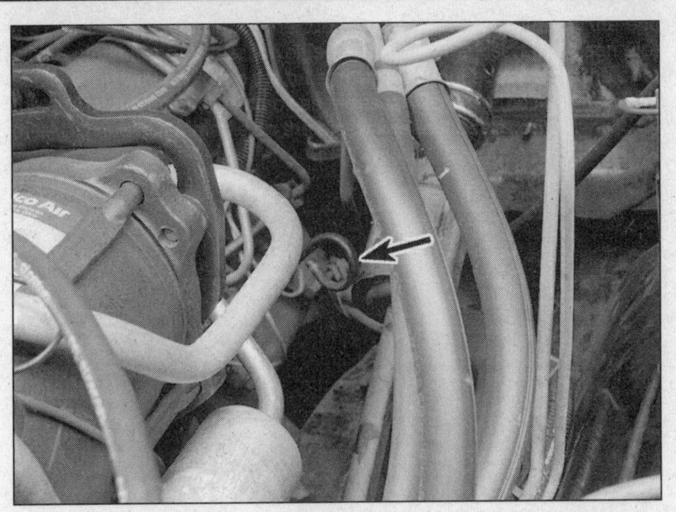

2.1a The dipstick (arrow) is located in a tube on the drivers aide of the engine

2.1b The oil level must be maintained between the marks at all times - It takes one quart of oil to raise the level from the ADD mark to the FULL mark

Fluid level checks

Note: *The following are fluid level checks to be done on a 250 mile or weekly basis. Additional fluid level checks can be found in specific maintenance procedures which follow. Regardless of intervals, be alert to fluid leaks under the vehicle which would indicate a fault to be corrected immediately.*

1 Fluids are an essential part of the lubrication and cooling systems. Because the fluids gradually become depleted and/or contaminated during normal operation of the vehicle, they must be periodically replenished. See the Specifications at the beginning of this Chapter before adding fluid to any of the following components. **Note:** *The vehicle must be on level ground when fluid levels are checked.*

Engine oil

2 The engine oil level is checked with a dipstick that extends through a tube and into the oil pan at the bottom of the engine.

3 The oil level should be checked before the vehicle has been driven, or about 15 minutes after the engine has been shut off. If the oil is checked immediately after driving the vehicle, some of the oil will remain in the upper engine components, resulting in an inaccurate reading on the dipstick.

4 Pull the dipstick out of the tube and wipe all the oil from the end with a clean rag or paper towel **(see illustration)**. Insert the clean dipstick all the way back into the tube, then pull it out again. Note the oil at the end of the dipstick. Add oil as necessary to keep the level between the ADD and FULL marks on the dipstick **(see illustration)**.

2.2 Oil is added to the engine after unscrewing the filler cap

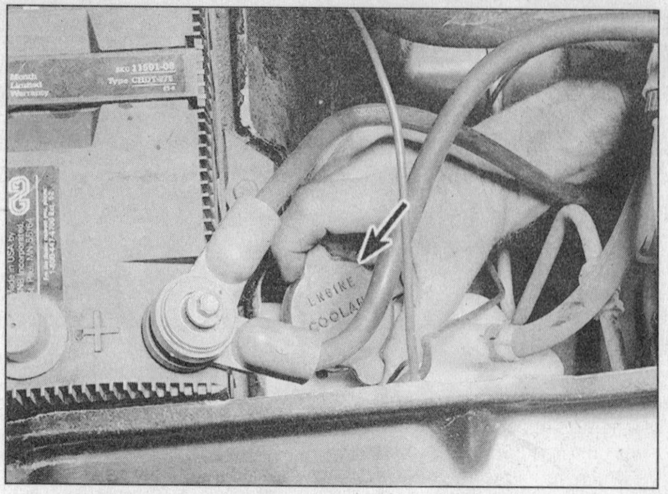

2.3 Check the coolant level in the coolant reservoir

5 Do not overfill the engine by adding too much oil since this may result in oil fouled spark plugs, oil leaks or oil seal failures.
6 Oil is added to the engine after unscrewing the filler cap **(see illustration)**. A funnel may help to reduce spills.
7 Checking the oil level is an important preventive maintenance step. A consistently low oil level indicates oil leakage through damaged seals, defective gaskets or past worn rings or valve guides. If the oil looks milky or has water droplets in it, the cylinder head gasket(s) may be blown or the head(s) or block may be cracked. The engine should be checked immediately. The condition of the oil should also be checked. Whenever you check the oil level, slide your thumb and index finger up the dipstick before wiping off the oil. If you see small dirt or metal particles clinging to the dipstick, the oil should be changed.

Engine coolant

Warning: *Do not allow antifreeze to come in contact with your skin or painted surfaces of the vehicle. Flush contaminated areas immediately with plenty of water. Don't store new coolant or leave old coolant lying around where it's accessible to children or pets - they're attracted by its sweet smell. Ingestion of even a small amount of coolant can be fatal! Wipe up garage floor and drip pan coolant spills immediately. Keep antifreeze containers covered and repair leaks in the cooling system as soon as they are noted.*
8 All vehicles covered by this manual are equipped with a pressurized coolant recovery system. A white plastic coolant reservoir located in the engine compartment is connected by a hose to the radiator filler neck **(see illustration)**. If the engine overheats, coolant escapes through a valve in the radiator cap and travels through the hose into the reservoir. As the engine cools, the coolant is automatically drawn back into the cooling system to maintain the correct level. **Warning:** *Do not remove the radiator cap to check the coolant level when the engine is warm.*
9 The coolant level in the reservoir should be checked regularly. The level in the reservoir varies with the temperature of the engine. When the engine is cold, the coolant level should be at or slightly above the FULL COLD mark on the reservoir. Once the engine has warmed up, the level should be at or near the FULL HOT mark. If it isn't, allow the engine to cool, then remove the cap from the reservoir and add a 50/50 mixture of ethylene glycol based antifreeze and water.
10 Drive the vehicle and recheck the coolant level. If only a small amount of coolant is required to bring the system up to the proper level, water can be used. However, repeated additions of water will dilute the antifreeze and water solution. In order to maintain the proper ratio of antifreeze and water, always top up the coolant level with the correct mixture. An empty plastic milk jug or bleach bottle makes an excellent container for mixing coolant. Do not use rust inhibitors or additives.
11 If the coolant level drops consistently, there may be a leak in the system. Inspect the radiator, hoses, filler cap, drain plugs and water pump (see the Cooling system section later in this Chapter). If no leaks are noted, have the radiator cap pressure tested by a service station.
12 If you have to remove the radiator cap, wait until the engine has cooled, then wrap a thick cloth around the cap and turn it to the first stop. If coolant or steam escapes, let the engine cool down longer, then remove the cap.
13 Check the condition of the coolant as well. It should be relatively clear. If it's brown or rust colored, the system should be drained, flushed and refilled. Even if the coolant appears to be normal, the corrosion inhibitors wear out, so it must be replaced at the specified intervals.

Battery electrolyte

14 Most vehicles with which this manual is concerned are equipped with batteries which are permanently sealed (except for vent holes) and have no filler caps. Water doesn't have to be added to these batteries at any time. If maintenance type batteries are installed, the caps on the top of the batteries should be removed periodically to check for a low water level. This check is most critical during the warm summer months.

Air filter replacement

1 At the specified intervals, the air filter should be replaced with a new one.
2 The filter is located on top of the air crossover manifold and is replaced by unscrewing the wing nut(s) from the top of the filter housing and lifting off the cover **(see illustration)**.

2.4 Remove the wingnuts; and raise the air cleaner cover

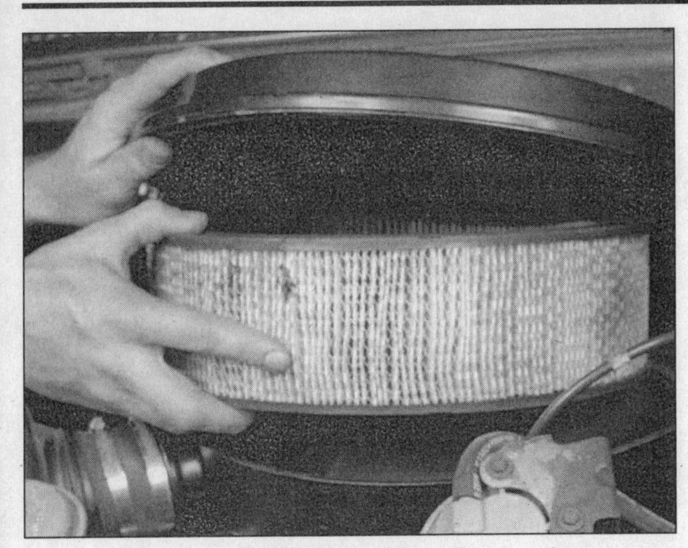

2.5 Lift the air filter element out of the housing

2.6 Open the petcock on the bottom on the primary filter, then drain the water by opening the petcock at the top

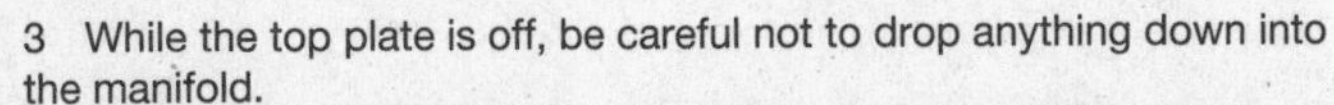

3 While the top plate is off, be careful not to drop anything down into the manifold.

4 Lift the air filter element out of the housing **(see illustration)** and wipe out the inside of the air cleaner housing with a clean rag.

5 Place the new filter in the air cleaner housing. Make sure it seats properly in the bottom of the housing.

6 Installation is the reverse of removal.

Fuel filter replacement

Warning: *Diesel fuel is flammable and may be hot which could cause burns, so take extra precautions when working on any part of the fuel system. If you spill fuel on your skin, rinse it off immediately with soap and water. Have a Class B fire extinguisher on hand.*

5.7L engine

1 The rectangular fuel filter used on these models is located behind the injection pump.

2 Remove the air cleaner assembly.

3 Remove the wingnut that fastens the fuel filter to the mounting bracket.

4 Use a flare nut wrench to disconnect the fuel lines and detach the filter from the bracket, tilting it back to prevent the residual fuel from spilling out.

5 Install the new filter by reversing the removal procedure. Tighten the fittings securely, but don't cross thread them.

6.2L and 6.5L engines

Early models - primary filter

6 The primary filter is located on the firewall. This filter also acts as a water separator and water can be drained by opening the petcock at the bottom, then the top of the filter **(see illustration)**. Once all the water is drained, close the bottom petcock.

7 Remove the primary filter using an oil filter wrench **(see illustration)**. This is a good time to replace the filter with a new one.

8 Fill the filter with clean diesel fuel and install it, turning it an additional 3/4 turn after it contact the filter gasket. Close the petcock tightly, then start the engine and let it run for a short time until the air is bled from the system. It may run roughly at first, until the air is purged.

Early models - secondary filter

9 Remove the air cleaner assembly.

10 The secondary filter is mounted on the back of the intake manifold **(see illustration)**.

2.7 Use an oil filter wrench to unscrew the primary filter

2.8 The secondary filter is mounted on the back of the intake manifold (arrow); disconnect the fuel lines, remove the adapter bolts and remove the adapter from the manifold

2.9 Unplug the pink wire (arrow) from the fuel injection pump so the engine won't start during the fuel 'filter purging process

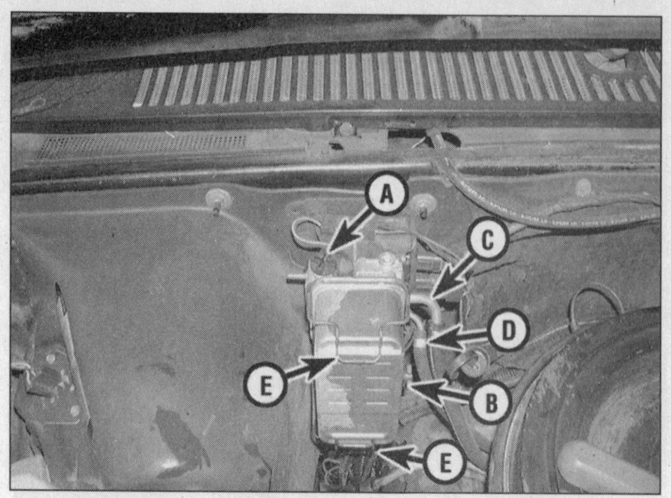

2.10 Details of the air bleed valve equipped fuel filter

A Air bleed valve
B Fuel filter
C Fuel inlet hose
D Fuel outlet hose
E Retaining clips

11 Use a flare nut wrench to disconnect the fuel lines from the adapter housing.
12 Unbolt the adapter, remove it and unscrew the filter element **(see illustration 2.8)**.
13 Screw the new filter onto the adapter and rotate it an additional 2/3 turn after contacting the gasket.
14 Install the adapter and bolts. Tighten the bolts securely.
15 Connect the inlet fuel line only and the fitting securely.
16 Air must be purged from the filter before connecting the outlet line. Unplug the pink wire lead from the fuel injection pump **(see illustration)** so the engine won't start, and then place a rag under the filter outlet. Crank the engine over for not more than 10 seconds, until fuel comes out the outlet port. If no fuel is observed, wait 15 seconds before repeating the procedure. Connect the outlet fuel line and tighten it securely.
17 Reconnect the pink wire, install the air cleaner, then start the engine and allow it to idle for several minutes to purge any remaining air from the system and check for leaks.

Later models

18 On these models the fuel filter is mounted on firewall or intake manifold and is held in place by clips. On some later models, the filter is equipped with an air bleed valve **(see illustration)**.
19 Remove the fuel filler cap, engine cover (van models) and air cleaner assembly.
20 On filters without an air bleed valve, place a rag under the filter, then detach the lower clip with a screwdriver to release the fuel pressure. Detach the upper clip and lift the filter off the bracket.
21 On air bleed valve equipped filters, drain the fuel from the filter by opening the air bleed valve and, if equipped, the engine water drain valve (at the front of the engine, near the thermostat housing). Detach the clips and lift the filter off.
22 Make sure the fuel filter sealing surfaces are clean before placing the filter in position. On non air bleed valve equipped filters, snap the upper clip securely into place. On filters with air bleed valves, secure both clips, then close the valve and connect one end of an 1/8 inch hose to the air bleed port, placing the other end in a container to catch the purged fuel.
23 The air must be purged from the filter after reinstallation or starting or stalling problems will result. Disconnect the pink wire lead from the fuel injection pump **(see illustration 2.9)** and crank the engine over for no more than 10 seconds. Then let the starter motor cool for 30 seconds and repeat the procedure until clear, bubble free fuel comes out of the lower fitting or air bleed. Connect the lower clip, close the bleed hose and plug in the fuel injection connector.
24 Start the engine and allow it to idle for several minutes to purge any remaining air from the system and check for leaks.

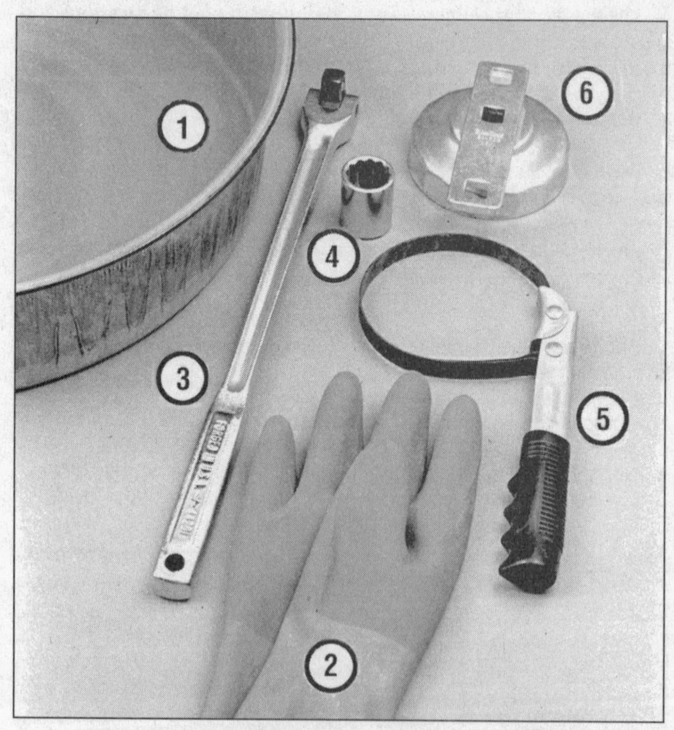

2.11 These tools are required when changing the engine oil and filter

1 ***Drain pan*** *– It should be fairly shallow in depth, but wide to prevent spills*
2 ***Rubber gloves*** *– When removing the drain plug and filter, you will get oil on your hands (the gloves will prevent burns)*
3 ***Breaker bar*** *– Sometimes the oil drain plug is tight and a long breaker bar is needed to loosen it*
4 ***Socket*** *– To be used with the breaker bar or a ratchet (must be the correct size to fit the drain plug – six-point preferred)*
5 ***Filter wrench*** *– This is a metal bank-type wrench, which requires clearance around the filter to be effective*
6 ***Filter wrench*** *– This type fits on the bottom of the filter and can be turned with a ratchet or breaker bar (different size wrenches are available for different types of filters)*

2.12 The oil drain plug is located at the bottom of the pan and should be removed with a socket or box-end wrench - DO NOT use an open-end wrench (the corners on the hex can be very easily rounded off)

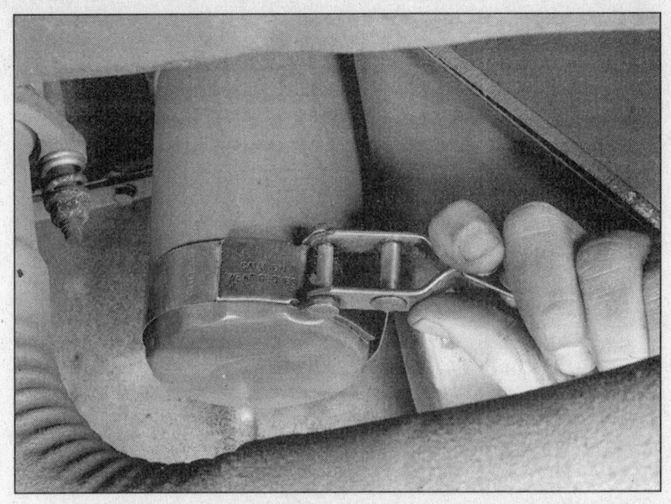

2.13 Use a strap-type oil filter wrench to loosen the filter- If access makes removal difficult, other types of filter wrenches are available

Engine oil and filter change

1 Frequent oil changes are the most important preventive maintenance procedures that can be done by the home mechanic. As engine oil ages, it becomes diluted and contaminated, which leads to premature engine wear.

2 Although some sources recommend oil filter changes every other oil change, we feel that the minimal cost of an oil filter and the relative ease with which it is installed dictate that a new filter be installed every time the oil is changed.

3 Gather together all necessary tools and materials before beginning this procedure **(see illustration)**.

4 You should have plenty of clean rags and newspapers handy to mop up any spills. Access to the underside of the vehicle may be improved if the vehicle can be lifted on a hoist, driven onto ramps or supported by jackstands. **Warning:** *Do not work under a vehicle that is supported only by a bumper, hydraulic or scissors type jack.*

5 If this is your first oil change, get under the vehicle and familiarize yourself with the locations of the oil drain plug and the oil filter. The engine and exhaust components will be warm during the actual work, so note how they are situated to avoid touching them when working under the vehicle.

6 Warm the engine to normal operating temperature. If the new oil or any tools are needed, use this warm up time to gather everything necessary for the job. The correct type of oil for your application can be found in the Specifications at the beginning of this Chapter.

2.14 Lubricate the oil filter gasket with clean engine oil before installing the filter on the engine

7 With the engine oil warm (warm engine oil will drain better and more built up sludge will be removed with it), raise and support the vehicle. Make sure it's safely supported!

8 Move all necessary tools, rags and newspapers under the vehicle. Set the drain pan under the drain plug. Keep in mind that the oil will initially flow from the pan with some force; position the pan accordingly.

9 Being careful not to touch any of the hot exhaust components, use a wrench to remove the drain plug near the bottom of the oil pan **(see illustration)**. Depending on how hot the oil is, you may want to wear gloves while unscrewing the plug the final few turns.

10 Allow the old oil to drain into the pan. It may be necessary to move the pan as the oil flow slows to a trickle.

11 After all the oil has drained, wipe off the drain plug with a clean rag. Small metal particles may cling to the plug and would immediately contaminate the new oil.

12 Clean the area around the drain plug opening and reinstall the plug. Tighten the plug securely with the wrench. If a torque wrench is available, use it to tighten the plug.

13 Move the drain pan into position under the oil filter.

14 Use the filter wrench to loosen the oil filter **(see illustration)**. Chain or metal band filter wrenches may distort the filter canister, but it doesn't matter since the filter will be discarded anyway.

15 Completely unscrew the old filter. Be careful - it's full of oil. Empty the oil inside the filter into the drain pan.

16 Compare the old filter with the new one to make sure they're the same type.

17 Use a clean rag to remove all oil, dirt and sludge from the area where the oil filter mounts to the engine. Check the old filter to make sure the rubber gasket isn't stuck to the engine. If the gasket is stuck to the engine (use a flashlight if necessary), remove it.

18 Apply a light coat of clean oil to the rubber gasket on the new oil filter **(see illustration)**.

19 Attach the new filter to the engine, following the tightening directions printed on the filter canister or packing box. Most filter manufacturers recommend against using a filter wrench due to the possibility of overtightening and damage to the seal.

20 Remove all tools, rags, etc. from under the vehicle, being careful not to spill the oil in the drain pan, then lower the vehicle.

21 Move to the engine compartment and locate the oil filler cap.

22 Pour the fresh oil through the filler opening. A funnel may be helpful.

23 Pour seven quarts of fresh oil into the engine. Wait a few minutes to allow the oil to drain into the pan, then check the level on the oil dipstick (see the *Fluid level checks* section earlier in this Chapter). If the oil level is above the ADD mark, start the engine and allow the new oil to circulate.

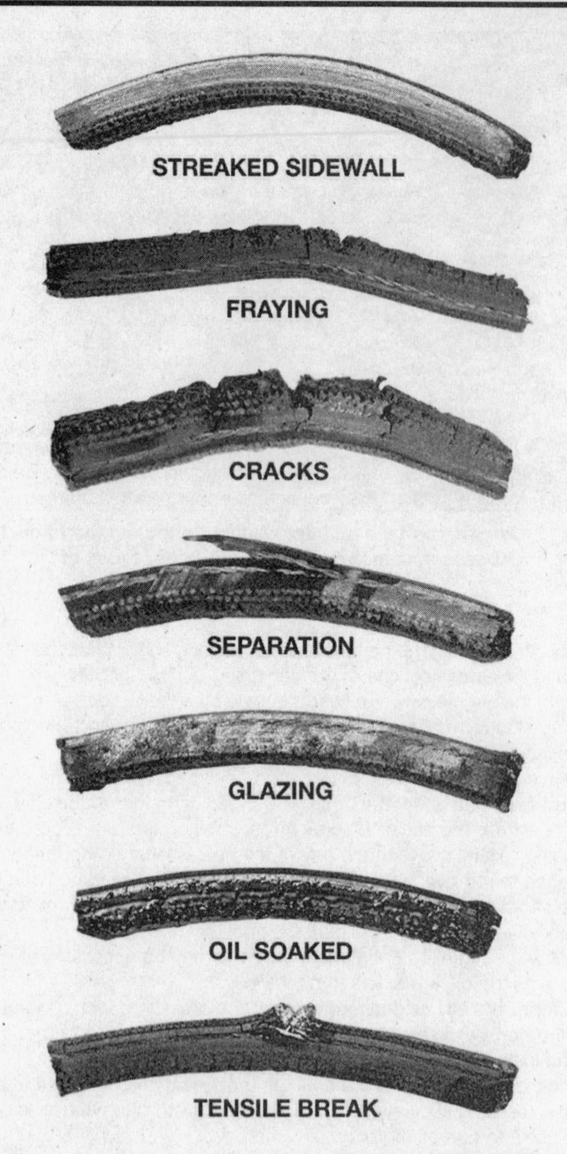

2.15 Here are some of the more common problems associated with drivebelts (check the belts very carefully to prevent an untimely breakdown)

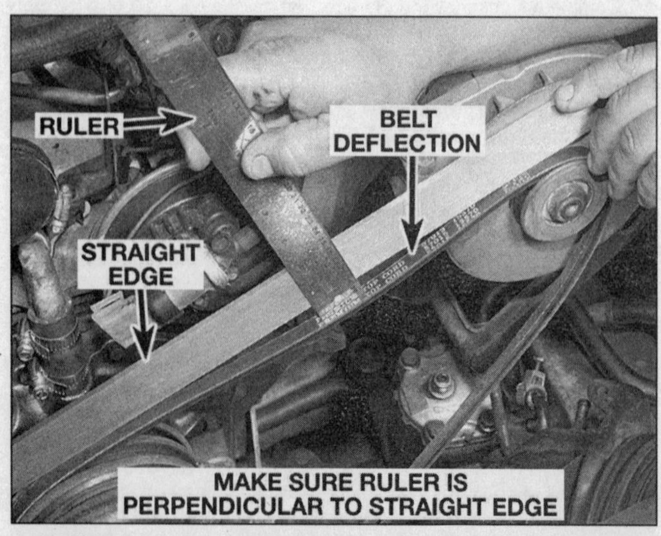

2.16 Measuring drivebelt deflection with a straightedge and ruler

24 Run the engine for only about a minute and then shut it off. Immediately look under the vehicle and check for leaks at the oil pan drain plug and around the oil filter. If either is leaking, tighten with a bit more force.

25 With the new oil circulated and the filter now completely full, recheck the level on the dipstick and add more oil as necessary.

26 During the first few trips after an oil change, make it a point to check frequently for leaks and proper oil level.

27 The old oil drained from the engine cannot be reused in its present state and should be disposed of. Oil reclamation centers, auto repair shops and gas stations will normally accept the oil, which can be refined and used again. After the oil has cooled it can be drained into a container (capped plastic jugs, topped bottles, milk cartons, etc.) for transport to one of these disposal sites. Don't dispose of the oil by pouring it on the ground or down a drain!

Drivebelt check, adjustment and replacement

1 The drivebelts, or V belts as they are sometimes called, at the front of the engine play an important role in the overall operation of the engine and its accessories. Due to their function and material makeup, the belts are prone to failure after a period of time and should be inspected periodically.

2 The number of belts used on a particular vehicle depends on the accessories installed. Drivebelts are used to turn the alternator, power steering pump, water pump and air conditioning compressor. Depending on the pulley arrangement, more than one of these components can be driven by a single belt. On some later models, a single serpentine drivebelt is located at the front of the engine and drives all of the components.

3 With the engine off, locate the drivebelts at the front of the engine. Using your fingers (and a flashlight, if necessary), move along the belts, checking for cracks and separation of the belt plies **(see illustration)**. Also check for fraying and glazing, which gives the belt a shiny appearance. Both sides of the belts should be inspected, which means you will have to twist the belt to check the underside. Check the pulleys for nicks, cracks, distortion and corrosion. On serpentine belts, check the ribs on the underside of the belt. All ribs should be the same depth, with no uneven surface.

4 The tension of each belt is checked by pushing on the belt at a distance halfway between the pulleys. Push firmly with your thumb and see how much the belt moves (deflects) **(see illustration)**. A rule of thumb is that if the distance from pulley center to pulley center is between 7 and 11 inches, the belt should deflect 1/4 inch. If the belt travels between pulleys spaced 12 to16 inches apart, the belt should deflect 1/2 inch. On single serpentine belts, the tension of the belt is automatically controlled by a tensioner, so the tension does not need to be adjusted.

5 If it is necessary to adjust the belt tension on non serpentine belts, either to make the belt tighter or looser, it is done by moving the belt driven accessory on the bracket.

6 For each component there will be an adjustment or strap bolt and a pivot bolt. Both bolts must be loosened slightly to enable you to move the component.

7 After the two bolts have been loosened, move the component away from the engine (to tighten the belt) or toward the engine (to loosen the belt). Hold the accessory in this position and check the belt tension. If it's correct, tighten the two bolts until snug, then recheck the tension. If it's okay, tighten the two bolts securely.

8 You may have to use some sort of pry bar to move the accessory while the belt is adjusted. If this must be done to gain the proper leverage, be very careful not to damage the component being moved or the part being pried against.

9 To replace a non serpentine belt, follow the above procedures for drivebelt adjustment but slip the belt off the pulleys and remove it. Since belts tend to wear out more or less at the same time, it's a good idea to replace all of them at the same time. Mark each belt and the corresponding pulley grooves so the replacement belts can be installed properly. To replace the serpentine belt, use a breaker bar and socket on the tensioner bolt and rotate the tensioner counterclockwise. This will release the tension so the belt can be removed.

When the belt is out of the way, release the tensioner slowly so you don't damage it.

10 Take the old belt with you when purchasing a new one to make a direct comparison for length, width and design.

11 When installing the new serpentine belt, make sure it is routed correctly (refer to the label in the engine compartment). Also, the belt must completely engage the grooves in the pulleys.

Cooling system check

1 Many major engine failures can be attributed to a faulty cooling system. If the vehicle is equipped with an automatic transmission, the cooling system also cools the transmission fluid and thus plays an important role in prolonging transmission life.

2 The cooling system should be checked with the engine cold. Do this before the vehicle is driven for the day or after it has been shut off for at least three hours.

3 Remove the radiator cap by turning it to the left until it reaches a stop. If you hear a hissing sound (indicating there is still pressure in the system), wait until this stops. Now press down on the cap with the palm of your hand and continue turning to the left until the cap can be removed. Thoroughly clean the cap, inside and out, with clean water. Also clean the filler neck on the radiator. All traces of corrosion should be removed. The coolant inside the radiator should be relatively transparent. If it is rust colored, the system should be drained and refilled (see the *Cooling system servicing* Section). If the coolant level is not up to the top, add additional antifreeze/coolant mixture (see the *Fluid level checks* Section).

4 Carefully check the large upper and lower radiator hoses along with the smaller diameter heater hoses that run from the engine to the firewall. Inspect each hose along its entire length, replacing any hose that is cracked, swollen or shows signs of deterioration. Cracks may become more apparent if the hose is squeezed **(see illustration)**. Regardless of condition, it's a good idea to replace hoses with new ones every two years.

5 Make sure all hose connections are tight. A leak in the cooling system will usually show up as white or rust colored deposits on the areas adjoining the leak. If wire type clamps are used at the ends of the hoses, it may be a good idea to replace them with more secure screw type clamps.

6 Use compressed air or a soft brush to remove bugs, leaves, etc. from the front of the radiator or air conditioning condenser. Be careful not to damage the delicate cooling fins or cut yourself on them.

7 Every other inspection, or at the first indication of cooling system problems, have the cap and system pressure tested. If you don't have a pressure tester, most gas stations and repair shops will do this for a minimal charge.

Check for a chafed area that could fail prematurely.

Check for a soft area indicating the hose has deteriorated inside.

Overtightening the clamp on a hardened hose will damage the hose and cause a leak.

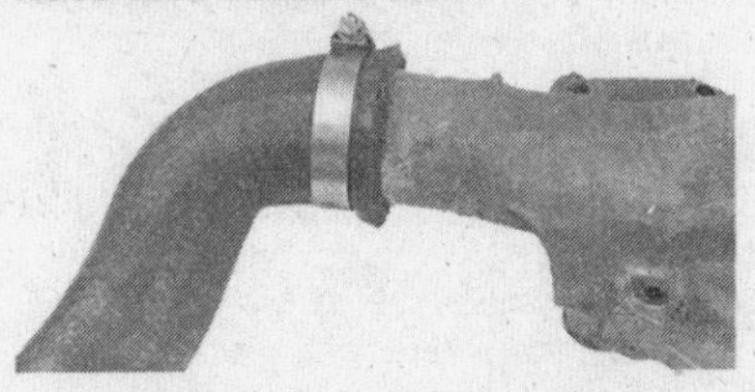

Check each hose for swelling and oil-soaked ends. Cracks and breaks can be located by squeezing the hose.

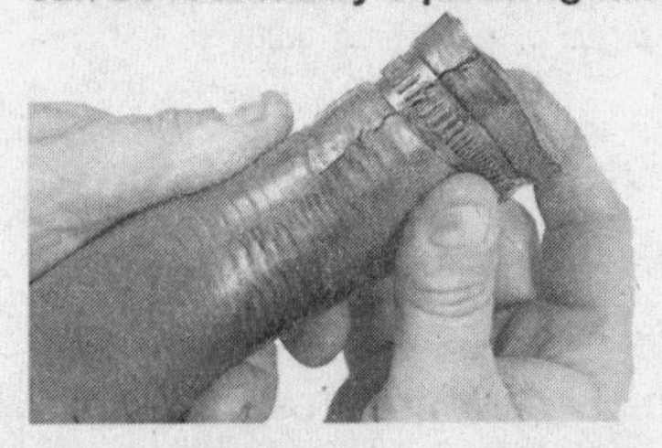

2.17 Hoses, like drivebelts, have a habit of failing at the worst possible time - to prevent the inconvenience of a blown radiator or heater hose, inspect them carefully as shown here

Cooling system servicing (draining, flushing and refilling)

Warning: *Do not allow antifreeze to come in contact with your skin or painted surfaces of the vehicle. Rinse off spills immediately with plenty of water. Antifreeze is highly toxic if ingested. Never leave antifreeze lying around in an open container or in puddles on the floor; children and pets are attracted by its sweet smell and may drink it. Check with local authorities regarding disposal of used antifreeze. Many communities have collection centers which will see that antifreeze is disposed of safely.*

1 Periodically, the cooling system should be drained, flushed and refilled to replenish the antifreeze mixture and prevent formation of rust and corrosion, which can impair the performance of the cooling system and cause engine damage. When the cooling system is serviced, all hoses and the radiator cap should be checked and replaced if necessary.

2 Apply the parking brake and block the wheels. If the vehicle has just been driven, wait several hours to allow the engine to cool down before beginning this procedure.

3 Once the engine is completely cool, remove the radiator cap.

4 Move a large container under the radiator drain to catch the coolant. Attach a 3/8 inch diameter hose to the drain fitting to direct the coolant into the container, then open the drain fitting **(see illustration)** (a pair of pliers may be required to turn it).

2.18 The radiator drain plug is located at the bottom of the radiator

5 After the coolant stops flowing out of the radiator, move the container under the engine block drain plug(s) on the side(s) of the block. Remove the plug(s) and allow the coolant in the block to drain.
6 While the coolant is draining, check the condition of the radiator hoses, heater hoses and clamps (see the previous Section).
7 Replace any damaged clamps or hoses.
8 Once the system is completely drained, flush the radiator with fresh water from a garden hose until it runs clear at the drain. The flushing action of the water will remove sediments from the radiator but will not remove rust and scale from the engine and cooling tube surfaces.
9 These deposits can be removed with a chemical cleaner. Follow the procedure outlined in the manufacturer's instructions. If the radiator is severely corroded, damaged or leaking, it should be removed and taken to a radiator repair shop.
10 Remove the overflow hose from the coolant recovery reservoir. Drain the reservoir and flush it with clean water, then reconnect the hose.
11 Close and tighten the radiator drain. Install and tighten the block drain plugs.
12 Place the heater temperature control in the maximum heat position.
13 Slowly add new coolant (a 50/50 mixture of water and antifreeze) to the radiator until it's full. Add coolant to the reservoir up to the lower mark.
14 Leave the radiator cap off and run the engine in a well-ventilated area until the thermostat opens (coolant will begin flowing through the radiator and the upper radiator hose will become hot).
15 Turn the engine off and let it cool. Add more coolant mixture to bring the level back up to the lip on the radiator filler neck.
16 Squeeze the upper radiator hose to expel air, then add more coolant mixture if necessary. Replace the radiator cap.
17 Start the engine, allow it to reach normal operating temperature and check for leaks.

Battery cable check and replacement

Check

1 Periodically inspect the entire length of each battery cable for damage, cracked or burned insulation and corrosion. Poor battery cable connections can cause starting problems and decreased engine performance.
2 Check the cable to terminal connections at the ends of the cables for cracks, loose wire strands and corrosion. The presence of white, fluffy deposits under the insulation at the cable terminal connection is a sign the cable is corroded and should be replaced. Check the terminals for distortion, missing mounting bolts and corrosion.
3 Too much resistance in the battery cables will result in poor starter performance. Disconnect the battery feed at the shutoff solenoid so the engine won't start.
4 With the transmission in Neutral and the parking brake On, connect one voltmeter probe on the negative battery terminal and the other on a good ground, such as the frame, operate the starter and check the voltage.
5 With the starter operating, check the voltage drop between the positive battery cable and the starter terminal stud.
6 Operate the starter again and check the voltage drop between the starter housing and the frame.
7 If there is a drop of over one volt during this check, disconnect the battery cables and clean the connectors. If the voltage drop is still excessive, replace the cables with new ones.

Replacement

8 When removing the cables, always disconnect the negative cable first and hook it up last or the battery may be shorted by the tool used to loosen the cable clamps. Even if only the positive cable is being replaced, be sure to disconnect the negative cable from the battery first.
9 Disconnect the old cables from the battery, then trace each of them to their opposite ends and detach them from the starter solenoid and ground terminals. Note the routing of each cable to ensure correct installation.

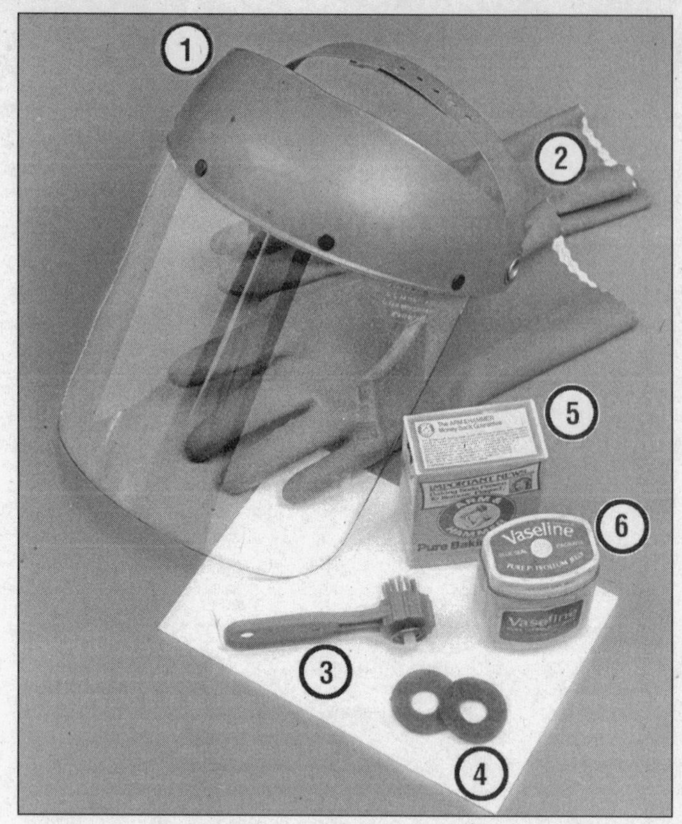

2.19 Tools and materials required for battery maintenance

1 ***Face shield/safety goggles*** – *When removing corrosion with a brush, the acidic particles can easily fly up into your eyes*
2 ***Rubber gloves*** – *Another safety item to consider when servicing the battery; remember that's acid inside the battery!*
3 ***Battery post/cable cleaner*** – *This wire brush cleaning tool will remove all traces of corrosion from the battery posts and cable clamps*
4 ***Treated felt washers*** – *Placing one of these on each post, directly under the cable clamps, will help prevent corrosion*
5 ***Baking soda*** – *A solution of baking soda and water can be used to neutralize corrosion*
6 ***Petroleum jelly*** – *A layer of this on the battery posts will help prevent corrosion*

10 If you're replacing either or both of the old cables, take them with you when buying new cables. It's vitally important that you replace the cables with identical parts. Cables have characteristics that make them easy to identify: Positive cables are usually red, larger in cross section and have a larger diameter battery post clamp; ground cables are usually black, smaller in cross section and have a slightly smaller diameter clamp for the negative post.
11 Clean the threads of the solenoid or ground connection with a wire brush to remove rust and corrosion. Apply a light coat of battery terminal corrosion inhibitor, or petroleum jelly, to the threads to prevent future corrosion.
12 Attach the cable to the solenoid or ground connection and tighten the mounting nut/bolt securely.
13 Before connecting a new cable to the battery, make sure it reaches the battery post without having to be stretched.
14 Connect the positive cable first, followed by the negative cable.

Battery check, maintenance and charging

Warning: *Certain precautions must be followed when checking and servicing the batteries. Hydrogen gas, which is highly flammable, is always present in the battery cells, so don't smoke and keep open flames and sparks away from the batteries. The electrolyte inside the*

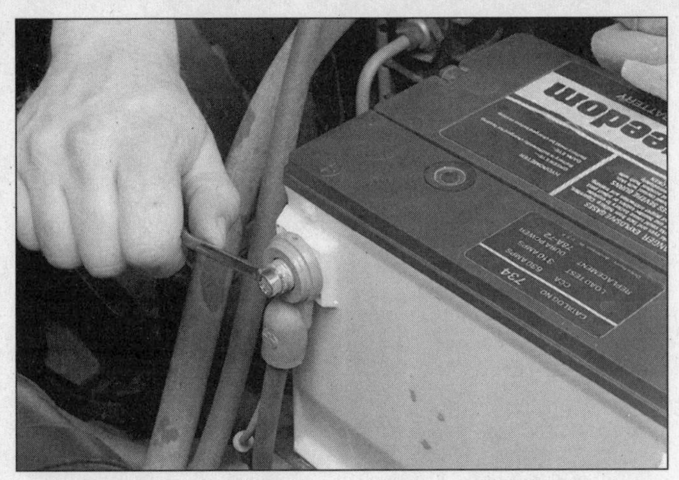

2.20 Make sure the battery terminals on both batteries are tight

battery is actually dilute sulfuric acid, which will cause injury if splashed on your skin or in your eyes. It will also ruin clothes and painted surfaces. When removing the battery cables, always detach the negative cable first and hook it up last!

Note: *These models use two batteries, connected in parallel. The batteries can be serviced and charged separately, but must be replaced in pairs.*

Check and maintenance

1 Battery maintenance is an important procedure which will help ensure that you are not stranded because of dead batteries. Several tools are required for this procedure **(see illustration)**.

2 When checking/servicing the batteries, always turn the engine and all accessories off.

3 Sealed (sometimes called maintenance free), side terminal batteries are standard equipment on these vehicles. The cell caps cannot be removed, no electrolyte checks are required and water cannot be added to the cells. However, if a maintenance type aftermarket battery has been installed, the following maintenance procedure can be used.

4 Remove the caps and check the electrolyte level in each of the battery cells. It must be above the plates. There's usually a split ring indicator in each cell to indicate the correct level. If the level is low, add distilled water only, and then reinstall the cell caps. **Caution:** *Overfilling the cells may cause electrolyte to spill over during periods of heavy charging, causing corrosion and damage to nearby components.*

5 The external condition of each battery should be checked periodically. Look for damage such as a cracked case.

6 Check the tightness of the battery cable bolts **(see illustration)** to ensure good electrical connections. Inspect the entire length of each cable, looking for cracked or abraded insulation and frayed conductors.

7 If corrosion (visible as white, fluffy deposits) is evident, remove the cables from the terminals, clean them with a battery brush and reinstall them. Corrosion can be kept to a minimum by applying a layer of petroleum jelly or grease to the bolt threads.

8 Make sure the battery carrier is in good condition and the hold down clamp is tight. If a battery is removed, make sure that no parts remain in the bottom of the carrier when it's reinstalled. When reinstalling the hold down clamp, don't overtighten the bolt.

9 Corrosion on the carrier, battery case and surrounding areas can be removed with a solution of water and baking soda. Apply the mixture with a small brush, let it work, then rinse it off with plenty of clean water.

10 Any metal parts of the vehicle damaged by corrosion should be coated with a zinc based primer, then painted.

Charging

Note: *Diesel models have two batteries connected in parallel - each battery should be charged separately, after disconnecting the cables.*

11 Remove all of the cell caps (if equipped) and cover the holes with a clean cloth to prevent spattering electrolyte. Disconnect the negative battery cable and hook the battery charger leads to the battery posts (positive to positive, negative to negative), then plug in the charger. Make sure it is set at 12 volts if it has a selector switch.

12 If you're using a charger with a rate higher than two amps, check the battery regularly during charging to make sure it doesn't overheat. If you're using a trickle charger, you can safely let the battery charge overnight after you've checked it regularly for the first couple of hours.

13 If the battery has removable cell caps, measure the specific gravity with a hydrometer every hour during the last few hours of the charging cycle. Hydrometers are available inexpensively from auto parts stores - follow the instructions that come with the hydrometer. Consider the battery charged when there's no change in the specific gravity reading for two hours and the electrolyte in the cells is gassing (bubbling) freely. The specific gravity reading from each cell should be very close to the others. If not, the battery probably has a bad cell(s).

14 Some batteries with sealed tops have built in hydrometers on the top that indicate the state of charge by the color displayed in the hydrometer window. Normally, a bright colored hydrometer indicates a full charge and a dark hydrometer indicates the battery still needs charging. Check the battery manufacturer's instructions to be sure you know what the colors mean.

15 If the battery has a sealed top and no built in hydrometer, you can hook up a digital voltmeter across the battery terminals to check the charge. A fully charged battery should read 12.6 volts or higher.

Underhood hose check and replacement

Caution: *Replacement of air conditioning hoses must be left to a dealer service department or air conditioning shop that has the equipment to depressurize the system safely. Never remove air conditioning components or hoses until the system has been depressurized.*

General

1 High temperatures in the engine compartment can cause the deterioration of the rubber and plastic hoses used for engine, accessory and emission systems operation. Periodic inspection should be made for cracks, loose clamps, material hardening and leaks.

2 Information specific to the cooling system hoses can be found in Section 8.

3 Some, but not all, hoses are secured to the fittings with clamps. Where clamps are used, check to be sure that they haven't lost their tension, allowing the hose to leak. If clamps aren't used, make sure the hose has not expanded and/or hardened where it slips over the fitting, allowing it to leak.

Vacuum hoses

4 It's quite common for vacuum hoses, especially those in the emissions system, to be color-coded or identified by colored stripes molded into them. Various systems require hoses with different wall thicknesses, collapse resistance and temperature resistance. When replacing hoses, be sure the new ones are made of the same material.

5 Often the only effective way to check a hose is to remove it completely from the vehicle. If more than one hose is removed, be sure to label the hoses and fittings to ensure correct installation.

6 When checking vacuum hoses, be sure to include any plastic T fittings in the check. Inspect the fittings for cracks and the hose where it fits over the fitting for distortion, which could cause leakage.

7 A small piece of vacuum hose (1/4 inch inside diameter) can be used as a stethoscope to detect vacuum leaks. Hold one end of the hose to your ear and probe around vacuum hoses and fittings, listening for the "hissing" sound characteristic of a vacuum leak. **Warning:** *When probing with the vacuum hose stethoscope, be very careful not to come into contact with moving engine components such as the drivebelts, cooling fan, etc.*

Fuel hose

Warning: *There are certain precautions that must be taken when inspecting or servicing fuel system components. Work in a well-ventilated area and do not allow open flames (cigarettes, appliance pilot lights, etc.) or bare light bulbs near the work area. Mop up any spills* *immediately and do not store fuel soaked rags where they could ignite.*

8 Check all rubber fuel lines for deterioration and chafing. Check especially for cracks in areas where the hose bends and just before fittings.

9 High quality fuel line, usually identified by the word *Fluroelastomer* printed on the hose, should be used for fuel line replacement. Never, under any circumstances, use unreinforced vacuum line, clear plastic tubing or water hose for fuel lines.

10 Spring type clamps are commonly used on fuel lines. These clamps often lose their tension over a period of time, and can be "sprung" during removal. Replace all spring type clamps with screw clamps whenever a hose is replaced.

Metal lines

Warning: *Use extreme care when inspecting the fuel lines. The fuel is under very high pressure and spray from a leak can actually penetrate the skin, causing serious personal injury.*

11 Sections of metal line are used for fuel line between the fuel pump and the fuel injection unit and injectors. Check carefully to be sure that the lines have not been bent or crimped, and that cracks have not started in the lines.

12 If a section of metal fuel line between the fuel injection pump and the injectors must be replaced, use only direct replacement steel tubing from a dealer, since each line is a specific length.

Cooling system

General information

The cooling system consists of a radiator, a thermostat and a crankshaft pulley driven water pump.

The radiator cooling fan is mounted on the front of the water pump and incorporates a fluid drive fan clutch, saving horsepower and reducing noise. A fan shroud is mounted on the rear of the radiator.

The system is pressurized by a spring-loaded radiator cap, which increases the boiling point of the coolant. If the coolant temperature goes above this increased boiling point, the extra pressure in the system forces the radiator cap valve off its seat and exposes the overflow pipe. The overflow pipe leads to a coolant recovery system. This consists of a plastic reservoir into which the coolant that normally escapes due to expansion is retained. When the engine cools, the excess coolant is drawn back into the radiator, maintaining the system at full capacity. This is a continuous process and provided the level in the reservoir is correctly maintained, it is not necessary to add coolant to the radiator.

Coolant in the radiator circulates up the lower radiator hose to the water pump, where it is forced through the water passages in the cylinder block. The coolant then travels up into the cylinder head, circulates around the combustion chambers and valve seats, travels out of the cylinder head past the open thermostat into the upper radiator hose and back into the radiator.

When the engine is cold the thermostat restricts the circulation of coolant to the engine. The thermostat is located in the front of the intake manifold. When the minimum operating temperature is reached, the thermostat begins to open, allowing coolant to return to the radiator.

Automatic transmission equipped models have a cooler element incorporated into the radiator to cool the transmission fluid.

The heating system works by directing air through the heater core mounted in the dash and then to the interior of the vehicle by a system of ducts. Temperature is controlled by mixing heated air with fresh air, using a system of flapper doors in the ducts, and a heater motor.

Air conditioning is an optional accessory, consisting of an evaporator core located under the dash, a condenser in front of the radiator, an accumulator in the engine compartment and a belt driven compressor mounted at the front of the engine.

Antifreeze - general information

Warning: *Do not allow antifreeze to come in contact with your skin or painted surfaces of the vehicle. Rinse off spills immediately with plenty of water. Antifreeze is highly toxic if ingested. Never leave antifreeze lying around in an open container or in puddles on the floor; children and pets are attracted by its sweet smell and may drink it. Check with local authorities regarding disposal of used antifreeze. Many communities have collection centers that can safely dispose of antifreeze.*

2.21 The 6.2L and 6.5L engine thermostat is located in the crossover pipe which goes between the cylinder heads; to remove it, loosen the hose clamp and remove the thermostat housing cover bolts (arrows) and remove the cover

The cooling system should be filled with a water/ethylene glycol based antifreeze solution, which will prevent freezing down to at least 20 degrees F (even lower in cold climates). It also provides protection against corrosion and increases the coolant boiling point.

The cooling system should be drained, flushed and refilled at least every other year (see page 2-10). The use of antifreeze solutions for periods of longer than two years is likely to cause damage and encourage the formation of rust and scale in the system.

Before adding antifreeze to the system, check all hose connections. Antifreeze can leak through very minute openings.

The exact mixture of antifreeze to water that you should use depends on the relative weather conditions. The mixture should contain at least 50 percent antifreeze, but should never contain more than 70 percent antifreeze.

Thermostat - check and replacement

Warning: *The engine must be completely cool when this procedure is performed.*

Note: *Don't drive the vehicle without a thermostat! Emissions and fuel economy will suffer.*

Check

1 Before condemning the thermostat, check the coolant level, drivebelt tension, drivebelt tension and temperature gauge (or light) operation.

2 If the engine takes a long time to warm up, the thermostat is probably stuck open. Replace the thermostat.

3 If the engine runs hot, check the temperature of the upper radiator hose. If the hose isn't hot, the thermostat is probably stuck shut. Replace the thermostat.

4 If the upper radiator hose is hot, it means the coolant is circulating and the thermostat is open. Refer to the *Diesel engine troubleshooting* section in Chapter 1 for the cause of overheating.

5 If an engine has been overheated, you may find damage such as leaking head gaskets, scuffed pistons and warped or cracked cylinder heads.

Replacement

6 Drain coolant (about 3 quarts) from the radiator, until the coolant level is below the thermostat housing.

7 Disconnect the upper radiator hose from the thermostat housing cover, which is located at the forward end of the intake manifold (5.7L engine) or in a crossover pipe between the cylinder heads (6.2L

2.22 Remove the thermostat housing cover gasket

2.23 Note how the thermostat is installed (with the spring facing into the housing) and then remove the thermostat

2.24a To detach the fan clutch hub, remove the four nuts (arrows on three nuts, fourth nut not visible)

and 6.5L engine).

8 Remove the bolts **(see illustration)** and lift the cover off. It may be necessary to tap the cover with a soft face hammer to break the gasket seal. Remove the gasket **(see illustration)**.

9 Note how it's installed, then remove the thermostat **(see illustration)**.

10 Use a scraper or putty knife to remove all traces of old gasket material and sealant from the mating surfaces. Make sure that no gasket material falls into the coolant passages; it is a good idea to stuff a rag in the passage. Wipe the mating surfaces with a rag saturated with lacquer thinner or acetone.

11 Apply a thin layer of RTV sealant to the gasket mating surfaces of the housing and cover, then install the new thermostat in the engine. Make sure the correct end faces up - the spring is directed into the housing.

12 Position a new gasket on the housing and make sure the gasket holes line up with the bolt holes in the housing.

13 Carefully position the cover on the housing and install the bolts. Tighten them to the torque listed in this Chapter's Specifications - do not overtighten them or the cover may be distorted.

14 Reattach the radiator hose to the cover and tighten the clamp - now may be a good time to check and replace the hoses and clamps.

15 Refer to the *Cooling system servicing* section earlier in this Chapter and refill the system, then run the engine and check carefully for leaks.

Coolant temperature switch - check and replacement

Temperature warning light system check

1 If the light doesn't come on when the ignition switch is turned on, check the bulb. If the light stays on even when the engine is cold, unplug the wire at the switch. If the light goes off, replace the switch. If the light stays on, the wire is grounded somewhere in the harness.

Temperature gauge system check

2 If the gauge is inoperative, check the fuse.

3 If the fuse is OK, unplug the wire connected to the switch and ground it with a jumper wire. Turn the ignition switch On. The gauge should now register at maximum. If it does, replace the temperature switch. If it still doesn't work, the gauge or wiring may be faulty.

Temperature switch replacement

4 With the engine cold, unplug the wire from the switch. Prepare the new switch by wrapping Teflon tape around the threads, or by applying a thin film of RTV sealant to them.

5 Unscrew the switch and quickly thread the new one in to prevent loss of coolant.

2.24b To remove the fan blade from the clutch, remove the four bolts

6 Connect the wire to the new switch. Check the coolant level and add some, if necessary.

Cooling fan and fan clutch - removal and installation

1 The cooling fan should be replaced if the blades become damaged or bent. The fluid drive fan clutch is disengaged when the engine is cold, or at high engine speeds, when the silicone fluid inside the clutch is contained in the reservoir section by centrifugal action. Symptoms of failure of the fan clutch are continuous noisy operation, looseness leading to vibration and evidence of silicone fluid leaks.

2 Disconnect the negative cable from the battery. Remove the upper fan shroud.

3 Loosen the water pump pulley bolts/nuts.

4 Remove the drivebelt.

5 Unbolt the fan assembly **(see illustration)** and detach it from the water pump.

6 The fan clutch can also be unbolted from the fan blade assembly for replacement **(see illustration)**.

7 Installation is the reverse of removal.

Water pump - check

1 Water pump failure can cause overheating and serious damage to the engine. There are three ways to check the operation of the water

2.25 Check the weep hole for leakage

2.26 Try to move the water pump pulley to check for bearing wear

pump while it is installed on the engine. If any one of the three following quick checks indicates water pump problems, it should be replaced immediately.

2 Start the engine and warm it up to normal operating temperature. Squeeze the upper radiator hose. If the water pump is working properly, you should feel a pressure surge as the hose is released.

3 A seal protects the water pump impeller shaft bearing from contamination by engine coolant. If this seal fails, a weep hole in the water pump snout will leak coolant **(see illustration)** (an inspection mirror can be used to look at the underside of the pump if the hole isn't on top). If the weep hole is leaking, shaft bearing failure will follow. Replace the water pump immediately.

4 Besides contamination by coolant after a seal failure, the water pump impeller shaft bearing can also be prematurely worn out by an improperly tensioned drivebelt. When the bearing wears out, it emits a high pitched squealing sound. If such a noise is coming from the water pump during engine operation, the shaft bearing has failed - replace the water pump immediately. **Note:** *Do not confuse belt noise with bearing noise.*

5 To identify excessive bearing wear before the bearing actually fails, grasp the water pump pulley and try to force it up and down or from side to side **(see illustration)**. If the pulley can be moved either horizontally or vertically, the bearing is nearing the end of its service life. Replace the water pump.

Water pump - removal and installation

Removal

1 Disconnect the negative cable at the battery.

2 Drain the coolant.

3 Loosen the fan pulley nuts/bolts and then remove the drivebelt.

4 Loosen the hose clamps for the lower radiator hose and disconnect the hose from the water pump.

5 Remove the fan assembly and water pump pulley.

6 On some models the air conditioning compressor, alternator, power steering pump and vacuum pump may be attached to the water pump. Remove the accessories and brackets from the water pump and set them aside. DO NOT disconnect the hoses from the air conditioning compressor or power steering pump.

7 Remove the water pump mounting bolts **(see illustrations)**. On 6.2L and 6.5L engines, remove the bolts from the mounting plate **(see illustration)** and separate the pump and mounting plate together. Remove the bolt from the backside of the mounting plate and separate the pump from the mounting plate **(see illustration)**. Note the location of each bolt so it can be installed in its original location.

Installation

8 Clean the sealing surfaces on both the block and the water pump. Wipe the mating surfaces with a rag saturated with lacquer thinner or

2.27a On 6.2L and 6.5L engines, remove the two oil filler neck retaining bolts (upper arrow) (left bolt shown, right bolt not visible in this photo), remove the left water pump plate bolt (lower arrow) . . .

2.27b . . . remove the water pump plate bolt and stud (A), remove the five upper water pump bolts (B) and, from underneath the vehicle, remove the lower water pump bolts (not visible)

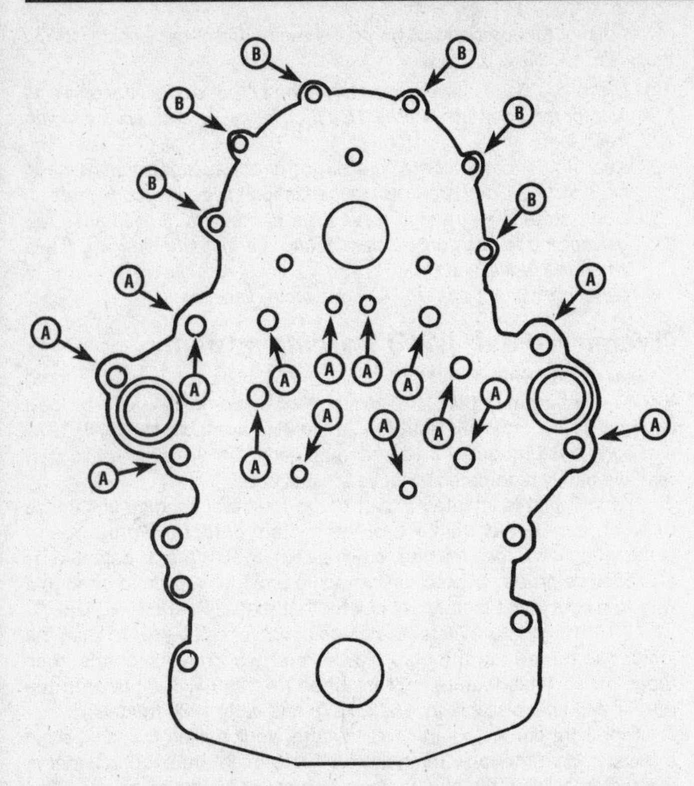

2.28 To remove the water pump and water pump plate, remove these bolts (6.2L and 6.5L engines)

A Water pump bolts *B Water pump plate bolts*

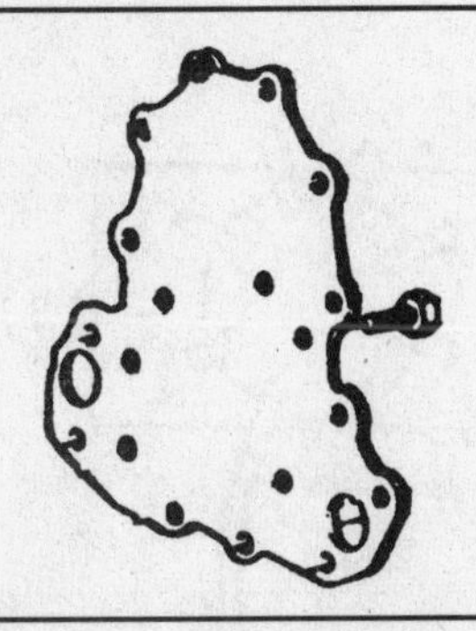

2.29 The water pump plate is also attached to the water pump by a bolt that can be removed only after the pump and plate have been detached from the engine

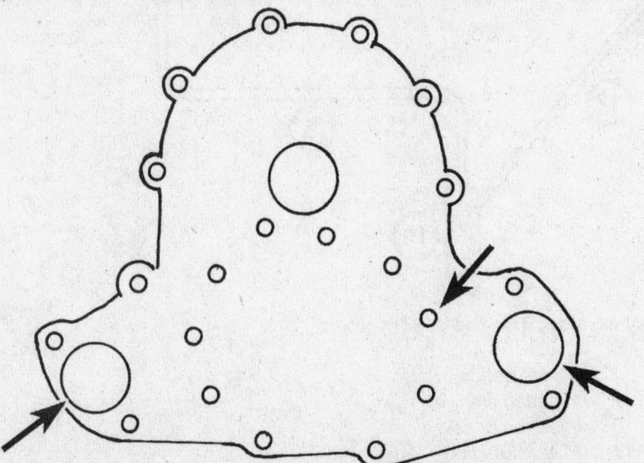

2.30 Apply anaerobic sealer to the points shown on the water pump plate - the sealer must be damp to the touch when the plate is installed

acetone.

9 Apply a thin layer of RTV sealant to the block mounting surfaces and install new water pump gaskets.

10 On 6.2L and 6.5L engines, Install the water pump to the mounting plate, tighten the bolt from the backside to the torque listed in this Chapter's Specifications. Apply a thin layer of anaerobic sealer to the mounting plate at the points shown **(see illustration)**. The sealer must be damp when the plate is installed.

11 Place the water pump in position and install the bolts and studs finger tight. Use caution to ensure that the gaskets do not slip out of position. Remember to reinstall any mounting brackets secured by the water pump mounting bolts. Tighten the bolts to the torque values listed in this Chapter's specifications.

12 Install the water pump pulley and fan assembly and tighten the pulley bolts by hand.

13 Install the lower radiator hose, heater hose and hose clamps. Tighten the hose clamps securely.

14 Install the drivebelt and tighten the fan mounting bolts securely.

15 Fill the cooling system with the recommended type of coolant, referring to the *Cooling system servicing* Section earlier in this Chapter.

16 Connect the cable to the negative terminal of the battery.

17 Start the engine and check the water pump and hoses for leaks.

Fuel system

Low pressure fuel delivery system - general information

The low pressure fuel delivery system **(see illustrations)** on the 5.7L and 6.2L diesel engines consists of the fuel tank, the lift pump, the fuel filter(s) and the fuel lines connecting all these components. Early versions of the low-pressure systems on both engines are fairly simple: They use mechanical fuel pumps similar to conventional gasoline fuel pumps and a pleated paper type filter element. However, as GM has refined the low-pressure system, it has become more and more complex. Later versions use an electric solenoid pump and a combination fuel filter/heater/water separator with a built-in Water-In-Fuel warning system.

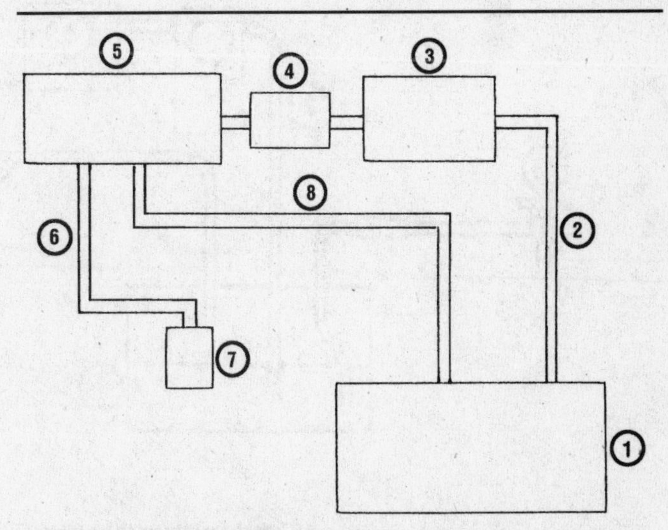

2.31 Schematic of typical "leak-off pencil nozzle type" fuel delivery system (all 1978 and 1979 and some 1980 and 1981 5.7L engines

1 *Fuel tank*
2 *Fuel inlet line*
3 *Mechanical fuel pump*
4 *Fuel filter*
5 *Fuel injection pump*
6 *Injection line*
7 *Pencil type injection nozzle*
8 *Fuel return line*

Fuel storage

Good quality diesel fuel contains rust inhibitors to stop the formation of rust in the fuel lines and the injectors, so as long as there are no leaks in the fuel system, it's generally safe from water contamination. Diesel fuel is usually contaminated by water as a result of careless storage. There's not much you can do about the storage practices of service stations where you buy diesel fuel, but if you keep a small supply of diesel fuel on hand at home, as many diesel owners do, follow these simple rules:

1) *Diesel fuel "ages" and goes stale. Don't store containers of diesel fuel for long periods of time. Use it up regularly and replace it with fresh fuel.*
2) *Keep fuel storage containers out of direct sunlight. Variations in heat and humidity promote condensation inside fuel containers.*
3) *Don't store diesel fuel in galvanized containers. It will cause the galvanizing to flake off, contaminating the fuel and clogging filters when the fuel is used.*
4) *Label containers properly as containing diesel fuel.*

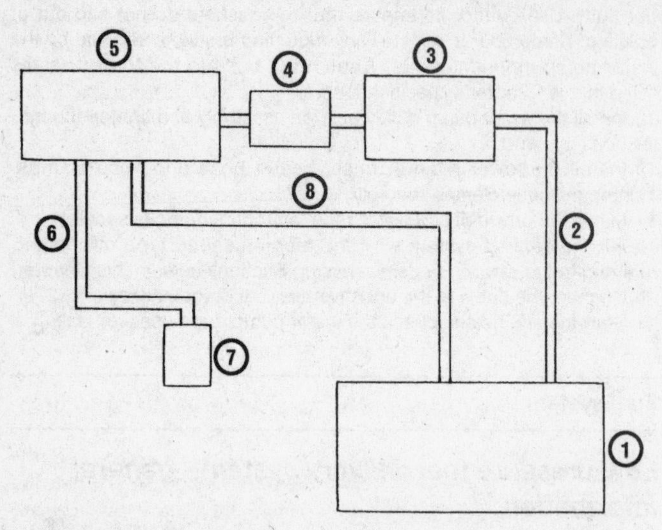

2.32 Schematic of typical later (poppet nozzle) fuel delivery system (all 5.7L engines except those listed in Figure 2.31)

1 *Fuel tank*
2 *Fuel inlet line*
3 *Mechanical fuel pump*
4 *Fuel filter*
5 *Fuel injection pump*
6 *Injection line*
7 *Poppet type injection nozzle*
8 *Fuel return line*

"Water-in-Fuel" (WIF) warning system

Since late 1980, a "Water-in-Fuel" (WIF) system has been offered by GM - first as an option, later as standard equipment - on the models covered by this manual. Retrofit kits are also available for 1978, 1979 and early 1980 models - if you have an early vehicle without this system, we highly recommend that you install one.

The WIF unit is an integral part of the fuel level sending unit inside the fuel tank. Earlier versions of the system detect the presence of water when it reaches the one to two gallon level. Water is detected by a capacitive probe; an electronic module provides a ground through a wire to a light in the instrument cluster that reads "WATER-IN-FUEL."

All 1981 and later WIF units include a bulb check feature: When the ignition is turned on, the bulb glows from two to five seconds, then fades away. (This feature isn't included on 1980 WIF units or in the retrofit WIF kits available for 1978, 1979 and early 1980 models.)

If the light comes on immediately after you've filled the tank, drain the water from the tank immediately. There might be enough water in the system to shut the engine down before you've driven even a short distance. If, however, the light comes on during a cornering or braking maneuver, there's less than a gallon of water in the system; the engine

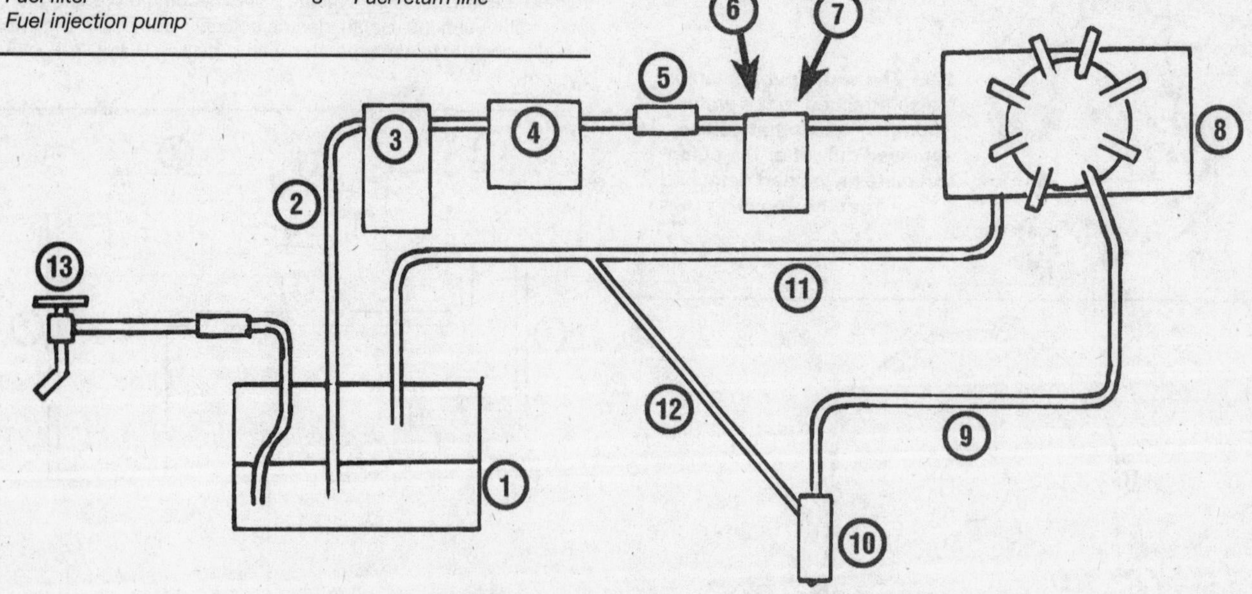

2.33 Schematic of the fuel delivery system used on the 6.2L engine

1 *Fuel tank*
2 *Fuel inlet line*
3 *Primary fuel filter (1982 and 1983 models)*
4 *Mechanical fuel pump*
5 *Fuel line heater (1982 and 1983 models)*
6 *Secondary fuel filter (1982 and 1983 models)*
7 *Two-stage Model 80 filter (1984 and later models)*
8 *Fuel injection pump*
9 *Injection line*
10 *Injection nozzle*
11 *Fuel return line*
12 *Fuel return line*
13 *Water drain siphon valve (1982 and 1983 models)*

probably won't shut down immediately, but you still should drain the water within a day or two.

Water is heavier than diesel fuel, so it sinks to the bottom of the fuel tank. An extended return pipe on the fuel tank sending unit, which reaches down into the bottom of the tank, enables you to siphon most of the water from the tank without removing the tank. But siphoning won't remove all of the water; you'll still need to remove the tank and thoroughly clean it.

The extended return pipe is used on all fuel tank sending units equipped with the WIF feature; it was also phased into production on some 1980 units that weren't equipped with the WIF feature. How can you tell whether your sending unit has the siphoning feature without re moving the sending unit? If it has got a purple ground wire, it has got the siphoning feature.

A check valve is installed at the upper end of the fuel sending unit to allow fuel to return in the event that frozen water plugs the end of the pipe; all 1981 and later units have a sock with a relief valve in the top end. This relief valve is designed to open up in case high cloud point fuel is used in cold weather and the sock gets plugged with wax crystals.

The sensor module in 1980 models turns on the WIF light as soon as water touches the probe, and remains on until the 12 volt signal is removed. If vehicle motion moves the water around inside the tank, the sensor can indicate smaller amounts of water present in the tank.

The module in the detector probe used with 1981 modules must remain submerged in water for about 15 to 20 seconds before it will indicate the presence of water. The indicator light comes on and stays on until the 12-volt signal is removed. This unit will accommodate larger amounts of water.

In 1983 and 1984, the WIF sensitivity was improved to trigger at .26 to .80 gallons instead of 1 to 2 gallons, and the time delay was changed from 15 to 20 seconds to 3 to 6 seconds.

Diagnosing the Water-In-Fuel detector

The Water-In-Fuel light stays on all the time

1 With the ignition turned to ON, unplug the two-wire connector (yellow wire and black/pink wire) at the rear of the fuel tank **(see illustration)** and check the Water-In-Fuel light.

2 **If the light comes on**, locate and repair a short to ground in the yellow/black wire from the two-wire connector to the Water-In-Fuel light on the instrument cluster.

3 **If the light doesn't come on**, purge the fuel tank with a pump or siphon out the contents. Remove the cap from the fuel tank filler neck. Hook up the pump or siphon hose to the 1/4-inch fuel return hose (the smaller of the two fuel hoses) above the rear axle or near the fuel pump in the engine compartment. Be sure to remove all water from the fuel tank. To determine when clear fuel begins to flow, use a clear plastic hose or watch the filter bowl on the siphoning equipment. Replace the filler cap. **Warning:** *Diesel fuel is flammable, so take extra precautions when you work on any part of the fuel system. Don't smoke or allow open flames or bare light bulbs near the work area, and don't work in a garage where a natural gas-type appliance (such as a water heater or clothes dryer) with a pilot light is present. Diesel fuel is carcinogenic, so wear latex gloves when there's any possibility of being exposed to fuel. If you spill any fuel on your skin, rinse it off immediately with soap and water. Mop up any spills immediately and do not store fuel-soaked rags where they could ignite. When you perform any kind of work on the fuel system, wear safety glasses and have a Class B fire extinguisher on hand.*

4 **If the light comes on now**, remove the fuel tank and check the wires for a short circuit. If the wires are okay, replace the Water-In-Fuel detector unit.

5 **If the light still doesn't come on**, it's operating correctly. (The fuel had water in it.)

The Water-In-Fuel light doesn't come on during the bulb check

1 With the ignition turned to ON, unplug the two-wire connector (yellow wire and black/pink wire) at the rear of the fuel tank, ground the yellow/black wire in the body harness and check the Water-In-Fuel light.

2 **If the light comes on**, remove the fuel gauge sending unit from the fuel tank and check the yellow/black wire for an open circuit. Check the connections to the Water-In-Fuel detector and make sure that the mounting screw is tight. If the screw is tight and there are no opens in the circuit, replace the Water-In-Fuel detector.

3 **If the light doesn't come on**, check the Water-In-Fuel bulb. If the bulb is okay, check for an open circuit in the yellow/black wire from the two-wire connector (at the rear of the tank) to the water in the Water-In-Fuel socket at the instrument cluster.

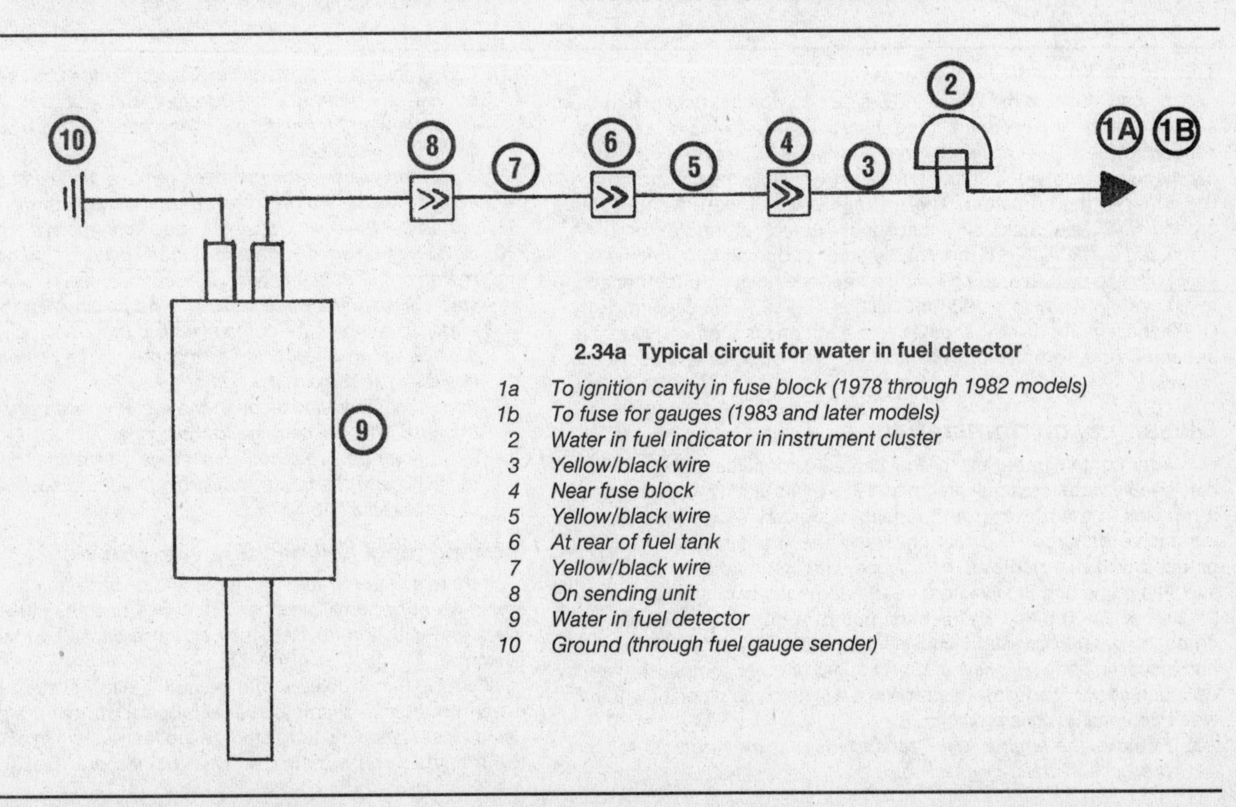

2.34a Typical circuit for water in fuel detector

1a *To ignition cavity in fuse block (1978 through 1982 models)*
1b *To fuse for gauges (1983 and later models)*
2 *Water in fuel indicator in instrument cluster*
3 *Yellow/black wire*
4 *Near fuse block*
5 *Yellow/black wire*
6 *At rear of fuel tank*
7 *Yellow/black wire*
8 *On sending unit*
9 *Water in fuel detector*
10 *Ground (through fuel gauge sender)*

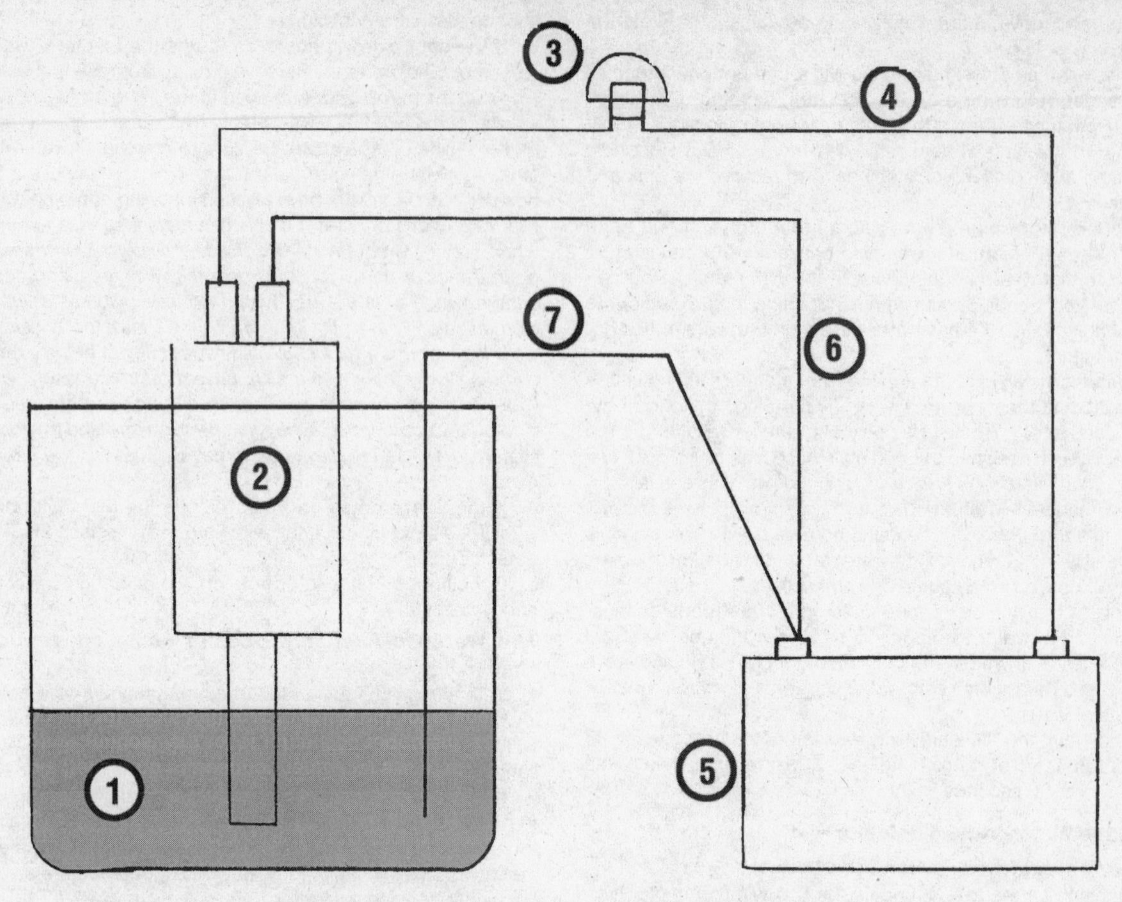

2.34b Test setup for water-in-fuel detector diagnosis

1 *Container of water (water must be grounded)*
2 *Water-in-fuel detector*
3 *12-volt bulb*
4 *12-volt wire*
5 *12-volt battery*
6 *Ground wire (to negative terminal of water in fuel detector)*
7 *Ground wire (to water in container)*

Testing the Water-In-Fuel detector

You can also test the Water-In-Fuel detector on the bench with the simple setup shown here **(see illustration)**. Remove the fuel sender/Water-In-Fuel assembly from the fuel tank. Using a 12-volt, 2 candlepower bulb, hook up the WIF as shown. Note the ground wire to the water in the container. The test setup won't work without this ground wire. Also, make sure that about 3/8-inch of the WIF probe is immersed in the water. If the WIF detector is operating correctly, the light bulb should turn on for two to five seconds, then go out. After a 15 to 20 second delay (pre-1983 models) or a three to six second delay (1983 and later models), it should come back on for another two to five seconds. And so on. If the WIF unit doesn't operate as described, replace it.

Diesel fuel contamination

Water contaminated diesel fuel can lead to major problems. You can identify water contaminated diesel fuel by its cloudy appearance. If it remains in the fuel system too long, water will cause serious and expensive damage. The fuel lines and the fuel filter can become plugged with rust particles, or clogged with ice in cold weather. The injection pump and the injectors can be damaged or ruined.

Before you replace an injection pump or some other expensive component, find out what's causing the problem. If water contamination is present, buying a new or rebuilt pump or other component won't do much good. The following procedure will help you pinpoint whether water contamination is present:

1) *Remove the engine fuel filter and inspect the contents for the presence of water or gasoline.*
2) *If the vehicle has been stalling, performance has been poor or the engine has been knocking loudly, suspect fuel contamination. Gasoline won't harm the injection system, but it must be removed by flushing (see below).*
3) *If you find water in the fuel filter, remove the injection pump cover (refer to the part of this chapter that deals with the high-pressure system). If the pump is full of water, flush the system (see below).*
4) *Small quantities of surface rust in the injection pump won't create a problem. If contamination is excessive, the vehicle will probably stall. Remove the metering valve and polish it lightly with 600-grit paper to remove the contaminant. If the advance piston is stuck (as evidenced by poor performance, smoke or noise), it may be necessary to remove the pump to free it up.*
5) *Sometimes contamination in the system becomes severe enough to cause physical damage to the springs and linkage in the pump. If the damage reaches this stage, have the damaged parts replaced and the pump rebuilt by an authorized Stanadyne shop, or buy a rebuilt pump.*

Fighting fungi and bacteria with biocides

If there's water in the fuel, fungi and/or bacteria can form in diesel fuel in warm or humid weather. Fungi and bacteria plug fuel lines, fuel filters and injection nozzles; they can also cause corrosion in the fuel system.

If you've had problems with water in the fuel system and you live in a warm or humid climate, have your dealer correct the problem. Then use a diesel fuel biocide to sterilize the fuel system in accordance with the manufacturer's instructions. Biocides are available from your dealer, service stations and auto parts stores. Consult your dealer for advice on using biocides in your area, and for recommendations on which ones to use.

Cleaning the low pressure fuel system

Water-In-Fuel system

1 Disconnect the cable from the negative battery cable.
2 Drain the fuel tank.
3 Remove the tank gauge sending unit.
4 Thoroughly clean the fuel tank. If it's rusted inside, replace it. Clean or replace the fuel pick up filter and check the valve assembly.
5 Reinstall the fuel tank but don't connect the fuel lines to the fuel tank.
6 Disconnect the main fuel line from the low pressure fuel pump. Using low air pressure, blow out the line toward the rear of the vehicle.
7 Disconnect the fuel return fuel line at the injection pump and again, using low air pressure, blow out the line toward the rear of the vehicle.
8 Reconnect the main fuel and return lines at the tank. Fill the tank to a fourth of its capacity with clean diesel fuel. Install the cap on the fuel filler neck.
9 Remove and discard the fuel filter.
10 Connect the fuel line to the fuel pump.
11 Reconnect the battery cable.
12 Purge the fuel pump and pump to filter line by cranking the engine until clean fuel is pumped out. Catch the fuel in a closed metal container.
13 Install a new fuel filter.
14 Install a hose from the fuel return line (from the injection pump) to a closed metal container with a capacity of at least two gallons.
15 On 1981 and later vehicles, if the engine temperature is above 125 degrees F, activate the injection pump Housing Pressure Cold Advance (HPCA): Detach the connectors for both leads at the engine temperature switch and bridge them with a jumper.
16 Crank the engine until clean fuel appears at the return line. Don't crank the engine for more than 30 seconds at a time. If it's necessary to crank it again, allow a three-minute interval before resuming.
17 Remove the jumper from the engine temperature switch connector and plug in the connectors to the engine temperature switch.
18 Using two wrenches to prevent damage to the nozzles, crack open each high pressure line at the nozzles.
19 On 1981 and later vehicles, disconnect the lead to the HPCA solenoid on the injection pump.
20 Crank the engine until clean fuel appears at each nozzle. Don't crank the engine for more than 15 seconds at a time. If it's necessary to crank it again, allow a two-minute interval before resuming.
21 On 1981 and later vehicles, reconnect the HPCA lead at the pump.

Gasoline in the fuel system

If gasoline has been accidentally pumped into the fuel tank, it won't hurt the fuel system or the engine, but the engine won't run either. Gasoline has a characteristic known as "octane," which is its ability to resist ignition under high temperatures. Gasoline has a high octane rating, i.e. it resists ignition under high heat. It must be ignited by a spark. Gasoline in the fuel in small amounts - from 0 to 30 percent - isn't noticeable. At higher ratios, the engine will make a knocking noise, which will get louder as the amount of gasoline goes up. Gasoline in any amount, however, will make the engine harder to start when it's hot. In the summertime, hot starting can be a big problem. Here's how to rid the fuel system of gasoline:

1 Drain the fuel tank and fill it with diesel fuel.
2 Remove the fuel line between the fuel filter and the injection pump.
3 Connect a short pipe and hose to the fuel filter outlet and run it to a closed metal container.
4 Crank the engine to purge gasoline out of the fuel pump and fuel filter. Don't crank the engine for more than 15 seconds at a time. Allow two minutes between cranking intervals.
5 Remove the short pipe and hose and install the fuel line between the fuel filter and the injection pump.
6 Try to start the engine. If it doesn't start, purge the injection pump and lines: Crack the fuel line fittings a little, just enough for fuel to leak out. Depress the accelerator pedal to the floor and, holding it there, crank the engine until all gasoline is gone, i.e. diesel fuel leaks out of the fittings. Tighten the fittings. Limit cranking to 15 seconds with two minutes between cranking intervals.
7 Start the engine and run it at idle for 15 minutes.

Low pressure pump - general information

Although the injection pump is capable of pulling fuel from the fuel tank, a low-pressure pump (also known as a delivery pump, lift pump, supply pump or transfer pump) is installed between the tank and the injection pump. The low-pressure pump prevents the injection pump from burning up if the tank runs dry or if a leak allows air into the fuel lines. And because the low-pressure pump raises the pressure level of the fuel entering the injection pump, it prolongs the service life of the injection pump significantly.

Mechanical low pressure pump

Some models use a mechanical, diaphragm type low-pressure pump to deliver fuel from the fuel tank to the injection pump. The mechanical lift pump is similar to a mechanical gasoline pump, except that a gasoline pump is operated by an eccentric on the camshaft. On the diesel version, the lift pump, which is bolted to the right side of the engine, is driven by an eccentric on the nose of the crankshaft.

Here's how it works: A rocker arm rides directly on an eccentric, or is actuated by a pushrod that rides on the eccentric. A spring holds the rocker arm in constant contact with the eccentric or pushrod. As the eccentric or pushrod moves the rocker arm, the other end of the arm pulls a diaphragm toward it. The vacuum action of this diaphragm enlarges the fuel chamber drawing fuel from the tank through the inlet valve and into the fuel chamber. The return stroke, which begins at the high point of the cam or eccentric, releases the compressed diaphragm spring, expelling fuel through the outlet valve. When there's enough fuel in the line between the low pressure pump and the injection pump, the pressure build up in this line forces the diaphragm and piston to make shorter and shorter strokes. As the pressure in the line goes down, the stroke length of the diaphragm and piston increases. And so on.

In 1989, the original pump (part number 6442199) was replaced by a newer design (part number 6443254). At first glance, both units look identical. But there's an important difference between the two: The inlet and outlet valves of the older pump each have a .007 inch diameter "drain back" hole. This design is known as a "vented" pump. The drain back holes allow fuel to drain back to the fuel tank when small air leaks are present. This allows the fuel injection system to become somewhat "aerated," and sometimes causes cold start problems. The newer design - known as a "non vented" pump - has no vent holes in the valves; it eliminates one of the "known" fuel drain back paths in the fuel injection system. The newer pump greatly reduces difficult cold start problems.

Testing the mechanical pump

There are two simple tests you can perform to determine whether a mechanical pump is operating satisfactorily. But first, do the following preliminary inspection:

1 Check the fuel line fittings and connections - make sure they're tight. If a fitting is loose, air and/or fuel leaks may occur.
2 Check for bends or kinks in the fuel lines.
3 Start the engine and let it warm up. With the engine idling, conduct the following checks before proceeding to the actual fuel pump tests:

a) *Look for leaks at the pressure (outlet) side of the pump.*
b) *Look for leaks on the suction (inlet) side of the pump. A leak on the suction side will suck air into the pump and reduce the volume of fuel on the pressure (outlet) side of the pump.*
c) *Inspect the steel cover and the fittings on the fuel pump for leaks. Tighten or replace the fittings as necessary.*
d) *Look for leaks around the diaphragm, the flange and the breather holes in the cast pump housing. If any of them are leaking, replace the pump.*

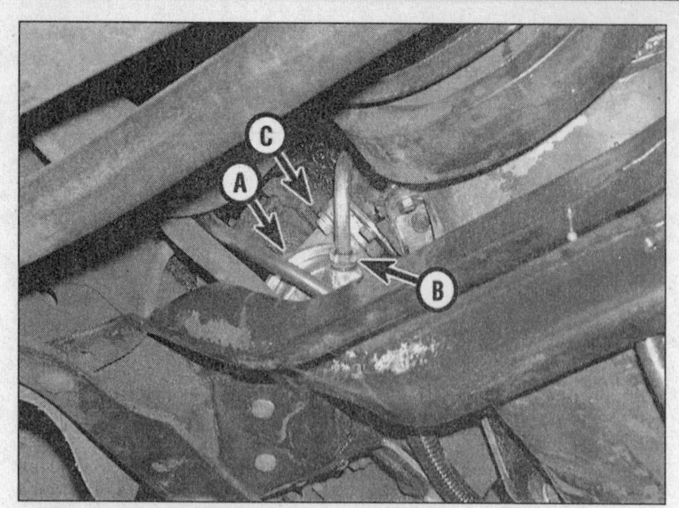

2.35 The mechanical lift pump (A) is located on the right side of the block, at the front of the engine; to remove the pump, disconnect the fuel inlet hose (not visible), the outlet fitting (B) and the two mounting bolts (C) (one bolt not visible) (6.2L pump shown, 5.7L. pump similar)

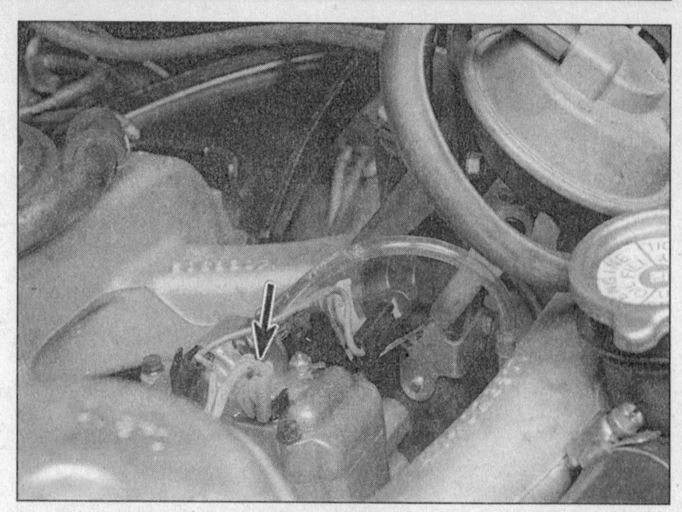

2.36 Fuel-injection pump electric shut-off (ESO) solenoid connector

Fuel flow test

1 Unscrew the threaded fittings and detach the metal fuel line from the filter inlet **(see illustration)**.

2 Disconnect the pink wire at the fuel injection pump electric shut off (ESO) solenoid **(see illustration)**.

3 Put the end of the metal line in a suitable container and crank over the engine a few times.

4 If little or no fuel flows from the open end of the fuel line, either the line is clogged or the fuel pump is dysfunctional. Remove the line between the fuel tank and the pump, blow it out and try this test again.

5 If there's still little fuel flowing from the pipe, replace the pump (see below).

Vacuum test

The vacuum test is the best indicator of how well a mechanical low pressure pump is working. If the low-pressure pump can only draw a low vacuum, or has a total loss of vacuum, it can't provide sufficient fuel to the injection pump to operate the engine throughout its normal speed range. The following vacuum test will help you determine whether the pump has the ability to pump fuel.

1 Detach the hose from the fuel tank to the fuel pump at the pump. Plug the hose to prevent contamination or fuel leakage.

2 Connect one end of a short hose to the fuel pump inlet and attach a vacuum gauge to the other end **(see illustration)**.

3 Start the engine. Note the vacuum gauge reading with the engine at idle. Shut off the engine. If the indicated vacuum is less than 12 inches Hg, replace the pump **(see illustration)**.

4 Remove the vacuum gauge and reattach the fuel inlet line.

Mechanical pump - removal and installation

Warning: *Diesel fuel isn't as volatile as gasoline, but it is flammable, so take extra precautions when you work on any part of the fuel system. Don't smoke or allow open flames or bare light bulbs near the work area. And don't work in a garage where a natural gas type appliance (such as a water heater or clothes dryer) with a pilot light is present. Finally, when you perform any kind of work on the fuel system, wear safety glasses and have a Class B type fire extinguisher on hand. If you spill any fuel on your skin, rinse it off immediately with soap and water.*

1 Detach the cable from the negative battery terminal.

2 Disconnect the fuel inlet and outlet lines from the low pressure pump.

3 Remove the pump mounting bolts **(see illustration 2.35)** and remove the pump, gasket(s) and spacer plate (if equipped) from the engine.

4 Remove all old gasket material from the pump to block mating surface with a gasket scraper and clean the surface thoroughly with solvent.

5 Installation is the reverse of removal. Be sure to use new gaskets.

2.37 Use a vacuum gauge, attached to the fuel pump inlet pipe (not visible in this photo), to check the ability of the pump to draw vacuum

2.38 To replace a mechanical fuel pump, disconnect the inlet and outlet hoses, remove the hold-down bolts and pull the pump straight out

Electric low pressure pump - general information

Some models use a solenoid type, electric low-pressure pump instead of a mechanical pump. The pump is usually located on the left frame rail; however, it may also be bolted onto the block. The basic operation of the electric pump is simple: A solenoid oscillated plunger pumps fuel in one check valve and out the other. This design has one big advantage over mechanical units: No rubber parts - like a diaphragm or bellows - are constantly flexed, so the pump is very reliable. The pump is completely sealed against fuel leakage. However, the pump won't last long if it's operated without fuel circulating through it because fuel cools the solenoid windings. Without it, they'll quickly overheat.

The electric pump operates on 12 volts, but will continue to operate during surges as high as 24 volts and as low as 8 volts. It's also protected against solenoid burnout if polarity is accidentally reversed.

Testing the electric pump

1 Check the fuel line fittings and connections - make sure they're tight. If a fitting is loose, air and/or fuel leaks may occur.
2 Check for bends or kinks in the fuel lines.
3 Start the engine and let it warm up. With the engine idling, conduct the following preliminary inspection before proceeding to the actual fuel pump tests:

a) *Listen to the pump. It should emit a steady purring sound. If it sounds shrill or metallic, it's probably about to fail.*
b) *Look for leaks at the pressure (outlet) side of the pump.*
c) *Look for leaks on the suction (inlet) side of the pump. A leak on the suction side will suck air into the pump and reduce the volume of fuel on the pressure (outlet) side of the pump.*
d) *Inspect the pump itself and the fittings on the pump for leaks. Tighten or replace the fittings as necessary.*

Voltage test

1 First, make sure the pump is getting battery voltage. Detach the B+ lead and, using a test light, make sure the pump is receiving voltage.
2 If the pump isn't receiving voltage, trace the lead back to the main wire harness, locate the problem (open, short, loose connector, etc.), fix it and recheck.
3 If the pump is receiving voltage, but isn't operating, the solenoid or transistor is burned out. Replace the pump.

Output tests

1 If is getting battery voltage and sounds like it's operating properly, check the output.
2 Disconnect the outlet hose from the pump, attach a short section of hose in its place and place the end of the hose in a suitable container.
3 Start the engine and note the quality of the squirts of fuel being emitted by the pump. They should be distinct and pronounced. If they're dribbling out, the pump is worn out. Replace it.
4 If the pump appears to be operating properly, remove the auxiliary hose, reattach the outlet line, detach the line at the inlet to the fuel filter and attach a conventional low pressure gauge to the line.
5 Start the engine and run it for 10 to 15 seconds. The indicated fuel pressure should be 5.8 to 8.7 psi. If it isn't, replace the pump.

Electric pump - removal and installation

1 Detach the cable from the negative battery terminal.
2 Detach the electrical lead to the B+ terminal on the pump.
3 Disconnect the fuel inlet and outlet lines from the pump and plug the lines to prevent contamination and leakage.
4 Loosen the pump mounting clamp and remove the pump.
5 Installation is the reverse of removal.

Inspecting the fuel lines

Air in the lines

The fuel lines, hoses and fittings in the fuel system must not allow air to enter the system. With the fuel tank filler cap screwed on and the low pressure and injection pumps pulling fuel through the lines, a slight vacuum of 0 to 1 pound Hg is created. During this vacuum condition, the smallest leak - it may not even leak fuel externally - can draw air into the system. Depending on how much air enters the lines, a variety of driveability problems - reduced mileage, smoking, poor performance and hard starting - can occur.

For instance, if the inlet fitting is slightly loose at the engine fuel filter, it will probably admit air into the lines. The symptoms are a fuel leak and/or a diesel fuel odor. The engine will start, then die and refuse to restart. Why? Because when the engine is shut down, the fuel siphons out of the lines and fuel pump, and drains back into the tank. Then it's replaced by air, which enters at the loose fitting, until the fuel system is emptied. As a result, the engine must be cranked until the lines are full again. The moral here is: Keep the fittings tight and the lines in good shape if you want to prevent air in the lines.

Diagnosing air in the fuel lines

The hydraulic advance mechanism won't work properly unless it's receiving transfer pump pressure and pump housing pressure. Even a small deviation below the factory adjusted operating pressure range will affect the advance mechanism and, therefore, the injection timing. Fuel pump delivery less than 5 1/2 to 6 1/2 pounds pressure reduces the total advance in direct proportion to the amount of the loss in pressure. Leaks, plugged filters, air ingestion, restricted lines, etc. - all reduce pressure delivery. Depending on the size of the obstruction, a return line restriction can elevate housing pressure to as high as transfer pump pressure, and will eventually stall the engine by upsetting the delicate balance between the transfer pump and housing pressures. Here's how to check for air in the lines:

1 Attach a short, clear plastic hose into the return line at the top of the injection pump **(see illustration)**.
2 Start the engine and watch for air bubbles or foam in the line.
3 If foam or bubbles are present, raise the vehicle and detach both fuel lines at the tank sending unit. Plug the disconnected return line (the one with the smaller diameter).
4 Attach a low-pressure air pressure source (a hand-operated pump will work fine) to the larger hose (the one with a 3/8-inch diameter) and apply 8 to 12 psi.
5 Note the indicated pressure reading: If the pressure quickly goes down, there's a leak in the line. The pressure you pumped into the line should force fuel out the leak point, enabling you to pinpoint its location. **Note:** *If you've got a vehicle with two fuel tanks, check the right fuel line with the dash switch at the right tank position, then check the left line with the switch pointing to the left tank. When you pressurize each line, watch the switching valve itself - that may be the source of the leak.*

2.39 Attach a short, clear plastic hose into the return line on top of the injection pump, start the engine and watch for air bubbles or foam in the line

2.40 On all 1981 through 1983 engines, the fuel heater (arrow) is installed on the fuel filter inlet line, under the right side of the intake manifold, along the upper edge of the right cylinder head (fuel heater not visible in this photo)

6 Repair the fuel line(s) as necessary. Make sure the right size hose clamps are installed. Look for burrs on the edge of metal lines - it could tear the inside of a rubber hose and allow air to enter. Pay particular attention to things like incorrect installation of fittings and defective or damaged auxiliary filters or water separators.

Be sure to use the correct replacement lines

If a fuel hose can't be repaired and you have to replace it, be sure you buy the right replacement hose. All 1981 and later models are equipped with GM SPEC. 6031 fuel hoses, which are made of "Viton" and contain a non-permeable tube inside the outer material. These hoses have a yellow stripe and the word *Fluroelastomer* imprinted on them. DO NOT use any other type of fuel hose material if you need to replace the fuel hoses.

Diesel fuel heater - general information and testing

A cold weather package, consisting of an in line diesel fuel heater and an engine block heater, is available on 1981 and later vehicles. Its purpose is to heat the fuel so the filter doesn't become clogged with wax crystals when the temperature drops below 20 degrees F. The use of a fuel heater allows the use of the (more efficient) No. 2 diesel fuel at temperatures substantially below its cloud point. The heater on all 1981 through 1983 models **(see illustration)**, which is electrically powered by the ignition circuit, is installed on the fuel filter inlet line a short distance upstream from the filter. (On later models, the fuel heater is located inside the fuel filter.) The heater is thermostatically controlled to operate when the temperature dips low enough that waxing of the fuel is likely to occur.

Basically, the heater can be divided into two components - the heater itself and the power control assembly. The heater is nothing more than an electric resistance strip, spiral wound and bonded to the fuel line. To minimize heat loss to the environment, the heating element is surrounded by a layer of insulation. The power control assembly senses fuel temperature and responds by closing an electrical circuit to the heater. The sensing element is a bimetal switch that turns on at 20 degrees F and shuts off at 50 degrees F. Power consumption is 100 watts. The heater will only remain on until the under hood temperature gets hot enough to warm the fuel. Thermal feedback from the heating element to the bimetal actuator protects the element from burning out in case there's no fuel flowing through the fuel heater.

The fuel tank filter sock has a bypass valve which opens when the filter is covered with wax, allowing fuel to flow to the heater. Without this bypass valve, the fuel line heater would be ineffective because the fuel would be trapped in the tank. The bypass valve is located at the upper end of the sock, so fuel will only be drawn into the bypass if the tank contains more than four gallons of fuel. Obviously, maintaining a minimum level of at least a quarter of a tank of fuel is critical when the temperature dips below 20 degrees F.

If you decide to retrofit a fuel heater, make sure you install the filter sock equipped with the bypass check valve. And since the connection for the fuel pipe is at a point in the side of the sock, you'll have to obtain the tank sending unit that's equipped with the "Water-In-Fuel" (WIF) option.

Because it's thermostatically controlled, the fuel heater can only be checked when the ambient temperature is 20 degrees F or colder. To check it, attach an ammeter in series. If it's operating properly, it should draw about 7 amps. If it isn't, replace it. The heater can't be serviced.

Fuel filters - general information

Particles larger than 10 microns in diameter can damage the internal components of the fuel injection pump. A micron is tiny - about .000039 inch in diameter. To put its size in the proper perspective, if you have perfect vision, the smallest object you can see is about 44 microns, or .0017 inch, in diameter; a human hair is about .003 inch across! Secondary fuel filters must be capable of removing nearly all particles bigger than 10 microns across, or they'll cause serious damage to the pump. A number of different filters have been used on the GM models covered in this manual. Familiarize yourself with the filter(s) on your vehicle so you'll get the right unit(s) when buying a replacement.

Roosa master Model 50 fuel filter (1978 through 1984 models with 5.7L engines)

The Roosa master Model 50 fuel filter is a two-stage, pleated paper type design designed to prevent particles in the fuel from damaging the pump. The first stage, which consists of about 400 square inches of filtering area, removes 94 percent of the 10-micron and larger particles in the fuel. The second stage, which is made of the same paper material and consists of about 200 square inches of filtering surface area, removes 98 percent of the stuff still left in the fuel filtered in the first stage.

The filter is impossible to install incorrectly: The inlet and outlet fittings are different sizes. Because the outlet is lower than the inlet, any water that enters the filter will pass through, instead of remaining trapped inside the bottom of the filter where it can freeze and hamper the operation of the injection pump, or rust and cause damage to the pump.

NB9 fuel filter system (1985 and later models with 5.7L engines)

The NB9 fuel filter system **(see illustration)** minimizes corrosion in the fuel injection pump and improves cold weather driveability. It continuously removes contaminants from the fuel tank and transports them through auxiliary lines to a water-separating filter. The NB9 system also alerts the driver of water in the fuel and provides a convenient way to drain contaminated fuel.

The NB9 offers a number of improvements over earlier low pressure filtering systems employed on the 5.7L engine. The AC filter assembly contains a water "coalescer" or "conditioner" that separates water from the fuel; a fuel filter; a 100-watt fuel heater; a water storage sump; a drain cock and a "Water-In-Fuel" (WIF) indicator. A sight glass (similar to the device used on the receiver drier in an air conditioning system) offers instant visual verification of contaminated fuel. An electric lift pump makes draining the fuel filter safer because it can be drained with the engine turned off. A water faucet type remote filter drain valve is mounted away from the engine fan and hot engine components for safe and clean draining of the fuel filter. It also serves as a back-up shut-off in case the push button drain valve mounted on the filter fails to open. O-ring fuel line connections virtually eliminate fuel system leaks. A relocated wax bypass insures that the vehicle won't run out of fuel with the gauge needle indicating above "Empty" during sub zero driving. The NB9 sending unit uses a tank sock with an anti collapsing stiffener and a controlled orifice to purge water from the bottom of the tank at a controlled rate.

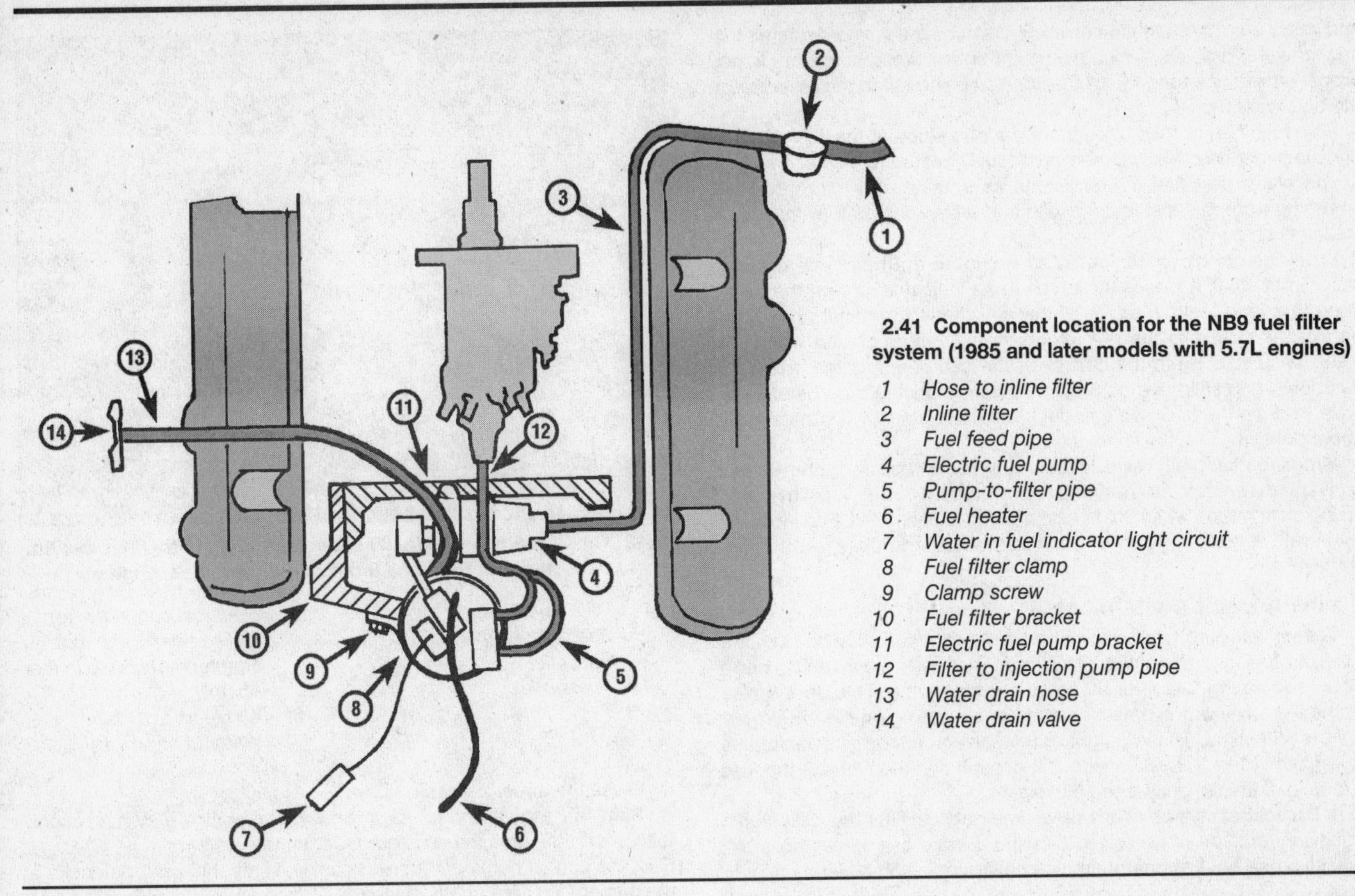

2.41 Component location for the NB9 fuel filter system (1985 and later models with 5.7L engines)

1 *Hose to inline filter*
2 *Inline filter*
3 *Fuel feed pipe*
4 *Electric fuel pump*
5 *Pump-to-filter pipe*
6 *Fuel heater*
7 *Water in fuel indicator light circuit*
8 *Fuel filter clamp*
9 *Clamp screw*
10 *Fuel filter bracket*
11 *Electric fuel pump bracket*
12 *Filter to injection pump pipe*
13 *Water drain hose*
14 *Water drain valve*

How does the system **(see illustration)** work? Water is pulled from the tank and separated at the filter assembly. A light on the dash comes on when 30 ml has been collected in the filter sump, which has a total capacity of 250 ml. To purge the system, the ignition key is turned on to activate the electric fuel pump, which pressurizes the system (engine off). The drain cock is closed once the fuel runs clear, indicating the purging is complete.

To check the WIF electrical circuit, note whether the WIF light on the dash comes on, then goes off, when you turn the ignition key to On. Then, if there's water in the fuel tank, you should be able to drive 28 miles before the light turns on again, and 200 miles before the sump is filled with water.

The AC engine mounted diesel fuel filter combines the following functions into one assembly: fuel heater (in the adapter), water coalescer and separator, fuel filter, water detector and water drain. The spin on filter assembly houses the coalescer and filter/separator; a die cast cover assembly contains the fuel heater, the water detector circuit and the water detector probe/siphon tube assembly.

Diesel fuel entering the adapter first passes through the heater, a Positive Temperature Coefficient (PTC) device which provides rapid warm up (it may approach 200 watts under certain conditions of cold fuel and high flow). A snap disc thermostat within the heater assembly turns the heater on and off as necessary. As the diesel fuel flows through the adapter, it actually contacts the ceramic heating element. Coupled with the heater's close proximity to the filter, this direct contact with the heat source provides excellent heating efficiency (remote mounted heaters lose heat through exposed fuel lines).

The first stage of the filter assembly is a water coalescer. When fuel

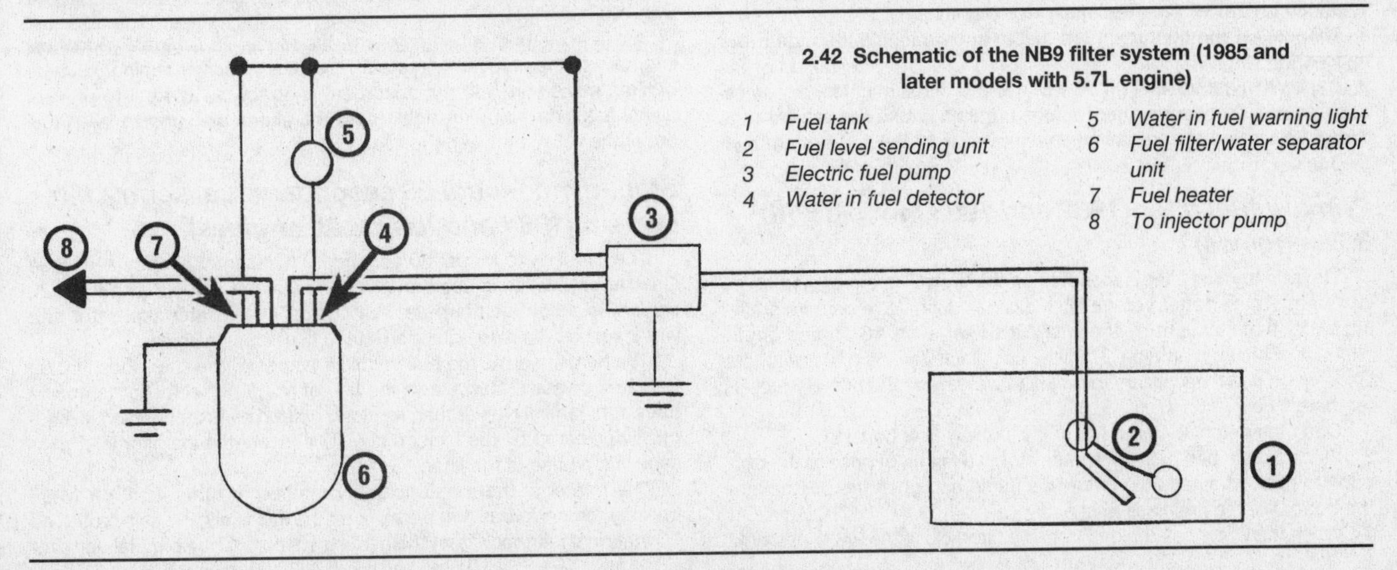

2.42 Schematic of the NB9 filter system (1985 and later models with 5.7L engine)

1 *Fuel tank*
2 *Fuel level sending unit*
3 *Electric fuel pump*
4 *Water in fuel detector*
5 *Water in fuel warning light*
6 *Fuel filter/water separator unit*
7 *Fuel heater*
8 *To injector pump*

and water pass through the coalescer element, the element retards the flow of water droplets, which merge with other droplets to form larger drops, until they emerge from the coalescer and fall to the reservoir at the bottom of the filter.

The fuel then passes through the second stage of the filter assembly, the paper filter. This element is specially treated to restrict passage of the water and fuel contaminants as small as 10 microns. Fuel emerging from this element is clean and water-free as it proceeds to the injection pump.

If the electronic water detector, mounted on the cover, detects water in the fuel, it closes the circuit to an indicator light on the dash. The other lead, held in place by the WIF ground terminal bracket, is part of the circuit to the WIF probe; this circuit is closed when the water accumulating in the bottom of the can contacts the probe. At this point, you still have a few more miles of driving time before you have to drain the filter system, but it's time to start thinking seriously about doing it.

A pushbutton drain valve, located in the cover assembly, provides a convenient spigot for draining accumulated water out the WIF probe/siphon tube when the solenoid pump is energized. The indicator lights will, of course, go off as soon as the water drops below the trigger level.

Draining water contaminated fuel

During refueling, water and other contamination may be pumped into your fuel tank along with diesel fuel. That's why your vehicle has a water-separating fuel filter mounted on the engine. This device also warns the driver of the presence of water by turning on the red Water-In-Fuel (WIF) light. The WIF light also comes on during engine starting to verify that the bulb is working. If it doesn't come on, check the fuse and bulb. If they're okay, see your dealer.

If the light comes on at any other time, the fuel filter has collected a significant amount of water and must be drained as soon as possible, or within one hour of operation, whichever comes first. Here's how to do it:

1 Turn off the engine, place the shift lever in P or N and apply the parking brake.
2 Turn the ignition key to the Run position, but don't start the engine. This energizes the electric fuel pump and activates the WIF light. Open the hood.
3 Place a container with a capacity of 12 fluid ounces under the drain valve. The valve is located on the left fender.
4 Open the drain valve two complete turns.
5 Reach over to the engine mounted fuel filter and push the drain button (indicated by a red arrow). Fuel and water will start flowing out the valve.
6 Continue draining until the color of the fuel indicates that all water is gone and clean fuel fills the filter. If visibility is poor, hold the drain button for 15 to 20 seconds, or stop pushing it when the container is filled.
7 Close the fender mounted drain valve tightly.
8 Dispose of the contaminated fuel in accordance with local ordinances.
9 If the WIF light doesn't go off within two or three minutes, or comes back on within seconds after repeated draining, take the vehicle to a dealer and have the contamination removed from the fuel system and the tank.

Primary fuel filter (1982 and 1983 models with 6.2L engines)

The primary fuel filter is mounted on the firewall in the engine compartment. At the bottom of the filter, a water drain valve permits draining water that's caught by the filter. This filter is an AC fibrous depth element of the spin on type. The filter case includes the drain petcock. Should you find it necessary to drain the water from the primary fuel filter, here's how to do it:

1 Open the petcock on the top of the primary filter housing.
2 Place a drain pan below the filter, attach a length of hose to the petcock to direct drained fluid below the frame and open the petcock on the bottom of the drain assembly.
3 When the water has drained from the filter, close the lower petcock.

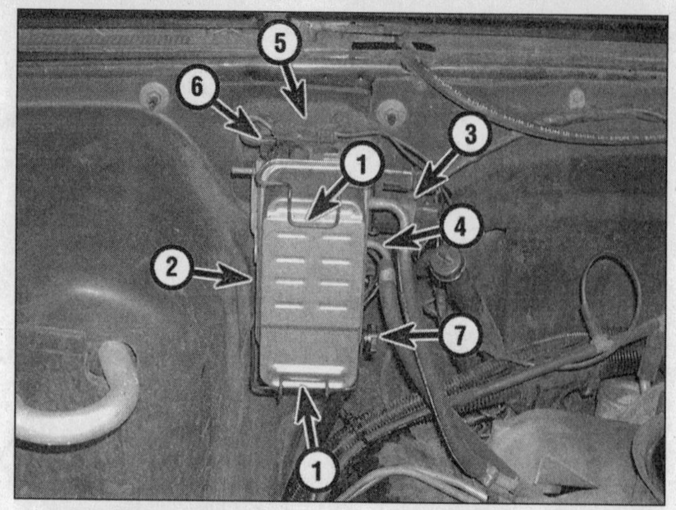

2.43 The Stanadyne Model 80 secondary fuel sentry filter system is used on all 1984 and later models with 6.2L engines

1 Wire retainers
2 Fuel filter element
3 Fuel inlet
4 Fuel outlet
5 Electrical connector for fuel heater (heater inside adapter, not visible in this photo)
6 Air bleed
7 Water drain valve

4 Close the upper petcock.
5 Start the engine and let it run briefly. The engine may run roughly for a short time until the air is purged from the system.
6 If the engine continues to run roughly, verify that both petcocks at the primary filter are tightly closed.

Secondary fuel filter (1982 C, K and P, and 1983 C/K models with 6.2L engine)

All 1982 C, K and P, and 1983 C/K models with the 6.2L engine use a spin on style secondary fuel filter mounted at the rear of the intake manifold. Though its internal design is different from the 5.7L unit discussed above, its 10 micron filtering capability is the same. Because of its spin on design, this unit is particularly easy to replace.

Stanadyne Model 75 secondary fuel filter (1983 G and P models with 6.2L engine)

1983 G and P models use a Stanadyne Model 75 secondary fuel filter. This unit is fastened to a bracket on the rear of the intake manifold with two bail clips.

Aside from its mounting system, the Model 75 is virtually identical internally to the unit used on the 5.7L engine. When changing this type of filter, make sure you put absorbent shop towels under it to prevent diesel fuel from dripping down onto the clutch disc and to keep fuel out of the valley between the heads.

Stanadyne Model 80 secondary fuel sentry filter system (1984 and later 6.2L engines)

The Stanadyne Model 80 secondary fuel sentry filter system **(see illustration)**, which is used on all 1984 and later models with the 6.2L engine, is a combination fuel heater, fuel filter, water separator and water sensor. It's also equipped with a filter change signal.

The heater warms the fuel in cold weather so the filter doesn't plug with wax crystals. This allows the use of No. 2 diesel fuel at temperatures substantially lower than its cloud point. The heater, which is electrically powered by the ignition circuit, is located in the filter inlet passage in the base of the filter.

The heater is thermostatically controlled: It turns itself on when freezing temperatures make waxing of the fuel likely. It's self-protected - by thermal feedback from the heating element to the bimetal actuator

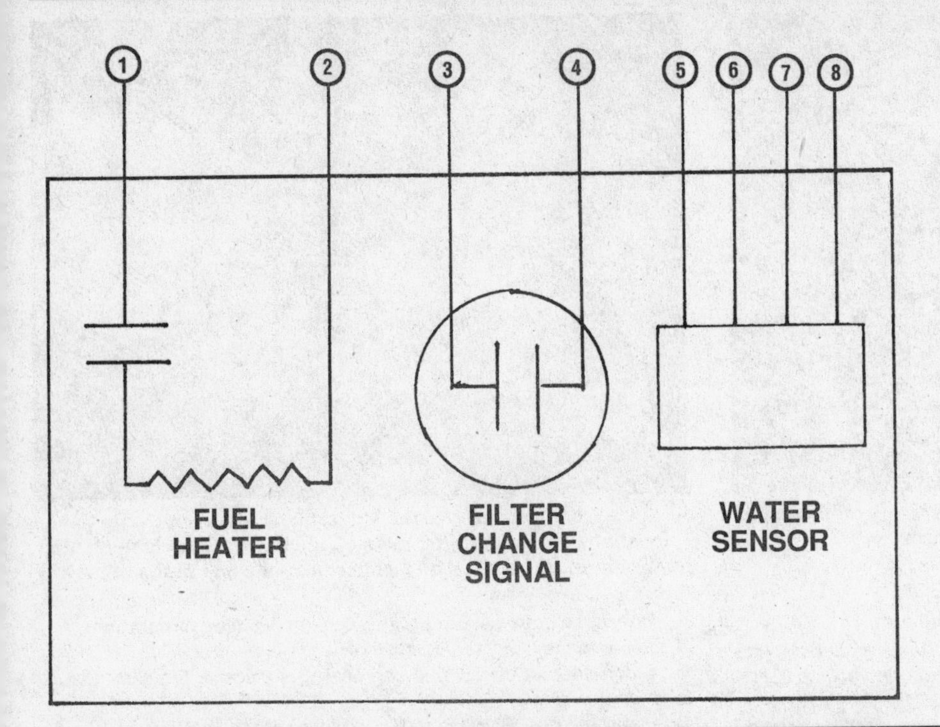

2.44 On 1984 models (shown), the Model 80 system tests its WIF circuit every time the ignition switch is turned to Start: Battery voltage at the test switch terminal grounds the WIF signal indicator bulb (on 1984-1/2 and later models, the circuit looks slightly different - the water sensor only has three wires instead of four and the self-check is also slightly different - see text)

1 *Ground*
2 *Battery*
3 *Signal indicator light*
4 *Battery*
5 *Test switch*
6 *Ground*
7 *Battery*
8 *Signal indicator light*

- against overheating which might result from poor fuel flow. Because the heater is located within the filter base, it's 50 percent more heat efficient than a line heater.

There are two main parts to the heater - the heater itself and the power control assembly. The heater is really nothing more than a spiral wound resistor strip, about 7/8 inches in diameter. The power control assembly senses fuel temperature and, when the temperature is cool enough, responds by closing an electrical circuit to the heater. The sensing element is a bimetal switch. The internal bimetal switch turns on at 20 degrees F and shuts off at 46 degrees F. The heater consumes about 110 watts at 14 volts DC. The heat remains on only until the under hood temperature gets hot enough to warm the fuel.

The fuel filter is a two stage, pleated paper element. The first stage, which has 350 square inches of filtering area, can remove 96 percent of the particles in the 5 to 6 micron and larger range. The second stage is made of the same paper material; it also contains glass particles and consists of about 100 square inches of filtering surface. This second stage is 98 percent effective in filtering the fuel already filtered by the first stage.

The water separator is a hollow water collector in the bottom of the fuel filter. Because of their greater density, water droplets separate from the fuel oil. The separator holds about 260 cc of water. A nylon/fiberglass coalescent promotes the formation of larger water droplets: One micron water droplets collect in the coalescent fibers; when the droplets become heavy enough and large enough, they drop into the filter bottom. The coalescing increases the water concentration from 20 to 30 parts per million (ppm), which promotes more efficient water collection.

The Water-In-Fuel system detects water with a capacitive probe located in the filter base. The probe itself is made from iron ferrite, which isn't susceptible to electrolysis. The sensor turns on at the 50-cc level. How does it work? The electronics within the probe connect the ground circuit with the ground side of the Water-In-Fuel warning light (circuit 508) on the dash. This light is in the center of the instrument cluster next to the glow plug light. On 1984 models with the four wire water sensor module, the bulb is checked every time the ignition switch is turned to the Start position. A B+ signal on the purple wire at the test switch **(see illustration)** causes the pin at the signal indicator light to pull low, grounding the Water-In-Fuel bulb. On 1984 1/2, and 1985 and later, models with the three wire water sensor module, the bulb check is a little different: The WIF light glows from two to five seconds when the ignition switch is turned on, then fades out. To distinguish a 1984 1/2 module from a 1985 and later module, note the number of wires going into the module and the number of wires going into the connector: The 1984 1/2 module is a three wire unit with a four wire male connector; the 1985 and later module is a three wire unit with a three wire female connector.

The pressure switch is incorporated into the filter base. It's used to indicate filter blockage. The pressure differential valve is set at 14 inches of mercury (Hg) $\pm$ 2 inches.

Water in the fuel and the Model 80 system

If water is inadvertently pumped into your fuel tank along with diesel fuel, the water separation system in the fuel filter will remove most of it. A low-pressure sensor will activate the Water-In-Fuel light in the instrument cluster if the water accumulates in the fuel filter and plugs it up. The light also comes on when you start the engine to let you know the bulb is working; if it doesn't come on, check the fuse and the bulb. If the sensor does come on, the first step (and usually the only step that's necessary) is to drain the water from the filter.

Draining water from the Model 80 fuel filter

The multifunction filter is mounted on the firewall. To drain water from the filter:

1 Remove the fuel tank filler cap.

2 Place a container below the filter drain hose (located below the filter).

3 With the engine off, open the water drain valve **(see illustration 2.43)** two to three turns. **Note:** *G and P models have a remote mounted water drain valve near the thermostat housing.*

4 Start the engine and allow it to idle for one to two minutes, or until you note clear fuel flowing from the drain valve.

5 Stop the engine and close the water drain valve.

6 Install the fuel tank filler cap.

If the Water-In-Fuel light comes on again after you've driven a short distance, or if the engine runs roughly or stalls, a large amount of water has probably been pumped into the fuel tank. Purge the tank.

Using the Model 80 Water-In-Fuel light to pinpoint a specific problem

1 If the Water-In-Fuel light comes on intermittently, drain the water from the fuel tank.

2.45 To remove the Model 80 fuel filter element, pry off the upper and lower bailing wire retainers

2.46 When you remove the Model 80 filter element, note the locations of the four grommets on the back of the element; the new filter must be installed just like this - it won't fit the other way

2 If the light stays on when the engine is running:

a) *And the temperature is above freezing, drain the fuel filter immediately. If no water comes out, and the light stays on, replace the filter.*
b) *And the temperature is below freezing, drain the fuel filter immediately. If no water comes out, it may be frozen. Open the air bleed to check for fuel pressure. If there's no fuel pressure, replace the filter.*

3 If the light comes on at high speed, or during heavy acceleration, the fuel filter is plugged. Replace it.

4 If the light stays on continuously, the engine stalls and it won't restart:

a) *After initial start up, the fuel filter or fuel lines may have become plugged. Inspect the system.*
b) *Large amounts of water may have been pumped into the tank during refueling. Purge the fuel tank.*

Purging the fuel tank

Warning: *Diesel fuel isn't as volatile as gasoline, but it is flammable, so take extra precautions when you work on any part of the fuel system. Don't smoke or allow open flames or bare light bulbs near the work area. And don't work in a garage where a gas type appliance (such as a water heater or clothes dryer) is present. Finally, when you perform any kind of work on the fuel system, wear safety glasses and have a Class B type fire extinguisher on hand. If you spill any fuel on your skin, rinse it off immediately with soap and water.*

We recommend having the tank purged by an authorized dealer, but if you wish to do it yourself, here's how :

1 Park the vehicle on level ground. The fuel tank sending unit/pick up is in the approximate center of the tank.

2 Place a large container under the filter drain hose. Open the drain about three or four turns.

3 Disconnect the fuel return hose at the injection pump.

4 With the fuel tank cap correctly installed, apply low pressure air (3 to 5 psi maximum) through the fuel return hose. The fuel tank cap is designed to retain 3 to 5 psi pressure, allowing water to be forced out of the fuel tank via the filter drain hose.

5 Continue to drain until only clear fuel comes out.

6 Close the drain valve tightly. Reinstall the fuel return hose.

Removing and installing the Model 80 fuel filter

1 To prevent fuel from spilling, open both the air bleed and the water drain valve **(see illustration 2.43)**, and drain the filter into a suitable container.

2 Remove the fuel tank filler cap to release any pressure or vacuum in the tank.

3 Pop loose both bail wires with a screwdriver **(see illustration)**.

4 Remove the filter **(see illustration)**.

5 Clean any dirt off the fuel port sealing surface of the filter adapter and the new filter.

6 Install the new filter and lock it in place with the bail wires.

7 Close the water drain valve and open the air bleed. Connect a 1/8-inch I.D. hose to the air bleed port and place the other end into a suitable container.

8 Disconnect the fuel injection pump shut off solenoid wire (the pink wire to the pump).

9 Crank the engine for 10 to 15 seconds, then wait one minute for the starter motor to cool. Repeat this process until you see clear fuel coming from the air bleed.

10 Close the air bleed, reconnect the injection pump solenoid wire and replace the fuel tank cap.

11 Start the engine and allow it to idle for five minutes.

12 Check the fuel filter for leaks.

Removing and installing the Model 80 fuel filter base

The Model 80 fuel filter base is mounted on the firewall of C and K models. It's mounted on the rear of the intake manifold on G and P models. To replace it:

1 Remove the air vent plug located at the top of the filter base.

2 On C and K models, loosen the drain plug and drain the fuel from the filter. On G and P models, open the remote drain valve to drain the filter.

3 Disconnect the fuel hoses **(see illustration 2.43)** and unplug the electrical connectors **(see illustrations)**.

4 Remove the three filter base mounting bracket bolts and remove the base and bracket assembly from the vehicle.

5 Installation is the reverse of removal.

Model 80 fuel filter seal leaks

Fuel and/or air can leak past the drain and/or vent seal(s) in the base of Model 80 fuel filters with part numbers 14071933 (C and K models) and 14071064 (G and P models). A new style seal (part number 15529641) - which has a slightly smaller outside diameter, enabling it to bottom in the bore and seal properly - eliminates this problem. Replace the leaky seal(s) with the new style seal if you note either of the following symptoms:

a) *External fuel leakage from the vent or drain plugs.*
b) *Engine starts normally, then stalls, and then is difficult to start.*

Replacing the Model 80 fuel filter seal

1 Remove the fuel filter assembly from the vehicle (see above).

2 Unclip and remove the filter/separator from the base, and then remove the filter base assembly from the firewall.

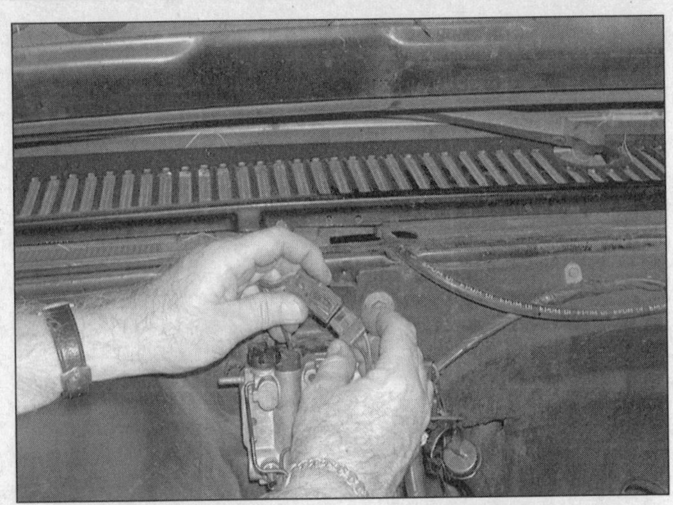

2.47 Before unbolting the Model 80 fuel filter base assembly, be sure to unplug the electrical connector for the fuel heater (the connector on top)

2.48 There are two other connectors on the left side of the filter base: the upper connector is for the vacuum switch, the lower one is for the water sensor (the locks for the connectors face toward the firewall)

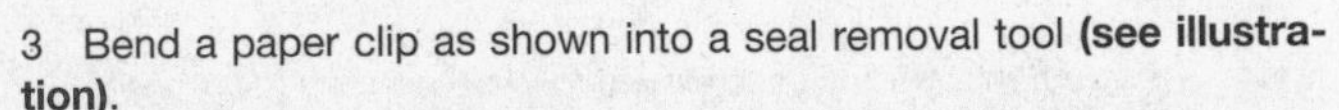

3 Bend a paper clip as shown into a seal removal tool **(see illustration)**.

4 Remove the drain valve **(see illustration)**. Then, using your seal removal tool, remove the drain plug seal from the filter base **(see illustration)**.

5 Visually inspect the bore(s) for evidence of seal particles. If particles are present, blow compressed air into the filter base outlet to remove them.

6 Use a short length of 1/4-inch bar stock to install the new seal in the bore. Apply a small amount of Synkut lubricant (or equivalent, such as STP) to one end of the rod, attach a new seal to the end of the rod and insert the seal into the bore until it seats firmly in the bottom of the bore. Visually inspect the seal to ensure it's squarely bottomed in the bore. Install the drain plug. **Note:** *Two styles of drain plugs have been used on the Model 80 filter. The earlier style has a conical end and uses a separate seal at the base of the plug bore. The later style plug has a flat end and a rubber like seal attached to the end of the plug. The newer plug is the only one used on later models. If you replace the flat end plug with a conical plug, measure the length of the boss of the threaded plug hole. If the boss is 1/4-inch long, install the drain plug without modification and tighten it until it bottoms. If it's 1/8-inch long, add a plain washer (GM part number 561890) to the drain plug, then thread the plug until it bottoms (the drain plug washer prevents you from overtightening the drain plug and damaging the seal).*

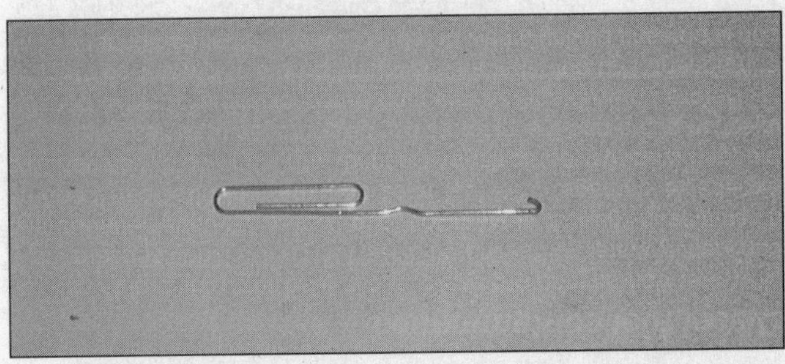

2.49 To remove the seal from a Model 80 water drain or air bleed bore, bend a paper clip as shown

7 Repeat the above procedure, if necessary, to replace the seal in the air vent bore.

8 Attach the element to the base.

9 Using a maximum of 10 psi, pressure check the filter assembly with compressed air by plugging the fuel inlet, outlet and drain outlet (if applicable). Open and close the air vent and drain plug (if equipped) several times to verify that the valves are sealing.

10 Reinstall the filter assembly.

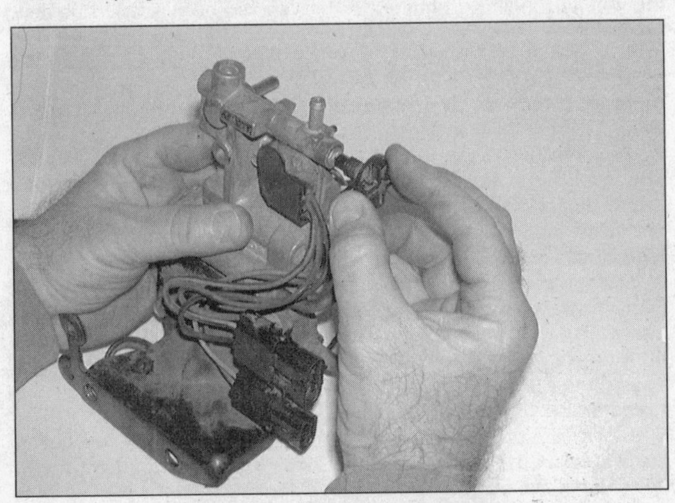

2.50 Unscrew the water drain plug

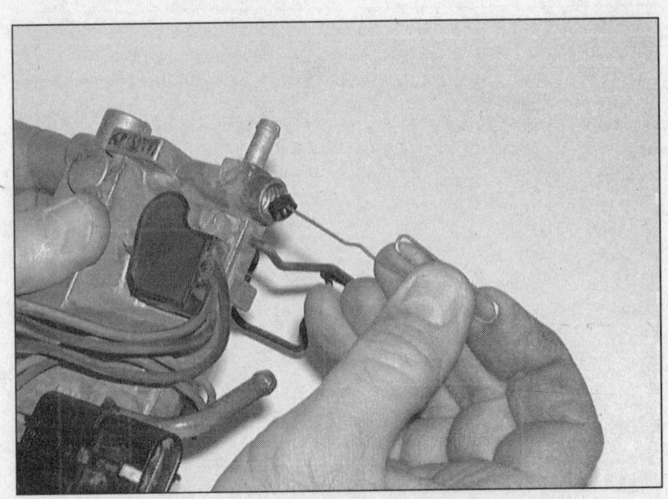

2.51 Carefully insert your special seal removal tool into the water drain bore, push it through the old seal, hook the seal and pull it out of the bore; be extremely careful not to damage the seal bore

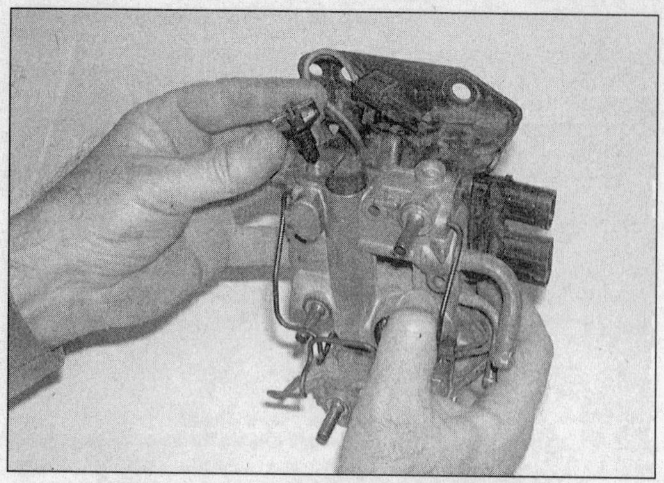

2.52 The air vent plug must be unscrewed before you can remove the fuel heater element

2.53 To remove the Model 80 fuel heater element, grasp the heater with a pair of slip-joint pliers, twist it counterclockwise only and pull straight up and out of the base assembly

Checking and replacing the Model 80 fuel heater

1 To check the heater, hook up an ammeter in series. Checking must take place below 20 degrees F ambient temperature. If the heater is operating properly, it should draw about 8.6 amps. If it isn't, remove it and test it again.
2 Remove the Model 80 filter base assembly.
3 Remove the vent plug from the filter base **(see illustration)**.
4 Grasp the heater with a pair of slip joint pliers **(see illustration)**, twist it counterclockwise and pull straight up and out of the adapter. Plug in the heater side of the electrical connector to the harness side of the connector. The heater must be connected for the next test.
5 Turn the ignition key to the On position.
6 Hold the heater by its lead wires and spray the heater head (in the area above the O ring seal) with Freon or some other type of quick freeze aerosol until the heater element begins to give off a vapor from the heated fuel. This confirms that the heater switch has closed and the element is heating. **Warning:** *Be sure to wear eye protection and don't get any on your skin.* **Caution:** *Do NOT allow the heater to remain on for longer than 30 seconds, or you may damage the heater element. Turn the key off or unplug the electrical connector.*
7 If the heater element still doesn't heat, unplug the connector, discard the faulty element and replace it.
8 Installation is the reverse of removal. Be sure to lubricate the O ring seal and position the flat opposite the vent plug hole by rotating the heater counterclockwise only.

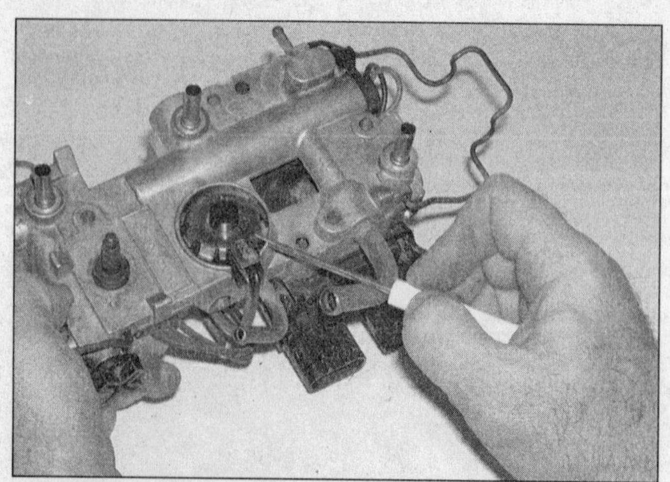

2.54 Place the base assembly flat on a workbench and pry the vacuum switch retaining clip from the base with a small screwdriver - initially prying the clip up at the wire lead protective tab will score the base bore the least

Checking the Model 80 vacuum switch and water sensor

Note: *The vacuum switch was removed in 1990 on the C/K model.*
1 If the WIF light is lit while the engine is running, shut off the engine, open the drain plug located at the bottom of the filter base and open the vent plug at the top of the base. Drain the fuel into a suitable container until clear fuel is detected.
2 Close the drain and vent plugs and then start the engine. The WIF light should go off.
3 If the WIF light remains on with the engine running, shut off the engine and disconnect the vacuum switch terminal.
4 If the WIF light goes out, the vacuum switch is faulty. Replace it.
5 If the WIF light remains lit, disconnect the water sensor terminal.
6 If the WIF light now goes out, the water sensor is faulty. Replace it.
7 The vacuum switch is normally open; it closes when a vacuum is applied.
8 All components of the Model 80 system must be installed to perform the vacuum switch test.
9 Plug off the inlet.
10 Hook up an ohmmeter to the vacuum switch terminals.
11 Connect a hand-operated vacuum pump to the fuel outlet connector.
12 Draw a vacuum of 10 inches Hg - the vacuum switch should still be open (no continuity reading on the meter).
13 Slowly increase vacuum to 18 inches Hg; the switch should close (continuity reading on meter). If the vacuum switch fails this test, replace it.

Replacing the Model 80 vacuum switch

1 Remove the filter element from the base and remove the filter base assembly from the firewall.
2 Place the base assembly flat on the workbench and pry the vacuum switch retaining clip from the base with a small screwdriver. Be careful - don't damage the bore. Initially prying the clip upward at the wire lead protective tab will cause the least amount of scoring to the base bore **(see illustrations)**.
3 Using a pair of small screwdrivers, remove the vacuum switch by gently prying up the switch at the wire leads and simultaneously prying under the switch opposite the leads **(see illustration)**.
4 After the switch is removed, check the bore by running your finger around the inside of the bore and visually inspect for sharp edges or raised metal burrs caused by removal of the retaining clip.
5 If burrs are present, prevent harm to the switch O ring seal during installation by using a fine (300) grit paper or round stone to remove any sharp metal burrs or scratches in the bore.
6 Using a lint free cloth and solvent, wipe off any debris from the bore area that may have been generated during the deburring process, then blow off the area with compressed air.

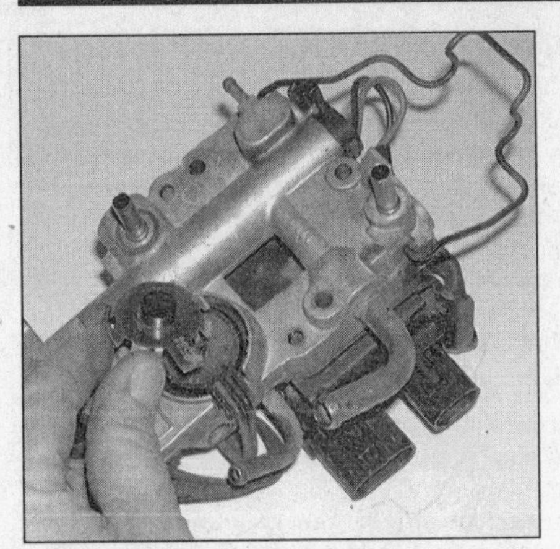
2.55 Remove the vacuum switch clip

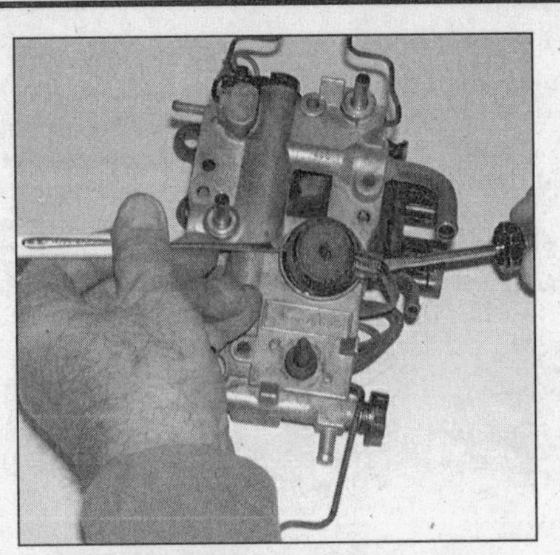
2.56 Using a pair of small screwdrivers, remove the vacuum switch by prying it out of the bore at the wire leads and simultaneously prying under the switch opposite the leads

2.57 To prevent cutting the O-ring seal, angle the new vacuum switch into the bore, the edge with the wire lead first

7 Pry off the retaining clip from the new switch with a knife blade or screwdriver (pry the clip upward at the wire lead area).

8 Apply a liberal amount of Vaseline, grease or STP to the switch O ring seal and apply a small amount of lubricant to the base switch bore.

9 To prevent cutting the O ring seal, insert the new switch - without the retaining clip - into the bore at an angle, inserting the wire lead portion first **(see illustration)**. Use finger pressure until the switch is seated in the bottom of the bore. **Note:** *Current versions of the vacuum switch have a rubber button at the center of the retaining clip. This button should be in place prior to installation of the switch retaining clip to the base and prior to installation of the filter/separator element.*

10 Make sure the cellular air filter is in place on top of the switch, then place the retaining clip in position with the protective tab over the wire lead. Place the vacuum switch installation tool (27752) on top of the retaining ring tabs; make sure the cutout portion of the tool is in line with the lead wires. Press it into place slowly and uniformly.

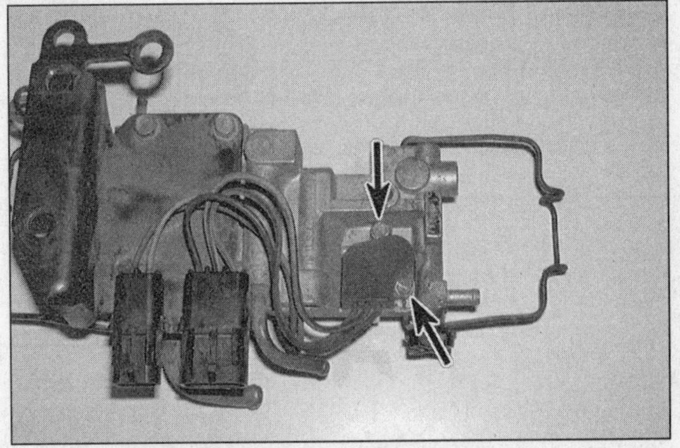
2.58 To detach the water sensor from the filter base, remove these two retaining screws (arrows)

Testing and replacing the water sensor

1 Remove the two screws holding the sensor to the base **(see illustration)**.

2 Connect the sensor to a variable voltage source and indicator light **(see illustration)**.

3 Switch on the variable voltage source and adjust it to 12 volts DC.

4 Close the switch in lead C (sensor tip not grounded). The indicator light should come on for about 10 to 15 seconds. If it doesn't, replace the sensor.

5 With the switch lead still closed in lead C, ground the sensor tip. The light should come on. If it doesn't, replace the sensor.

6 Installation is the reverse of removal. Be sure to use two new sensor hold-down screws. If you're using the old base, tighten the screws to about 10 or 15 ft-lbs; if you're using a new base, initially tighten the screws to about 20 or 25 ft-lbs, then loosen them and retorque them to 10 or 15 ft-lbs.

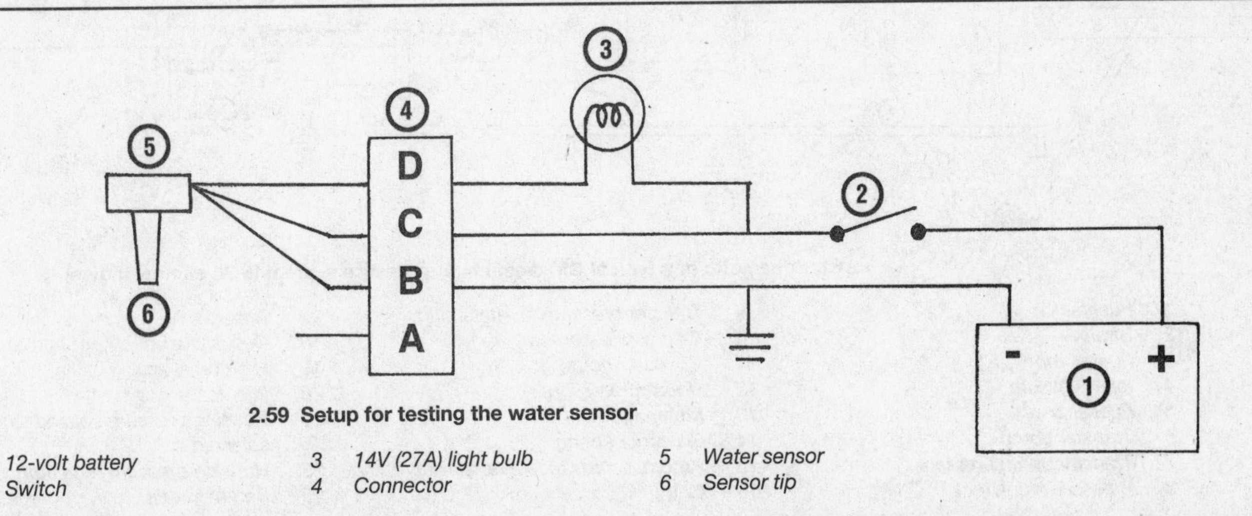

2.59 Setup for testing the water sensor

1 *12-volt battery*
2 *Switch*
3 *14V (27A) light bulb*
4 *Connector*
5 *Water sensor*
6 *Sensor tip*

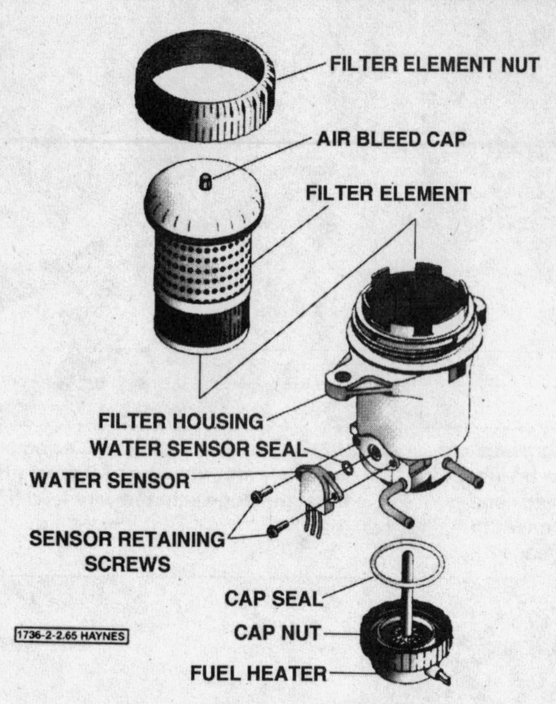

2.60 Exploded view of the fuel filter on 6.5L engines

Fixing a blinking WIF light (1984 C, K, G and P models with 6.2L engine)

A blinking instrument panel light can be isolated to the WIF sensor or the vacuum sensor by disconnecting the sensor connectors one at a time and operating the engine. If the WIF sensor is causing the light to blink or come on after water has been drained from the filter, install a three wire sensor (C, K part number 15588775; G, P part number 15588774) to rectify the condition.

Model FM 100 fuel manager/filter (1993 and later 6.5L engines) - general information

The FM 100 fuel filter **(see illustration)** is an in line type filter which acts as a two stage fuel filter, water detector, water separator, water drain and fuel heater.

The filter acts as a water detector by turning on the "WATER-IN-FUEL" light on the instrument panel. When the light comes on, the filter should be drained immediately, but definitely within one to two hours maximum. A water drain is located in the bottom of the filter assembly.

Draining water or fuel from the Model FM 100 fuel filter

The model FM 100 fuel filter/water separator follows the same procedure as the Model 80, previously described in this Chapter, for draining accumulated water from the filter. **Note:** *When draining fuel from the housing, in order to replace the filter element, once the fuel begins flow, open the air bleed valve on top of the fuel filter (see illustration 2.65), then turn off the engine. With the air bleed open and the engine turned off the remaining fuel will siphon out of the fuel filter housing.*

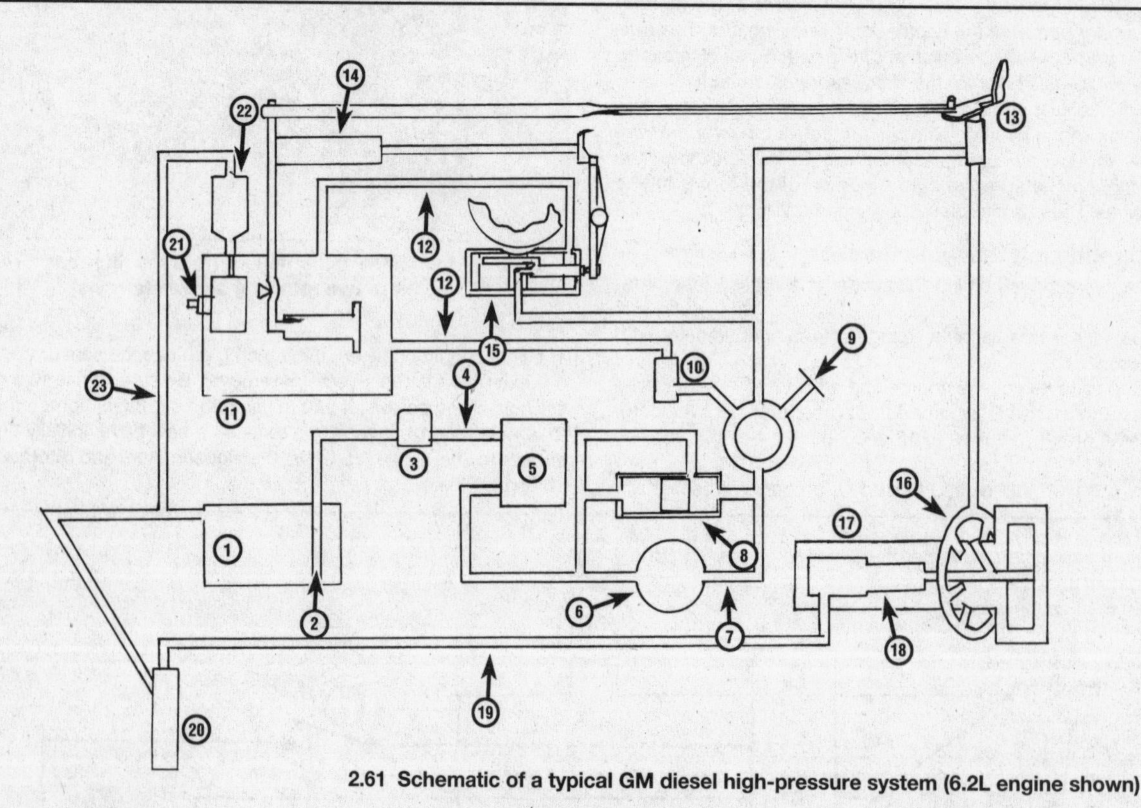

2.61 Schematic of a typical GM diesel high-pressure system (6.2L engine shown)

1 *Fuel tank*
2 *Strainer*
3 *Fuel pump*
4 *Inlet pressure*
5 *Fuel filter*
6 *Transfer pump*
7 *Transfer pump pressure*
8 *Pressure regulator*
9 *Transfer pressure tap hole plug*
10 *Vent wire assembly*
11 *Governor housing*
12 *Housing pressure*
13 *Metering valve*
14 *Governor spring*
15 *Automatic advance mechanism*
16 *Charging passage*
17 *Rotor*
18 *Delivery valve*
19 *High pressure*
20 *Injector*
21 *Housing pressure cold advance solenoid*
22 *Housing pressure regulator*
23 *Return circuit*

Model FM 100 filter element replacement

1 Unscrew the element nut **(see illustration 2.60)** counterclockwise (left). **Note:** *It may take an oil filter strap wrench to break the nut loose.*
2 Lift the element straight up and out of the housing.
3 When placing a new filter element into the housing align the widest key slot on the element with the slot in the housing assembly, and push the filter into the housing.
4 Install the threaded nut on the housing and tighten it.
5 Disconnect the fuel injection pump shutdown solenoid wire.
6 Crank the engine in short (15 second) intervals until clear fuel is seen at the air bleed hose.
7 Reconnect the solenoid wire.
8 Start and run the engine at least 5 minutes at idle.
9 Check for leaks.

High pressure system - general information

The heart of GM's 5.7L, 6.2L, and 6.5L diesel engines is the high-pressure system which delivers fuel to the combustion chambers. The high-pressure system **(see illustration)** consists of the injection pump, the injection nozzles and the metal lines between the pump and the nozzles.

The high-pressure system sounds pretty simple in principle, but in fact it's anything but simple.

The high-pressure system performs many functions, many more than a carburetor or fuel injection system on a gasoline-powered vehicle. Not only must it meter - in accordance with the load and speed of the engine - the quantity of fuel required for each cycle of the engine. It must also develop the high pressure necessary to inject fuel into each cylinder at a precisely determined instant in its operating cycle (firing order). And it must control the rate at which the fuel is injected, atomize and distribute the fuel throughout the combustion chamber, and start and end each injection cycle abruptly. Obviously, the high-pressure system is fairly complex.

To inject fuel into an engine with a compression ratio as high as 22.5:1 (5.7L), 21.5:1 (6.2L), or 21.0:1 (6.5L), the injection pump must operate at very high pressure. The injection pump used on GM diesels has an operating pressure of about 1800 psi, and can produce up to 5000 peak psi.

Fuel metering must be absolutely accurate. The same amount of fuel must be delivered to each cylinder for each power stroke; should the quantity of fuel vary from cylinder to cylinder, the power produced in each cylinder will vary, and the engine will run roughly. The quantity of fuel injected must be varied in accordance with the load on the engine and its speed.

Fuel must be injected at just the right moment. If fuel is injected too early, the piston will not yet have reached top dead center, compression won't yet have reached its maximum, the temperature will still be low and ignition will be delayed. If the fuel is late, the piston will be past top dead center, compression will have passed its maximum, the temperature will be going down and the burning fuel won't have time to reach its maximum expansion. In either case, a loss of power is the result. Injection must therefore begin instantly, continue for the prescribed time, and then cease instantly.

Fuel isn't injected in one single spurt, but continues for a precise period of time. If the fuel is injected too quickly, the effect is similar to early injection. If the fuel is injected too slowly, and continues for too long a period of time, the effect is similar to late injection. The injection rate varies with different engines and is affected largely by the type and contour of the combustion chambers, the engine speed and the fuel characteristics.

Fuel is spurted into the combustion chamber as a spray. How well the fuel atomizes is determined by the shape of the combustion chamber and the spray pattern itself. Good atomization increases the surface area of the fuel that is exposed to the oxygen of the air, and produces better combustion and power.

To prevent all the droplets of sprayed fuel from igniting at once, the spray pattern is usually formed of some fine droplets to start ignition, and larger droplets for further combustion. The ratio of fine to large droplets is determined mainly by the diameter and shape of the nozzle orifice, injection pressure and the density of the air into which the fuel is injected.

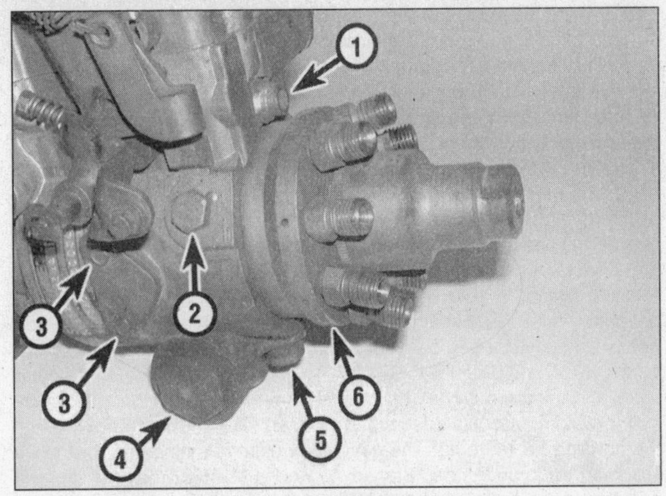

2.62a Details of Stanadyne Roosa-Master DB2 injection pump (left side)

1 Guide stud
2 Head locking screw
3 Timing line cover screws
4 Piston hole plug (power side)
5 Head locating screw
6 Head and rotor assembly

Stanadyne Roosa Master DB2 injection pump

1993 and earlier GM diesel engines use a Stanadyne Roosa Master DB2 distributor type injection pump **(see illustrations)** located in the valley between the cylinder banks on both engines.

The DB2 is a rotary type pump, gear driven by the camshaft on both engines. However, the means by which the pump driven gear meshes with the cam drive gear are different. On 5.7L engines, the pump drive shaft, which faces forward, is angled downward toward the front of the cam. A spiral bevel gear on the end of the driveshaft meshes with another spiral bevel gear on the front of the cam. On 6.2L engines, the pump driveshaft also faces forward, but it's level. On the

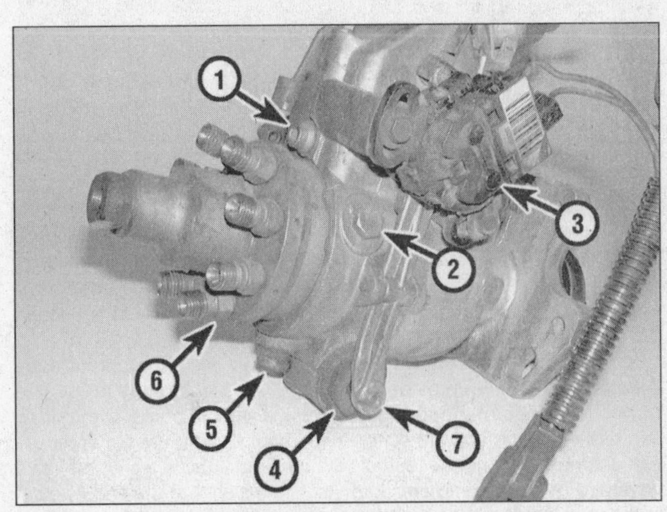

2.62b Details of Stanadyne Roosa-Master DB2 injection pump (right side)

1 Guide stud
2 Head locking screw
3 Throttle position sensor (TPS)
4 Piston hole plug (trimmer side)
5 Head locating screw
6 Head and rotor assembly
7 Advance cam actuating arm

end of the driveshaft, a straight cut gear meshes with another straight cut gear on the front end of the camshaft.

Unlike many older pump designs, which employ a separate pumping element for each engine cylinder, the DB2 employs one pump barrel and plunger to supply all cylinders. Because of its compact size, this design lends itself to the limited space available between the intake manifold and the "valley" (the space between the cylinder heads) of a V8 engine.

There isn't a lot you can service on the DB2 at home. Most repairs require the use of sophisticated calibration equipment, expensive tools and specialized training. Testing the electric solenoid pull in voltage, housing pressure cold advance, face cam positioning, the mini max governor, return fuel volume and automatic advance must be carried out by an authorized Roosa Master technician.

DO NOT ATTEMPT TO REPAIR THE INJECTION PUMP! Leaking seals and a worn out governor weight retaining ring are the most common problems associated with the DB2, so we'll show you how to test the housing pressure and the governor weight retaining ring, but leave servicing to an authorized Stanadyne Roosa Master repair facility. The number of special tools and skills required to service the pump make home repair a difficult - and expensive - proposition. And because of the Federally mandated extended warranty offered by GM on emissions related components, it just doesn't make sense to repair it yourself.

Stanadyne electronic fuel injection pump

1994 and later models use a Stanadyne electronic fuel injection pump. The electronic pump is a high-pressure rotary type pump similar in operation to the Roosa Master BD2, but it's controlled by the Powertrain Control Module (PCM). This system uses a unique accelerator control system. Attached to the accelerator pedal is a pedal position module, which sends an electrical signal to the PCM. The PCM controls fuel metering electronically at the fuel solenoid; there is no accelerator cable.

Removal and installation procedures for the electronic pump are identical to the earlier DB2 pump. Procedures for servicing the high-pressure lines and fuel injection nozzles are identical as well. One major difference to note concerns fuel injection timing. Timing is controlled by the PCM; therefore an expensive electronic SCAN tool is needed to set fuel injection timing correctly. If the pump is removed, align the original factory timing marks on reinstallation, then take the vehicle to a dealer service department and have the timing adjusted with the proper equipment.

High pressure lines

Eight metal lines carry the pressurized fuel from the pump to the injection nozzles. These lines must be of equal length and inside diameter to prevent uneven injection timing. For example, most lines are 26 inches in length, but their inside diameters vary, according to year of production. 1978 and 1980 lines have an I.D. of approximately .073 inch. 1981 lines have an I.D. of .075 inch. 1982 through 1984 lines have an I.D. of .078 inch. And so on. If you have to replace some of the high-pressure lines, make sure you buy lines of the same length and I.D. as the lines already on your engine. These dimensions are absolutely essential for proper engine timing: If one line is longer or shorter than the others are, the cylinder fed by that line will fire out of time.

Only a small amount of fuel is forced into each line during each injection. Each tiny pulse of fuel pushes the fuel already in the line toward its respective nozzle, forcing a small amount of fuel at the other end of the line through the spring-loaded nozzle and into the combustion chamber. The nozzle restricts the fuel and acts as a shut off valve governed by its opening pressure. As it slams shut at the end of each injection pulse, the already highly pressurized fuel is compressed a little more; depending on the volume inside the line, a little more or a little less compression occurs inside the line. The larger the volume in the line, the less the pressurization, reducing the volume of fuel which passes through the nozzle in a measured time period. The line also expands slightly when subjected to high pressure. The larger the inside diameter of the line, the larger the expansion and, consequently, the less fuel injected. Again, uneven delivery of fuel can result in a loss of power and/or poor running.

Pencil type nozzles

Two different types of nozzles are used - the pencil type and the poppet type. All 1978 and 1979 models, and some 1980 and 1981 models, use a pencil type nozzle. The nozzle body consists of the inlet fitting, the tip and the valve guide. An edge filter, which is located in the inlet fitting, is a final screen for catching debris which may have entered the line when the system was last opened. The inward opening valve is spring loaded. The amount of pressure required to open it is controlled by the lift and pressure adjusting screws, both of which are secured by locknuts. **Caution:** *These adjustments are extremely critical - they can only be adjusted on a flow meter. Do not attempt to adjust these screws at home.*

A nylon seal beneath the inlet banjo fitting prevents leakage of engine compression, and a Teflon carbon dam prevents carbon accumulation in the bore in the cylinder head for the nozzle. Between injections, positive sealing is maintained by the interference angle between the valve and its seal. During injection, a small amount of fuel leaks through the clearance between the nozzle and its guide, lubricating and cooling all moving parts.

The fuel flows through a leak off boot at the top of the nozzle body and returns to the fuel tank.

You can buy replacement pencil type nozzles from an authorized GM dealer, or you can obtain them directly from Stanadyne. Make sure you get the right nozzles! Most applications call for the nozzle with two .017 inch diameter holes. However, GM models sold in California have three .014 inch holes; the two-hole type doesn't conform to California's more stringent emission control requirements.

Fuel return lines return unused fuel to the fuel tank. Keep these return lines open and unobstructed at all times. If you remove the plastic nozzle return line connectors from the nozzles during servicing, make sure you align the connectors with the holes in the return lines when you reinstall them.

Flare type connectors for pencil type nozzles

Beginning in November 1978, all 5.7L diesel engines, except those produced for high altitude service, were equipped with a "flare" type connection. Make sure you don't accidentally interchange the two types of pencil nozzles and/or their respective high-pressure lines.

If you do, a severe leak - and permanent damage to the parts - could occur. Refer to the accompanying nozzle application chart when buying lines and/or nozzles.

Poppet nozzles

Poppet nozzles, or micro injectors, are used in all other GM diesel engines covered in this chapter. The poppet nozzle **(see illustration)** is

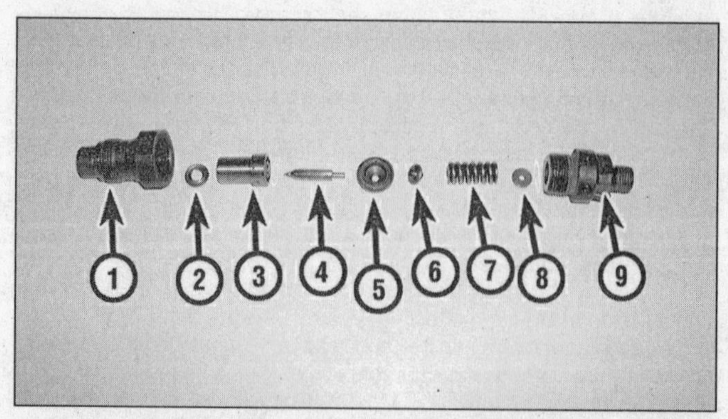

2.63 An exploded view of a typical poppet-type nozzle, or micro-injector (don't attempt to disassemble an injector at home - it will never work correctly again!)

1. *Nozzle nut*
2. *Heat shield*
3. *Pintle nozzle*
4. *Needle valve*
5. *Intermediate plate*
6. *Pressure spindle*
7. *Pressure spring*
8. *Spring shim*
9. *High pressure inlet housing*

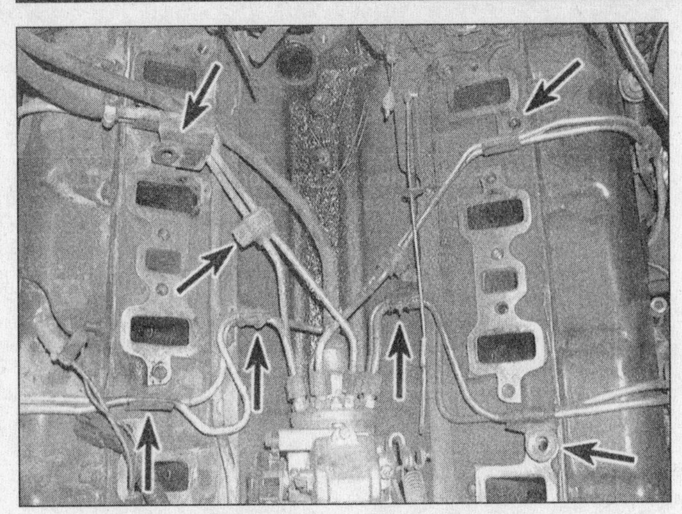

2.64a Typical injection line clips (arrows) (6.2L engine shown, other engines similar)

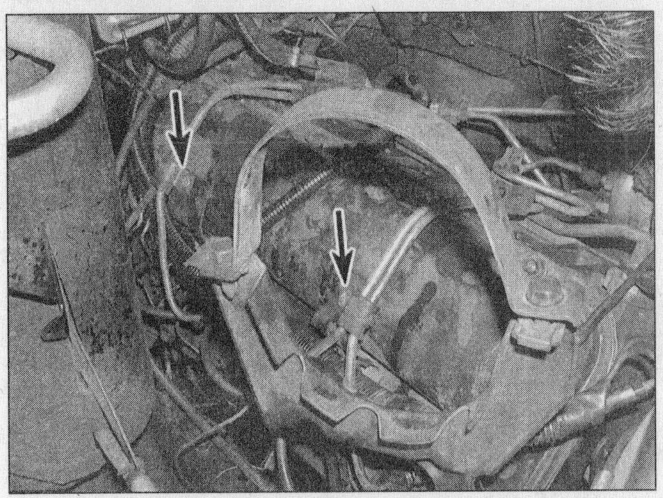

2.64b Typical injection line clips (arrows) (6.2L engine shown, other engines similar)

similar in principle to a gasoline fuel injector, but it's been modified for the indirect injection design of GM diesels. But unlike a conventional fuel injector, which sprays fuel into the intake port, the poppet nozzle sprays fuel into a precombustion chamber. And a regular fuel injector is pressed into the head and sealed with one or two nylon or plastic sealing rings, while a poppet nozzle is screwed directly into the cylinder head to prevent combustion leaks.

As fuel is injected from the poppet nozzle, it's swirled just before it passes around the head of the valve, producing a high velocity, narrow cone, atomized spray which promotes more efficient combustion in the prechamber. The nozzle assembly is a precisely machined assembly, matched and preset at the factory, and isn't serviceable.

Do not disassemble poppet nozzles! They're assembled to precise tolerances and won't work properly once you've taken them apart. When you replace worn nozzles, make sure you get the correct units for your engine. Though they may look identical, poppet nozzles are not interchangeable. For instance, nozzles for V6 engines in passenger vehicles look absolutely identical to the nozzles used in V8s, but they have a lower opening pressure than those used on V8s.

Another example: The nozzles employed on 1984 and later vehicles have a "semi guided" valve pintle. The diameter of the mid-way slotted guide has been reduced to minimize carbon formation around the pintle area. The extra clearance causes the pintle to oscillate slightly, breaking up the carbon. When you go to a GM dealer to buy new nozzles, specify the nozzle part number, which is stamped on the side of the nozzle.

Servicing the fuel lines and nozzles

Removing and installing the fuel lines

1 Before starting this procedure, clean the lines and fittings thoroughly to prevent contamination of the fuel system.

2 Remove the air cleaner.

3 Remove the filters and pipes from the valve covers and air crossover (5.7L engines) or intake manifold (6.2L engines).

4 If you're working on a 5.7L engine, remove the air crossover and cover the intake manifold opening. If you're working on a 6.2L engine, remove the intake manifold (see the *In vehicle engine repairs* Section later in this Chapter).

5 Remove the injection pump line clips **(see illustrations)**.

6 Clearly label the fuel line(s) at the injection pump then, using a flare-nut wrench, disconnect the lines from the pump **(see illustration)** and from the injectors. Plug the pump, lines and injectors to prevent contamination of the fuel system.

7 Installation is the reverse of removal. Be sure to securely tighten the fuel line fittings at the pump and the nozzle(s).

Removing and installing the injection nozzle(s)

1 Remove and plug the fuel line(s) (see above).

2 Remove the nozzle(s) by applying torque to the larger of the two hexes **(see illustration)**.

3 If you're going to install the old nozzle(s), remove and discard the old copper washer and install a new washer. **Note:** *Poppet nozzles*

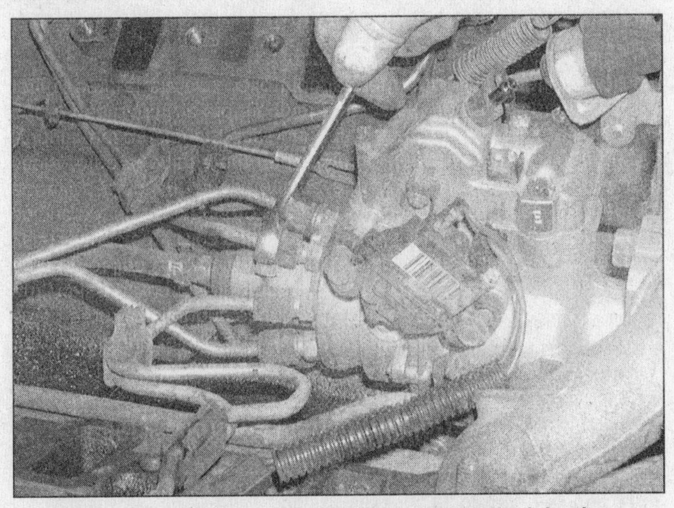

2.65 Use a flare-nut wrench to disconnect the injection lines from the injection pump

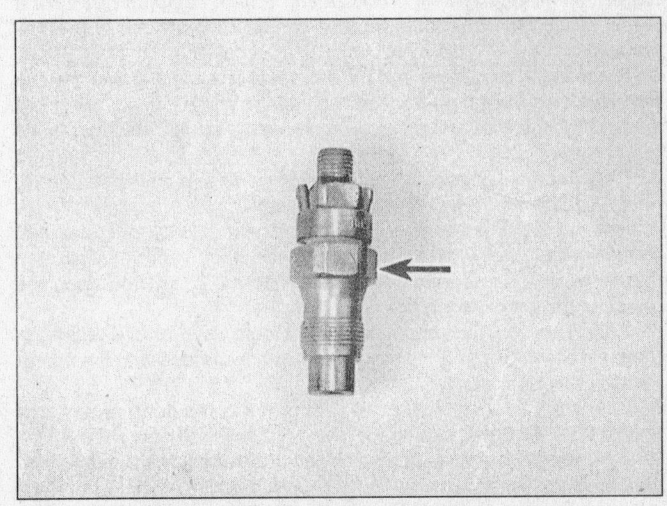

2.66 When removing an injector nozzle, put the wrench on the big hex (arrow)

2.67 Location of housing pressure cold advance terminal (arrow) (5.7L engines)

2.68 Governor weight retainer ring (arrow)

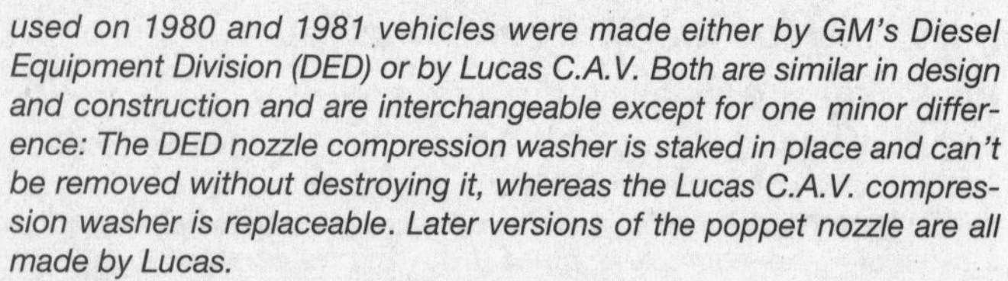

used on 1980 and 1981 vehicles were made either by GM's Diesel Equipment Division (DED) or by Lucas C.A.V. Both are similar in design and construction and are interchangeable except for one minor difference: The DED nozzle compression washer is staked in place and can't be removed without destroying it, whereas the Lucas C.A.V. compression washer is replaceable. Later versions of the poppet nozzle are all made by Lucas.

4 If you're going to install the old nozzle(s), clean all carbon from the tip of the nozzle with a soft brass wire brush.

5 Expensive specialized equipment is needed for testing injection nozzles. If you suspect trouble with any of the nozzles, have them tested by a dealer or an authorized Stanadyne Roosa Master repair facility.

6 Remove the protective caps from the new nozzle(s).

7 Make sure the copper washer is installed on the nozzle.

8 If the old nozzle was a Lucas C.A.V unit, make sure the old copper washer isn't sitting on top of the nozzle hole.

9 Install the new nozzle and washer and tighten the nozzle to 25 ft-lbs (make sure you torque the nozzle using the larger hex).

10 Using a back-up wrench, connect the fuel line fittings and tighten to 25 ft-lbs.

11 Installation is otherwise the reverse of removal.

On vehicle testing of the injection pump

Pump housing pressure (5.7L models)

Note: *The injection pumps on 6.2L and 6.5L models do not have a pressure tap plug; the following test cannot be performed on those models.*

1 Remove the air cleaner and the crossover assembly. Cover the two air intakes to prevent dirt from getting into the pump.

2 Wipe off the area around the injection pump so dirt won't get in the pump.

3 Remove the pressure tap plug from the pump housing (if equipped), on the front of the pump housing.

4 Hook up a low pressure gauge (Kent Moore J 28552 or equivalent) to the pressure tap hole (depending on the gauge you're using, you may need an adapter such as a Kent Moore J 28526 to attach the gauge to the pressure tap hole).

5 Attach a magnetic tachometer (Kent Moore J 26925 or J 33300, or Snap On MT 480 or 1480, or equivalent) in accordance with the manufacturer's instructions.

6 With the engine running at 1000 rpm, the pump housing pressure should be 8 to 12 psi, with no more than a 2 psi fluctuation.

7 If the pressure is zero, check the operation of the housing pressure cold advance as follows:

a) *Unplug the electrical connector from the housing pressure cold advance terminal* **(see illustration)**. *If the pressure remains at zero, remove the injection pump cover and check the operation of the advance solenoid. If it's binding, free up or replace parts as needed.*

b) *If the pressure returns to normal when the lead is disconnected, check the operation of the temperature switch on the cylinder head bolt.*

8 If the pressure is still low, replace the fuel return line connector assembly, then recheck. If the pressure is too high, the fuel return system may be restricted. Remove the fuel return line at the injection pump, install a fitting and short piece of hose to allow the return flow to empty into a small container. If the fuel return line connector assembly is replaced, check and, if necessary, reset the pump timing.

9 If the pressure is lower than before, correct the restriction in the fuel line.

10 If it's still high, replace the fuel return line connector assembly.

11 If it remains too high after replacing the return line connector, remove the pump and have it tested and repaired by an authorized Stanadyne Roosa Master repair facility.

12 Remove the gauge and tachometer.

13 Install the pressure tap plug. Use a new seal.

14 Install the air crossover and air cleaner assembly.

Testing the governor weight retainer ring

The governor weight retainer ring **(see illustration)** is mounted on the governor weight retainer. When subjected to water in the fuel, or alcohol (an additive in some diesel fuels not recommended by GM), this ring can become brittle and break off in small black particles. These particles can plug the fuel return check valve. When this happens, the engine will idle roughly and might not run at all.

Stanadyne repair facilities now install an elastomer insert drive (EID) in place of the older pellathane ring (which, in turn, replaced the original elasticast unit). The new EID unit, which is standard on 1986 and later vehicles, is the standard replacement part for older pumps, so once it's replaced, the governor shouldn't give you any more trouble.

To check for a governor ring failure:

1 Remove the air cleaner and (on 5.7L engines) the crossover assembly.

2 Cover the air intakes.

3 Wipe off any dirt from around the pump.

4 Disconnect the fuel return line at the top of the pump.

5 Unplug the electrical connectors from the pump.

6 Remove the three screws holding the governor cover **(see illustration)** and lift off the cover.

7 Using a screwdriver, rotate the governor weight retainer in both directions. It shouldn't move more than 1/16 inch in either direction, and it should return to its original position.

8 If the ring is within these limits, it's still got some service life left.

2.69 To remove the governor cover assembly, remove these three screws (arrows) (6.2L shown, 5.7L and 6.5L similar)

2.70 Disconnect the electrical leads and fuel return line hose (arrows) from the injection pump

9 If the ring moves more than 1/16 inch in either direction, or if it doesn't return, it has failed. Remove the pump and have a new ring installed by an authorized Stanadyne repair facility. **Caution:** *If you're going to drive the vehicle to a dealer to have the pump removed, inspect the fuel return check valve and the inside of the pump housing for signs of the small black particles from the disintegrating ring and remove them before installing the governor cover.*

10 Install the governor cover. Tighten the cover screws to 35 to 45 in lbs.

11 Test the fuel cutoff solenoid with a 12-volt power source. You should hear a click when the connector is powered up - this indicates the governor cover is installed correctly. **Caution:** *If you don't perform this check, you won't know if the metering valve is jammed open, which will cause the engine to run wide open (full throttle) when started.*

12 Connect the fuel return line and solenoid wire.

13 Install the air crossover and air cleaner assembly.

Injection pump - removal and installation

Once you've determined that the injection pump has a bad governor ring, or the driveshaft or hydraulic seals are leaking (or that perhaps there's a more serious problem) you'll have to remove the pump and have it serviced by an authorized Stanadyne Roosa Master repair facility.

Removal (5.7L engines)

1 Remove the air cleaner. Disconnect the cables from the negative terminals of the batteries.

2 Remove the ventilation filters from the valve covers and the pipes from the ventilation filters in the air intake crossover.

3 Disconnect all electrical leads and hoses from the top of the injection pump **(see illustration)**.

4 Remove the air crossover and cover the openings.

5 Disconnect the throttle rod and return spring and remove the bellcrank.

6 Detach the throttle cable from the intake manifold brackets and set it aside, away from the engine.

7 If the fuel filter is mounted on the intake manifold, disconnect the lines to the fuel filter, then remove the fuel filter and bracket.

8 Disconnect the fuel line at the fuel pump.

9 Disconnect the fuel return line from the top of the injection pump.

10 Using two wrenches, disconnect the injection pump lines at the nozzles.

11 Using Kent Moore J 26987, or an equivalent tool, remove the three nuts retaining the injection pump.

12 Remove the pump and cap all open lines and nozzles.

Installation (5.7L engines)

Note: *Before you install the pump, make sure that the T on the injection pump driveshaft is facing up, and that the number one cylinder is at TDC.*

1 Remove all protective caps, then line up the offset tang on the pump driveshaft with the pump driven gear and install the pump. Make sure the slotted end of the driveshaft mates with the slot in the injection pump drive gear. Push the pump into place by hand until it's fully seated.

2 Install the three pump mounting nuts and lock washers, but don't tighten them yet.

3 Connect the injection pump lines at the nozzles, then tighten the pump mounting nuts to 25 ft-lbs.

4 Connect the fuel return lines to the injection pump.

5 Using a 3/4-inch open-end wrench on the boss at the front of the injection pump to help you rotate the pump, align the mark on the injection pump with the line on the adapter and tighten the nuts to 18 ft-lbs.

6 To adjust the throttle rod:

a) Remove the transmission vacuum regulator valve.
b) Loosen the locknut on the throttle rod and shorten the rod several turns.
c) Rotate the bellcrank to the full throttle stop and lengthen the throttle rod until the injection pump lever contacts the injection pump full throttle stop.
d) Release the bellcrank and tighten the throttle rod locknut.

7 Install the bellcrank and hairpin clip.

8 Attach the throttle cable to the intake manifold and to the bellcrank.

9 Connect the throttle rod and return spring.

10 Start the engine and check for fuel leaks.

11 Remove the rags from the intake manifold and install the air crossover.

12 Install the pipes in the flow control valve in the crossover and the ventilation filters in the valve covers.

13 Install the air cleaner.

Removal (6.2L and 6.5L engines)

Caution: *DO NOT rotate the engine with the fuel injection pump removed.*

1 Disconnect the cables from the negative terminals of the batteries.

2 Remove the fan and the fan shroud.

3 Remove the air cleaner.

4 Remove the crankcase ventilator tubes.

5 Loosen the vacuum pump hold down clamp and rotate the pump in order to gain access to the manifold bolts.

6 Remove the EPR/EGR valve bracket (if equipped).

7 Remove the rear bracket for the air conditioning compressor (if equipped).

8 Remove the intake manifold bolts (the injection line clips are also retained by the same bolts).

9 Remove the intake manifold.

10 Remove the injection line clips at the loom brackets.

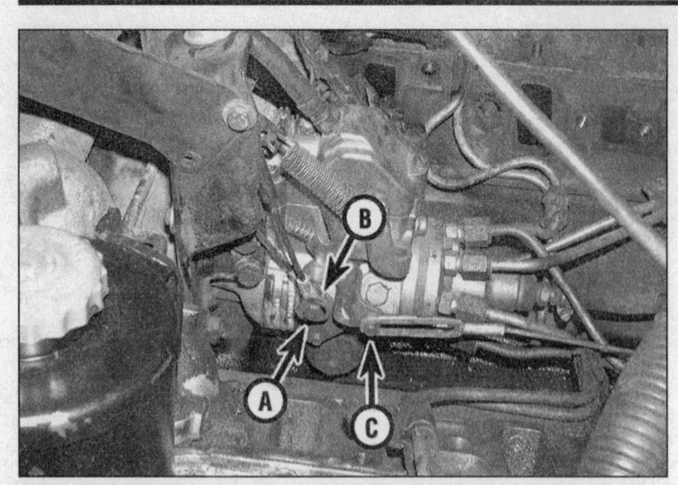

2.71 Remove the retaining clip (A) to detach the accelerator cable and cruise cable (B) from the throttle bellcrank; to detach the detent cable (C) on models with an automatic transmission, unhook the rear end of the detent rod at the bracket (not visible in this photo), slide the cable forward and disengage it from the pin (6.2L shown, 6.5L similar)

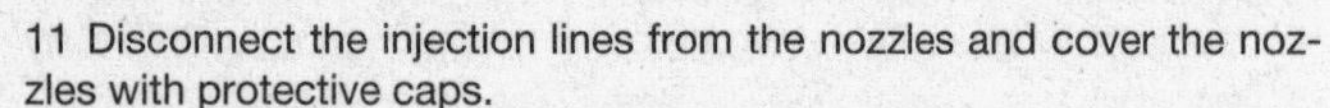

11 Disconnect the injection lines from the nozzles and cover the nozzles with protective caps.
12 Disconnect the injection lines from the pump and tag the lines for reinstallation.
13 Disconnect the accelerator cable at the injection pump and, if applicable, disconnect the detent cable **(see illustration)**.
14 Disconnect all hoses and wires from the injection pump.
15 Disconnect the fuel feed and return lines from the top of the injection pump.
16 Detach the air-conditioning hose retainer bracket (if equipped). **Warning:** *Don't disconnect the hoses.*
17 Remove the oil fill tube and the CDRV vent hose assembly.
18 Remove the grommet.
19 Scribe or paint an alignment mark on the front cover and injection pump flange.
20 Rotate the engine to gain access to the three driven gear-to-injection pump bolts through the oil filler neck hole **(see illustration)**.
21 Remove the injection pump to front cover nuts **(see illustration)**.
22 Remove the pump, remove the pump gasket **(see illustration)** and cap all open lines and nozzles.

Installation (6.2L and 6.5L engines)

1 Replace the gasket.
2 Align the locating pin on the pump hub with the slot in the injection pump driven gear, and line up the aligning marks **(see illustrations)**.
3 Install the injection pump to front cover nuts and tighten them to 30 ft-lbs. Make sure the timing marks are aligned before fully tightening the nuts **(see illustrations)**.
4 Install the drive gear to injection pump bolts and tighten the bolts to 20 ft-lbs.

2.72 Rotate the engine to gain access to each of the three driven gear-to-injection pump bolts through the oil filler neck (6.2L shown, 6.5L similar)

2.73a To detach the injection pump, remove these three nuts (arrows) (6.2L shown, 6.5L similar)

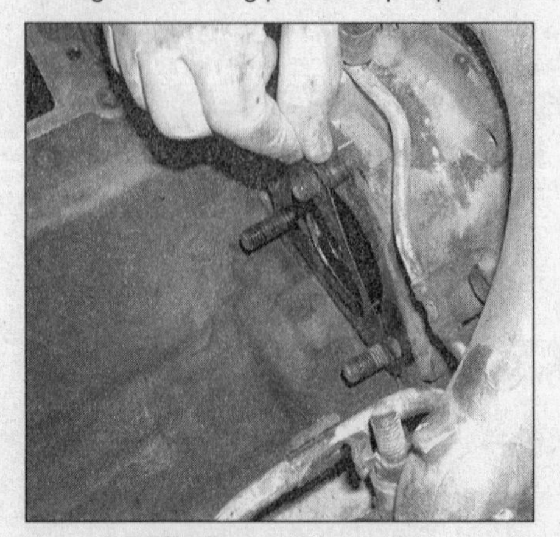

2.73b Remove the gasket

2.74a When installing the injection pump, align the locating pin (arrow) on the pump hub . . .

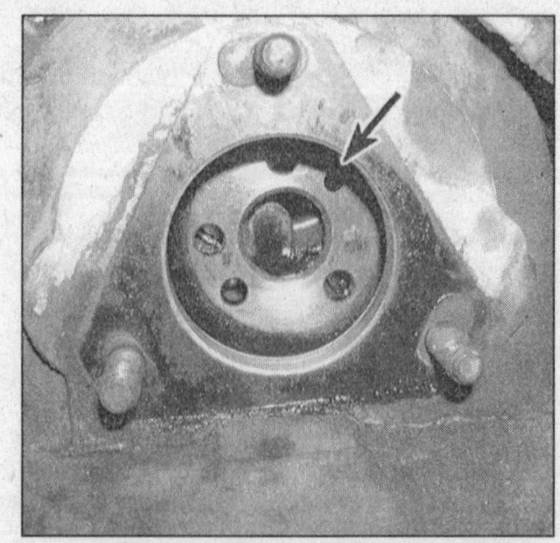

2.74b . . . with the slot (arrow) in the injection pump driven gear (don't confuse the slot with one of the three holes for the injection pump gear bolts

2.74c Line up the alignment marks, which may be either a pair of lines, like this . . .

2.74d . . . or a zero stamped across the mating line between the pump flange and the front cover, like this

5 Install the oil fill tube and the CDRV vent hose assembly.
6 Install the grommet.
7 Install the air-conditioning hose retainer bracket (if equipped).
8 Connect the fuel feed line at the injection pump and tighten it to 20 ft-lbs. Connect the fuel return line to the top of the injection pump.
9 Connect the necessary wires and hoses.
10 Attach the accelerator cable.
11 Install the intake manifold.
12 Install the fan shroud and the fan.
13 Connect the batteries.

Initial check and adjustment of injection pump timing (static timing)

Before the engine will run right, the pump must be properly timed. First, time it statically (engine not running), and then time it dynamically (engine running). If you don't have the equipment to time the engine dynamically, drive the vehicle to a dealer after you've timed the engine statically and have it done there.

One more thing: If the engine has been overhauled, or the pump adapter was removed during servicing, you'll need a special static timing gauge to check and remark initial timing. For 5.7L engines, get a Kent-Moore J-26896 static timing gauge; for 6.2L and 6.5L engines, ask for J-33042.

Static timing (5.7L engines)

The marks on the top of the injection pump adapter and the flange of the injection pump must be aligned. If they're not, here's how to do it:

1 Note whether the marks on the pump flange and the pump adapter are aligned.
2 Loosen the three pump retaining bolts or nuts.
3 Using a wrench on the boss at the front of the injection pump, rotate the pump to the left to advance the timing and to the right to retard it. The width of the mark on the adapter is equal to about one degree. Move the pump the necessary amount and tighten the bolts or nuts to 18 ft-lbs.
4 Start the engine and recheck the timing dynamically (see below) - or drive the vehicle straight to a dealer and have it done there. Reset if necessary.
5 Adjust the throttle linkage, and reset the fast idle and curb idle speeds (as described in this Chapter).

Static timing (6.2L and 6.5L engines)

1 Note whether the marks on pump flange and the engine front cover are aligned.
2 If they're not, loosen the three pump retaining nuts.
3 Using an injection pump adjusting tool (Kent-Moore J-29872, or equivalent) to rotate the pump, align the mark on the injection pump with the mark on the cover and tighten the nuts to 30 ft-lbs. **Note:** *All 1988 and later units and all LH6 California applications have a stamped "O" on both the engine front cover and the injection pump flange. Align the two semicircles to make a complete "O." This is the static timing mark.*

Dynamic timing

Note: *On 1994 and later models, the fuel injection timing is controlled by the Powertrain Control Module (computer). An electronic SCAN tool is required to perform the necessary adjustments to the injection timing. This is an expensive device that communicates directly with the PCM. Take the vehicle to a dealership or other properly equipped repair facility if it becomes necessary to adjust injection pump timing.*

Measuring the timing of diesel engines when they're running is a little more difficult than it is on gasoline engines. There's no spark from an ignition system to tell you when combustion starts.

Since you can't use a timing light, you'll need a special kind of timing indicator, such as a Kent-Moore (J-26925 or J-33300) or Snap-On (MT-480 or 1480) that employs a magnetic pick-up mounted to the crankshaft vibration damper. The pick-up measures engine speed and degrees of crankshaft rotation. A special luminosity probe mounted in one of the glow plug holes can "see" the point at which combustion occurs. A microprocessor in the timing meter receives the information from the magnetic and luminosity probes and converts it into a timing reading on the meter (indicated in degrees of crankshaft rotation).

Because this luminosity probe depends on actual combustion in the cylinder, a miss in that cylinder will give an erroneous timing reading. So make sure you have corrected any engine malfunction - like a sticking injector, bent fuel delivery line or other problem - before you attempt to measure or adjust the timing. And before you install the probe, make sure it's clean - a dirty or sooty probe will retard the timing reading. It will also soot up rapidly in a cold engine, so be prepared to remove and clean it if necessary.

Diesel timing is also affected by altitude. For example, if the engine timing were set at 4-1/2 degrees ATDC in a low-altitude area, the same setting would be 5-1/2 degrees ATDC in a high-altitude area, because combustion occurs later as air density decreases at higher elevations. The rule-of-thumb guide for engines at high altitude is to retard the timing by one degree.

1 Place the transmission in Park, apply the parking brake and block the drive wheels.
2 Start the engine and warm it to operating temperature. Failure to do this will result in incorrect timing readings.
3 Turn off the engine and remove the air cleaner.
4 Clean all dirt from the magnetic probe holder and the edge of the crankshaft vibration damper.

5 Look through the end of the luminosity probe - you should be able to see through it. If it's dirty, use a toothpick to scrape the carbon and soot from the combustion side of the probe.
6 Install the magnetic probe into the engine-mounted holder so it lightly contacts the crankshaft balancer rim.
7 Remove the glow plug from the number one cylinder and install the luminosity probe.
8 On the Snap-On MT-480 timing meter, set the offset selector to the B position (on other models, set the meter in accordance with the manufacturer's instructions).
9 Connect the timing meter battery leads to the battery.
10 Start the engine. Check and, if necessary, adjust the idle speed (see below). **Note:** *If the idle speed given on the Vehicle Emission Control Information (VECI) label under the hood differs from the specified idle speed, go with the specification on the VECI label.*
11 Wait for the idle to stabilize, then note the indicated timing. Repeat this procedure until the timing reading is the same at least twice. Compare your reading to the VECI label. If the timing is satisfactory, proceed to Step 19; if it isn't, go to the next Step.
12 Shut off the engine.
13 Note the relative position of the mark on the pump and the mark on the pump adapter (5.7L) or engine front cover (6.2L and 6.5L).
14 Loosen the fasteners holding the pump to the adapter, or front cover, enough to allow the pump to be rotated.
15 To adjust the timing, use a 3/4-inch open end wrench on the boss at the front of the injection pump. Rotate the pump to the left to advance the timing, or to the right to retard it. The width of the timing mark is equal to about one degree.
16 Tighten the pump mounting fasteners to 18 ft-lbs.
17 Start the engine and recheck the timing (see Step 11).
18 Adjust the injection pump throttle rod.
19 Reset the slow and fast idle.
20 Shut off the engine. Remove the two probes and install the glow plug. Remove the timing meter connections from the battery.
21 Install the air cleaner and connect the EGR vacuum hose (if so equipped).

Adjusting the idle speed

GM diesel engines have two idle-speed settings - slow and fast. Both have to be adjusted in accordance with GM specifications to prevent the engine from stalling. If the slow-idle speed is incorrect, the engine will start, but will stall when accelerated. If the fast-idle speed is wrong, the engine will start, but will stall at idle.

If you've got a 5.7L engine, first adjust the throttle linkage and the vacuum regulator valve before adjusting the idle speed.

Adjusting the throttle linkage (5.7L engines)

1 Detach the transmission throttle valve (TV) cable (or detent cable) from the throttle assembly and, if equipped, the cruise control servo rod.
2 Loosen the locknut on the pump rod and shorten it several turns.
3 Rotate the bellcrank lever assembly to the full throttle position and hold it in that position.
4 Lengthen the pump rod until the injection pump lever just contacts the full throttle stop.
5 Release the bellcrank assembly and tighten the pump rod locknut.
6 Depress and hold the metal lock on the upper end of the TV (or detent) cable. Move the slider through the fitting in the direction away from the bellcrank lever assembly until the slider stops against the metal fitting. Release the metal tab.
7 Install the cruise control servo rod (if equipped) and connect the transmission TV (or detent) cable.
8 Rotate the bellcrank lever assembly to the full throttle stop and release the lever assembly.
9 Adjust the vacuum regulator valve (see below).

Adjusting the vacuum regulator valve (5.7L engines)

The injection pump on the 5.7L engine has a vacuum regulator valve (VRV) that sends a variable vacuum signal to the transmission, causing it to shift at the proper time. After servicing the injection pump, you must reinstall this valve and adjust it to provide the correct amount of vacuum to the transmission.

1 Remove the air crossover and stuff clean, lint-free shop rags into the intake manifold openings.
2 Disconnect the throttle rod from the pump.
3 Loosen the vacuum regulator valve-to-injection pump mounting bolts.
4 Install a carburetor angle gauge vacuum valve adapter (Kent-Moore J-26701-15, J-26701-20 or equivalent) in accordance with the manufacturer's instructions.
5 Install a carburetor angle gauge (Kent-Moore J-26701, J-26701-A or equivalent) on the adapter.
6 Rotate the throttle lever to the wide-open position and set the angle gauge to zero degrees.
7 Center the bubble level in the gauge.
8 Set the angle gauge to 49-degrees (1978 and 1979 models), 50-degrees (1980 models) or 58-degrees (1981 through 1984 models).
9 Rotate the throttle lever so the bubble is centered.
10 Attach a vacuum source to port A on the VRV; attach a vacuum gauge to port B on the VRV.
11 Apply 18 to 24 in-Hg of vacuum to port A.
12 Rotate the VRV until the vacuum indicates 8.5 to 9 in-Hg (1978 and 1979 models), 7 to 8 in-Hg (1980 models), 8.6 to 9.2 in-Hg (1981 models) or 10.6 in-Hg (1982 through 1984 models).
13 Tighten the VRV mounting bolts.
14 Remove the vacuum source and vacuum gauge.
15 Connect the throttle rod to the pump throttle lever.
16 Remove the rags from the intake manifold openings.
17 Install the air crossover.

Adjusting the idle speed on 5.7L engines

Note: *Before you adjust the idle speed, check and, if necessary, adjust the throttle linkage (see above). If throttle linkage adjustment is necessary, you'll also need to adjust the vacuum regulator valve (see above).*
1 Apply the parking brake, place the transmission selector lever in Park and block the drive wheels.
2 Adjust the throttle linkage and vacuum regulator valve as just described.
3 Start the engine and allow it to warm up for 10 to 15 minutes.
4 Shut off the engine and remove the air cleaner assembly.
5 Clean the front cover probe holder and the rim of the crankshaft vibration damper.
6 Insert the magnetic pick-up probe of a diesel timing meter (Kent-Moore J-26925, or equivalent) all the way into the probe holder. Connect the battery leads, red to positive and black to negative.
7 Disconnect the two lead connectors at the generator.
8 Turn off all electrical accessories.
9 DO NOT TOUCH the steering wheel or the brake pedal during this procedure.
10 Start the engine and place the transmission selector lever in Drive.
11 Check the slow-idle speed reading against the one specified on the Vehicle Emission Control Information (VECI) label. If necessary, reset the slow idle speed.
12 To reset the slow-idle speed, adjust the slow idle screw **(see illustration 2.75)** on the injection pump to 575 rpm. **Note:** *If the slow-idle speed provided on the Vehicle Emission Control Information (VECI) label under the hood differs from this figure, go with the specification on the VECI label.*
13 Before checking the fast idle speed:

a) *On 1978 and 1979 models with the slow-glow plug system, unplug the A/C compressor clutch 2-pin connector and turn on the A/C. On vehicles without A/C, unplug the solenoid connector, connect the jumper wires to the solenoid terminals and ground one jumper wire. Connect the other jumper to 12 volts and energize it.*
b) *On 1979 and 1980 models with the fast-glow plug system, unplug the coolant temperature switch connector, at the right rear of the intake manifold.*
c) *On 1981 through 1983 models, unplug the connector from the fast-idle cold advance engine temperature switch and install a*

jumper between the connector terminals. Don't let the jumper touch ground.

d) *On 1984 models, unplug the electrical connector from the top of the TVS and install a jumper between the connector terminals. Don't let the jumper touch ground.*

14 Compare the fast-idle speed against the one provided on the VECI label. Reset it if necessary.

15 To reset the fast-idle speed, make sure the fast-idle solenoid is energized by attaching a jumper wire across the fast-idle temperature switch connector terminals. This temperature switch is located in the left rear corner of the intake manifold. It shouldn't be necessary to remove the connector from the temperature switch. Advance the throttle momentarily to verify that the fast-idle solenoid is fully extended (and check to make sure it's energized). Adjust the extended solenoid until the engine is idling at 650 rpm in Drive gear. Remove the jumper wire. **Note:** *If the slow-idle speed provided on the Vehicle Emission Control Information (VECI) label under the hood differs from this figure, go with the specification on the VECI label.*

16 After checking the fast idle speed:

a) *On 1978 and 1979 slow-glow plug models with air conditioning, reconnect the air conditioning compressor clutch and turn off the air conditioning. On models without air conditioning, remove the jumpers from the fast idle solenoid.*

b) *On 1979 and 1980 fast-glow plug models, reconnect the coolant temperature switch.*

c) *On 1981 through 1983 models, remove the jumper and reconnect it to the temperature switch.*

17 Recheck and reset the slow-idle speed if necessary.

18 Shut off the engine. Reconnect the lead at the generator (and air-conditioning compressor, if equipped). Disconnect the tachometer.

19 If the vehicle is equipped with cruise control, adjust the servo throttle rod to minimum slack, then put the clip in the first free hole closest to the bellcrank or throttle lever, but within the servo bail.

20 Install the air cleaner assembly and connect the EGR valve hose, if equipped.

Setting the slow-idle speed (6.2L and 6.5L engines)

1 Set the parking brake and block the drive wheels.

2 Start the engine and let the engine warm up to its normal operating temperature.

3 Turn off all accessories.

4 Hook up a Kent-Moore J-26925, a Snap-On MT-480, or any other suitable diesel timing meter, in accordance with the manufacturer's instructions.

5 Adjust the slow-idle screw **(see illustration)** on the fuel injection pump to an engine speed of 650 rpm, with the transmission in Neutral or Park. **Note:** *If the fast-idle speed provided on the Vehicle Emission Control Information (VECI) label under the hood differs from this figure, go with the specification on the VECI label.*

2.75 Location of the slow-idle adjustment screw (arrow) on the injection pump (6.2L shown, other engines similar)

Setting the fast-idle speed (6.2L engines)

1 Unplug the connector from the fast-idle solenoid. Use an insulated jumper wire from a 12-volt terminal to energize the solenoid.

2 Open the throttle momentarily to ensure that the fast-idle solenoid plunger is energized and fully extended.

3 Adjust the extended plunger by turning the hex head until the engine has a fast-idle speed of 800 rpm in Neutral. **Note:** *If the fast-idle speed provided on the Vehicle Emission Control Information (VECI) label under the hood differs from this figure, go with the specification on the VECI label.*

4 Remove the jumper wire and reattach the electrical connector to the fast-idle solenoid.

5 Remove the tachometer.

Adjusting the vacuum regulator valve (6.2L and 6.5L engines)

1 Attach vacuum regulator valve to fuel injection pump. Valve must be free to rotate on pump.

2 Attach vacuum source to bottom nipple and apply 9.0 to 10.5 in/hg for 1982 model, 18.5 to 21.5 in/hg for 1983-86 models, or 18.5 to 21.5 in/hg for 1987-91 models. Attach vacuum gauge to top nipple.

3 Insert vacuum regulator valve gage bar (which can be fabricated from a piece of stock 0.646 in. thick) between gage boss on injection pump and wide open stop screw on throttle lever.

4 Rotate and hold throttle shaft against gage bar.

5 Slowly rotate vacuum regulator valve clockwise until vacuum gauge reads 5.3 to 5.9 in/hg for 1982 model, 7.4 to 8.6 in/hg for 1983-86 models, or 10.9 to 12.1 for 1987-91 models. Hold the valve in this position and tighten mounting screws to 4-5 ft/lbs. **Note:** *The valve must be set while rotating valve body in a clockwise direction only.*

6 Check by releasing the throttle shaft, allowing it to return to idle stop position. Then rotate throttle shaft back against gage bar, if vacuum gauge does not read within specifications, reset valve.

Turbocharger - general information

The turbocharger increases power by using an exhaust gas-driven turbine to pressurize the fuel/air mixture before it enters the combustion chambers. The amount of boost (intake manifold pressure) is controlled by the wastegate (exhaust bypass valve). The wastegate is operated by a spring-loaded actuator assembly, which controls the maximum boost level by allowing some of the exhaust gas to bypass the turbine. The wastegate is controlled by the computer. Only the 6.5L engine is equipped with the turbocharger option.

Turbocharger - check

General checks

Warning: *Never feel any components of the turbocharger system while the engine is running. The engine must be stopped and the turbocharger and engine must be cooled down before handling. Rotating parts and extremely high temperatures can cause serious injury if handled improperly.* **Note:** *The turbocharger system is normally covered by a Federally mandated extended warranty (5 years or 50,000 miles at the time of publication). Check with a dealer service department concerning coverage.*

1 While it is a relatively simple device, the turbocharger is also a precision component, which can be severely damaged by an interrupted oil supply or loose or damaged ducts.

2 Due to the special techniques and equipment required, checking and diagnosis of suspected problems dealing with the turbocharger should be left to a dealer service department. The home mechanic can, however, check the connections and linkages for security, damage and other obvious problems.

3 Because each turbocharger has its own distinctive sound, a change in the noise level can be a sign of potential problems.

4 A high-pitched or whistling sound is a symptom of an inlet air or exhaust gas leak.

5 A cycling up or down in pitch often indicates a blockage in the air inlet duct, a restricted air cleaner or a build up of dirt on the compressor wheel.
6 A high-pitched scream may indicate that the one or more of the bearings has deteriorated and possibly one, or both of the wheels is rubbing on the housing.
7 Check for excessive oil leaks into the turbocharger housing. **Note:** *It is considered normal that the inside of the air intake duct, rubber connector hose, compressor wheel and housing be quite oily due to the venting of crankcase vapors into the intake system.* Some possibilities that could create excessive oil in the turbocharger are:

- a) *Damage to turbocharger bearings.*
- b) *Too much idling.*
- c) *Plugged or kinked oil drain tube.*
- d) *Blocked or obstructed air intake or filter element.*

8 If an unusual sound comes from the vicinity of the turbine, the turbocharger can be removed and the turbine wheel inspected. **Warning:** *All checks must be made with the engine off and cool to the touch and the turbocharger stopped or personal injury could result. Operating the engine without all the turbocharger ducts and filters installed is also dangerous and can result in damage to the turbine wheel blades.*
9 With the engine turned off and completely cool, reach inside the housing and turn the turbine wheel to make sure it spins freely. If it doesn't, it's possible the cooling oil has sludged or cooked from overheating. Push in on the turbine wheel and check for binding. The turbine should rotate freely with no binding or rubbing on the housing. If it does, the turbine bearing is worn out.
10 Check the exhaust manifold for cracks and loose connections.
11 Because the turbine wheel rotates at speeds up to 140,000 rpm, severe damage can result from the interruption of coolant or contamination of the oil supply to the turbine bearings. Check for leaks in the oil inlet lines and obstructions in the oil drain-back line, as this can cause severe oil loss through the turbocharger seals. Burned oil on the turbine housing is a sign of this. **Caution:** *Whenever a major engine bearing such as a main or connecting rod bearing is replaced, the turbocharger should be flushed with clean engine oil.*

Turbocharger - removal and installation

Removal

1 Disconnect the cable from the negative terminal of the battery.
2 Drain the cooling system (see Chapter 1).
3 Remove the upper intake manifold cover.
4 Remove the screw holding the valve tube to the top of the turbocharger.
5 Remove the valve and the tube assembly from the right valve cover and air cleaner assembly.

2.76 A typical glow-plug installation (arrow) (6.2L shown, other engines similar)

6 Remove the air cleaner extension hose from the compressor inlet.
7 Remove the two bolts at the wheel well and the air cleaner assembly from the vehicle for easier access to turbocharger flange nuts. **Note:** *The right front wheel and inner fender splash shield can be removed for easier access to the rear flange nuts.*
8 Remove the vacuum hose-to-waste gate actuator.
9 Remove the short and long braces.
10 Remove the exhaust clamp at the turbocharger.
11 Disconnect the oil feed hose.
12 Disconnect the oil return pipe.
13 Remove the four turbocharger flange nuts.
14 Remove the turbocharger assembly from the exhaust manifold.

Installation

15 Carefully clean the mating surfaces of the turbocharger and exhaust manifold. **Caution:** *Check for any debris in the intake and exhaust systems. Even very small debris can destroy a turbocharger operating at high speeds.*
16 Place the turbocharger in position on the manifold studs.
17 Apply anti-seize compound to the studs and install the nuts. Tighten the nuts to the torque listed in this Chapter's Specifications.
18 The remainder of the installation is the reverse of the removal procedure.
19 Change the engine oil.

Electrical system

General information

The charging system on a diesel engine is virtually identical to the charging system on a gasoline engine. The starting system is also similar, but there are some important differences. Starting a diesel engine isn't quite as easy as starting a gasoline engine. Diesels have higher compression ratios, which impose a higher load on the starter motor during starting. Yet cranking speeds must be faster because the compression heat needed to fire a diesel dissipates (heat is lost to the cylinder walls) too quickly at the slower cranking speeds of a conventional starter. So, diesel engines use two 12-volt batteries connected in parallel, and a heavy-duty, high-torque starter motor to overcome the higher compression and to crank the engine faster.

Cold weather is a more serious problem. Starting a diesel in cold weather is almost impossible without the right cold-weather starting accessories. As the temperature drops, battery efficiency is reduced and cranking loads become higher. Cold cylinder walls chill the incoming air, preventing it from reaching the temperature required for combustion.

GM diesels use several starting system devices to aid cold starting. A fuel heater warms up the fuel to prevent the formation of wax crystals in cold weather (see the section on the fuel heater in the low-pressure part of this Chapter). "Glow plugs," low voltage heating elements installed next to the injection nozzles, heat the combustion chamber to help the engine start. A block heater, located in the center freeze plug of the left or right side of the block, also helps the engine to start by maintaining the engine coolant and the block at a warm temperature in extremely cold weather.

Glow plugs - general information

Note: *The glow plug system, however, is normally covered by the Federally mandated extended emissions warranty (5 years or 50,000 miles at the time of publication). Check with a Dealer service department concerning coverage.*

Just before, during and after start-up, a glow plug **(see illustration)** in each combustion chamber is energized to preheat the air in the combustion chamber. The period during which the glow plug is energized prior to cranking is referred to as "pre-glow." A light on the dash tells you when pre-glow is over and the engine can be started. The glow plug also remains on for up to a minute after the engine starts to maintain ignition in all cylinders, improve throttle response and reduce

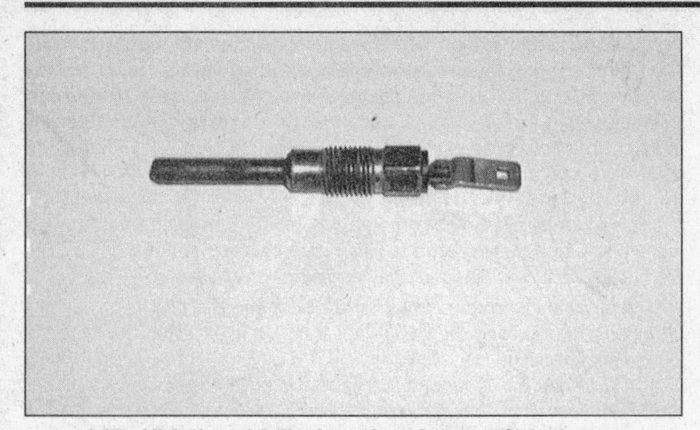

2.77 All 6.2L and 6.5L glow plugs have a 1/4-inch tang and are stamped "9G"

exhaust smoke. This period is known as "after-glow." How long the glow plugs remain on depends on the coolant temperature and how long it takes the engine to warm up enough to run on heat compression alone. As soon as the engine can run smoothly on its own, the glow plugs are turned off.

Three basic types of glow plugs are used on GM engines:

Slow-glow 12-volt
Fast-glow 6-volt
Positive Temperature Coefficient (PTC) 6-volt

If you have to replace the glow plug(s), make sure you buy the correct replacement plug(s) for the system you're servicing. If you install a 12-volt slow-glow plug in a 6-volt system, it won't get hot enough to do its job. Conversely, if you install a 6-volt fast-glow plug in an early 12-volt system, it will quickly burn out and the tip may break off and drop into the prechamber.

5.7L glow plugs

How do you know you've got the right glow plug for your 5.7L engine? Look at the color of the insulator, the width of the spade terminal and the plug type stamped on the body:

a) *The slow-glow 12-volt plug has a black insulator, a 1/4-inch spade terminal and is stamped "7G"*
b) *The fast-glow 6-volt plug has a black insulator, a 5/16-inch spade terminal and is stamped "8G"*
c) *The Positive Temperature Coefficient (PTC) plug has a tan insulator, a 1/4-inch spade terminal and is stamped "12G"*

Refer to the following chart when replacing the glow plug(s) in a 5.7L engine:

6.2L and 6.5L glow plugs

All 6.2L and 6.5L glow plugs have a 1/4-inch tang and are stamped "9G" **(see illustration)**.

Testing the glow plugs

Glow-plug system problems are usually at the plug itself. Here's a simple way to test each plug and quickly determine its condition:

1 Unplug the electrical connector from each glow plug.

2 Hook up one lead of an ohmmeter to the glow plug terminal and the other lead to a good ground such as the engine block. Note the indicated resistance.

3 The resistance of a fast-glow plug is 0.8 ohms; the resistance of an older slow-glow plug is 1.8 ohms.

4 Resistance readings of 1 to 2 ohms for slow-glow plugs, or below 1 ohm for fast-glow plugs, are acceptable. If the indicated resistance for either type of plug is infinity, or approaching infinity, the plug is failing. Replace it.

Diagnosing the glow-plug system - a basic check

Several different glow-plug systems have been used by GM. Many of the problems they develop are subtle and can't be tracked down unless you know the specific troubleshooting procedure for the system on your vehicle. However, if your engine refuses to start when it's cold, there are a few quick tests you can use to check out most systems (they won't work on early 12-volt, slow-glow, module-control systems).

Turn the ignition switch to the Run position, and then listen closely to the glow-plug relay(s) (Most 5.7L systems have two relays; most 6.2L systems have one).

If the relay(s) is/are clicking:

Note: *This test can't be used on the early 12-volt, slow-glow, module-control type system.*

1 Turn off the ignition switch and unplug the electrical connectors from all eight glow plugs.

2 Hook up a self-powered test light between the rear post of the glow-plug relay and ground.

3 At each glow plug, touch the harness connector to the glow-plug terminal.

a) *If the test light comes on, the glow plug and harness leads are okay.* **Caution:** *After you have tested each glow plug, set the connector aside so you don't accidentally energize one plug while you're testing the other plugs.*
b) *If the test light doesn't come on, touch the connector to a metal part of the engine.*
   1) *If the test light now comes on, the glow plug is bad. Replace it.*
   2) *If the test light still doesn't come on, the wire to the glow plug is bad. Replace it.*

Glow plug application Chart

Series	1978	1979	1980	1981	1982	1983	1984 - 85
A	N/A	7G	8G	8G	8G	8G	12G
B	7G	7G	8G	8G	8G	8G	12G
C	7G	7G	8G	8G	8G	8G	12G
E	N/A	8G	8G	8G	8G	8G	12G
G	N/A	N/A	N/A	N/A	8G	8G	12G
K	N/A	N/A	8G	8G	8G	8G	12G
C truck	7G	7G	8G	8G	9G (6.2L)	9G (6.2L)	9G (6.2L)
K truck	N/A	N/A	N/A	N/A	9G (6.2L)	9G (6.2L)	9G (6.2L)

N/A Diesel engine not available

c) *If the test light doesn't come on at ANY of the glow plugs, the glow plugs aren't the problem. The controller or something else in the circuit is bad. Further testing is called for. Study the description of the system on your engine and, using the accompanying diagnostic charts (beginning on page 2-54), troubleshoot the system.*

d) *If the test light comes on at all of the glow plugs, your starting problem isn't in the glow-plug system. It's somewhere else.*

If the relay(s) isn't/aren't clicking:

Further testing is called for. Study the description of the system on your engine and, using the accompanying diagnostic charts, troubleshoot the system.

Glow-plug systems

Unfortunately, the above tests don't always solve the problem. Each major type of glow-plug system is coordinated and controlled by a net work of electrical components - various relays, sensors and modules or controllers. Before you can really diagnose a system, you need to know the type of system used on your engine. The following comprehensive description and troubleshooting sections for each of the major glow-plug systems will help you find the more subtle problems that sometimes occur in a GM glow-plug system.

Several different glow plug systems are used on 5.7L engines; 6.2L engines all use basically the same system. Briefly, these are the principal different types of systems found on GM vehicles:

System No. 1: Slow-glow, module-controlled (5.7L engine)

System No. 2: Fast-glow, module-controlled (5.7L engine)

System No. 3: Fast-glow, thermal controller (5.7L engine)

System No. 4 (early): Fast-glow, thermal controller (6.2L engine)

System No. 4 (later): Fast-glow, electronic controller (6.2L engine)

System No. 5: Positive Temperature Coefficient (PTC) glow plug (5.7L engine)

System No. 6: Powertrain Control Module (PCM) controlled glow plug system (1994 and later models)

Before trying to track down a problem in the glow-plug system, make sure you know which system you've got on your engine and how it works.

System No. 1: Slow-glow, module-controlled (5.7L models)

System No. 1 is a 12-volt "slow-glow" system. It was used on 1978 Oldsmobile 88s and 98s (Series B and C), the 1979 Olds-mobile Cutlass (Series A) and some early production 1979 Oldsmobile 88s and 98s (Series B and C).

System No. 1 sends a steady 12-volt current to 12-volt glow plugs, which require 30 to 60 seconds to provide sufficient warm-up for starting; Wait and Start lights on the dash tell you how long to wait and when to start, respectively. The slow-glow system is controlled by the Diesel Electronic Control Module.

The slow-glow system needs a constant 12-volt current to the glow plugs to reach its normal operating temperature. When you turn the ignition to "Run," power is supplied to the WAIT light and glow plugs. When the plugs reach the proper temperature, the module turns off the WAIT light and turns on the Start light. If the ignition remains in the "Run" position from two to five minutes, the glow plugs and indicator lights are shut off to prevent battery discharge.

Once the engine starts, the Start light goes off, but the glow plugs stay on for about the same length of time as the pre-glow period. This is called the "after-glow" period. After-glow is needed to maintain enough heat in the prechambers to keep the engine running smoothly. During the after-glow period, charging voltage is sent to the module; nine or more volts tells the module that the engine is running - the charging light then goes off.

If the engine doesn't start - and the ignition switch is still in the "Run" position with the charging light still on - the module starts the entire process over again after a two to four second delay.

When you turn on the ignition switch, it directs battery current to the starter solenoid, the control module and the fuel solenoid. Voltage to the module tells the module that the ignition switch has been turned to the Start position; voltage to the fuel solenoid starts fuel flowing inside the injection pump. A two-amp diode in the charging circuit (behind the instrument cluster) prevents charging-system "feedback" to the fuel solenoid when the engine is turned off, which prevents engine run-on. When you turn on the ignition switch, two glow-plug relays (mounted on the firewall) direct current to the plugs. The Wait and Start lights tell you when to start the engine.

The module is an electronic device located under the instrument panel; it's usually taped to the main wire harness about eight inches from the fuse block. The module has several purposes:

a) *It provides a ground circuit for the Wait and Start lights.*

b) *It times how long the WAIT light should stay on after receiving the signal from the thermistor.*

c) *It provides the ground circuit for the glow-plug relays.*

d) *It provides the timing circuit for the glow plugs' after-glow period after the WAIT light goes off.*

e) *It provides a ground circuit for the fast-idle solenoid.*

As a general rule, never replace the module until you've systematically eliminated all other circuits and components as potential problems.

The thermistor senses engine coolant temperature. It's located in the right (passenger's side) front corner of the intake manifold. The thermistor enables the module to measure the length of the preglow period and energizes/de-energizes the fast-idle solenoid.

In summary, glow-plug "on" time is determined by the following sequence (use the accompanying electrical schematic to follow the explanation):

a) *When the ignition switch is off, no voltage goes to the fuel solenoid, which cuts off fuel to the injection pump.*

b) *When you turn the switch to the Run position, voltage is directed to the fuel solenoid, alternator, both relays, the module, the fast-idle solenoid, and the Wait and Start lights. The ground circuit for the relays - which turns on the glow plugs - is supplied through the module. The ground circuit which turns on the WAIT light and fast-idle solenoid is also supplied through the module.* **Note:** *On light trucks, the WAIT light says Don't Start and the Start light says Glow Plugs.*

c) *While the ignition switch is in the Run position, the WAIT light goes out after a period determined by the engine coolant temperature, and the Start light - which is grounded through the module - comes on.*

d) *Once the ignition switch is in the Start position, the Start light goes out. Voltage from the ignition switch is directed to the starter motor and the module, which signals the module that the ignition switch has been turned to the Start position. The glow plugs are still on.*

e) *After the engine has started and the ignition switch is still in the Run position, the Wait and Start lights are off. The glow plugs remain on for the same length of time as it took for the Start light to come on. After that period, the module opens the ground circuit for the relays, which disables the glow plugs.*

System No. 2: Fast-glow, module-controlled (5.7L models)

System No. 2 is a 6-volt fast-glow system. It was used on 1979 through 1983 Oldsmobile Toronados and Trofeos; Buick Rivieras and Cadillac Eldorados (Series E); some 1979 Oldsmobile 88s and 98s (Series B and C); 1980 through 1983 Oldsmobile 88s and 98s (Series B and C); the 1982 and 1983 Oldsmobile Cutlass (Series A); and the 1980 through 1983 Chevrolet and GMC van (Series G).

The fast-glow system sends a pulsing 12-volt current to 6-volt glow plugs, which require five to seven seconds to provide sufficient warm-up for starting. 1979 models have Wait and Start lights on the dash tell you how long to wait and when to start; all other models have only a WAIT light. Series E, B, C and A models have an electro-mechanical thermal controller; series G models have a Diesel Electronics Module.

When you turn the ignition switch to Run, the WAIT light goes on and the glow plugs are energized. When the engine is cold, the WAIT light stays on for about six seconds (at 0-degrees F), then goes out, indicating the engine is ready to start. There is no Start light used with this system. On 1980 and 1981 trucks, the WAIT light says "Don't start/glow plugs." The glow plugs cycle on and off for about 25 seconds - as many as ten times - to provide after-glow once the engine starts. You can tell if the glow plugs are cycling by the way various electrical accessories waiver - the headlights dim, then brighten, then dim; the blower runs fast, then slow, etc.

These are normal characteristics of a fast-glow system, so don't take them as signs that something's wrong with the system.

If there's a problem somewhere in the system, the WAIT light warns you by staying on, or coming back on. You can usually still start the engine, but the system should be repaired.

While the ignition switch is in the Run position, the Electronics Module sends a pulsing voltage to the single glow-plug relay to maintain the correct glow-plug temperature and prevent the plugs from overheating and breaking. The control sensor senses engine coolant temperature and relays this information to the module as an electrical signal; this signal tells the module when to turn off the glow plugs and the WAIT light. The module also has a circuit to monitor the system for malfunctions, and a circuit that keeps the WAIT light on if a problem develops somewhere in the system. **Caution:** *NEVER bypass the glow-plug relay on a fast-glow system. The relay is a heavy-duty design, which controls the temperature relationship at the glow plugs. Bypassing the internal circuitry of the relay can damage the plugs.*

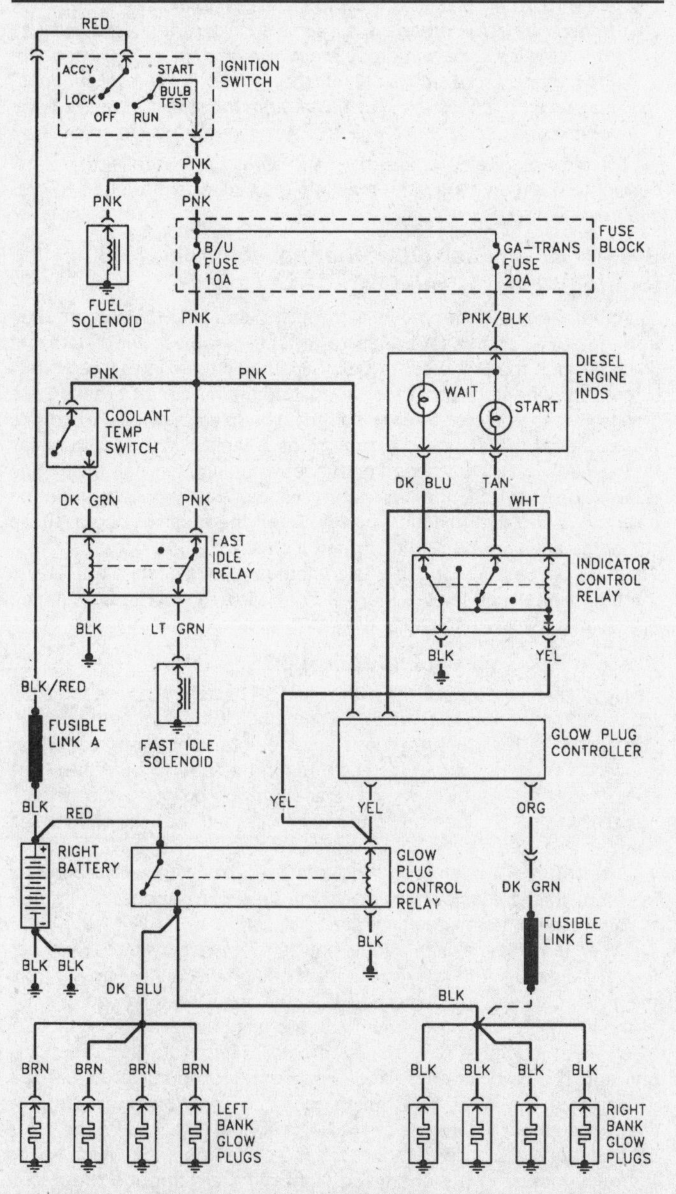

2.78 Typical electrical circuit for 1979 version of System No. 3 (fast-glow, thermal controller); System No. 2 is similar, except that it uses an electronic module to control glow-plug temperature, pre-glow and after-glow, whereas this system uses an electro-mechanical controller (1979 Cadillac shown)

Let's look at the operation of a typical System No. 2. Follow the accompanying electrical schematic (which is for 1980 and 1981 C model light trucks, but is basically the same on other models).

When you turn the key on:

a) *The fuel solenoid, the electronic module, the fast-idle circuit, the glow plug relay, the glow plugs, the temperature sensor circuit, the dash light and the charging circuit are energized.*

b) *The fast-idle circuit activates the coolant temperature switch (which is closed when cold on this system). If there's no air conditioning, it directly energizes the fast-idle solenoid; if there is air conditioning, it indirectly energizes it through the fast-idle control circuit.*

c) *The glow-plug relay is energized by a wire from the module, and is grounded through another wire to the module. The module determines pulse time based on input from the control sensor.*

d) *The glow plugs and temperature circuit are energized by a feed from the batteries, which goes through the relay, then through left and right splices to the two banks of plugs; fusible links are installed in both wires between the relay and the splices. There are two temperature-sensing circuits - one for each bank of plugs - between the splices and terminals C and L at the module; both sensing circuits are routed through the control sensor. The control sensor varies the voltage signal with a thermistor-resistor type voltage divider, which detects unequal voltage at the glow-plug banks by comparing the two voltages.*

e) *Another sensor circuit between the left bank splice and terminal E on the module feeds the "Don't Start" or "Glow Plugs" light on the dash. The dash light pulses off and on as the glow plug relay pulses off and on.*

f) *The sensing circuits are hot only when the relay is pulsing on. If the engine is warmed up, there shouldn't be any voltage in any of these circuits and the relay shouldn't be pulsing. The module needs a glow-plug relay "on" pulse to determine the pulse time and temperature. The dash light should flash on momentarily each time the ignition is turned on when starting or testing on a warmed-up engine.*

g) *The charging circuit is fed by the ignition switch, through the GEN light and a parallel resistance wire in the instrument panel harness, and to the generator field and the module. The module is like a high-resistance connection, and uses this input as an engine-start signal when the generator output starts. A couple of other leads go to the automatic starter disengagement-lockout relay, which senses engine start and automatically opens the solenoid feed wire from the ignition switch. This prevents re-engagement of the starter once the engine has been started.*

System No. 3: Fast-glow, thermal controller (1979 5.7L models)

The big difference between system nos. 2 and 3 is the means by which glow-plug temperature, pre-glow and after-glow are controlled. System no. 2 uses an electronic module; system no. 3 **(see illustration)** uses an electro-mechanical thermal controller.

When you turn the ignition switch to the Run position, the WAIT light and the glow plugs go on. When the plugs reach the correct temperature, the thermal controller signals the relay to turn out the WAIT light. The relay also turns on the Start light, which stays on until the engine is started. If the engine doesn't start for some reason, the Start light stays on as long as the ignition switch remains on.

Once the engine has started, the glow plugs cycle on and off during the after-glow period; they'll continue cycling for up to 30 seconds, then go off and remain off as long as the engine temperature remains above 120-degrees F.

If you leave the ignition switch in the Run position without turning it to Start, the glow plugs will continue to pulse on and off until the batteries run down - this takes about four hours if the coolant switch is open.

In certain situations - when there's no WAIT light with a cold engine, when there's a flashing Start light or when both the Wait and Start lights come on at the same time - the engine can be started cold, even though the glow plug system needs repairs. If the engine is starting under the conditions, the system needs to be repaired.

The ignition switch directs battery current to the starter solenoid and the electro-mechanical controller. Voltage to the fuel solenoid allows fuel to flow through the injection pump. A diode in the charging circuit prevents "feedback" that would cause the engine to run-on when the engine is off.

The thermal controller:

a) *Signals the light control relay for the WAIT and START lights;*
b) *Senses engine coolant temperature and glow-plug voltage;*
c) *Controls current to the glow plugs and maintains glow-plug temperature by pulsing the relay on and off.*

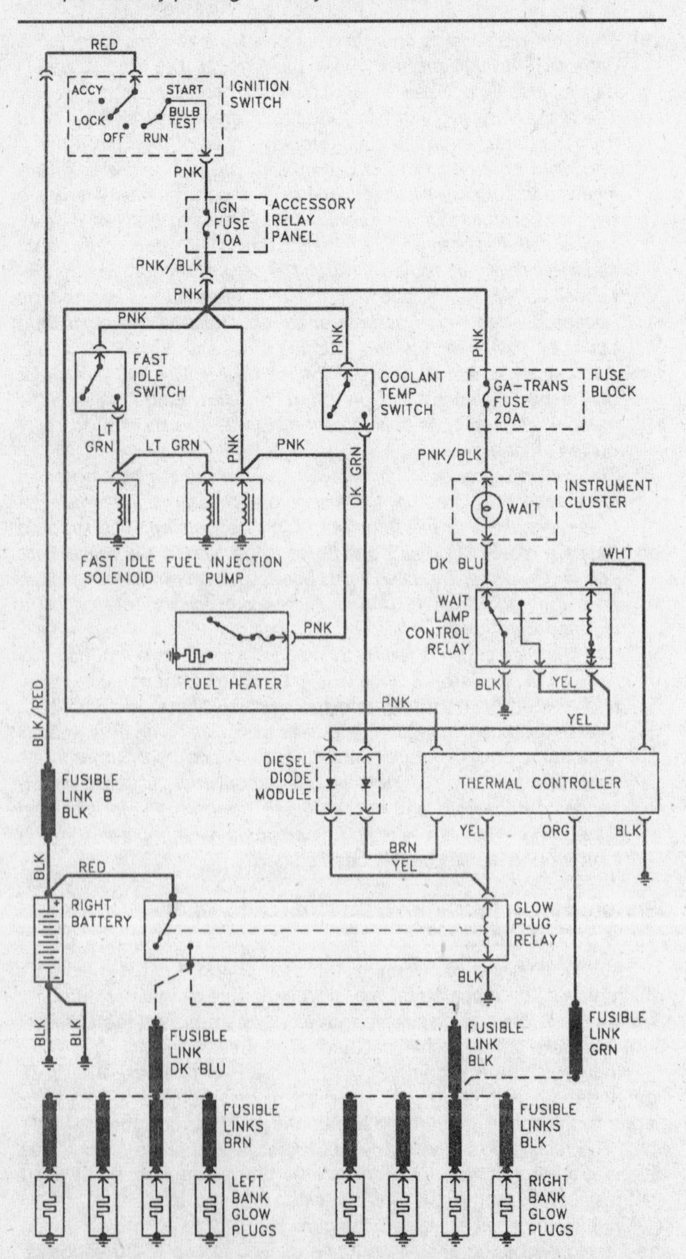

2.79a Typical electrical circuit for 1980 through 1983 version of System No. 3 (fast-glow, thermal controller) (1982 Cadillac shown)

The light control relay responds to the controller signal to turn the Wait (or Start) light on or off.

When it closes, the coolant temperature switch completes the circuit, energizing the fast-idle relay. When the engine coolant temperature reaches about 120-degrees F, the energized fast-idle relay opens the circuit to the fast-idle solenoid to reduce the idle speed.

The WAIT light on some engines may remain on a couple of minutes before it cycles to Start. Then it recycles back to Wait. Check for reversed leads at the coolant temperature switch and the glow-plug system thermistor. The lead to the temperature switch should be green and the lead to the thermistor should be green with a white stripe. If the start light isn't coming on, diagnose it as follows:

Unplug the connector with the tan wire at the control relay. Turn the ignition to ON and touch the tan wire with a jumper wire connected to ground.

If the start light doesn't come on, *check the start light bulb. If the start light bulb is okay, locate and repair the open circuit in the tan wire from the control relay to the start light bulb.*

If the start light comes on, *turn off the ignition, hook up a test light to a 12-volt source and probe the brown wires in the relay connector. If the test light comes on, replace the relay. If the test light doesn't come on, check the brown wire from the relay to the alternator for an open circuit. If this wire is okay, check for a reversed diode.*

The glow plug relay is pulsed on and off by the controller to control current to the glow plugs and maintain glow-plug temperature to prevent overheating.

System No. 3: Fast-glow, thermal controller (1980 through 1983 5.7L models)

Another (more common) version of System No. 3 is used on 1980 through 1983 models **(see illustrations)**. The power supply current flows through the DSL/VAC or DSL/ECM fuse and follows three paths to ground. One is through the thermal controller circuit breaker, the time limiter, the pulser switches and the glow-plug relay coil to ground. The same voltage is applied to both sides of the coil in the control relay for the WAIT light. The second path goes to ground through the engine temperature switch, the fast idle solenoid and the cold-advance solenoid in the fuel injection pump. When the engine is cold, these solenoids advance pump timing and increase the idle speed slightly. The third path is through the coolant temperature switch, diode C and the time limiter coil for the thermal controller; this starts time limiter operation early when the engine is warm.

System No. 3 glow plug operation

How are the glow plugs controlled? The following description applies to both types of System No. 3 circuits. When you turn the ignition switch to Run, the Wait and Charge lights come on. Voltage is applied to the glow-plug relay coil through the thermal controller. The contacts on the glow-plug relay close, applying power to the left and right glow-plug banks, and to the pulser in the controller. The plugs heat up, and so does the controller pulser coil.

After about six seconds (cold start, engine temperature below freezing), the pulser switch inside the controller opens. In warmer ambient temperatures, the time period is less.

When the pulser switch opens, power is removed from the relay coil for the glow plug. The relay contacts open and disconnect power from the glow plugs and the pulser coil. With the pulser switch open, the coil of the WAIT light control relay no longer has battery voltage on both sides. With a voltage drop across the coil, current flows through the coil and the relay is energized. The contacts open, removing the ground from the WAIT light, which goes out. The glow plugs are now warm enough for the engine to start. The control relay coil for the WAIT light has a new ground path through the closed contact, which keeps the relay locked in the energized (light off) state through the circuit breaker in the thermal controller.

If the engine isn't started immediately, the pulser switch closes again, and applies voltage again to the relay coil for the glow plug. Power is again applied to the glow plugs and the pulser. After a short time (1/2 to 2 seconds) the pulser switch opens and removes voltage from the glow plug relay coil. The pulser will continue pulsing power to the glow plugs. **Caution:** *If the ignition switch is turned to Run - and the engine isn't started - the glow plugs will continue to pulse on and off until the batteries run down (about four hours).*

When the engine is started, voltage from the charging circuit is applied through diode A to the charge indicator. This balances the voltage on the fuse side of the indicator, and the light goes out.

Even after the WAIT light goes out, the glow plugs must remain on for awhile. The combustion chambers probably aren't yet hot enough to keep the engine running smoothly on compression alone. This pulsing on and off of the glow plugs is the after-glow period; it lasts for a minute or less. Voltage is also applied through diode B to the time limiter coil in the thermal controller. When the time limiter switch in the controller opens, the glow plug pulsing stops, and so does the after-glow period.

If the pulser circuit fails, the pulser switch can't pulse the glow plug relay on and off. If the glow plugs remain on too long (over eight seconds), the circuit breaker in the controller opens, removing voltage from the glow-plug relay and the glow plugs.

Diagnosing the cause of failure of the glow-plug controller

Glow-plug controller failures can be caused by a short to ground in the white or yellow wire going to the controller. It can also be caused by a diode failure in the control relay for the WAIT light. To determine if the controller failure was caused by the above, measure the resistance between pins No. 3 (pink wire) and No. 2 (white wire) on the controller. If the resistance is around 30 ohms, one of the failures described above is the cause. If it's 10 or fewer ohms, the failure was caused by some other condition.

Diagnosing the thermal controller glow-plug system

Note: *If a problem occurs when the engine is cold, then the engine must be cold during the following troubleshooting procedure(s).*

Condition 1: Engine is cold and there's no WAIT light

1 With the ignition in the Run position, and the engine stopped, not whether the Charge light is on. If it isn't, check the Gages fuse. Check the operation of the glow-plug relay by listening for a click at the relay. If the relay operates, look for a break in the pink and black wire between the fuse block and splice S287 (S221). If the relay doesn't operate, check the pink wires between the ignition switch and the fuse block.

2 Disconnect the connector at the control relay for the WAIT light near the wiper motor. With the jumper wire, connect the blue wire at the connector to ground. If the WAIT light doesn't go on, check to see if the bulb is burned out. Look for a break in the blue wire between the connector at the WAIT light control relay and the WAIT light itself; also check the pink and black wire between the WAIT light and splice S287 (S221).

3 Check the connection at ground G178 (near the wiper motor). Touch one end of the test light to the red wire terminal of the relay for the glow plug, and the other end to ground. Then touch the lead to the black wire at the connector for the WAIT light control relay. If the light comes on, replace the WAIT light control relay.

Condition 2: The WAIT light stays on more than 10 seconds

1 Check the operation of the glow-plug relay. Unplug the connector at the diode module. Touch one end of the test light to ground and the other end to the red (front) post of the relay for the glow plug. Then touch the test lead to the rear post (blue and black wires) of the glow-plug relay. If the relay operates, the test light should come. If the relay doesn't operate, go to Step 2; if it does operate, go to Step 3.

2 If the relay doesn't operate, check the DSL/VAC (DSL/ECM) fuse and ground G178. Unplug the connector at the thermal controller. Hook up a test light between pins 3 (pink and black wire) and 6 (yellow wire) at the harness connector. If the test light comes on with the ignition turned on, check the continuity between pins 3 and 6 of the thermal controller. If there's no continuity, replace the controller.

3 If the relay operates, unplug the connector to the control relay for the WAIT light. The WAIT light should go out. Connect the test light between the white and yellow wires at the connector. If the test light comes on, replace the control relay for the WAIT light. The engine should start. If the test light doesn't come on, turn off the power and check for 30 ohms resistance at pins 4 and 5 of the thermal controller. Replace the controller if the resistance is high. Check the continuity of the black, dark green, orange, white, yellow and black wires.

Condition 3: The engine runs roughly during cold starts

1 With the engine off, turn the ignition to Run. Unplug the connector at the engine temperature switch. Touch the jumper to the two terminals on the connector and note whether the fast idle solenoid extends. If it doesn't, check the throttle linkage or the solenoid plunger for binding. If the linkage is okay, replace the solenoid.

2 If the fast idle solenoid extends, hook up a magnetic tach (Kent-Moore J-26925 or equivalent) to the engine. Start the engine. Leave the jumper attached to the connector at the engine temperature switch. Unplug the connector at the cold advance solenoid of the fuel injection pump. The engine speed should change 30 rpm when the connector is removed. Check the solenoid and pump if there's no change.

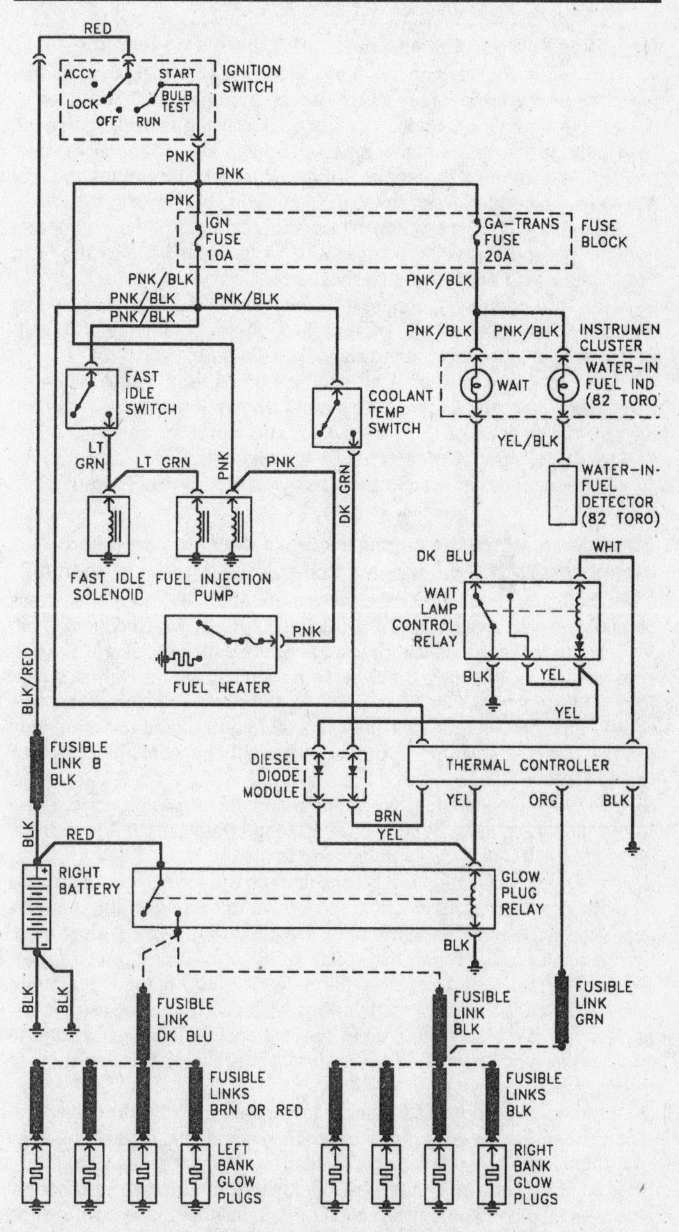

2.79b Typical electrical circuit for 1980 through 1983 version of System No. 3 (fast-glow, thermal controller) (1983 Buick Riviera shown)

3 Turn the engine off. Check the continuity of the engine temperature switch with a self-powered test light. There should be continuity below 120-degrees F; there shouldn't be continuity above 120-degrees F. Replace the switch if it's bad.
4 Attach one end of the test light to ground. With the engine off and the ignition in Run, touch the other end of the test light to the green wire pin at the diode module. If the light comes on, and the engine is cold, replace the coolant temperature switch.
5 Turn the ignition to Off. Unplug all eight connectors for the glow plugs. Attach a self-powered test light between the rear post (blue and black wires) of the glow-plug relay and ground. At each glow plug, touch the harness connector to the spade terminal of the glow plug. If the test light comes on, the glow plug and the harness lead are good. Be sure to set each harness connector aside after testing its respective plug.
6 If the test light doesn't come on, touch the harness connector to the engine block or some other ground. If the light then comes on, replace the glow plug. If the light doesn't come on, replace the wire to the glow plug.
7 Check the fuel system for correct fuel delivery.

Condition 4: When the engine is cold, there's no fast idle

1 Turn off the engine, then turn the ignition to Run. Unplug the connector at the engine temperature switch. Attach a jumper wire to both terminals of the connector. Unplug the connector at the fast-idle solenoid, then reconnect it. Note whether the solenoid operates. If it does, readjust the fast idle (refer to the section of this chapter that details the idle adjustment procedures and to the VECI label on your vehicle).
2 If the fast idle solenoid doesn't operate, connect one end of a test light to ground. Touch the other end of the test light to the light green/black lead at the fast-idle solenoid.
3 If the light comes on, unplug the connector at the fast idle solenoid and check the continuity to ground through the solenoid with a self-powered test light. Replace the solenoid if it's bad.
4 If the light doesn't come on, touch the test lead to the pink and black terminal of the engine temperature switch. If the light comes on, remove the connector from the switch and test it for continuity. The switch should now be closed below 120-degrees F, and open above 125-degrees F. If the switch opens below 120-degrees F, remove and replace it.

Condition 5: When the engine is cold, it doesn't start (even though the WAIT light appears to be okay - it goes on, then off)

1 Is the engine cranking speed at least 100 rpm? If it isn't, check battery voltage with a voltmeter - it should be 12.4 volts with the ignition off.
2 Using a test light, check for voltage at the pink wire lead at the fuel solenoid of the injection pump with the ignition in the Run position. Repair the pink wire if there's no voltage. If the test light comes on, turn off the ignition. Using a self-powered test light, check for continuity through the fuel solenoid to ground. If there is no continuity, replace the fuel solenoid.
3 With the ignition in the Run position, and the engine stopped, listen to the glow plug relay. It should be clicking on and off. If it's clicking, go to step 4; if it isn't clicking, proceed to step 7.
4 If the relay is clicking, turn off the ignition. Unplug the connectors to all eight glow plugs. Connect a self-powered test light between the rear post (blue and black wires) of the glow plug relay and ground. Touch each connector to the spade terminal of its respective glow plug. If the test light comes on, the glow plug and harness leads are good. Test all the glow plugs this way. Make sure you set aside each connector after testing its plug to prevent accidentally energizing the plug. When you're finished testing, reconnect all the connectors to their respective glow plugs.
5 If the test light doesn't come on, touch the connector to the engine block or some other ground. If the light then comes on, replace the glow plug. If the light doesn't come on, replace the wire to the glow plug.
6 If all eight glow plugs are open-circuited, then replace the thermal controller. If a glow plug has broken off and fallen into the combustion chamber, remove the cylinder head and clean out the debris.
7 If the relay isn't clicking, check the thermal controller and glow plug relay circuit. Unplug the connector at the thermal controller. Connect one end of the test light to ground. With the ignition in Run, check for voltage as follows:
8 Check for voltage through the brown wire. If the light comes on, note whether the coolant temperature switch is closed. If the engine is cold, replace the switch.
9 Check for voltage through the orange wire. If the light comes on, check for shorted contacts in the glow plug relay (red to blue and black wires). If the contacts are shorted, replace the relay. Also replace all glow plugs and the thermal controller.
10 Connect one end of the test light to 12-volts at the battery. Touch the yellow wire. If the light doesn't come on, repair the yellow wire between the controller and the glow plug relay. If the light comes on, replace the thermal controller.

Condition 6: The engine stays on fast idle

1 With the engine off, turn the ignition to Run. Unplug the connector at the engine temperature switch. Touch the jumper to the two terminals on the connector and note whether the fast idle solenoid extends. If the solenoid doesn't extend, check the throttle linkage or the solenoid plunger for binding. If the linkage is okay, replace the solenoid.
2 Check the continuity of the engine temperature switch with a self-powered test light. There should be continuity below 120-degrees F and no continuity above 120-degrees F. Replace the switch if it's bad.

Condition 7: The WAIT light pulses slowly on and off

1 Hook up a test light between the white and yellow wires at the WAIT light control relay. If the test light is on when the WAIT light is off, and off when the WAIT light is on, replace the WAIT light control relay.
2 If the test light doesn't come on, hook it up between ground and the orange wire at the thermal controller. With the WAIT light pulsing, note whether the test light pulses on and off with the WAIT light. If the test light doesn't come on, touch it to the front (red wire) and rear (blue and black wires) posts of the glow plug relay. The fault may be in the red wire between the batteries and the glow plug relay, in the black, green or orange wires between the relay and the thermal controller or in the relay itself.
3 If the test light flashes on and off with the WAIT light when connected to the orange wire at the thermal controller, hook up the test light between the black wire at the controller and ground. If the test light again flashes on and off, repair the black wire between the controller and ground. If the test light doesn't flash, replace the thermal controller.

Condition 8: The engine continues to run with the ignition off

1 With the ignition turned off, and the engine still running, unplug the connector at the diode unit. If the engine stops, diode C is shorted. Replace the diode unit.
2 If the engine continues to run, unplug the pink wire connector at the fuel solenoid in the fuel injection pump. If the engine still continues to run, stop it by crimping the flexible fuel return line near the fuel supply pump (lower right side of engine). Then repair or replace the fuel solenoid.

Condition 9: The Water-In-Fuel (WIF) light stays on

Drain the contaminated fuel. Refer to the part of this chapter that details how to service the low-pressure side of the fuel system.

Diagnostic flow charts for glow-plug systems used on 5.7L engines

Before using the diagnostic flow charts to troubleshoot the glow-plug system, always:

1 Make sure all the glow plugs are working.
2 Check all fuses, bulbs and grounds before testing or replacing any components.
3 Make sure no wires are exposed and all connections are clean and dry.
4 Perform an ammeter test **(see illustration)** *to pinpoint the cause of the problem.*

If the WAIT light circuit operates correctly, but then comes on after the engine is started, the system is malfunctioning. Check the glow plug system **(see illustration 2.80)**. Make sure that module terminals H and M are correctly grounded. Check the diode trio in the alternator.

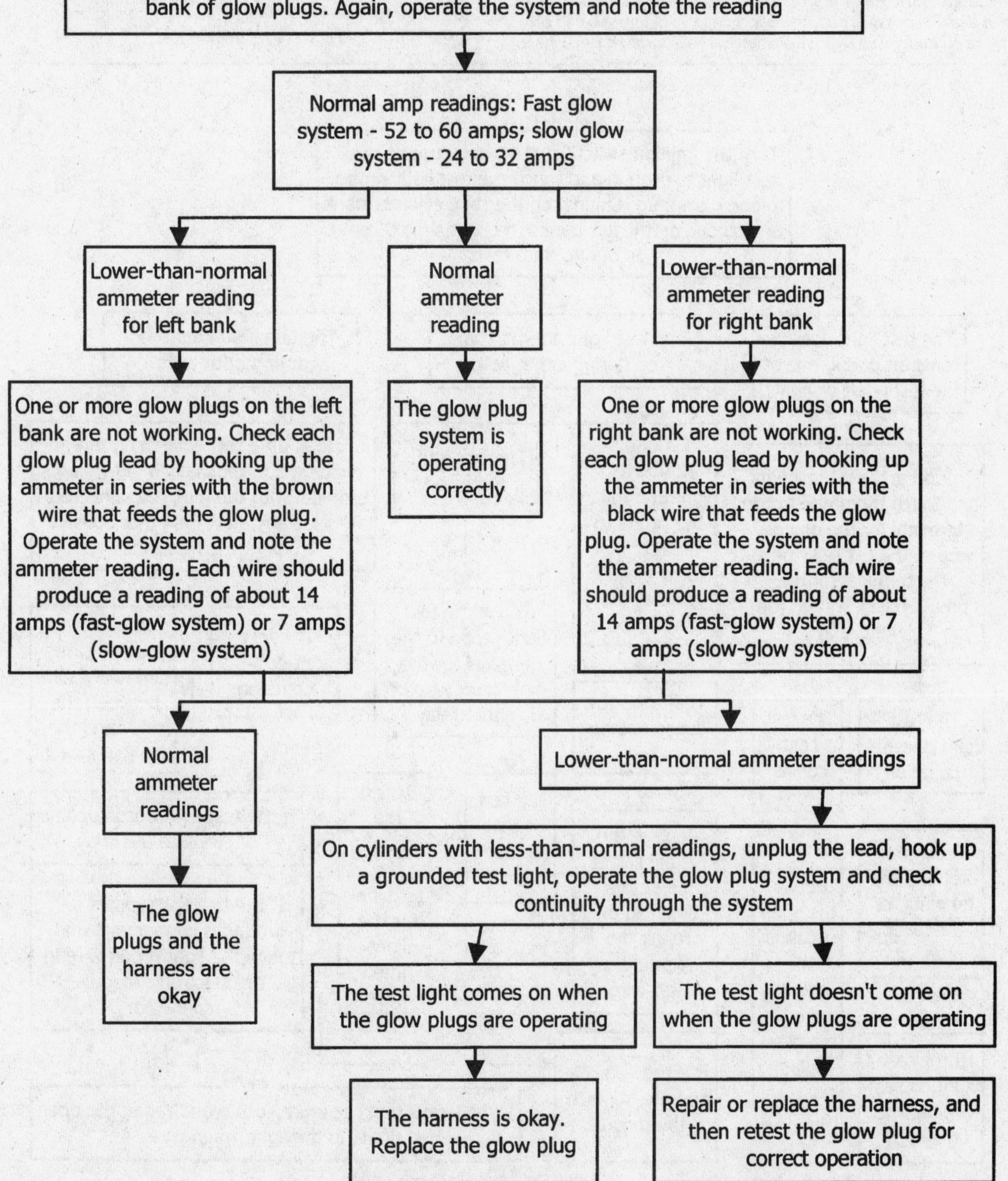

2.80 Preliminary glow plug system testing with an ammeter (5.7L engines)

And check the following circuits from the module for an open, a short or a bad connection:

1) *the purple, gray, black/white, black/pink and orange wires to the control sensor.*
2) *The brown wire to the alternator.*
3) *The orange, dark green, dark blue, orange/black and black wires to the glow plug relay.*

The accompanying flow charts **(see illustrations)** for the various glow-plug systems used on 5.7L engines will help you pinpoint the cause of a problem. You'll need the following tools before you can perform these procedures: a digital multimeter, a self-powered test light, a 12-volt test light and a volt-amp tester. The digital multimeter is essential for testing fast-glow systems because it can measure extremely low voltages without burning out the delicate circuits in the electronic control module. You can also use it as an ohmmeter instead of a conventional ohmmeter when diagnosing hard-to-start or no-start conditions. The self-powered test light is essential for checking glow-plug continuity and opens and grounds in circuit checks where no voltage can be applied. The 12-volt test light is used to check for opens and grounds in circuit checks that require 12 volts for proper diagnosis. The volt-amp tester can measure amperage (current flow) draw between the left and right glow-plug banks.

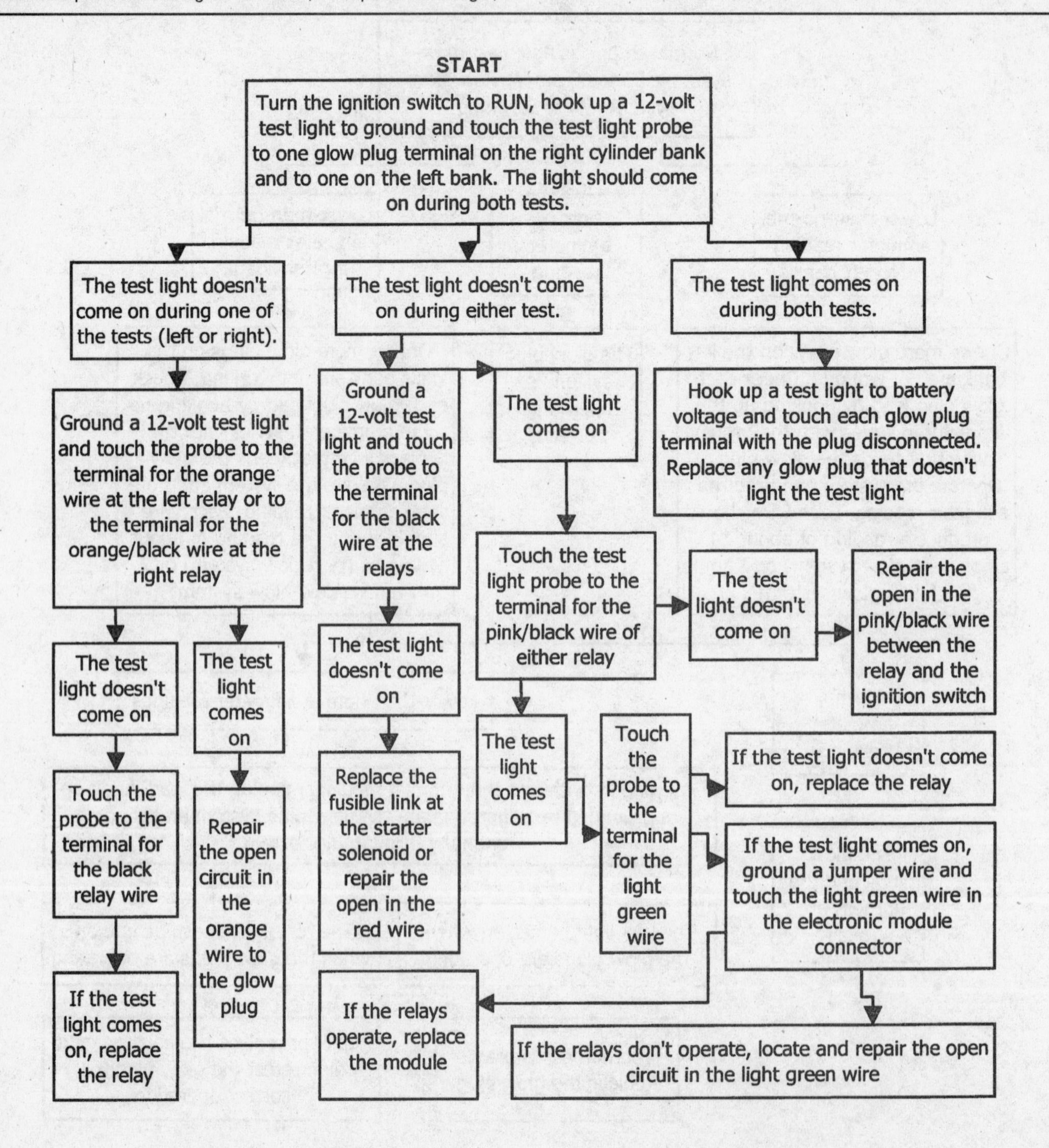

2.81a Flowchart No. 1 (5.7L engines) - The engine doesn't start and the glow plugs are inoperative

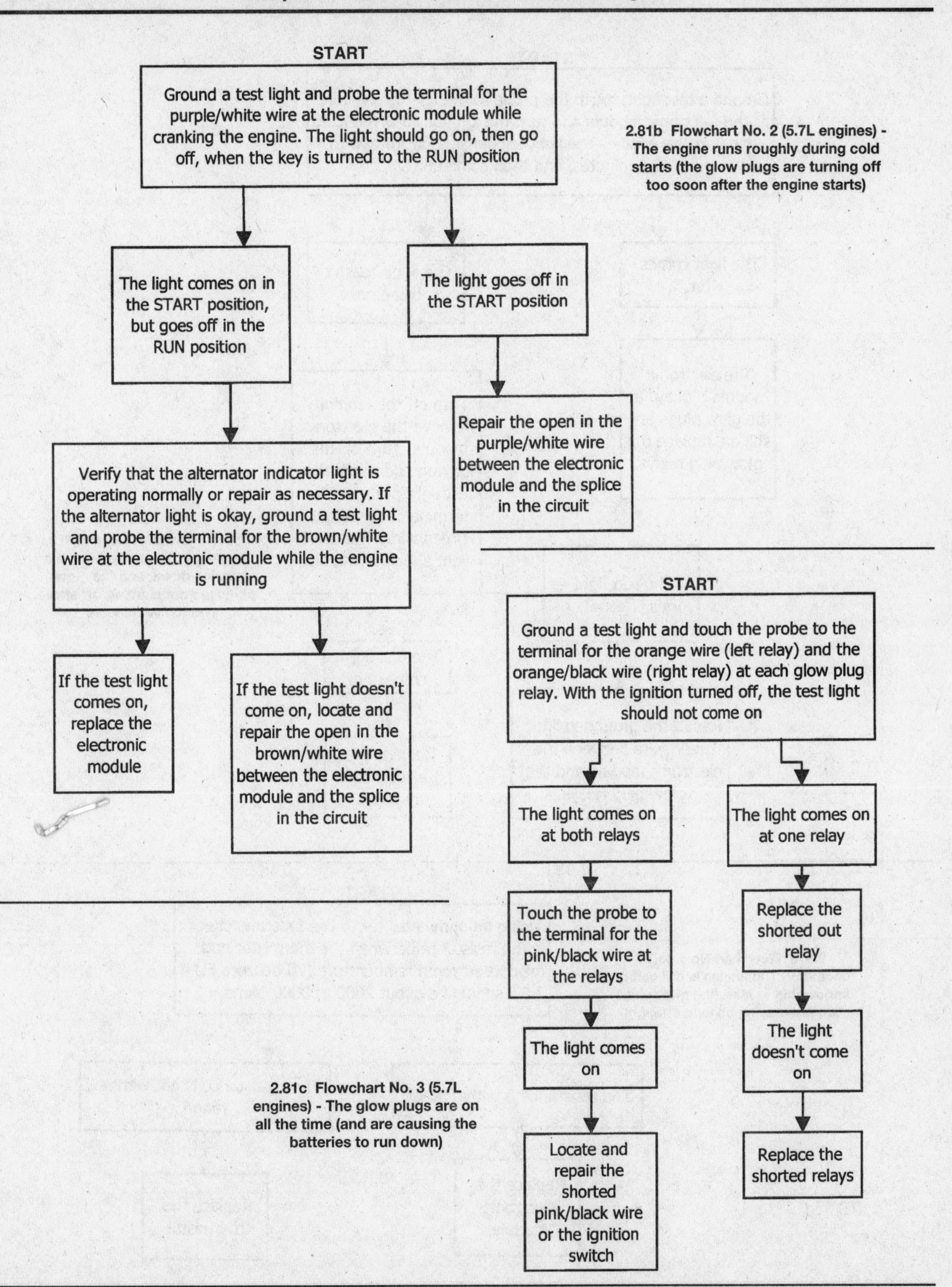

2.81b Flowchart No. 2 (5.7L engines) - The engine runs roughly during cold starts (the glow plugs are turning off too soon after the engine starts)

2.81c Flowchart No. 3 (5.7L engines) - The glow plugs are on all the time (and are causing the batteries to run down)

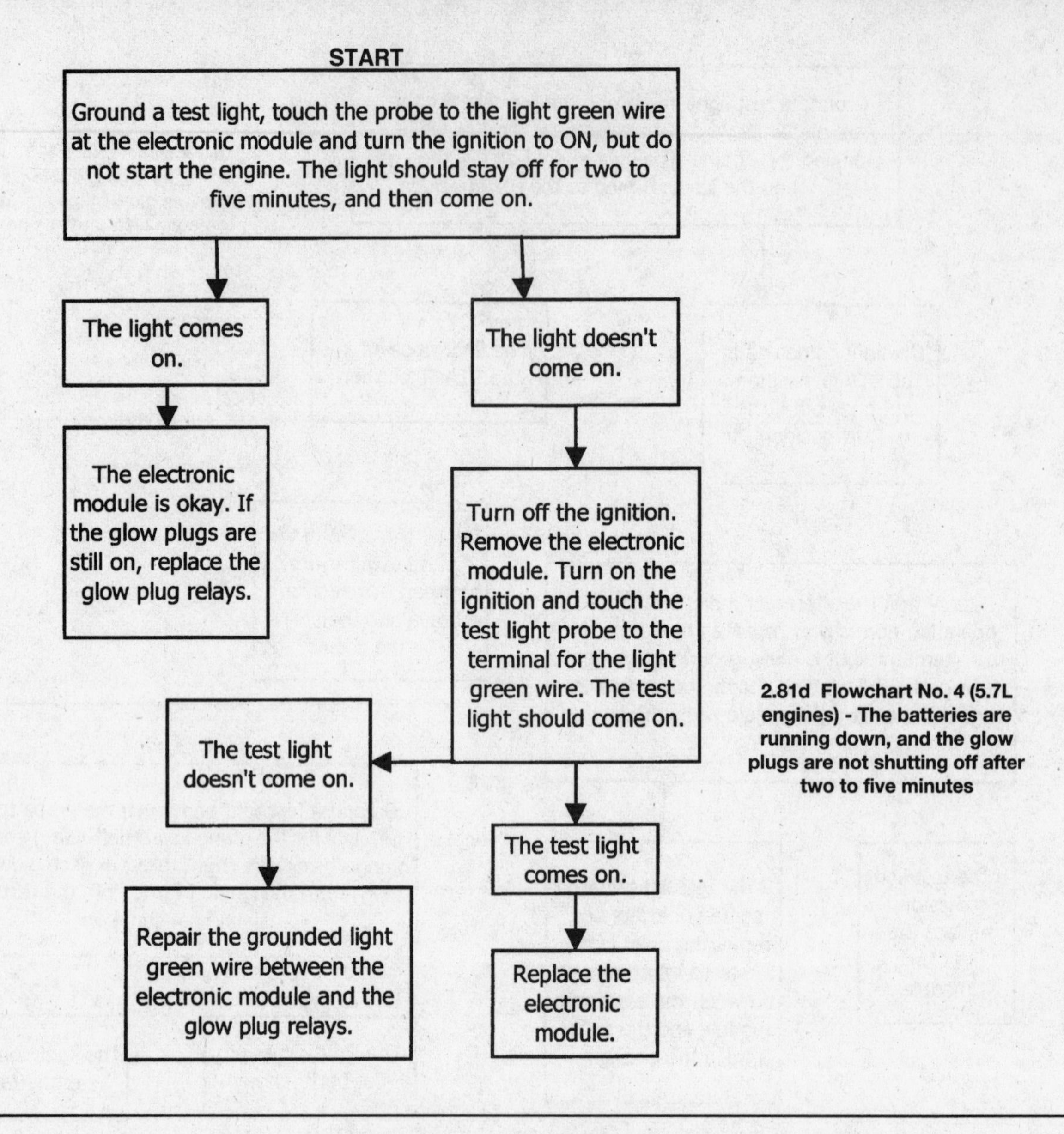

2.81d Flowchart No. 4 (5.7L engines) - The batteries are running down, and the glow plugs are not shutting off after two to five minutes

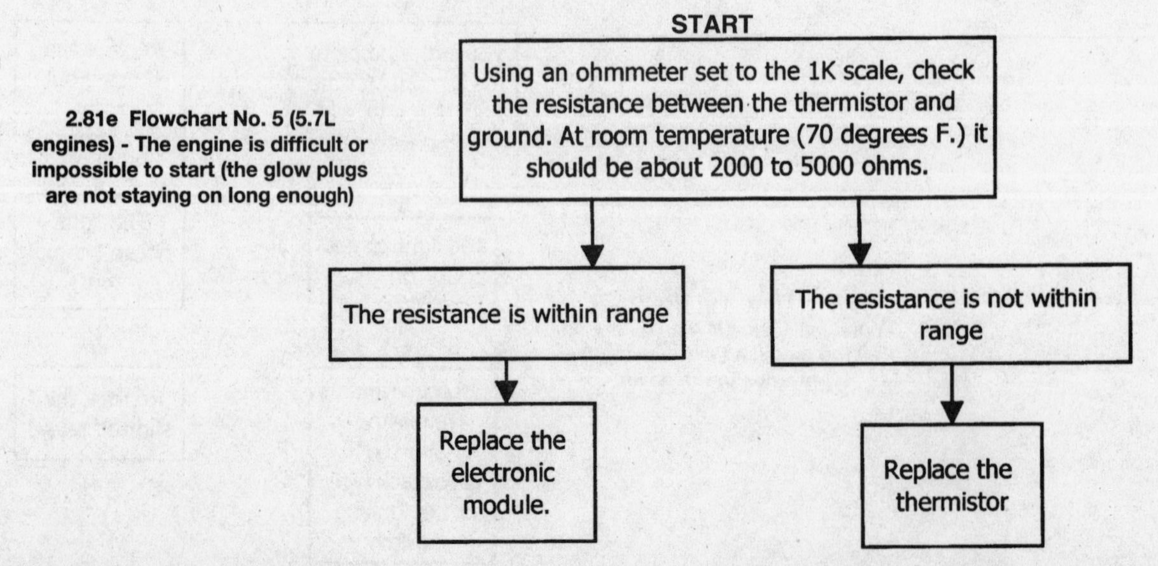

2.81e Flowchart No. 5 (5.7L engines) - The engine is difficult or impossible to start (the glow plugs are not staying on long enough)

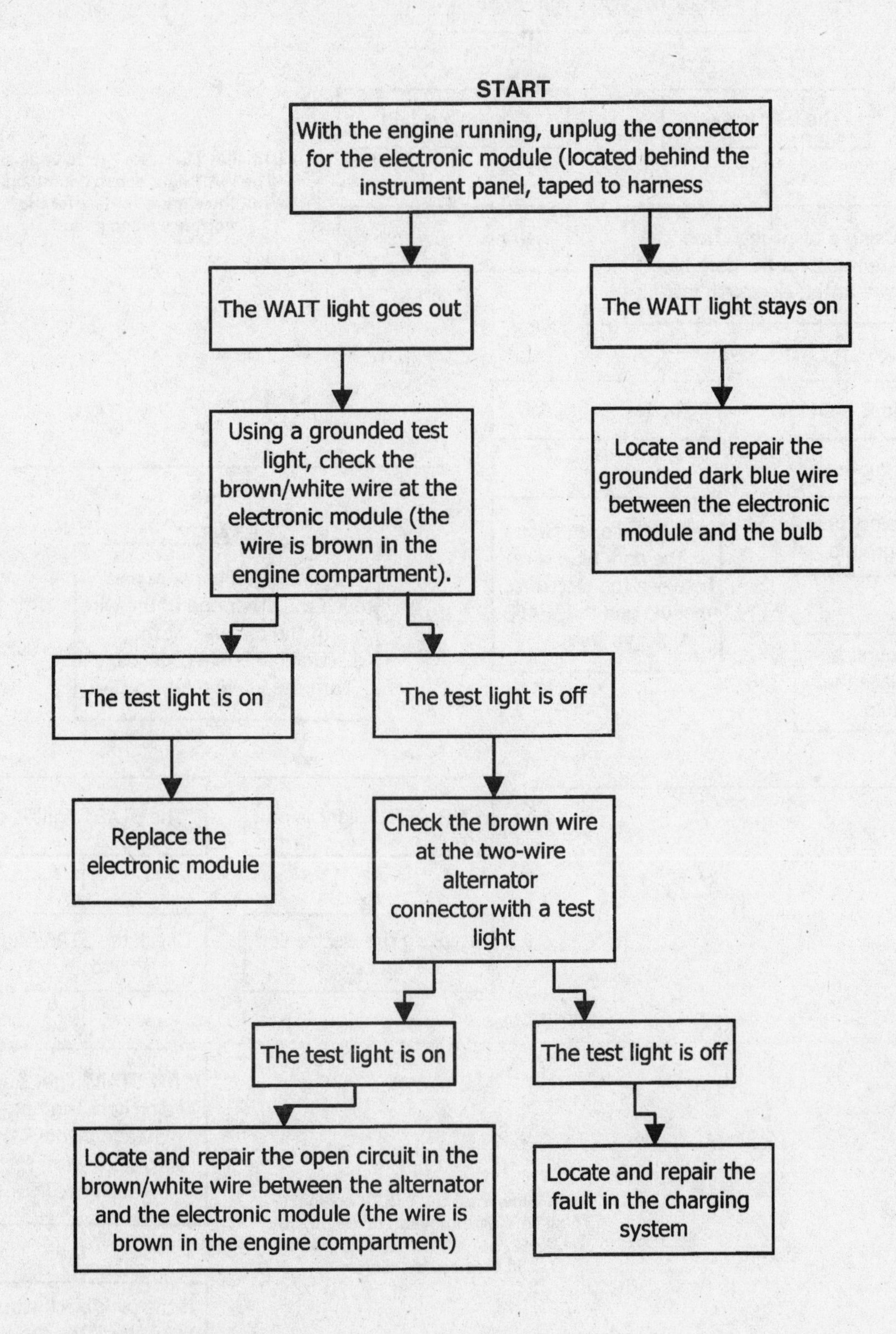

2.81f Flowchart No. 6 (5.7L engines) - The WAIT light remains on after the engine starts

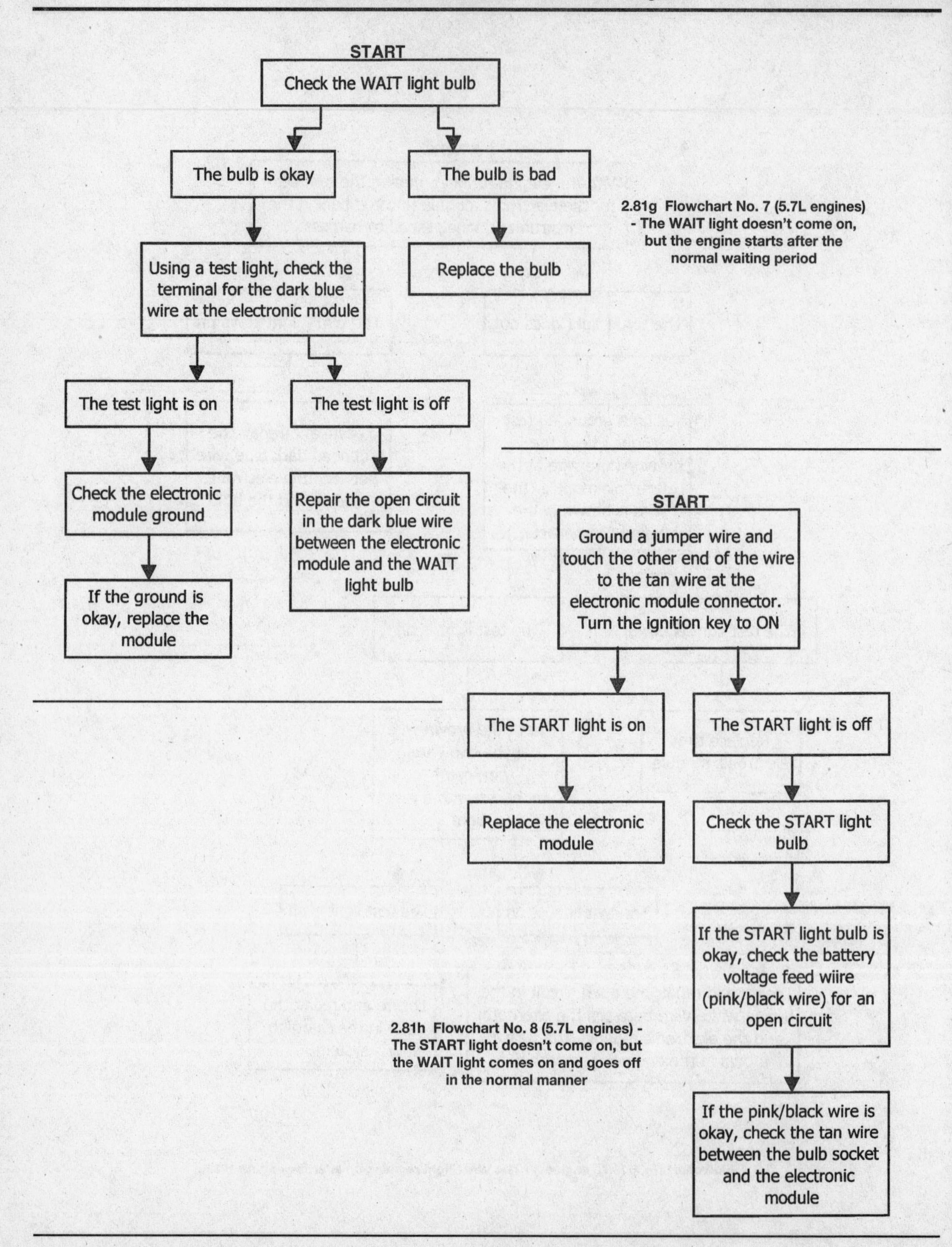

2.81g Flowchart No. 7 (5.7L engines) - The WAIT light doesn't come on, but the engine starts after the normal waiting period

2.81h Flowchart No. 8 (5.7L engines) - The START light doesn't come on, but the WAIT light comes on and goes off in the normal manner

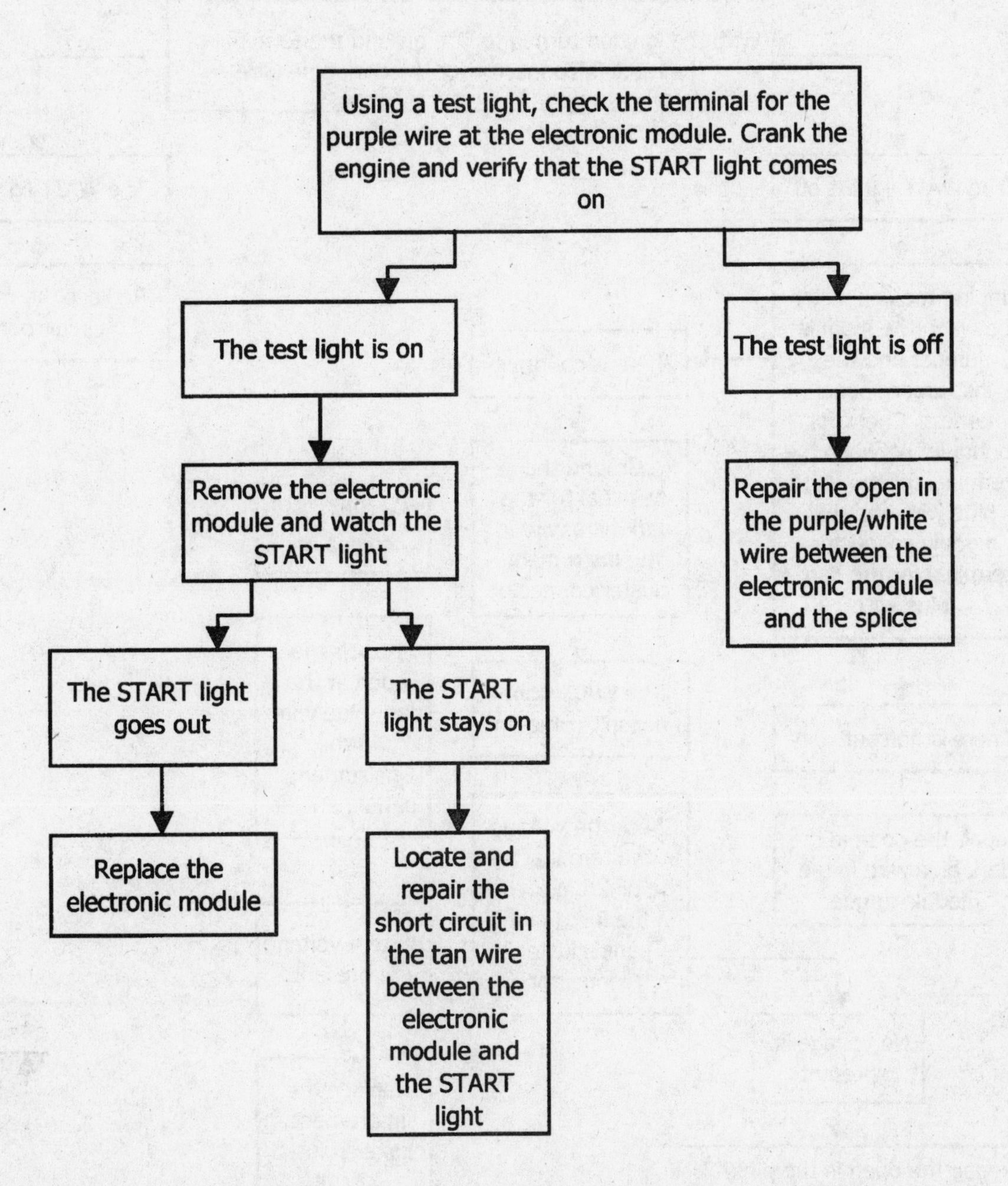

2.81i Flowchart No. 9 (5.7L engines) - The START light remains on after the engine starts, or after the two to five minute emergency shutdown period

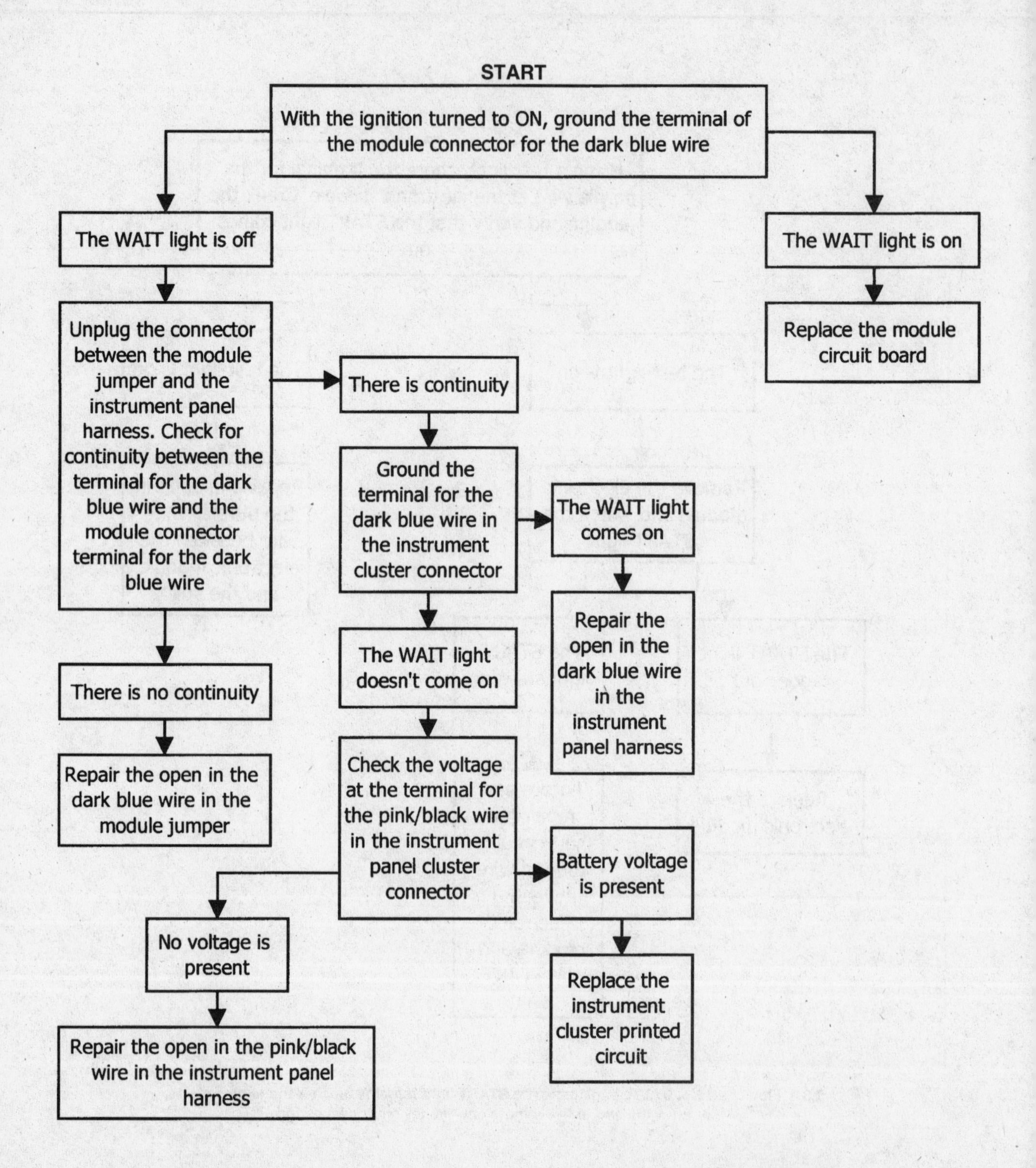

2.81j Flowchart No. 10 (5.7L engines) - The WAIT light doesn't come on (the bulb is okay), but the glow plug relay is okay and the engine starts

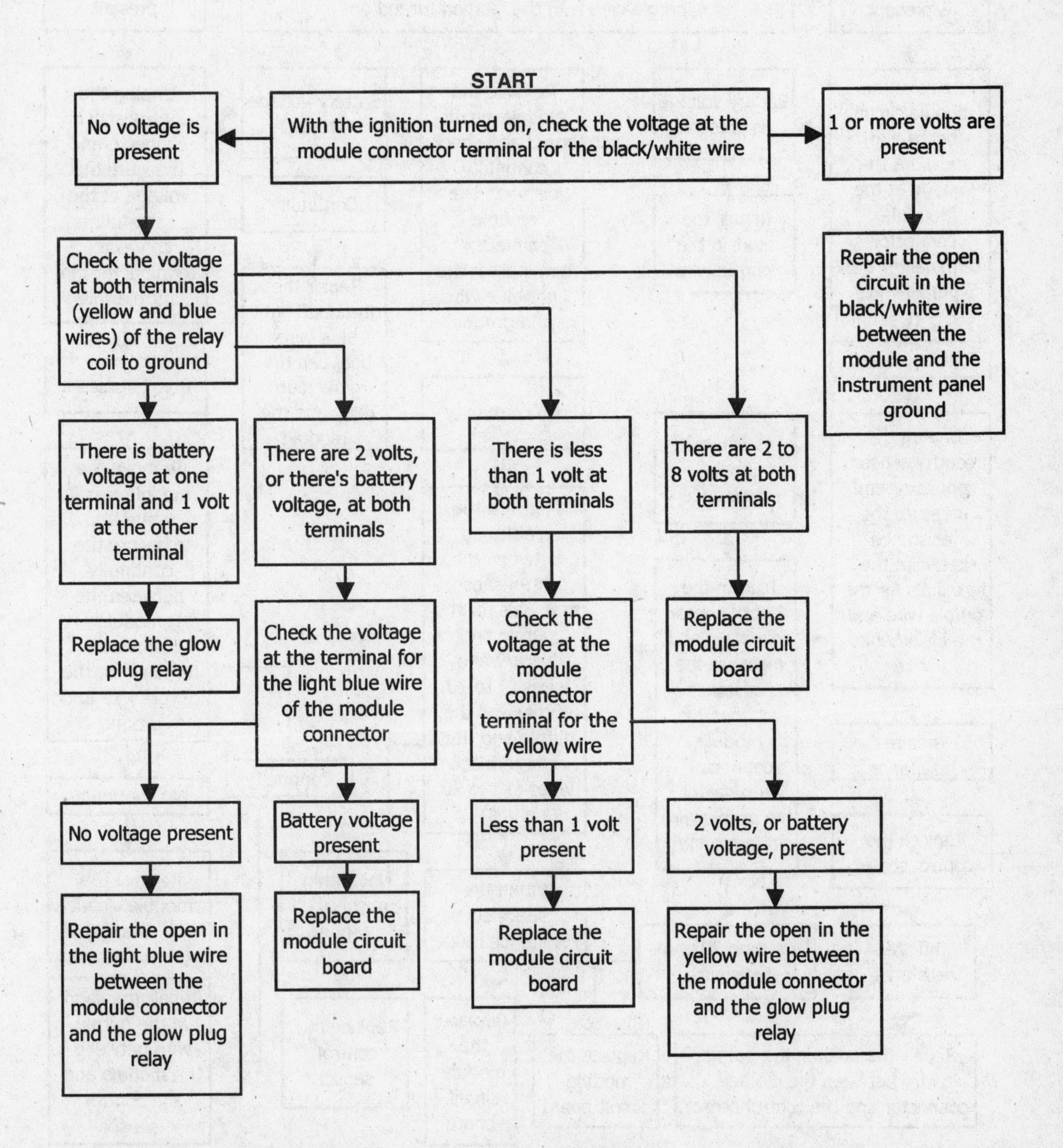

2.81k Flowchart No. 11 (5.7L engines) - The engine won't start, the WAIT light won't go off and the glow plug relay doesn't operate

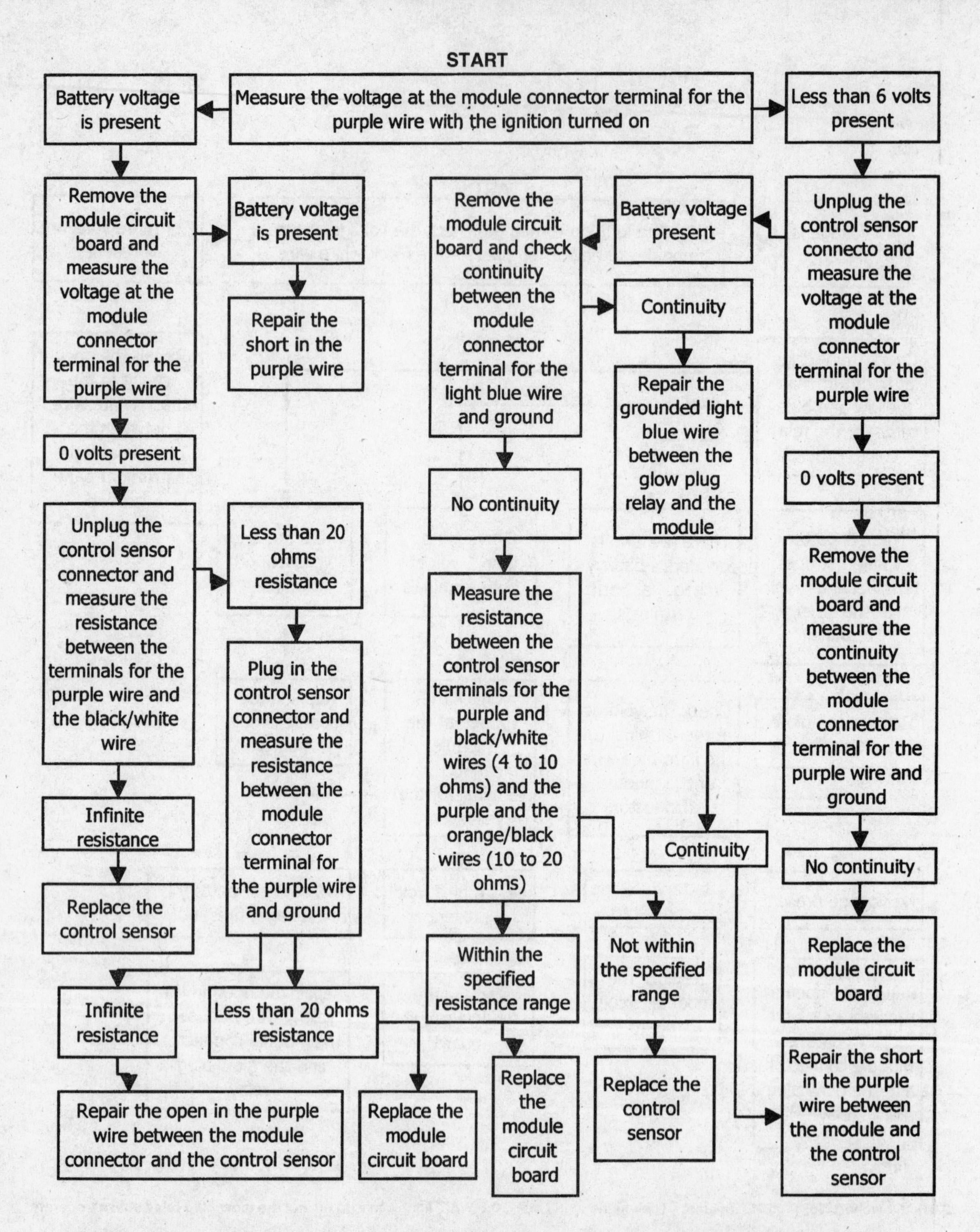

2.81l Flowchart No. 12 (5.7L engines) - The WAIT light remains on constantly and the glow plug relay comes on and then quickly goes off; OR, the engine is hard to start, or won't start, when cold, and the WAIT light is operating normally

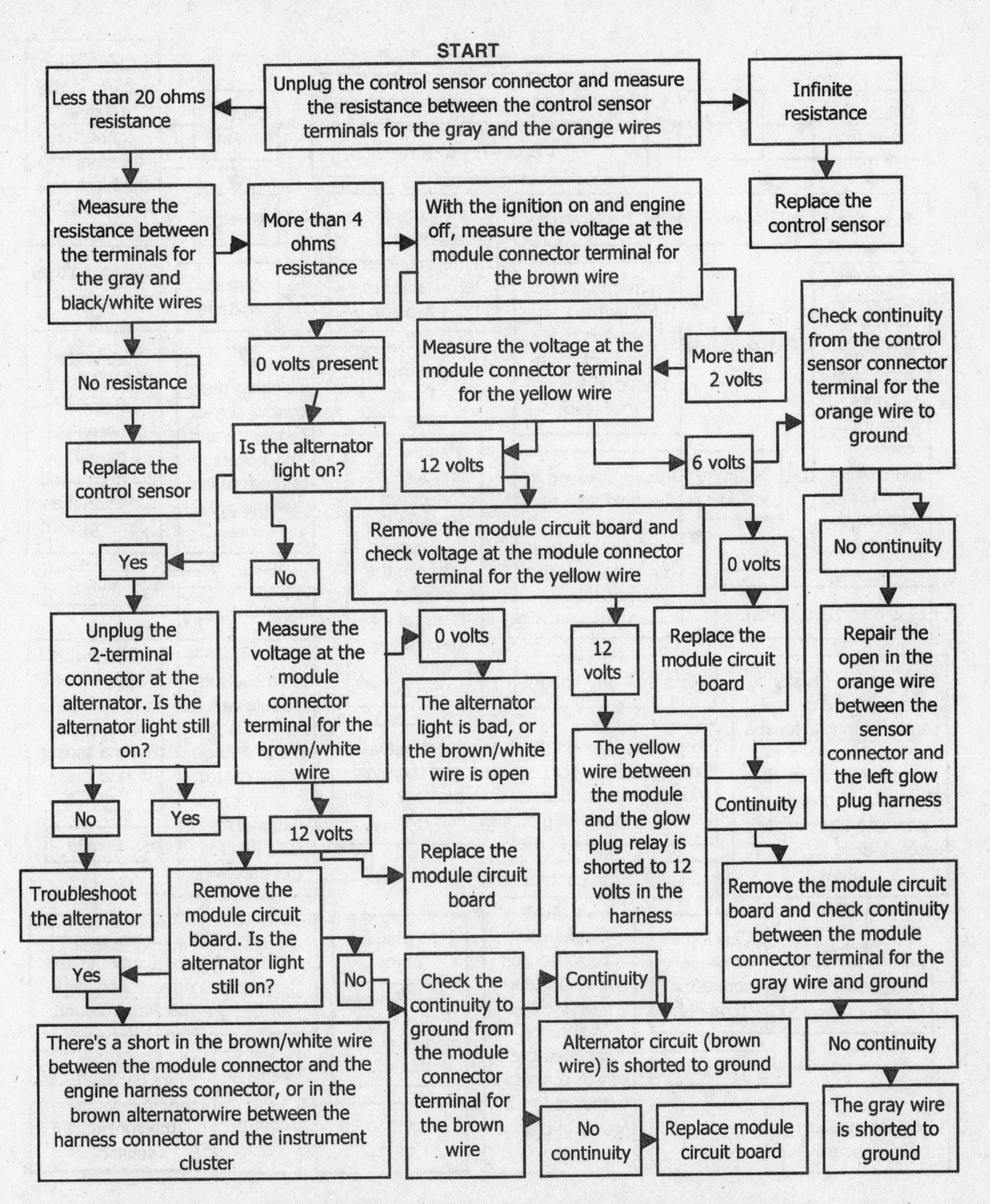

2.81m Flowchart No. 13 (5.7L engines) - The engine is difficult to start when it's cold; OR, the WAIT light is constantly on, and the glow plug relay comes on the first time, and then goes off

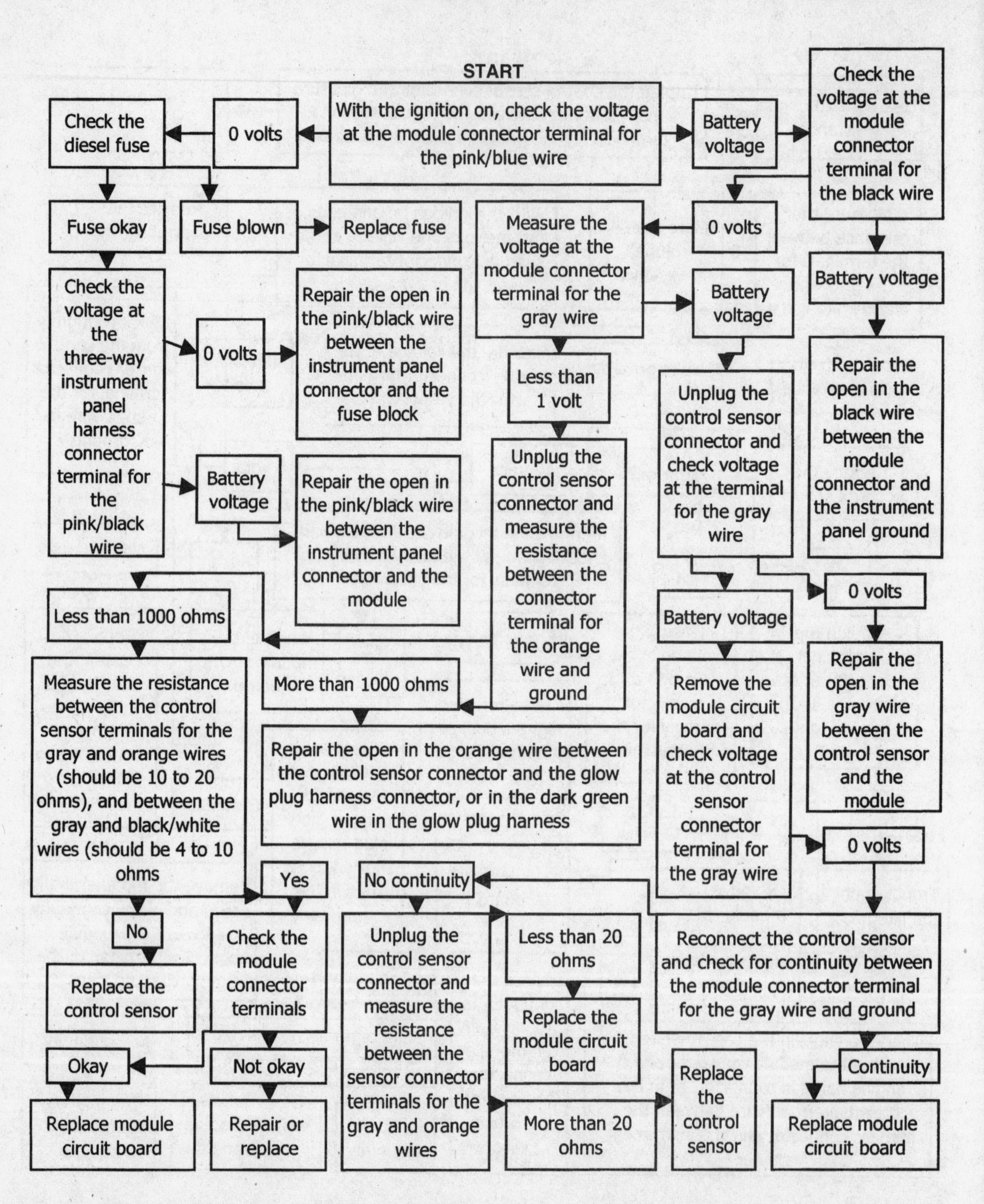

2.81n Flowchart No. 14 (5.7L engines) - The WAIT light doesn't come on and the glow plug relay doesn't operate; OR, the engine won't start when it's cold

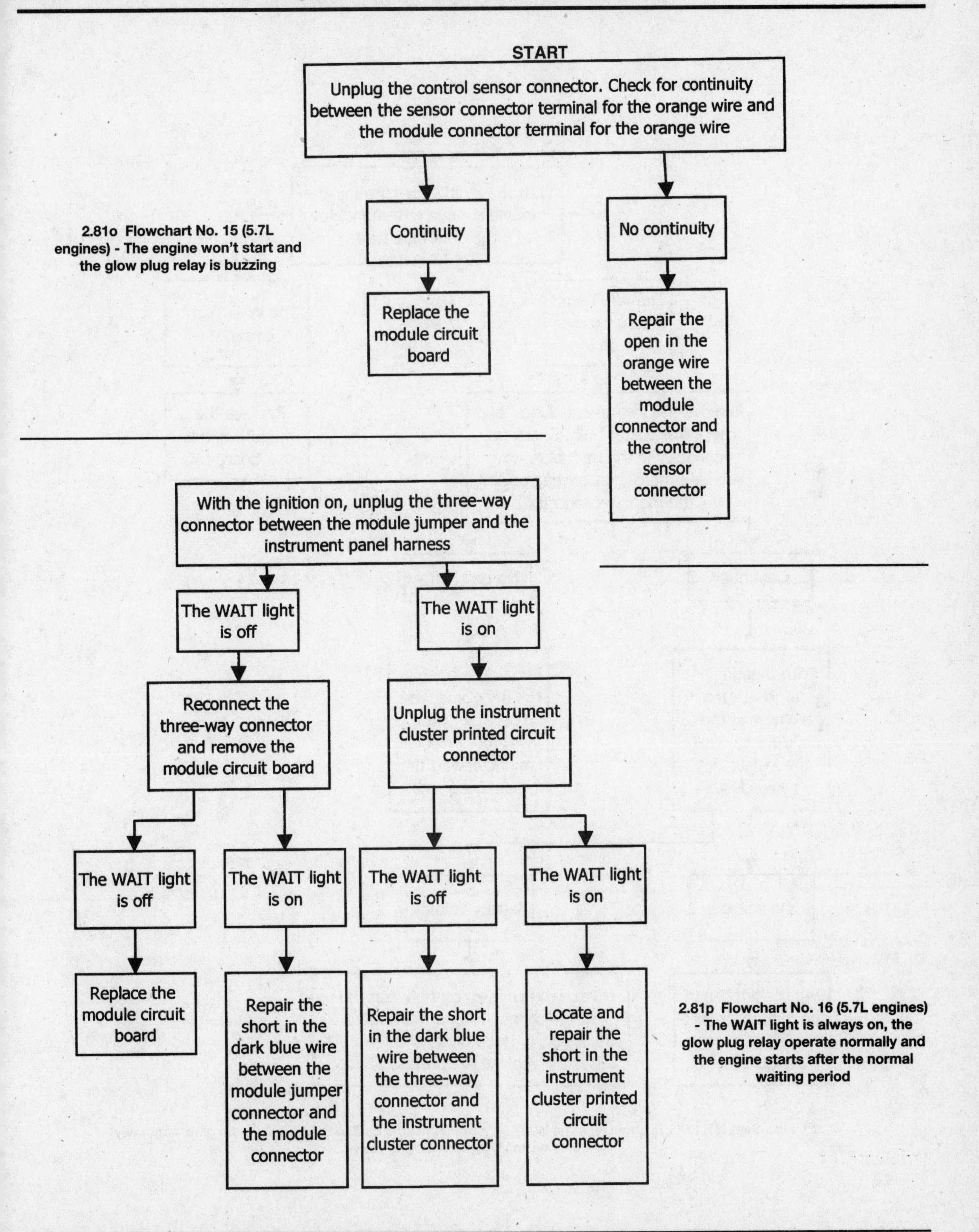

2.81o Flowchart No. 15 (5.7L engines) - The engine won't start and the glow plug relay is buzzing

2.81p Flowchart No. 16 (5.7L engines) - The WAIT light is always on, the glow plug relay operate normally and the engine starts after the normal waiting period

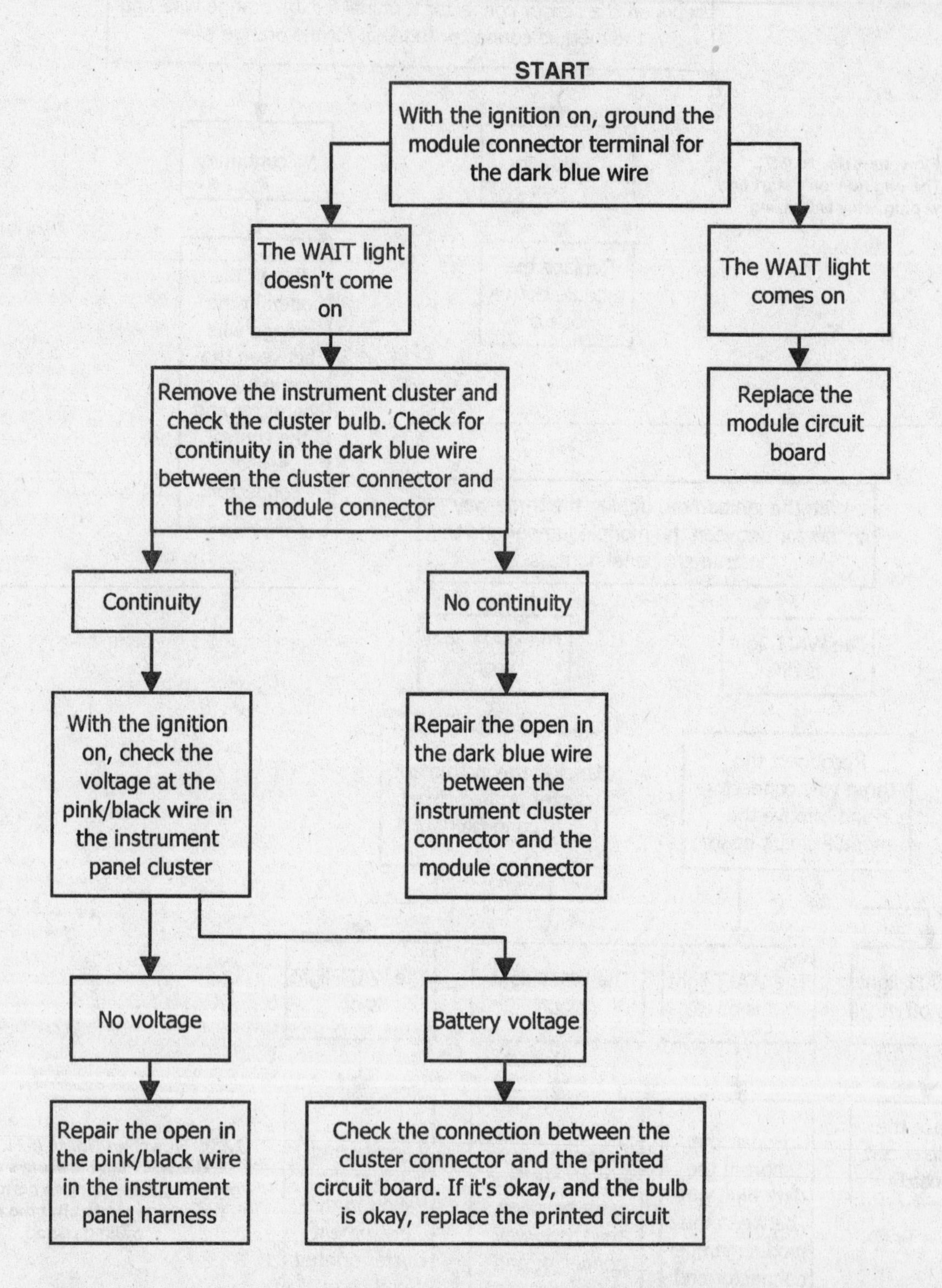

2.81q Flowchart No. 17 (5.7L engines) - The WAIT light doesn't come on (the bulb is okay), the glow plug relay operates normally and the engine starts

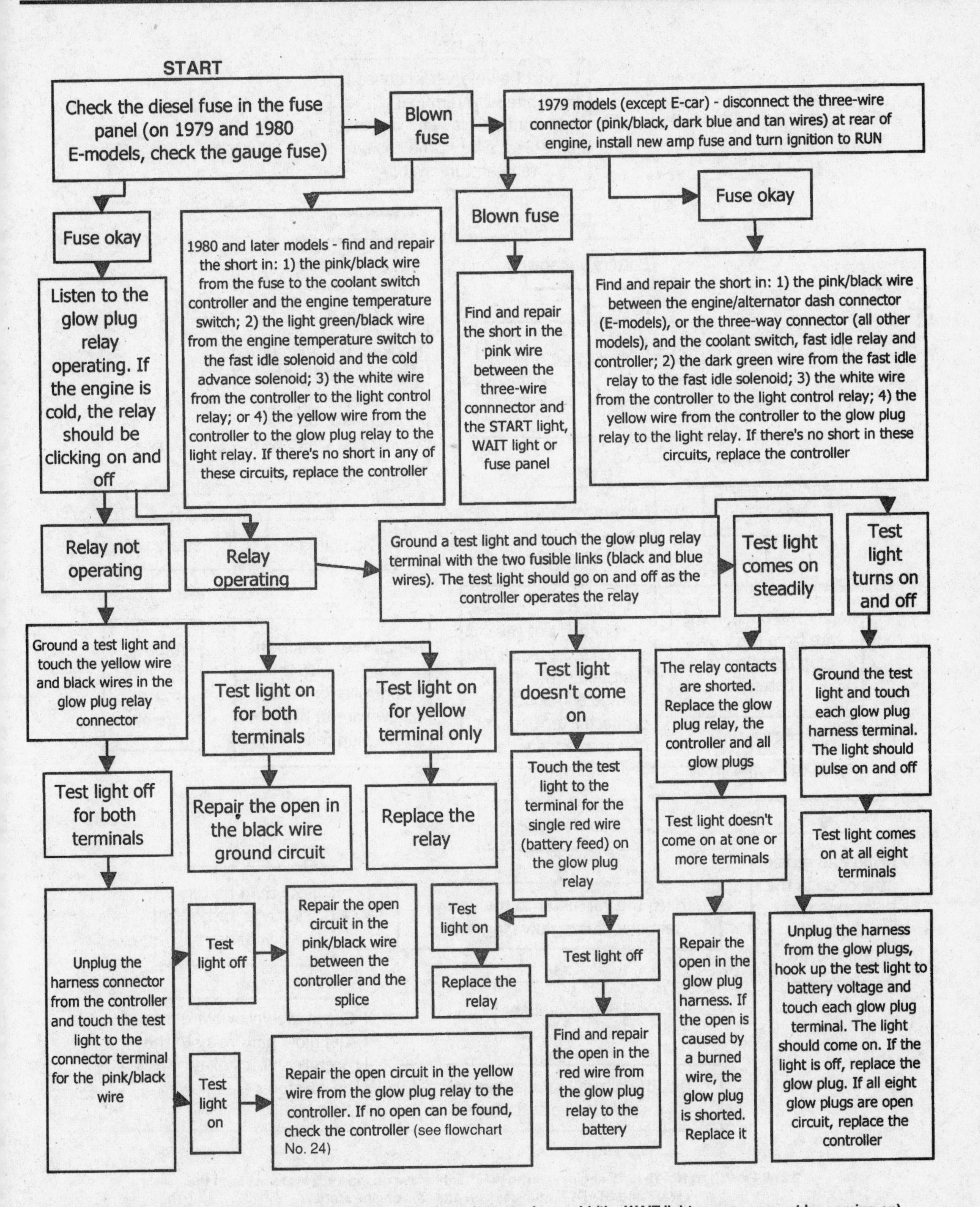

2.81r Flowchart No. 18 (5.7L engines) - The engine doesn't start when cold (the WAIT light may or may not be coming on)

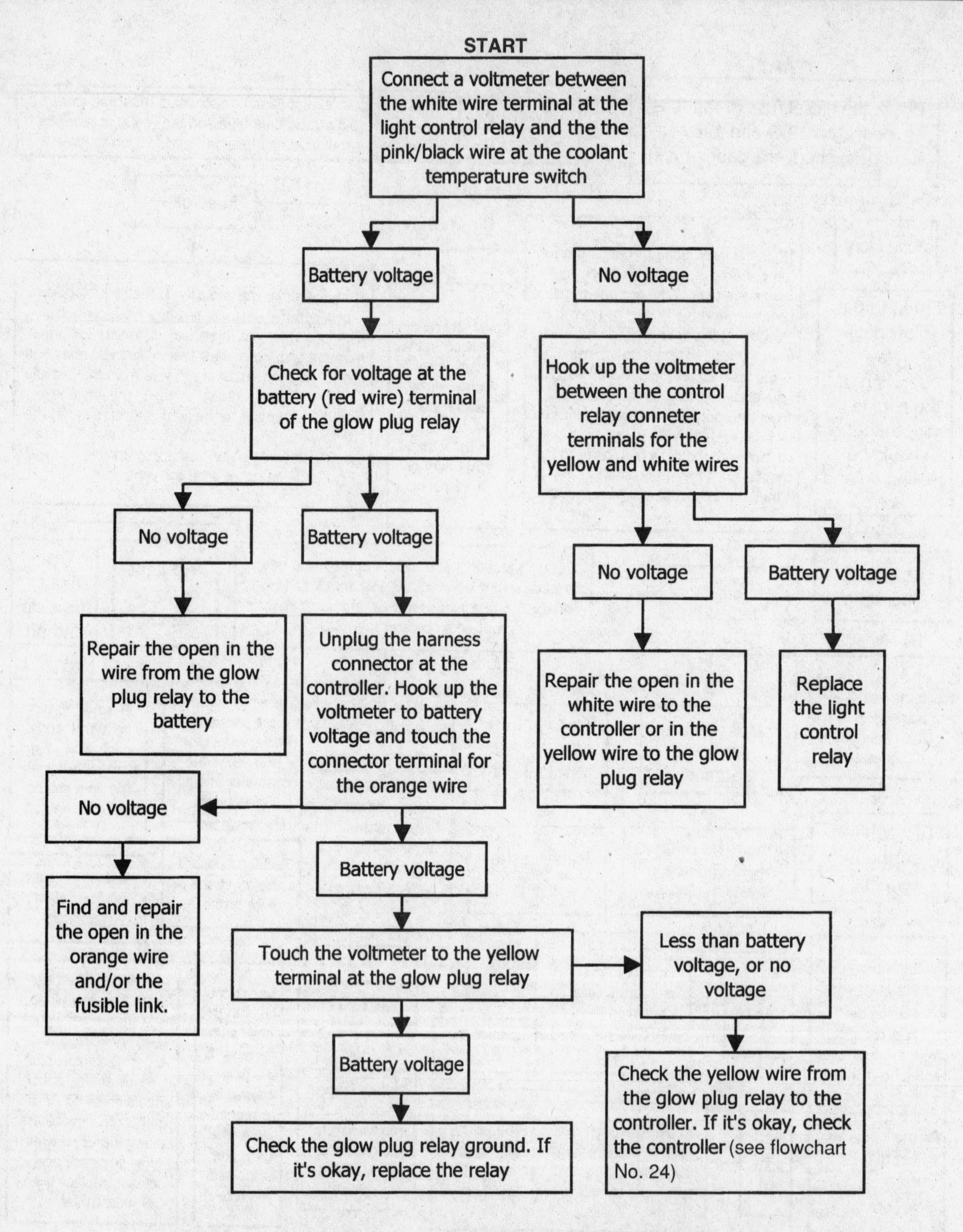

2.81s Flowchart No. 19 (5.7L engines) - The WAIT light stays on (on 1979 systems, both the WAIT and START lights stays on after the engine starts)

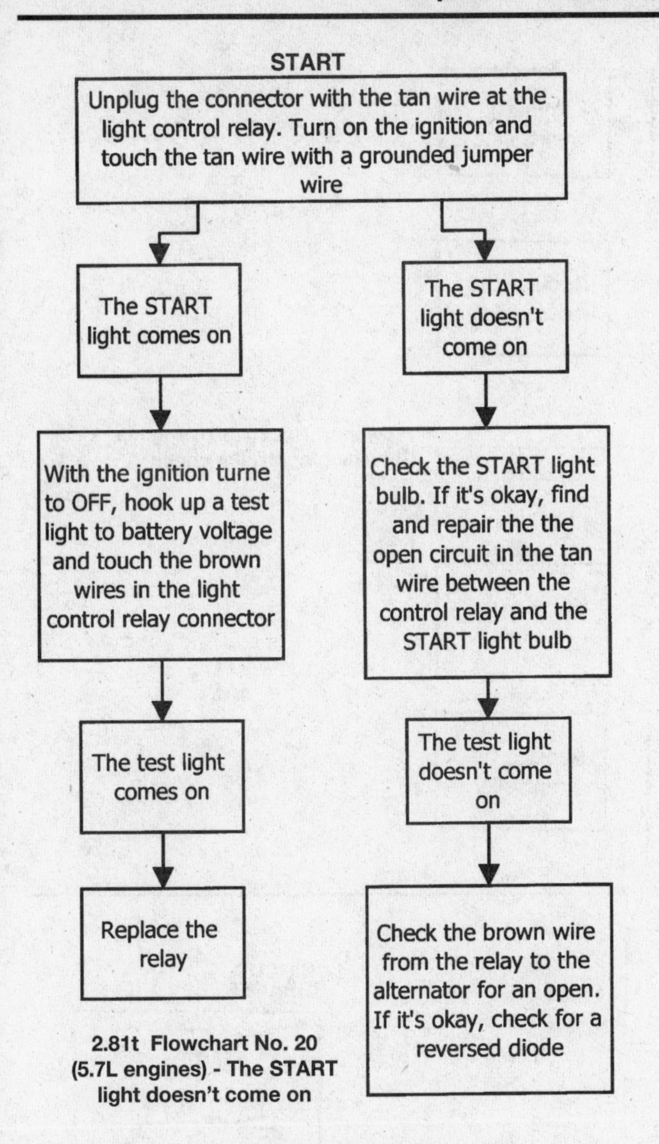

2.81t Flowchart No. 20 (5.7L engines) - The START light doesn't come on

START

Ground a test light and touch the terminal for the yellow wire at the light control relay

The test light doesn't come on

The test light comes on

Repair the open in the yellow wire between the light control relay and the glow plug relay

Replace the light control relay

2.81w Flowchart No. 23 (5.7L engines) - There is no START light, the WAIT light stays on, and the glow plugs operate correctly

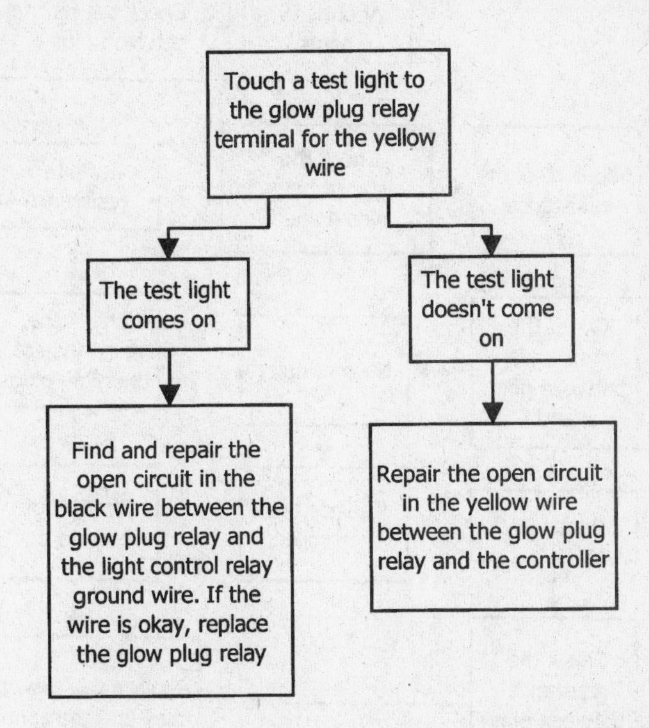

2.81u Flowchart No. 21 (5.7L engines) - The START light comes on immediately on a cold engine, and the glow plugs are off

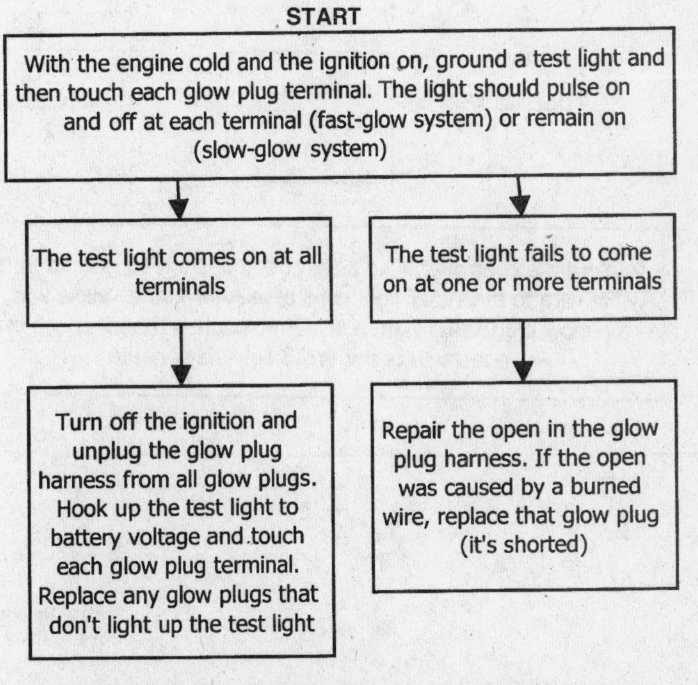

2.81v Flowchart No. 22 (5.7L engines) - The engine runs roughly when started cold (the glow plugs are not on after the engine starts and the fuse is okay)

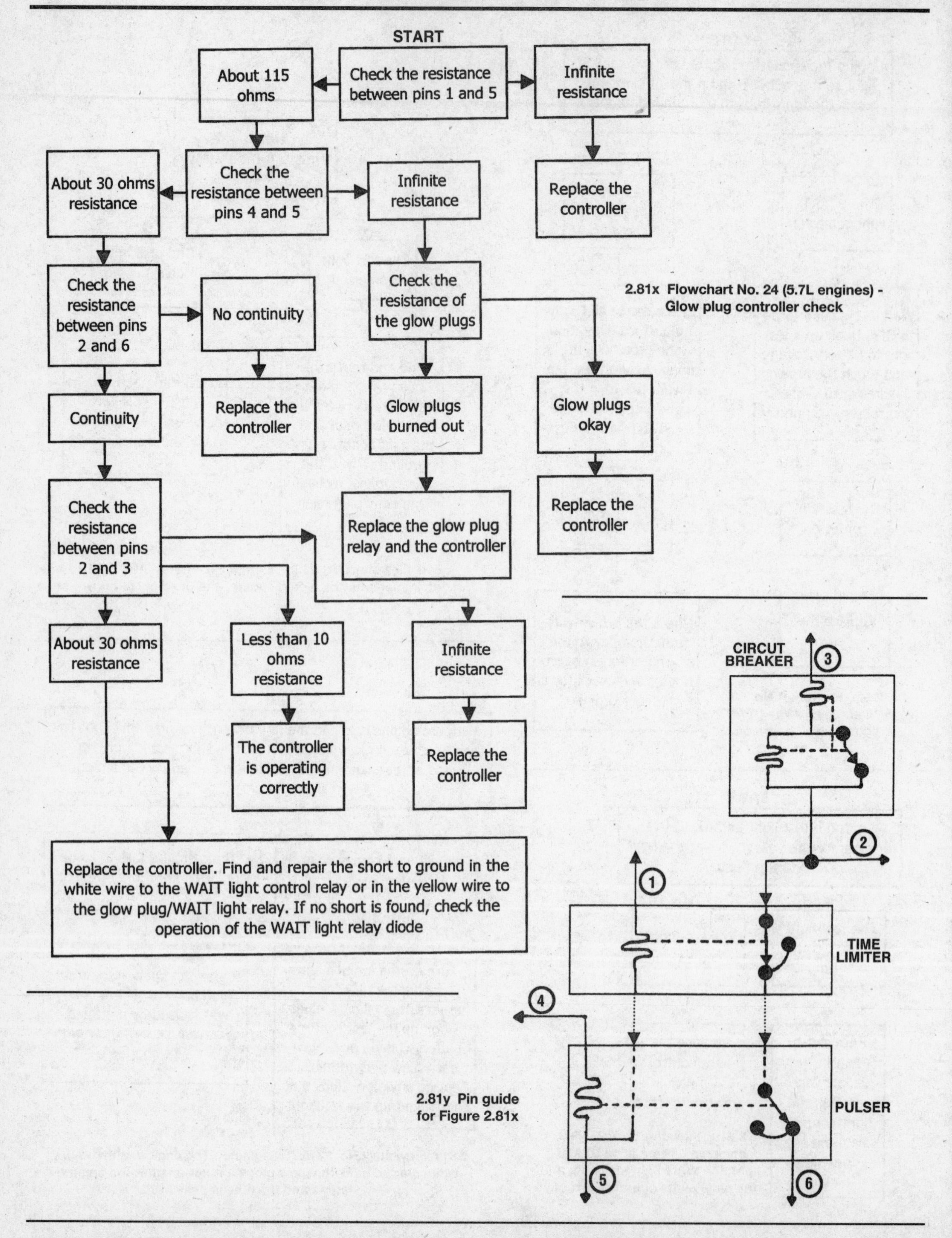

2.81x Flowchart No. 24 (5.7L engines) - Glow plug controller check

2.81y Pin guide for Figure 2.81x

Is the glow-plug system controlled by an electronic module or by an electro-mechanical thermal controller?

Trying to determine what kind of system you've got on your vehicle can be confusing. One quick way to figure out what you're dealing with is to note whether the system is controlled by an electronic module or by an electro-mechanical thermal controller. Look at the right (passenger's side) front corner of the intake manifold and note whether the engine has a single-wire thermistor (module-controlled) or a six-wire glow plug control switch (thermal controller).

System No. 5: Positive Temperature Coefficient (PTC) glow plug system (all 1984 and 1985 car models with 5.7L engines)

The 1984 and 1985 glow plug control circuit **(see illustrations)** uses glow plugs that are protected from burnout by increasing resistance as they get hotter. The circuit operates the glow plug system in the same three steps as other systems: pre-glow, after-glow and off. During pre-glow, the circuit turns on the WAIT light and heats the glow plugs until they're hot enough to start the engine. During after-glow, the circuit turns off the WAIT light and keeps power applied to the glow plugs to help the cold engine run. Once the engine is hot enough to run on its own, the circuit cuts the power to the glow plugs and stays off until the engine is re-started.

The glow-plug controller/relay, or glow-plug module, controls all circuit functions - pre-glow and after-glow operation, the WAIT light, over-voltage production and system reset.

Thermal controls, which open the pre-glow switch to end pre-glow and the after-glow switch to end after-glow, respond to engine temperature. When the system is energized with the engine cold, both switches are in the "cold" position. As the engine heats up, current flow heats the thermal controls, and both switches move toward their "hot" positions. How much time it takes for each switch to move to its hot position depends on how cold the engine and the controller were when the system was first energized. The colder the engine and controller are, the longer the pre-glow time period.

As the pre-glow switch reaches its hot position, the WAIT light goes out. The engine can now be started. Whether the engine is started or not, the controller continues to operate. The after-glow switch hasn't yet reached its hot position. Current continues to flow, heating the thermal controls. When the thermal controls reach their maximum temperature, the after-glow switch moves to its hot position.

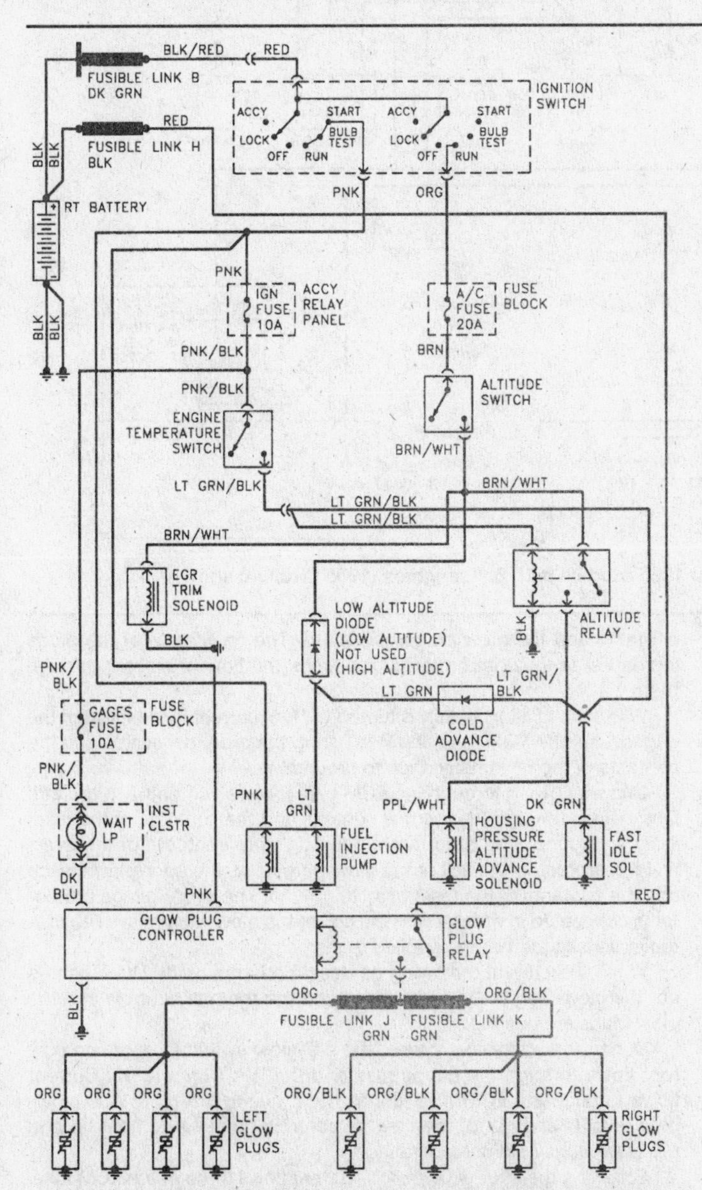

2.82a Typical electrical circuit for System No. 5 used on 1984 models with 5.7L engines (1984 Cadillac shown)

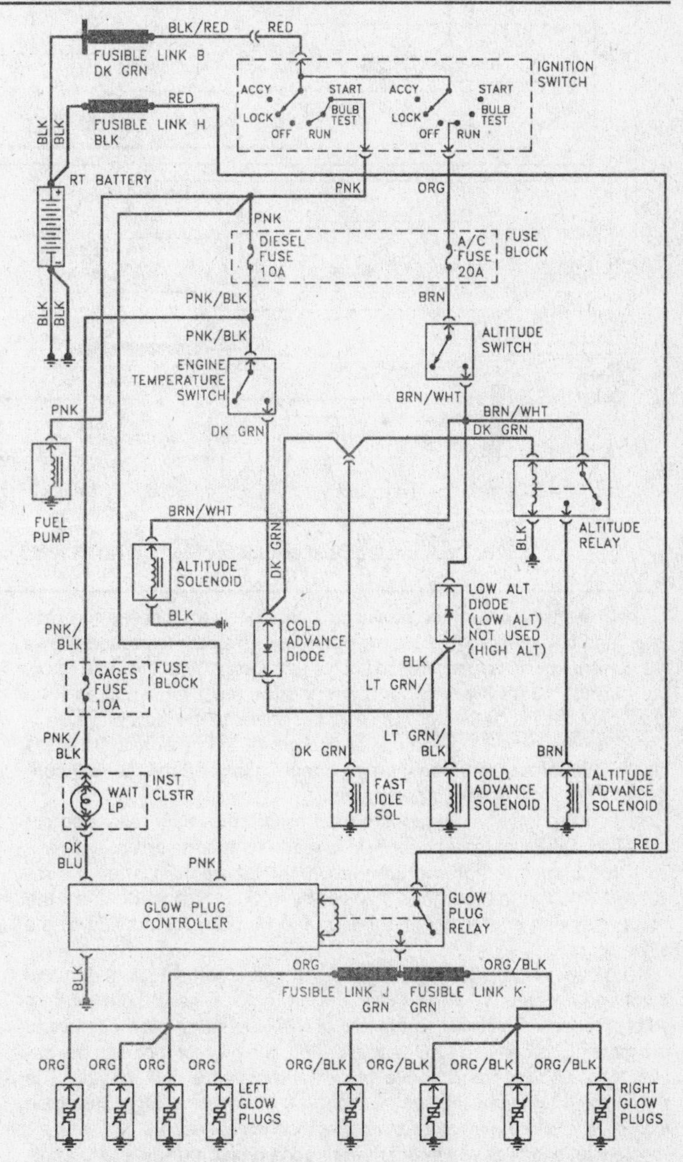

2.82b Typical electrical circuit for System No. 5 used on 1984 models with 5.7L engines (1984 Buick Riviera shown)

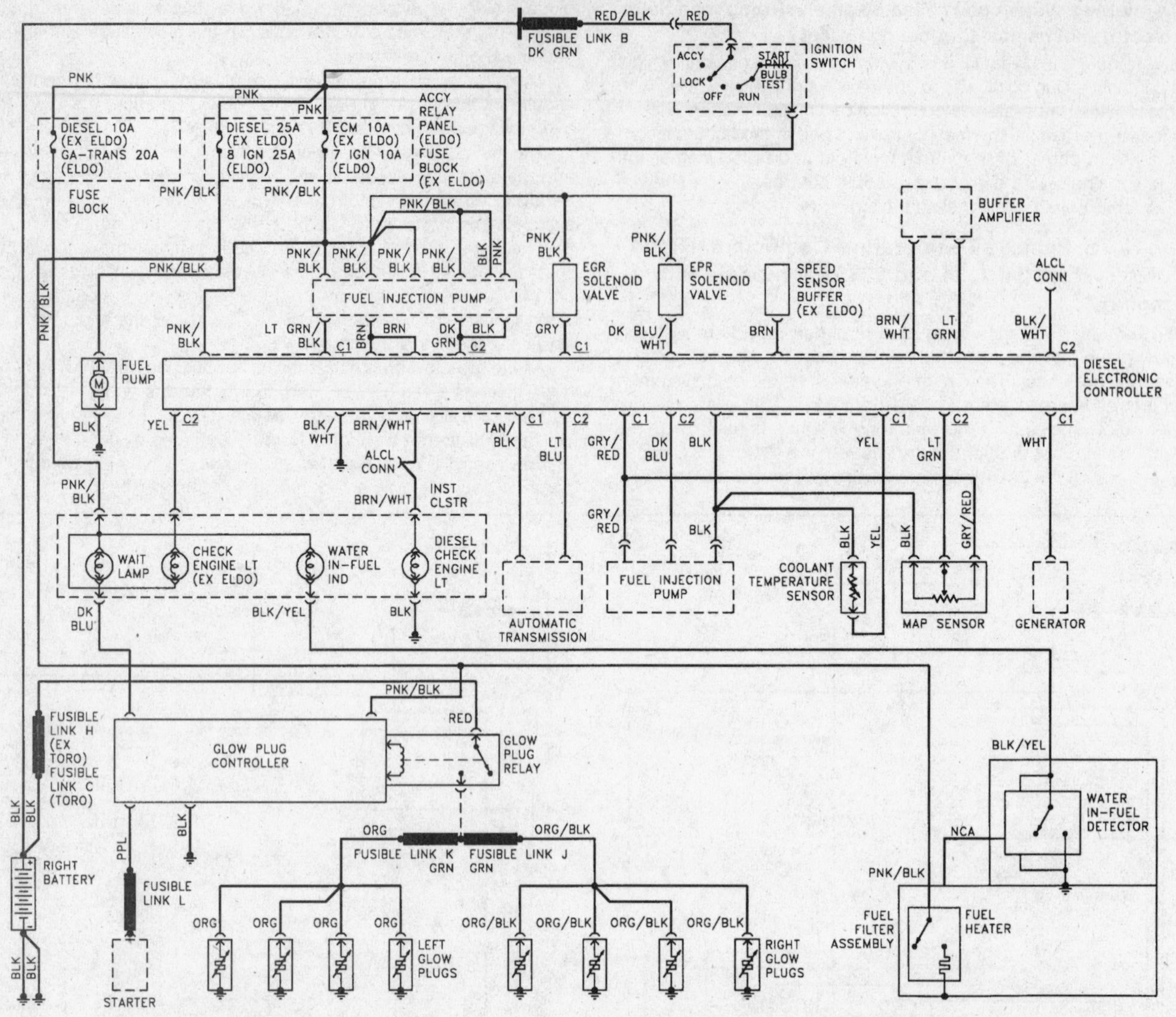

2.82c Typical electrical circuit for System No. 5 used on 1985 models with 5.7L engines (1985 Cadillac shown)

As the after-glow switch moves to its hot position, it opens the path to ground from the coil of the glow plug relay. This path to ground allows the current flow to bypass the coil of the reset relay. With this path open, the current must flow to ground through the reset relay coil and the glow-plug relay is de-energized, removing power to the glow plugs.

When the glow-plug relay is de-energized, the reset relay re-energizes. The contact of the reset relay opens to lock off the thermal controller. This ends glow-plug operation.

When the ignition switch is turned to Off, the reset relay contact closes and the glow-plug controller is ready to operate again.

If the engine is hot enough (above 140-degrees F) when it's restarted, the thermal controls will still be in their hot position, so the reset relay can energize immediately, and no power is applied to the glow plugs.

The over-voltage protector protects the glow plugs against damage if voltage exceeds 14 volts. When the protector senses more than 14 volts, it opens the circuit to the coil of the glow-plug relay. Current to the glow plugs ceases. After a brief period, the protector closes the circuit. If the voltage is still above 14 volts, it opens the circuit again. The protector may cycle this way as long as the over-voltage condition exists - and as long as glow-plug operation is needed.

Now let's look at where the current goes during pre-glow and after-glow and following after-glow in a PTC system. During pre-glow, voltage is applied at all times from the battery to the contacts of the glow-plug relay. When the relay is energized, current flows through the relay contacts and the glow plugs to ground. The resistance of the plugs increases as they get hotter, which limits the flow of current through them.

When the ignition switch is turned to Run, current flows through the gages fuse, the ECM fuse, the WAIT light, the pre-glow switch and the contacts of the reset relay, then to ground.

Current flowing through the ECM fuse follows two paths to ground: One path is through the thermal controls and the contacts of the reset relay to ground. The other path is through the contacts of the over-voltage protector, the coil of the glow-plug relay, the after-glow switch and the contacts of the reset relay to ground. The over-voltage protector is connected in parallel with the coil of the glow-plug relay. The protector senses the voltage applied to the coil.

When the current is flowing as described above, the WAIT light is on, the glow-plug relay is energized and the thermal controls and the glow plugs are heating.

During the after-glow phase, the pre-glow switch is open, no current flows through the gauge fuse and the WAIT light is off. Current flows through the ECM fuse just as it did during pre-glow. The glow-plug relay is energized, the thermal controls continue to heat up and the glow plugs continue operating.

Following the after-glow period, current flows through the ECM fuse, the contacts of the over-voltage protector, the coil of the glow plug relay (which doesn't energize) and the coil of the reset relay (which does energize) and to ground. No current flows through any of the glow plugs.

Diagnosing the PTC glow-plug system

Condition 1: When the engine is cold, there's no WAIT light

1 With the ignition in the Run position and the engine off, note whether the Charge indicator light is on. If it isn't, check the gages fuse. If the gages fuse is good, check the operation of the glow-plug relay by listening for a clicking sound. If the relay is clicking, check for a break in the pink and black wire between the fuse block and splice S204 **(see illustrations 2.82a and 2.82b)**. If the relay doesn't operate, check the pink wires between the ignition switch and the fuse block.
2 Unplug the connector of the thermal controller. Connect the dark blue wire (pin A) at the connector to ground with a jumper wire. If the WAIT light doesn't go on, note whether the bulb is burned out. Check for a break in the dark blue wire between the thermal controller connector and the WAIT light, and in the pink and black wire between the WAIT light and splice S204.
3 Check the connection at ground G104. Hook up one lead of a test light to the red wire terminal of the glow-plug relay and the other lead to ground. Then touch the lead to the black wire (pin C) of the connector at the thermal controller. If the light comes on, replace the thermal controller assembly.

Condition 2: The WAIT light remains on more than 15 seconds

1 With the ignition turned off, check the battery voltage. Charge the batteries if they're below 10.5 volts (the temperature must be above 0-degrees F to charge the batteries).
2 Turn the ignition switch to Run, then to Off. Listen for a clicking sound from the glow-plug relay. If the glow-plug relay is operating, remove the thermal controller assembly and replace it.
3 If the glow-plug relay doesn't operate, check the ECM fuse. Unplug the connector at the thermal controller. Hook up a test light between the black and pink wire (pin B) and the black wire (pin C) at the harness connector. Turn the ignition switch to Run. If the test light doesn't come on, check the continuity of the pink and black wires.
4 If the test light comes on when the ignition is turned on, check the continuity between pins B and C of the thermal controller. It should be less than 75 ohms. If there's no continuity, remove the thermal controller assembly from the vehicle, disconnect the glow-plug relay and replace the thermal controller.

Condition 3: The engine doesn't start when it's cold (even though the WAIT light is okay - it goes on, then off)

1 Is the cranking speed at least 100 rpm? If it isn't, check battery voltage with a voltmeter. It should be 12.4 volts with the ignition off.
2 Using a test light, check for voltage at the pink wire lead at the fuel solenoid with the ignition in Run. If there's no voltage, repair the pink wire. If the test light comes on, turn off the ignition and check for continuity through the fuel solenoid to ground with a self-powered test light. If there's no continuity, replace the fuel solenoid.
3 Check the operation of the glow-plug relay. Touch one probe of a test light to ground and the other probe to the red wire post of the glow-plug relay. With the ignition on, touch the probe to the green wire post of the glow-plug relay. The test light should come on if the relay operates. If it does, go to the next step; if it doesn't, go to step 7.
4 If the glow-plug relay operates, turn the ignition off. Unplug all eight connectors to the glow plugs. Hook up a self-powered test light between the green wire post of the glow-plug relay and ground. At each glow plug, touch the harness connector to the spade terminal of the glow plug. If the test light comes on, the glow plug and harness lead are good. Test all eight glow plugs this way. Make sure you set each connector aside after testing its respective glow plug to prevent accidentally energizing the plug. When you're finished, reconnect all eight connectors to their respective glow plugs.
5 If the test light doesn't come on, touch the harness connector to the engine block or some other suitable ground. If the light then comes on, replace the glow plug. If the light doesn't come on, replace the wire to the glow plug.
6 If all eight glow plugs are open-circuited, replace the thermal controller.
7 If the glow-plug relay doesn't operate, and the WAIT light is going on and off, check the circuit between the thermal controller and the glow-plug relay.
8 Remove the thermal controller assembly and the glow-plug relay and have them bench-tested by a dealer service department. Replace as necessary.

System No. 4 (early): Fast-glow, thermal controller (1982 through 1984 6.2L models)

The circuit for this system **(see illustration)** includes an electro-

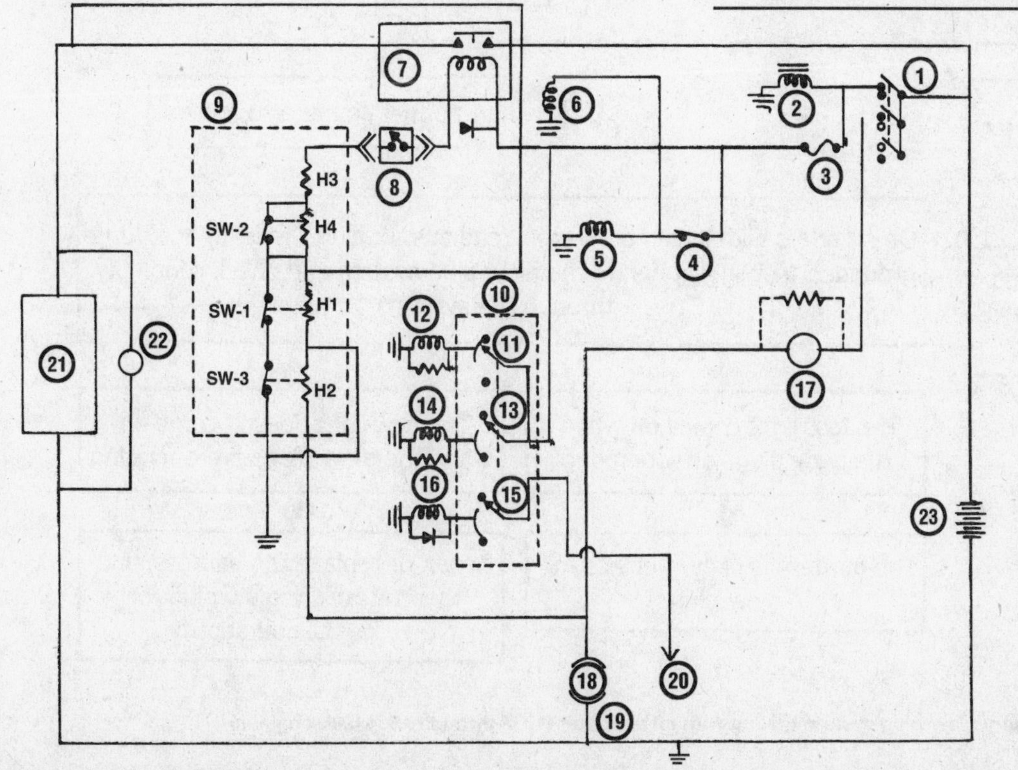

2.83 Typical electrical circuit for System No. 4 (early): Fast-glow, thermal controller (1982 through 1984 6.2L engines)

1 *Ignition switch*
2 *Fuel solenoid*
3 *Amp gauge idle fuse*
4 *Fast idle and cold advance temperature switch*
5 *Cold advance solenoid*
6 *Fast idle solenoid*
7 *Glow plug relay*
8 *Temperature inhibit switch*
9 *Glow plug controller*
10 *Throttle switch (light-duty vehicles only)*
11 *Switch (closed)*
12 *EPR solenoid*
13 *WOT switch*
14 *EGR solenoid*
15 *Switch*
16 *TCC solenoid*
17 *Alternator indicator light*
18 *Alternator light output*
19 *Alternator*
20 *To brake switch*
21 *Glow plugs*
22 *Glow plug indicator light*
23 *Battery*

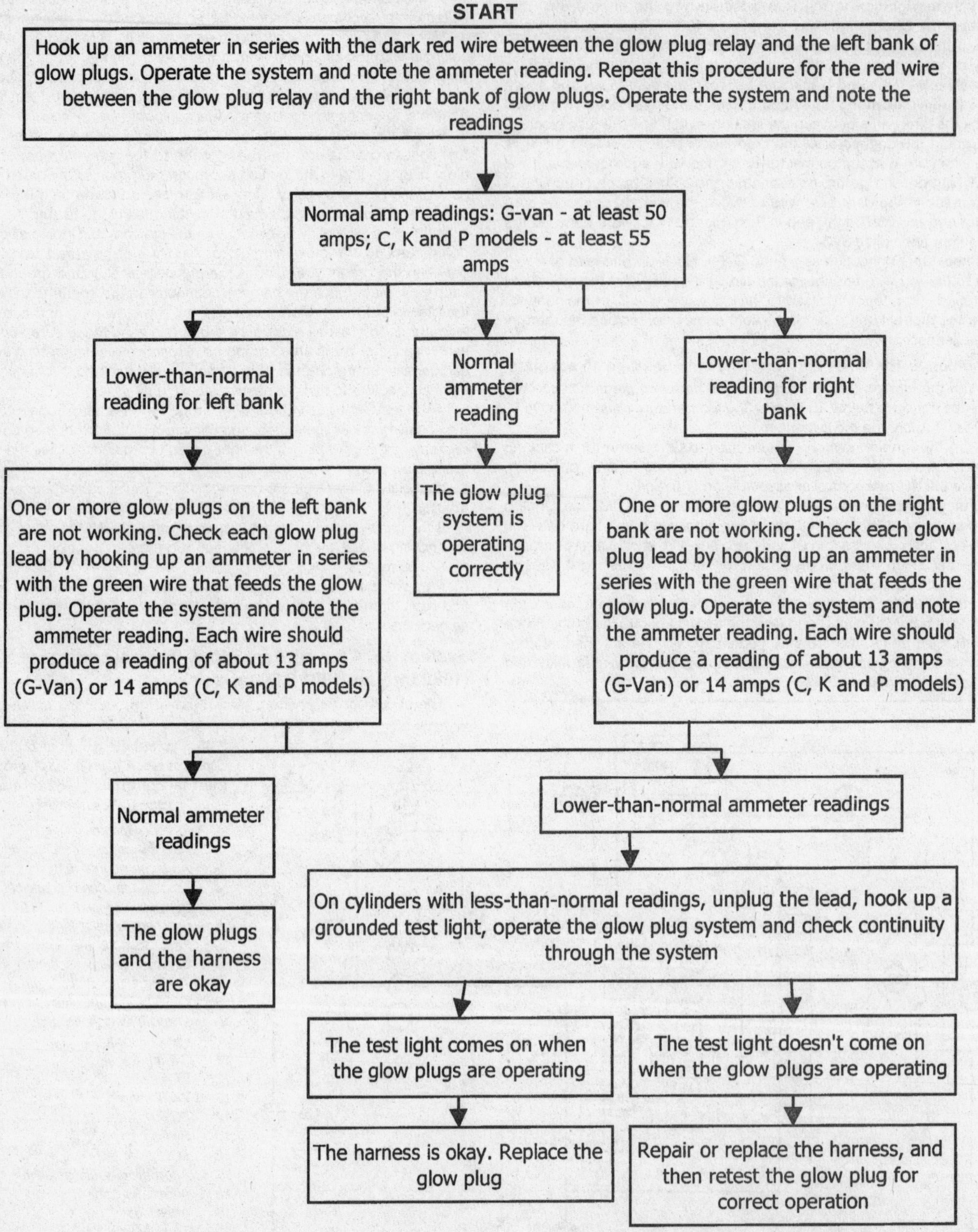

START

Hook up an ammeter in series with the dark red wire between the glow plug relay and the left bank of glow plugs. Operate the system and note the ammeter reading. Repeat this procedure for the red wire between the glow plug relay and the right bank of glow plugs. Operate the system and note the readings

Normal amp readings: G-van - at least 50 amps; C, K and P models - at least 55 amps

Lower-than-normal reading for left bank

One or more glow plugs on the left bank are not working. Check each glow plug lead by hooking up an ammeter in series with the green wire that feeds the glow plug. Operate the system and note the ammeter reading. Each wire should produce a reading of about 13 amps (G-Van) or 14 amps (C, K and P models)

Normal ammeter reading

The glow plug system is operating correctly

Lower-than-normal reading for right bank

One or more glow plugs on the right bank are not working. Check each glow plug lead by hooking up an ammeter in series with the green wire that feeds the glow plug. Operate the system and note the ammeter reading. Each wire should produce a reading of about 13 amps (G-Van) or 14 amps (C, K and P models)

Normal ammeter readings

The glow plugs and the harness are okay

Lower-than-normal ammeter readings

On cylinders with less-than-normal readings, unplug the lead, hook up a grounded test light, operate the glow plug system and check continuity through the system

The test light comes on when the glow plugs are operating

The harness is okay. Replace the glow plug

The test light doesn't come on when the glow plugs are operating

Repair or replace the harness, and then retest the glow plug for correct operation

2.84 Preliminary glow-plug system testing with an ammeter (1982 through 1984 6.2L engines)

thermal controller, a glow-plug relay and a glow-plug temperature inhibit switch. The glow-plug relay is pulsed on and off by the controller to control current flow to the glow plugs and maintain glow-plug temperature to prevent overheating.

The electro-thermal controller, which threads into the water jacket, senses engine coolant pressure and contains small electric heaters to operate three bimetal switches. Four electrical heaters inside the controller alternately heat up and cool down, causing the bimetal switches to open and close the glow-plug relay circuit.

The circuit for heater H-1 is fed by the glow-plug side of the glow plug relay, and gets hot when the glow plugs get hot. It has a resistance of 300 ohms. This causes switch S-1 to open at 180-degrees F, then close, to pulse the glow-plug relay coil and the glow plugs.

Heater H-2 is energized by alternator output and gets hot when the engine runs. It has a resistance of 115 ohms. Heater H-2 opens switch S-3 at 160-degrees F. and keeps the glow-plug relay e-energized when the glow plugs are no longer needed. Switch S-2 is activated by heat from heaters H-3 and H-4. H-3 has a resistance of 45 ohms, H-4 a resistance of 32 ohms.

The operating temperature of switch S-2 is about 300-degrees F. Heater H-3 is a low-heat unit and heater H-4 only supplies heat when switch S-2 is open. Switch S-2 is also affected by engine coolant temperature: The engine operating temperature, in conjunction with the heat from heater H-3 and water temperature will open switch S-2. This causes heater H-4 to warm up and maintain switch S-2 in its open position.

When the ignition switch is turned to the On position, current flows through the gage-idle fuse, the 3 ohm glow-plug relay coil, the closed temperature switch and closed controller switches S-2, S-1 and S-3 to ground. This energizes the glow-plug relay, which connects the battery to the eight glow plugs. The "glow plugs" light is wired in parallel with the glow plugs - when they cycle on and off, so does the glow plugs light.

When heater H-1 of the controller reaches 180-degrees F, switch S-1 opens, de-energizing the glow plug relay, and turns off the glow plugs. As H-1 cools off, switch S-1 will close again if the engine isn't started, and turn on the glow plugs. The glow plugs and light will continue to cycle on and off in this controlled heating mode until the engine is started.

Once the engine is running, the glow plugs light may pulse for a brief period, which is determined by alternator output. The total length of the after-glow period is controlled by a signal from the alternator to the control system. Current from the alternator entering the controller at pin 1 will cause heater H-2 to get hot and open switch S-3 at 160-degrees F. This is the end of the after-glow period.

Controller contamination on early 1982 models with 6.2L engines

On early 1982 models with 6.2L engines, there's a hole at the no. 2 pin of the engine wire harness connector for the controller. Dirt and moisture can enter this hole and corrode the pin connections, causing the controller to malfunction.

If you've got an early 1982 6.2L model, and have experienced poor starts, burned glow plugs, etc., remove the connector from the controller and inspect it for moisture and dirt. Clean the pin areas on the controller and the connector and reinstall the connector. Apply a small amount of RTV sealant over the hole at the No. 2 pin to prevent any more dirt or water from entering.

If the controller now recycles correctly, the problem is solved; if it doesn't, corrosion of the pins was probably excessive. Replace the connector. The two wires from the No. 5 and No. 6 pin positions are ground wires. The ground connection is at the rear of the right head on a stud that also grounds the body ground strap. The stud is on the opposite side of the engine from the controller. Make sure the ground connection is secure.

Diagnosing System No. 4 problems (1982 through 1984 models with 6.2L engines)

You'll need the following tools to perform the diagnostic procedures contained below - a digital multimeter, a self-powered test light, a 12-volt test light and a volt-ampere tester with an inductive pick-up. You'll need the digital multimeter to measure extremely low voltages without burning out the delicate circuits in the electronic controller. The self-powered test light is necessary for checking glow-plug continuity in circuit checks where no voltage can be applied. The 12-volt test light is handy for checking normal 12-volt circuits. You need the volt-amp tester for measuring amperage draw between left and right glow-plug banks. **Note:** *To get a good grasp of how System No. 4 works, read through the entire diagnostic section before tackling any troubleshooting procedures. It's essential that you understand the interrelationship of the various system components before trying to track down a problem.*

Preliminary checks

Note: *Make sure you read this entire section, steps 1 through 5, before proceeding.*

1 Check the operation of the glow plugs. With the ignition switch in the Run position, the glow plugs light should come on, and stay on, for 8 to 10 seconds. When it goes out, start the engine. The light should then cycle a few times and go out. If the system doesn't cycle, or continues to cycle, repair the glow-plug system. **Note:** *All 1984 through 1989 systems have a temperature inhibit switch, which prevents glow-plug operation when the engine coolant temperature is above 125-degrees F.*

2 Verify that all the glow plugs are working correctly by performing a preliminary diagnosis with an ammeter **(see illustration)**. If the glow plugs are working correctly, the problem is somewhere else; if they're not working correctly, refer to the accompanying troubleshooting charts **(see illustrations)**.

3 Check all fuses, bulbs and grounds before you replace any components.

4 Make sure that no wires are exposed and all connections are clean and dry.

5 Finally, there are a few other things to look at before you get into the troubleshooting procedures: Is the pre-glow timer functioning correctly? Does the system have the right controller? Is the engine timing advanced? These questions must be answered satisfactorily before tackling the accompanying diagnostic procedures.

Is the pre-glow timer functioning correctly?

The glow-plug controller has three internal circuits - a pre-glow timer, an after-glow timer and a circuit breaker. Failure of the pre-glow timer will cause the circuit breaker to take over. During a cold start, you can determine whether the pre-glow timer or the circuit breaker is operating the glow plugs by watching what the glow plugs light on the dash does.

If the pre-glow timer is working properly, the glow plugs light will continue cycling on and off as long as the ignition key it turned to On and the engine is off. If the pre-glow timer isn't working, the glow plugs light will ONLY CYCLE ONCE with the ignition turned to On and the engine off. If the circuit breaker is controlling the glow plugs, replace the controller.

Caution: *Prolonged operation of the glow plugs with the circuit breaker will result in premature failure of the plugs.*

Does the system have the right controller?

The controller for 6.2L engines has a light gray connector and a silver label; the 5.7L controller has a black connector and a gold label. These two controllers are NOT interchangeable. They have different resistance values.

Is the engine timing advanced?

Advanced fuel injection pump timing causes higher-than-normal cylinder temperatures, which will cause the glow plugs to fail. When the timing is advanced, some of the glow plugs will not operate. A normal looking glow plug that is electrically open, or has a small blister, could be due to a glow-plug or system default. But when a glow plug tip has been burned off, it's most likely the result of advanced engine timing.

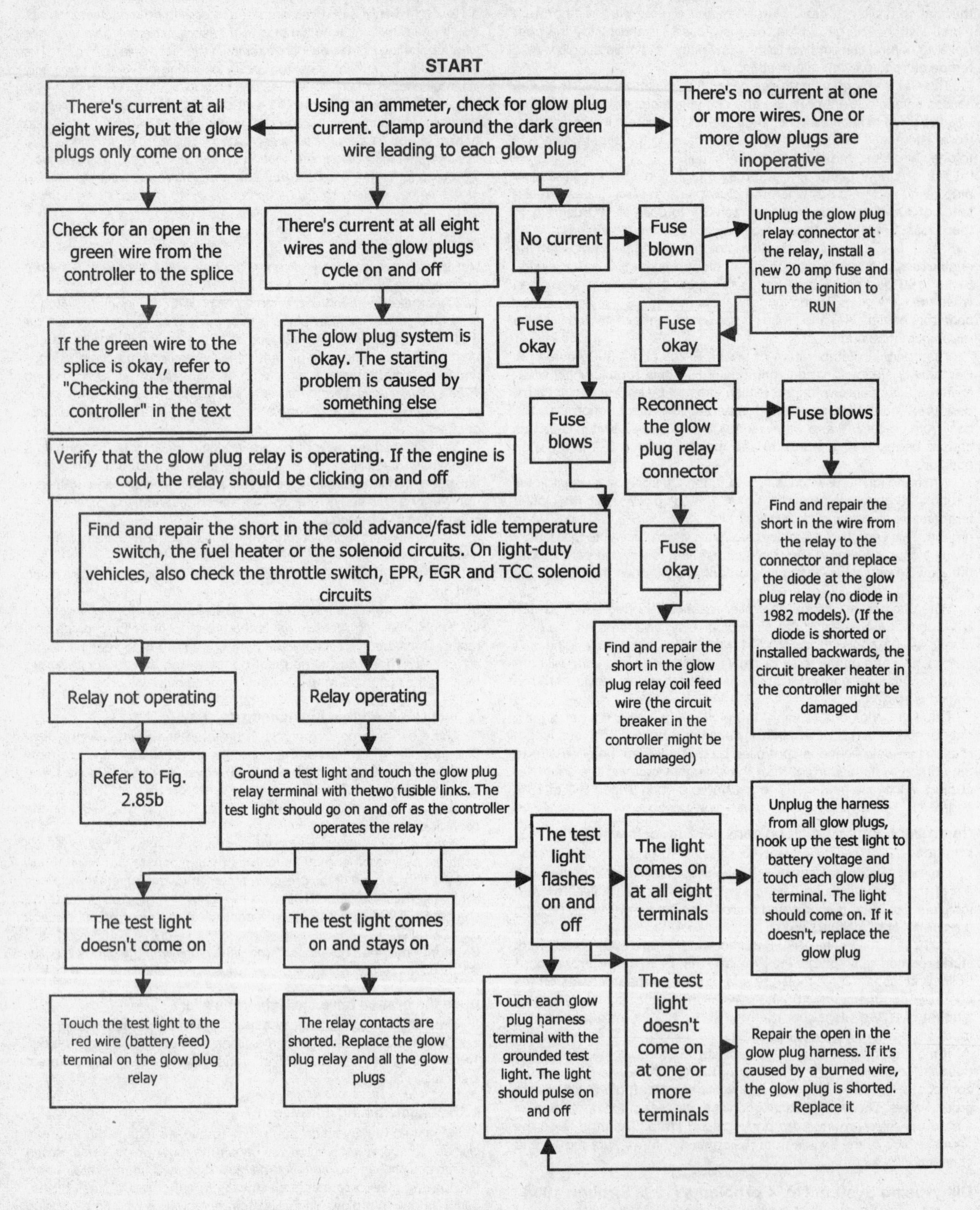

2.85a Glow-plug system troubleshooting (1982 through 1984 6.2L engines)

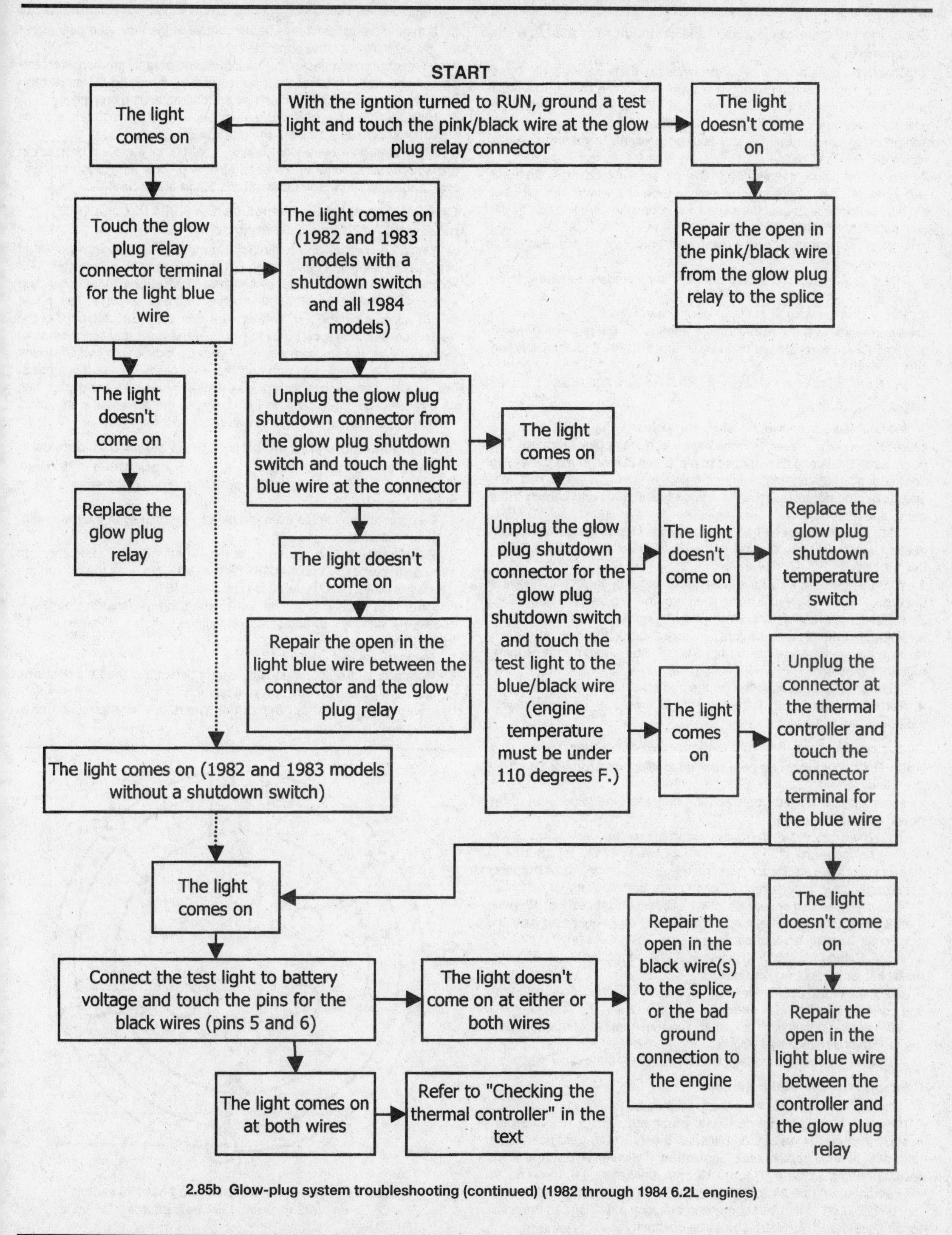

2.85b Glow-plug system troubleshooting (continued) (1982 through 1984 6.2L engines)

Checking the system circuits at the connector for the glow-plug controller

1 Check the resistance at pin 1 (brown wire). If the engine's off, there should be zero resistance in the brown wire to the alternator. Now check the voltage at pin 1 with the engine running. There should be alternator voltage at the pin. Now check the voltage at pin 1 with the engine off. If there's still battery voltage present, the alternator has blown a diode (see below).
2 Pin 2: There's no connection to this pin on 1982 vehicles. On 1983 and 1984 vehicles, there may be a white tube or a wire to this pin, but it's only purpose is to seal the connector in position. It's hooked to the wire harness. DO NOT GROUND THIS PIN FOR ANY REASON, or you will upset the operation of the controller and may even damage the glow plugs.
3 The blue wire at Pin 3 should have battery voltage when the ignition key is turned on.
4 Pin 4 (orange wire 82MY and dark green wire 83, 84MY) should have continuity with the twin-red-wire terminal on the glow-plug relay.
5 Pin 5 (black wire) should have continuity with Pin 6 (black wire) and with ground.
6 Pin 6 should have continuity with Pin 5 (see previous step).

When a diode dies

When the diode fails in the alternator, battery voltage can get to the glow-plug controller even with the key off and the engine stopped! This battery voltage is usually supplied to pin 1 (see above) in the glow-plug controller from the alternator after the engine starts. In effect, this voltage "tells" the controller that the engine has started and shuts off the glow-plug heating cycle.

If the engine has been hard to start, and no glow plugs have been heating, and you have verified that a diode failed (see above), follow this procedure:
1 With the ignition key off and the engine stopped, check for voltage in the brown wire that goes from the alternator to the glow-plug controller.
2 If voltage is present, disconnect the wire from the alternator. Check for voltage in the brown wire again - there shouldn't be any. (If the key is on, a low voltage - about three volts - will be present in a properly functioning system).
3 Wait for at least 15 minutes (so the controller can cool off).
4 Turn the key on again. If the glow plugs now heat, the problem was a diode in the alternator.

Checking the operation of the glow-plug controller

Note: *The engine must be cold and the ignition switch must be off for the following test.*
1 Hook up a multimeter between the negative side (blue wire) of the power relay and ground.
2 Turn the ignition switch on and watch the meter. Depending on the coolant temperature, it should read about two volts DC for about 4 to 10 seconds. If the engine isn't completely cooled down, you'll probably get a continuous voltage reading of 12 volts (battery voltage).
3 This initial reading should be followed by an on-off cycle of 12-volts, then 2-volts, then 12-volts, then 2-volts, etc. You should also hear the power relay making an audible clicking sound as it turns on and off.
4 If the voltage reading is as specified, the controller is working normally. If it isn't, replace the controller.
5 Disconnect the positive lead of the meter from the power relay coil and connect it to the relay output.
6 With the ignition switch still on, the meter should continue to repeat the on-off cycle described above.
7 If it does, the correct voltage is being applied to the glow plugs.

Checking the thermal controller

1 Unplug the connector from the glow-plug controller.
2 Using a high impedance digital ohmmeter set to the 200-ohm scale, measure the resistance between the indicated terminals of the controller bimetal heaters **(see illustration)**. Between pins 2 and 3, the resistance should be 0.40 to 0.75 ohms; between pins 4 and 5, the resistance should be 24 to 30 ohms; between pins 1 and 5, the resistance should be 117 to 143 ohms; and between pins 2 and 6, the resistance should be zero (there should be continuity).
3 If the indicated resistance values between the indicated pins aren't as specified, replace the controller.
4 If the controller checks out satisfactorily, plug in the harness connector and note whether the controller cycles on and off more than once when you turn the ignition key on (don't start the engine).
5 If the controller cycles more than once, and the resistance measurements were as specified, the controller is good.
6 If the controller cycles only once, it's bad, or there's a problem in the wire harness. Refer to the accompanying Glow Plug Electrical System Diagnosis chart **(see illustrations 2.85a and 2.85b).**

Glow-plug temperature inhibit switch (1984 through 1989 models with 6.2L and 6.5L engines)

Diesel engines which are already warmed up, and engines operated in warm climates, don't always require glow plugs to help them start. GM installed a temperature inhibit switch **(see illustrations)** in the glow-plug systems on 1984 through 1989 vehicles with 6.2L engines. The inhibit switch is located in the wire between the glow plug relay and pin 3 of the controller, which is located in the rear of the right (passenger's side) cylinder head, across from the glow-plug controller. The switch opens above 125-degrees F and turns off the glow plugs. This reduces needless glow-plug cycling, which gives the glow plugs a longer service life. As a result of improvements in the design of the controller in the latest version of the system, the switch was dropped after 1989.

Installing the glow-plug inhibit switch on earlier 6.2L diesels

If you own a 1982 or 1983 C, K, G or P truck with a 6.2L diesel engine, you can retrofit an inhibit switch to prolong the service life of the glow plugs.
1 Disconnect the cables from the negative terminals of the batteries.
2 Drain some of the coolant.
3 Replace the cover on the right rear head with the 1984 through 1989-style cover (part no. 14028949, the same cover as the one on the left head). Use a new gasket (part no. 14028951).
4 Coat the threads of the new temperature switch (part no. 15599010) with pipe thread sealer, install the switch and tighten it to 13 to 20 ft-lbs.
5 Replace any lost coolant.
6 Referring to the accompanying wiring schematic **(see illustration)**, connect the glow-plug inhibit wire assembly to the switch with the light blue wire (the one without the stripe), where the wire terminal mates with the switch terminal.

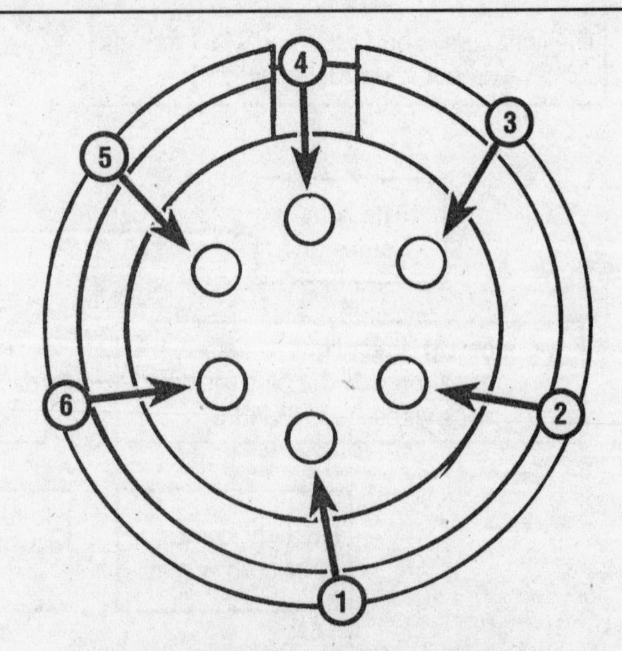

2.86 Terminal guide for glow plug control switch on 1982 through 1984 6.2L engines

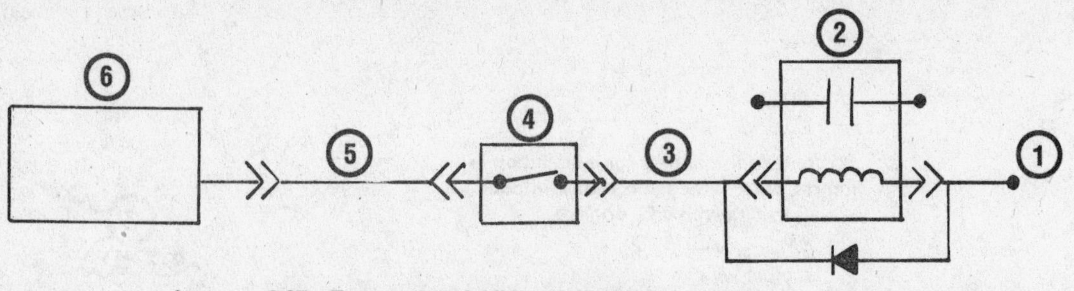

1 System voltage source
2 Glow plug relay
3 Light blue wire
4 Temperature switch
5 Light blue/black wire (the new wire)
6 Glow plug controller

2.87a Temperature inhibit switch circuit (1984 through 1989 6.2L and 6.5L engines)

2.87b The temperature inhibit switch (arrow) is located at the rear of the right cylinder head (1984 through 1989 6.2L and 6.5L engines)

7 Route the wire assembly toward the left side of the vehicle and strap it to the engine harness as required.

8 Unplug the four-way engine harness connector **(see illustration)** from its mate on the glow-plug relay extension harness (part no. 12031 1493).

9 Remove the female terminal of the light blue wire from the engine harness four-way connector and install connector body (part no. 2977253) on this terminal. Plug the mating connector on the new jumper wire assembly (light blue wire with the black stripe) into 2977253.

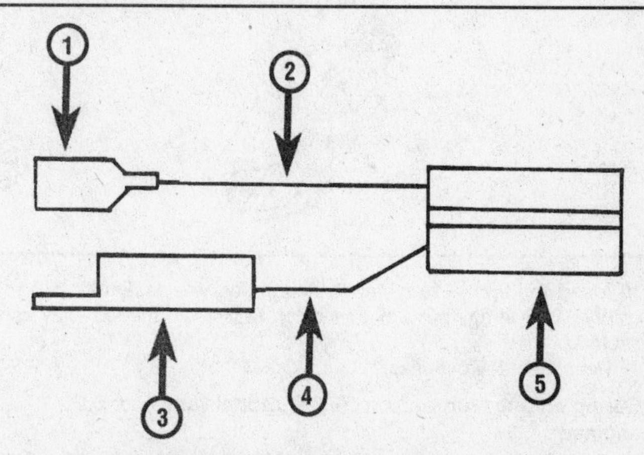

2.88 Here's how the now circuit looks with the inhibit switch installed (1982 and 1983 C, K, G and P trucks with 6.2L and 6.6L engine)

1 Terminal (2965867)
2 Light blue wire (39.4 inches)
3 Connector (6294828)
4 Light blue/black wire (37.4 inches)
5 Connector (2984528)

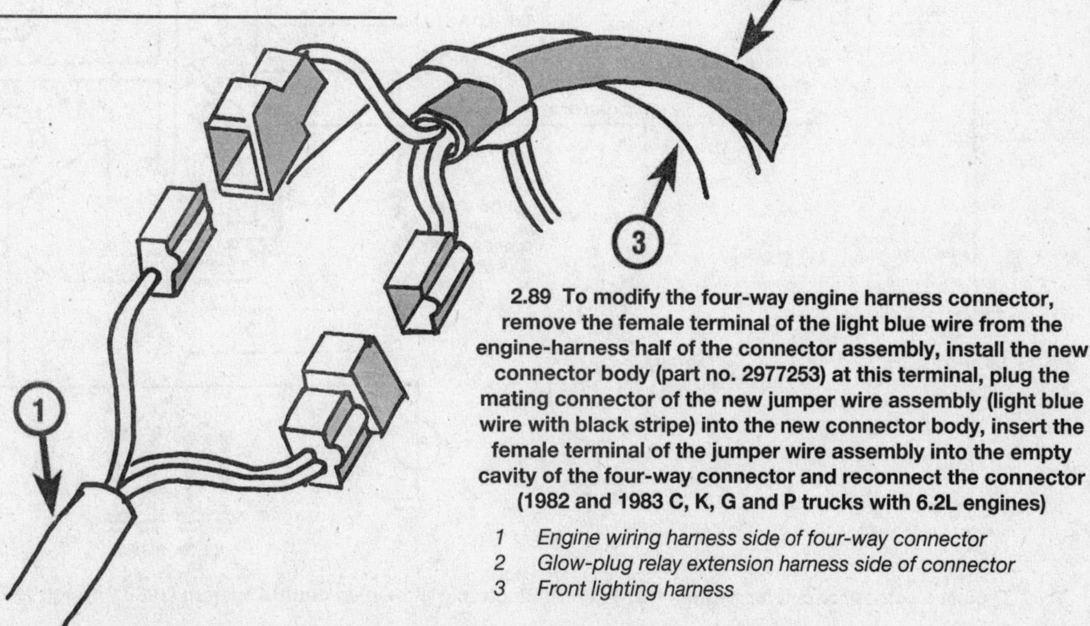

2.89 To modify the four-way engine harness connector, remove the female terminal of the light blue wire from the engine-harness half of the connector assembly, install the new connector body (part no. 2977253) at this terminal, plug the mating connector of the new jumper wire assembly (light blue wire with black stripe) into the new connector body, insert the female terminal of the jumper wire assembly into the empty cavity of the four-way connector and reconnect the connector (1982 and 1983 C, K, G and P trucks with 6.2L engines)

1 Engine wiring harness side of four-way connector
2 Glow-plug relay extension harness side of connector
3 Front lighting harness

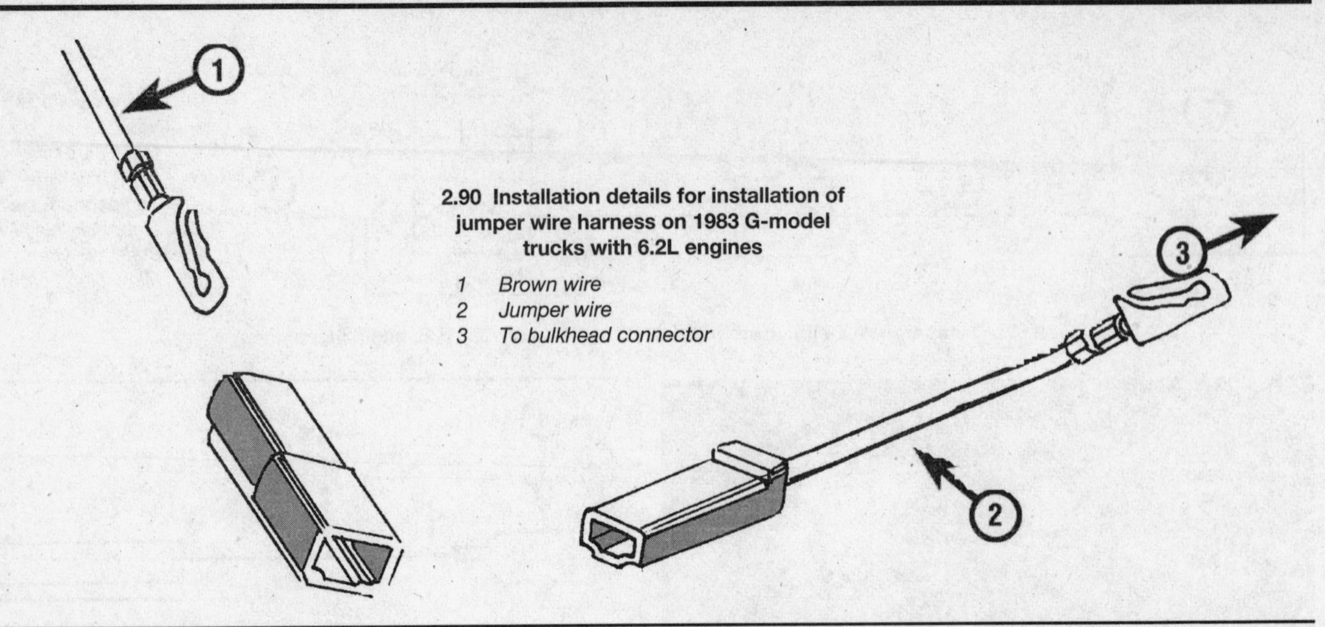

2.90 Installation details for installation of jumper wire harness on 1983 G-model trucks with 6.2L engines

1. *Brown wire*
2. *Jumper wire*
3. *To bulkhead connector*

10 Insert the female terminal of the jumper wire assembly into the empty cavity of the four-way connector. Reconnect the four-way connectors.

11 Reconnect the batteries.

Curing engine "run-on" on 1983 G-model vans with 6.2L engines

On 1983 G-model vans, an electrical feedback signal from the alternator can cause the engine to continue running with the key in the Off position. This signal prevents the injection pump solenoid from shutting off the fuel supply. To correct this condition, you'll need to install a jumper wire kit (part no. 12038051) as shown **(see illustration)**. (This jumper assembly was installed by the factory in G-trucks manufactured after March 1983).

1 Detach the cable from the negative battery terminal.

2 Remove the engine harness connector at the firewall (located right above the Hydra-Booster assembly).

3 Looking into the terminal end of the bulkhead connector, locate the brown wire and remove it.

4 Insert the end of the jumper harness into the bulkhead connector.

5 Using the terminal supplied with the jumper wire kit, connect the other end of the brown wire to the jumper harness.

6 Attach the connector to the firewall.

7 Reattach the cable to the negative battery terminal.

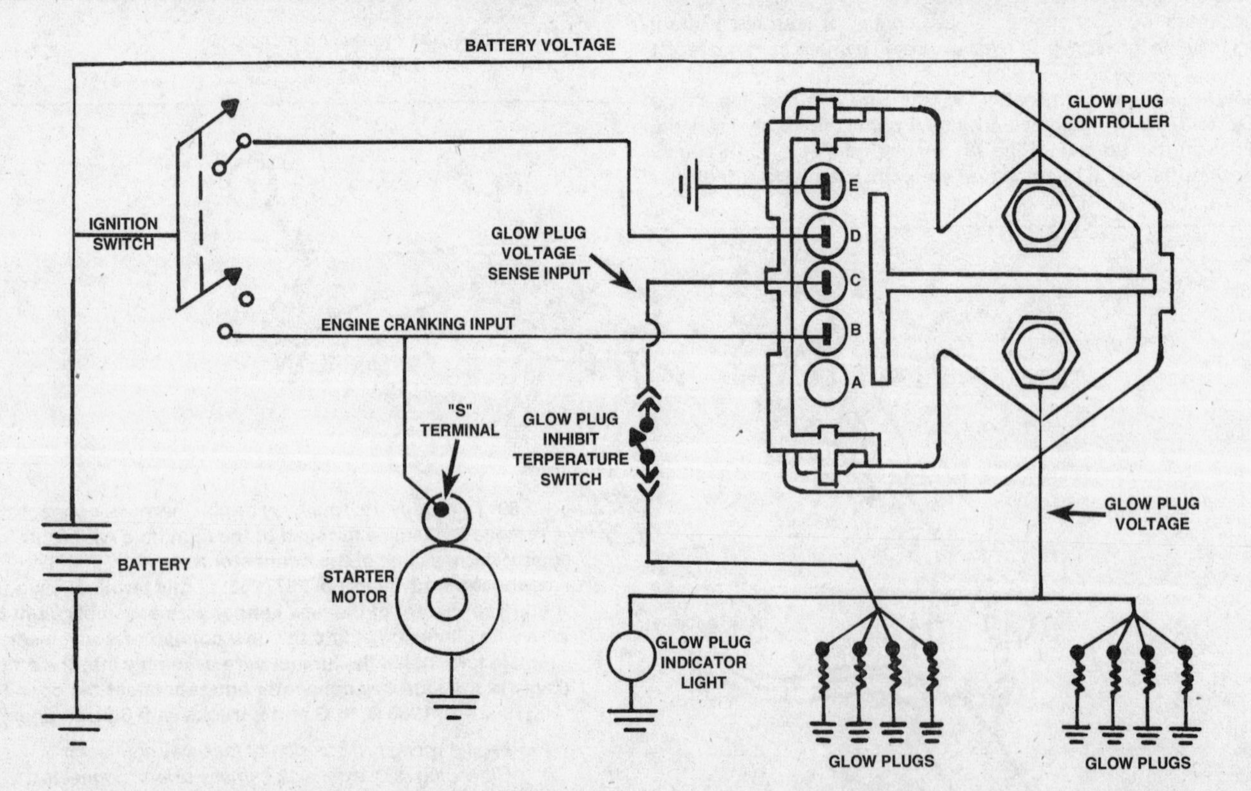

2.91 Typical electrical circuit for System No. 4 (later): Electronic glow-plug control system (1985 through 1993 6.2L and 6.5L engines)

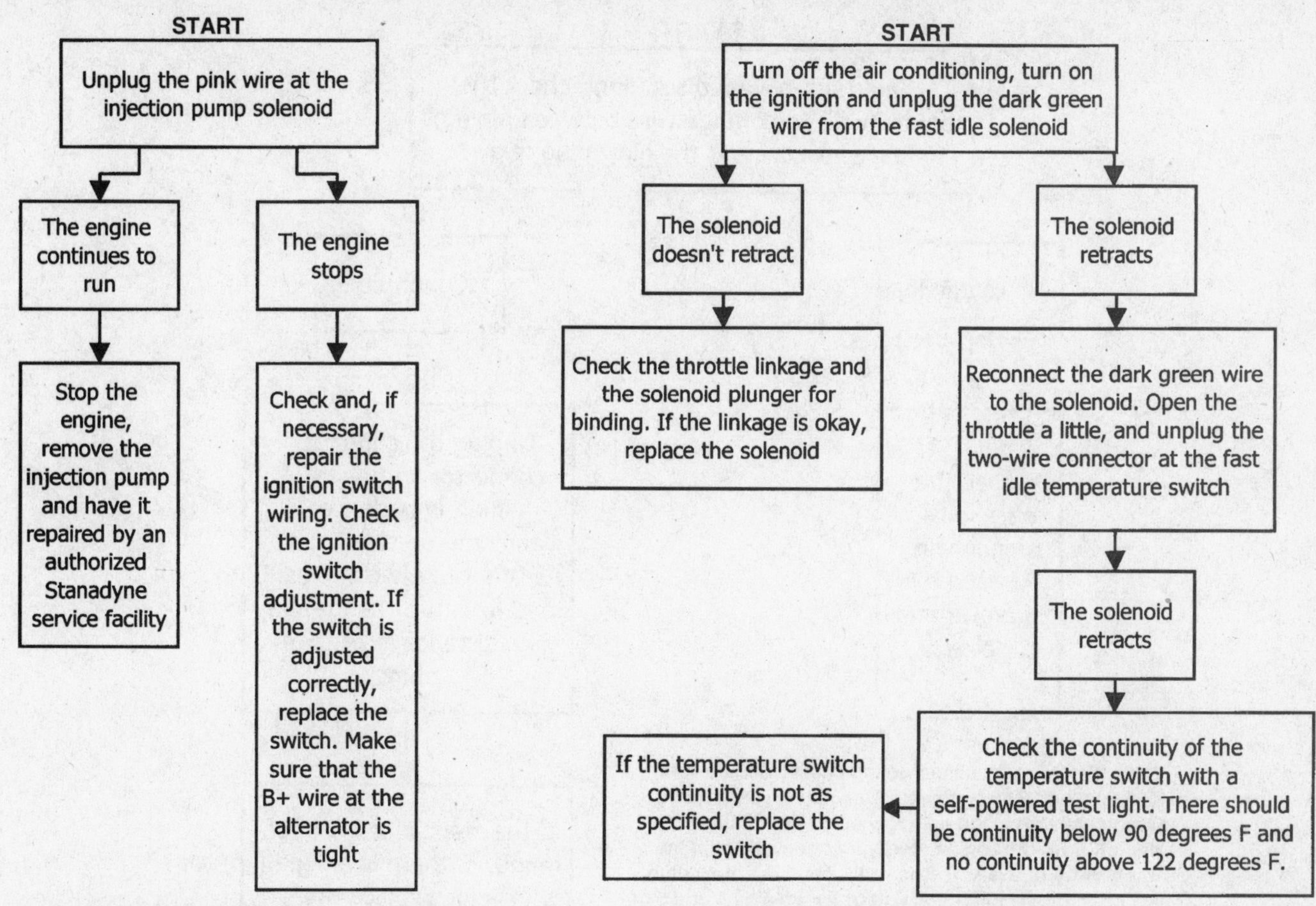

2.92a Electrical troubleshooting chart (1985 through 1993 6.2L and 6.5L engines) - The engine continues to run with the ignition key turned to OFF

2.92b Electrical troubleshooting chart (1985 through 1993 6.2L and 6.5L engines) - The engine stays on fast idle all the time

System No. 4 (later): Electronic glow-plug control system (1985 through 1993 6.2L and 6.5L models)

The glow-plug system on 1985 and later vehicles with 6.2L engines consists of an integral electronic controller/glow-plug relay assembly, eight six-volt glow plugs, a glow-plug light and a glow-plug inhibit temperature switch.

The electronic controller/glow-plug relay assembly contains the circuitry that monitors and controls the operation of the glow-plug relay. Inputs at pins B and C of the controller **(see illustration)** provide the information that it needs to determine the operating requirements of the glow plugs. Pin B senses "engine cranking." Pin C senses glow-plug voltage through the glow-plug inhibit switch, which is wired in series with the glow-plug voltage sensor lead to the glow plugs. The electronic controller is mounted on the rear of the left head.

The glow-plug inhibit switch opens at about 125-degrees F to prevent operation of the glow plugs above this temperature. There must be battery voltage at pin C of the controller to prevent an operational amplifier inside from inverting to ground. If this occurs, the transistor that turns on the glow plug relay opens, preventing the glow plug from operating.

Here's how the circuit works. When you turn on the ignition key (engine not running and at room temperature), the glow plugs go on for 4 to 6 seconds, then off for about 4.5 seconds. After that, they cycle on for about 1.5 seconds, then go off for about 4.5 seconds, then on for another 1.5 seconds, then 4.5 off, etc., for a total duration of about 25 seconds (including the initial 4 to 6 seconds).

If you crank the engine during, or after, the above sequence, the glow plugs will cycle on and off for a total duration of 25 seconds AFTER the ignition switch returns from its cranking position, whether the engine starts or not. The engine doesn't have to be running to terminate glow-plug cycling.

The cycling times used here are approximations - they vary somewhat in accordance with engine temperature. The initial "on" time and the on and off cycling time also vary, depending on system voltage. Longer "on" times are produced by lower voltage and/or temperature. The duration of the cycling time is lengthened by a lower voltage and/or temperature. The inhibit switch is calibrated to 125-degrees F; above this temperature, the switch opens and the glow plugs aren't energized. The "always closed" switch is always on, and the glow plugs will energize unless the engine temperature is above 110-degrees F.

If the system doesn't operate as described here:

1 Check all connectors and verify that they're a tight fit. The ground connection for the engine harness is critical on all 6.2L diesels. Make sure the nut is tightened securely and the ground ring terminal is tight.

2 Check the four-wire connector on the controller. If it isn't fully seated and latched, the glow plugs may not function.

3 Check both stud nuts on the controller. Make sure they're tight.

4 Check the temperature switch connector at the glow-plug inhibit switch, or the "always closed" (always on) switch. If this connection isn't good, the glow plug won't come on (and engine temperature must be below 110-degrees F for the plugs to operate).

5 If the glow plugs function normally, but the "glow plugs" light doesn't, check the connections in the harness behind the instrument panel and the bulb in the dash.

6 If all connections are intact, but the glow plug system doesn't operate as described, proceed to the general electrical diagnosis charts **(see illustrations)**.

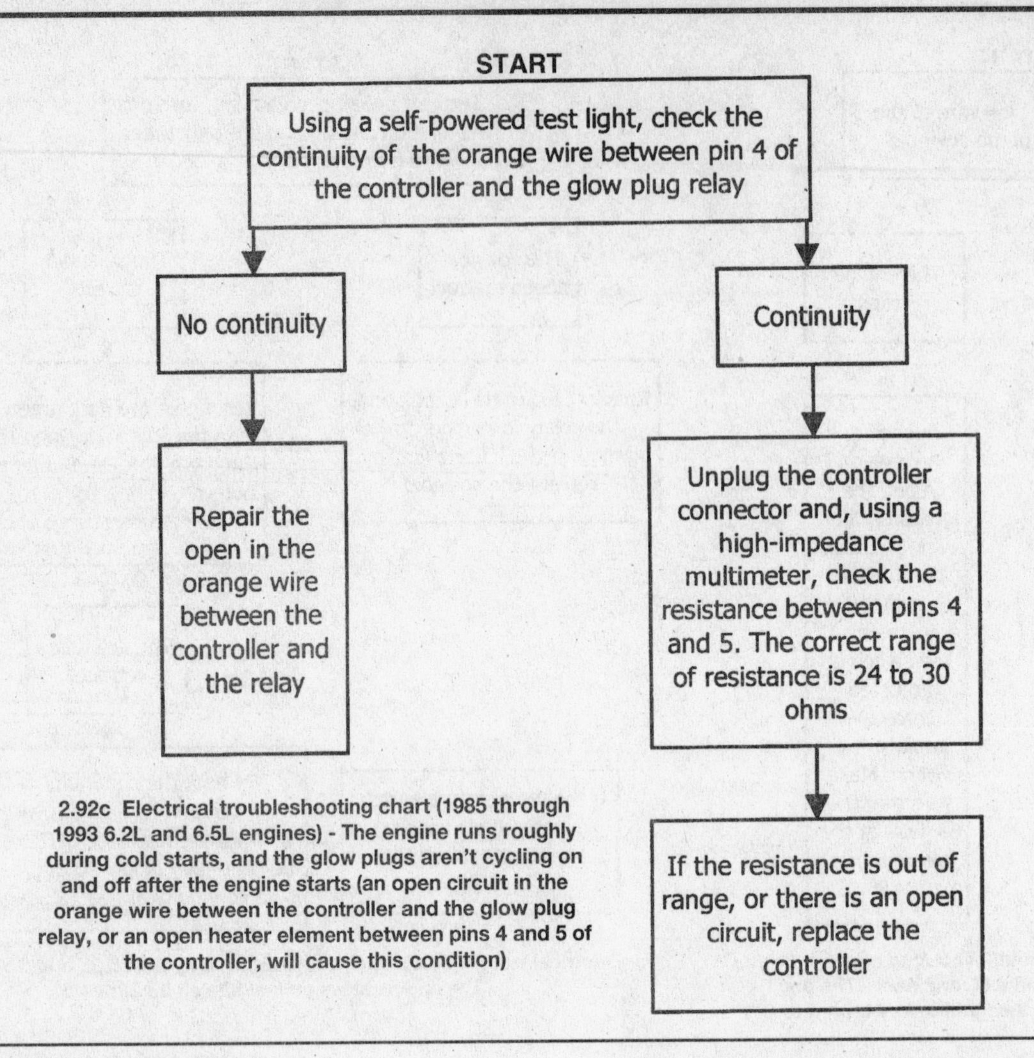

2.92c Electrical troubleshooting chart (1985 through 1993 6.2L and 6.5L engines) - The engine runs roughly during cold starts, and the glow plugs aren't cycling on and off after the engine starts (an open circuit in the orange wire between the controller and the glow plug relay, or an open heater element between pins 4 and 5 of the controller, will cause this condition)

Diagnosing the glow-plug system

If the glow-plug system still fails to work correctly, follow the troubleshooting procedure outlined in the accompanying diagnostic flow chart **(see illustration)** to troubleshoot a problem with the glow plug system.

Diagnosing a "no after-glow" problem

The glow-plug controller provides glow-plug operation after starting a cold engine. This after-start operation is initiated when the ignition switch is returned to Run from the Crank position. Loss of the after-start operation can result in excessive white smoke and/or poor idle quality after starting. To check for proper operation of this circuit:

1 Turn the ignition switch to the Run position and allow the glow plugs to cycle.

2 Wait two minutes, then turn the ignition switch to Crank for about one second (the engine need not actually start) and return to Run. The glow plugs should cycle on at least once.

3 If the glow plugs don't turn on, unplug the controller connector and check the B terminal of the harness connector with a grounded 12-volt test light. The light should be off with the ignition switch in the Run position, and on with the switch in the Crank position.

4 If the light doesn't operate correctly, repair a short or open in the purple wire of the engine harness.

5 If the light operates correctly, but the after-glow feature doesn't, replace the controller.

System No. 6: 1994 and later PCM controlled glow plug control system

On 1994 and later models, the PCM controls the on-off time of the glow plugs through the glow plug relay. The design of the computer controlled glow plug system is fairly simple and some diagnosis can be performed using the previous methods described for the System 4 (later) electronic system and the accompanying troubleshooting chart **(see illustration)**. Refer to the circuit illustration for wire color codes and components used **(see illustration)**. To properly diagnose an unusual problem, a SCAN tool must be used to retrieve diagnostic trouble codes (DTCs) from the PCM.

Block heater - check and replacement

If the temperature drops below 32-degrees F for engines with 30W oil, or below 0-degrees F for engines with 15W-30 or 10W-30, the block heater must be used to warm up the block so the engine can turn over fast enough to start. The block heater is located on the left or right side of the engine block, in the center freeze plug hole in the block. The block heater is designed to plug into a three-prong (grounded) 110-volt AC outlet. If you have to use an extension cord between the heater cord and the wall outlet, make sure you use a heavy-duty cord with a grounded three-prong plug; a smaller household-type cord will overheat and could cause a fire.

Leave the heater on for at least two hours to warm up an engine with 30W oil when the temperature is between 0 and 32-degrees F. Leave it on for at least eight hours to warm up engines with 30W, 15W-40 or 10W-30 oil when the temperature drops below 0-degrees F. After using the block heater, always store the cord properly to keep it away from hot or moving parts.

Check

1 Check the cord for continuity. Unplug both ends and test the wires with an ohmmeter or continuity tester. There are three separate wires, the center one is a ground and the two outer ones are "live". Look for

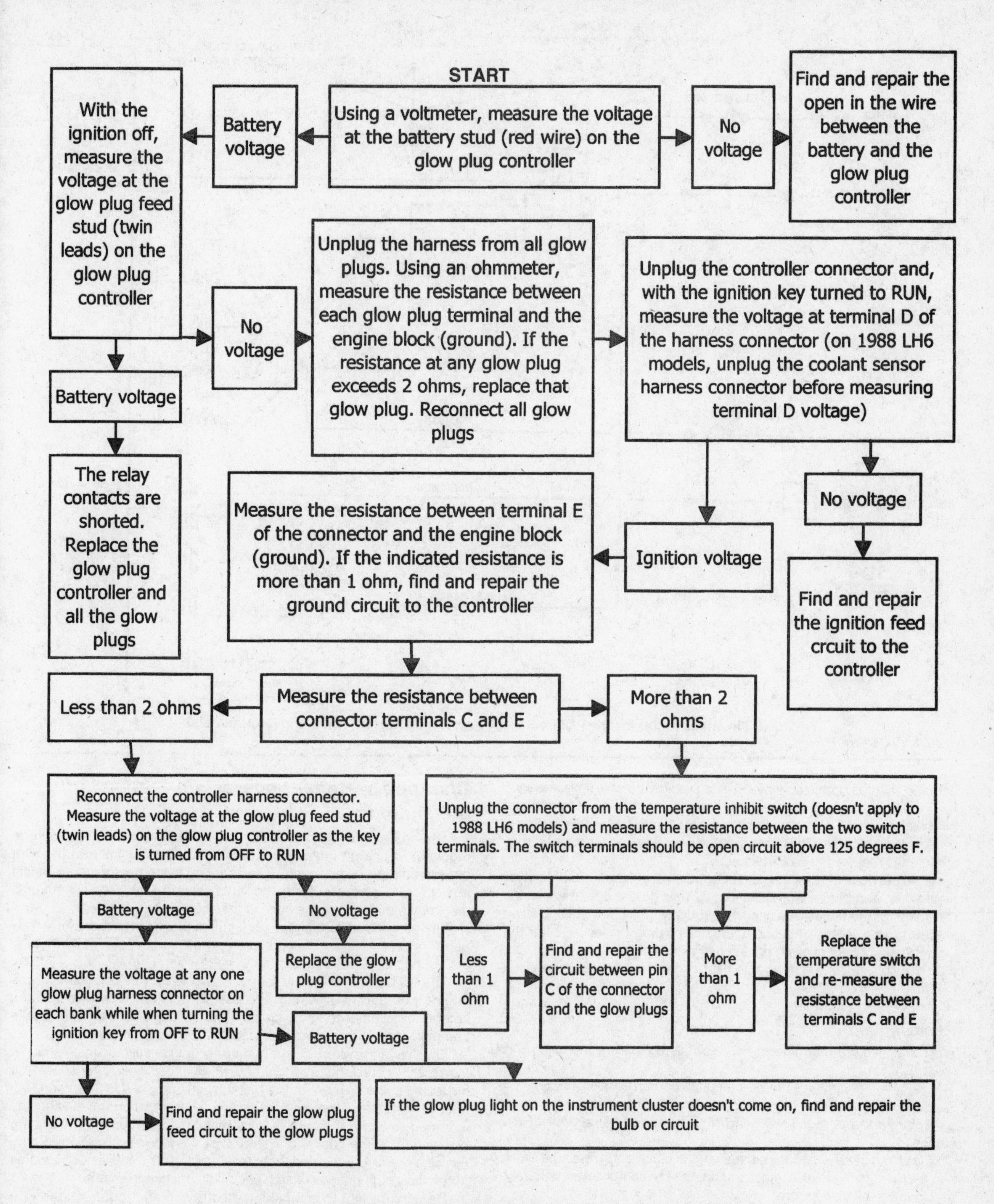

2.93 Glow-plug system troubleshooting chart (1994 and later 6.5L engines)

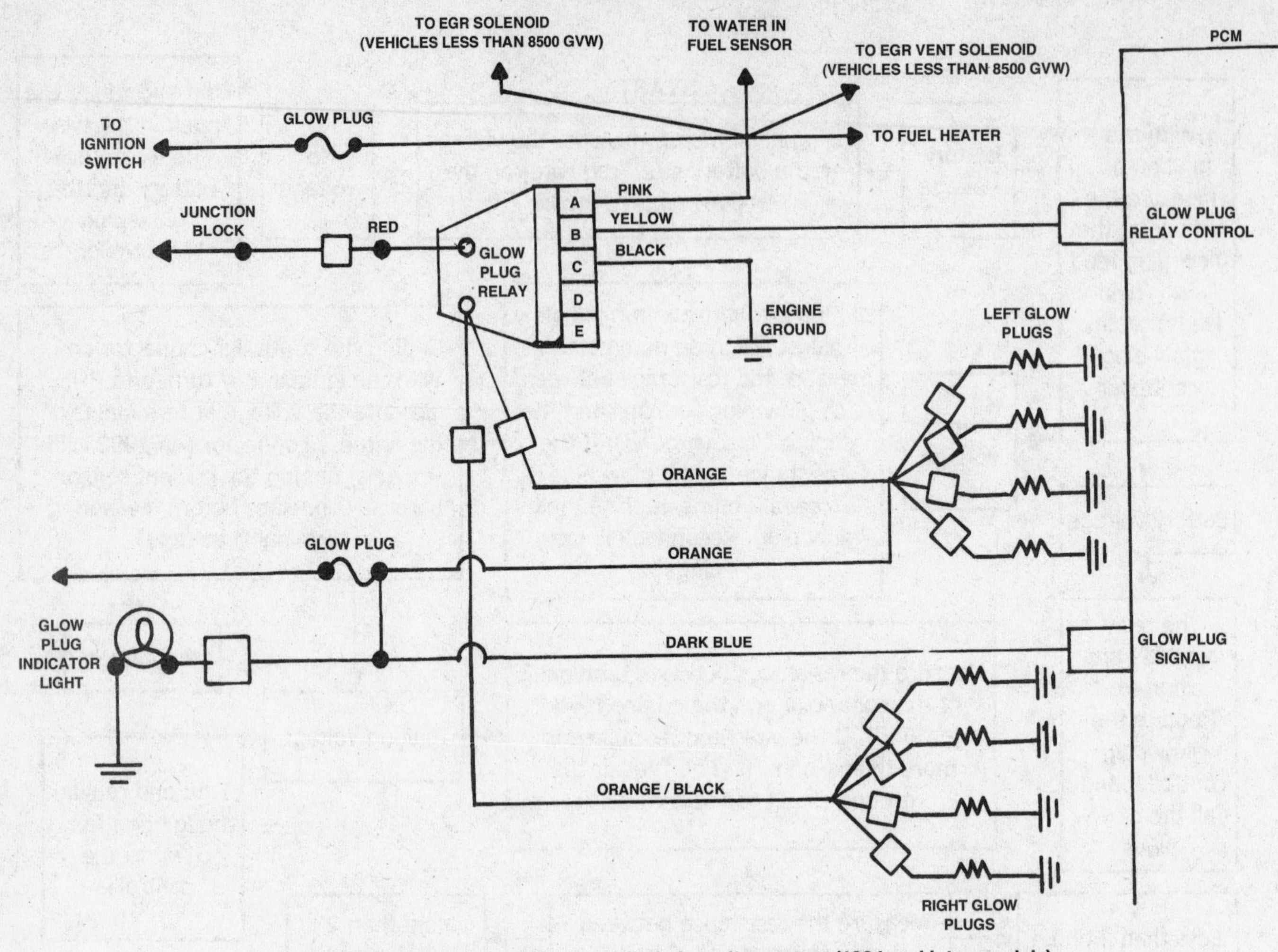

2.94 Electrical circuit for the PCM controlled glow plug system (1994 and later models)

damage such as opens and shorts, and replace the wire if necessary.

2 If the wires are OK, connect an ohmmeter to the two terminals of the block heater. There should be a very low resistance reading (near zero). If the ohmmeter indicates that the heating element is open (infinite ohms), replace the block heater.

3 If the previous checks indicate no problems, choose a double electrical outlet and plug a work light into it. Then plug the heater into the other side of the outlet. The light should dim slightly. If it does, current is flowing to the block heater. Check for heating by feeling near the block heater. If heat is given off, the unit is working.

4 If the light doesn't dim and no heat is given off, replace the block heater.

Replacement

Warning: *Allow the engine to cool completely before following this procedure.*

1 Drain the coolant (see the Maintenance portion of this Chapter).

2 Unplug the electrical connector from the block heater.

3 Remove the center screw and pull the block heater out of the engine.

4 Clean the sealing surfaces in the engine block and install the block heater. Be sure the seal is intact and tighten the center screw securely.

5 Connect the power cord.

6 Refill and bleed the cooling system.

7 Test the block heater as described previously.

8 Run the engine and check for coolant leaks.

Charging system - general information and precautions

The charging system includes the alternator, an internal voltage regulator, a charge indicator, the battery, a fusible link and the wire between all the components. The charging system supplies electrical power for the ignition system, the lights, the radio, etc. The alternator is driven by a drivebelt at the front of the engine.

The purpose of the voltage regulator is to limit the alternator's voltage to a preset value. This prevents power surges, circuit overloads, etc., during peak voltage output.

The fusible link is a short length of insulated wire integral with the engine compartment wiring harness. The link is four wire gauges smaller in diameter than the circuit it protects. Production fusible links and their identification flags are identified by the flag color.

The charging system doesn't ordinarily require periodic maintenance. However, the drivebelt, battery and wires and connections should be inspected at the intervals outlined at the beginning of this Chapter.

The dashboard warning light should come on when the ignition key is turned to Start, then go off immediately. If it remains on, there's a malfunction in the charging system (see the following Section). Some vehicles are also equipped with a voltmeter. If the voltmeter indicates abnormally high or low voltage, check the charging system.

Be very careful when making electrical circuit connections to a vehicle equipped with an alternator and note the following:

a) *When reconnecting wires to the alternator from the battery, be sure to note the polarity.*

b) *Before using arc welding equipment on the vehicle, disconnect the wires from the alternator and the battery terminals.*
c) *Never start the engine with a battery charger connected.*
d) *Always disconnect both battery cables before using a battery charger.*
e) *The alternator is turned by an engine drivebelt which could cause serious injury if your hands, hair or clothes become entangled in it with the engine running.*
f) *Since the alternator is connected directly to the battery, it could arc or cause a fire if overloaded or shorted out.*
g) *Wrap a plastic bag over the alternator and secure it with rubber bands before steam cleaning the engine.*

Charging system - check

1 If a charging system malfunction occurs, don't immediately assume the alternator is causing the problem. First check the following items:

a) *Check the drivebelt tension and condition. Replace it if it's worn or deteriorated.*
b) *Make sure the alternator mounting and adjustment bolts are tight.*
c) *Inspect the alternator wiring harness and the connectors at the alternator and voltage regulator. They must be in good condition and tight.*
d) *Check the fusible link (if equipped) located between the starter solenoid and the alternator. If it's burned, determine the cause, repair the circuit and replace the link (the engine won't start and/or the accessories won't work if the fusible link blows). Sometimes a fusible link may look good, but still be bad. If in doubt, remove it and check for continuity.*
e) *Start the engine and check the alternator for abnormal noises (a shrieking or squealing sound indicates a bad bearing).*
f) *Check the specific gravity of the battery electrolyte. If it's low, charge the battery (doesn't apply to maintenance free batteries).*
g) *Make sure the battery is fully charged (one bad cell in a battery can cause overcharging by the alternator).*
h) *Disconnect the battery cables (negative first, then positive). Inspect the battery posts and the cable clamps for corrosion. Clean them thoroughly if necessary. Reconnect the cable to the positive terminal.*
i) *With the key off, connect a test light between the negative battery post and the disconnected negative cable clamp.*
    1) *If the test light does not come on, reattach the clamp and proceed to the next step.*
    2) *If the test light comes on, there is a short (drain) in the electrical system of the vehicle. The short must be repaired before the charging system can be checked.*
    3) *Disconnect the alternator wiring harness.*
        (a) *If the light goes out, the alternator is bad.*
        (b) *If the light stays on, pull each fuse until the light goes out (this will tell you which component is shorted).*

2.96 Typical alternator mounting details

2.95 A voltmeter will be needed to check the charging system

2 Using a voltmeter, check the battery voltage with the engine off. It should be approximately 12-volts **(see illustration)**.

3 Start the engine and check the battery voltage again. It should now be approximately 14-to-15 volts.

4 Turn on the headlights. The voltage should drop, and then come back up, if the charging system is working properly.

5 If the voltage reading is more than specified, the voltage regulator must be replaced. If the voltage is less, the alternator diodes, stator or rectifier may be bad or the voltage regulator may be malfunctioning.

Alternator - removal and installation

1 Detach the cable from the negative terminal of the battery.

2 On van models, remove the air cleaner and hoses and oil dipstick, for access. On all models, detach the wires from the alternator.

3 Remove the alternator drivebelt.

4 Remove the mounting bolts and separate the alternator from the engine **(see illustration)**.

5 If you're replacing the alternator, take the old one with you when purchasing the replacement. Make sure the new/rebuilt unit looks identical to the old one. Look at the terminals - they should be the same in number, size and location as the terminals on the old alternator. Finally, look at the identification numbers stamped into the housing or printed on a tag attached to the housing. Make sure the numbers are the same on both alternators.

6 Many new/rebuilt alternators DO NOT have a pulley installed, so you may have to switch the pulley from the old unit to the new/rebuilt one. When buying an alternator, find out the shop's policy regarding pulleys - some shops will perform this service free of charge.

7 Installation is the reverse of removal.

8 After the alternator is installed, check the drivebelt tension.

Starting system - general information and precautions

The sole function of the starting system is to turn over the engine quickly enough to allow it to start.

The solenoid/starter motor assembly is installed on the lower part of the engine, next to the transmission bellhousing.

The starting system consists of the battery, the starter motor, the starter solenoid and the wires connecting them. The solenoid is mounted directly on the starter motor.

When the ignition key is turned to the Start position, the starter solenoid is actuated through the starter control circuit. The starter solenoid then connects the battery to the starter. The battery supplies the electrical energy to the starter motor, which does the actual work of cranking the engine.

The starter motor on a vehicle equipped with a manual transmission can only be operated when the clutch pedal is depressed; the starter

2.97 There are three terminals on the end of a typical starter solenoid

1 Battery terminal
2 Battery terminal (S)
3 Motor terminal (M)

on a vehicle equipped with an automatic transmission can only be operated when the shift lever is in Park or Neutral.

Always observe the following precautions when working on the starting system:

a) Excessive cranking of the starter motor can overheat it and cause serious damage. Never operate the starter motor for more than 15 seconds at a time without pausing to allow it to cool for at least two minutes.

b) The starter is connected directly to the battery and could arc or cause a fire if mishandled, overloaded or shorted out.

c) Always detach the cable from the negative terminal of the battery before working on the starting system.

Starter motor - check

1 If the starter motor doesn't turn at all when the switch is operated, make sure the shift lever is in Neutral or Park (automatic transmission) or the clutch pedal is depressed (manual transmission).

2 Make sure the battery is charged and all cables, both at the battery and starter solenoid terminals, are clean and secure.

3 If the starter motor spins but the engine doesn't turn, the overrunning clutch in the starter motor is slipping and the starter motor must be replaced.

4 If, when the switch is actuated, the starter motor doesn't operate at all but the solenoid clicks, then the problem lies with either the battery, the main solenoid contacts or the starter motor itself (or the engine is seized).

5 If the solenoid plunger cannot be heard when the switch is actuated, the battery is bad, the fusible link is burned (the circuit is open) or the solenoid is defective.

6 To check the solenoid, connect a jumper wire between the battery (+) and the ignition switch wire terminal (the small terminal) on the solenoid. If the starter motor now operates, the solenoid is okay and the problem is in the ignition switch, neutral start switch or the wires.

7 If the starter motor still does not operate, remove the starter/solenoid assembly for disassembly, testing and repair.

8 If the starter motor cranks the engine at an abnormally slow speed, first make sure the battery is charged and all terminal connections are tight. If the engine is partially seized, or has the wrong viscosity oil in it, it'll crank slowly.

9 Run the engine until normal operating temperature is reached, then

2.98 Starter motor mounting details

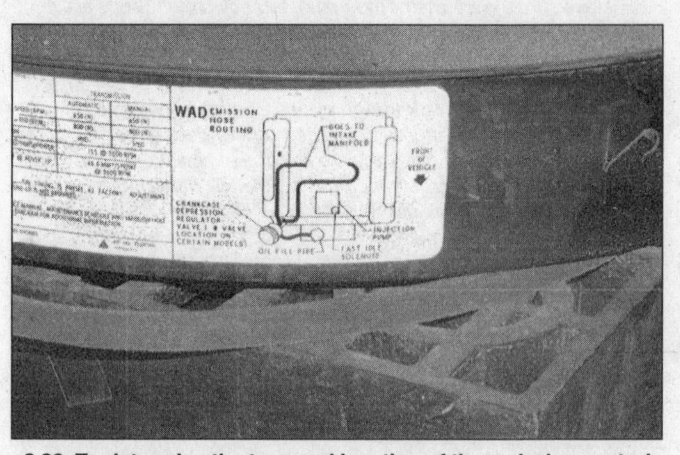

2.99 To determine the type and location of the emission control components on your vehicle, refer to the Vehicle Emission Control Information (VECI) label in the engine compartment (1985 6.2L truck shown)

stop the engine and disconnect the pink wire from the fuel injection pump. Reconnect the wire after completing the test.

10 Connect a voltmeter positive lead to the positive battery post and connect the negative lead to the negative post.

11 Crank the engine and take the voltmeter readings as soon as a steady figure is indicated. Do not allow the starter motor to turn for more than 30 seconds at a time. A reading of 9-volts or more, with the starter motor turning at normal cranking speed, is normal. If the reading is 9-volts or more but the cranking speed is slow, the motor is faulty. If the reading is less than 9-volts and the cranking speed is slow, the solenoid contacts are probably burned, the starter motor is bad, the battery is discharged or there's a bad connection.

Starter motor - removal and installation

1 Detach the cable from the negative terminal of the battery.

2 Raise the vehicle and support it securely on jackstands.

3 Clearly label, then disconnect the wires from the terminals on the starter motor and solenoid **(see illustration)**.

4 Remove the mounting bolts and detach the starter, taking care to note the number and location of any shims **(see illustration)**.

5 Installation is the reverse of removal.

Emission control systems

General information

GM diesels are equipped with several emission control devices **(see illustration)**. All engines have some sort of crankcase ventilation

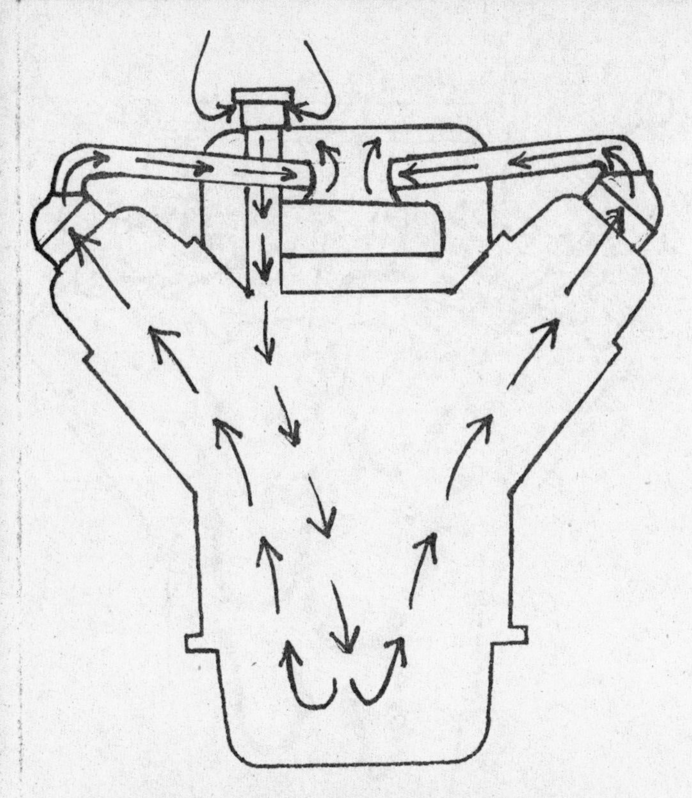

2.100 Typical Positive Crankcase Ventilation (PCV) system (all 1978 through 1980 5.7L engines and 1981 California-spec 5.7L engines)

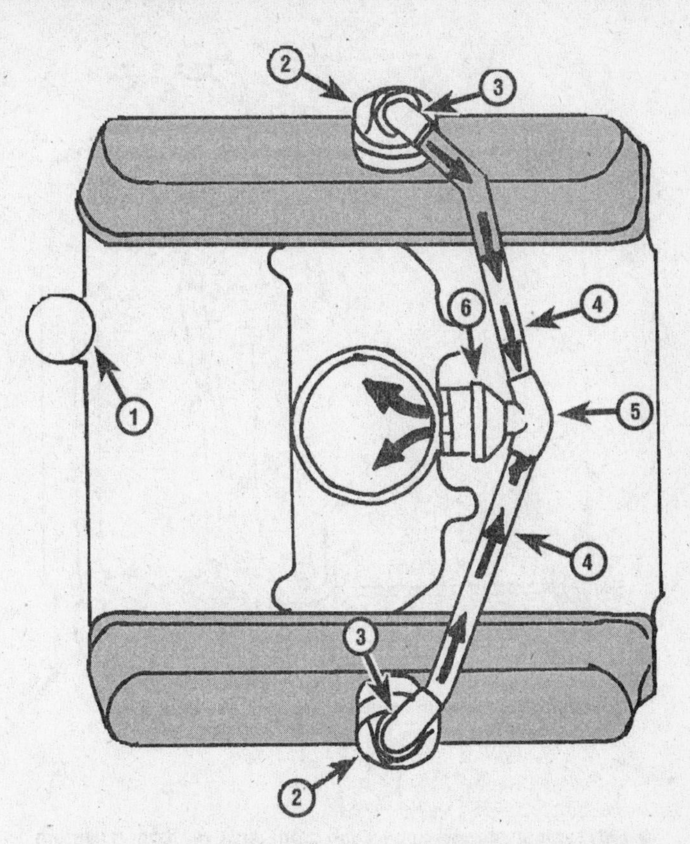

2.101 Typical flow control valve and crankcase ventilation system (1978 through 1980 5.7L engines)

1 *Breather cap*
2 *Ventilation filters*
3 *Ventilation elbows*
4 *Ventilation pipes*
5 *Ventilation-to-pipe nipple connector*
6 *Flow control valve*

system to reduce crankcase pressure at idle. A Positive Crankcase Ventilation (PCV) system is used on 1978 through 1980 5.7L engines; these engines, and 1981 California models, also use a flow control valve. A Crankcase Depression Regulator (CDR) replaces the PCV system on 1981 through 1984 5.7L engines; the CDR system is also used on all 6.2L engines.

All 1980 through 1984 5.7L engines are equipped with an Exhaust Gas Recirculation (EGR) system to reduce oxides of nitrogen (NOx) at idle and during the partial loads of around-town driving; all LH6 versions of the 6.2L engine also use an EGR system. An Exhaust Pressure Regulator (EPR) is used on 1981 5.7L EGR systems and on LH6 versions of the 6.2L EGR system.

The injection pumps on 5.7L engines and LL4 versions of the 6.2L engine are equipped with a Vacuum Regulator Valve (VRV); the pumps on LH6 versions of the 6.2L engine use a Throttle Position Switch (TPS) instead.

GM emission systems are controlled by the Diesel Electronic Control System (DECS). The brain of the DECS system is the Electronic Control Module (ECM). The ECM monitors inputs such as engine rpm, vehicle speed, manifold absolute pressure, the metering valve sensor, the transmission gear position, coolant temperature, the brake switch and the A/C system. The ECM uses this information to control the Exhaust Gas Recirculation (EGR) system, the Exhaust Pressure Regulator (EPR), the Transmission Control Clutch (TCC), the Housing Pressure Cold Advance (HPCA), the Altitude Fuel Limiter (AFL) and the Check Engine Light (CEL). On 6.2L engines, there are two versions of DECS - one is for Federal models and the other is for California.

Positive Crankcase Ventilation (PCV) system (5.7L engines only)

Early 5.7L engines (all 1978 through 1980 models and 1981 California models) use a Positive Crankcase Ventilation (PCV) system **(see illustration)** similar to those found on conventional gasoline engines. Fresh air enters the crankcase through a filter/check valve located in the oil fill tube cap at the front of the engine. Ventilation vapors are drawn from the valve covers into the rear of the air intake crossover. If the air cleaner becomes clogged, the normally low level of purging can increase dramatically and suck oil into the system; a ventilation flow control valve in the air crossover prevents this from happening.

The PCV systems on all but the earliest 1978 5.7L engines, and all 1979 and 1980 engines, are equipped with flow control valves to further limit crankcase pressure at idle. If you have one of these early models, and the engine isn't already equipped with a flow control valve, install one into the air crossover and modify the plumbing as necessary **(see illustration)**.

Testing the flow control valve (5.7L engines only)

To test the flow control valve, you'll need a water manometer (Kent-Moore J-23951, or equivalent). The manometer indicates pressure or vacuum by the difference in the height of two columns of liquid.

1 Connect one end of the manometer in the oil dipstick hole; leave the other end open, so it's vented to the atmosphere. Don't remove the air cleaner.

2 Run the engine at idle. If the valve is good, it should produce three inches of water pressure at idle; if it's stuck or closed, it will produce as much as 10-inches. If the valve indicates excessive water pressure at idle, remove it, clean it and retest it. If that doesn't work, replace the valve. **Note:** *On 1978 models, the original flow control valve had a metered orifice that was too small. This caused a pressure build-up in the crankcase which resulted in oil leaks. On 1979 models, the orifice was enlarged by .012-inch to improve flow. Whenever you replace the flow control valve on a 1978 model, make sure you get the newer valve with the larger orifice.*

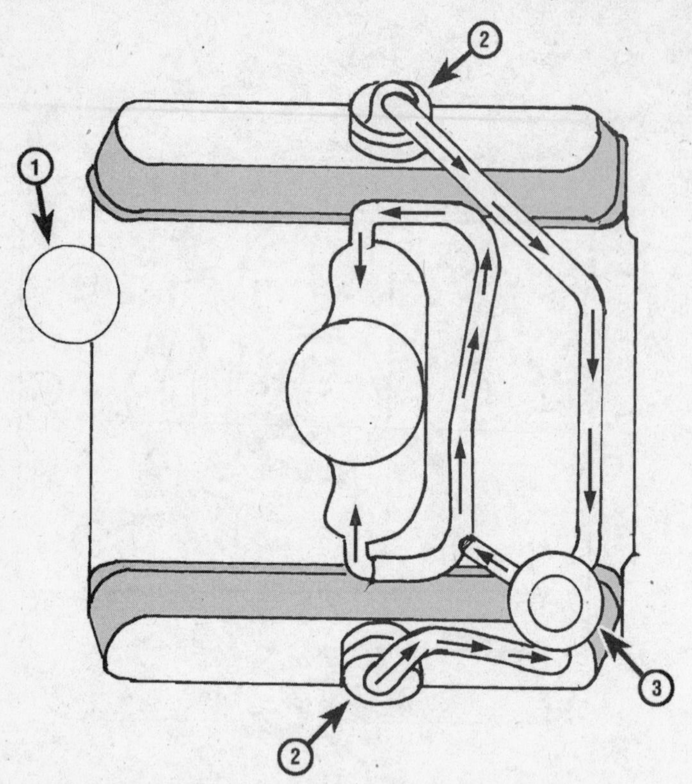

2.102 Typical second-generation crankcase ventilation system with Crankcase Depression Regulator (CDR) valve (1981 through 1984 6.7L engines, except 1981 California models)

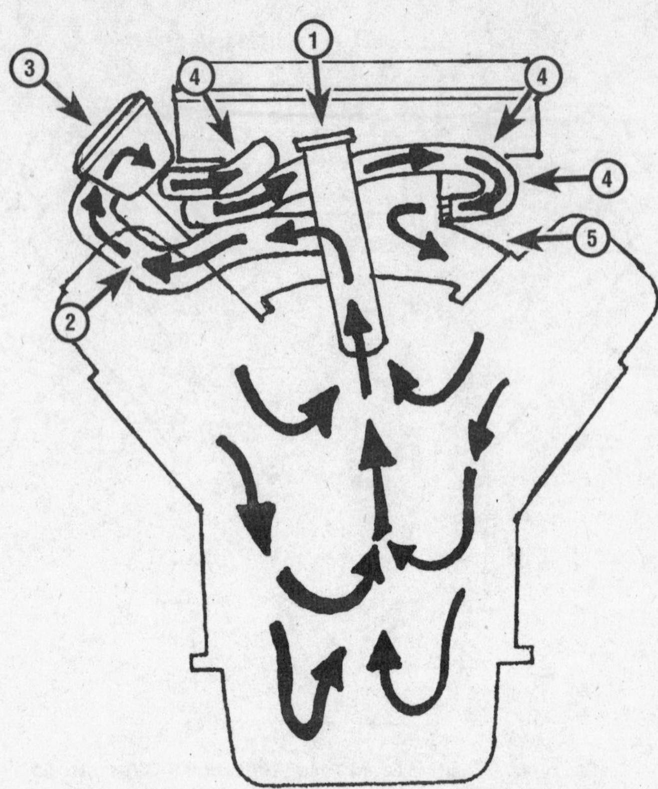

2.103 Typical crankcase ventilation system with Crankcase Depression Regulator (CDR) valve on (6.2L and 6.5L engines)

1 *Breather*
2 *Breather-to-CDR hose*
3 *Crankcase Depression Regulator (CDR) valve*
4 *CDR-to-intake manifold hoses*
5 *Intake manifold*

Crankcase Depression Regulator (CDR) valve (1981 through 1984 5.7L engines, except 1981 California models, and all 6.2L and 6.5L engines)

In 1981, the PCV system on early 5.7L engines was replaced by an improved ventilation system **(see illustration)**, which employs a Crankcase Depression Regulator (CDR) to reduce pumping pressure more effectively at idle. Early (1978 through 1980) 5.7L diesel engines are prone to oil leaks at idle because of high crankcase pressure. The redesigned crankcase ventilation system reduces crankcase pressure - and the likelihood of engine oil leaks - at idle.

The main difference between the new system and the earlier system is the addition of a Crankcase Depression Regulator (CDR) valve **(see illustration)**, located at the rear of the left (driver's side) cylinder head, and larger ventilation system tubing. Crankcase gases enter the air crossover on each side, below the EGR valve. The breather cap includes a check valve which prevents blow-by gases from entering the atmosphere. As the gases (blow-by and fresh air) are drawn from the valve covers, through the CDR valve and into the intake manifold, the CDR limits crankcase vacuum. Too little vacuum can force an oil leak; too much can pull oil into the crossover.

The CDR system is also used on all 6.2L and 6.5L engines **(see illustration)**. On some early models, the valve is located at the front of the right cylinder head. The early valve has a tendency to collect condensation, which can freeze during cold spells, causing the crankcase to pressurize and the engine to lose oil.

On earlier models, the CDR valve is integrated into the oil filler cap, which is located between the intake manifold and the right (passenger's side) valve cover. On later models, the CDR valve is located in the same place, but isn't part of the oil filler cap. The two later versions of the valve eliminate the condensate problem by draining it back into the engine crankcase. You can retrofit the later-style CDR valve to an earlier crankcase ventilation system.

2.104 Typical Crankcase Depression Regulator (CDR) valve installation (1982 through 1987 C, K and P trucks and RVs and 1983 through 1987 G vans)

1 *Breather-to-CDR hose*
2 *CDR-to-intake manifold hoses*
3 *CDR mounting bolts*
4 *Crankcase Depression Regulator (CDR) valve*

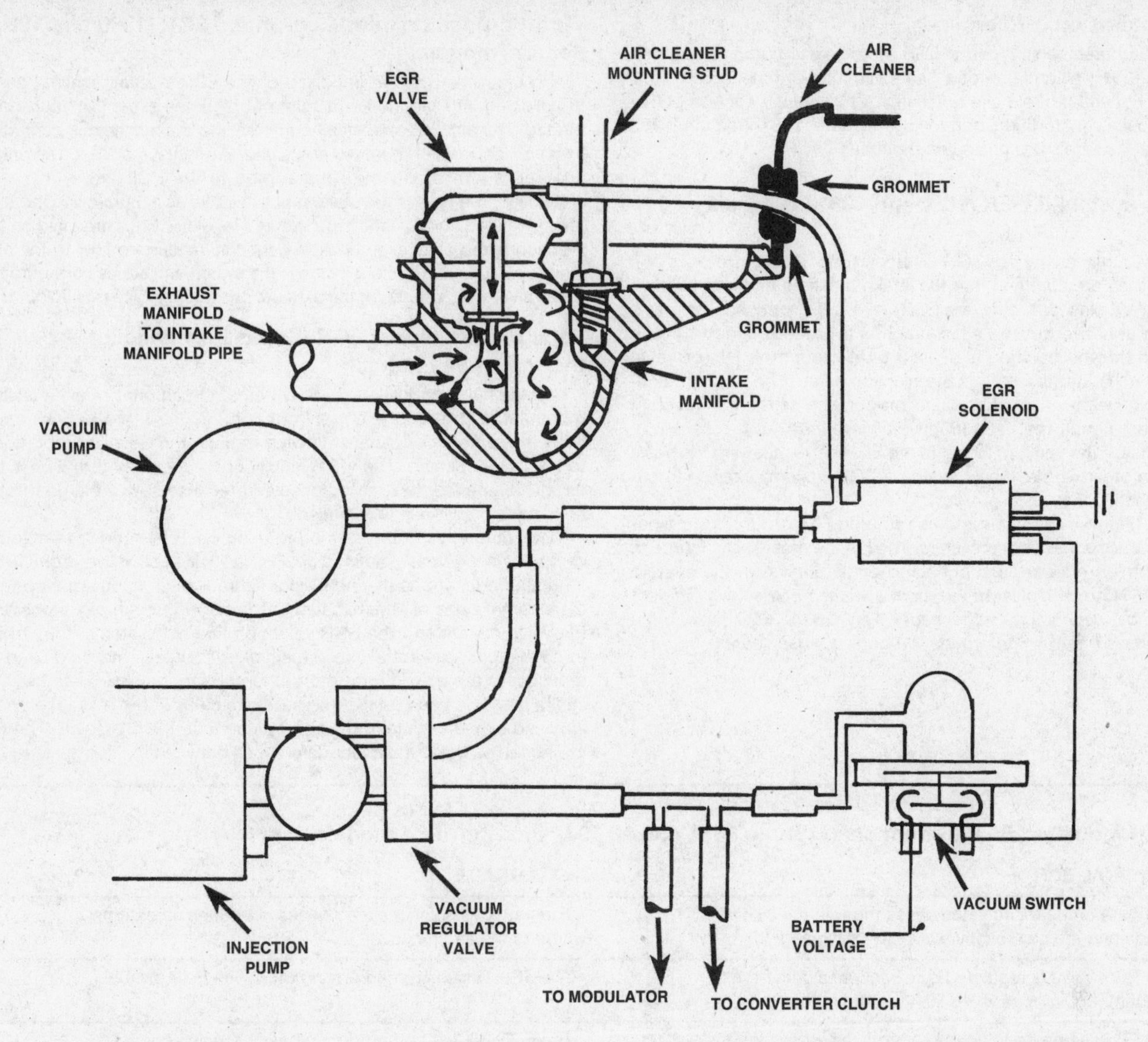

2.105 Schematic for a typical vacuum-switched EGR system (1980 and 1981 California model with 5.7L engine shown)

Servicing the Crankcase Depression Regulator (CDR) system

Periodically, clean the ventilation pipes, the valve cover filters and the valve cover openings. Replace the crankcase depression regulator and oil cap every 30,000 miles.

Checking the CDR valve

1 Hook up a U-tube water manometer (Kent-Moore J23951, or equivalent). Connect one end of the manometer to the oil dipstick tube hole and vent the other end to atmosphere. If equipped with a turbocharger, unplug the rubber vent tube from the turbo inlet elbow.
2 Start the engine and let it idle. If the CDR is functioning properly, it should limit water pressure at idle to one-inch, and no more than 3 to 4-inches water at full load (about 2000 rpm).
3 If the CDR doesn't perform as described, replace it.

Replacing the CDR valve

1 Disconnect the ventilation pipe(s) from the CDR valve.
2 Remove the CDR valve mounting bracket bolt (earlier models).
3 Remove the CDR valve.
4 Clean the ventilation pipes and connectors with solvent, but make sure they're dry before installing them; if solvent enters the crankcase depression regulator, it will destroy the diaphragm.
5 Place the new CDR valve in position, install the mounting bolt (if applicable) and attach the ventilation pipes.

Exhaust Gas Recirculation (EGR) systems

All 1980 and later GM diesel engines use an Exhaust Gas Recirculation (EGR) system. EGR systems have been widely used on gasoline engines since 1975 to reduce pollutants known as oxides of nitrogen, or NOx. The only way to reduce NOx is to reduce the temperature inside the combustion chamber. The EGR system accomplishes this by admitting exhaust gas into the intake manifold, which reduces the concentration of oxygen and soaks up some of the heat generated during combustion, lowering the temperature. The amount of gas admitted into the intake stream is determined at the EGR valve, which is installed at a point where the gases are admitted into the intake manifold **(see illustration)**.

There are many different types of EGR systems used on GM diesel engines. Some EGR valves are open all the time except at wide open throttle settings, some are switched on and off by a calibrated throttle switch and some are gradually opened and closed by a vacuum regulator valve at the throttle. The EGR valve on 1994 and later models is controlled by the Powertrain Control Module. Before you can diagnose an EGR system, you should have a basic idea of how the specific system on your engine works.

EGR systems on 5.7L engines

There are three basic types of EGR systems used on 5.7L engines: 1980 and 1981 California models use a vacuum-switched EGR; 1980 and 1981 Federal models use a variation of vacuum-switched EGR - throttle-position, modified-orifice EGR; some 1981 and all 1982 through 1984 models use a vacuum-modulated EGR.

Vacuum-switched EGR (1980 and 1981 California models)

In a vacuum-switched EGR system **(see illustration 2.105)**, exhaust gases are admitted into the intake manifold either at full flow or not at all. At closed throttle, the EGR valve is fully opened. It remains fully open until the throttle shaft reaches a predetermined throttle angle. The throttle position is sensed by a vacuum regulator valve mounted on the throttle shaft on the injection pump. The vacuum regulator valve generates a vacuum signal proportional to throttle travel. At a calibrated signal, the vacuum switch de-energizes the solenoid, which shuts off the vacuum signal to the EGR valve, allowing the valve to close. In other words, this is an on-off EGR system; it doesn't control the quantity of exhaust gas flow.

Always inspect the vacuum hose routings before trying to troubleshoot a vacuum-switched system. They're the most likely source of trouble; if they're loose, disconnected or deteriorated, the system won't work. **Note:** *Vehicles with vacuum-switched EGRs have different hose routings, depending on the model. Refer to the VECI label in the engine compartment to determine the correct hose routing for your vehicle.*

Throttle-position, modified-orifice EGR (1980 and 1981 Federal models)

On a throttle-position, modified-orifice EGR system, exhaust gases are allowed into the intake manifold at all times except at wide-open throttle. The amount of exhaust admitted into the manifold is controlled by an orifice adapter mounted at the end of exhaust manifold. The orifice adapter is connected by a metal pipe to the EGR valve in the air crossover. The rest of the system is similar to a vacuum-switched system described above. When the vacuum regulator valve mounted on the injection pump throttle shaft allows enough vacuum to flow to the normally open vacuum switch, it closes the switch and sends current to the EGR valve, which shuts off the exhaust gas to the intake manifold.

Vacuum-Modulated EGR (Some 1981 and all 1982 through 1984 models)

On vacuum-modulated EGR systems, the amount of exhaust gas recirculated to the intake manifold is modulated by the EGR valve itself. The amount of exhaust flowing through the valve is proportional to the valve opening. The valve is fully open at closed throttle. As the throttle is opened, the valve closes and reduces flow. At large throttle openings, the valve is fully closed.

Two different EGR systems are used on 1981 models: California cars use an external system, which is basically a carryover from 1980; 49-state Federal models and California trucks use an internal system. On external systems, the exhaust gases travel through pipes between the valve covers and the EGR valve; on internal systems, they travel through a passage in the intake manifold. The intake manifolds and air crossovers are also different on external and internal systems. The intake manifold used on the internal system has larger air intake openings and exhaust inlets from each cylinder into the crossover. The following tables detail the major differences between the two systems.

Vacuum-Modulated EGR system chart (Some 1981 and all 1982 through 1984 models)

INTERNAL EGR	EXTERNAL EGR
The path of the recirculated gases through the cylinder heads and the air crossover is inside the manifold.	The path of the recirculated gases is through an external pipe to the air crossover.
The EGR valve is regulated by a vacuum signal from the vacuum regulator valve on the injection pump.	The EGR valve is turned on and off by the EGR switch.
The EGR is boosted during idle mode by increasing the exhaust back-pressure (on 1981 models, back-pressure	The EGR is switched off during heavy acceleration; this is controlled by the EGR switch.
is boosted by closing the EGR valve, which is controlled by the EPR switch.	
The amount of EGR is reduced or eliminated when the Torque Converter Clutch (TCC) is activated; this is controlled by the EGR control (trimmer).	

ITEM	FEDERAL	CALIFORNIA	CANADA	FEDERAL TRUCK	CALIFORNIA TRUCK
Internal or external EGR	Internal	External	None	None	Internal
EGR valve type	Modulated	On/Off	None	None	Modulated
EGR switch	No	Yes	No	No	No
EPR switch	Yes	No	No	No	No
EGR control (trimmer)	Yes	No	No	No	Yes
EPR valve	Yes	No	No	No	No
Air cleaner	Two studs	C/O	Two studs, no hose	Two studs, no hose	Two studs
Low vacuum switch	No	Yes	No	No	No

On vacuum-switched or vacuum-modulated 1981 models, exhaust gases are also modulated by the Exhaust Pressure Regulator (EPR) mounted in the exhaust manifold. When the throttle is closed, the EPR valve is closed by the EPR switch; this increases backpressure at idle and recirculates more exhaust to the EGR valve. As engine speed increases, the EPR valve opens and exhaust backpressure is reduced, sending less exhaust to the EGR valve. The EPR valve also opens when the Torque Converter Clutch (TCC) is engaged. This is controlled by the EGR control, or trimmer.

Throttle position on 1981 models is sensed by a vacuum control valve on the throttle shaft on the injection pump. The vacuum control valve generates a vacuum signal proportional to throttle travel. When vacuum regulated by the Vacuum Regulator Valve (VRV) is above 12 in-Hg, the EPR switch closes the circuit; this energizes a solenoid, which allows the full vacuum signal from the vacuum pump to the EPR valve, closing the valve and boosting backpressure. When vacuum is below 12 in-Hg, the switch opens and de-energizes the solenoid, blocking vacuum to the solenoid and venting the EPR valve.

The EGR control assembly on 1981 models is located between the VRV and the EGR valve. When the TCC is disengaged, the solenoid in the EGR control assembly is de-energized, and VRV-regulated vacuum is directed to the EGR valve. When the TCC is engaged, it energizes the solenoid, allowing the vent to open. The reducer valve then reduces VRV-regulated vacuum by 2.5 in-Hg. The EGR valve varies the exhaust flow in accordance with the strength of the vacuum signal it receives from the VRV and the EGR control assembly.

Two systems are used on 1982 models - one on B-body station wagons, and the other on everything else. On B-body station wagons, vacuum from the vacuum pump is modulated by the VRV on the injection pump. Vacuum is highest at idle and decreases to zero by the time the throttle is wide open. So the EGR valve is fully open at idle, and fully closed at wide open throttle. A response vacuum reducer valve is used between the VRV and the EGR valve to allow the EGR to react quickly to changes in throttle position. The VRV is a calibrated bleed - it prevents a trapped signal so the EGR valve can close more quickly.

On all other 1982 models (except the B-body station wagon) and on all 1983 models, a solenoid in the system shuts off vacuum to the EGR valve when the TCC is engaged. This solenoid gets battery voltage from the TCC switch and is grounded through the transmission's governor pressure switch.

Diagnosing the 5.7L EGR system

Now that you have a good idea how the 5.7L EGR systems work, you can test them. In this section, we'll show you how to test the EGR valve, the Exhaust Pressure Regulator (EPR), the Vacuum Regulator Valve (VRV), the Response Vacuum Reducer (RVR), the Vacuum Modulator Valve (VMV), the Torque Converter Clutch (TCC) and the EGR vacuum solenoid. The following diagnostic procedure will help you pinpoint the trouble:

1 If the EGR valve won't open, it's either binding, or stuck, or it's not getting any, or enough, vacuum. Check the EGR vacuum solenoid, the VRV, the RVR, the TCC, the vacuum pump and the hoses. If all of them are okay, replace the EGR valve.

2 If the EGR valve won't close (heavy smoke on acceleration), it's either binding or stuck. Replace the EGR valve.

3 If the EGR valve closes late on systems with a VRV, either the vacuum switch is opening the circuit at below 8 in-Hg vacuum or the VRV is out of adjustment. If the switch is faulty, replace it. If the VRV is the problem, adjust it.

4 If the EGR valve only opens partially, either the EGR valve is binding, or there's low vacuum at the EGR valve. If the valve is binding, replace it. If there's low vacuum at the valve, check the VRV, RVR, vacuum solenoid, vacuum pump and hoses.

Removing, testing and installing the EGR valve

1 Remove the air cleaner.

2 Remove the air crossover.

3 Bend the tabs away from the EGR valve mounting bolts.

4 Remove the EGR valve mounting bolts and remove the valve from the air crossover.

5 Apply a vacuum source to the port on the EGR valve and watch the underside of the valve. It should be fully open at 9.5 to 10.5 in-Hg of vacuum. The valve should close by the time vacuum drops below 5 to 6 in-Hg.

6 Inspect the valve sealing area and verify that it seals tightly when closed.

7 Check the diaphragm by opening the EGR valve by hand and covering the vacuum port with your finger. The valve should remain open for at least 20 seconds. If it doesn't, the diaphragm has a leak. Replace the EGR valve.

8 Clean the gasket area on the air crossover before installing the EGR valve.

9 Install the EGR valve with a new gasket.

10 Install the EGR valve mounting bolts.

11 Bend the locking tabs up against the mounting bolts.

12 Install the air crossover and air cleaner.

Testing the Exhaust Pressure Regulator (EPR) valve

The EPR valve can be tested either on or off the engine. To verify that it's opening and closing correctly, hook up a vacuum pump and gauge to the EPR vacuum port, apply vacuum to the port and watch the lever on the side of the valve under the vacuum diaphragm. It should be fully open at 6 in-Hg vacuum and fully closed at 12 in-Hg.

Testing the Vacuum Regulator Valve (VRV)

The VRV, which is attached to the side of the injection pump, regulates the amount of vacuum in proportion to throttle angle. Vacuum from the vacuum pump is supplied to port A and vacuum at port B is reduced as the throttle is opened. At closed throttle, the vacuum at port B should be about 15 in-Hg; at half-throttle, it should be about 5 or 6 in-Hg; at wide open throttle, it should be zero. If the VRV doesn't operate as specified, replace it.

Testing the Response Vacuum Reducer (RVR)

1 Trace the vacuum hose that connects the TCC solenoid to the RVR. Disconnect this hose from the RVR.

2 Connect a hand-operated vacuum pump to the RVR port.

3 Apply 15 in-Hg to the RVR. The gauge should indicate 0.75 in-Hg.

4 If the RVR doesn't operate as specified, replace it.

Testing the Torque Converter Clutch (TCC)-operated solenoid

The VRV mounted on the side of the injection pump contains a switch which - depending on throttle position - engages or disengages the TCC. This switch also operates a TCC-operated solenoid which turns the EGR valve on and off as the TCC is engaged and disengaged. When the TCC is engaged, the TCC-operated solenoid is energized, which shuts off port 3 and connects port 1 to port 2, shutting off vacuum to the EGR; when the TCC is disengaged, the TCC-operated solenoid is de-energized, and port 1 (the vent port) is shut off while ports 2 and 3 are interconnected, which applies vacuum to the EGR valve. If the TCC-operated solenoid doesn't operate as described, replace it.

Vacuum-modulated EGR system (1984 models with 5.7L engines)

The 1984 system is designed to meet emission standards at low AND high altitude. It compensates for changes in altitude by altering timing and controlling the EGR valve with an altitude sensitive switch and a TCC-EGR cut-off solenoid. The altitude switch, which is located in the position used by the glow plug control relay on 1983 models, controls the HPCA, HPAA and EGR valve. The altitude switch is open at its intended design altitude and closes when the altitude changes. The TCC-EGR cut-off solenoid is normally closed at low altitude and normally open at high altitude. When battery voltage is applied to the HPCA, it has zero pressure; when battery voltage is applied to the HPAA, it has 5 psi.

The Housing Pressure Cold Advance (HPCA) system

The HPCA system is an emission control device designed to improve cold starts and idling, and reduce white smoke and noise when the engine is cold. The HPCA solenoid allows more engine advance during engine warm-up. It consists of a solenoid assembly and a ball check return connector installed inside the governor cover of the injection pump. On 1981 through 1983 models, a sensing unit mounted in the cylinder head generates the electrical signal which controls the operation of the solenoid; on 1984 models, it's part of a thermal vacuum switch in the intake manifold. The switch is calibrated to open the circuit at 120-degrees F (89-degrees on 1984 models). Below the switching point, housing pressure is decreased from 8 to 12 psi to zero, which advances the timing 3-degrees. Above 120-degrees F. (89-degrees on 1984 models), the switch opens, de-energizing the solenoid and returning the housing pressure to 8 to 12 psi. The fast-idle solenoid is also energized by the same switch. The switch again closes when the temperature falls below 95-degrees F (79-degrees F. on 1984 models).

During cold warm-ups, the solenoid plunger moves up and the rod contacts the return connector ball. When the ball is moved off its seat, housing pressure is reduced by the increased flow through the connector. Because of lower housing pressure, resistance to the movement of the advance piston is less, so the piston can move further in the advance direction.

When the engine reaches its normal operating temperature, the electrical signal to the solenoid is terminated and the plunger returns to its initial position.

On a hot engine, whenever you change the fuel filter or the injection pump, or when the vehicle runs out of fuel, unplug the connector from the temperature switch and bridge the connector terminals. This will help you purge air from the pump by allowing more fuel to pass through the return line. On a cold engine, the circuit is already closed, so this step is unnecessary.

The Housing Pressure Altitude Advance (HPAA) system

The HPAA solenoid is located in the fuel return line; it regulates housing pressure in accordance with altitude. When the HPAA solenoid is "on," the glass check ball is seated and regulates pressure at 5 psi; when the HPAA solenoid is "off," the ball is moved of its seat, the return line is open and there is no regulation. It's possible to have both the HPCA off and the HPAA regulating housing pressure at the same time. It's also possible to have just the HPCA off while the HPAA regulates housing pressure. The HPCA must be energized, i.e. the plunger extended, holding the housing pressure regulating check ball off its seat, and not regulating to allow the HPAA solenoid to regulate at its calibrated value. Higher housing pressures retard timing. To understand the relationship between the altitude, the engine coolant temperature, the HPAA, the HPCA and the timing, refer to the table below.

The electrical signal which controls the operation of the HPAA solenoid is generated by an altitude sensing switch. Two different switches are used: One is for Federal vehicles intended for use below 4000 feet and the other is for vehicles which will be operated above 4000 feet. The altitude switch is open at its design altitude and closes when the altitude changes.

An altitude relay with normally closed contacts completes the path for battery voltage to the HPAA solenoid. The contacts are open when the relay is energized by battery voltage, which comes through the TVS electrical switch when it's closed at less than 89-degrees F. When the engine temperature is less than 89-degrees F., the altitude relay contacts are open, preventing the HPAA from operating. The TVS electrical switch also provides the current which energizes the fast-idle solenoid.

A diode in the circuit between the TVS electrical switch and the EGR trim solenoid blocks the current path from the TVS switch to prevent operation of the EGR trim solenoid when cold. It also provides a path for the altitude switch current to energize the HPCA solenoid for operation of a Federal vehicle at higher altitude. This diode is removed from the high-altitude version of the system, and the circuit is open, so no current is provided to the HPCA solenoid and current can't get to the EGR trim solenoid from the TVS electrical path.

How the circuit works on a Federal (NA5) low-altitude version of the system

When the engine is cold (temperature is less than 89-degrees F.) and the altitude is less than 4000 feet, battery voltage comes from the diesel fuse out pin 139 and through the closed contacts of the TVS electrical switch. It goes through a diode in the diode module and energizes the fast-idle solenoid and passes through an external diode and energizes the HPCA solenoid. The HPCA plunger moves up and the rod contacts the check ball and moves off its seat. With the check ball off its seat, housing pressure is reduced because of the increased flow through the connector. Reduced housing pressure advances the timing. Current from the TVS switch also energizes the altitude relay, opening the normally closed contacts, which prevents the HPAA from operating when the engine is cold. When the HPAA solenoid is de-energized, the check-ball is held off its seat so it's open and not regulating housing pressure. The diode at pin 939 blocks current to the EGR trim solenoid.

When the engine is hot (temperature is above 89-degrees F.) and the altitude is less than 4000 feet, the TVS switch opens and de-energizes the fast-idle and HPCA solenoids and altitude relay. The altitude relay contacts are again normally closed.

When the engine is cold (temperature is less than 100-degrees F.) and the altitude is above 4000 feet, the operation is the same as low altitude, except the altitude switch is closed, which energizes the EGR trim solenoid. This connects ports 1 and 2, which allows air to bleed into the VRV and cause a 2 in-Hg pressure drop, lowering the EGR valve position at altitude.

When the engine is hot (temperature is above 100-degrees F.), the TVS head bolt switch is open, so the normally closed altitude relay contacts are closed. The altitude switch is closed, so current energizes the EGR trim solenoid, reducing EGR. Current also passes through the normally closed contacts of the altitude relay to the HPAA solenoid. The HPAA is on, allowing the regulating check ball to seat and regulate the housing pressure at its calibrated value. Altitude switch current must also energize the HPCA solenoid. The HPCA plunger is extended, holding the housing pressure regulating check ball off its seat. With the HPCA open, the HPAA can regulate at its calibrated value of 5 psi alone. This advances timing at altitude by 2-degrees.

Housing Pressure Altitude Advance (HPAA) system chart

	Altitude	Coolant	HPCA	HPAA	HSG pressure	Timing change
Federal (NA5)	Sea level	Cold	On	Off	0 psi	+4 degrees
Federal (NA5)	Sea level	Hot	Off	Off	10 psi	0 degrees
Federal (NA5)	4000 feet	Cold	On	Off	0 psi	+4 degrees
Federal (NA5)	4000 feet	Hot	On	On	5 psi	+2 degrees
Altitude (NA6)	4000 feet	Cold	On	Off	0 psi	+4 degrees
Altitude (NA6)	4000 feet	Hot	Off	Off	10 psi	0 degrees
Altitude (NA6)	Sea level	Cold	On	Off	0 psi	+4 degrees
Altitude (NA6)	Sea level	Hot	Off	On	15 psi	-2 degrees

How the circuit works on a high-altitude version of the system

When the engine is cold (temperature below 89-degrees F.) and the altitude is above 4000 feet, operation is the same as a Federal low-altitude system.

When the engine is hot (temperature above 89-degrees F.) and the altitude is above 4000 feet, operation is the same as a Federal low-altitude system.

When the engine is cold (temperature below 100-degrees F.) and the altitude is below 4000 feet, operation is the same as at high altitude, except the altitude switch is now closed, which energizes the EGR trim solenoid and connects ports 1 and 2 on the solenoid. This blocks off the atmospheric bleed to the VRV, which increases the EGR valve opening.

When the engine is hot (temperature above 89-degrees F.) and the altitude is below 4000 feet, the TVS electrical switch is open, so the normally closed altitude relay contacts are closed. The altitude switch is closed, so current energizes the EGR trim solenoid, which connects ports 1 and 2 on the solenoid and blocks off atmospheric bleed to the VRV, increasing the EGR valve opening.

Altitude switch current also flows through the normally closed contacts of the altitude relay, energizing the HPAA solenoid. When the HPAA solenoid is energized, the plunger is pulled in and the check ball is seated, regulating pressure at about 5 psi. The HPCA is still de-energized and its check ball seated at about 10 psi. The HPAA regulation of 5 psi plus the HPCA regulation of 10 psi produces a total housing pressure of 15 psi. This higher housing pressure causes the timing to retard about 2-degrees.

Checking the 1984 EGR system components

Thermostatic Vacuum Switch (TVS)

The 1984 EGR uses a Thermostatic Vacuum Switch (TVS), mounted in the coolant passage in the intake manifold, as a cold override. The TVS has two roles: Its upper portion is an electrical switch that energizes the HPCA and fast-idle solenoids. This occurs when the coolant temperature is below 89-degrees F; its lower portion is a vacuum switch that prevents the EGR from operating below the calibration point. When the engine coolant temperature is less than 100-degrees F., ports 1 and 2 are connected and the switch allows the atmosphere to vent the EGR valve, which prevents the system from operating. When the coolant temperature exceeds 100-degrees F., the TVS piston moves up, which shuts off the vent and connects ports 2 and 3, allowing vacuum modulated by the VRV to operate the EGR valve through the system. Port 1, the vent, is sealed off.

Vacuum Regulator Valve (VRV)

The VRV is mounted on the side of the injection pump. Vacuum is highest at idle and decreases to zero at wide-open throttle. So the EGR valve is fully opened at idle and closed at wide-open throttle. Vacuum from the vacuum pump is supplied to port A and vacuum at port B is reduced as the throttle is opened. At closed throttle, there should be 15 in-Hg present; at half-throttle, there should be about 6 in-Hg; at wide-open throttle, there should be no vacuum.

Torque Converter Clutch (TCC) solenoid

When the TCC is engaged, the solenoid vents the EGR system. When the solenoid is energized, ports 1 and 2 are connected; when it's de-energized, ports 2 and 3 are connected.

Altitude solenoid

This solenoid changes the amount of EGR at low or high altitude. When the solenoid is energized, ports 1 and 2 are connected; when it's de-energized, ports 2 and 3 are connected.

Checking the Vacuum Reducer Valve

The VRV is located between the Vacuum Regulator Valve (VRV) and the EGR valve. When used, it allows a calibrated atmospheric bleed to the EGR valve. To check it, connect a vacuum gauge to the port marked "To EGR valve or TCC solenoid." Connect a hand-operated vacuum pump to the VRV port. Apply 15 in-Hg. The reading on the gauge should be 2.0 in-Hg lower than the pump reading at any altitude.

Quick Vacuum Response (QVR) valve

The QVR valve is located between the vacuum reducer and the EGR valve. It's designed to vent the EGR valve to the atmosphere quickly when the vacuum drops on the vacuum reducer side. This improves throttle response. When there's vacuum on the vacuum reducer side, the external vent is closed. When there's no vacuum on the reducer side, the vent is open, allowing atmospheric pressure to close the EGR valve.

Checking the EGR valve

Apply vacuum to the vacuum port. The valve should be closed below 6 in-Hg and fully open at 10.6 in-Hg.

The altitude switch

The altitude switch controls the altitude solenoid, the HPCA and the HPAA. It's usually located in the position used by the glow plug relay on earlier systems. On Federal vehicles, the switch directs battery voltage to the altitude solenoid, the HPAA and the HPCA at high altitude. The switch is open at its design altitude, but closes when the altitude changes. Below 4000 feet, the low altitude version of the switch is open; above 4000 feet, it's closed. The high altitude version is just the opposite - below 4000 feet, it's closed; above 4000 feet, it's open.

The altitude relay

The altitude relay contacts are normally closed. When the engine coolant temperature is below 89-degrees F., the TVS cold override switch (the upper part of the TVS) is closed. This energizes the altitude relay, opening the contacts and preventing the HPAA from operating when cold.

Troubleshooting the 1984 EGR system

Federal (low-altitude) emissions

When the engine is cold (temperature less than 100-degrees F.) at low or high altitude (below or above 4000 feet), and the TCC is not applied, modulated vacuum from the VRV goes to port 3 of the TCC solenoid and ports 2 and 3 of the TCC solenoid are connected when it's de-energized. Vacuum then goes from port 2 of the TCC solenoid to port 3 of the TVS. Ports 1 and 2 on the TVS are connected when coolant temperature is less than 100-degrees F. Port 1 is a vent to the atmosphere, so the EGR system is open and can't function.

When the engine is hot (temperature above 100-degrees F.) at low altitude (below 4000 feet) and the TCC is not applied, ports 2 and 3 on the TVS are connected. This permits modulated vacuum to get to port 3 at the altitude solenoid and the vacuum reducer valve. Ports 2 and 3 on the altitude solenoid are connected when it's de-energized. Port 2 goes to the VRV, so the vacuum is the same on both sides of the VRV. Vacuum then acts on the QRV, closing it to atmosphere. The modulated vacuum then opens the EGR valve in proportion to the throttle position. When the TCC is applied, the TCC solenoid is energized by a ground from the governor pressure switch. Ports 1 and 2 on the TCC solenoid are then connected. Port 1 is an atmospheric vent, so the EGR system is vented and inoperative. The QVR valve is used to quickly vent the EGR valve when vacuum drops to zero on the VRV side.

When the engine is hot (temperature above 100-degrees F.) at high altitude (above 4000 feet) and the TCC is not applied, the altitude switch closes, which sends battery voltage to the altitude solenoid and energizes it. Ports 1 and 2 on the altitude solenoid are then connected, allowing atmosphere to bleed down the vacuum through an orifice in the vacuum reducer valve by about 2 in-Hg.

The vacuum at the EGR valve is two inches lower than at the VRV, so it closes slightly. This reduces EGR at altitude on a Federal emission vehicle. When the TCC is applied, the TCC solenoid is energized by a ground from the transmission governor pressure switch. Ports 1 and 2 on the TCC solenoid are then connected. Port 1 is an atmospheric vent, so the EGR system is vented and inoperative.

High-altitude emissions

When the engine is cold (temperature below 100-degrees F.) at low altitude (below 4000 feet) and the TCC is not applied, the system operates the same way as a Federal (low-altitude) emission vehicle. The

EGR doesn't operate on an engine unless the temperature is above 100-degrees F.

When the engine is hot (temperature above 100-degrees F.) at high altitude (above 4000 feet) and the TCC is not applied, modulated vacuum from the VRV goes through ports 3 and 2 on the TCC solenoid, to port 3 on the.

TVS and out port 2 on the TVS. Then it goes to port 1 on the altitude solenoid, to the vacuum reducer, then the QVR and finally to the EGR valve. The altitude solenoid is de-energized, so ports 2 and 3 are connected. Port 3 is an atmospheric vent that allows atmosphere to bleed down the vacuum through an orifice in the vacuum reducer valve by 2 in-Hg. The vacuum at the EGR valve is 2 in-Hg lower than at the VRV, so it closes slightly. This reduces the amount of exhaust gases admitted into the intake manifold through the EGR at high altitude. When the TCC is applied, the TCC solenoid is energized by a ground from the transmission governor pressure switch. Ports 1 and 2 on the TCC solenoid are now connected. Port 1 is an atmospheric vent, so the EGR system is vented and inoperative.

When the engine is hot (temperature above 100-degrees F.) at low altitude (below 4000 feet) and the TCC is not applied, the altitude switch closes, sending battery voltage to the altitude solenoid and energizing it. Ports 1 and 2 on the altitude solenoid are then connected, which removes the calibrated 2 in-Hg vacuum drop at the altitude solenoid. Vacuum at the EGR valve is now the same as vacuum at the VRV. The EGR valve opens more, increasing the amount of exhaust admitted into the intake by the EGR valve. The QVR valve is used to quickly vent the EGR valve when vacuum drops to zero on the VRV side.

Vacuum-switched EGR system (LH6 engine)

EGR systems on 6.2L and 6.5L engines

Note: *The operation of the EGR valve on 1994 and later systems is controlled entirely by the Powertrain Control Module. The PCM receives inputs from accelerator pedal position, baro sensor reading and engine speed. The computer then sends signals to the EGR solenoid and EGR vent solenoid which in turn operate the EGR valve. The valve receives vacuum from the engine vacuum pump. The EGR valve is located on top of the air intake pleneum and the solenoid is located on the rear of the left valve cover.*

On this type of EGR system **(see illustration)**, exhaust gas is admitted at full-flow or not at all. At closed throttle, the EGR valve is fully open and remains open until a calibrated throttle position is reached, at which point it closes. A throttle position switch (TPS) senses this throttle position. As the throttle is opened, the TPS closes when the calibration point is reached and de-energizes a solenoid which shuts the vacuum signal to the EGR valve.

The system also uses an Exhaust Pressure Regulator (EPR) valve in the exhaust system to restrict flow and increase exhaust backpressure. The EPR valve is used in conjunction with the vacuum-switched EGR valve at the intake manifold. When the throttle is closed, the EPR valve is closed, increasing the recirculation of exhaust gas. As the throttle is opened, the valve opens, decreasing the amount of exhaust backpressure.

Vacuum isn't modulated to the EGR or EPR valves. The TPS mounted on the injection pump throttle shaft operates a pair of on-off vacuum solenoid switches controlling flow to the EGR and EPR valves.

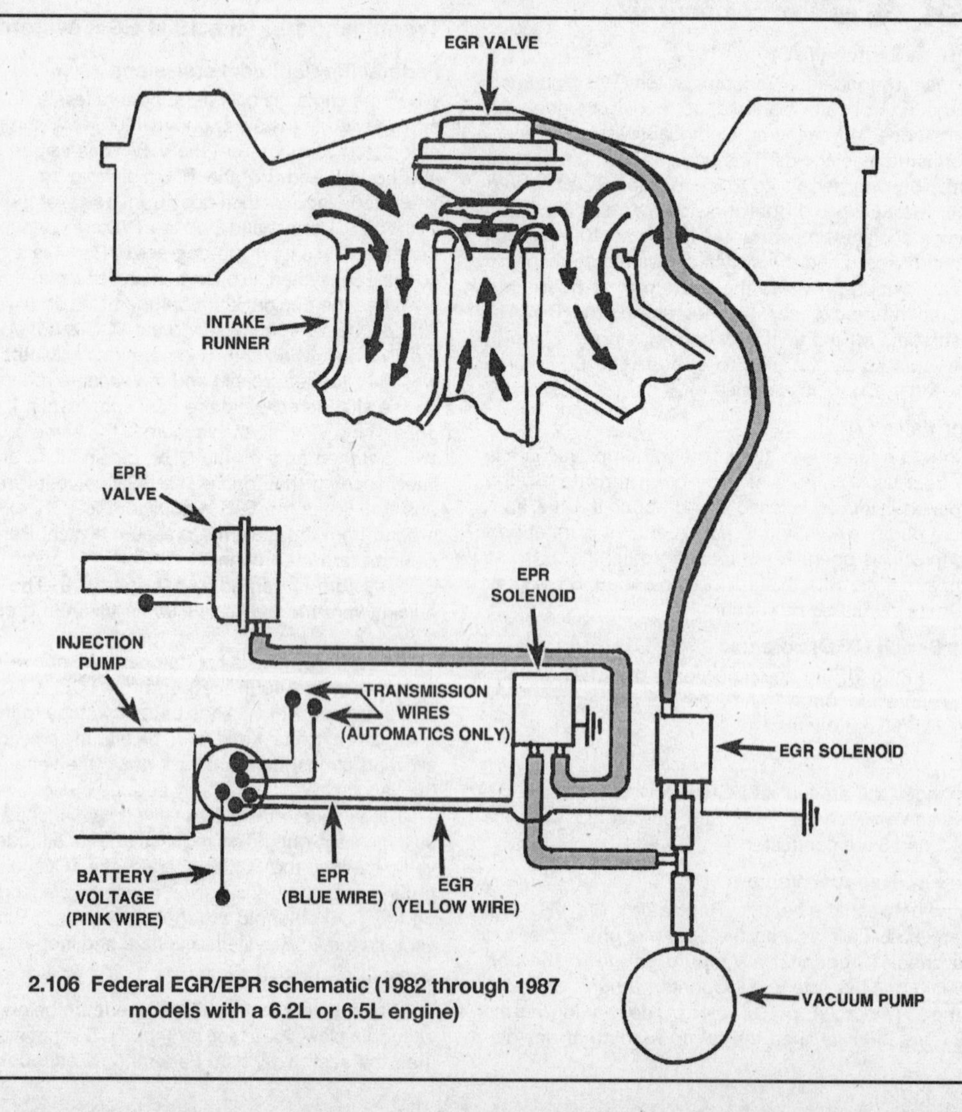

2.106 Federal EGR/EPR schematic (1982 through 1987 models with a 6.2L or 6.5L engine)

At about 15-degrees of throttle angle, current is broken to the normally closed EPR vacuum solenoid, and the EPR valve opens when the solenoid cuts off vacuum. The normally open EGR vacuum solenoid gets current from the TPS at a calibrated throttle position, vacuum to the EGR valve is cut off and the EGR valve closes, preventing the flow of exhaust into the intake manifold.

The TPS has two contacts inside it: One contact (blue or purple wire) sends battery voltage to the normally closed EPR solenoid at idle; this opens the solenoid valve and vacuum closes the EPR valve. The other contact (yellow wire) sends battery voltage to the normally open EGR solenoid at a calibrated throttle angle. This current energizes, and closes, the EGR solenoid, cutting off vacuum and closing the EGR valve. On some switches, there's a slight delay in the time between when the EPR opens and the EGR closes. You can identify the switch type by looking at the color of the cam used to change the EPR/EGR switch points: A blue cam has no difference; a black cam has a five-degree difference; a red cam has a 10-degree difference.

On some models equipped with 700-R4 (MD8) automatic transmissions, a third switch inside the TPS operates the Transmission Converter Clutch (TCC). This switch, which controls battery voltage coming from the brake switch to the TCC solenoid, disengages the converter clutch at a throttle angle of less than 8-degrees. On 4WD models, there's a normally closed relay in the parallel path of current to the TCC solenoid at pin B of the connector at the transmission. It's energized by grounding pin 4 of the relay. Pin 4 is grounded through the 4WD indicator switch in the transfer case. It's grounded in either "4 high" or "4 low." When the relay is grounded, the normally closed relay is switched to open, interrupting the battery voltage path to pin B of the connector at the transmission. The only way to get the converter to operate in 4WD is to close the fourth-gear switch in the transmission, allowing battery voltage from the other parallel path at pin A to pass through the now closed contacts of the fourth gear switch to the TCC solenoid.

Diagnosing the EGR/EPR system on 6.2L LH6 engines

1 Start the engine and warm it up until the thermostat opens.
2 Remove the air cleaner so you can observe the operation of the EGR valve. Using the table below, compare the operation of the EGR valve, EGR solenoid, EPR valve and EPR solenoid.
3 With the engine at idle, the EGR valve should be open (the valve head is in its "up" position and there's exhaust noise at the intake). If the EGR valve isn't open, check and correct any electrical and hose connections which may be loose and/or disconnected.
4 Remove the vacuum hose from the EGR valve. The valve head should drop and the exhaust noise should diminish significantly. Reattach the hose.
5 At idle, the hose to the EGR valve should have about 20 inches of vacuum. If vacuum isn't present, check the output of the vacuum pump at the pump. It should produce at least 20 in-Hg.
6 If vacuum is present at the EGR valve, but the valve doesn't open and close as the hose is attached and detached, the EGR valve is stuck. Try to "unstick" it; if you can't get it working, replace it.
7 Manually operate the throttle lever at the injection pump through 15 to 20 degrees of travel. The EGR valve should close when the TPS reaches the calibrated point.
8 Verify that battery voltage is present in the pink wire to the TPS when the ignition switch is turned on. If battery voltage isn't present, check for any loose connections, an open in the wire or a blown 20-amp gauge idle fuse.
9 Check the wiring connections and make sure nothing's loose. Check the fuse and replace it if it's blown. Turn the key on and make sure there's battery voltage in the blue wire from the TPS. This blue wire feeds the EPR solenoid. At idle, if the pink wire has battery voltage, but the blue one doesn't, the TPS is bad. Replace it.
10 With the engine off, and the ignition switch turned on, operate the throttle through 20-degrees of travel. At about 15-degrees, the TPS will cut out battery voltage to the EPR (blue wire); at about 20-degrees, the TPS will cut in battery voltage to the EGR (yellow wire). If it doesn't operate as described, the TPS is bad. Replace it.
11 Verify that the electrical connections are good at the EGR/EPR solenoid assembly and that the hoses are routed correctly and connected to the solenoids.
12 If there's vacuum present at the solenoid assembly and the solenoids are receiving an electrical signal as described above, but the TPS doesn't operate the EGR and/or EPR valves, the solenoid assembly itself is faulty. Replace it.

Identifying a bad diode in the solenoid assembly (EGR system on 1982 6.2L LH6 engine)

On 1982 LH6 engines, a condition exists that can allow excessive electrical feedback to short out a diode in the EGR and/or EPR solenoids. The diodes are for radio noise suppression only. When the diode shorts, it usually blows the 20 amp gauge fuse. If the fuse blows, there won't be any power to electrical components such as the glow plugs, cold advance, fast idle, etc. AND the EGR system won't operate either. When the EGR system malfunctions, full vacuum is supplied to the EGR valve at all speeds, resulting in heavy black exhaust smoke and low power.

To prevent the diodes from shorting out as a result of heavy feedback loads, a special jumper harness with a built-in diode to reduce feedback load (part no. 14048052) is available. If you've noticed heavy black exhaust smoke and traced the condition to a diode in the EGR/EPR solenoid assembly, here's how to install the jumper:

1 Stop the engine and turn the key to the Off position.
2 Disconnect the vacuum and electrical connections to the EGR/EPR solenoid assembly.
3 Remove the solenoid assembly mounting bolt and remove the solenoid assembly.
4 Install the new solenoid assembly.
5 Reattach the vacuum and electrical connections to the new solenoid assembly.
6 Disconnect the wire to the TPS and install the new jumper harness between the connectors. Make sure the blue, pink and yellow wires line up, i.e. are at matching terminals of the connector.
7 Install a new 20-amp gauge idle fuse.

Adjusting the TPS

The TPS must be adjusted on the throttle shaft of the injection pump. It's not necessary to remove the pump. You'll need a special "go-no go" gauge and the correct throttle position switch gage block to adjust it. Different blocks are needed for different applications. You'll find the correct block for your particular application specified on the Vehicle Emission Control Information (VECI) label in the engine compartment. (Gage block dimensions weren't listed on the VECI label until the 1983 model year.) The accompanying tables specify the correct gage blocks for 1982 through 1986 models. If the gage block dimension isn't specified here, or the VECI label is missing, check with a dealer service department for the correct block dimensions. You'll also need a self-powered test light or ohmmeter.

EGR/EPR system chart for 6.2L LH6 engine

Engine speed	EGR valve	EGR solenoid	EPR valve	EPR solenoid
Idle to 15 degrees of throttle opening	Open	Not energized (vacuum to valve)	Closed	Energized (vacuum to valve)
15 to 20 degrees of throttle opening	Open	Not energized (vacuum to valve)	Open	Not energized (no vacuum to valve)
20 degrees to wide open throttle	Closed	Energized (no vacuum to valve)	Open	Not energized (no vacuum to valve)

TPS chart for 1982 LH6 engines

Usage	Altitude	Transmission	Models	Throttle Position Switch part number	Gage bar switch closed	Gage bar switch open	Gage tool number
Nationwide	All	Manual	All	14050405	0.646 inch	0.668 inch	J-33043-2
Federal	All	Automatic	All	14033943	0.646 inch	0.668 inch	J-33043-2
California	All	Automatic	C and K	14033943	0.646 inch	0.668 inch	J-33047-2
California	All	Automatic	C and K	14050408	0.602 inch	0.624 inch	J-33043-4

TPS chart for 1983 LH6 engines

Engine assembly part number	Broadcast code	Throttle position switch part number	Gage bar switch closed	Gage bar switch open	Gage tool number
14061529	UHB	14050405	0.646 inch	0.668 inch	J-33043-2
14061531	UHC	14050405	0.646 inch	0.668 inch	J-33043-2
14061545	UHD	14050405	0.602 inch	0.624 inch	J-33043-4
14061549	UHF	14050405	0.646 inch	0.668 inch	J-33043-2
14061550	UHH	14066239	0.646 inch	0.668 inch	J-33043-2
14061552	UHJ	14066238	0.602 inch	0.624 inch	J-33043-4
14061560	UHN	14066239	0.646 inch	0.668 inch	J-33043-2
14060581	UHA	14050405	0.646 inch	0.668 inch	J-33043-2
14061573	UHS	14050405	0.646 inch	0.668 inch	J-33043-2
14066299	UHZ	14050405	0.602 inch	0.624 inch	J-33043-4
14061571	UHR	14050405	0.646 inch	0.668 inch	J-33043-2
14061576	UHT	14066239	0.646 inch	0.668 inch	J-33043-2
14061578	UHU	14066238	0.602 inch	0.624 inch	J-33043-4
14061580	UHW	14066239	0.646 inch	0.668 inch	J-33043-2

TPS chart for 1984 through 1986 LH6 engines

Engine assembly broadcast codes	Throttle Position Switch part number	Gage bar switch closed	Gage bar switch open	Gage tool number
HHB, DHB and FHB	14050405	0.602 inch	0.624 inch	J-33043-4
HHF, DHF and FHF	14050405	0.602 inch	0.624 inch	J-33043-4
HHD, DHD and FHD	14066239	0.646 inch	0.668 inch	J-33043-2
HHJ, DHJ and FHJ	14066239	0.646 inch	0.668 inch	J-33043-2
HHK, DHK and FHK	14050405	0.602 inch	0.624 inch	J-33043-4
HHN, DHN and FHN	14050405	0.602 inch	0.624 inch	J-33043-4
HHW, DHW and FHW	14066239	0.646 inch	0.668 inch	J-33043-2
HHY, DHY and FHY	14066239	0.646 inch	0.668 inch	J-33043-2
HHR, — and —	23500164	0.624 inch	0.646 inch	J-33043-4, J-33043-2

TPS chart for 1984 and later LL4 engines

Throttle position switch part number	Gage bar switch closed	Gage bar switch open	Gage tool number
14066207	0.751 inch	0.773 inch	J-33043-5

1 With the throttle lever in the closed position, loosen the TPS retaining screws.
2 On LH6 engines, attach the leads of an ohmmeter or self-powered test light to the IGN terminal (pink wire) and the EGR terminal (yellow wire); on an LL4 engine with a 700 R4 transmission, hook up the leads across the connector terminals.
3 Insert the correct "switch-on" gage block between the gage boss on the injection pump and the wide-open stop screw on the throttle shaft.
4 Rotate and hold the throttle lever against the gage block.
5 Facing the TPS, rotate it clockwise until continuity is indicated (low meter reading) across the IGN and EGR terminals. Hold the switch body at this position and tighten the mounting screws to 4 to 5 ft-lbs.
6 Release the throttle lever and allow it to return to the idle position. Remove the "switch-on" gage bar and insert the "switch-off" gage bar. Rotate the throttle lever against the "switch-off" gage bar. There shouldn't be any continuity (infinite resistance reading) across the IGN and EGR terminals. If no continuity exists, the TPS is now correctly adjusted. If there's continuity, the switch must be readjusted. Return to Step 1 and repeat this procedure.

LL4 engines with a Vacuum Regulator Valve (VRV)

LL4 engines equipped with the M40 THM 400 automatic transmission use a Vacuum Regulator Valve (VRV) instead of a TPS. The VRV supplies a vacuum signal to the transmission vacuum modulator when the engine is under a load. This signal is proportional to throttle travel: At idle, maximum vacuum is present, at wide-open-throttle, zero vacuum is present. This enables the vacuum modulator to regulate the transmission shift points and line pressure.

Adjusting the VRV on LL4 engines

1 Loosen the VRV retaining screws so it can rotate freely.
2 Attach a vacuum source to the inner vacuum nipple and a vacuum gauge to the outer nipple. Apply 20 in-Hg to the inner nipple.
3 Insert the "switch-on" end of the VRV gage bar (Kent-Moore J-33043-2) between the gage boss on the injection pump and the wide-open stop screw on the throttle lever.
4 Rotate and hold the throttle shaft against the gage bar.
5 Facing the VRV, slowly rotate it clockwise until the vacuum gauge reads 8 in-Hg (1982 through 1986 models) or 11.5 in-Hg (1987 and later models). Hold the VRV at this position and tighten the retaining screws to 4 to 5 ft-lbs.
6 Check your work by releasing the throttle shaft and allowing it to return to the idle stop position, then rotate the throttle shaft back against the gage bar to determine whether the vacuum gauge reads with 8 in-Hg (1982 through 1986 models) or 11.5 in-Hg (1987 and later models). If the indicated vacuum reading is outside specification, readjust the VRV. If it's within specification, and the engine is a 1982 through 1986 model, you're done; if it's a 1987 or later model, proceed to the next Step.
7 With the VRV correctly adjusted, the detent switch point must be confirmed. Attach the leads of an ohmmeter or self-powered test light to the ignition terminal (pink wire with black stripe) and detent terminal (orange wire) of the pigtail connector.
8 Insert the "switch-open" end of the gage tool (Kent-Moore J-36142, 0.175-in) between the wide-open stop pin on the injection pump and the wide-open stop screw on the throttle lever.
9 Rotate the throttle lever against the gage tool. There shouldn't be any continuity (infinite resistance reading).
10 Release the throttle lever and remove the switch gage tool. Rotate the throttle lever to the wide-open throttle position (lever stop screw touching the pump stop-pin).
11 The switch must make continuity before wide-open-throttle (meter reading of 0 to 1 ohm). If switch closure is outside of specification, readjust it. Return to Step 1 and repeat the procedure.

Adjusting the VRV on some early 1982 C, K or P trucks with the THM 400 transmission

As a result of incorrect VRV calibration, these models have had problems with high or late upshifts. The problem was corrected on models manufactured after March 1982 by installing a different VRV

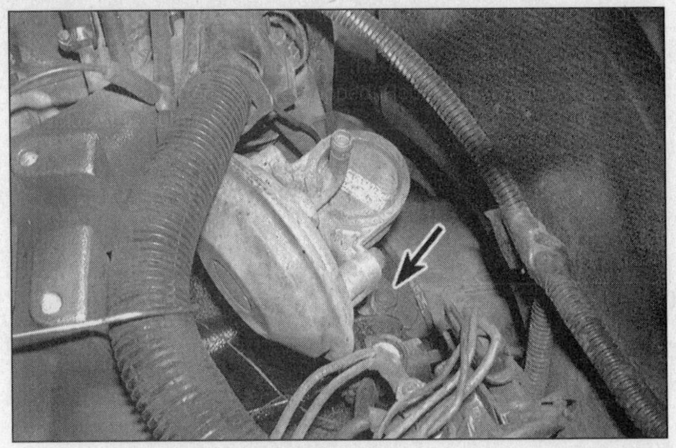

2.107 The gear-driven vacuum pump is mounted in the same location as the distributor on a gasoline engine; to remove it, simply remove the hold-down bolt (arrow) and pull the pump straight up

(part no. 14057219). If you've had this type of problem with an early 1982 truck, here's how to verify that the VRV is the cause:
1 Identify which VRV is on the engine. The old VRV rotating cam is green and the part number, 14033982, is cast into the body; the new VRV rotating cam is orange and the part number, 14057219, is white-lettered on the face of the VRV.
2 If the VRV is the old design, remove it; if it's the new design, go to Step 5.
3 Install a new VRV (part no. 14057219).
4 Adjust the new VRV (see previous procedure).
5 If the new VRV is already installed, verify that it produces a vacuum of 8 in-Hg at the outer nipple when a vacuum of 20 in-Hg is applied to the inner nipple (see Steps 1 through 6 in the previous procedure).
6 Check the vacuum output of the vacuum pump, which should be 20 or 21 in-Hg.

Vacuum pump - general information

Unlike a gasoline engine, a diesel engine doesn't develop intake manifold vacuum - its air crossover and intake manifold are unrestricted. The vacuum needed to run vacuum-powered accessories (air conditioning servos, cruise control servo, transmission vacuum modulator, etc.) is provided by a diaphragm type vacuum pump mounted in the same location as the distributor on a gasoline engine **(see illustration)**. (Some 6.2L and 6.5L engines equipped with the Stanadyne Model 80 fuel filter use a belt-driven pump - see below). The pump is driven by a cam inside the drive assembly to which it's attached. The pump's diaphragm moves back and forth, pulling air through the inlet port on the front, through the pump and out the rear port.

The drive housing assembly has a drive gear on the lower end which meshes with a drive gear on the rear end of the camshaft. This drive gear rotates the cam in the drive housing; it also powers the oil pump. **Caution:** *Because the vacuum pump drive assembly is also the drive for the engine oil pump, NEVER operate the engine without the drive and vacuum pump assembly installed.*

Some 1984 and later 6.2L models with the Stanadyne Model 80 fuel filter system, and all 6.5L models, use a belt-driven vacuum pump. The oil pump drive on these models is also different - an oil pump drive replaces the gear driven vacuum pump.

Diagnosing the vacuum pump

1 If the pump makes a loud clattering noise, the screws between the pump assembly and the drive assembly are probably loose. Tighten them.
2 If the pump makes a hooting noise, the valve isn't functioning properly. Replace the pump.
3 If pump assembly is loose on the drive assembly - and won't tighten down - the threads are stripped. Repair the threads, if possible, or replace the pump.

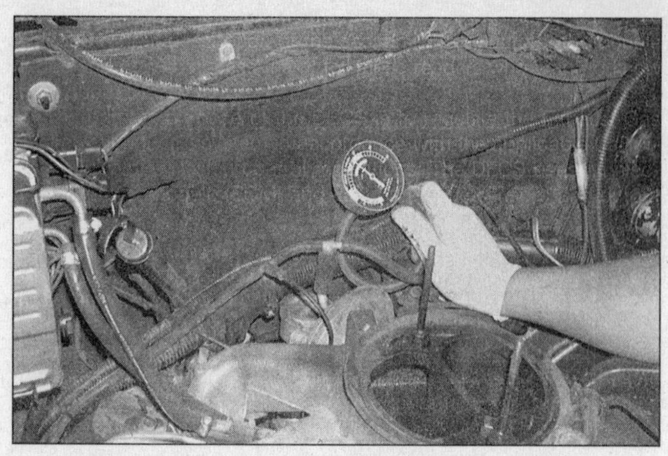

2.108 Test setup for checking the vacuum pump at idle

4 If there's oil around the drive assembly end plug, the plug is loose. Seat the plug or replace the drive assembly.
5 If there's oil around the crimped bead at the circumference of the pump, the crimp is leaking. Replace the pump.
6 If the pump appears to be okay so far, check its operation (see below).

Testing the vacuum pump

1 Disconnect the vacuum hose from the pump inlet pipe.
2 Attach a hose and vacuum gauge to the inlet pipe.
3 With the engine running, the pump should reach at least 20 to 21 in-Hg within 30 seconds; with the engine off, the pump shouldn't lose more than 1 in-Hg in less than 1-1/2 seconds.
4 If the pump checks out okay, go to the next Step. If it doesn't, i.e. if there's a low reading or a fluctuating reading, check the gauge and connections for leaks. Also check the idle rpm. If the pump is belt-driven, check the belt tension and the fit of the pulley on its shaft.
5 Reattach the vacuum hose to the pump. Insert a tee-fitting between the vacuum hose and accessories. Place the tee as close as possible to the pump inlet. Attach the vacuum gauge to the tee **(see illustration)**. Start the engine. At idle, the indicated vacuum should be no more than 3 in-Hg less than the vacuum measured previously after one minute. If it isn't, the problem is not in the vacuum system; if it is, recheck all hoses for leaks and repair as necessary. If the indicated vacuum is still low, check all vacuum accessories for out-of-specification leaks. Repair or replace as required.

Diesel Electronic Control System (DECS) - general information

Note: *At the time of publication, both the California and Federal versions of the Diesel Electronic Control System (DECS) - and a lot of other fuel system components as well (see your dealer for a complete description of the coverage on your vehicle) - are protected by a Federally-mandated, extended warranty of 5 years/50,000 miles; recent Federal legislation will have doubled this figure by the time you read this. During the warranty period, every emission-related device in the DECS that malfunctions can be tested, serviced and/or replaced under the terms of the emissions warranty. If a failure occurs after the terms of the warranty have expired, we don't recommend that you attempt to service either system. Troubleshooting them is fairly complex and requires considerable experience with electrical diagnosis.*

GM has equipped 5.7L, 6.2L and 6.5L engines with a trouble code system to help you pinpoint the source of a problem. So we've included a brief description of the sensors, the actuators, the trouble codes and how to access them, for those readers who want to know more about DECS.

The primary purpose of the DECS is to regulate the Exhaust Gas Recirculation (EGR) system in accordance with the applicable California or Federal emission control regulations. The DECS uses an electronic control module (ECM) and various engine sensors and output actuators to accomplish this task.

Though they operate within different parameters, the California and Federal versions of the DECS are similar in operation. The ECM controls the amount of EGR introduced into the intake by turning an EGR solenoid - which regulates the vacuum signal to the EGR valve - on and off. The ECM computes the amount of EGR based on various inputs to the ECM from the engine sensors. The three main sensor inputs to the ECM are engine rpm, vehicle speed and the position of the fuel metering valve on the injection pump.

The ECM varies the on and off time of the EGR solenoid in accordance with the information it receives from these three inputs. A Manifold Absolute Pressure (MAP) sensor monitors the ECM control of EGR by measuring the amount of absolute pressure in the EGR vacuum line. Even a minor variance between calculated EGR and actual EGR causes the ECM to alter EGR to the correct amount. If the variance exceeds the amount of correction possible, an error is detected by the ECM and a Check Engine Light is displayed on the dash. A self-diagnostic feature is also included on 1985 and later California models with 6.2L engines.

Electronic Control Module (ECM)

The Electronic Control Module (ECM) **(see illustration)** is the brain of the DECS. It constantly monitors the engine sensors and controls the EGR and other systems that affect engine performance. In order to use the same ECM unit on different vehicles, GM manufactures a Programmable Read-Only Memory (PROM) for each specific model. If you have to replace the ECM, it won't have a PROM. You will have to either swap the PROM from the old ECM unit or install a new PROM. If you have to replace a PROM, make absolutely sure you get the right unit for your vehicle. It's a good idea to take the PROM with you when buying a new one to prevent the possibility of buying the wrong one.

Information sensors (inputs)

A typical DECS uses several inputs - an engine rpm signal, a vehicle speed sensor, a Manifold Absolute Pressure (MAP) sensor, a Metering Valve Sensor (MVS), a transmission gear position indicator (not used on THM 200-C units), an engine coolant temperature sensor and a brake switch signal.

The engine rpm signal is supplied from the alternator terminal R to the ECM. As engine rpm increases, the frequency of voltage "pulses" from the alternator also increases, providing the ECM with engine speed data. The ECM uses this rpm signal to compute various output functions and to compare it with other input signals.

The ECM applies 12 volts to its own terminal N and monitors the voltage at this terminal. Terminal N is connected to the VSS circuit, which alternately grounds and ungrounds when the drive wheels are turning. This pulsing action takes place about 2000 times per mile; the ECM calculates vehicle speed based on the time between "pulses." The speed of vehicle is sensed by a speed sensor, which consists of a light-emitting diode (LED) and a phototransistor, both of which are enclosed in the connector. The connector is located in the back of the speedometer cluster, next to the speedometer cable. A wiring harness connects the sensor to the ECM. When a voltmeter is hooked up between terminal N of the ECM and ground, and the speedometer cable is turned, the voltage swings between 8 and less-than-1 volt.

The MAP sensor monitors vacuum in the EGR vacuum circuit. It also sends the ECM a barometric (BARO) pressure (altitude) signal when the ignition switch is on with the engine not running, or with the engine running when the EGR isn't functioning. The BARO signal allows the ECM to control altitude-related injection pump timing and metering valve travel at wide-open throttle. The BARO signal is also used by the ECM - along with the EGR control pressure, coolant temperature, metering valve position, TCC state, vehicle speed and engine rpm - to determine and control the EGR pulse-width modulated duty cycle (the length of the on and off times for the EGR solenoid). During normal operation, the ECM compares its EGR duty cycle signal with the MAP signal and makes appropriate corrections.

If the EGR control pressure (line vacuum) varies more than 8.11 kPa from what the ECM has determined the pressure should be - and if this variance lasts for more than 10 seconds - a Code 53 is set and the

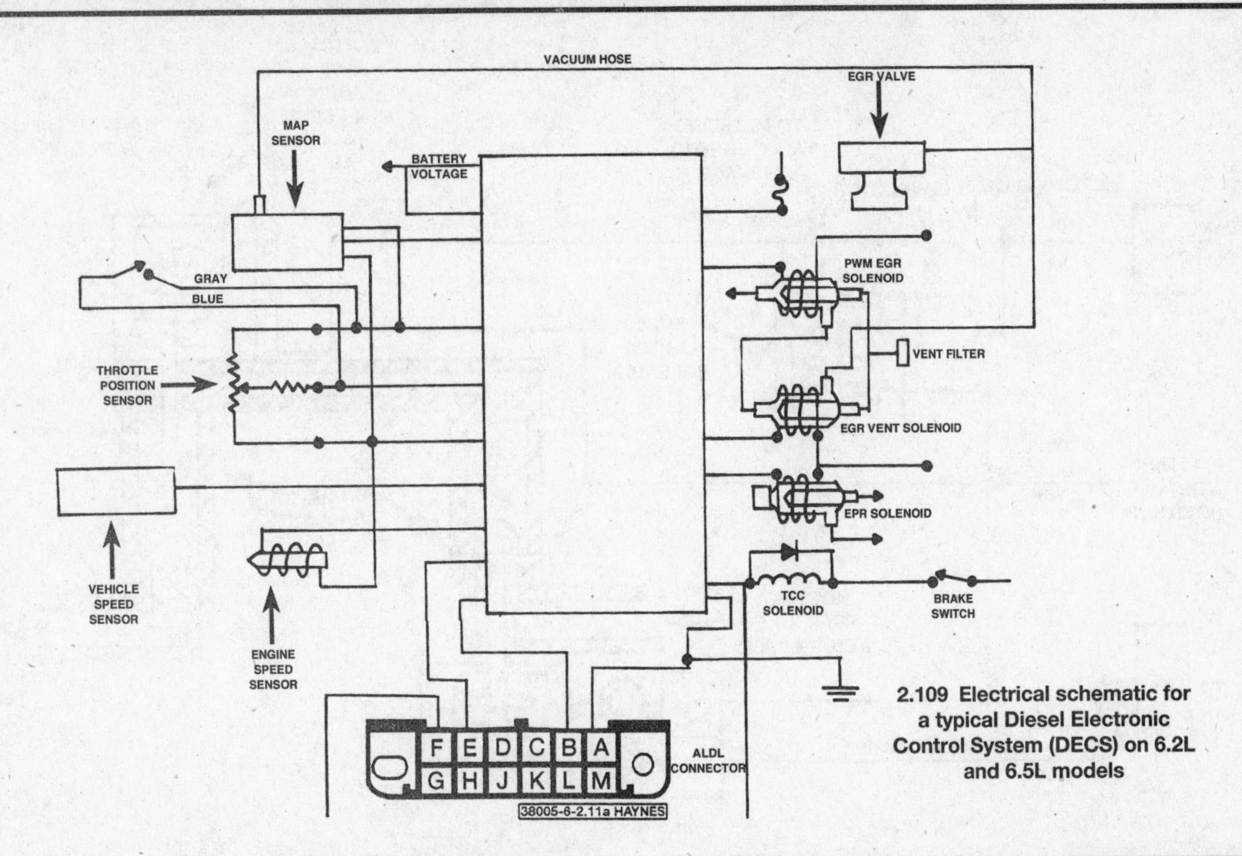

2.109 Electrical schematic for a typical Diesel Electronic Control System (DECS) on 6.2L and 6.5L models

ECM shuts down the EGR. The Check Engine light (CEL) goes on, and stays on, and the EGR remains off, until the ignition switch is cycled off. But the CEL turns off if the malfunction is cleared on all codes except Code 53. In either case, the code remains in memory while the ignition switch is on and voltage is applied to ECM terminal A.

If the throttle goes to its wide-open position, the EGR vacuum signal has zero vacuum (atmospheric pressure). The MAP sensor reads the atmospheric pressure, which is proportional to vehicle altitude. The pressure data is stored in the ECM and is used to modify the EGR vacuum signal. Thus, the EGR vacuum signal is altered for changes in altitude.

The Metering Valve Sensor (MVS) is a variable resistor that tells the ECM the position of the metering valve. The sensor is connected to V-REF and has the highest resistance at closed throttle. At wide open throttle, the resistance is lowest and output is about five volts. A failure in the sensor turns on the CEL, setting a Code 21 for a low MVS signal, or a Code 22 for a high MVS signal. The MVS output is used by the ECM to determine the duty cycle of the EGR.

Some transmissions have a 3rd/4th gear switch (2nd/3rd on THM 125-C units) to send a signal to the ECM telling it what gear the transmission is in. The ECM uses this information to vary the conditions under which the clutch applies or releases. However, the transmission doesn't have to be in high gear for the ECM to turn on the TCC. Transmissions using gear select switches can be identified by three or four wires coming out of the TCC connector.

The brake switch signal is 12 volts to the ECM from the brake light switch. It tells the ECM that the brakes are applied. The ECM uses this signal to delay the EPR to improve driveability.

The engine coolant temperature sensor is a thermistor (a resistor which changes its resistance value in accordance with the temperature) mounted in the engine coolant stream. Low coolant temperature produces a high resistance (100,000 ohms at -40-degrees F); conversely, a high temperature causes a low resistance (70 ohms at 266-degrees F). The ECM supplies a 5-volt signal to the coolant sensor through a resistor in the ECM and measures the voltage. The voltage is high when the engine is cold, and low when the engine is hot. By measuring the voltage, the ECM knows the engine coolant temperature, which affects most of the systems controlled by the ECM. A failure in this circuit sets either a Code 14 or a Code 15. Remember, these codes indicate a failure in the coolant temperature circuit, so service will consist of either repairing a wire or replacing the sensor.

Outputs

A typical DECS also uses several outputs - an EGR solenoid, an Exhaust Pressure Regulator (EPR) solenoid, a Transmission Converter Clutch (TCC), a Housing Pressure Cold Advance (HPCA) and a Housing Pressure Altitude Advance (HPAA). The EGR solenoid controls vacuum to the EGR valve. The EPR solenoid controls vacuum to the EPR valve. Both of these solenoids are described earlier in this chapter.

The TCC system has a solenoid-operated valve in the automatic transmission which couples the flywheel to the output shaft of the transmission through the torque converter to reduce slippage losses in the converter, which increases fuel economy.

The fast idle solenoid **(see illustration)** is an ECM-controlled

2.110 Typical fast Idle solenoid installation

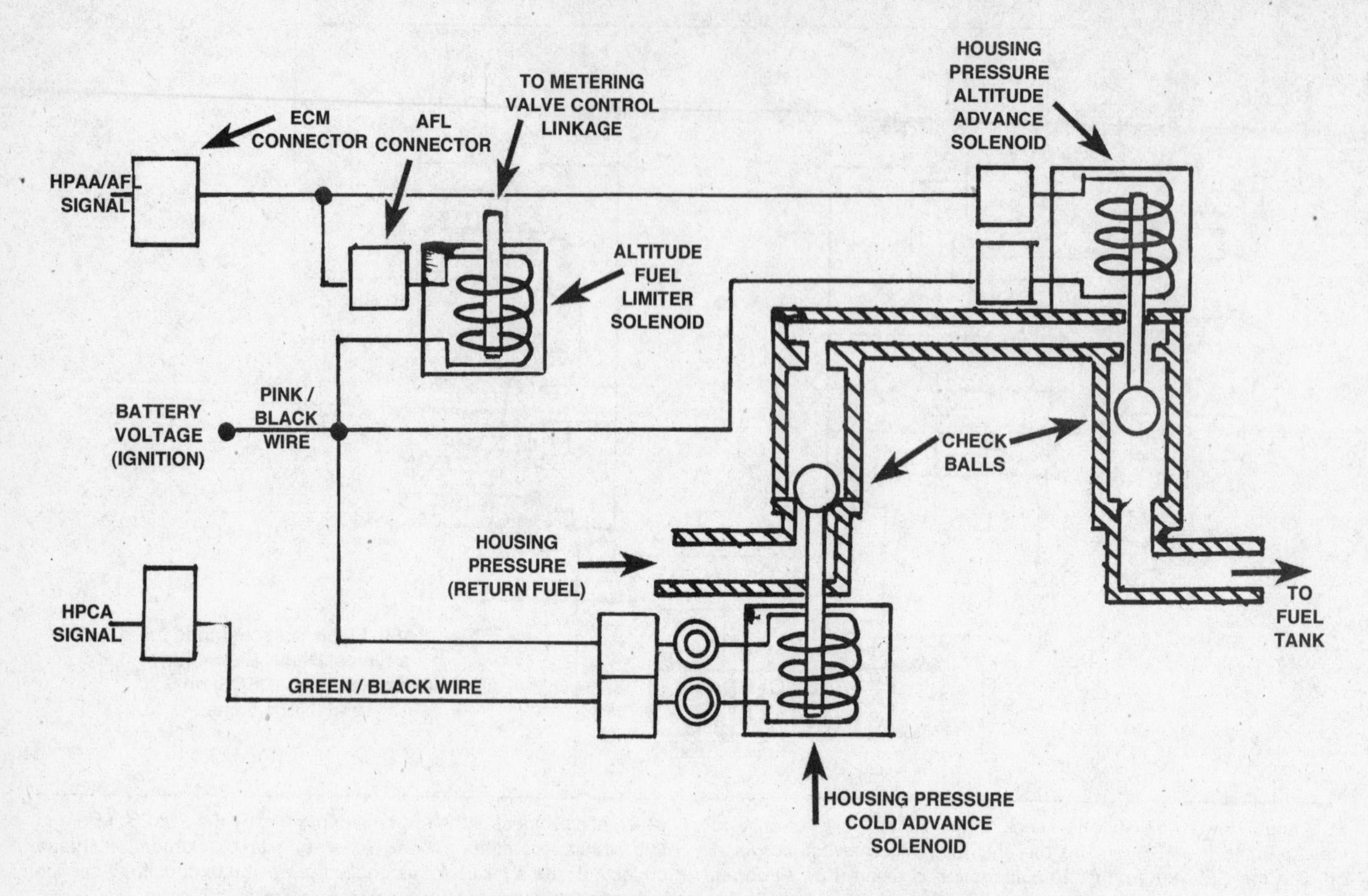

2.111 Schematic for a typical Housing Pressure Cold Advance (HPCA)-Housing Pressure Altitude Advance (HPAA) and Altitude Fuel Limiter (AFL) circuit

plunger-type solenoid which is normally retracted (de-energized). It's extended (energized) when engine coolant temperature is below 100-degrees F (cold engine) or above 248-degrees F (overheated engine).

The HPCA system advances engine timing when the engine is cold. The HPCA solenoid is described earlier in this chapter. The HPAA solenoid is located in the fuel return line between the fuel injection pump and the fuel tank **(see illustration)**. It regulates pump housing pressure by altering pump timing in response to changes in altitude.

The trouble codes

Note: *Diagnostic trouble codes apply to 1995 and earlier models equipped with the DECS only. As of 1996, GM has adopted the OBD II computer system. (On Board Diagnostics - version 2). This system uses a different 16 pin Assembly Line Diagnostic Link (ALDL) connector and an expensive electronic SCAN tool must be used to access the trouble codes. If the CHECK ENGINE light comes on while driving, on a 1996 or later model, take the vehicle to a dealer service department for diagnosis.*

Trouble in the DECS is indicated by a Check Engine Light (CEL) that doesn't go off after the engine is started. Normally, the CEL comes on when the key is turned to On, then goes off as soon as the engine starts. If the CEL stays on, you've got trouble somewhere. To access the trouble codes, you must bridge the diagnostic terminal of the Assembly Line Data Link (ALDL) connector to the ground terminal **(see illustration)**. The ALDL connector is wired to the ECM and is located under the dash.

When the ALDL is grounded, the CEL flashes its usual Code 12 and then follows that with any stored codes. Each code appears as a series of blinking flashes. Here are a few helpful things to know about the way the system works:

1 Generally, it takes at least 10 seconds for the CEL to come on when a problem occurs. If the malfunction disappears before that threshold, the light will usually go out - but a trouble code will be set in the memory of the ECM.

2 The exception is Code 53, which will lock the CEL on until the ignition is turned off.

3 Code 12 doesn't store in memory.

4 If the CEL comes on intermittently, but no code is stored, check all connections for a tight fit. If the problem persists, take the vehicle to a dealer service department or other repair shop.

5 Any codes stored will be erased when the ignition switch is turned off.

2.112 Details of the 12 pin Assembly Line Data Link (ALDL) connector (1995 and earlier)

A Ground
B Diagnostic test terminal

Trouble codes (5.7L engines)

Code 12	No engine rpm signal pulses to the ECM while cranking the engine. This code isn't stored in memory and only flashes while the fault is present. With the ignition on and the engine off, this is a normal code.
Code 14	Grounded coolant sensor circuit - low voltage at ECM terminal W.
Code 15	Open coolant sensor circuit - high voltage at ECM terminal W.
Code 21	Metering Valve Sensor (MVS) circuit voltage low at ECM terminal R (grounded circuit or misadjusted MVS)
Code 22	Metering Valve Sensor (MVS) circuit voltage high at ECM terminal R (open circuit or misadjusted MVS).
Code 24	Vehicle Speed Sensor (VSS) circuit - rpm and MVS signals indicate vehicle should be in motion with inadequate VSS signal.
Code 41	No engine rpm signal pulses to the ECM while the vehicle is in motion. Code will store in memory.
Code 51	Faulty or improperly installed calibration unit (PROM).
Code 53	Exhaust Gas Recirculation (EGR) valve MAP sensor has incorrect EGR vacuum.
Code 55	Grounded 5-volt reference circuit and/or faulty ECM. Low voltage at terminal C of ECM.

*Trouble codes (6.2 and 6.5L engines)**

Code 12	No codes present
Code 13	Engine shutoff solenoid circuit
Code 14	Engine Coolant Temperature (ECT) sensor circuit (high temperature indicated)
Code 15	Engine Coolant Temperature sensor (ECT) circuit (low temperature indicated)
Code 16	Vehicle Speed Sensor (VSS) output signal low
Code 17	Fuel temperature sensor
Code 18	Fuel temperature sensor/cam reference problem
Code 19	Crankshaft position sensor
Code 21 (1993 and earlier)	Throttle Position Sensor (TPS) sensor voltage high
Code 21 (1994 and later)	Accelerator pedal position sensor (circuit 1 high)
Code 22 (1993 and earlier)	Throttle Position Sensor (TPS) sensor voltage low
Code 22 (1994 and later)	Accelerator pedal position sensor (circuit 1 low)
Code 23 (1993 and earlier)	Throttle Position Sensor (TPS) sensor voltage out-of-range
Code 23 (1994 and later)	Accelerator pedal position sensor (circuit 1 out-of-range)
Code 24	Vehicle Speed Sensor (VSS) open or grounded circuit
Code 25	Accelerator pedal position sensor (circuit 2 high)
Code 26	Accelerator pedal position sensor (circuit 2 low)
Code 27	Accelerator pedal position sensor (circuit 2 out-of-range)
Code 28	Transmission range pressure switch fault
Code 29	Glow plug relay circuit
Code 31 (1993 and earlier)	Manifold Absolute Pressure(MAP) sensor signal voltage low
Code 31 (1994 and later)	EGR control vacuum too high
Code 32	EGR circuit fault
Code 33 (1993 and earlier)	Manifold Absolute Pressure(MAP) sensor signal voltage high
Code 33 (1994 and later)	EGR control sensor circuit high (low vacuum)
Code 34	Ignition timing stepper motor fault

** Not all codes will set in all models*

Trouble codes (6.2 and 6.5L engines) (continued)*

Code 35	Ignition pulse width fault
Code 36	Ignition pulse width fault
Code 37	Brake pedal switch stuck on
Code 38	Brake pedal switch stuck off
Code 41	Brake pedal switch circuit fault
Code 42	Fuel temperature circuit problem (high temperature)
Code 43	Fuel temperature circuit problem (low temperature)
Code 44	EGR pulse width fault
Code 45	EGR vent fault
Code 46	"Service Engine Soon" lamp circuit problem
Code 47	Intake Air Temperature (IAT) sensor indicates high temperature
Code 48	Intake Air Temperature (IAT) sensor indicates low temperature
Code 49	"Service Throttle Soon" lamp problem
Code 51	Programmable Read Only Memory (PROM) malfunction or incorrect PROM installed
Code 52	Fault in Powertrain Control Module (PCM) circuit or system voltage high
Code 53	Fault in Powertrain Control Module (PCM) circuit or 5-volt reference circuit overloaded
Code 54	Powertrain Control Module (PCM) fuel circuit problem
Code 56	Injection pump driver fault
Code 57	Fault in Powertrain Control Module (PCM) circuit or 5-volt reference circuit grounded
Code 58	Transmission fluid temperature sensor circuit low
Code 59	Transmission fluid temperature sensor circuit high
Code 63	Accelerator pedal position sensor (circuit 3 high)
Code 64	Accelerator pedal position sensor (circuit 3 low)
Code 65	Accelerator pedal position sensor (circuit 3 out-of-range)
Code 66	Transmission 3-2 shift solenoid circuit
Code 67	Torque Converter Clutch (TCC) circuit
Code 68	Transmission slipping
Code 69	Torque Converter Clutch (TCC) stuck on
Code 71	Cruise control set switch circuit
Code 72	Vehicle Speed Sensor (VSS) circuit
Code 73	Transmission pressure control solenoid circuit
Code 75	System voltage low
Code 76	Cruise control Resume/Accel switch circuit
Code 79	Transmission fluid over temperature
Code 81	Transmission 2-3 shift solenoid circuit
Code 82	Transmission 1-2 shift solenoid circuit
Code 88	TDC offset problem
Code 91 thru 98	Cylinder balance
Code 99	Accelerator pedal position sensor 5-volt reference problem

** Not all codes will set in all models*

2.113 Remove the air conditioning compressor from the bracket, but do not disconnect the refrigerant lines

2.114 On the 6.2L and 6.5L diesel engines, disconnect the fuel lines from the injectors and lift the lines up only three to four inches to allow just enough clearance for the valve cover. Do not bond or twist the lines in any way that will damage them

In-vehicle engine repairs

General information

This Part of Chapter 2 is devoted to in-vehicle repair procedures for the 5.7L and 6.2L GM diesel engine. All information concerning engine removal and installation and engine block and cylinder head overhaul can be found in Chapter 4.

The following repair procedures are based on the assumption that the engine is installed in the vehicle. If the engine has been removed from the vehicle and mounted on a stand, many of the steps outlined in this Part of Chapter 2 will not apply.

The Specifications included in this Part of Chapter 2 apply only to the procedures contained in this Part. Chapter 4 contains the Specifications necessary for cylinder head and engine block rebuilding.

Repair operations possible with the engine in the vehicle

Many major repair operations can be accomplished without removing the engine from the vehicle.

Clean the engine compartment and the exterior of the engine with some type of degreaser before any work is done. It will make the job easier and help keep dirt out of the internal areas of the engine.

Depending on the components involved, it may be helpful to remove the hood to improve access to the engine as repairs are performed. Cover the fenders to prevent damage to the paint. Special pads are available, but an old bedspread or blanket will also work.

If vacuum, exhaust, oil or coolant leaks develop, indicating a need for gasket or seal replacement, the repairs can generally be made with the engine in the vehicle. The intake and exhaust manifold gaskets, timing cover gasket, oil pan gasket, crankshaft oil seals and cylinder head gasket are all accessible with the engine in place.

Exterior engine components, such as the intake and exhaust manifolds, the oil pan (and the oil pump), the water pump, the starter motor, the alternator, the vacuum pump and the fuel system components can be removed for repair with the engine in place.

Since the cylinder head can be removed without pulling the engine, valve component servicing can also be accomplished with the engine in the vehicle. Replacement of the timing chain and sprockets is also possible with the engine in the vehicle.

In extreme cases caused by a lack of necessary equipment, repair or replacement of piston rings, pistons, connecting rods and rod bearings is possible with the engine in the vehicle. However, this practice is not recommended because of the cleaning and preparation work that must be done to the components involved.

Valve covers - removal and installation

Note: *When removing the valve covers on the 5.7L diesel engine, it is not necessary to remove the injection pump and the fuel lines as an assembly. Remove the injection pump lines from the left or right side, depending on which cover is being removed. On 6.2L and 6.5L engines, carefully lift the fuel lines off the valve cover and make sure they do not bend - bent lines will cause poor performance.*

Removal

1 Disconnect the negative cable from the battery.

2 Remove the air cleaner assembly. On 6.2 and 6.5L engines, remove the intake manifold and cover the intake ports in the cylinder head. Remove the CDR valve and hose, turbocharger brace (if equipped), dipstick tube and other components as necessary. If equipped with air conditioning, remove the compressor, without disconnecting the refrigerant lines, and secure it aside **(see illustration)**.

3 Disconnect the injection lines **(see illustration)** (refer to the *Fuel System* Section in this Chapter, if necessary).

4 Remove the valve cover mounting bolts/nuts **(see illustration)**. Remember to make a note of the locations of the stud bolts that are used to attach the injector line brackets.

5 A special tool (Burroughs BT-8315) will simplify valve cover

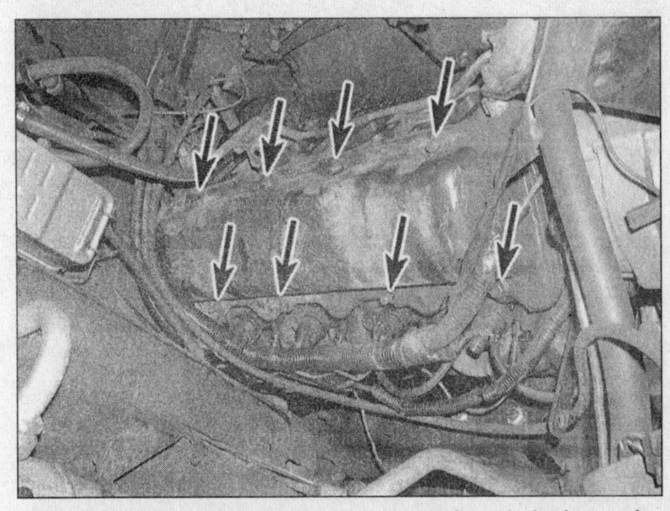

2.115 To remove the valve cover, remove these bolts (arrows) (6.2L engine shown, other engines similar)

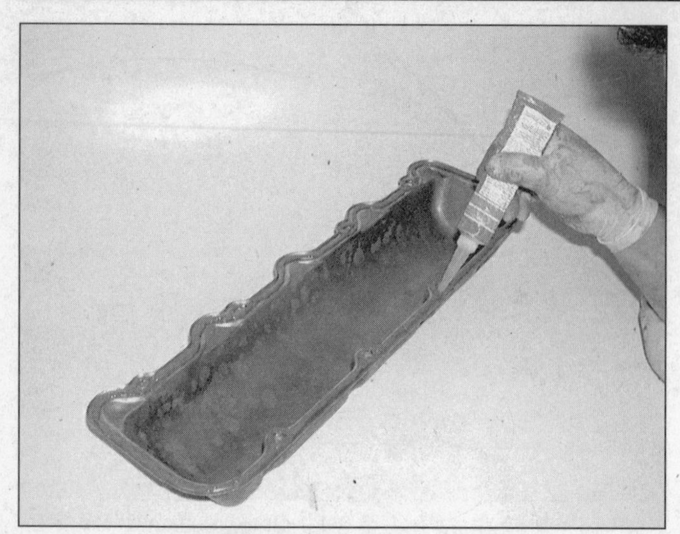

2.116 Apply a thin (3/32-inch) bead of sealant to the valve cover flange, and then allow the RTV sealer to "set up" (slightly harden) before installing the valve cover

removal. This tool will allow the flanges to separate from the cylinder head without damaging the valve cover. Engage the tabs on the special tool under the valve cover flange and tighten the tool, forcing it against the intake manifold. If the flange starts to bend, stop and leave the tool in place. Wrap a narrow board in a shop rag and place it against the side of the valve cover. Strike the board with a mallet or hammer. This will dislodge the valve cover from the cylinder head. Remove the valve cover.

Installation

6 The mating surfaces of each cylinder head and valve cover must be perfectly clean when the covers are installed. Use a gasket scraper to remove all traces of sealant or old gasket, then wipe the mating surfaces with a cloth saturated with lacquer thinner or acetone. If there is sealant or oil on the mating surfaces when the cover is installed, oil leaks may develop. Also, check the surface of the valve cover to make sure it is level and not bent or damaged.

7 Make sure the threaded holes are clean. Run a tap into them to remove corrosion and restore damaged threads.

8 Apply a thin bead (3/32 inch) of RTV sealant to the cover flange **(see illustration)**, then position the gasket inside the cover lip and allow the sealant to set up so the gasket adheres to the cover (if the sealant is not allowed to set, the gasket may fall out of the cover as it is installed on the engine).

9 Carefully position the cover on the head and install the bolts.

10 Tighten the nuts/bolts in three steps to the torque listed in this Chapter's Specifications.

11 The remaining installation steps are the reverse of removal.

12 Start the engine and check carefully for oil leaks as the engine warms up.

Rocker arms and pushrods - removal, inspection and installation

Note: *Because the diesel engine has such a high compression ratio, there is minimal valve-to-piston clearance. It is important to have a rigid valve train that insures precise valve train motion throughout the speed range.*

The rocker arms consist of nodular iron with a steel backed bushing. The shafts are bolted to the case stanchions on the cylinder head. Oil is supplied to the rocker arm through the hollow pushrod and the rocker arm has drilled passages that provide a path for the oil to the bushings, etc.

The 1982 6.2L engine uses hardened steel spacers and a metric washers at the rocker shaft attachments. 1983 and 1984 models use a non-hardened spacer and steel cleat (the steel cleat has a 90-degree gap to the bolt to prevent closure). 1985 and later models are equipped with steel stamped rocker arms, a large diameter steel shaft bolted directly to the cylinder head pedestal, individual plastic locators and an open top which permits splash lubrication of the bearing surface **(see illustration)**.

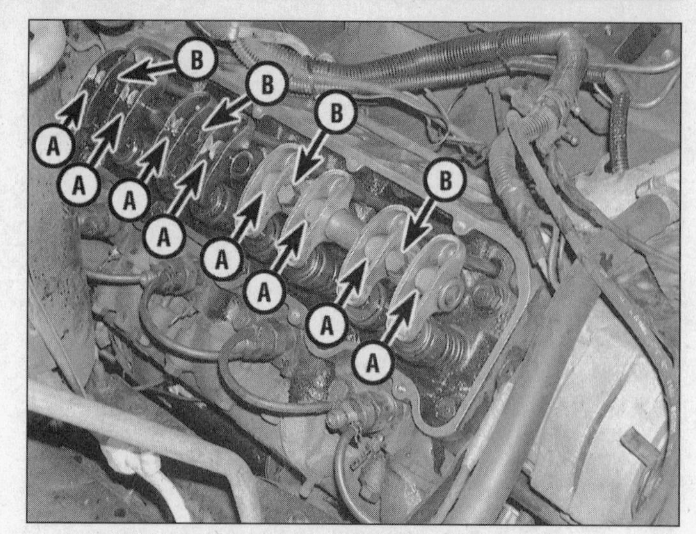

2.117 Typical rocker arm assembly on 1985 6.2L engine (all 1985 and later 6.2L and 6.5L diesel engines similar)

A Nylon rocker arm retainers
B Rocker shaft bolts

Removal

1 Detach the valve covers from the cylinder heads.

2 Beginning at the front of one cylinder head, loosen and remove the rocker arm nuts or bolts **(see illustration)**. Store them separately in marked containers to ensure that they will be reinstalled in their original locations. **Note:** *On 1985 and later 6.2L and 6.5L diesel engines, insert a screwdriver into the bore of the rocker shaft and break off the ends of the nylon rocker arm retainers. Use a pair of pliers and remove the remaining piece of the retainer* **(see illustration)** *by prying up.*

3 Lift off the rocker arms and pivots or shafts and store them in the marked containers with the nuts or bolts (they must be reinstalled in their original locations).

4 Remove the pushrods and store them in order to make sure they will not get mixed up during installation **(see illustration)**.

2.118 Instead of loosening each bolt completely, alternate the bolt pattern to allow the assembly to remain level as it is loosened

2.119 To remove the nylon rocker arm retainers on 1985 and later 6.2L and 6.5L engines, insert a screwdriver into the bore of the rocker shaft, break off the ends of the retainers and use a pair of pliers to pry out the remaining pieces

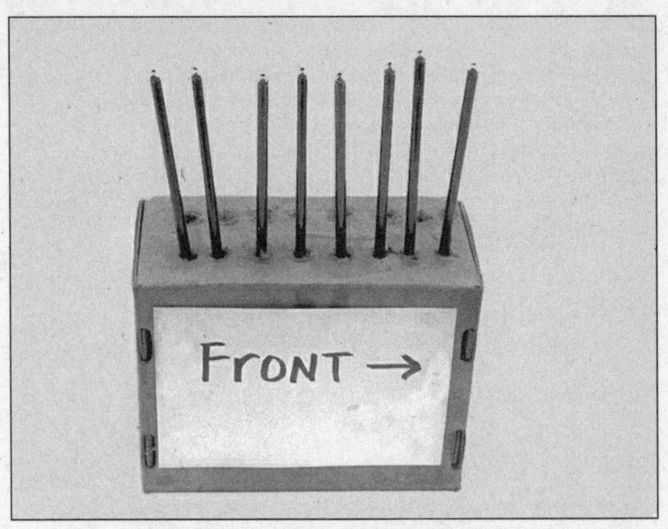

2.120 A perforated cardboard box can be used to store the pushrods to ensure that they are reinstalled in their original locations

Inspection

5 Check each rocker arm for wear, cracks and other damage, especially where the pushrods and valve stems contact the rocker arm faces.

6 Make sure the hole at the pushrod end of each rocker arm is open.

7 Check each rocker arm pivot area for wear, cracks and galling. If the rocker arms are worn or damaged, replace them with new ones and use new pivots or shafts as well.

8 Inspect the pushrods for cracks and excessive wear at the ends. Roll each pushrod across a piece of plate glass to see if it is bent (if it wobbles, it is bent).

Installation

9 Lubricate the lower end of each pushrod with clean engine oil or moly-base grease and install them in their original locations. Make sure each pushrod seats completely in the lifter socket.

10 Apply moly-base grease to the ends of the valve stems and the upper ends of the pushrods before installing the rocker arms **(see illustration)**.

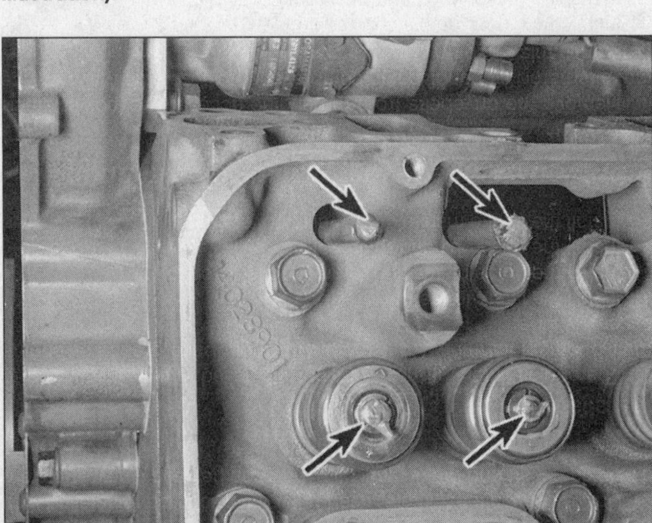

2.121 The ends of the pushrods and the valve stems should be lubricated with moly-base grease prior to installation of the rocker arms

11 Before installing the bolts through the shafts on 1982 models, be certain the ring around the shaft is installed with the split at the bottom. On 1983 and later, the split is 90-degrees to the right. On 1985 and later, there is NO split used. On 1985 and later 6.2L and 6.5L diesel engines, install a new nylon rocker arm retainer (Part #2350076) in each 1/4-inch hole using a drift of approximately 1/2-inch in diameter.

12 Set the rocker arms in place.

Adjustment

13 Bring the number one piston to top dead center (TDC) on the compression stroke (see the *Timing chain and sprockets - inspection, removal and installation* procedure later in this Chapter).

14 Rotate the engine 30-degrees BTDC or 3-1/2 inches counterclockwise measured on the balancer. This will position the engine so that NO valves will be close to the piston heads.

15 Tighten the bolts evenly. Make several passes to insure that the rocker arm assembly does not bind at any point.

16 Tighten each bolt to the torque listed in this Chapter's Specifications.

17 Install the valve covers. Start the engine, listen for unusual valve train noises and check for oil leaks at the valve cover joints.

Valve springs, retainers and seals - replacement in vehicle

Note: *Broken valve springs and defective valve stem seals can be replaced without removing the cylinder head. Two special tools and a compressed air source are normally required to perform this operation, so read through this Section carefully and rent or buy the tools before beginning the job. If compressed air is not available, a length of nylon rope can be used to keep the valves from falling into the cylinder during this procedure.*

1 Remove the valve cover from the affected cylinder head. If all of the valve stem seals are being replaced, remove both valve covers.

2 Remove the glow plug from the cylinder which has the defective component. If all of the valve stem seals are being replaced, all of the glow plugs should be removed.

3 Turn the crankshaft until the piston in the affected cylinder is at top dead center (TDC) on the compression stroke (refer to the Timing chain removal and installation procedure later in this Chapter). If you are replacing all of the valve stem seals, begin with cylinder number one and work on the valves for one cylinder at a time. Move from cylinder-to-cylinder, following the firing order sequence: 1-8-4-3-6-5-7-2 on 5.7L; 1-8-7-2-6-5-4-3 on 6.2L and 6.5L engines.

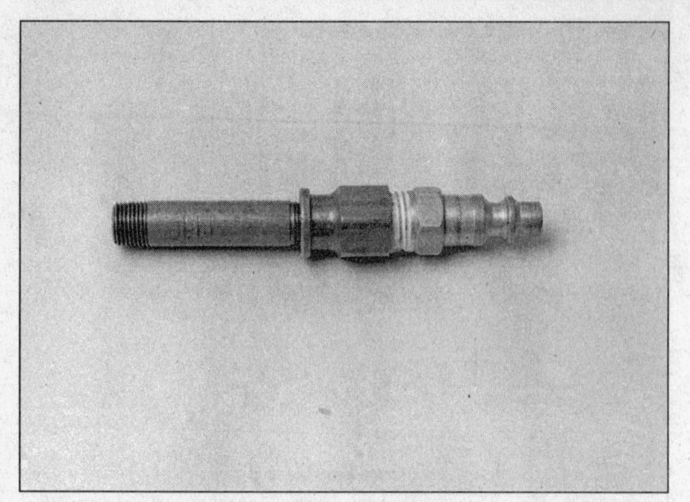

2.122 This air hose adapter is a combination of a special pipe fitting and an air hose fitting - both were purchased at a hardware store

2.123 With the compressed air holding the intake and exhaust valve closed, compress the springs with a valve spring compressor

4 Thread an adapter into the glow plug hole **(see illustration)** and connect an air hose from a compressed air source to it. Most auto parts stores can supply an air hose adapter. **Note:** *Many cylinder compression gauges utilize a screw-in fitting that may work with your air hose quick-disconnect fitting.*

5 Remove the rocker arm(s) (or rocker arm shaft assembly) for the valve with the defective part and pull out the pushrod. If all of the valve stem seals are being replaced, all of the rocker arms and pushrods should be removed.

6 Apply compressed air to the cylinder. The valves should be held in place by the air pressure. If the valve faces or seats are in poor condition, leaks may prevent the air pressure from retaining the valves - refer to the alternative procedure below.

7 If you do not have access to compressed air, an alternative method can be used. Position the piston at a point 45-degrees before TDC on the compression stroke, then feed a long piece of nylon rope through the glow plug hole until it fills the combustion chamber. Be sure to leave the end of the rope hanging out of the engine so it can be removed easily. Use a large breaker bar and socket to rotate the crankshaft in the normal direction of rotation until slight resistance is felt as the piston comes up against the rope in the combustion chamber.

8 Stuff shop rags into the cylinder head holes above and below the valves to prevent parts and tools from falling into the engine, then use a valve spring compressor to compress the spring/damper assembly **(see illustration)**. Remove the keepers with a pair of small needle-nose pliers or a magnet. **Note:** *A couple of different types of tools are available for compressing the valve springs with the head in place. One type (shown here) grips the lower spring coils and presses on the retainer as the knob is turned, while the other type utilizes the rocker arm shaft for leverage. Both types work very well, although the lever type is usually less expensive.*

9 Remove the spring retainer or rotator, oil shield and valve spring assembly, then remove the valve stem seal. Several different sizes of valve stem oil seals are used on these engines, depending on intake or exhaust valves and oversize valves. Consult the chart in Chapter 4, Valves - servicing. **Note:** *If air pressure fails to hold the valve in the closed position during this operation, the valve face or seat is probably damaged. If so, the cylinder head will have to be removed for additional repair operations.*

10 Wrap a rubber band or tape around the top of the valve stem so the valve will not fall into the combustion chamber, then release the air pressure. **Note:** *If a rope was used instead of air pressure, turn the crankshaft slightly in the direction opposite normal rotation.*

11 Inspect the valve stem for damage. Rotate the valve in the guide and check the end for eccentric movement, which would indicate that the valve is bent.

12 Move the valve up-and-down in the guide and make sure it doesn't bind. If the valve stem binds, either the valve is bent or the guide is damaged. In either case, the head will have to be removed for repair.

13 Inspect the rocker arms for wear (see the previous Sub-section).

14 Reapply air pressure to the cylinder to retain the valve in the closed position, then remove the tape or rubber band from the valve stem. If a rope was used instead of air pressure, rotate the crankshaft in the normal direction of rotation until slight resistance is felt.

15 Lubricate the valve stem with engine oil and install a new oil seal of the type originally used on the engine. If the engine has O-ring type seals, compress the spring first, then install the seal.

16 Install the spring/damper assembly and shield in position over the valve.

17 Install the valve spring retainer or rotator and compress the valve spring assembly.

18 Position the keepers in the upper groove. Apply a small dab of grease to the inside of each keeper to hold it in place if necessary **(see illustration)**. Remove the pressure from the spring tool and make sure the keepers are seated.

19 Disconnect the air hose and remove the adapter from the glow plug hole. If a rope was used in place of air pressure, turn the crankshaft in the direction opposite normal rotation and pull it out of the cylinder.

20 Install the rocker arms and pushrods.

21 Install the glow plugs.

22 Install the valve covers.

23 Start and run the engine, then check for oil leaks and unusual sounds coming from the valve cover area.

2.124 Apply a small dab of grease to each keeper as shown here before installation - it'll hold them in place on the valve stem as the spring is released

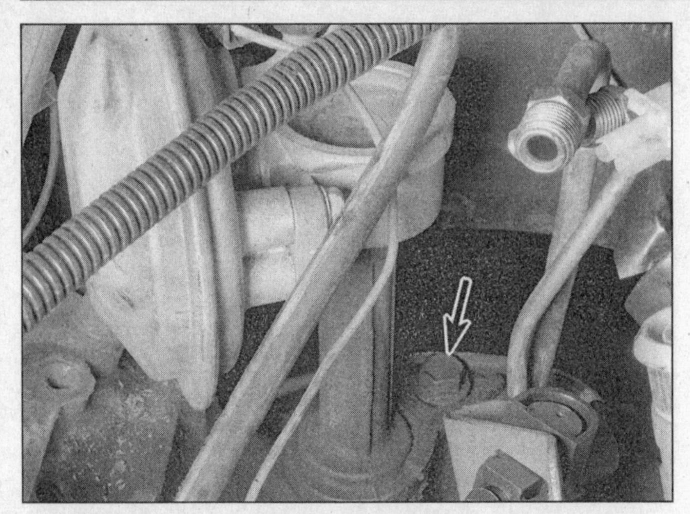

2.125 Remove the bolt (arrow) and bracket and lift the pump from the block (intake manifold removed for clarity)

2.126 To disassemble the vacuum pump, remove these two bolts (arrows)

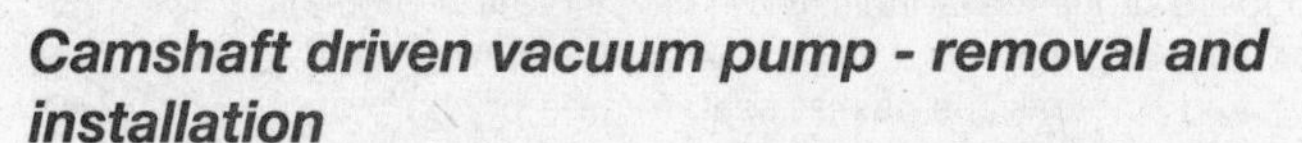

Camshaft driven vacuum pump - removal and installation

Caution: *Do not operate the engine without the vacuum pump in place or the oil pump will not operate, leading to serious engine damage.*
Note: *Further information on the vacuum pump, including testing procedures, are in the Emissions control Section of this Chapter.*

Removal

1 Remove the hose from the outlet on the vacuum pump.
2 Remove the bolt and the bracket that retain the pump to the intake manifold **(see illustration)**.
3 Lift the pump out of the manifold. The pump can be disassembled, if necessary; be sure to replace the O-ring if you disassemble the pump **(see illustrations)**.

Installation

4 Insert the pump into the engine. Make sure the driven gear on the shaft of the pump meshes with the gear on the camshaft while inserting the oil pump intermediate shaft into the bottom of the vacuum pump. Sometimes making the connection takes some patience and a slight turn until the pump drops down and "clicks" into position.
5 Rotate the pump until the bracket and the bolt can be installed into the intake manifold.
6 Install the vacuum hose.

Belt driven vacuum pump - removal and installation

1 Remove the drive belt.
2 Remove the hose from the vacuum pipe. Leave it attached to the pump.
3 Remove the mounting bolts.
4 If required, a special puller (available at auto parts stores) must be used to remove the pulley. Do not attempt to remove the pulley any other way. Use a pulley installer to install the pulley.
5 Installation is the reverse of removal.

Intake manifold - removal and installation

The intake manifold on the 5.7L engine has a separate air crossover assembly bolted to it. For information on removing and installing this assembly, refer to the *Fuel system* Section of this Chapter. The manifold on 1980 through 1984 models has a manifold drain tube which is used to drain diesel fuel that accumulates in the area.

The intake manifold on the 6.2L and 6.5L is a combination of an intake manifold and an air crossover assembly **(see illustration)**. This "spider" air plenum type is completely separated from the coolant system. This allows the intake manifold to be removed without draining the cooling system. One important detail to remember is that the intake manifold can be installed backwards and still operate the engine. However, the mounting boss for the secondary fuel filter would be in the wrong position for filter installation.

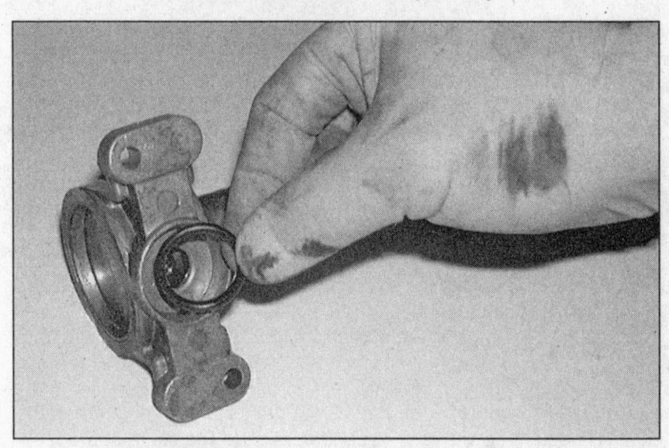

2.127 If you disassemble the vacuum pump, be sure to replace the O-ring

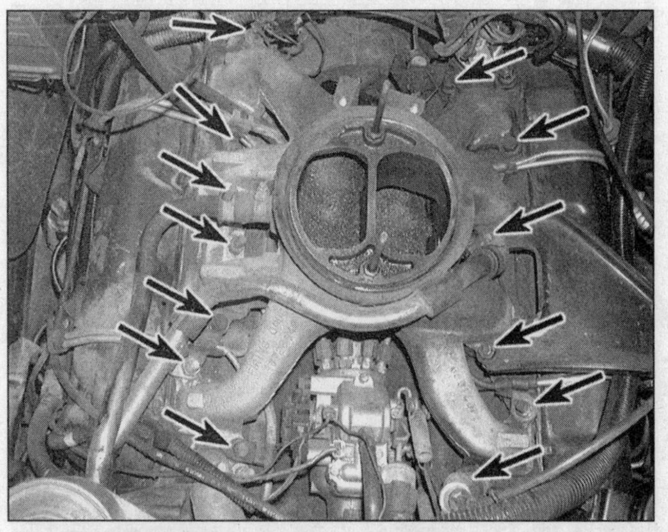

2.128 The intake manifold on 6.2L and 6.5L engines has 16 mounting bolts (arrows), eight on each side (not all bolts visible in this photo)

2.129 Remove the clamps from the crankcase ventilation lines and remove them from the intake manifold

2.130 Remove the secondary fuel filter from the intake manifold

Removal

1 Disconnect the negative cable from the battery. Drain the cooling system (5.7L engines only).

2 Remove the air cleaner assembly. On 6.2L and 6.5L engines, remove the crankcase ventilation lines from the intake manifold **(see illustration)**.

3 On 5.7L engines only disconnect the upper radiator hose and the thermostat bypass hose from the water outlet. Also, disconnect the heater hose and the vacuum hose from the water control valve.

4 On 5.7L engines only, remove the air crossover assembly and stuff shop rags into the openings to prevent any objects from falling inside.

5 Label and then disconnect any fuel lines, wires and vacuum hoses from the vehicle to the intake manifold. On 6.2L and 6.5L engines, disconnect the secondary fuel filter adapter **(see illustration)** and remove the adapter and filter as one unit.

6 Detach the throttle rod or cable from the fuel injection pump. If the engine is equipped with cruise control, remove the servo and accessories.

7 Remove the alternator and brackets, if necessary.

8 On 5.7L engines only, disconnect the fuel line from the fuel pump and the fuel filter and remove the fuel filter with the bracket.

9 On 5.7L engines only, disconnect the fuel lines at the nozzles using a back-up wrench to prevent twisting the lines. Remove the injection pump and lines. (See the *Fuel system* Section in this chapter). Be very careful not to bend the injection pump lines. Remove the fuel injection pump mounting adapter and intake manifold drain tube.

10 Loosen the intake manifold mounting bolts **(see illustration 2.128)** in 1/4-turn increments until they can be removed by hand. Follow the opposite sequence as indicated in the installation procedure. The manifold will probably be stuck to the cylinder heads and force may be required to break the gasket seal. A large pry bar can be positioned under the cast-in lug near the thermostat housing to pry up the front of the manifold. **Caution:** *Do not pry between the block and manifold or the heads and manifold or damage to the gasket sealing surfaces may result and vacuum leaks could develop.* Lift the intake manifold from the engine block **(see illustration)**.

2.131 Lift the intake manifold from the engine block

Installation

Note: *The mating surfaces of the cylinder heads, block and manifold must be perfectly clean when the manifold is installed. Gasket removal solvents in aerosol cans are available at most auto parts stores and may be helpful when removing old gasket material that is stuck to the heads and manifold. Be sure to follow the directions printed on the container.*

11 Use a gasket scraper to remove all traces of sealant and old gasket material, then wipe the mating surfaces with a cloth saturated with lacquer thinner or acetone. If there is old sealant or oil on the mating surfaces when the manifold is installed, oil or vacuum leaks may develop. Cover the lifter valley with shop rags to keep debris out of the engine. Use a vacuum cleaner to remove any gasket material that falls into the intake ports in the heads.

12 Use a tap of the correct size to chase the threads in the bolt holes, then use compressed air (if available) to remove the debris from the holes. **Warning:** Wear safety glasses or a face shield to protect your eyes when using compressed air.

13 Apply a thin coat of RTV sealant around the coolant passage holes (5.7L engine only) on the cylinder head side of the new intake manifold gaskets (there is normally one hole at each end).

14 Position the gaskets on the cylinder heads. Make sure all intake port openings, coolant passage holes (5.7L engines only) and bolt holes are aligned correctly and the THIS SIDE UP is visible.

15 Connect the thermostat bypass hose to the water pump.

16 Install the end seals onto the block (5.7L only). Apply a bead of RTV sealant to the edges to insure against oil leaks in the corners of the intake manifold.

17 Carefully set the manifold in place. Do not disturb the gaskets and do not move the manifold fore-and-aft after it contacts the front and rear seals.

18 Dip each manifold bolt into regular engine oil up to the bolt threads, then install the bolts. Tighten the bolts to 15 ft-lbs in the sequence shown **(see illustrations)**. Work up to the final torque (see the Specifications) in three steps.

19 On 5.7L engines only, install the fuel injection pump mounting adapter and seal. **Note:** *A special seal installation tool may be needed to install the seal. Install the intake manifold drain tube with RTV sealant.*

20 Install the injection pump as described in the Fuel System Section

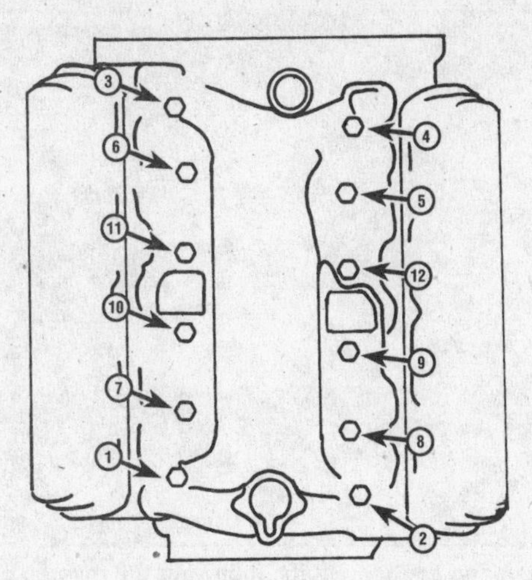

2.132 Intake manifold tightening sequence on the 5.7L engine

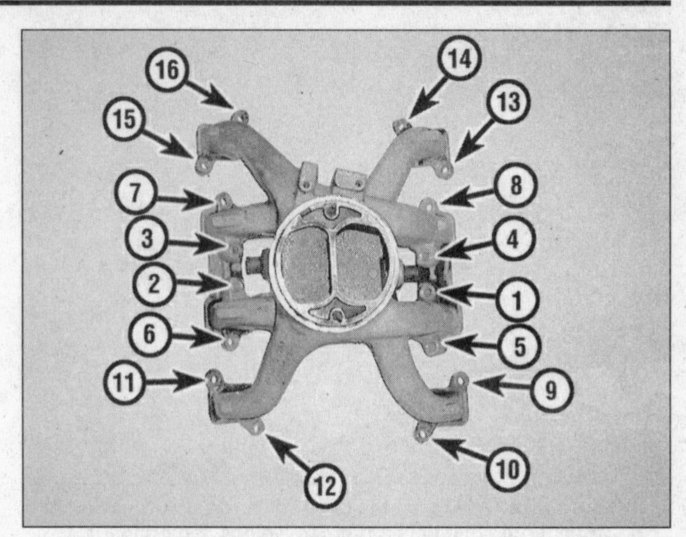

2.133 Intake manifold tightening sequence on the 6.2L and 6.5L engine

(5.7L engines only). Follow the sequence and use the proper alignment marks.

21 The remaining installation steps are the reverse of removal.

22 Start the engine and check carefully for oil, vacuum and coolant leaks at the intake manifold joints.

Exhaust manifolds - removal and installation

Warning: *Allow the engine to cool completely before following this procedure.*

Removal

1 Disconnect the negative cable from the battery.

2 Remove the air cleaner assembly.

3 Remove the alternator and power steering braces (if equipped) from the right exhaust manifold.

4 Disconnect the glow plug wires and the glow plugs (see the *Electrical system* Section earlier in this Chapter).

5 Unbolt the dipstick tube.

6 Remove the heat shields, if equipped.

7 Set the parking brake and block the rear wheels. Raise the front of the vehicle and support it securely on jackstands. Disconnect the exhaust crossover pipe from the manifold outlet **(see illustration)**. **Note:** *Often a short period of soaking with penetrating oil is necessary to remove frozen exhaust pipe attaching nuts. Use caution not to apply excessive force to frozen nuts, which could shear off the exhaust manifold studs.*

8 Remove the two front and two rear manifold mounting bolts first, then the center bolt(s) to separate the manifold from the head. Some models use locking tabs under the manifold bolts to keep the bolts from vibrating loose. On these models the tabs will have to be straightened before the bolts can be removed.

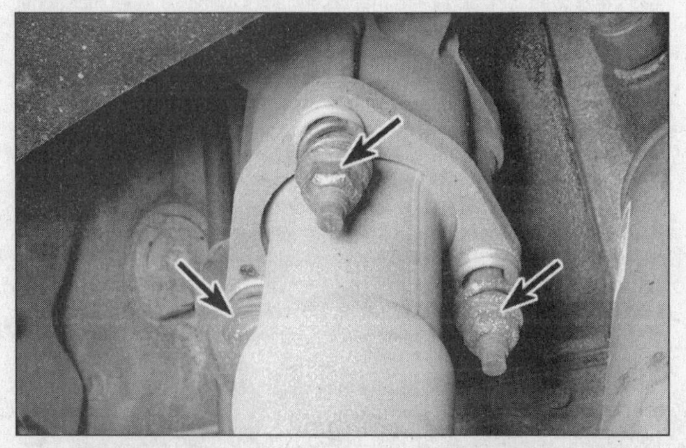
2.134 Use penetrating lubricant to soak the threads before removing the nuts (arrows) from the studs

Installation

9 Installation is basically the reverse of the removal procedure. Clean the manifold and head gasket surfaces of old gasket material, then install new gaskets. Do not use any gasket cement or sealer on exhaust system gaskets.

10 Install all the manifold bolts and tighten them to the torque listed at the beginning of the Chapter. Work from the center to the ends and approach the final torque in three steps.

11 Apply anti-seize compound to the exhaust manifold-to-exhaust pipe studs and nuts, and use a new exhaust "doughnut" gasket, if equipped.

Cylinder heads - removal and installation

Note: *The 5.7L diesel engine uses updated cylinder head bolts that must be replaced when the cylinder head is removed and serviced. The old style head bolts may stretch without breaking, resulting in insufficient torque at the bolt. The new style bolt is identified by a dash (-) on the bolt head. If the cylinder head is equipped with the new-style head bolt(s), it is not necessary to replace them.*

The 6.2L diesel engine uses updated head bolts on 1982 and 1983 models only. The updated head bolts (part number 14077193) are available at a dealership parts department **(see illustration)**.

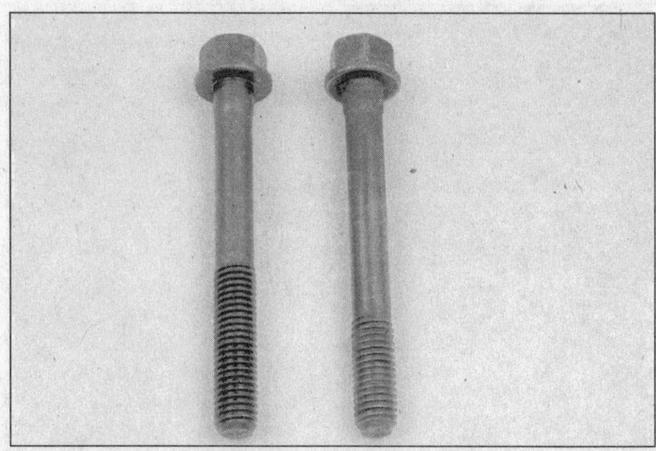
2.135 The updated head bolts on the 6.2L engine (left) are equipped with more threads to maintain the proper torque over a long period of time

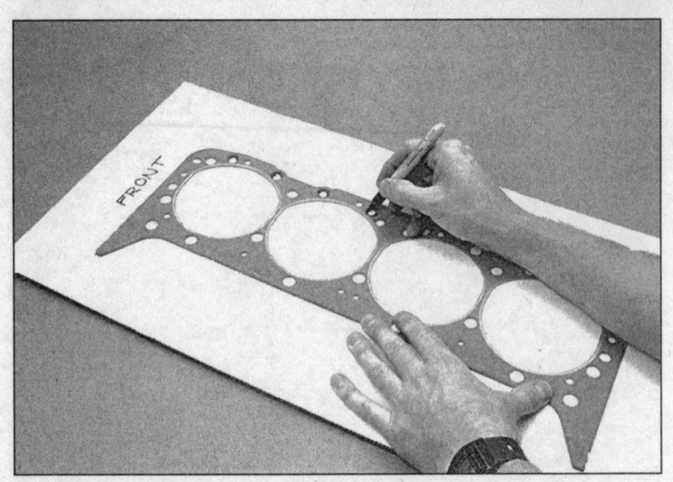

2.136 To avoid mixing up the head bolts, use a now gasket to transfer the bolt hole pattern to a piece of cardboard, then punch holes to accept the bolts

2.137 Loosen the head bolts, following the reverse pattern of removal (see illustration 2.140 or 2.141)

Removal

1 Disconnect the negative cable from the battery. Drain the cooling system.

2 On 5.7L engines only, remove the fuel injection pump and lines, the intake manifold and the valve covers.

3 On 6.2L and 6.5L engines, remove accessory brackets, dipstick tube, hoses, turbocharger (if equipped) and wiring harness as required. Remove the glow plug wires. Pull only on the wire-end connectors. A special tool (GM tool no. J-3083) must be used to remove the wires from the right side of the 6.5L Turbo-diesel as they cannot be reached by hand.

4 On 5.7L and 6.5L engines, detach both exhaust manifolds. On 6.2L engines, leave the manifolds attached until the heads are removed from the engine.

5 Remove the rocker arms and pushrods.

6 Using a new head gasket, outline the cylinders and bolt pattern on a piece of cardboard **(see illustration)**. Be sure to indicate the front of the engine for reference. Punch holes at the bolt locations.

7 Loosen the head bolts in 1/4-turn increments until they can be removed by hand **(see illustration)**. Work from bolt-to-bolt in a pattern that is the reverse of the tightening sequence. **Note:** *Don't overlook the row of bolts on the lower edge of each head, near the glow plug holes. Store the bolts in the cardboard holder as they are removed. This will insure that the bolts are reinstalled in their original holes.*

8 Lift the heads off the engine. If resistance is felt, do not pry between the head and block as damage to the mating surfaces will result **(see illustration)**. To dislodge the head, place a block of wood against the end of it and strike the wood block with a hammer. Store the heads on blocks of wood to prevent damage to the gasket sealing surfaces. **Note:** *On 6.2L and 6.5L engines, remove the exhaust manifolds. Also, it is sometimes necessary to leave the corner head bolt in the cylinder head when removing the head because of the lack of clearance to the firewall. Remember to install the head bolt in the same hole before reinstalling the cylinder head.*

2.138 Pry the cylinder head, using a prybar on a bolt that is threaded into a cylinder bolt hole

9 Cylinder head disassembly and inspection procedures are covered in detail in Chapter 4.

Installation

Note: *The head gaskets on the 5.7L engine have been replaced with another type of head gasket normally found on engines equipped with 0.030 inch oversize pistons. This head gasket is designed to prevent the possibility of a sealing ring "falling" into the combustion chamber. The updated head gasket can be purchased at a dealership parts department.*

10 The mating surfaces of the cylinder heads and block must be perfectly clean when the heads are installed.

11 Use a gasket scraper to remove all traces of carbon and old gasket material, then wipe the mating surfaces with a cloth saturated with lacquer thinner or acetone. If there is oil on the mating surfaces when the heads are installed, the gaskets may not seal correctly and leaks may develop. When working on the block, cover the lifter valley with shop rags to keep debris out of the engine. Use a vacuum cleaner to remove any debris that falls into the cylinders.

12 Check the block and head mating surfaces for nicks, deep scratches and other damage. If damage is slight, it can be removed with emery cloth. If it is excessive, machining may be the only alternative.

13 Use a tap of the correct size to chase the threads in the head bolt holes in the block. Mount each bolt in a vise and run a die down the threads to remove corrosion and restore the threads **(see illustration)**. Dirt, corrosion, sealant and damaged threads will affect torque readings.

14 Position the new gaskets over the dowel pins in the block. New cylinder heads should not be installed without checking the dowel pin clearance. To test the dowel pins, install the cylinder head onto the engine WITHOUT the head gasket. Measure the clearance with a 0.005 inch feeler gauge. There should be NO clearance, meaning the dowel pins are NOT keeping the cylinder head from resting completely on the block.

15 Install the cylinder head gaskets and carefully position the heads on the block without disturbing the gaskets.

16 On 5.7L engines, coat the threads of the head bolts with clean engine oil. On 6.2L and 6.5L engines, coat the threads of the head bolts with Teflon thread-sealant.

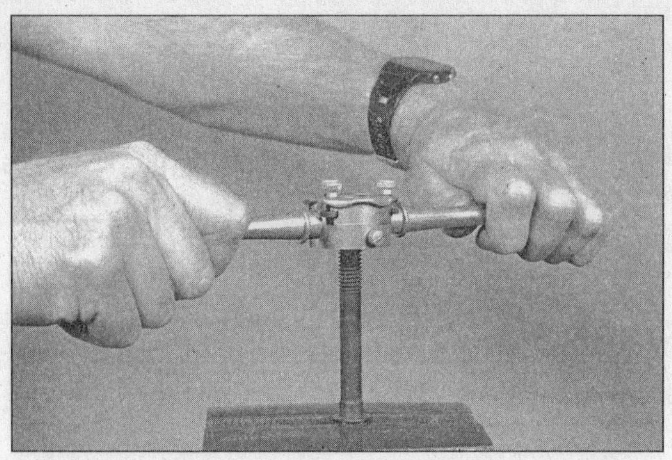

2.139 A tap should be used to remove sealant and corrosion from the head bolt threads prior to installation

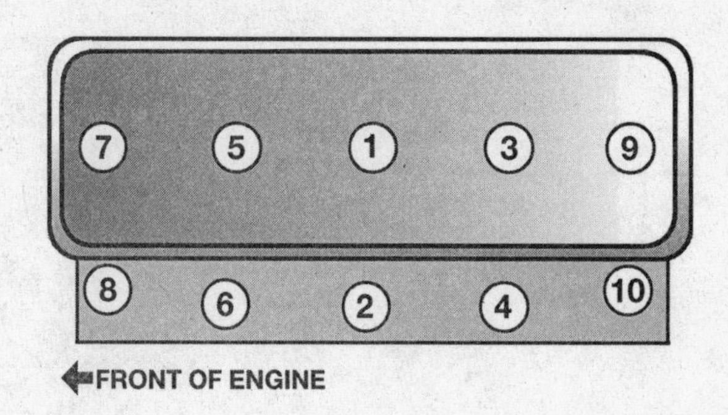

2.140 Cylinder head torque sequence on a 5.7L diesel engine

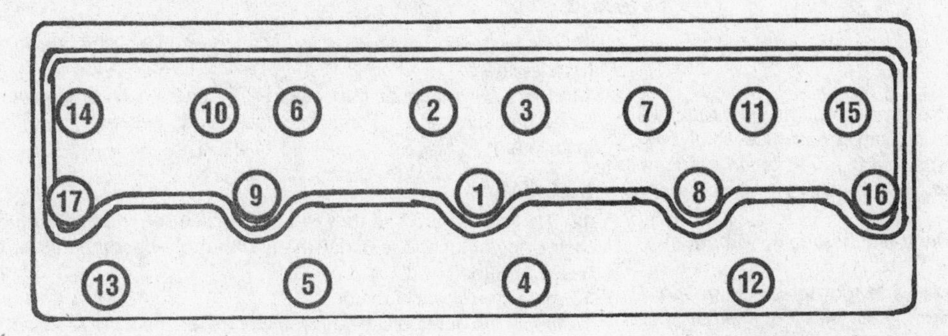

2.141 Cylinder head torque sequence on 6.2L and 6.5L diesel engines

17 Install the bolts in their original locations and tighten them finger tight. Following the recommended sequence, tighten the bolts in several steps to the torque listed in this Chapter's Specifications. **Note:** *On 6.2L and 6.5L diesel engines, use the torque and angle tightening procedure to tighten the head bolts. Also be sure to replace the headbolts (1982 and 1983) with the updated versions. First, torque all the bolts to 20 ft-lbs in the correct sequence. Second, retorque all the bolts to 50 ft-lbs, following the same sequence. Finally, turn each bolt an additional 90-degrees (1/4 turn). This is to insure a uniform bolt tension.*
18 The remaining installation steps are the reverse of removal.

Hydraulic lifters - removal, inspection and installation

GM diesel engines use two types of hydraulic lifters: disc type (1978 through 1980 5.7L engines) and roller type (all other models). The diesel type lifters are not the same as the gasoline type lifters and they should NOT be interchanged at any time.

A noisy valve lifter can be isolated when the engine is idling. Hold a mechanic's stethoscope or a length of hose near the location of each valve while listening at the other end.

Removal

1 Remove the intake manifold.
2 On 6.2L and 6.5L engines, remove the cylinder heads.
3 Remove the rocker arms and pushrods. There are several ways to extract the lifters from the bores. Special tools designed to grip and remove lifters are manufactured by many tool companies and are widely available, but may not be needed in every case. On newer engines without a lot of varnish buildup, the lifters can often be removed with a small magnet or even with your fingers. A machinist's scribe with a bent end can be used to pull the lifters out by positioning the point under the retainer ring in the top of each lifter. **Caution:** *Don't use pliers to remove the lifters unless you intend to replace them with new ones (along with the camshaft). The pliers may damage the precision machined and hardened lifters, rendering them useless. On engines with a lot of sludge and varnish, work the lifters up and down, using carburetor cleaner spray to loosen the deposits.* **Note:** *Be sure to check the casting on the block for an oversize designation. If the block is stamped with a "O", then the lifter is oversize. Make a note for proper parts identification.*

2.142 To remove roller lifters, remove the retainer bolts (arrows) and remove the retainers and lifters as an assembly; make sure that the lifters are installed in their original locations if they're reused

4 Before removing the lifters, arrange to store them in a clearly labeled box to insure that they're reinstalled in their original locations. **Note:** *On engines equipped with roller lifters, the guide retainer and guide plates must be removed before the lifters are withdrawn* **(see illustration)**. *Remove the lifters and store them where they won't get dirty.*

2.143 If the lifters are pitted or rough, they shouldn't be reused

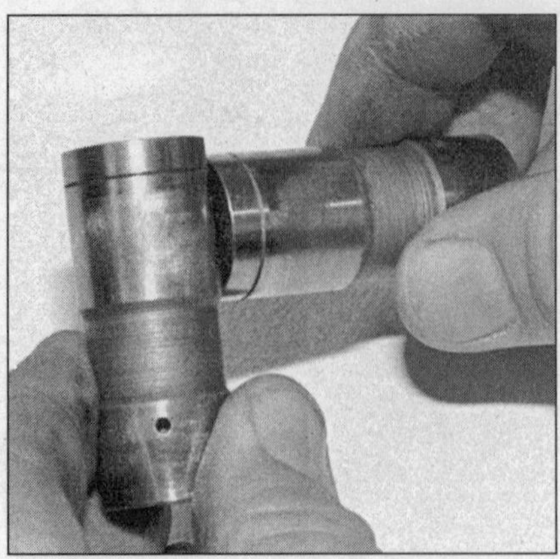

2.144 The foot of each lifter should be slightly convex - the side of another lifter can be used as a straightedge to check it; if It appears flat, It is worn and must not be reused

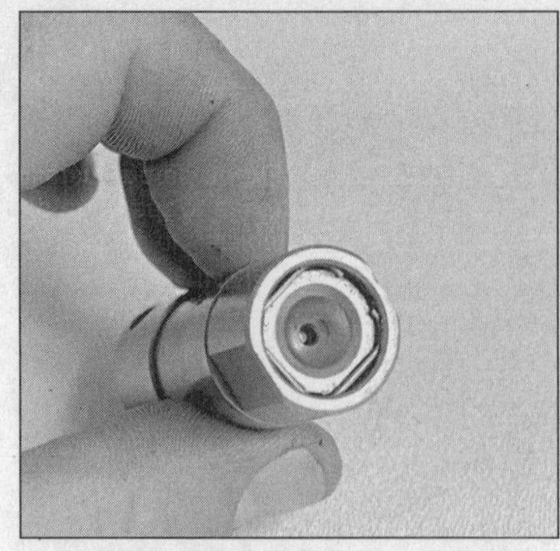

2.145 Check the pushrod seat (arrow) in the top of each lifter for wear

Inspection

Disc lifters

5 Parts for valve lifters are not available separately. The work required to remove them from the engine again, if cleaning is unsuccessful, outweighs any potential savings from repairing them. In other words, if you suspect there might be a problem with the lifters - replace them with new parts.

6 Clean the lifters with solvent and dry them thoroughly without mixing them up.

7 Check each lifter wall, pushrod seat and foot for scuffing, score marks and uneven wear **(see illustration)**. Each lifter foot (the part that rides on the camshaft) must be slightly convex, although this can be difficult to determine by eye **(see illustration)**. If the base of the lifter is concave, the lifters and camshaft must be replaced. If the lifter walls are damaged or worn (which is not very likely), inspect the lifter bores in the engine block as well. If the pushrod seats **(see illustration)** are worn, check the pushrod ends.

8 If new lifters are being installed, a new camshaft must also be installed. If a new camshaft is being installed, then install new lifters as well. Never install used lifters unless the original camshaft is used and the lifters can be installed in their original locations. When installing lifters, make sure they're coated with moly-based grease or engine assembly lube. Soak the new lifters in oil to remove the trapped air.

Roller lifters

9 Check the rollers carefully for wear and damage and make sure they turn freely without excessive play **(see illustration)**.

10 The inspection procedure for disc (conventional) lifters also applies to roller lifters.

11 Unlike conventional lifters, used roller lifters can be reinstalled with a new camshaft and the original camshaft can be used if new lifters are installed.

Installation

12 The original lifters, if they're being reinstalled, must be returned to their original locations. Coat them with moly-base grease or engine assembly lube.

13 Install the lifters in the bores.

14 Install the guide plates and retainer (roller lifters only).

15 Install the pushrods and rocker arms.

16 Install the intake manifold and valve covers.

Vibration damper - removal and installation

Removal

1 Remove the drivebelts from the engine accessories (power steering, alternator, air conditioning compressor etc. - see the *Maintenance* Section in this Chapter)

2 Remove the pulley from the face of the vibration damper **(see illustration)**.

3 Use a special puller **(see illustration)** and remove the vibration damper from the crankshaft. Watch carefully to make sure the pulley remains even with the timing chain cover as it is "backed-out". Any unnecessary force will damage the crankshaft.

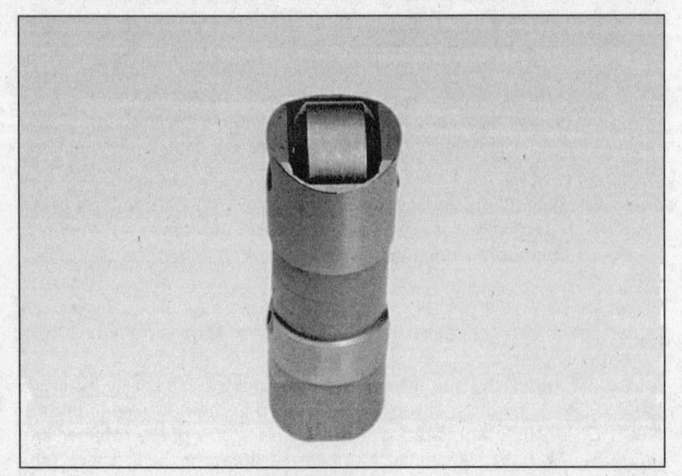

2.146 The roller in a roller lifter must turn freely - check for wear and excessive play as well

2.147 Remove the bolts from the front pulley

2.148 Attach the puller to the vibration damper tighten the center bolt against the crankshaft

2.149 Attach the installation tool to the damper and slowly turn the nut until the damper rests directly against the crankshaft

Installation

4 Use a special tool to install the vibration damper **(see illustration)**. Watch carefully to make sure the damper is perfectly flush with the crankshaft and not binding or offset.

Crankshaft front oil seal - replacement

Timing cover in place

1 Remove the vibration damper.

2 Carefully force the seal out of the cover with a seal removal tool **(see illustration)**. Be careful not to distort the cover or scratch the wall of the seal bore.

3 Clean the bore to remove any old seal material and corrosion. Position the new seal in the bore with the open end of the seal facing IN. A small amount of oil applied to the outer edge of the new seal will make installation easier - don't overdo it!

4 Drive the seal into the bore with GM tool no. J-25264 or a large socket and hammer **(see illustration)** until it's completely seated. Select a socket that's the same outside diameter as the seal.

5 Lubricate the seal lips with engine oil and reinstall the vibration damper.

2.150 Pry out the old seal with a seal removal tool (shown) or a screwdriver

Timing cover removed

6 Use a punch or screwdriver and hammer to drive the seal out of the cover from the backside. Support the cover as close to the seal bore as possible **(see illustration)**. Be careful not to distort the cover or scratch the wall of the seal bore. If the engine has accumulated a lot of miles, apply penetrating oil to the seal-to-cover joint on each side and allow it to soak in before attempting to drive the seal out.

7 Clean the bore to remove any old seal material and corrosion. Sup-

2.151 Carefully drive the new seal into place with a hammer and large socket

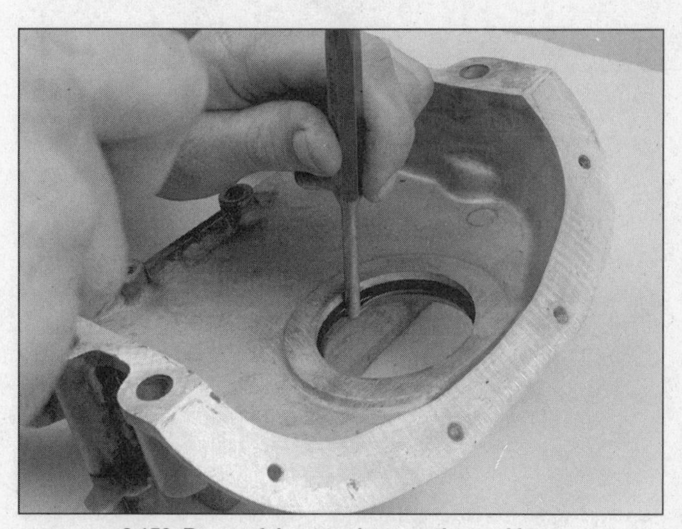

2.152 Be careful not to damage the seal bore when removing the old seal

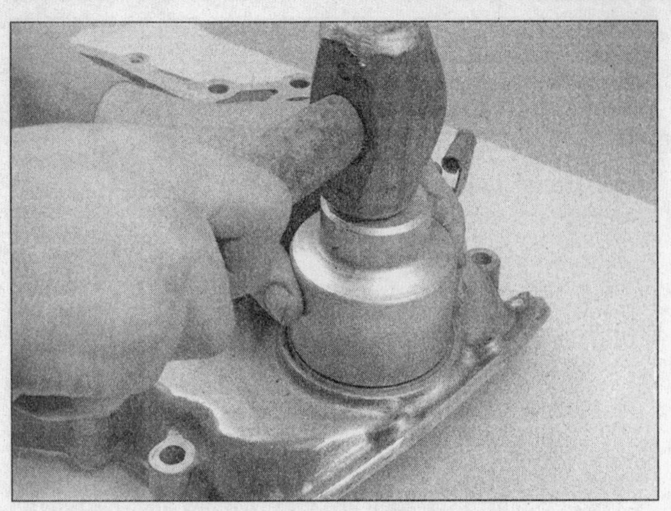

2.153 Use a large socket that is slightly smaller than the outside diameter of the now seal

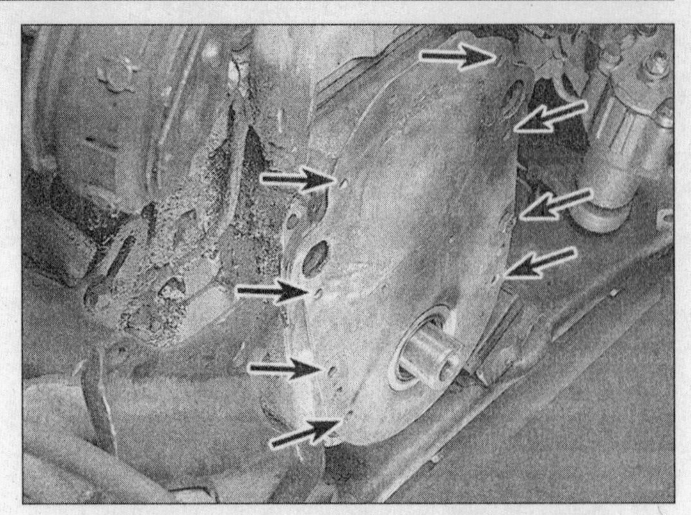

2.154 Engine front cover bolts (arrows) (5.7L engine)

port the cover on blocks of wood and position the new seal in the bore with the open end of the seal facing in. A small amount of oil applied to the outer edge of the new seal will make installation easier.
8 Drive the seal into the bore with a large socket and hammer until it is completely seated **(see illustration)**. Select a socket that is the same outside diameter as the seal (a section of pipe can be used if a socket is not available).
9 Reinstall the timing chain cover.

Timing chain cover - removal and installation

Note: *On 1994 and later models, whenever the front cover, timing chain, timing gears, crankshaft position sensor or other parts affecting timing are replaced, it is necessary to reprogram TDC offset into the PCM (computer). This requires an expensive electronic SCAN tool, take the vehicle to a dealership service department for the reprogramming procedure.*

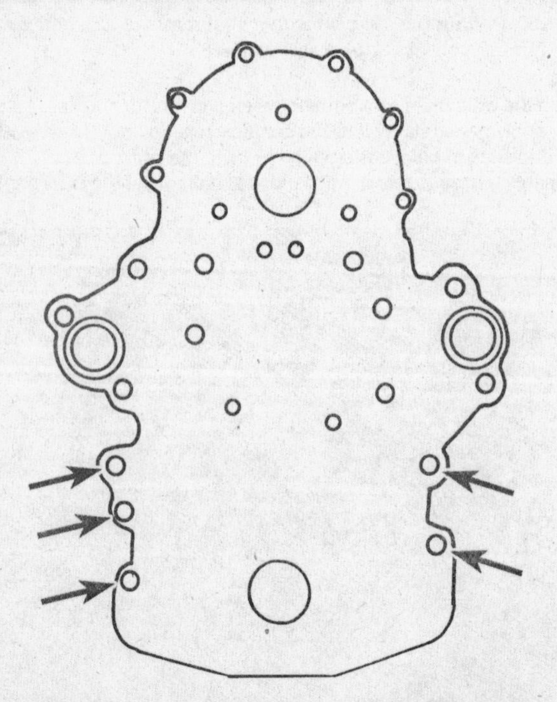

2.155 Engine front cover bolts (arrows) on 6.2L and 6.5L engines (water pump and water pump plate removed)

The timing cover on the 5.7L is a one-piece unit that can be removed with the water pump attached **(see illustration)**. The timing cover on the 6.2L and 6.5L is a two-piece unit which separates from the water pump. The baffle is removed after the water pump is off to allow the cover to clear the injection pump gears **(see illustration)**. Also, it is necessary to remove the intake manifold to gain access to the injection pump.

Removal

1 Remove the air conditioning compressor from the engine but do not disconnect the lines. Set the compressor off to the side. Remove the power steering pump from its brackets and set the assembly off to the side. Do not disconnect the power steering fluid lines from the pump. On 6.2L engines, refer to the *Cooling system* Section and remove the water pump.
2 Remove the bolts and separate the crankshaft drivebelt pulley from the vibration damper.
3 Refer to the next Sub-section *(Timing chain and sprockets - inspection, removal and installation)* and position the number one piston at TDC on the compression stroke. **Caution:** *This step is essential if the timing cover or timing probe is to be replaced or removed for any purpose. Once this has been done, do not turn the crankshaft until the timing chain and sprockets have been reinstalled.*
4 Most engines use a large bolt threaded into the nose of the crankshaft to secure the vibration damper in position. If your engine has a bolt, remove it from the front of the crankshaft, then use a puller to detach the vibration damper. **Caution:** *Do not use a puller with jaws that grip the outer edge of the damper. The puller must be the type that utilizes bolts to apply force to the damper hub only.*
5 On 6.2L and 6.5L engines, remove the intake manifold.
6 On 6.2L and 6.5L engines, remove the injection pump (refer to the *Fuel system* Section in this Chapter).
7 On 6.2L and 6.5L engines, remove the bolts that retain the baffle to the inside of the cover and remove the baffle **(see illustration)**.
8 On some models, the timing cover cannot be removed with the oil pan in place. The pan bolts will have to be loosened and the pan lowered slightly for the timing cover to be removed. If the pan has been in place for an extended period of time it is likely the pan gasket will break when the pan is lowered. In this case the pan should be removed and a new gasket installed.
9 Remove the bolts from the timing cover. It is important that the bolts are marked or a diagram is constructed designating the exact size and location of each of the timing cover bolts to prevent mix-ups when the cover is reassembled. Be sure to note the position of the stud bolts.
10 Remove the timing cover. Be careful not to damage the gasket sealing surface of the cover by prying the cover on the wrong areas.

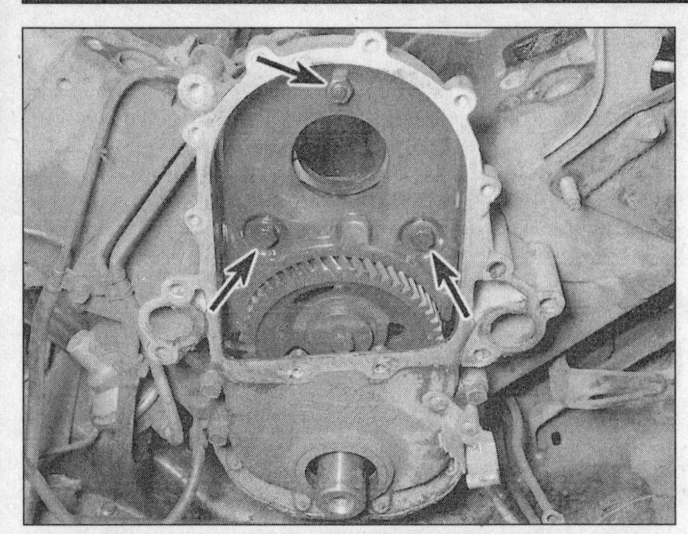

2.156 Remove the three bolts (arrows) from the baffle to release it from the timing chain cover

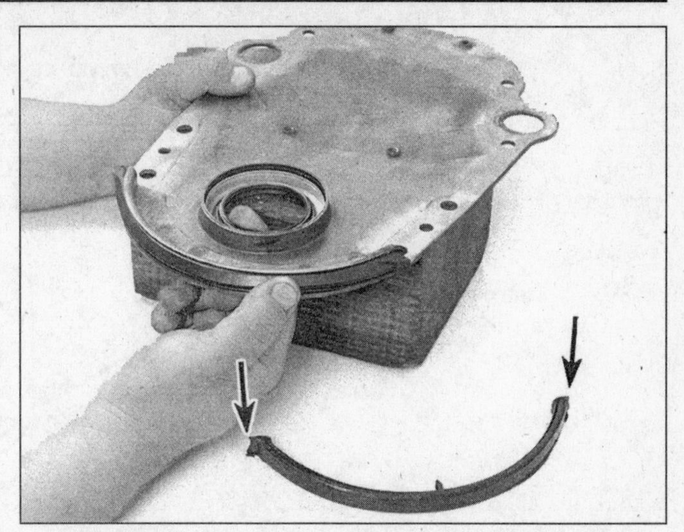

2.157 On the 5.7L timing cover, install the front seal into the timing cover first, then apply RTV sealer into the corners of the seal

Installation

11 On 5.7L engines, grind a chamfer on the end of each dowel pin to aid in reassembly.

12 If a new oil pan gasket was installed, cut the excess material from the front end of the gasket on each side of the block.

13 On 5.7L engines, trim about 1/8-inch from each end of a new front pan seal.

14 Install a new front cover gasket on the engine block and a new seal **(see illustration)** in the front cover (5.7L engine only).

15 Apply RTV to the gasket surface and allow the sealer to "set up" (slightly harden). Apply a more generous portion to the corners.

16 On 5.7L engines, place the cover on the front of the block and press down to compress the seal. Rotate the cover left and right and guide the pan seal into the cavity using a small screwdriver **(see illustration)**.

17 On 5.7L engines, install the two dowel pins (chamfered side first).

18 The remaining installation is the reverse of removal.

Marking TDC on the front housing (6.2L and 6.5L engine)

19 Turn the engine so the number one cylinder is at TDC.

20 Install the timing fixture (Kent-Moore J-33042, or equivalent) in the fuel injection pump location. Don't use a gasket.

21 The slot of the fuel injection pump gear should be vertical, i.e. in the six o'clock position **(see illustration)**. If it isn't, remove the fixture and rotate the crankshaft 360-degrees. The timing marks on the gears should now be aligned.

22 Fasten the timing fixture with one 8 mm bolt and tighten it snugly.

23 Install a 10 mm nut to the upper housing stud to hold the fixture. The flange nut should be finger tight.

24 Tighten the large bolt (18 mm head) counterclockwise (toward the left cylinder bank) to 50 ft-lbs. Tighten the 10 mm nut securely.

25 Make sure the crankshaft hasn't rotated (and the fixture didn't bind on the 10 mm nut).

26 Strike the scriber with a mallet to mark "TDC" on the front housing.

27 Remove the timing fixture.

28 Install the fuel injection pump with the gasket.

29 Install one 8 mm bolt to attach the gear to the pump hub and tighten it to 16.5 ft-lbs.

30 Align the timing mark on the fuel injection pump to the front housing mark. Tighten the three 10 mm attaching nuts to 31 ft-lbs.

31 Rotate the engine and install the remaining two pump gear attaching bolts and tighten to 16.5 ft-lbs.

Marking TDC on the injection pump adapter (5.7L engines)

32 On the 5.7L engine, it is necessary to mark the TDC mark on the injection pump adapter instead of the front cover if the adapter is replaced or damaged.

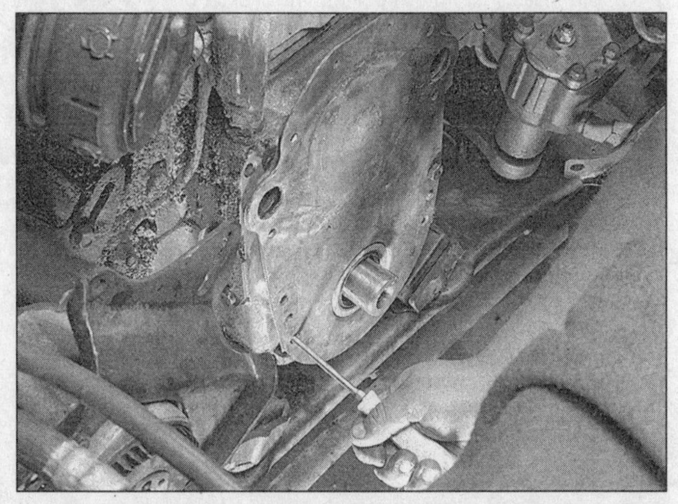

2.158 Guide the seal into the pan while pushing down

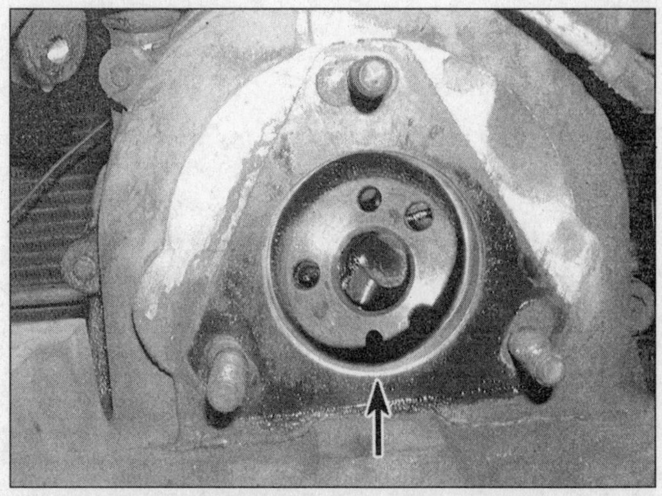

2.159 On 6.2L and 6.5L engines, the slot on the fuel injection pump gear must be at the six o'clock position

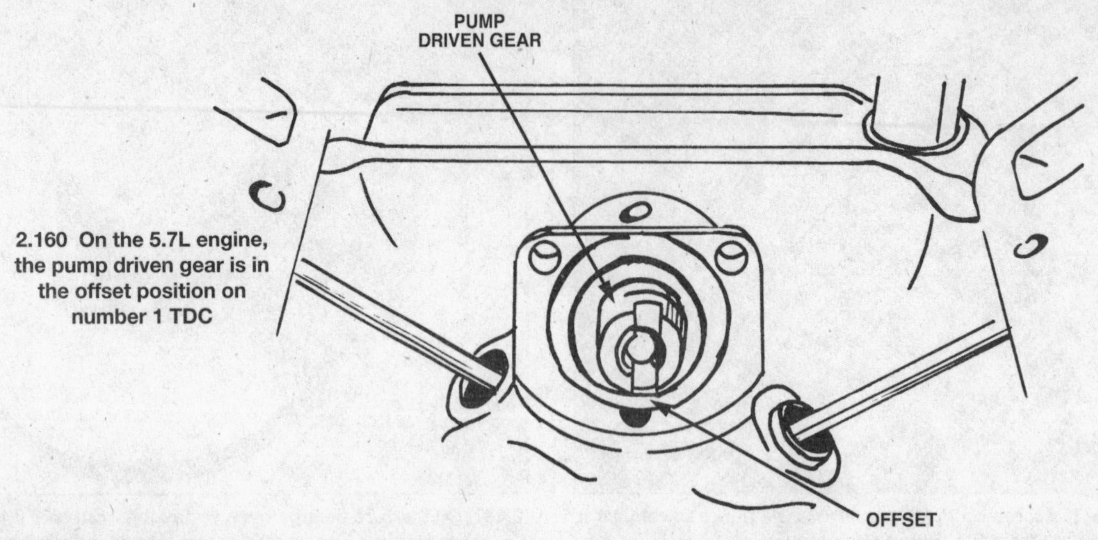

2.160 On the 5.7L engine, the pump driven gear is in the offset position on number 1 TDC

2.161 Align the timing mark on the vibration damper with the zero mark on the timing probe

2.162 If the exhaust valve (arrow) starts to close as the mark on the vibration damper nears the zero mark on the timing probe, then the cylinder is on the exhaust stroke. Rotate the crankshaft 360-degrees to TDC

33 Position the engine on number 1 TDC (see the procedure later in this Chapter).

34 With the mark on the vibration damper on number 1 TDC, the index on the pump driven gear is offset to the right **(see illustration)**.

35 Install the timing tool (Kent-Moore J-26896) into the pump adapter and tighten the tool to 50 ft-lbs. Use a small hammer and strike the tool to make the mark in the adapter.

Timing chain and sprockets - inspection, removal and installation

Note: *On 1994 and later models, whenever the front cover, timing chain, timing gears, crankshaft position sensor or other parts affecting timing are replaced, it is necessary to reprogram TDC offset into the PCM (computer). This requires an expensive electronic SCAN tool, take the vehicle to a dealership service department for the reprogramming procedure.*

Inspection

1 Position the number one piston at TDC on the compression stroke (see Step 2). **Caution:** *Once this has been done, do not turn the crankshaft until the timing chain and sprockets have been reinstalled.* To inspect the timing chain, remove the timing chain cover and temporarily reinstall the vibration damper bolt. Using this bolt, rotate the crankshaft in a counterclockwise direction to take up the slack in the right (passenger's) side of the chain. Establish a reference point on the block and measure from that point to the chain. Rotate the crankshaft in a clockwise direction to take up the slack in the left (driver's) side of the chain. Force the left side of the chain out with your fingers and measure the distance between the reference points and the chain. The difference between the two measurements is the slack. If the slack exceeds 1/2-inch, install a new chain and sprockets.

Locating Top Dead Center (TDC)

2 It is very important to correctly identify TDC if the timing chain and sprockets are replaced or if the engine is overhauled. Also, if the timing cover is replaced or the timing probe holder has been moved or replaced, the TDC mark must be adjusted.

3 Remove the number 1 (left bank) valve cover. **Note:** *Check the diagrams in the Chapter 4 Specifications to verify the location of the number 1 cylinder.*

4 Set the timing line or pointer on the vibration damper on the 0-degree mark **(see illustration)**. This will set the cam timing to number 1 TDC and number 6 (companion cylinder).

5 To verify which cylinder is on the compression stroke it is important to know the position of the rocker arm assembly. Use a large breaker bar and a socket on the vibration damper bolt and move the damper back and forth approximately 30-degrees before TDC (BTDC) and after

2.163 With the intake valve resting on top of the number 1 piston, zero the dial caliper

2.164 If the cylinder heads have been removed, install the dial indicator so that it reads directly off the top of the piston

2.165 Note very carefully the position of the alignment marks on the injection timing gears before the gears are removed - make sure they're aligned the some way on installation

2.166a Use two prybars to remove the camshaft gear from the camshaft

TDC (ATDC) and observe the number one cylinder rocker arms **(see illustration)**. If the exhaust valve moves from its open position to a closed position the cylinder is on the exhaust stroke which means the engine is set up for number 6 (companion cylinder). Rotate the engine 360-degrees until the mark comes around again onto the timing probe and carefully observe the rocker arms for the number one cylinder. The engine is on number 1 TDC when both valves (intake and exhaust) are closed and there is NO movement on the rocker arms when the crankshaft is rotated slightly BTDC and ATDC. Now you are ready to check the exact location of TDC using a dial caliper.

6 Remove the rocker arm assembly. Apply compressed air to the number 1 cylinder to force the valves up against the cylinder head (see the *Valve springs, retainers and seals - replacement* procedure earlier in this Chapter).

7 Install the valve spring removal tool onto the intake valve and compress the spring and remove the retainers from the end of the valve stem. Remove the valve spring and install a dial caliper.

8 Set the dial caliper against the tip of the valve stem so that the dial caliper can measure the vertical travel of the intake valve **(see illustration)**.

9 Remove the compressed air line from the adapter fitting and allow the valve to rest against the top of the piston.

10 Zero the dial caliper and, using a breaker bar and a large socket, move the vibration damper very slightly clockwise. Watch the dial caliper very carefully as the needle drops from its initial setting. Move the vibration damper counter-clockwise and watch carefully as the dial indicator comes back to the zero setting. Now keep moving the vibration damper past the initial setting - this time counter-clockwise. The timing mark should simultaneously align with the zero mark on the timing pointer and the dial caliper should read zero as the top of the piston reaches its highest point. Make several passes and record the results onto a piece of paper if the mark does not quite center on the zero mark on the timing probe. Calculate the average of four or five passes to arrive at the most perfect TDC location (if the setting varies).

11 Adjust the timing probe if the TDC mark is not perfectly aligned with the mark on the vibration damper. Loosen the bolts on the timing probe and move it to the location of true TDC (as indicated on the dial caliper). Recheck the TDC location to insure accuracy **(see illustration)**.

Removal

12 Remove the timing chain cover. Make sure the timing is set for number 6 TDC (number 1 cylinder companion). Refer to the beginning of this Sub-section.

13 On 6.2L engines, it is necessary to remove the injection pump timing gears to be able to remove the timing chain and gears **(see illustration)** (see *Fuel system* in this Chapter). Remove the bolts from the end of the camshaft, then detach the camshaft sprocket and chain as

2.166b If the crankshaft gear fits loosely on the crankshaft, remove the gears and chain as one assembly

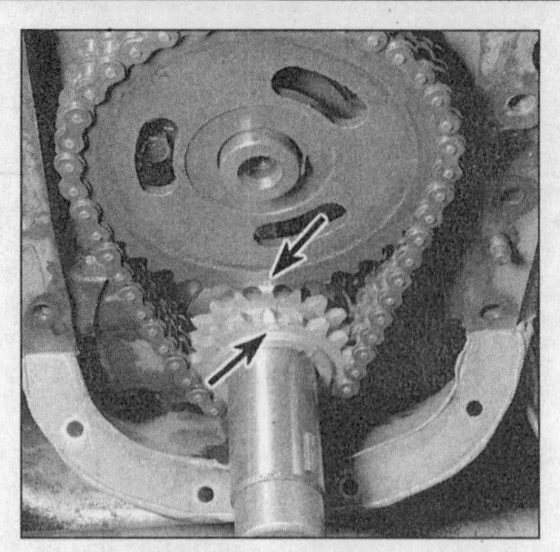

2.167 Install the timing gears with the timing marks aligned (arrows)

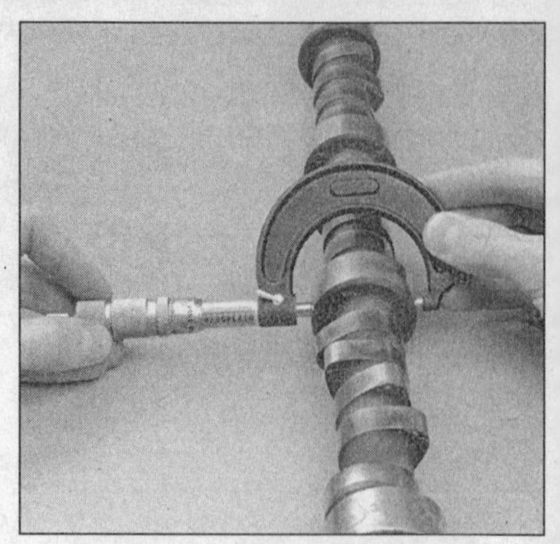

2.168 Check the diameter of each camshaft bearing journal to pinpoint excessive wear and out-of-round conditions

an assembly **(see illustrations)**. The sprocket on the crankshaft can be removed with a two or three-jaw puller, but be careful not to damage the threads in the end of the crankshaft. **Note:** *If the timing chain cover oil seal has been leaking, refer to the procedure earlier in this Section and install a new one.*

Installation

14 Use a gasket scraper to remove all traces of old gasket material and sealant from the cover and engine block. Stuff a shop rag into the opening at the front of the oil pan to keep debris out of the engine. Wipe the cover and block sealing surfaces with a cloth saturated with lacquer thinner or acetone.

15 On steel covers, check the cover flange for distortion, particularly around the bolt holes. If necessary, place the cover on a block of wood and use a hammer to flatten and restore the gasket surface.

16 If new parts are being installed, be sure to align the keyway in the crankshaft sprocket with the Woodruff key in the end of the crankshaft. **Note:** *Timing chains must be replaced as a set with the camshaft and crankshaft gears. Never put a new chain on old gears. Align the sprocket with the Woodruff key and press the sprocket onto the crankshaft with the vibration damper bolt, a large socket and some washers or tap it gently into place until it is completely seated.* **Caution:** *If resistance is encountered, do not hammer the sprocket onto the crankshaft. It may eventually move onto the shaft, but it may be cracked in the process and fail later, causing extensive engine damage.*

17 Loop the new chain over the camshaft sprocket. Make sure the timing mark is in the six o'clock position **(see illustration)**. Mesh the chain with the crankshaft sprocket and position the camshaft sprocket on the end of the cam. If necessary, turn the camshaft so the dowel pin fits into the sprocket hole with the timing mark in the 6 o'clock position. **Note:** *The number six piston must be at TDC on the compression stroke as the chain and sprockets are installed (see Step 2 above).*

18 Apply a thread locking compound to the camshaft sprocket bolt threads, then install and tighten them to the torque listed in this Chapter's Specifications. Lubricate the chain with clean engine oil.

19 Install the timing chain cover on the block, tightening the bolts finger tight (see the timing chain cover removal and installation procedure earlier in this Chapter).

20 Lubricate the oil seal contact surface of the vibration damper hub with moly-base grease or clean engine oil, then install the damper on the end of the crankshaft (see the procedure earlier in this Chapter). The keyway in the damper must be aligned with the Woodruff key in the crankshaft nose.

21 The remaining installation steps are the reverse of removal.

Camshaft and bearings - removal, inspection and installation

Note: *On 1994 and later models, whenever the front cover, timing chain, timing gears, crankshaft position sensor or other parts affecting timing are replaced, it is necessary to reprogram TDC offset into the PCM (computer). This requires an expensive electronic SCAN tool, take the vehicle to a dealership service department for the reprogramming procedure.*

Removal

1 Refer to the appropriate procedures and remove the intake manifold, the rocker arms, the pushrods and the timing chain and camshaft sprocket. The radiator should be removed as well. **Note:** *If the vehicle is equipped with air conditioning it may be necessary to remove the air conditioning condenser to remove the camshaft. If the condenser must be removed the system must first be depressurized by a dealer service department or air conditioning shop. Do not disconnect any air conditioning lines until the system has been properly depressurized.*

2 Before removing the lifters, arrange to store them in a clearly labeled box to ensure that they are reinstalled in their original locations. Remove the lifters from the engine block.

3 Obtain long bolts from the auto parts store and thread them into the camshaft sprocket bolt holes to use as a handle when removing the camshaft from the block.

4 Carefully pull the camshaft out. Support the cam near the block so the lobes do not nick or gouge the bearings as it is withdrawn.

Inspection

Camshaft and bearings

5 After the camshaft has been removed from the engine, cleaned with solvent and dried, inspect the bearing journals for uneven wear, pitting and evidence of seizure. If the journals are damaged, the bearing inserts in the block are probably damaged as well. Both the camshaft and bearings will have to be replaced. Replacement of the camshaft bearings requires special tools and techniques which place it beyond the scope of the home mechanic. The engine block will have to be removed from the vehicle and taken to an automotive machine shop for this procedure.

6 Measure the bearing journals with a micrometer to determine if they are excessively worn or out-of-round **(see illustration)**. Compare your measurements with this Chapter's Specifications.

7 Check the camshaft lobes for heat discoloration, score marks, chipped areas, pitting and uneven wear. If the lobes are in good condition and not worn excessively, the camshaft can be reused.

Installation

8 Lubricate the camshaft bearing journals and cam lobes with moly-base grease or engine assembly lube **(see illustration)**.

9 Slide the camshaft into the engine. Support the cam near the block and be careful not to scrape or nick the bearings.

10 Turn the camshaft until the dowel pin is in the 3 o'clock position.

11 Install the timing chain and sprockets.

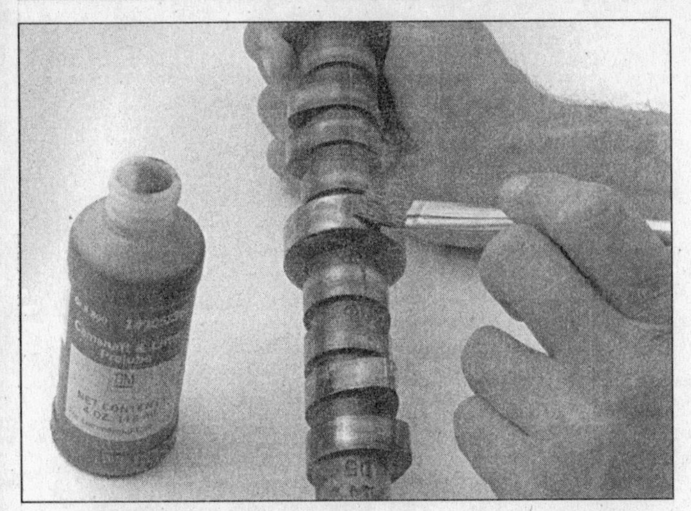

2.169 Coat the lobes and journals with moly-based grease or engine assembly lube

2.170 Tilt the pan down at the rear as shown to clear the crank throws, the oil pump and the crossmember

12 Lubricate the lifters with clean engine oil and install them in the block. If the original lifters are being reinstalled, be sure to return them to their original locations. If a new camshaft was installed, be sure to install new lifters as well.
13 The remaining installation steps are the reverse of removal.
14 Before starting and running the engine, change the oil and install a new oil filter.

Oil pan - removal and installation

Removal

1 Disconnect the negative cable from the battery.
2 Raise the vehicle and support it securely on jackstands.
3 Drain the engine oil and remove the oil filter.
4 Unbolt the crossover pipe at the exhaust manifolds.
5 Remove the vacuum pump and the oil pump drive.
6 Remove the lower bellhousing cover.
7 Remove the starter, if necessary for clearance.
8 Disconnect the oil cooler lines at the base of the oil filter.
9 Unbolt the fan shroud and move it back over the fan. If clearance is tight, insert a piece of heavy cardboard between the fan and radiator to protect the radiator fins from the fan when the engine is raised.
10 Remove the engine mount through bolts.
11 Use an engine hoist to lift the engine approximately three inches. If a hoist isn't available, you might try a floor jack under the oil pan with a wood block between the jack head and the pan. **Caution:** *On most engines the oil pump pickup is very close to the bottom of the oil pan, and it can be damaged easily if concentrated pressure from a jack is applied to the pan. When lifting the engine check to make sure the injection pump isn't hitting the firewall (5.7L) and the fan isn't hitting the radiator.*
12 Place blocks of wood between the crossmember and the engine block in the area of the motor mounts to hold the engine in the raised position, then remove the engine hoist.
13 Remove the oil pan bolts and reinforcements. Note that some models use studs and nuts in some positions.
14 Remove the pan by tilting the back downward and working it free of the crankshaft throws, oil pump pickup and the front crossmember **(see illustration)**.

Installation

15 Thoroughly clean the mounting surfaces of the oil pan and engine block of old gasket material and sealer.
16 Apply a thin layer of RTV-type sealer to the oil pan and install a new oil pan gasket.
17 Install the oil pan end seal(s). Apply RTV sealant to the ends of the oil pan gaskets to hold these seal(s) in place.
18 Lift the pan into position, being careful not to disturb the gasket, and install the bolts/nuts finger tight.
19 Starting at the ends and alternating from side-to-side towards the center, tighten the bolts to the torque listed in this Chapter's Specifications.
20 The remainder of the installation procedure is the reverse of removal. Fill the pan with new oil, start the engine and check for leaks before placing the vehicle back in service.

2.171 Remove the oil pump mounting bolt (shown with engine mounted on an engine stand)

Oil pump - removal and installation

1 Remove the oil pan as described in the previous procedure.
2 While supporting the oil pump, remove the pump-to-rear main bearing cap bolts. Remove the pump **(see illustration)** and, if equipped, the driveshaft extension.
3 Lower the pump and remove it.
4 Position the pump on the engine and make sure the slot in the upper end of the pump (or driveshaft extension) is aligned with the tang on the lower end of the driveshaft or injection pump.
5 Install the mounting bolt and tighten it to the torque listed in this Chapter's Specifications.
6 Install the oil pan.

Clutch and flywheel/driveplate - removal and installation

Note: *Some 1994 and later models may be equipped with a Dual Mass flywheel. Removal and installation of the Dual Mass flywheel is identical*

2.172 Typical clutch assembly

1 *Pressure plate* 2 *Flywheel*

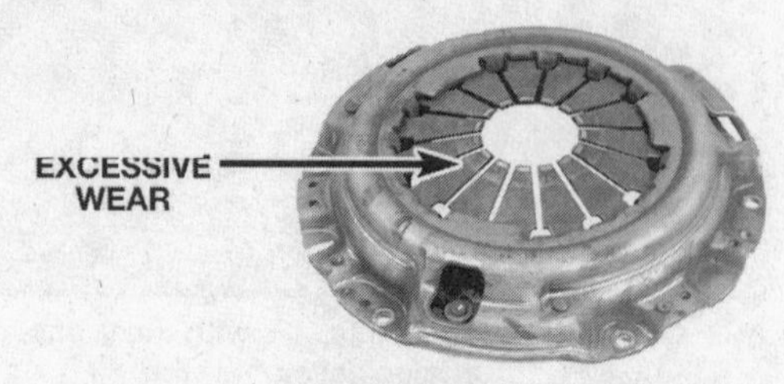

2.173 Replace the pressure plate if damage or excessive wear is noted

to the standard flywheel. Do not attempt to disassemble a Dual Mass flywheel, special tools are necessary to reassemble the unit correctly.

1 Raise the vehicle and support it securely on jackstands, then remove the transmission. If it's leaking, now would be a very good time to replace the front pump seal/O-ring (automatic transmission only).

2 If the vehicle has a manual transmission, remove the pressure plate and clutch disc **(see illustration)**. Now is a good time to check/replace the clutch components and pilot bearing **(see illustrations)**.

3 Make alignment marks on the flywheel/driveplate and crankshaft to ensure correct alignment during reinstallation **(see illustration)**.

4 Remove the bolts that secure the flywheel/driveplate to the crankshaft **(see illustration)**. If the crankshaft turns, wedge a screwdriver through the starter opening to jam the flywheel.

5 Remove the flywheel/driveplate from the crankshaft. Since the flywheel is fairly heavy, be sure to support it while removing the last bolt.

6 Clean the flywheel to remove grease and oil. Inspect the surface for cracks, rivet grooves, burned areas and score marks. Light scoring can be removed with emery cloth. Check for cracked and broken ring gear teeth. Lay the flywheel on a flat surface and use a straightedge to check for warpage.

7 Clean and inspect the mating surfaces of the flywheel/driveplate and the crankshaft. If the crankshaft rear seal is leaking, replace it before reinstalling the flywheel/driveplate.

8 Position the flywheel/driveplate against the crankshaft. Be sure to

2.174 The machined face of the pressure plate must be inspected for score marks and other damage; if the damage is slight, a machine shop can make the surface smooth again

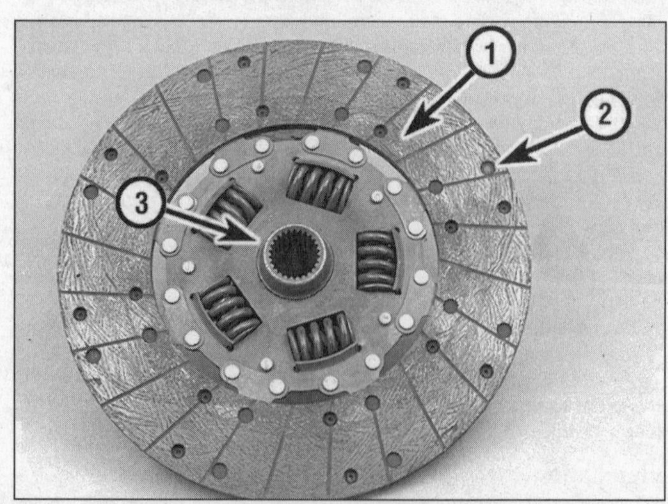

2.175 The clutch plate

1 ***Lining*** *- the lining will wear down in use*
2 ***Rivets*** *- the rivets secure the lining and will damage the flywheel or pressure plate is allowed to contact the surfaces*
3 ***Markings*** *- usually indicated by "Flywheel side" or something similar*

2.176 To insure correct balance, mark the flywheel's relationship to the crankshaft

2.177 A large screwdriver wedged in the starter ring gear teeth or one of the holes in the flywheel/driveplate can be used to keep the flywheel/driveplate from turning as the mounting bolts are removed

align the marks made during removal. Note that some engines have an alignment dowel or staggered bolt holes to ensure correct installation. Before installing the bolts, apply thread locking compound to the threads.

9 Wedge a screwdriver through the starter motor opening to keep the flywheel/driveplate from turning as you tighten the bolts to the torque listed in this Chapter's Specifications.

10 Position the clutch disc and pressure plate against the flywheel with the clutch held in place with an alignment tool **(see illustration)**. Make sure it's installed properly (most replacement clutch plates will be marked "flywheel side" or something similar - it not marked, install the clutch disc with the damper springs toward the transmission). Tighten the pressure plate-to-flywheel bolts only finger tight, working around the pressure plate.

11 Center the clutch disc by ensuring the alignment tool extends through the splined hub and into the pilot bearing in the crankshaft. Wiggle the tool up, down or side-to-side as needed to bottom the tool in the pilot bearing. Tighten the pressure plate-to-flywheel bolts a little at a time, working in a criss-cross pattern to prevent distorting the cover.

12 The remainder of installation is the reverse of the removal procedure.

2.178 A clutch alignment tool can be purchased at most auto parts stores and eliminates all guesswork when centering the clutch plate in the pressure plate

Crankshaft rear oil seal - replacement

Two-piece seals (1982 through 1991 models)

"Rope" seals

1 The crankshaft rear main seal is a "rope" seal. The rear main seal can be replaced with the engine in the vehicle. Refer to the appropriate Sub-sections and remove the oil pan and oil pump.

2 Remove the bolts and detach the rear main bearing cap from the engine.

3 The old seal section in the bearing cap can be packed tighter in the block housing and additional sections of the rope seal can be inserted to fill the excess space.

4 To compact the seal section in the block, tap on one end using the special tool (J-33154-2) or with a hammer and an aluminum, brass or wood dowel until the old seal is packed tight. This will vary 1/4 to 3/4-inch, depending on the condition of the old seal. **(see illustrations)**.

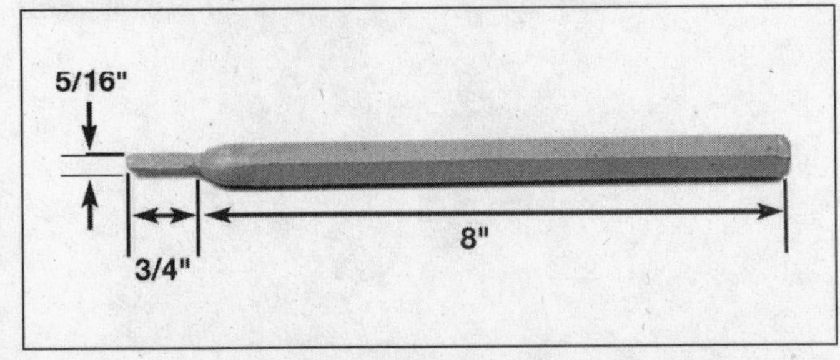

2.179 Grind a piece of aluminum or brass rod to these dimensions as a rear seal driver

2.180 Drive each end of the seal into the groove until it feels tightly packed

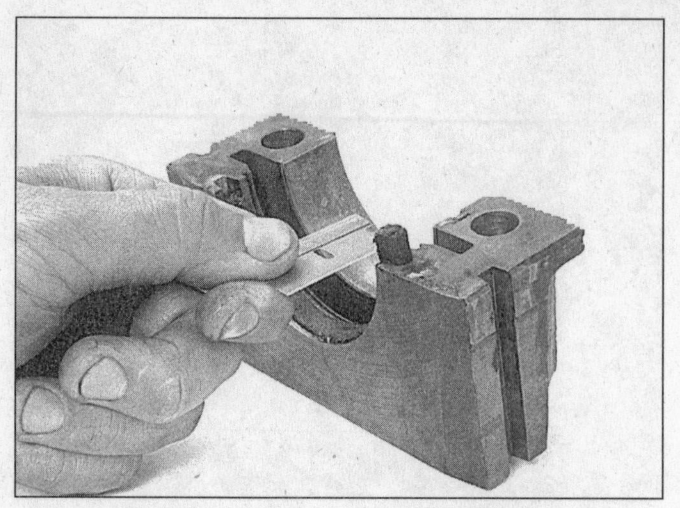

2.181 Use the bearing cap as a holding fixture when cutting short sections of the old seal

2.182 The new seal material used in the cap can be settled into the groove with a hammer handle or by tapping it in with a hammer and large-diameter socket

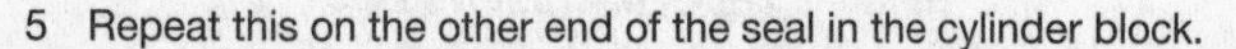

5 Repeat this on the other end of the seal in the cylinder block.

6 Measure the length the seal was driven up on one side, add 1/16 inch and cut a section that length off the new seal with a single edged razor blade. Repeat the procedure for the other side. Use the bearing cap as a holding fixture **(see illustration)**. Apply a sealant (GM 1052621 or equivalent) to each end of the rope seal.

7 Using the tool (J-33154-2), install both of these two small pieces into the vacated ends of the old seal. The ends should be flush with the mating surface of the cap. Make sure it is completely seated.

8 Trim the seal flush with the block.

9 Form a new rope seal in the rear main bearing cap.

10 Push the seal into place, using the special tool (J-33153), or a wooden hammer handle, like a "press" **(see illustration)**. When both ends of the seal are flush with the block surface, remove the tool.

11 Lubricate the cap bolts with clean engine oil.

12 Carefully position the bearing cap on the block, install the bolts and tighten them to the torque listed in the Chapter 4 Specifications. Tap the crankshaft forward and backward with a lead or brass hammer to line up the main bearing and crankshaft thrust surfaces, then tighten the rear bearing cap bolts to the torque listed in this Chapter's Specifications.

13 Install the oil pump and oil pan.

Neoprene seals

1 Always replace both halves of the rear main oil seal as a unit. While the replacement of this seal is much easier with the engine removed from the car, the job can be done with engine in place.

2 Remove the oil pan and oil pump as described previously in this Chapter.

3 Remove the rear main bearing cap from the engine.

4 Using a screwdriver, pry the lower half of the oil seal from the bearing cap.

5 To remove the upper half of the seal, use a small hammer and a brass pin punch to roll the seal around the crankshaft journal. Tap one end of the seal with the hammer and punch (be careful not to strike the crankshaft) until the other end of the seal protrudes enough to pull the seal out with a pair of pliers **(see illustration)**.

6 Clean all seal and and foreign material from the bearing cap and block. Do not use an abrasive cleaner for this.

7 Inspect components for nicks, scratches or burrs at all sealing surfaces.

8 Coat the seal lips of the new seal with light engine oil. Do not get oil on the seal mating ends.

9 Included in the purchase of the rear main oil seal should be a small plastic installation tool. If not included, make your own by cutting an old feeler gauge blade or shim stock **(see illustration)**.

10 Position the narrow end of this installation tool between the crankshaft and the seal seat. The idea is to protect the new seal from being damaged by the sharp edge of the seal seat.

2.183 Once enough of the upper half of the seal has been pushed out, pull it out with a pair of pliers

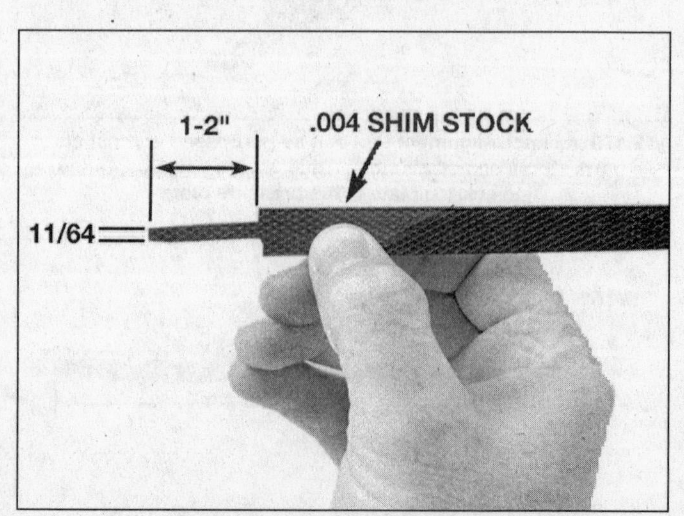

2.184 Installation tool for neoprene rear main seal

2.185 Use the protector tool (arrow) when pushing the seal into place

2.186 Sealant should be used where the rear main cap touches the engine block

11 Raise the new upper half of the seal into position with the seal lips facing towards the front of the engine. Push the seal onto its seat, using the installation tool as a protector against the seal contacting the sharp edge.
12 Roll the seal around the crankshaft, all the time using the tool as a "shoehorn" for protection. When both ends of the seal are flush with the engine block, remove the installation tool, being careful not to withdraw the seal as well.
13 Install the lower half of the oil seal in the bearing cap, again using the installation tool to protect the seal against the sharp edge **(see illustration)**. Make sure the seal is firmly seated, then withdraw the installation tools.
14 Smear a bit of sealant on the bearing cap areas immediately adjacent to the seal ends **(see illustration)**.
15 Install the bearing cap (with seal) and torque the attaching bolts to about 10 to 12 ft-lbs only. Now tap the end of the crankshaft first rearward, then forward to line up the thrust surfaces. Retorque the bearing cap bolts to the proper Specification (see Chapter 2B).

One-piece seal (1992 and later models)

Caution: *We recommend using a special seal installation tool (GM no. J39084) for this procedure. If the tool is not available, you may be able to install the seal using a piece of pipe with a diameter slightly smaller than the outer diameter of the seal or a blunt punch and a hammer.*
14 These engines use a one-piece rear main oil seal. The seal is pressed into a bore machined into the rear main bearing cap and engine block. Remove the transmission, clutch components (if equipped) and flywheel or driveplate.
15 Pry out the old seal with a hooked tool or large screwdriver **(see illustration)**. **Caution:** *To prevent an oil leak after the new seal is installed, be very careful not to scratch or otherwise damage the crankshaft sealing surface or the bore in the bearing cap.*
16 Clean the crankshaft and seal bore in the block/bearing cap thoroughly and de-grease these by wiping them with a rag soaked in lacquer thinner or acetone. Lubricate the lip and outer diameter of the new seal with engine oil. **Note:** *When installing the new seal, the lip of the seal must face the front of the engine.*
17 If tool J-39084 is available, install the new seal on the tool and position the tool against the crankshaft. Thread the attaching screws into the crankshaft, then tighten the screws securely with a screwdriver. Turn the tool handle until it bottoms, then remove the tool.
18 If tool J-39084 is not available, tap the new seal into place using a hammer and a piece of pipe or a blunt punch. Work around the seal, tapping it evenly into place until it bottoms. **Caution:** *The seal must be flush with the rear surface of the engine block and rear main bearing cap.*
19 The remainder of installation is the reverse of removal.

2.187 To remove the seal from the housing, insert the tip of the screwdriver into each notch and pry out the seal

Engine mounts - check and replacement

1 Engine mounts seldom require attention, but broken or deteriorated mounts should be replaced immediately or the added strain placed on the driveline components may cause damage or wear.

Check

2 During the check, the engine must be raised slightly to remove the weight from the mounts.
3 Raise the vehicle and support it securely on jackstands, then position a jack under the engine oil pan. Place a large block of wood between the jack head and the oil pan, then carefully raise the engine just enough to take the weight off the mounts. **Warning:** *DO NOT place any part of your body under the engine when it's supported only by a jack!*
4 Check the mounts to see if the rubber is cracked, hardened or separated from the metal plates. Sometimes the rubber will split right down the center.
5 Check for relative movement between the mount plates and the engine or frame (use a large screwdriver or prybar to attempt to move the mounts). If movement is noted, lower the engine and tighten the mount fasteners.
6 Rubber preservative should be applied to the mounts to slow deterioration.

2.188 To detach the engine block from the left engine mount, remove the through-bolt and nut (upper arrow); to detach the rubber insulator from the bracket, remove these three nuts (lower arrows; one nut not visible in this photo) (6.2L pick-up shown, other models similar)

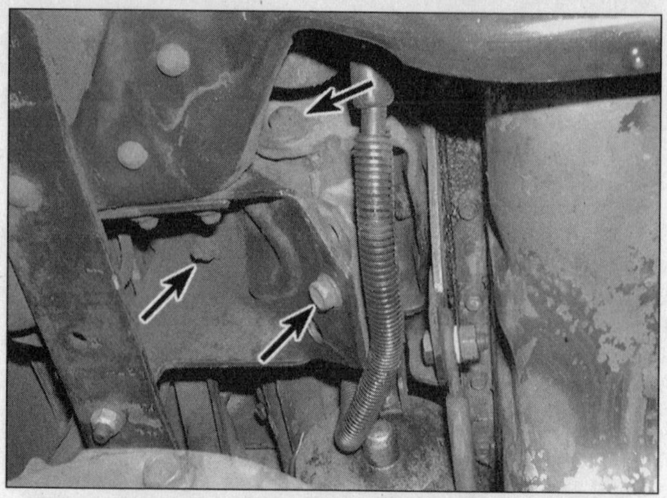

2.189 To detach the engine block from the right engine mount, remove the through-bolt and nut (upper arrow); to detach the rubber insulator from the bracket, remove these three nuts (lower arrows; one nut not visible in this photo) (6.2L pick-up shown, other models similar)

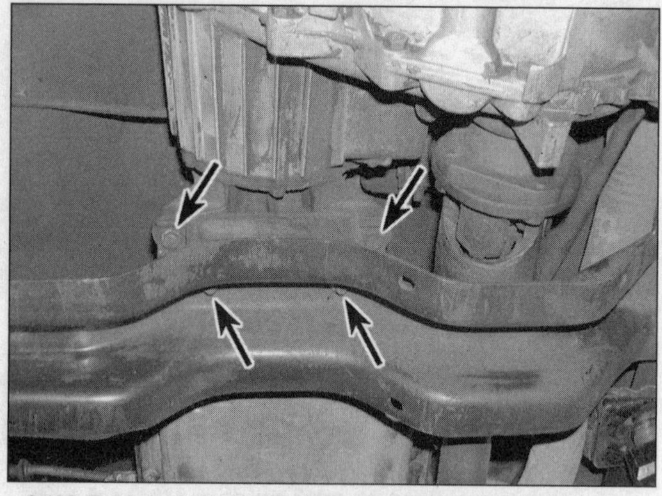

2.190 To detach the transmission from the rear crossmember, remove these four bolts (arrows) (6.2L pick-up shown, other models similar)

Replacement

7 Disconnect the negative battery cable from the battery, then raise the vehicle and support it securely on jackstands (if not already done).
8 Remove the fasteners and detach the mount from the frame bracket **(see illustrations)**.
9 Raise the engine slightly with a jack or hoist (make sure the fan doesn't hit the radiator or shroud). Remove the mount-to-block bolts and detach the mount.
10 Installation is the reverse of removal. Use thread locking compound on the mount bolts and be sure to tighten them securely.

Chapter 3 Ford 6.9L and 7.3L V8 engines

Specifications

Recommended lubricants and fluids

Engine oil	
Type	API grade SF/CD or SF/CE
Viscosity	See accompanying chart
Capacity	10.0 us qts
Engine coolant	50/50 mixture of water and ethylene glycol- based antifreeze

General

Cylinder numbering (front-to-rear)	
Right (passenger's) side	1-3-5-7
Left side	2-4-6-8
Firing order	1-2-7-3-4-5-6-8

Drivebelt tension

	New	Used
Vacuum pump	110 to 130 lbs	90 to 110 lbs
Power steering with A/C	140 to 180 lbs	110 to 130 lbs
All others (3/8 inch)	120 to 160 lbs	110 to 130 lbs

Engine idle speed

Warm	650 to 700 rpm (most models - check VECI label underhood)
Cold	
1983 through 1988	850 to 900 rpm
1989 on	800 to 850 rpm

Camshaft and timing gears

Camshaft bearing journal diameter	
6.9L	
1983	2.0990 to 2.100 in
1984	2.1010 to 2.1045 in
1985 and later	2.1020 to 2.1025 in
7.3L	2.1015 to 2.1025 in
Camshaft bearing oil clearance	
6.9L	0.0010 to 0.0055 in
7.3L	
1983 thru 1993 and 1994 non-DI Turbo	0.0015 to 0.0035 in
1994 and later DI Turbo	0.002 to 0.006 in
Camshaft end play	
6.9L	0.001 to 0.009 in
7.3L	
1983 thru 1993 and 1994 non-DI Turbo	0.002 to 0.009 in
1994 and later DI Turbo	0.002 to 0.008 in
Timing gear backlash	
6.9L	0.0055 to 0.0100 in
7.3L	
1983 thru 1993 and 1994 non-DI Turbo	0.0015 to 0.0130 in
1994 and later DI Turbo	0.0055 to 0.0100 in

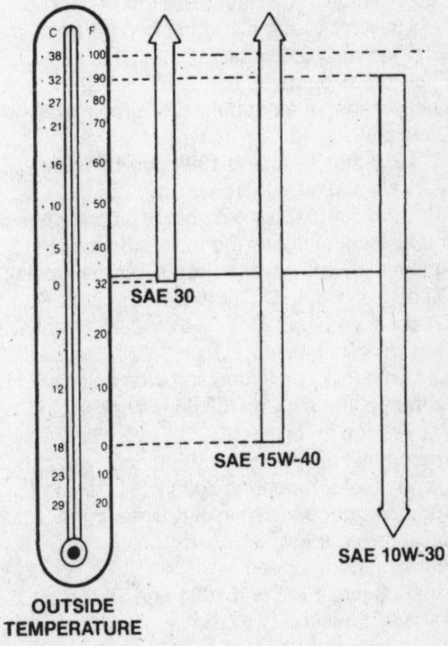

SAE 30
SAE 15W-40
SAE 10W-30
OUTSIDE TEMPERATURE

Engine oil viscosity chart

Torque specifications

Note: *If no separate listing is provided, use torque values based on fastener size.*

	Ft-lbs
1/4 - 20 UNC	7
5/16 - 18 UNC	14
3/8 - 16 UNC	24
7/16 - 14 UNC	38
1/2 - 13 UNC	60

Torque specifications (continued)

	Ft-lbs
Cylinder head	
1983 and 1984 models	
First step	40
Second step	65
Third step	75
Fourth step	75
1985 through 1987 models	
First step	40
Second step	70
Third step	80
Fourth step	80
1988 through 1991 models	
First step	65
Second step	85
Third step	100
Fourth step	100
1992, 1993 and 1994 non-DI Turbo	
First step	65
Second step	90
Third step	110
1994 and later DI Turbo	
First step	65
Second step	85
Third step	105
Driveplate-to-crankshaft bolts (auto trans)	47
Exhaust manifold bolts	
1983 thru 1992	35
1993 and 1994 non-DI Turbo	20
1994 and later DI Turbo	45
Fan-to-fan clutch nuts	12 to 18
Fan clutch-to-water pump bolts	
1983 thru 1993 and 1994 non-DI Turbo	45 to 120
1994 and later DI Turbo	83 to 113
Flywheel-to-crankshaft bolts (manual trans)	
1983 thru 1993 and 1994 non-DI Turbo	47
1994 and later DI Turbo	89
Flywheel, dual mass, secondary to primary flywheel bolts	47
Fuel injector-to-cylinder head (non-DI Turbo)	35
Fuel injector clamp mounting bolt (DI Turbo models)	108 in-lbs
Fuel line-to-nozzle fittings	22
Fuel line-to-pump fittings	22
Fuel pump mounting bolts (DI Turbo models)	24
Glow plugs	
1983 thru 1993 and 1994 non-DI Turbo	12
1994 and later DI Turbo	14
Injection pump drive gear cover (adapter housing) bolts	14
Intake manifold bolts (non-DI Turbo models)	24
Intake manifold cover bolts (DI Turbo models)	18
Oil deflector bolt (DI Turbo models)	106 in-lbs
Oil rail drain plugs (DI Turbo models) (install with thread lock)	53
Oil pan drain plug	28
Oil pump (high pressure) sprocket bolt (DI Turbo models)	95
Oil pump and pick-up (5/16 inch bolts)	14
Rocker arm bolts	20
Thermostat cover bolts	20
Turbocharger mounting bolts	18
Turbocharger exhaust inlet adapter bolts	36
Valve cover bolts	72 in-lbs
Vibration damper bolt	
1983 thru 1993 and 1994 non-DI Turbo	90
1994 and later DI Turbo	212
Water pump to front cover bolt	15

Maintenance schedule

Every 250 miles or weekly, whichever comes first

Check the engine oil level
Check the engine coolant level

Every 5000 miles or 6 months, whichever comes first

All items listed above, plus
Change the engine oil and filter*
Drain the water from the fuel separator*
Check and lubricate the throttle linkage

Every 15,000 miles or 12 months, whichever comes first

Replace the air filter
Check and adjust the engine drivebelts
Check the cooling system
Check and service the batteries
Inspect and replace, if necessary, all underhood hoses

Every 30,000 miles or 24 months, whichever comes first

Drain and flush the cooling system

Every 60,000 miles or 48 months, whichever comes first

Replace the fuel filter

** This item is affected by "severe" operating conditions as described below. If the vehicle is operated under severe conditions, perform all maintenance indicated with an asterisk (*) at 3000 mile/3 month intervals. Severe conditions exist if you mainly operate the vehicle . . .*

In dusty areas
Towing a trailer
Idling for extended periods and/or driving at low speeds
When outside temperatures remain below freezing and most trips are less than four miles long

Introduction

This Chapter contains information on the Ford 6.9L and 7.3L diesel engines. In 1994 Ford introduced the 7.3L Direct Injection (DI) turbocharged diesel, coined as the "Power-stroke" diesel. The naturally-aspirated 7.3L diesel was also produced in 1994 and discontinued for 1995. The two 7.3L diesel engines share nothing in common; if you have a 1994 model, make sure you know exactly which version your dealing with.

We've included sections on routine maintenance, on servicing the cooling, fuel, electrical and emission control systems, and even the engine repairs you can perform with the engine still installed in the vehicle (you'll find general engine overhaul procedures - those jobs that require engine removal - in Chapter 4).

Look at the mileage/time master maintenance schedule. It tells you what to do, when to do it and how to do it. Each recommended service or maintenance procedure - a visual check and/or adjustment, replacement of a component, etc. - is explained later in this Chapter.

Servicing your engine in accordance with this maintenance schedule will significantly prolong its service life. Keep in mind that this is a comprehensive plan: servicing selected items - but skipping others - will not produce the same results

When you service your engine, you'll find that many of the maintenance procedures can be grouped together because they're logically related, or because they're located next to each other **(see illustrations)**.

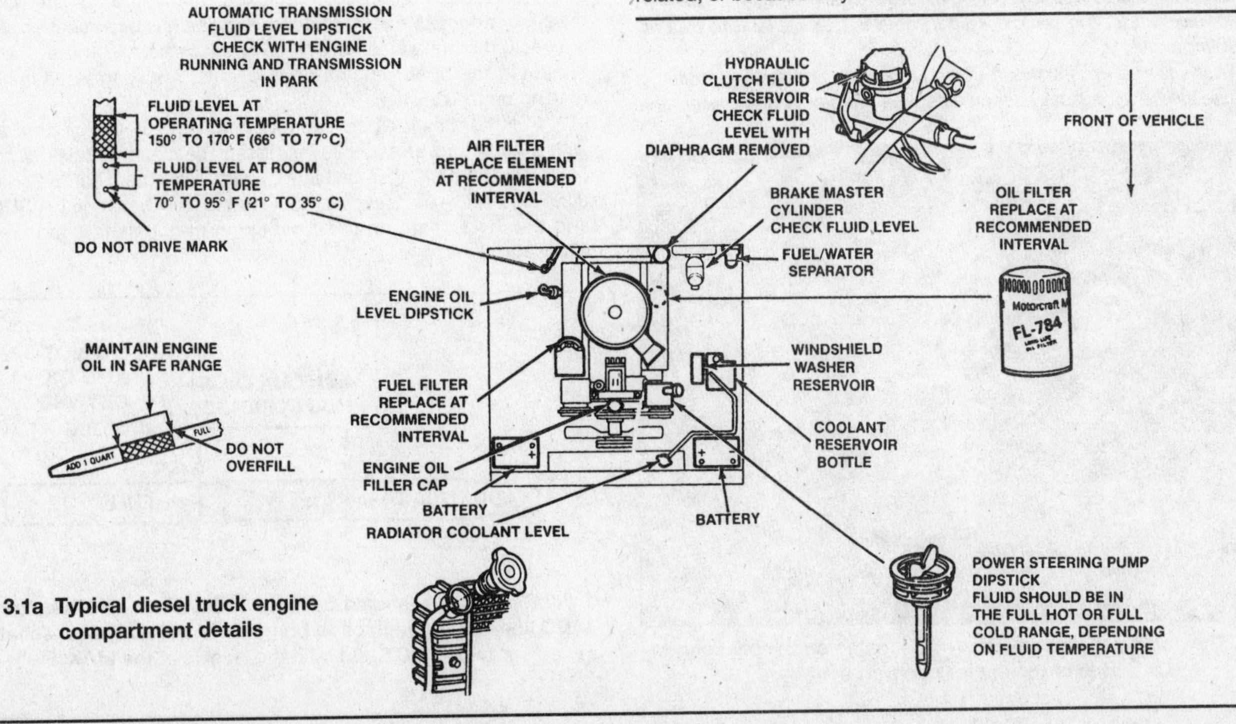

3.1a Typical diesel truck engine compartment details

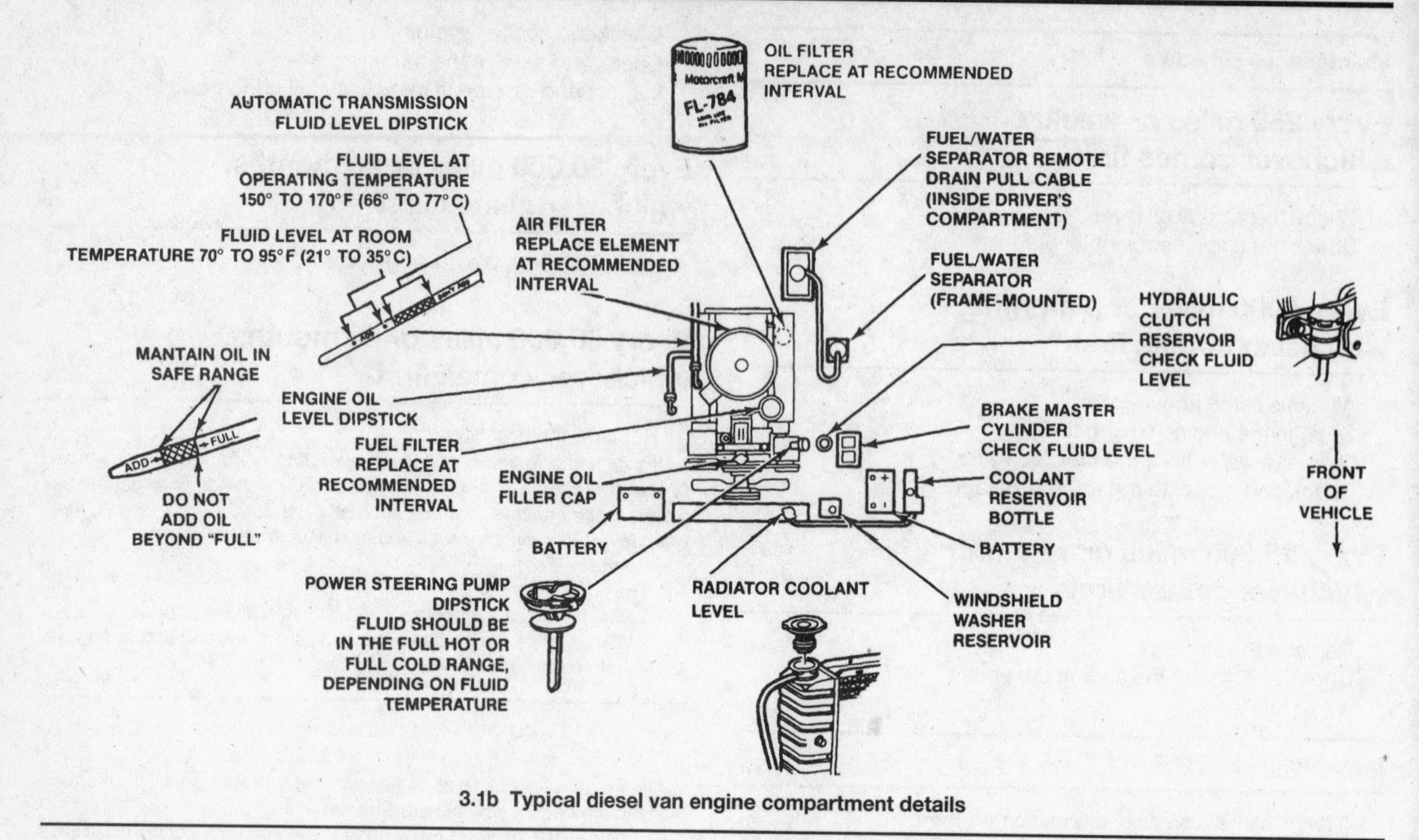

3.1b Typical diesel van engine compartment details

Before you get started, read through the service items you're planning to do, familiarize yourself with the procedures and gather up all the parts and tools you'll need. If it looks like you might run into problems during a particular job, seek advice from a mechanic or an experienced do-it-yourselfer.

Fluid level checks

Note: *The following are fluid level checks to be done on a 250 mile or weekly basis. Additional fluid level checks can be found in specific maintenance procedures which follow. Regardless of intervals, be alert to fluid leaks under the vehicle which would indicate a fault to be corrected immediately.*

1 Fluids are an essential part of the lubrication and cooling systems. Because the fluids gradually become depleted and/or contaminated during normal operation of the vehicle, they must be periodically replenished. See Recommended lubricants and fluids at the beginning of this Chapter before adding fluid to any of the following components. **Note:** *The vehicle must be on level ground when fluid levels are checked.*

Engine oil

2 The engine oil level is checked with a dipstick **(see illustration)** that extends through a tube and into the oil pan at the bottom of the engine.

3 The oil level should be checked before the vehicle has been driven, or about 15 minutes after the engine has been shut off. If the oil is checked immediately after driving the vehicle, some of the oil will remain in the upper engine components, resulting in an inaccurate reading on the dipstick.

4 Pull the dipstick from the tube and wipe all the oil from the end with a clean rag or paper towel. Insert the clean dipstick all the way back into the tube, then pull it out again. Note the oil at the end of the dipstick. Add oil as necessary to keep the level between the ADD mark and the FULL mark on the dipstick, in the crosshatched area **(see illustration)**.

3.2 The engine oil dipstick (arrow) is located on the passenger side of the engine compartment

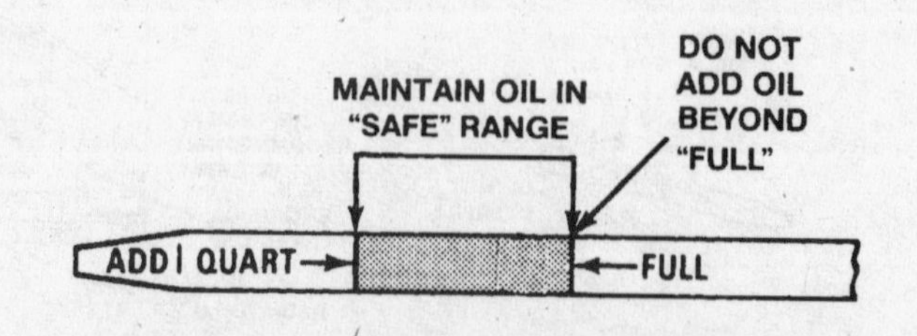

3.3 The oil level should be in the SAFE range - if it's below the ADD line, add enough oil to bring the level into the crosshatched area (DO NOT add oil if the level is at the MAX line!)

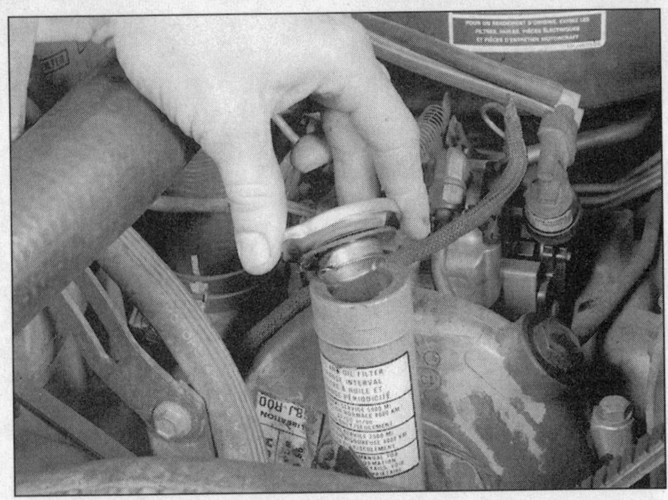

3.4 Rotate the oil filler cap counterclockwise to remove it

3.5 The combined windshield washer/coolant reservoir used on some models features separate tanks for each fluid - DO NOT add engine coolant to the windshield washer reservoir)

5 Don't overfill the engine by adding too much oil - it may result in oil leaks or oil seal failures.

6 Oil is added to the engine after removing the filler cap **(see illustration)**. An oil can spout or funnel may help to reduce spills.

7 Checking the oil level is an important preventive maintenance step. A consistently low oil level indicates oil leakage through damaged seals, defective gaskets or past worn rings or valve guides. If the oil looks milky in color or has water droplets in it, the cylinder head gasket(s) may be blown or the head(s) or block may be cracked. The engine should be checked immediately. The condition of the oil should also be checked. Whenever you check the oil level, slide your thumb and index finger up the dipstick before wiping off the oil. If you see small dirt or metal particles clinging to the dipstick, the oil should be changed (see Section 6).

Engine coolant

Warning: *Do not allow antifreeze to come in contact with your skin or painted surfaces of the vehicle. Rinse off spills immediately with plenty of water. Antifreeze is highly toxic if ingested. Never leave antifreeze lying around in an open container or in puddles on the floor; children and pets are attracted by it's sweet smell and may drink it. Check with local authorities about disposing of used antifreeze. Many communities have collection centers which will see that antifreeze is disposed of safely.*

8 All vehicles covered by this manual are equipped with a coolant recovery reservoir located in the left front corner of the engine compartment. The reservoir is connected by a hose to the radiator filler neck. If the engine overheats, coolant escapes through a valve in the radiator cap and travels through the hose into the reservoir. As the engine cools, the coolant is automatically drawn back into the system to maintain the correct level.

3.6 Remove the cell caps to check the water level in the battery - if the level is low, add distilled water only

9 The coolant level in the reservoir should be checked regularly. **Warning:** *Do not remove the radiator cap to check the coolant level when the engine is warm! The level in the reservoir varies with the temperature of the engine.* When the engine is cold, the coolant level should be at or slightly above the COLD mark on the reservoir. Once the engine has warmed up, the level should be at or near the HOT mark. If it isn't, allow the engine to cool, then remove the small cap from the reservoir and add a 50/50 mixture of ethylene glycol-based antifreeze and water **(see illustration)**. **Caution:** *Do not add coolant to the windshield washer reservoir!*

10 Drive the vehicle and recheck the coolant level. If only a small amount of coolant is required to bring the system up to the proper level, water can be used. However, repeated additions of water will dilute the antifreeze and water solution. In order to maintain the proper ratio of antifreeze and water, always top up the coolant level with the correct mixture. An empty plastic milk jug or bleach bottle makes an excellent container for mixing coolant. Don't use rust inhibitors or additives.

11 If the coolant level drops consistently, there may be a leak in the system. Inspect the radiator, hoses, filler cap, drain plugs and water pump. If no leaks are noted, have the radiator cap pressure tested by a service station.

12 If you have to remove the radiator cap, wait until the engine has cooled, then wrap a thick cloth around the cap and turn it to the first stop. If coolant or steam escapes, let the engine cool down longer, then remove the cap.

13 Check the condition of the coolant as well. It should be relatively clear. If it's brown or rust colored, the system should be drained, flushed and refilled. Even if the coolant appears to be normal, the corrosion inhibitors wear out, so it must be replaced at the specified intervals.

Battery electrolyte

14 Most vehicles with which this manual is concerned are equipped with a battery which is permanently sealed (except for vent holes) and has no filler caps. Water doesn't have to be added to these batteries at any time. If an aftermarket battery has been installed, the caps on the top of the battery should be removed periodically to check for a low water level **(see illustration)**. This check is most critical during the warm summer months.

Air filter replacement

1 At the specified intervals, the air filter should be replaced with a new one.

2 The filter is located on top of the intake manifold and is replaced

3.7 Remove the wing bolt and lift the cover up

3.8 Raise the cover and lift out the air filter element

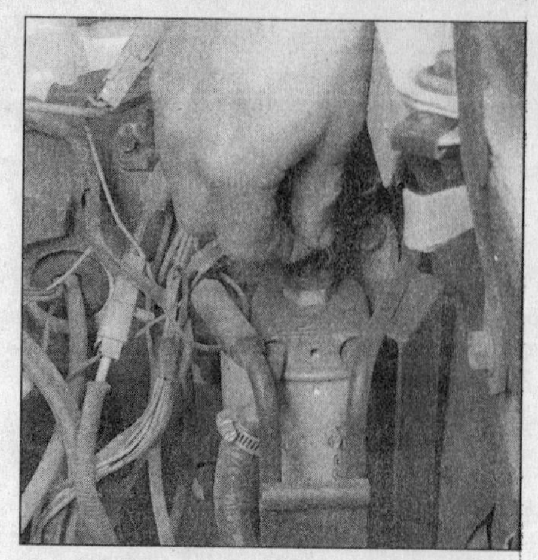

3.9a On earlier model trucks, the separator is located at the left rear comer of the engine compartment - pull up on the ring to drain the water

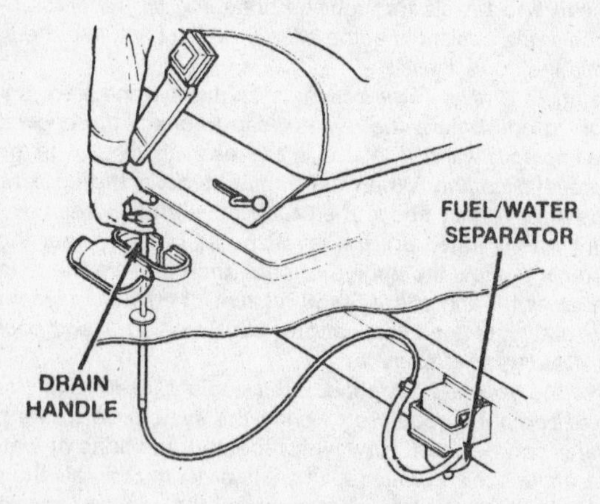

3.9b On earlier model vans, the separator can be drained from inside the vehicle

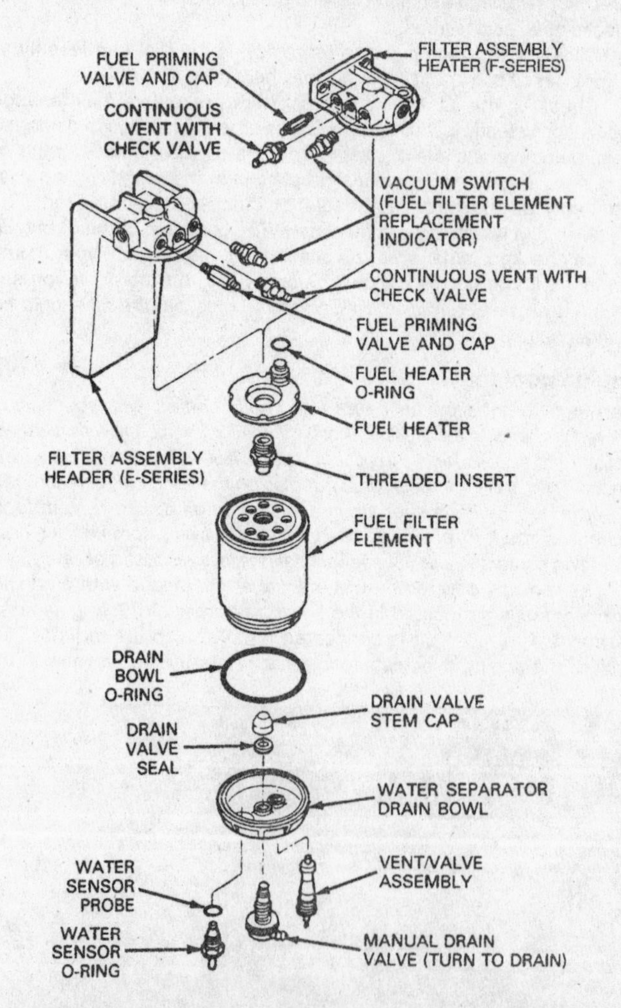

3.9c On later models the water separator is part of the filter - unscrew the manual drain valve at the bottom until only water-free diesel fuel comes out, then screw it back in

by unscrewing the wing bolt from the top of the filter housing and lifting off the cover **(see illustration).**

3 While the top plate is off, be careful not to drop anything down into the manifold.

4 Lift the air filter element out of the housing **(see illustration)** and wipe out the inside of the housing with a clean rag.

5 Place the new filter in the air filter housing. Make sure it seats properly in the bottom of the housing.

6 Installation is the reverse of removal.

Fuel filter replacement

Warning: *Diesel fuel is flammable and may be hot, so take extra precautions when working on any part of the fuel system. If you spill fuel on your skin, rinse it off immediately with soap and water. Have a Class B fire extinguisher on hand.*

1994 and earlier (except Direct Injection Turbo) models

1 The fuel system on these models has a fuel/water separator and a replaceable fuel filter. The water should be drained from the separator whenever the "Water in Fuel" light on the dash goes on, or before replacing the fuel filter. On 1988 and earlier models the separator is mounted separately, while on later models it's incorporated into the base of the filter. Place a container under the separator drain tube. With the engine off, drain the water by pulling up on the ring or handle (early models) or turning the manual drain valve (later models) **(see illustrations).**

2 Disconnect the negative battery cables.

3 Place rags or newspapers under the filter housing to catch the fuel that will drain out when the filter is removed.

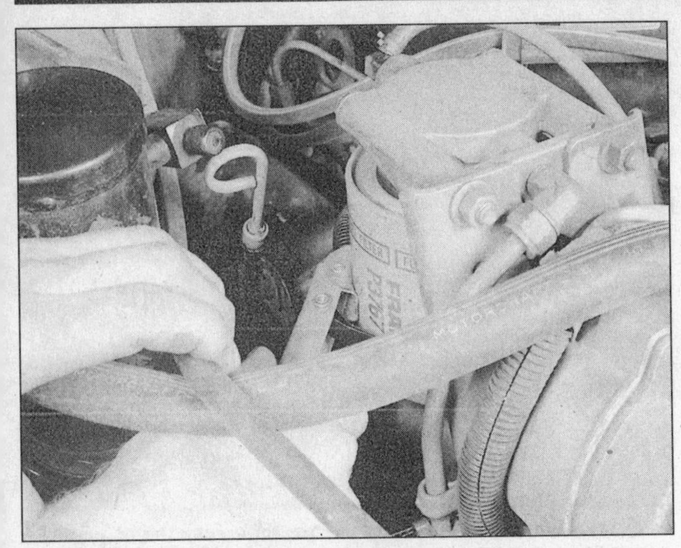

3.10a Unscrew the fuel filter with a large oil filter wrench (non-Direct Injection models)

4 Use a large oil filter wrench to unscrew the filter element **(see illustration).**
5 Screw the new filter onto the housing and rotate it an additional 2/3 turn after contacting the gasket.

1994 and later Direct Injection Turbo models

Note: *The filter housing must be drained of water whenever the water-in -fuel light comes on or at least every 5,000 miles, just as in earlier models.*
Refer to the illustration for operation of the drain valve lever.
1 Disconnect the cables from the negative battery terminals.
2 Place a container under the drain tube.
3 Open the valve by rotating the lever away from the housing. Allow it to remain open for 15 seconds or until clear fuel comes out.
4 Close the drain valve lever tightly. Dispose of the liquid in an approved manner.
5 To replace the fuel filter, unscrew the cap on top of the housing and replace the filter element **(see illustration).**

Engine oil and filter change

1 Frequent oil changes are the most important preventive maintenance procedure that can be done by the home mechanic. As engine oil ages, it becomes diluted and contaminated, which leads to premature engine wear.
2 Although some sources recommend oil filter changes every other oil change, the minimal cost of an oil filter and the fact that it's not hard to change dictate that a new filter be used every time the oil is changed.
3 Gather together all necessary tools and materials before beginning this procedure **(see illustration).**
4 You should have plenty of clean rags and newspapers handy to mop up any spills. Access to the underside of the vehicle is greatly improved if the vehicle can be lifted on a hoist, driven onto ramps or supported by jackstands. **Warning:** *Do not work under a vehicle which is supported only by a bumper, hydraulic or scissors-type jack!*
5 If this is your first oil change, get under the vehicle and familiarize yourself with the locations of the oil drain plug and the oil filter. The engine and exhaust components will be warm during the actual work, so note how they are situated to avoid touching them when working under the vehicle.
6 Warm the engine to normal operating temperature. If the new oil or any tools are needed, use this warm-up time to gather everything necessary for the job. The correct type of oil for your application can be found in *Recommended lubricants and fluids* at the beginning of this Chapter.

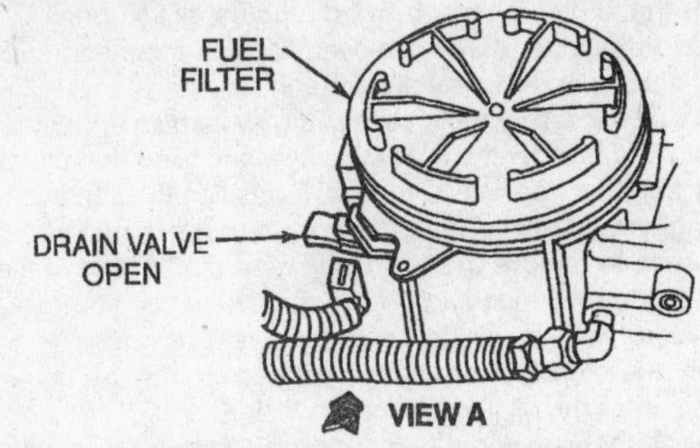

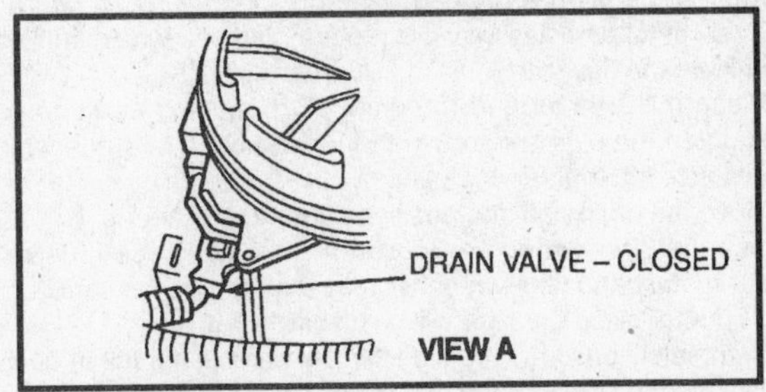

3.10b On Direct Injection Turbo models, unscrew the cap from the filter housing and remove the filter

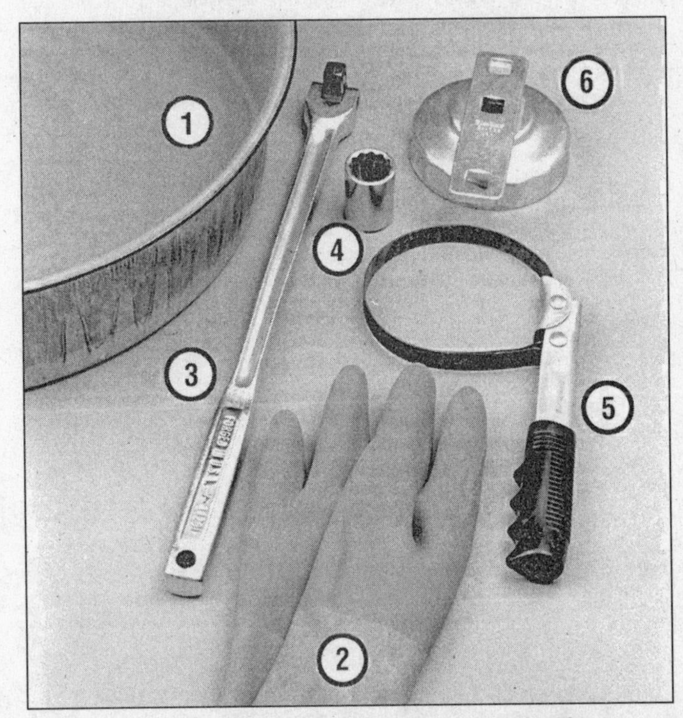

3.11 These tools are required when changing the engine oil and filter

1 ***Drain pan*** *- It should be fairly shallow in depth, but wide to prevent spills*
2 ***Rubber gloves*** *- When removing the drain plug and filter, you will get oil on your hands (the gloves will prevent burns)*
3 ***Breaker bar*** *- Sometimes the oil drain plug is tight and a long breaker bar is needed to loosen it*
4 ***Socket*** *- To be used with the breaker bar or a ratchet (must be the correct size to fit the drain plug - six-point preferred)*
5 ***Filter wrench*** *- This is a metal band-type wrench, with requires clearance around the filter to be effective*
6 ***Filter wrench*** *- This type fits on the bottom of the filter and can be turned with a ratchet or breaker bar (different size wrenches are available for different types of filters)*

7 With the engine oil warm (warm engine oil will drain better and more built-up sludge will be removed with it), raise and support the vehicle - make sure it's safely supported!
8 Move all necessary tools, rags and newspapers under the vehicle. Set the drain pan under the drain plug. Keep in mind that the oil will initially flow from the pan with some force; position the pan accordingly.
9 Being careful not to touch any of the hot exhaust components, use a wrench to remove the drain plug near the bottom of the oil pan **(see illustration)**. Depending on how hot the oil is, you may want to wear gloves while unscrewing the plug the final few turns.
10 Allow the old oil to drain into the pan. It may be necessary to move the pan as the oil flow slows to a trickle.
11 After all the oil has drained, wipe off the drain plug with a clean rag. Small metal particles may cling to the plug and would immediately contaminate the new oil.
12 Clean the area around the drain plug opening and reinstall the plug. Tighten the plug securely with the wrench. If a torque wrench is available, use it to tighten the plug.
13 Move the drain pan into position under the oil filter.
14 Use the filter wrench to loosen the oil filter **(see illustration)**. Chain or metal band filter wrenches may distort the filter canister, but it doesn't matter since the filter will be discarded anyway.
15 Completely unscrew the old filter. Be careful; it's full of oil. Empty the oil inside the filter into the drain pan.
16 Compare the old filter with the new one to make sure they're the same type.
17 Use a clean rag to remove all oil, dirt and sludge from the area where the oil filter mounts to the engine. Check the old filter to make sure the rubber gasket isn't stuck to the engine. If the gasket is stuck to the engine, remove it.
18 Apply a light coat of clean oil to the rubber gasket on the new oil filter **(see illustration)**.
19 Attach the new filter to the engine. Overtightening the filter will damage the gasket, so don't use a filter wrench. Most filter manufacturers recommend tightening the filter by hand only. Normally they should be tightened 3/4-turn after the gasket contacts the block, but be sure to follow the directions on the filter or container.
20 Remove all tools, rags, etc. from under the vehicle, being careful not to spill the oil in the drain pan, then lower the vehicle.
21 Move to the engine compartment and locate the oil filler cap.
22 Remove the oil filler cap and pour the fresh oil through the filler opening. A funnel may also be used.
23 Pour ten quarts of fresh oil into the engine. Wait a few minutes to allow the oil to drain into the pan, then check the level on the oil dipstick (see Section 3 if necessary). If the oil level is above the ADD mark, start the engine and allow the new oil to circulate.
24 Run the engine for only about a minute and then shut it off. Immediately look under the vehicle and check for leaks at the oil pan drain plug and around the oil filter. If either is leaking, tighten with a bit more force.
25 With the new oil circulated and the filter now completely full, recheck the level on the dipstick and add more oil as necessary.
26 During the first few trips after an oil change, make it a point to check frequently for leaks and proper oil level.
27 The old oil drained from the engine cannot be reused in its present state and should be disposed of. Oil reclamation centers, auto repair shops and gas stations will normally accept the oil, which can be refined and used again. After the oil has cooled it can be drained into a suitable container (capped plastic jugs, topped bottles, milk cartons, etc.) for transport to one of these disposal sites.

3.12 Use a socket or box-and wrench to remove the engine oil drain Plug - DO NOT use an open end wench, as the comers on the plug head can be easily rounded off!

Drivebelt check, adjustment and replacement

1 The accessory drivebelts, also referred to as V-belts or simply fan belts, are located at the front of the engine. The condition and tension of the drivebelts are critical to the operation of the engine and accessories. Excessive tension causes bearing wear, while insufficient tension produces slippage, noise, component vibration and belt failure. Because of their composition and the high stresses to which they are subjected, drivebelts stretch and deteriorate as they get older. As a result, they must be periodically checked and adjusted.

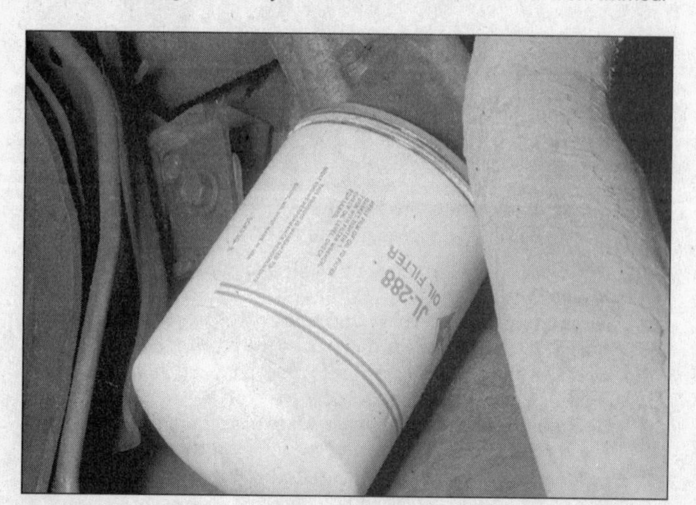

3.13 The oil filter is usually on very tight and will require an oil filter wrench for removal - DO NOT use the wrench to tighten the filter

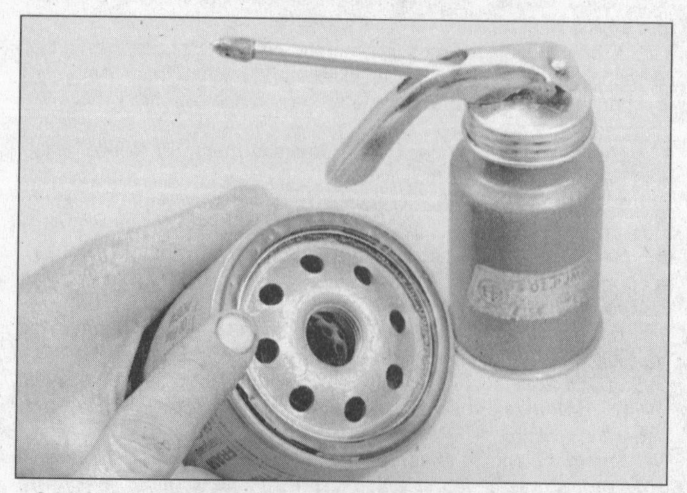

3.14 Lubricate the oil filter gasket with clean engine oil before installing the filter on the engine

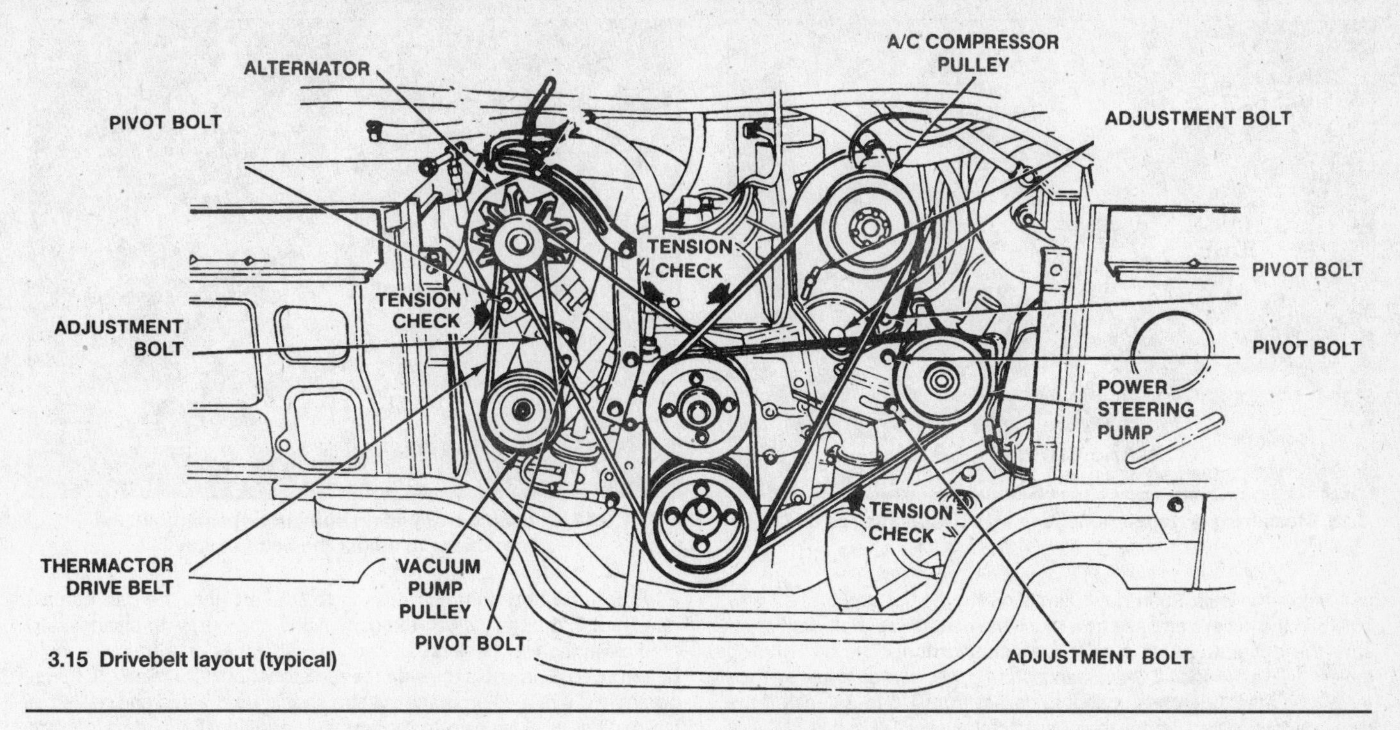

3.15 Drivebelt layout (typical)

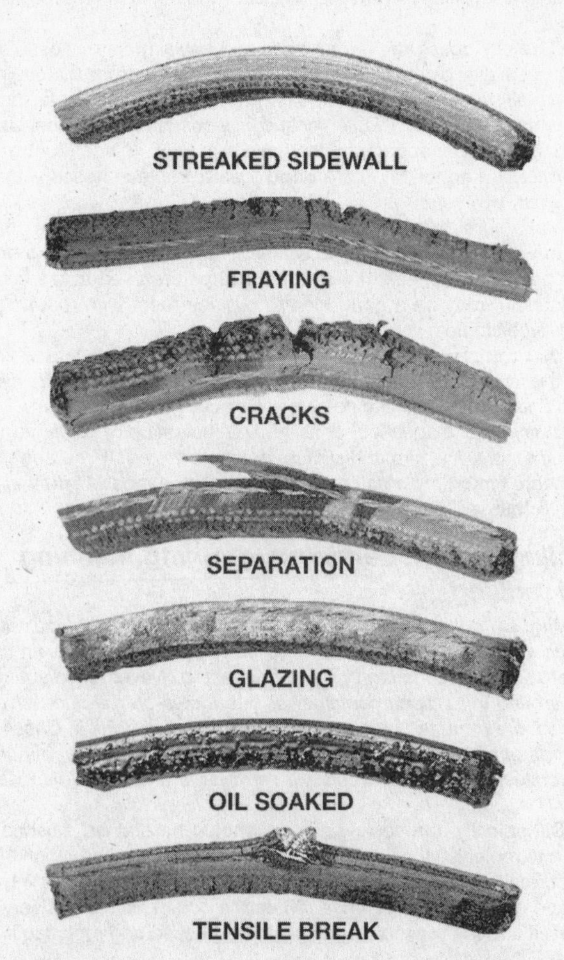

3.16 Here are some of the more common problems associated with drivebelts (check the belts very carefully to prevent an untimely breakdown)

Check

2 The number and type of belts used on a particular vehicle depends on the accessories installed. Various types of drivebelts are used on these models **(see illustration)**.

3 With the engine off, open the hood and locate the drivebelts. Using a flashlight, check each belt for separation of the rubber plies from each side of the core, a severed core, separation of the ribs from the rubber, cracks, torn or worn ribs and cracks in the inner ridges of the ribs. Also check for fraying and glazing, which gives the belt a shiny appearance **(see illustration)**. Both sides of each belt should be inspected, which means you'll have to twist them to check the undersides. Use your fingers to feel a belt where you can't see it. If any of the above conditions are evident, replace the belt as described below.

4 To check the tension of each belt in accordance with factory recommendations, install a drivebelt tension gauge (Ford tool no. T63L-8620-A, or equivalent) **(see illustration)**. Measure the tension in accordance with the tension gauge instructions and compare your measurement to the specified drivebelt tension for either a used or new belt. **Note:** *A "new" belt is defined as any belt which has not been run; a "used" belt is one that has been run for more than ten minutes.*

5 The special gauge is the most accurate way to check belt tension. However, if you don't have a gauge, and can't borrow one, the following "rule-of-thumb" method is recommended as an alternative. Lay a straightedge across the longest free span (the distance between two

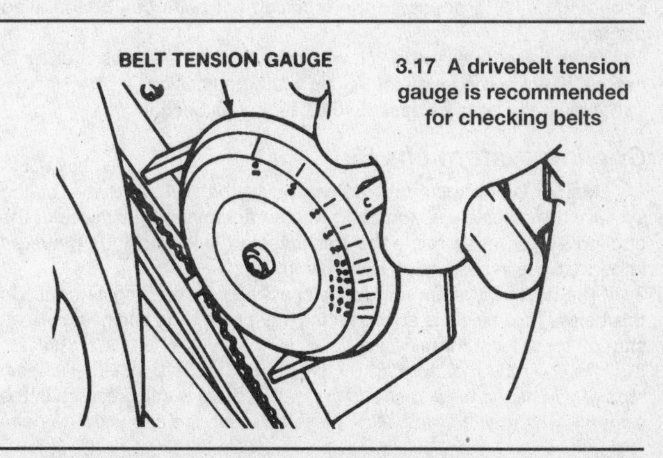

3.17 A drivebelt tension gauge is recommended for checking belts

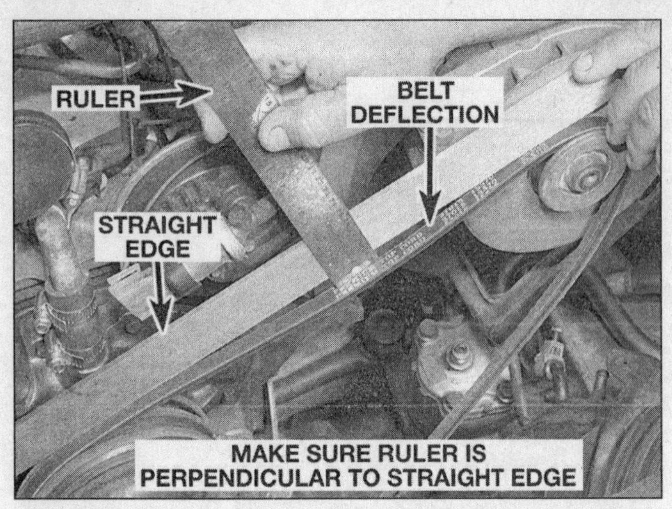

3.18 Measuring drivebelt deflection with a straightedge and ruler

3.19 Loosen the adjusting bolt (arrow) and move the alternator to adjust the belt tension

pulleys) of the belt. Push down firmly on the belt at a point half way between the pulleys and see how much the belt moves (deflects). Measure the deflection with a ruler **(see illustration)**. The belt should deflect 1/8 to 1/4-inch if the distance from pulley center-to-pulley center is less than 12-inches; it should deflect from 1/8 to 3/8-inch if the distance from pulley center-to-pulley center is over 12-inches.

Adjustment

6 If adjustment is required to make the drivebelt tighter or looser, it's done by moving the belt driven accessory on the bracket.

7 For each component, there will be a locking bolt and pivot bolt or nut. Both must be loosened slightly to enable you to move the component **(see illustration)**.

8 After the bolts have been loosened, move the component away from the engine to tighten the belt or toward the engine to loosen the belt. Hold the accessory in position and check the belt tension. If it's correct, tighten the two bolts until snug, then recheck the tension. If it's still correct, tighten the two bolts completely.

9 It will often be necessary to use some sort of pry bar to move the accessory while the belt is adjusted. If this must be done to gain the proper leverage, be very careful not to damage the component being moved, or the part being pried against.

10 Run the engine for about 15 minutes, then recheck the belt tension.

Replacement

11 To replace a belt, follow the above procedures for drivebelt adjustment but slip the belt off the pulleys and remove it. Since belts tend to wear out more or less at the same time, it's a good idea to replace all of them at the same time. Mark each belt and the corresponding pulley grooves so the replacement belts can be installed properly.

12 Take the old belts with you when purchasing new ones in order to make a direct comparison for length, width and design.

13 Adjust the belts as described earlier in this Section.

Cooling system check

1 Many major engine failures can be attributed to a faulty cooling system. If the vehicle is equipped with an automatic transmission, the cooling system also cools the transmission fluid and thus plays an important role in prolonging transmission life.

2 The cooling system should be checked with the engine cold. Do this before the vehicle is driven for the day or after the engine has been shut off for at least three hours.

3 Remove the radiator cap by turning it to the left until it reaches a stop. If you hear a hissing sound (indicating there is still pressure in the system), wait until it stops. Now press down on the cap with the palm of your hand and continue turning to the left until the cap can be removed. Thoroughly clean the cap, inside and out, with clean water. Also clean the filler neck on the radiator. All traces of corrosion should be removed. The coolant inside the radiator should be relatively transparent. If it's rust colored, the system should be drained and refilled. If the coolant level isn't up to the top, add additional antifreeze/coolant mixture.

4 Carefully check the large upper and lower radiator hoses along with the smaller diameter heater hoses which run from the engine to the firewall. Inspect each hose along its entire length, replacing any hose which is cracked, swollen or shows signs of deterioration. Cracks may become more apparent if the hose is squeezed **(see illustration)**. Regardless of condition, it's a good idea to replace hoses with new ones every two years.

5 Make sure all hose connections are tight. A leak in the cooling system will usually show up as white or rust colored deposits on the areas adjoining the leak. If wire-type clamps are used at the ends of the hoses, it may be a good idea to replace them with more secure screw-type clamps.

6 Use compressed air or a soft brush to remove bugs, leaves, etc. from the front of the radiator or air conditioning condenser. Be careful not to damage the delicate cooling fins or cut yourself on them.

7 Every other inspection, or at the first indication of cooling system problems, have the cap and system pressure tested. If you don't have a pressure tester, most gas stations and repair shops will do this for a minimal charge.

Cooling system servicing, draining, flushing and refilling

Warning: *Do not allow antifreeze to come in contact with your skin or painted surfaces of the vehicle. Rinse off spills immediately with plenty of water. Antifreeze is highly toxic if ingested. Never leave antifreeze lying around in an open container or in puddles on the floor; children and pets are attracted by it's sweet smell and may drink it. Check with local authorities about disposing of used antifreeze. Many communities have collection centers which will see that antifreeze is disposed of safely.*

1 Periodically, the cooling system should be drained, flushed and refilled to replenish the antifreeze mixture and prevent formation of rust and corrosion, which can impair the performance of the cooling system and cause engine damage. When the cooling system is serviced, all hoses and the radiator cap should be checked and replaced if necessary.

Draining

2 Apply the parking brake and block the wheels. If the vehicle has just been driven, wait several hours to allow the engine to cool down

Check for a chafed area that could fail prematurely.

Check for a soft area indicating the hose has deteriorated inside.

Overtightening the clamp on a hardened hose will damage the hose and cause a leak.

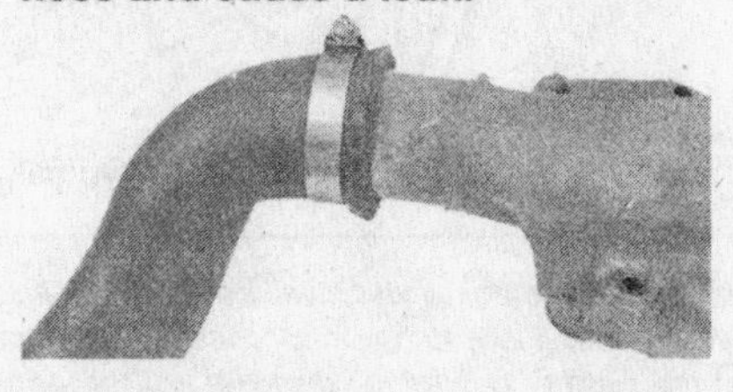

Check each hose for swelling and oil-soaked ends. Cracks and breaks can be located by squeezing the hose.

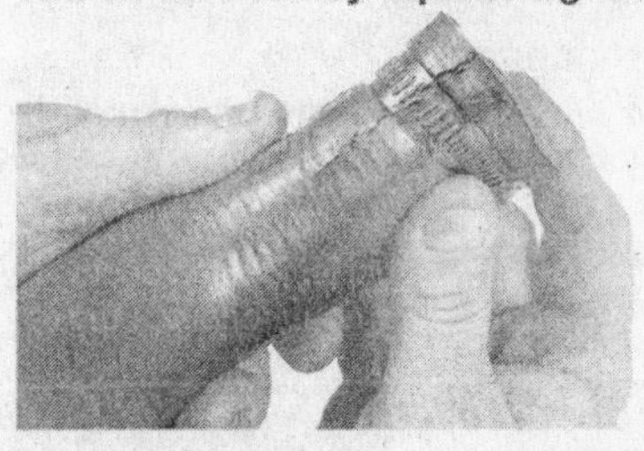

3.20 Hoses, like drivebelts, have a habit of failing at the worst possible time - to prevent the inconvenience of a blown radiator or heater hose, inspect them carefully as shown here

before beginning this procedure.

3 Once the engine is completely cool, remove the radiator cap.

4 Move a large container under the radiator drain to catch the coolant **(see illustration)**. Attach a 3/8-inch diameter hose to the drain fitting to direct the coolant into the container, then open the drain fitting (a pair of pliers may be required to turn it).

5 After the coolant stops flowing out of the radiator, move the container under the engine block drain plug. Remove the plug and allow the coolant in the block to drain **(see illustration)**.

6 While the coolant is draining, check the condition of the radiator hoses, heater hoses and clamps.

7 Replace any damaged clamps or hoses.

Flushing

8 Once the system is completely drained, flush the radiator with fresh water from a garden hose until water runs clear at the drain. The flushing action of the water will remove sediments from the radiator but will not remove rust and scale from the engine and cooling tube surfaces.

9 These deposits can be removed by the chemical action of a cleaner such as Ford Cooling System Fast Flush. Follow the procedure outlined in the manufacturer's instructions. If the radiator is severely corroded, damaged or leaking, it should be removed and taken to a radiator repair shop.

10 Remove the overflow hose from the coolant recovery reservoir. Drain the reservoir and flush it with clean water, then reconnect the hose.

Refilling

11 Close and tighten the radiator drain. Install and tighten the block drain plug.

12 Place the heater temperature control in the maximum heat position.

13 Slowly add new coolant (a 50/50 mixture of water and antifreeze) to the radiator until it's full. Add coolant to the reservoir up to the lower mark.

14 Leave the radiator cap off and run the engine in a well-ventilated area until the thermostat opens (coolant will begin flowing through the radiator and the upper radiator hose will become hot).

15 Turn the engine off and let it cool. Add more coolant mixture to bring the level back up to the lip on the radiator filler neck.

16 Squeeze the upper radiator hose to expel air, then add more coolant mixture if necessary. Replace the radiator cap.

17 Start the engine, allow it to reach normal operating temperature and check for leaks.

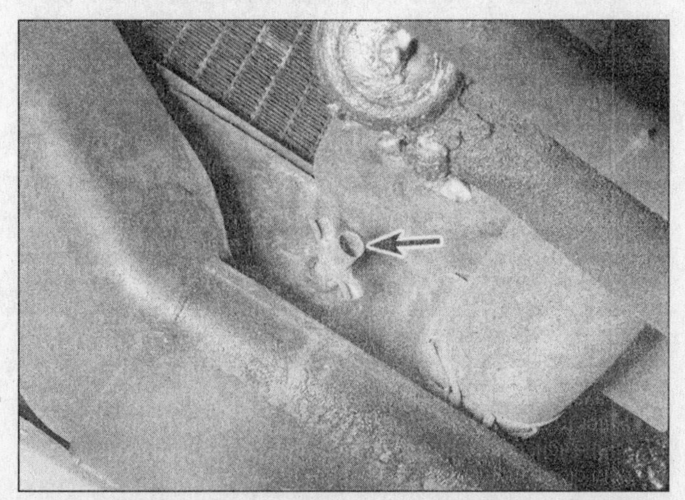

3.21 Before opening the radiator drain fitting (arrow), push a short section of 3/8-inch diameter rubber hose onto the fitting to prevent the coolant from splashing as it drains

3.22 The block drain plug (arrow) is located on the passenger side of the block, near the starter motor

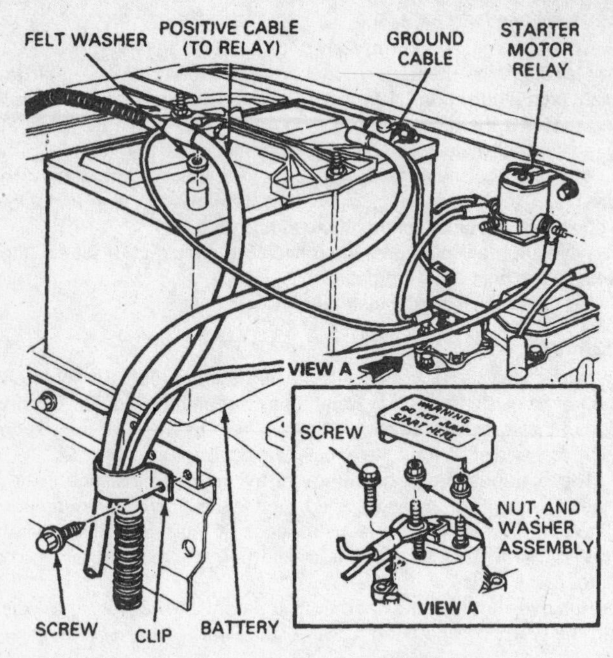

3.23a Battery cable details - right side (typical)

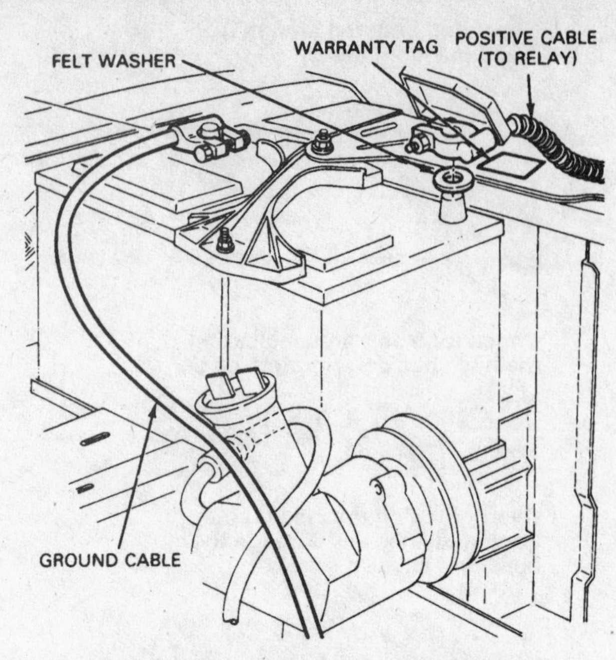

3.23b Battery cable details - left side (typical)

Battery cable check and replacement

1 Periodically inspect the entire length of each battery cable for damage, cracked or burned insulation and corrosion. Poor battery cable connections can cause starting problems and decreased engine performance.

2 Check the cable-to-terminal connections at the ends of the cables for cracks, loose wire strands and corrosion. The presence of white, fluffy deposits under the insulation at the cable terminal connection is a sign the cable is corroded and should be replaced. Check the terminals for distortion, missing mounting bolts and corrosion.

3 When removing the cables, always disconnect the negative cables first and hook them up last or the battery may be shorted by the tool used to loosen the cable clamps. Even if only the positive cable is being replaced, be sure to disconnect the negative cable from the battery first.

4 Disconnect the old cables from the batteries, then trace each of them to their opposite ends and detach them from the starter solenoid and ground terminals. Note the routing of each cable to ensure correct installation **(see illustrations)**.

5 If you're replacing some or all of the old cables, take them with you when buying new cables. It's vitally important that you replace the cables with identical parts. Cables have characteristics that make them easy to identify: Positive cables are usually red, larger in cross-section and have a larger diameter battery post clamp; ground cables are usually black, smaller in cross-section and have a slightly smaller diameter clamp for the negative post.

6 Clean the threads of the solenoid or ground connection with a wire brush to remove rust and corrosion. Apply a light coat of battery terminal corrosion inhibitor, or petroleum jelly, to the threads to prevent future corrosion.

7 Attach the cable to the solenoid or ground connection and tighten the mounting nut/bolt securely.

8 Before connecting a new cable to the battery, make sure it reaches the battery post without having to be stretched.

9 Connect the positive cables first, followed by the negative cables.

Battery check, maintenance and charging

Warning: *Certain precautions must be followed when checking and servicing the battery. Hydrogen gas, which is highly flammable, is always present in the battery cells, so don't smoke and keep open flames and sparks away from the battery. The electrolyte inside the battery is actually dilute sulfuric acid, which will cause injury if splashed on your skin or in your eyes. It will also ruin clothes and painted surfaces. When removing the battery cables, always detach the negative cable first and hook it up last!*

Note: *These models use two batteries, connected in parallel. The batteries must be checked and charged separately, but replaced in pairs.*

Check and maintenance

1 Battery maintenance is an important procedure which will help ensure that you're not stranded because of dead batteries. Several tools are required for this procedure **(see illustration)**.

2 Before servicing the battery, always turn the engine and all accessories off and disconnect the cables from the negative terminals.

3 A sealed (sometimes called maintenance-free) battery is standard equipment on these vehicles. The cell caps can't be removed, no electrolyte checks are required and water can't be added to the cells. However, if an aftermarket battery that requires regular maintenance has been installed, the following procedure can be used.

4 Check the electrolyte level in each of the battery cells. It must be above the plates. There's usually a split-ring indicator in each cell to indicate the correct level. If the level is low, add distilled water only, then install the cell caps. **Caution:** *Overfilling the cells may cause electrolyte to spill over during periods of heavy charging, causing corrosion and damage to nearby components.*

5 If the positive terminal and cable clamp on your vehicle's battery is equipped with a rubber protector, make sure that it's not torn or damaged. It should completely cover the terminal.

6 The external condition of the battery should be checked periodically. Look for damage, such as a cracked case.

7 Check the tightness of the battery cable clamps to ensure good electrical connections and inspect the entire length of each cable, looking for cracked or abraded insulation and frayed conductors.

8 If corrosion (visible as white, fluffy deposits) is evident, remove the cables from the terminals, clean them with a battery brush and reinstall them **(see illustrations)**. Corrosion can be kept to a minimum by installing specially treated washers available at auto parts stores or by applying a layer of petroleum jelly or grease to the terminals and cable clamps after they are assembled.

9 Make sure that the battery carrier is in good condition and that the hold-down clamp bolt is tight. If the battery is removed, make sure that no parts remain in the bottom of the carrier when it's reinstalled. When reinstalling the hold-down clamp, don't overtighten the bolt.

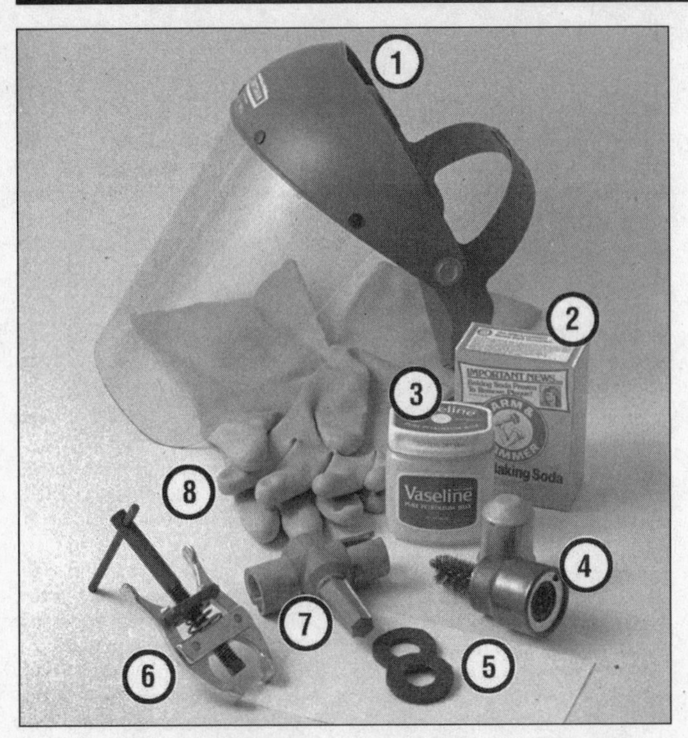

3.24 Tools and materials required for battery maintenance

1. ***Face shield/safety goggles*** - *When removing corrosion with a brush, the acidic particles can easily fly up into your eyes*
2. ***Baking soda*** - *A solution of baking soda and water can be used to neutralize corrosion*
3. ***Petroleum jolly*** - *A layer of this on the battery posts will help prevent corrosion*
4. ***Battery post/cable cleaner*** - *This wire brush cleaning tool will remove all traces of corrosion from the battery posts and cable clamps*
5. ***Treated felt washers*** - *Placing one of these on each post, directly under the cable clamps, will help prevent corrosion*
6. ***Puller*** - *Sometimes the cable clamps are very difficult to pull off the posts, even after the nut/bolt has been completely loosened. This tool pulls the clamp straight up and off the post without damage.*
7. ***Battery post/cable cleaner*** - *Here is another cleaning tool which is a slightly different version of number 4 above, but it does the same thing*
8. ***Rubber gloves*** - *Another safety item to consider when servicing the battery; remember that's acid inside the battery!*

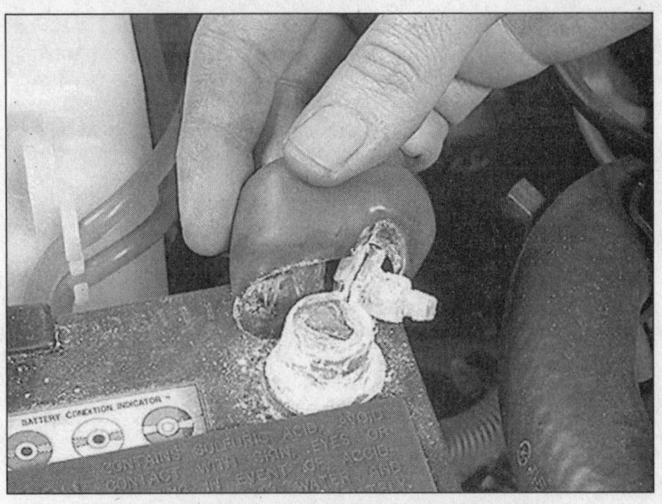

3.25a Battery terminal corrosion usually appears as light, fluffy powder

3.25b Removing the cable from a battery post with a wrench - sometimes a special battery pliers is required for this procedure if corrosion has caused deterioration of the nut hex (always remove the ground cable first and hook it up last!)

3.25c Regardless of the type of tool used on the battery posts, a clean, shiny surface should be the result

3.25d When cleaning the cable clamps, all corrosion must be removed (the inside of the clamp is tapered to match the taper on the post, so don't remove too much material)

10 Corrosion on the carrier, battery case and surrounding areas can be removed with a solution of water and baking soda. Apply the mixture with a small brush, let it work, then rinse it off with plenty of clean water.

11 Any metal parts of the vehicle damaged by corrosion should be coated with a zinc-based primer, then painted.

Charging

Note: *Diesel models have two batteries connected in parallel and each battery should be charged separately, after disconnecting the negative cables.*

12 Remove all of the cell caps (if equipped) and cover the holes with a clean cloth to prevent spattering electrolyte. Disconnect the negative battery cable and hook the battery charger leads to the battery posts (positive to positive, negative to negative), then plug in the charger. Make sure it is set at 12 volts if it has a selector switch.

13 If you're using a charger with a rate higher than two amps, check the battery regularly during charging to make sure it doesn't overheat. If you're using a trickle charger, you can safely let the battery charge overnight after you've checked it regularly for the first couple of hours.

14 If the battery has removable cell caps, measure the specific gravity with a hydrometer every hour during the last few hours of the charging cycle. Hydrometers are available inexpensively from auto parts stores - follow the instructions that come with the hydrometer. Consider the battery charged when there's no change in the specific gravity reading for two hours and the electrolyte in the cells is gassing (bubbling) freely. The specific gravity reading from each cell should be very close to the others. If not, the battery probably has a bad cell(s).

15 Some batteries with sealed tops have built-in hydrometers on the top that indicate the state of charge by the color displayed in the hydrometer window. Normally, a bright-colored hydrometer indicates a full charge and a dark hydrometer indicates the battery still needs charging. Check the battery manufacturer's instructions to be sure you know what the colors mean.

16 If the battery has a sealed top and no built-in hydrometer, you can hook up a digital voltmeter across the battery terminals to check the charge. A fully charged battery should read 12.6 volts or higher.

Throttle cable check and lubrication

1 Inspect the throttle linkage for damage and missing parts as well as for binding and interference when the throttle pedal is depressed.

2 Lubricate the linkage pivot points with engine oil.

Underhood hose check and replacement

Warning: *Replacement of air conditioning hoses must be left to a dealer service department or air conditioning shop that has the equipment to depressurize the system safely. Never remove air conditioning components or hoses until the system has been depressurized.*

General

1 High temperatures in the engine compartment can cause the deterioration of the rubber and plastic hoses used for engine, accessory and emission systems operation. Periodic inspection should be made for cracks, loose clamps, material hardening and leaks.

2 Information specific to the cooling system hoses can be found in Section 8.

3 Some, but not all, hoses are secured to the fittings with clamps. Where clamps are used, check to be sure they haven't lost their tension, allowing the hose to leak. If clamps aren't used, make sure the hose has not expanded and/or hardened where it slips over the fitting, allowing it to leak.

Vacuum hoses

4 It's quite common for vacuum hoses, especially those in the emissions system, to be color coded or identified by colored stripes molded into them. Various systems require hoses with different wall thicknesses, collapse resistance and temperature resistance. When replacing hoses, be sure the new ones are made of the same material.

5 Often the only effective way to check a hose is to remove it completely from the vehicle. If more than one hose is removed, be sure to label the hoses and fittings to ensure correct installation.

6 When checking vacuum hoses, be sure to include any plastic T-fittings in the check. Inspect the fittings for cracks and the hose where it fits over the fitting for distortion, which could cause leakage.

7 A small piece of vacuum hose (1/4-inch inside diameter) can be used as a stethoscope to detect vacuum leaks. Hold one end of the hose to your ear and probe around vacuum hoses and fittings, listening for the "hissing" sound characteristic of a vacuum leak. **Warning:** *When probing with the vacuum hose stethoscope, be very careful not to come into contact with moving engine components such as the drivebelts, cooling fan, etc.*

Fuel hose

Warning: *There are certain precautions which must be taken when inspecting or servicing fuel system components. Work in a well ventilated area and do not allow open flames (cigarettes, appliance pilot lights, etc.) or bare light bulbs near the work area. Mop up any spills immediately and do not store fuel soaked rags where they could ignite.*

8 Check all rubber fuel lines for deterioration and chafing. Check especially for cracks in areas where the hose bends and just before fittings.

9 High quality fuel line, usually identified by the word Fluroelastomer printed on the hose, should be used for fuel line replacement. Never, under any circumstances, use unreinforced vacuum line, clear plastic tubing or water hose for fuel lines.

10 Spring-type clamps are commonly used on fuel lines. These clamps often lose their tension over a period of time, and can be "sprung" during removal. Replace all spring-type clamps with screw clamps whenever a hose is replaced.

Metal lines

Warning: *Use extreme care when inspecting the fuel lines, particularly when the engine is running. The fuel is under very high pressure and can penetrate the skin causing injury.*

11 Sections of metal line are used for fuel line between the fuel pump and the fuel injection unit and injectors. Check carefully to be sure the lines have not been bent or crimped and that cracks have not started in the lines.

12 If a section of metal fuel line between the fuel injection pump and the injectors in must be replaced, use only direct replacement steel tubing from a dealer, since each line is a specific shape.

Cooling System

General information

The cooling system basically consists of a radiator, a thermostat and a crankshaft pulley-driven water pump.

The radiator cooling fan is mounted on the front of the water pump and incorporates a fluid drive fan clutch, saving horsepower and reducing noise. A fan shroud is mounted on the rear of the radiator.

The system is pressurized by a spring-loaded radiator cap, which increases the boiling point of the coolant. If the coolant temperature goes above this increased boiling point, the extra pressure in the system forces the radiator cap valve off its seat and exposes the overflow pipe. The overflow pipe leads to a coolant recovery system. This consists of a plastic reservoir into which the coolant which normally escapes due to expansion is retained. When the engine cools, the excess coolant is drawn back into the radiator, maintaining the system at full capacity. This is a continuous process and, provided the level in the reservoir is correctly maintained, it is not necessary to add coolant to the radiator.

Coolant from the radiator circulates into the lower radiator hose to the water pump, where it is forced through the water passages in the cylinder block. The coolant then travels up into the cylinder head, circulates around the combustion chambers and valve seats, travels out of the cylinder heads past the open thermostat into the upper radiator hose and back into the radiator.

When the engine is cold the thermostat restricts the circulation of coolant to the engine. The thermostat is located near the front of the engine where the upper radiator hose connects to the engine. When

3.26 Remove the thermostat cover bolts (arrows)

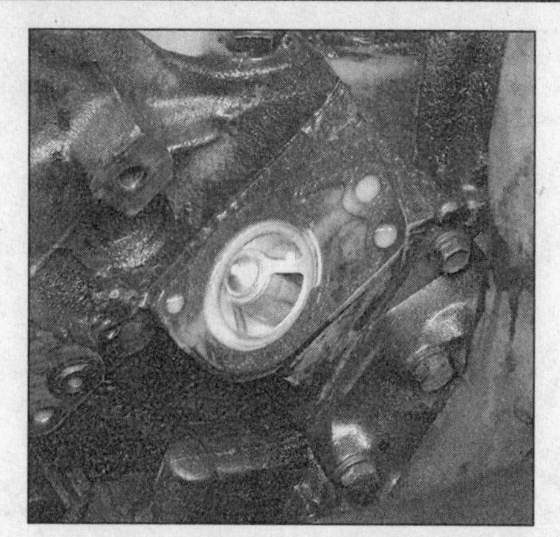

3.27 Install the new gasket with the spring side in

3.28 Position a new gasket so the holes align

the minimum operating temperature is reached, the thermostat begins to open, allowing coolant to return to the radiator.

Automatic transmission-equipped models have a cooler element incorporated into the radiator to cool the transmission fluid.

The heating system works by directing air through the heater core mounted in the dash and then to the interior of the vehicle by a system of ducts. Temperature is controlled by mixing heated air with fresh air, using a system of flapper doors in the ducts, and a heater motor.

Air conditioning is an optional accessory, consisting of an evaporator core located under the dash, a condenser in front of the radiator, a receiver-drier in the engine compartment and a belt-driven compressor mounted at the front of the engine.

Antifreeze - general information

Warning: *Do not allow antifreeze to come in contact with your skin or painted surfaces of the vehicle. Rinse off spills immediately with plenty of water. NEVER leave antifreeze lying around in an open container or in a puddle in the driveway or on the garage floor. Children and pets are attracted by it's sweet smell. Antifreeze is fatal if ingested.*

The cooling system should be filled with a water/ethylene glycol based antifreeze solution which will prevent freezing down to at least -20-degrees F (even lower in cold climates). It also provides protection against corrosion and increases the coolant boiling point.

The cooling system should be drained, flushed and refilled at least every other year. The use of antifreeze solutions for periods of longer than two years is likely to cause damage and encourage the formation of rust and scale in the system.

Before adding antifreeze to the system, check all hose connections. Antifreeze can leak through very minute openings.

The exact mixture of antifreeze to water which you should use depends on the relative weather conditions. The mixture should contain at least 50 percent antifreeze, but should never contain more than 70 percent antifreeze.

Thermostat - check and replacement

Warning: *The engine must be completely cool when this procedure is performed.*

Note: *Don't drive the vehicle without a thermostat!*

Check

1 Before condemning the thermostat, check the coolant level, drivebelt tension, drivebelt tension and temperature gauge (or light) operation.

2 If the engine takes a long time to warm up, the thermostat is probably stuck open. Replace the thermostat.

3 If the engine runs hot, check the temperature of the upper radiator hose. If the hose isn't hot, the thermostat is probably stuck shut. Replace the thermostat.

4 If the upper radiator hose is hot, it means the coolant is circulating and the thermostat is open. Check for other causes of overheating.

5 If an engine has been overheated, you may find damage such as leaking head gaskets, scuffed pistons and warped or cracked cylinder heads.

Replacement

6 Drain coolant from the radiator until the coolant level is below the thermostat housing.

7 Remove the alternator and vacuum pump.

8 Disconnect the upper radiator hose from the thermostat housing cover, which is located near the front of the engine.

9 Remove the bolts and lift the thermostat cover off **(see illustration)**. It may be necessary to tap the cover with a soft-face hammer to break the gasket seal.

10 Note how it's installed, then remove the thermostat.

11 Use a scraper or putty knife to remove all traces of old gasket material and sealant from the mating surfaces. Make sure that no gasket material falls into the coolant passages; it is a good idea to stuff a rag in the passage. Wipe the mating surfaces with a rag saturated with lacquer thinner or acetone.

12 The original thermostat does not have a bypass. The wrong replacement could cause overheating damage. Use only a Motorcraft E5TZ-8575-C or Navistar 1807945-C1 or equivalent thermostat. Apply a thin layer of RTV sealant to the gasket mating surfaces of the housing and cover, then install the new thermostat in the engine **(see illustration)**. Make sure the correct end faces up - the spring is directed into the housing.

13 Position a new gasket on the housing and make sure the gasket holes line up with the bolt holes in the housing **(see illustration)**.

14 Carefully position the cover on the housing and install the bolts. Tighten them to the torque listed in this Chapter's Specifications - do not overtighten them or the cover may be damaged.

15 Reattach the radiator hose to the cover and tighten the clamp - now may be a good time to check and replace the hoses and clamps.

16 Refer to the Cooling System portion of this Chapter and refill the system, then run the engine and check carefully for leaks.

Coolant temperature sending unit – check and replacement

Warning: *The engine must be completely cool before removing the sending unit.*

Check

1 If the coolant temperature gauge or light is inoperative, check the fuses first.

2 If either of the temperature indicators show excessive temperature after running awhile, check the cooling system for the possible source of overheating.

3 If the temperature gauge or light indicates Hot shortly after the engine is started cold, disconnect the wire at the coolant temperature sending unit **(see illustrations)**. If the light goes out or the gauge read-

3.29 The sending unit for the light (arrow) is located near the front of the cylinder head on the left (driver's) side of the engine

3.30 The sending unit for the gauge is located near the front of the intake manifold (arrow)

ing drops, replace the sending unit. If the reading remains high, the wire to the gauge may be shorted to ground or the gauge is faulty.

4 If the coolant temperature gauge fails to indicate after the engine has been warmed up (approximately 10 minutes) and the fuses checked out okay, shut off the engine. Disconnect the wire at the sending unit and, using a jumper wire, connect it to a clean ground on the engine. Turn on the ignition without starting the engine. If the gauge now indicates Hot, replace the sending unit.

5 If the gauge still does not work, the circuit may be open or the gauge may be faulty.

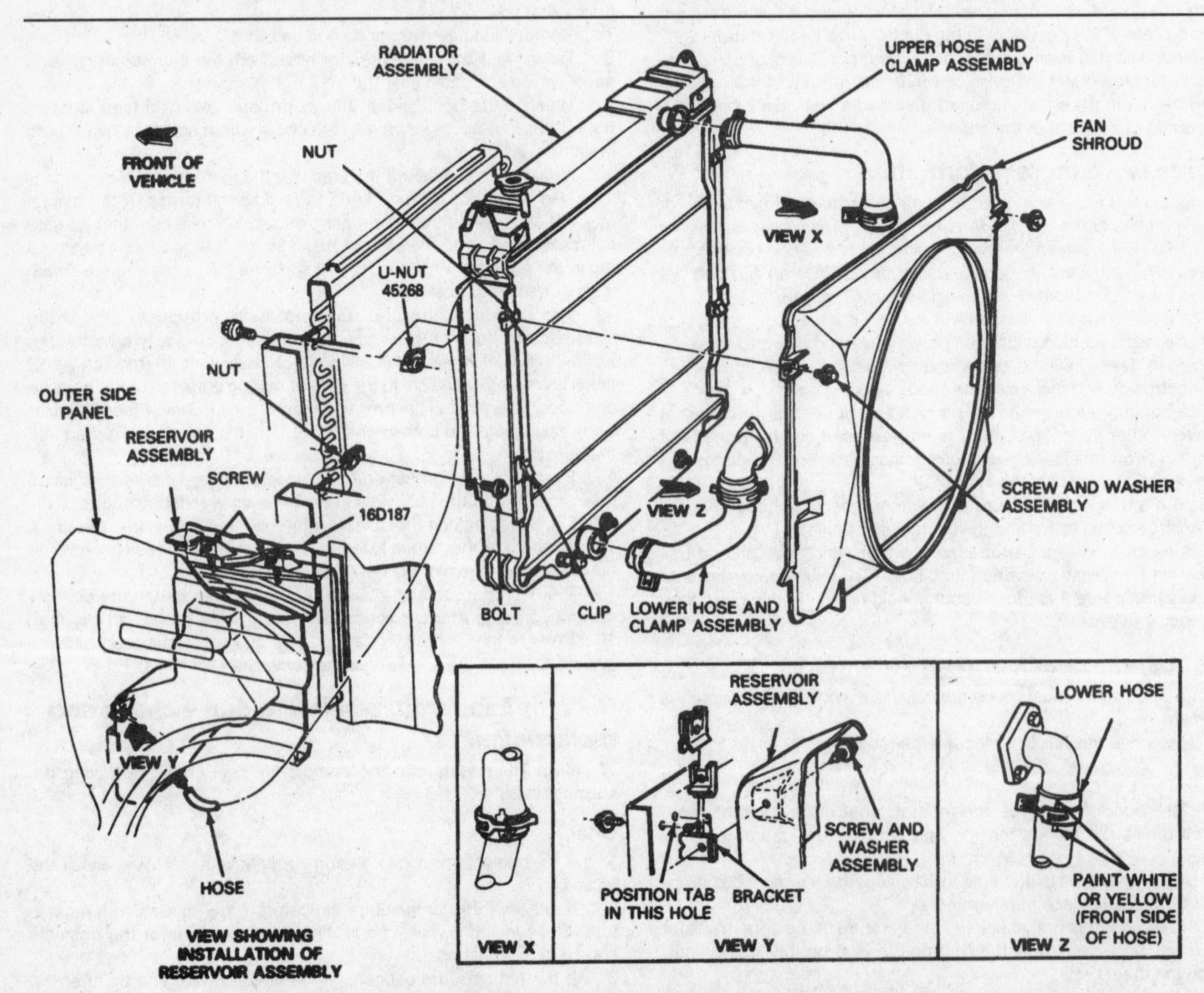

3.31a Typical E-Series van fan shroud and radiator details

Replacement

6 With the engine completely cool, remove the cap from the radiator to release any pressure, then reinstall the cap. This reduces coolant loss during sending unit replacement.
7 Disconnect the wiring harness from the sending unit.
8 Prepare the new sending unit for installation by applying a light coat of sealant to the threads.
9 Unscrew the sending unit from the engine and quickly install the new one to prevent coolant loss.
10 Tighten the sending unit securely and connect the wiring harness.
11 Refill the cooling system and run the engine. Check for leaks and proper gauge operation.

Cooling fan and fan clutch - removal and installation

1 The cooling fan should be replaced if the blades become damaged or bent. The viscous drive fan clutch is disengaged when the engine is cold, or at high engine speeds, when the silicone fluid inside the clutch is contained in the reservoir section by centrifugal action. Symptoms of failure of the fan clutch include continuous noisy operation (fan clutch seized), lack of cooling (clutch freewheels), looseness leading to vibration (worn bearings) and evidence of silicone fluid leaks (worn seals).
2 Disconnect the ground cables from the negative terminals of the batteries.
3 Partially drain the radiator. Unbolt the fan shroud **(see illustrations).**

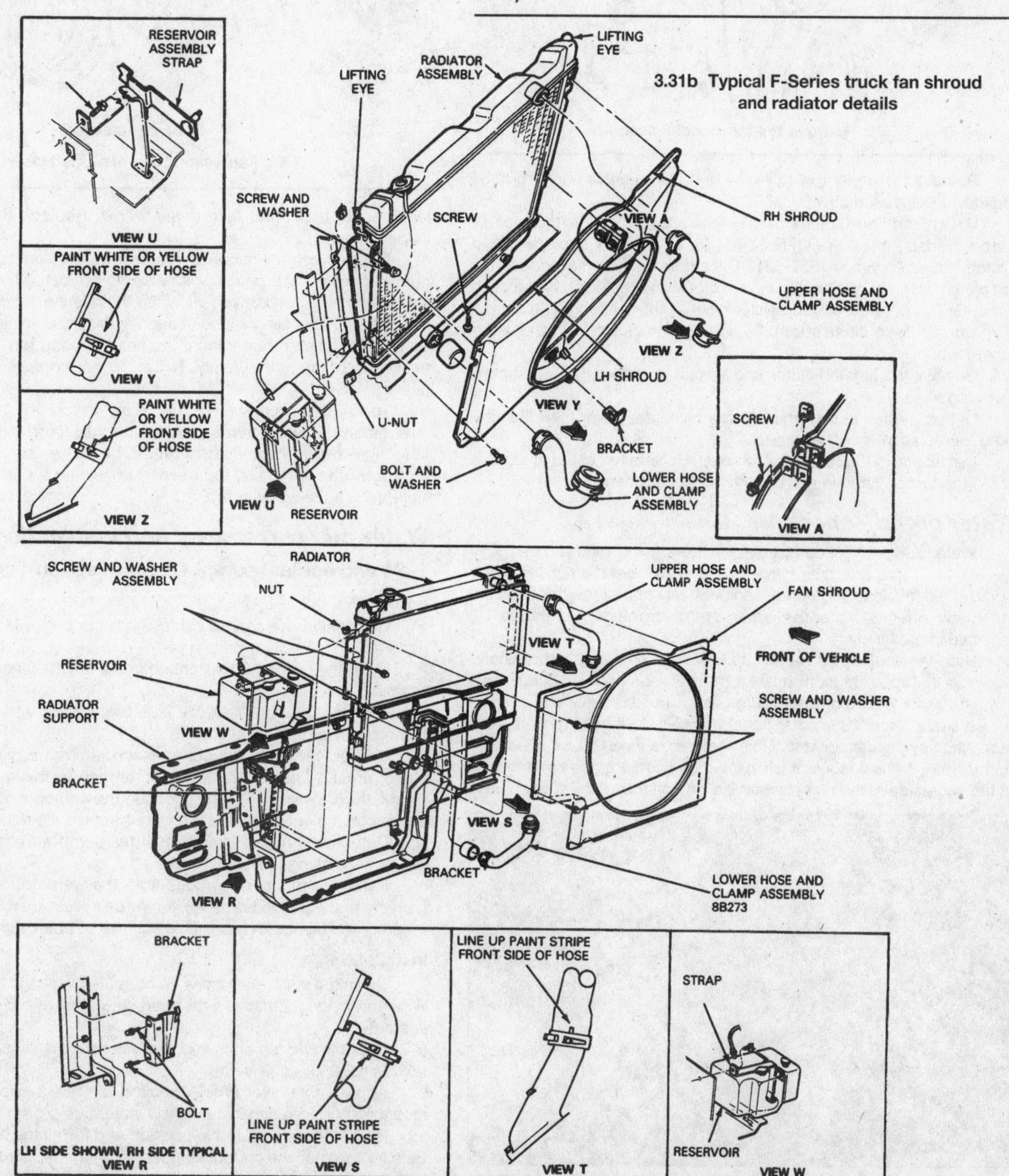

3.31b Typical F-Series truck fan shroud and radiator details

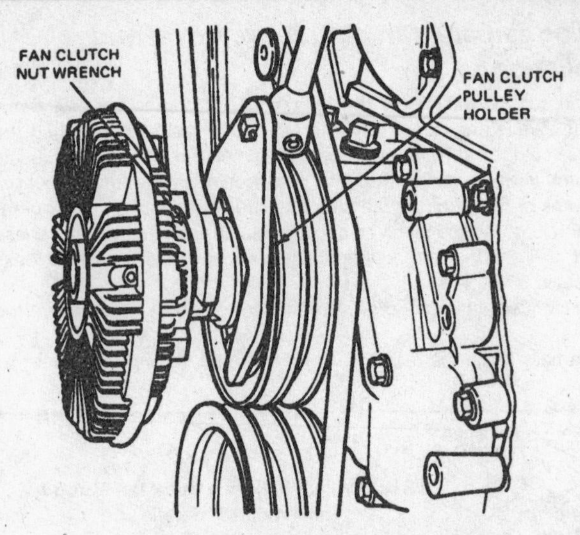

3.32 Remove the fan clutch assembly

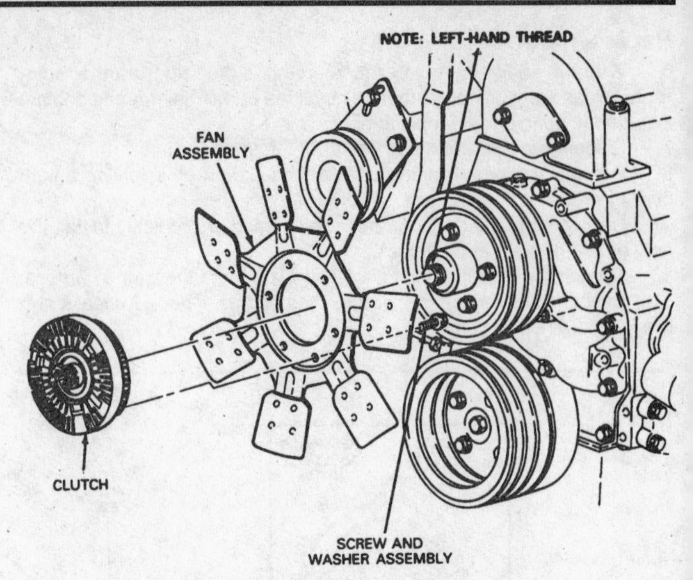

3.33 Fan assembly - exploded view

4 Remove the upper radiator hose and the reservoir hose from the radiator. Remove the drivebelts.

5 Using Ford Fan Clutch Pulley Holder no. T83T-6312-A for non-DI Turbo models, or no. T94T-6312-AH for DI Turbo models and Fan Clutch Nut Wrench no. T83T-6312-B (or their equivalents); loosen the fan clutch assembly by turning the nut clockwise (left-hand threads) on a non-DI Turbo model or counterclockwise (right-hand threads) on a DI Turbo model **(see illustration)**. Detach the fan clutch from the water pump hub.

6 Remove the fan, fan clutch and shroud from the engine compartment together.

7 Unbolt the fan clutch from the fan blade assembly **(see illustration)** for replacement, if necessary.

8 Installation is the reverse of removal. Be sure to tighten all fasteners to the torque listed in this Chapter's Specifications.

Water pump - check

1 Water pump failure can cause overheating and serious damage to the engine. There are three ways to check the operation of the water pump while it is installed on the engine. If any one of the three following quick checks indicates water pump problems, it should be replaced immediately.

2 Start the engine and warm it up to normal operating temperature. Squeeze the upper radiator hose. If the water pump is working properly, you should feel a pressure surge as the hose is released.

3 A seal protects the water pump impeller shaft bearing from contamination by engine coolant. If this seal fails, a weep hole in the water pump snout will leak coolant (an inspection mirror can be used to look at the underside of the pump if the hole isn't on top). If the weep hole is leaking, shaft bearing failure will follow. Replace the water pump immediately.

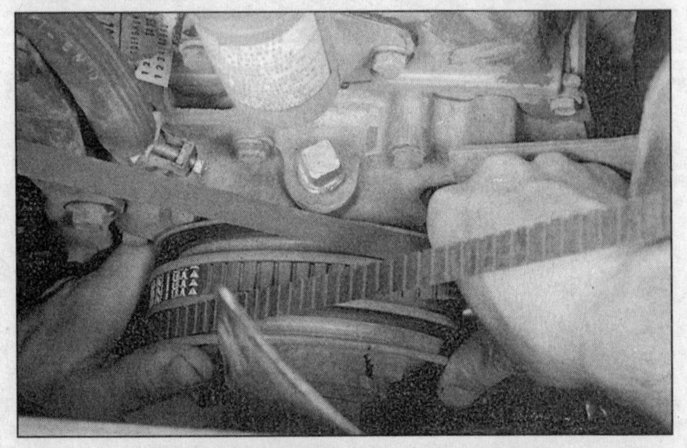
3.34 Try to rock the pulley back and forth

4 Besides contamination by coolant after a seal failure, the water pump impeller shaft bearing can also be prematurely worn out by an improperly tensioned drivebelt. When the bearing wears out, it emits a high pitched squealing sound. If such a noise is coming from the water pump during engine operation, the shaft bearing has failed - replace the water pump immediately. **Note:** *Do not confuse belt noise with bearing noise.*

5 To identify excessive bearing wear before the bearing actually fails, grasp the water pump pulley and try to force it up-and-down or from side-to-side **(see illustration)**. If the pulley can be moved either horizontally or vertically, the bearing is nearing the end of its service life. Replace the water pump.

Water pump - removal and installation

1994 and earlier (except Direct Injection Turbo) models

Removal

1 Disconnect the ground cables from the negative terminals of the batteries.

2 Drain the coolant as described in the Maintenance portion of this Chapter.

3 Remove the fan assembly (see the appropriate Section), water pump pulley and drivebelts.

4 On some models the compressor, alternator, vacuum pump or power steering pump brackets may attach to the water pump **(see illustration)**. Where necessary, loosen the components so the brackets can be moved aside and the water pump bolts removed.

5 Detach the hoses from the water pump and then remove the heater hose fitting from the pump.

6 Remove the mounting bolts from the water pump and detach it from the block **(see illustration)**. It may be necessary to grasp the pump securely and tap it with a soft-face hammer to break the gasket seal.

Installation

7 Clean the sealing surfaces on both the block and the water pump. Wipe the mating surfaces with a rag saturated with lacquer thinner or acetone.

8 Coat the two top bolts and two bottom bolts with Aviation Permatex TM no. 3 (or equivalent).

9 Apply a thin layer of RTV sealant to the water pump and block mounting surfaces and install a new water pump gasket.

10 Place the water pump in position and install the bolts finger tight. Be sure to install the correct length bolts in the proper holes **(see illustration)**. Use caution to ensure that the gaskets do not slip out of posi-

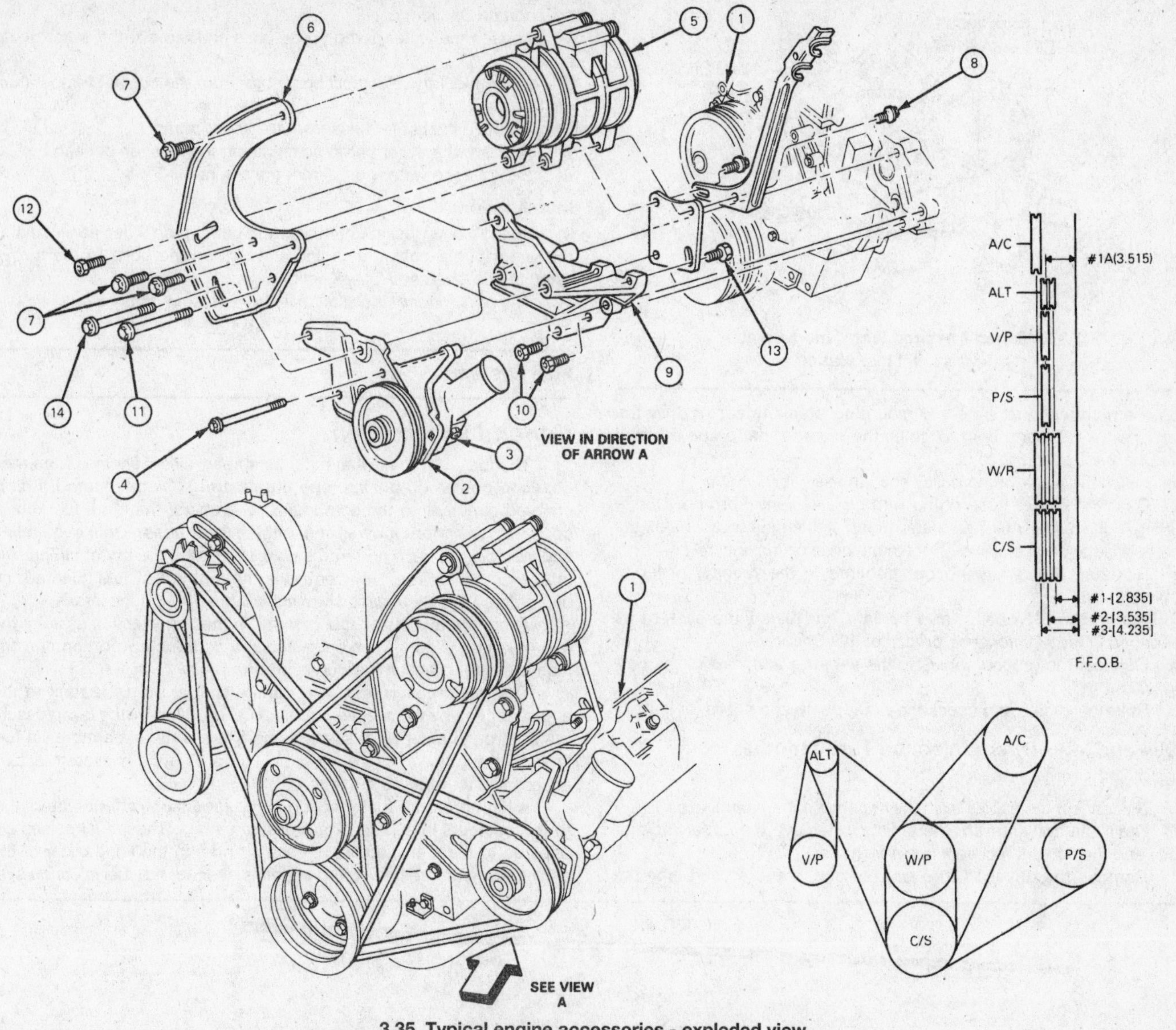

3.35 Typical engine accessories - exploded view

1 Engine
2 Power steering mounting bracket
3 Power steering pump
4 Bolt
5 Compressor and clutch
6 Bracket
7 Bolt
8 Bolt
9 Compressor mounting bracket
10 Bolt
11 Bolt
12 Bolt
13 Bolt
14 Bolt
15 Compressor rear bracket

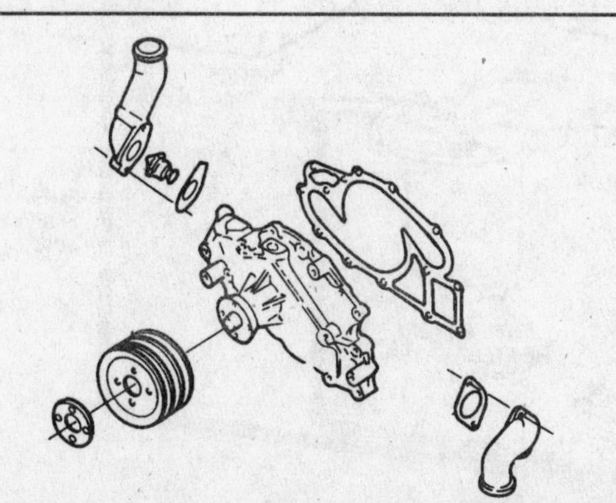

3.36 Water pump components (non-DI Turbo - exploded view)

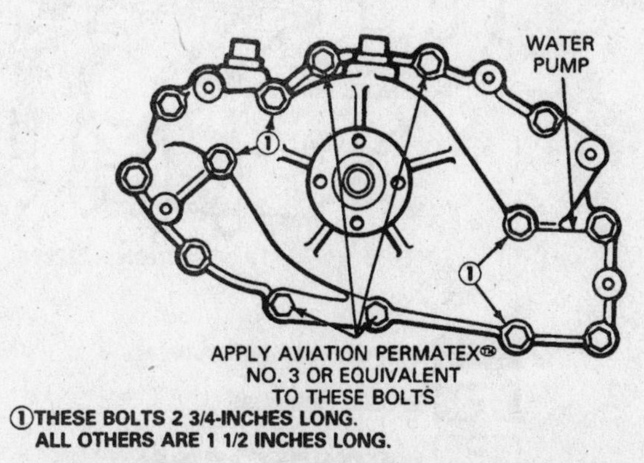

3.37a Be sure to install the bolts in the proper holes (there are two different lengths of bolts)

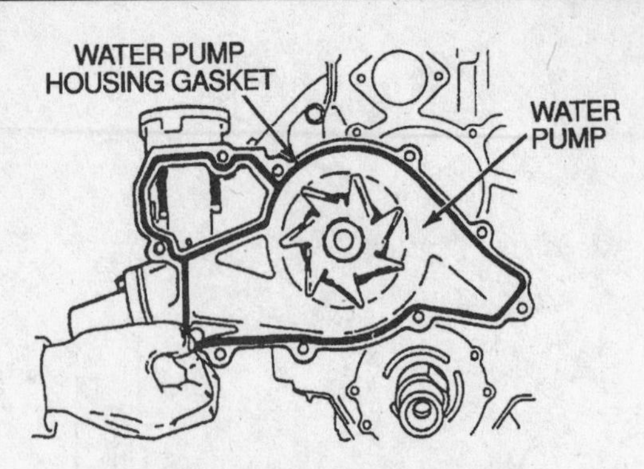

3.37b Direct Injection Turbo model water pump gasket installation

tion. Remember to replace any mounting brackets secured by the water pump mounting bolts. Tighten the bolts to the torque listed in this Chapter's specifications.

11 Install the water pump pulley and fan assembly.

12 Coat the heater hose fitting with pipe sealant (Ford no. D8AZ-19554-A or equivalent), then install it and tighten securely. Install the heater hose and hose clamp. Tighten the hose clamp securely.

13 Reinstall the remaining components in the reverse order of removal.

14 Adjust the drivebelts, add coolant and bleed the system as described in the *Maintenance* portion of this Chapter.

15 Connect the ground cables to the negative terminals of the batteries.

16 Start the engine and check the water pump and hoses for leaks.

1994 and later Direct Injection Turbo models

Removal

1 Disconnect the cables from the negative battery terminals.

2 Drain the coolant from the radiator. Remove the upper radiator hose and the coolant recovery reservoir hose.

3 Remove the drivebelt, the fan, fan clutch and shroud (see the appropriate Sections).

4 Loosen the water pump pulley bolts and remove the water pump pulley.

5 Disconnect the electrical connector from the coolant temperature sensor.

6 Remove the heater hose from the water pump.

7 Remove the water pump bolts, noting the location of each bolt.

8 Remove the water pump from the engine.

Installation

9 Install a new gasket onto the water pump **(see illustration)**. Install the water pump onto the engine and tighten the bolts to the torque listed in this Chapter's Specifications.

10 The remainder of installation is the reverse of removal.

Fuel System

General information

The fuel system is what most sets apart diesel engines from their gasoline powered cousins **(see illustration)**. Simply stated, fuel is sprayed directly into the combustion chambers; the more fuel that is sprayed, the more power the engine produces. Unlike gasoline engines, diesels have no throttle plates to limit the entry of air into the intake manifold. The only control is the amount of fuel injected; an unrestricted supply of air is always available through the intake.

There are two major sub-systems in the fuel injection system; the low pressure (also known as the supply or transfer) portion and the high pressure injection (delivery) portion.

The low pressure system "transfers" fuel from the tank to the engine for use by the injection pump. Fuel is drawn by the low pressure pump through a series of screens and filters/water separators via fuel lines to the high pressure injection pump. A bypass system allows excess fuel to return to the tank.

A high pressure fuel injection pump **(see illustration)** meters fuel to the cylinders in minute high pressure squirts. These fuel pulses are directed to the fuel injectors of each cylinder in the firing order of the engine. When more power and speed is desired, the fuel injection sys-

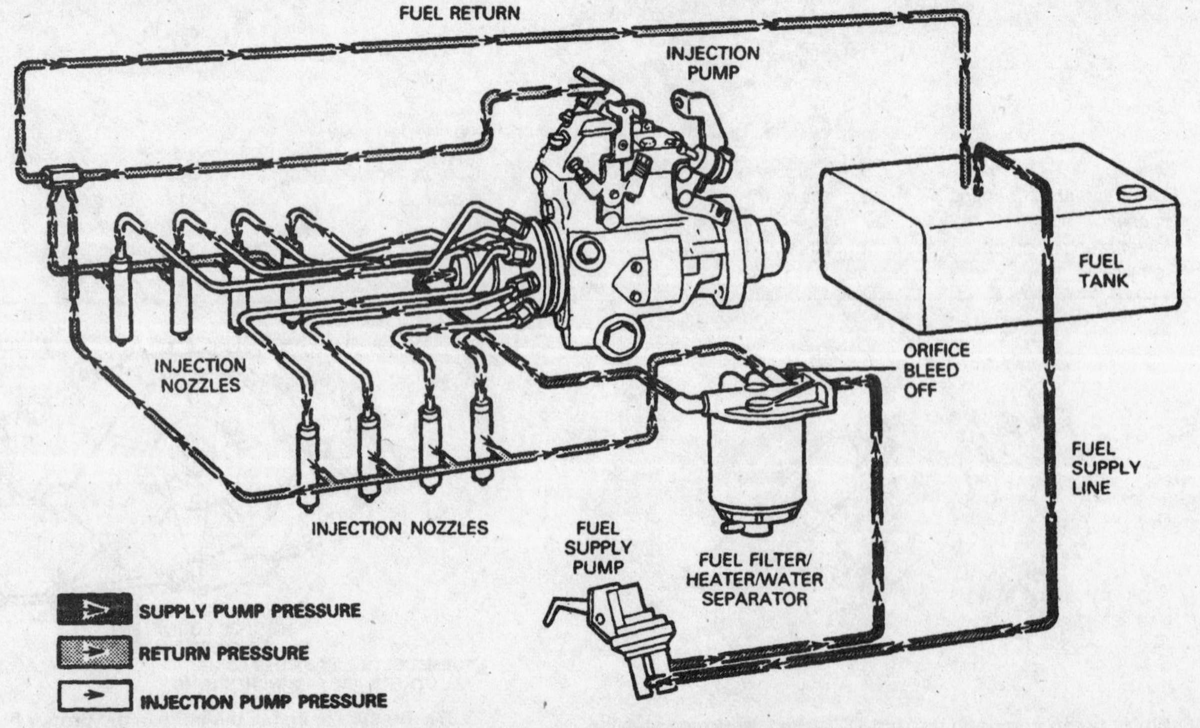

3.38 Typical Ford diesel fuel system (non-DI Turbo)

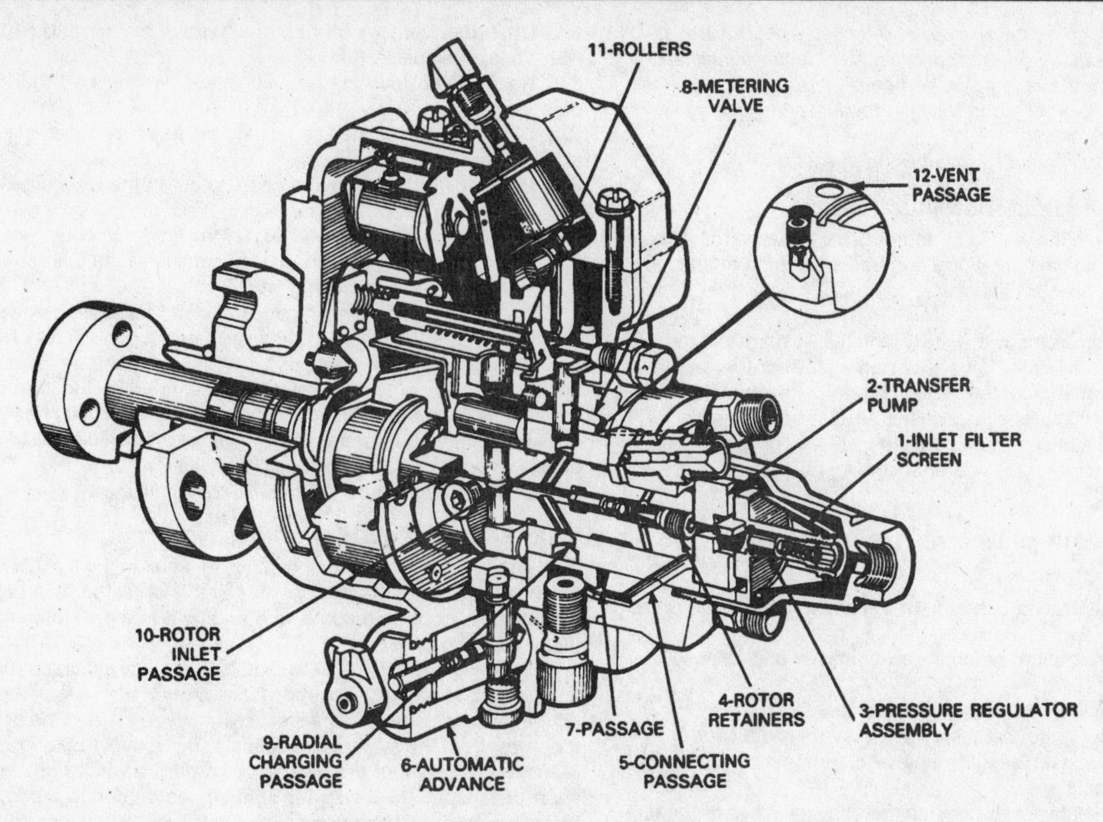

3.39 Cutaway view of the fuel injection pump (non-DI Turbo)

tem simply sprays more fuel into the cylinders. When a preset maximum engine speed is reached, a governor limits the delivery of fuel, thereby limiting speed.

The injection of fuel also determines the timing of combustion, similar to spark timing in a gasoline engine. As engine speed increases, timing must be advanced.

In 1994 Ford introduced an improved "Direct Injection" fuel injection system for their new turbocharged 7.3L engine. The fuel injector location was moved from the conventional location above the valve covers to beneath the valve covers, placing them near the center of the combustion chambers. The operation of the injectors is electronically controlled by the Powertrain Control Module (PCM). A mechanical timed injection pump is no longer used. A single supply pump provides a supply of fuel at high pressure to the injectors.

The fuel injectors are driven by extremely high pressure engine oil (as much as 3,000 psi) which is supplied by a special pump. This pump is driven from the camshaft and is located above the front cover. The oil circulates through a reservoir, to the pump and regulator, through hoses to the oil galleries in the redesigned heads and then to the injectors.

Fuel system problems are by far the most frequent cause of breakdowns and loss of power in diesel-powered vehicles. Whenever a diesel engine quits running or loses power for no apparent reason, check the fuel system first. Begin with the most obvious items first, such as the fuel filter and damaged fuel lines. If the vehicle has multiple fuel tanks, suspect a faulty tank switchover valve.

The fuel system on diesel engines is extremely sensitive to contamination. Fuel contamination is a serious problem due to the very small clearances in the injection pump and the minute orifices in the injection nozzles. The injection pump and the injectors can be damaged or ruined.

Water-contaminated diesel fuel is a major problem. If it remains in the fuel system too long, water will cause serious and expensive damage. The fuel lines and the fuel filter can also become plugged with rust particles, or clogged with ice in cold weather.

Diesel fuel contamination

Before you replace an injection pump or some other expensive component, find out what caused the failure. If water contamination is present, buying a new or rebuilt pump or other component won't do much good. The following procedure will help you pinpoint whether water contamination is present:

1) *Remove the engine fuel filter and inspect the contents for the presence of water or gasoline.*
2) *If the vehicle has been stalling, performance has been poor or the engine has been knocking loudly, suspect fuel contamination. Gasoline or water must be removed by flushing (see below).*
3) *If you find a lot of water in the fuel filter, remove the injection pump fuel return line and check for water there. If the pump has water in it, flush the system (see below).*
4) *Small quantities of surface rust won't create a problem. If contamination is excessive, the vehicle will probably stall.*
5) *Sometimes contamination in the system becomes severe enough to cause physical damage to the internal parts in the pump. If the damage reaches this stage, have the damaged parts replaced and the pump rebuilt by an authorized fuel injection shop, or buy a rebuilt pump.*

Storage

Good quality diesel fuel contains inhibitors to stop the formation of rust in the fuel lines and the injectors, so as long as there are no leaks in the fuel system, it's generally safe from water contamination. Diesel fuel is usually contaminated by water as a result of careless storage. There's not much you can do about the storage practices of service stations where you buy diesel fuel, but if you keep a small supply of diesel fuel on hand at home, as many diesel owners do, follow these simple rules:

1) *Diesel fuel "ages" and goes stale. Don't store containers of diesel fuel for long periods of time. Use it up regularly and replace it with fresh fuel.*

2) Keep fuel storage containers out of direct sunlight. Variations in heat and humidity promote condensation inside fuel containers.
3) Don't store diesel fuel in galvanized containers. It may cause the galvanizing to flake off, contaminating the fuel and clogging filters when the fuel is used.
4) Label containers properly, as containing diesel fuel.

Fighting fungi and bacteria with biocides

If there's water in the fuel, fungi and/or bacteria can form in diesel fuel in warm or humid weather. Fungi and bacteria plug fuel lines, fuel filters and injection nozzles; they can also cause corrosion in the fuel system.

If you've had problems with water in the fuel system and you live in a warm or humid climate, have your dealer correct the problem. Then, use a diesel fuel biocide to sterilize the fuel system in accordance with the manufacturer's instructions. Biocides are available from your dealer, service stations and auto parts stores. Consult your dealer for advice on using biocides in your area, and for recommendations on which ones to use.

Cleaning the low-pressure fuel system

Water in fuel system

1 Disconnect the ground cables from the negative terminals of the batteries.
2 Drain the fuel tank into an approved container and dispose of it properly.
3 Remove the tank gauge sending unit.
4 Thoroughly clean the fuel tank. If it's rusted inside, send it to a repair shop or replace it. Clean or replace the fuel pick-up filter and check the valve assembly.
5 Reinstall the fuel tank but don't connect the fuel lines to the fuel tank yet.
6 Disconnect the main fuel line from the low-pressure fuel pump. Using low air pressure, blow out the line toward the rear of the vehicle. **Warning:** *Wear eye protection.*
7 Temporarily disconnect the fuel return fuel line at the injection pump and again, using low air pressure, blow out the line toward the rear of the vehicle.
8 Reconnect the main fuel and return lines at the tank. Fill the tank to a fourth of its capacity with clean diesel fuel. Install the cap on the fuel filler neck.
9 Remove and discard the fuel filter.
10 Connect the fuel line to the fuel pump.
11 Reconnect the battery cables.
12 Purge the fuel pump and pump-to-filter line by cranking the engine until clean fuel is pumped out. Catch the fuel in a closed metal container.
13 Install a new fuel filter.
14 Install a hose from the fuel return line (from the injection pump) to a closed metal container with a capacity of at least two gallons.
15 Crank the engine until clean fuel appears at the return line. Don't crank the engine for more than 30 seconds at a time. If it's necessary to crank it again, allow a three-minute interval before resuming.
16 Using a line wrench, loosen each high pressure line at the injector nozzles.
17 Crank the engine until clean fuel appears at each nozzle. Don't crank the engine for more than 30 seconds at a time. If it's necessary to crank it again, allow a three-minute interval before resuming.

Gasoline in the fuel system

Warning: *Gasoline is extremely flammable, so take extra precautions when working with it. Don't smoke or allow open flames or bare light bulbs in the work area, and don't work in a garage where a gas-type appliance is present. If you spill fuel on your skin, wash it off immediately with soap and water. Have a Class B fire extinguisher handy.*

If gasoline has been accidentally pumped into the fuel tank, it should be drained immediately. Gasoline in the fuel in small amounts - up to 30 percent - isn't usually noticeable. At higher ratios, the engine may make a knocking noise, which will get louder as the amount of gasoline goes up. Here's how to rid the fuel system of gasoline:

1 Drain the fuel tank into an approved container and fill the tank with clean, fresh diesel fuel.
2 Remove the fuel line between the fuel filter and the injection pump.
3 Connect a short pipe and hose to the fuel filter outlet and run it to a closed metal container.
4 Crank the engine to purge gasoline out of the fuel pump and fuel filter. Don't crank the engine more than 30 seconds. Allow two or three minutes between cranking intervals for the starter to cool.
5 Remove the short pipe and hose and install the fuel line between the fuel filter and the injection pump.
6 Try to start the engine. If it doesn't start, purge the injection pump and lines: Crack the fuel line fittings a little, just enough for fuel to leak out. Depress the accelerator pedal to the floor and, holding it there, crank the engine until all gasoline is gone, i.e. diesel fuel leaks out of the fittings. Tighten the fittings. Limit cranking to 30 seconds with two or three minutes between cranking intervals. **Warning:** Avoid sources of ignition and have a fire extinguisher handy.
7 Start the engine and run it at idle for 15 minutes.

"Water-In-Fuel" (WIF) warning system

The system detects the presence of water in the fuel filter when it reaches the one-tenth gallon level. Water is detected by a probe that completes a circuit through a wire to a light in the instrument cluster that reads "Water In Fuel."

Units include a bulb-check feature: When the ignition is turned on, the bulb glows momentarily, then fades away.

If the light comes on immediately after you've filled the tank, drain the water from the system immediately. There might be enough water in the system to shut the engine down before you've driven even a short distance. If, however, the light comes on during a cornering or braking maneuver, there's less water in the system; the engine probably won't shut down immediately, but you still should drain the water soon.

Water is heavier than diesel fuel, so it sinks to the bottom of the fuel tank. An extended return pipe on the fuel tank sending unit, which reaches down into the bottom of the tank, enables you to siphon most of the water from the tank without having to remove the tank. But siphoning won't remove all of the water; you'll still need to remove the tank and thoroughly clean it. **Warning:** *Do not start a siphon by mouth - use a siphoning kit (available at most auto parts stores).*

If the water sensor probe needs to be replaced, you can service it separately from the fuel filter unit **(see illustration)**. Unscrew the old unit and install a new one.

Low pressure pump - check

There are some simple tests you can perform to determine whether a mechanical low pressure fuel pump is operating satisfactorily. But first, do the following preliminary inspection:

Preliminary inspection

1 Check the fuel line fittings and connections - make sure they're tight. If a fitting is loose, air and/or fuel leaks may occur.
2 Check for bends or kinks in the fuel lines.
3 Start the engine and let it warm up. With the engine idling, conduct the following preliminary inspection before proceeding to the actual fuel pump tests:

a) Look for leaks at the pressure (outlet) side of the pump.
b) Look for leaks on the suction (inlet) side of the pump. A leak on the suction side will let air into the pump and reduce the volume of fuel on the pressure (outlet) side of the pump.
c) Inspect the cover and the fittings on the fuel pump for leaks. Tighten or replace the fittings as necessary.
d) Look for leaks around the diaphragm, the flange and the breather holes in the pump housing. If any of them are leaking, replace the pump.

Mechanical pump fuel flow test

1 Unscrew the threaded fittings and detach the fuel line from the filter inlet.

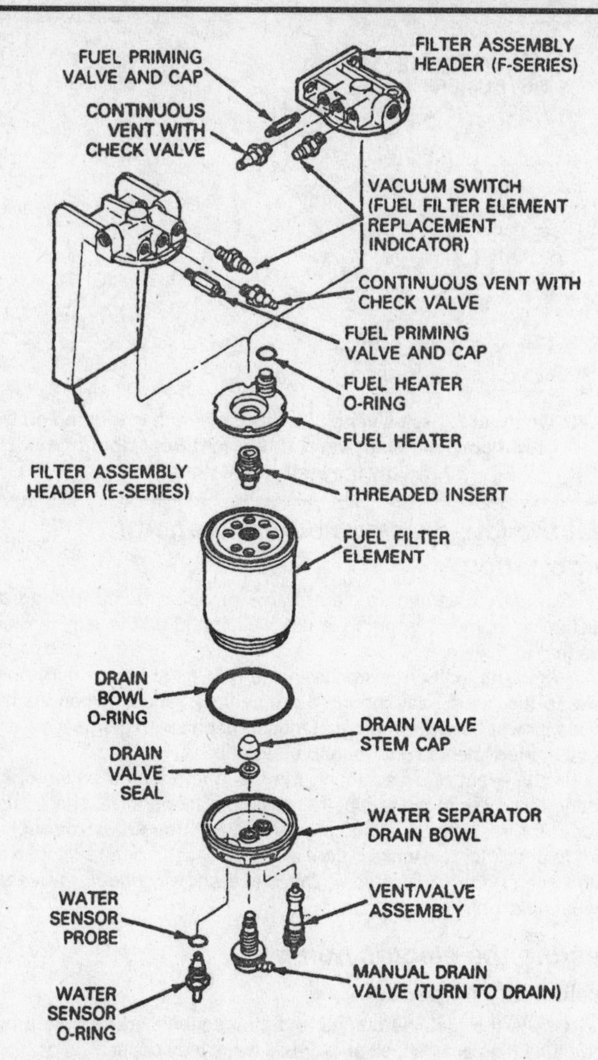

3.40 In this typical Ford fuel filter assembly, the water sensor probe is at the bottom

2 Disconnect the wire at the fuel injection pump electric shut-off solenoid.
3 Put the end of the metal line in a container and crank over the engine a few times.
4 If little or no fuel flows from the open end of the fuel line, either the line is clogged or the fuel pump is faulty. Remove the line between the fuel tank and the pump, blow it out and try this test again.
5 If there's still little fuel flowing from the pipe, replace the pump (see below).

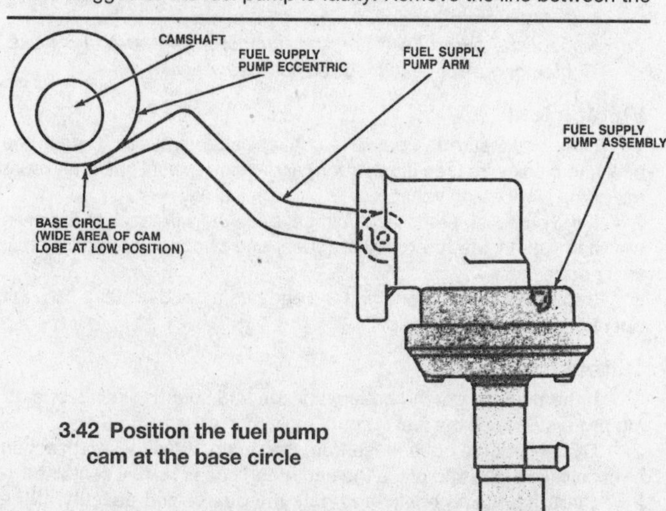

3.42 Position the fuel pump cam at the base circle

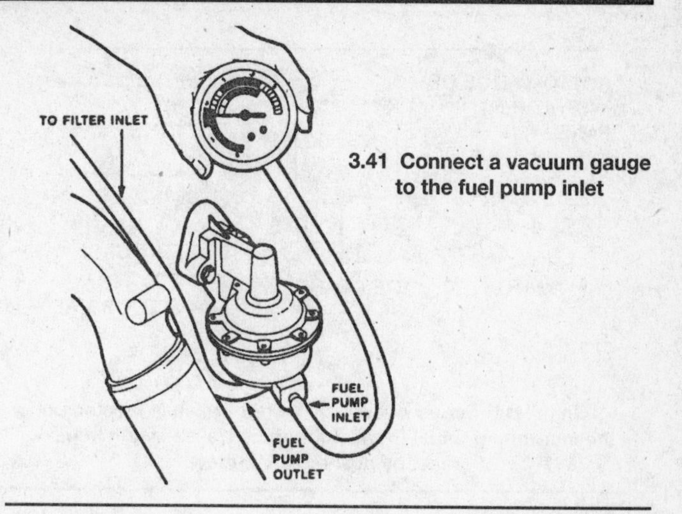

3.41 Connect a vacuum gauge to the fuel pump inlet

Vacuum test

The vacuum test is the best indicator of how well a mechanical low-pressure pump is working. If the low-pressure pump can only draw a low vacuum, or has a total loss of vacuum, it can't provide sufficient fuel to the injection pump to operate the engine throughout its normal speed range. The following vacuum test will help you determine whether the pump has the ability to pump fuel.
1 Detach the hose from the fuel tank to the fuel pump at the pump. Plug the hose to prevent contamination or fuel leakage.
2 Connect one end of a short hose to the fuel pump inlet and attach a vacuum gauge to the other end **(see illustration)**.
3 Start the engine. Note the vacuum gauge reading with the engine at idle. Shut off the engine. If the indicated vacuum is less than 12 inches Hg, replace the pump.
4 Remove the vacuum gauge and reattach the fuel inlet line.

Mechanical low pressure pump (except Direct Injection Turbo models)

Removal

1 The low pressure (also known as supply or transfer) mechanical fuel pump is located on the right (passenger's) side of the engine, low and near the front.
2 Using a flare nut wrench, loosen and then retighten the threaded fuel lines.
3 Loosen the pump mounting bolts one or two turns. Move the pump by hand to break the gasket seal. Using a remote starter switch, operate the starter to rotate the crankshaft in short bursts until the fuel pump cam is at its lowest position **(see illustration)**. **Warning:** *Stay clear of the fan and drivebelts.*
4 Disconnect the fuel lines from the fuel pump **(see illustration)**.
5 Remove the fuel pump mounting bolts and remove the pump.

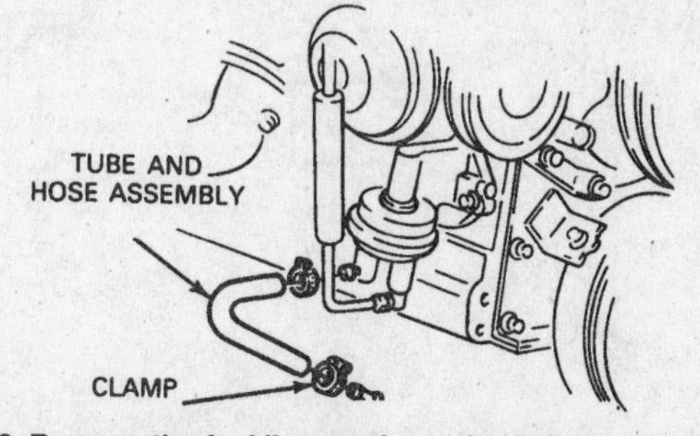

3.43 Remove the fuel lines and cap the fittings

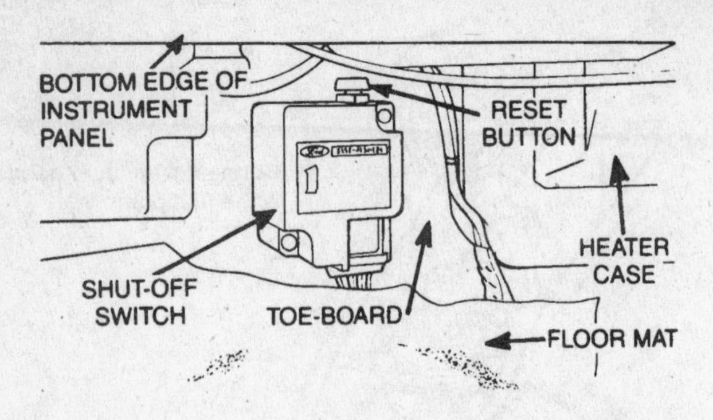

3.44 On most F-Series trucks, the inertia switch is located under the instrument panel to the right of the transmission hump - reset by pushing the button.

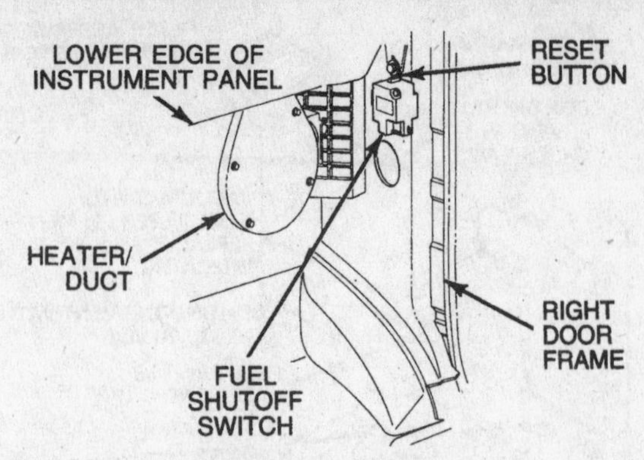

3.45 On most E-Series vans, the inertia switch is located on the right cowl panel forward of the right front door and below the instrument panel

Installation

6 Remove all traces of old gasket material.

7 Insert the fuel pump mounting bolts into the pump and position a new gasket over the bolts.

8 Apply a coat of assembly lube onto the fuel pump lever-to-cam contact surface. Hold the pump in place on the engine and start the bolts by hand to avoid cross-threading. **Note:** *The fuel pump cam must be in its low position during pump installation. If it is difficult to install the pump due to lever interference, turn the crankshaft as described in Step 3 until the pump seats properly.*

9 Tighten the mounting bolts securely.

10 Install the fuel lines. Start the outlet line by hand to avoid cross-threading.

11 Start the engine and check for fuel and oil leaks.

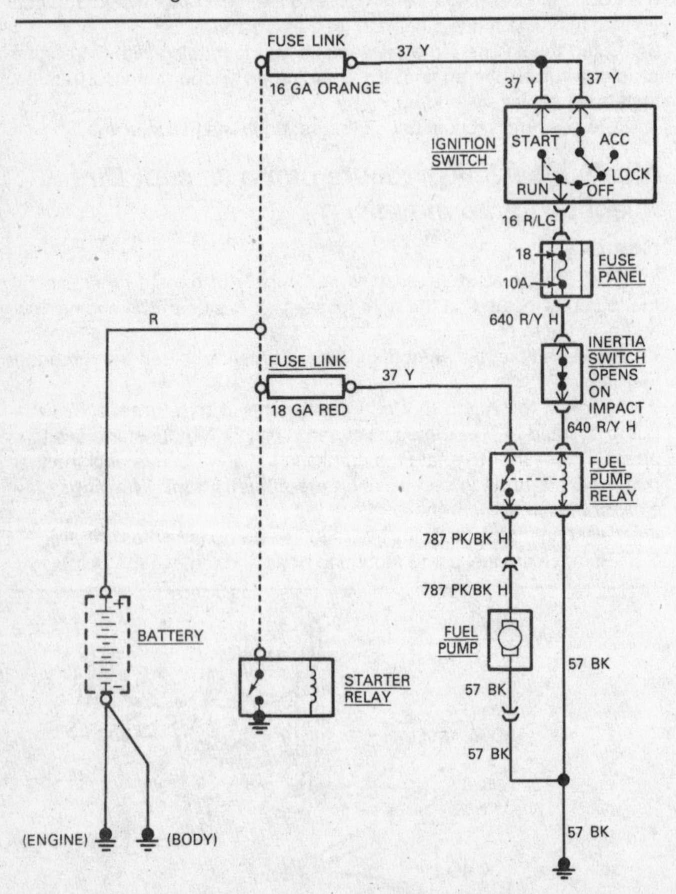

3.46 Typical electric fuel pump schematic wiring diagram

Electric low-pressure pump - general information

Some models use an electric low-pressure pump instead of a mechanical pump. The pump is usually located on the engine mount adjacent to the starter.

When the ignition switch is in the Run or Start position, power flows to the pump relay mounted on the inner fender. When the relay closes, power flows from the batteries through an 18 gauge fusible link at the starter solenoid and then to the pump.

In the event of an accident, the fuel pump inertia switch opens, cutting off power to the pump. If the pump is inoperative, check and, if necessary, reset the fuel pump inertia switch **(see illustrations)**. Be sure to check for fuel leaks before and after resetting the switch. **Note:** *If the switch is not located in the place shown, check the vehicle owner's manual.*

Testing the electric pump

Preliminary inspection

1 Check the fuel line fittings and connections -make sure they're tight. If a fitting is loose, air and/or fuel leaks may occur.

2 Check for bends or kinks in the fuel lines.

3 Start the engine and let it warm up. With the engine idling, conduct the following preliminary inspection before proceeding to the actual fuel pump tests:

a) *Listen to the pump. It should emit a steady purring sound. If it sounds shrill or metallic, it's probably about to fail.*
b) *Look for leaks at the pressure (outlet) side of the pump.*
c) *Look for leaks on the suction (inlet) side of the pump. A leak on the suction side will suck air into the pump and reduce the volume of fuel on the pressure (outlet) side of the pump.*
d) *Inspect the pump itself and the fittings on the pump for leaks. Tighten or replace the fittings as necessary.*

Voltage test

1 First, make sure the pump is getting battery voltage. Detach the pink and black wire **(see illustration)** and, using a test light, make sure the pump is receiving voltage.

2 If the pump isn't receiving voltage, trace the wire back to the main wire harness, locate the problem (open, short, loose connector, etc.), fix it and recheck.

3 If the pump is receiving voltage and has a good ground, but isn't operating, replace the pump.

Output tests

1 If the pump is getting battery voltage and sounds like it's operating properly, check the fuel output.

2 Disconnect the outlet hose from the pump, attach a short section of hose in its place and place the end of the hose in a fuel container.

3 Start the engine briefly and note the quality and quantity of the

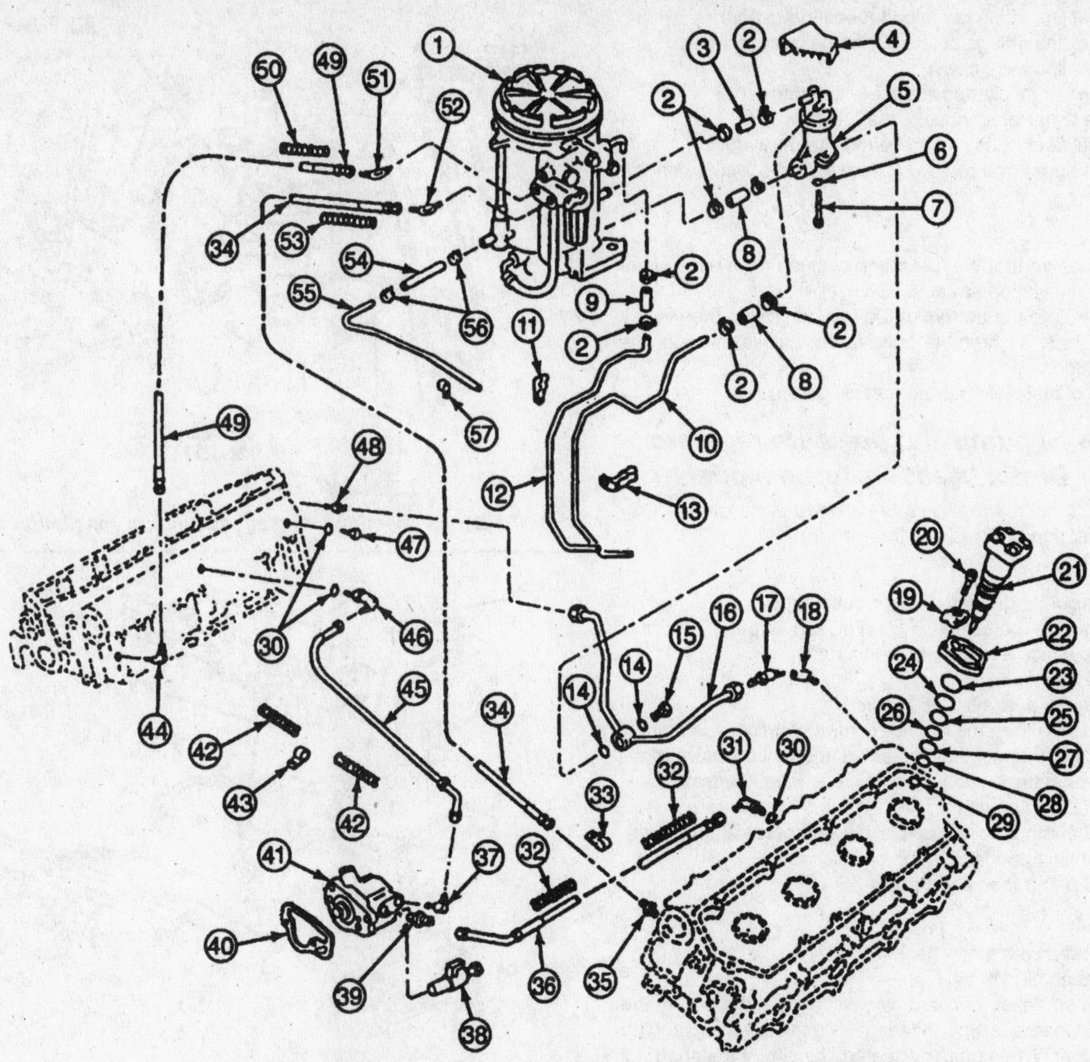

3.47a Fuel system components of the Direct Injection system

1 Fuel filter
2 Clamp
3 Hose
4 Shield
5 Fuel pump
6 Seal
7 Tappet
8 Hose, return
9 Hose, return
10 Tube, supply
11 Clamp
12 Tube, return
13 Clamp
14 Gasket
15 Bolt
16 Tube
17 Elbow
18 Elbow
19 Oil deflector
20 Bolt
21 Injector
22 Clamp
23 Backup ring
24 Cushion ring
25 Injector upper seal
26 Injector middle seal
27 Cushion ring
28 Injector lower seal
29 Gasket
30 O-ring
31 Elbow, high pressure
32 Sleeve
33 Clamp
34 Hose
35 Connector
36 Hose
37 Elbow
38 Pressure regulator
39 Connector
40 Gasket
41 Oil pump, high pressure
42 Sleeve
43 Clamp
44 Elbow
45 Hose
46 Elbow
47 Fuel rail plug with O-ring
48 Connector
49 Hose
50 Sleeve
51 Elbow
52 Elbow
53 Sleeve
51 Elbow
52 Elbow
53 Sleeve
54 Water drain hose
55 Water drain tube
56 Clamp
57 Clip

squirts of fuel being emitted by the pump. They should be distinct and pronounced. If they're dribbling out, the pump is worn out. Replace it. **Warning:** *Wear eye protection, don't spill fuel, and keep a fire extinguisher handy.*

Electric pump - removal and installation

1 Detach the ground cables from the negative battery terminals.
2 Detach the wires from the pump.
3 Disconnect the fuel inlet and outlet lines from the pump and plug the lines to prevent contamination and leakage.
4 Loosen the pump mounting bracket and remove the pump.
5 Installation is the reverse of removal.

Mechanical two stage fuel pump, 1994 and later Direct Injection Turbo models

The Direct Injection system does not use a conventional mechanical fuel injection pump assembly as in previous years. A single supply pump pulls fuel from the tank, routing it through the filter to the injectors **(see illustration)**. The injectors are electronically triggered by the PCM to meter fuel to the engine as required. The fuel pressure regulator is located on the side of the fuel filter assembly and is set to maintain a fuel pressure of 40 psi.

Removal

1 Remove the turbocharger.

2 Remove the fuel line fitting banjo bolt **(see illustration)**.
3 Remove the fuel line fittings at the rear of the cylinder heads.
4 Remove the fuel line assembly.
5 Loosen the three hose clamps at the pump fittings.
6 Disconnect the water drain hose at the fuel filter.
7 Remove the fuel filter bolts and move the filter forward.
8 Unbolt the fuel pump and remove it from it's bore along with the tappet.

Installation

9 Rotate the engine so that the fuel pump cam is on it's base circle (low point) to make installation easier.
10 Place the tappet into the bottom of the fuel pump and then install the pump into the block. Tighten the bolts to the torque listed in this Chapter's Specifications.
11 The remainder of installation is the reverse of removal.

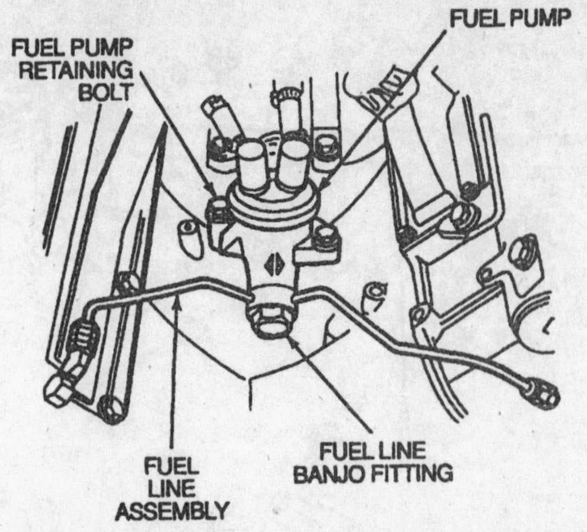

3.47b Direct Injection two stage fuel pump installation details

High pressure oil pump and pressure regulator, 1994 and later Direct Injection Turbo models

Note: *This oil pump does not provide lubrication to the engine. It only operates the fuel injection system* **(see illustration)**.

Removal

1 Remove the turbocharger compressor outlet pipe.
2 Remove the oil from the reservoir using a vacuum pump. There is a hole in the top into which a hose can be inserted.
3 Remove the fuel filter.
4 Remove the supply hoses from the pump.
5 Disconnect the wire from the injector control pressure solenoid.
6 Remove the nut and remove the solenoid and fuel pressure regulator if you are replacing the pump with a new one **(see illustration)**.
7 Remove the oil pump cover from the engine front cover.
8 Remove the oil pump sprocket bolt and washer **(see illustration)**.
9 Remove the retaining bolts from the oil reservoir.
10 Remove the oil pump **(see illustration)**.

Installation

11 Install the oil pump with a new gasket.
12 Install the oil reservoir.
13 Install the drive sprocket, bolt and washer. Tighten the bolt to the torque listed in this Chapter's Specifications. Make sure the sprocket is seated onto the shaft. When properly seated, the end of the shaft will be flush with the edge of the sprocket.
14 Apply a bead of RTV sealant to the cover sealing surface and install the cover.
15 If removed, install the fuel pressure regulator.
16 Install the solenoid and nut.
17 The remainder of installation is the reverse of the removal procure. Refill the reservoir with fresh engine oil.

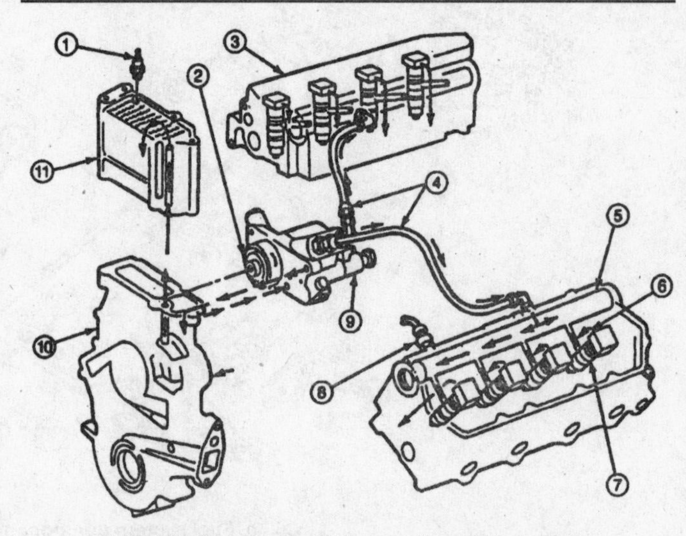

3.47c High pressure oil flow for the 1994 and later Direct Injection system

1 Oil pressure sensor
2 High pressure oil pump
3 Cylinder head
4 Hoses, high pressure
5 Oil rail
6 Oil galleys
7 Injector
8 Pressure sensor
9 Pressure regulator
10 Front cover
11 High pressure oil reservoir

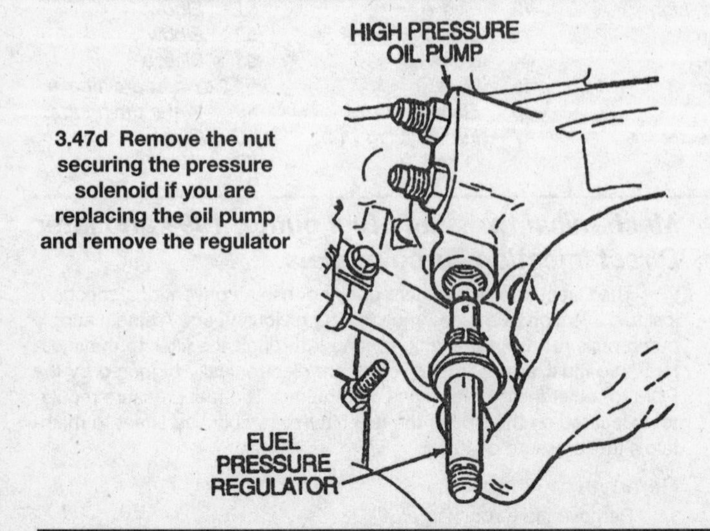

3.47d Remove the nut securing the pressure solenoid if you are replacing the oil pump and remove the regulator

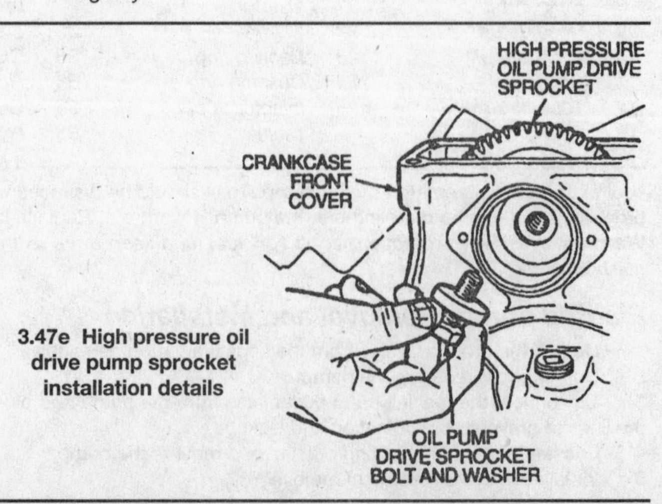

3.47e High pressure oil drive pump sprocket installation details

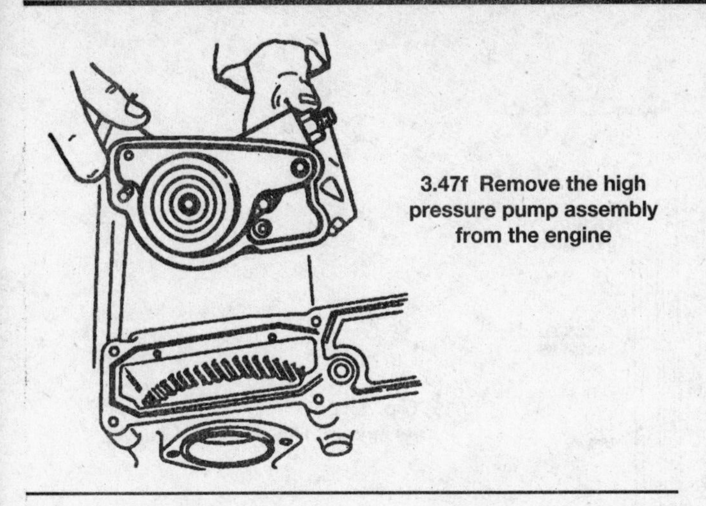

3.47f Remove the high pressure pump assembly from the engine

Injection lines - servicing

Low pressure lines

Special "push connect" fittings are used on some parts of the fuel system. To detach such fittings **(see illustration)**, remove the "hairpin" type clip by first bending the shipping tab down so it will clear the body. Using hands only, spread the two clip legs about 1/8-inch to disengage the body and push the legs into the fitting. Lightly pull from the triangular end of the clip and work it clear of the tube and fitting. Separate the fitting.

To install the fitting, use a new clip. Insert it into any two adjacent openings with the triangular portion pointing away from the fitting opening. Guide the clip with an index finger **(see illustration)** until it locks on the outside.

Some fuel supply and return lines used on these engines have spring lock couplings at some connections **(see illustration)**. The male end of the spring lock coupling, which is girded by two O-rings, is inserted into a female flared end fitting. The coupling is secured by a garter spring which prevents disengagement by gripping the flared end

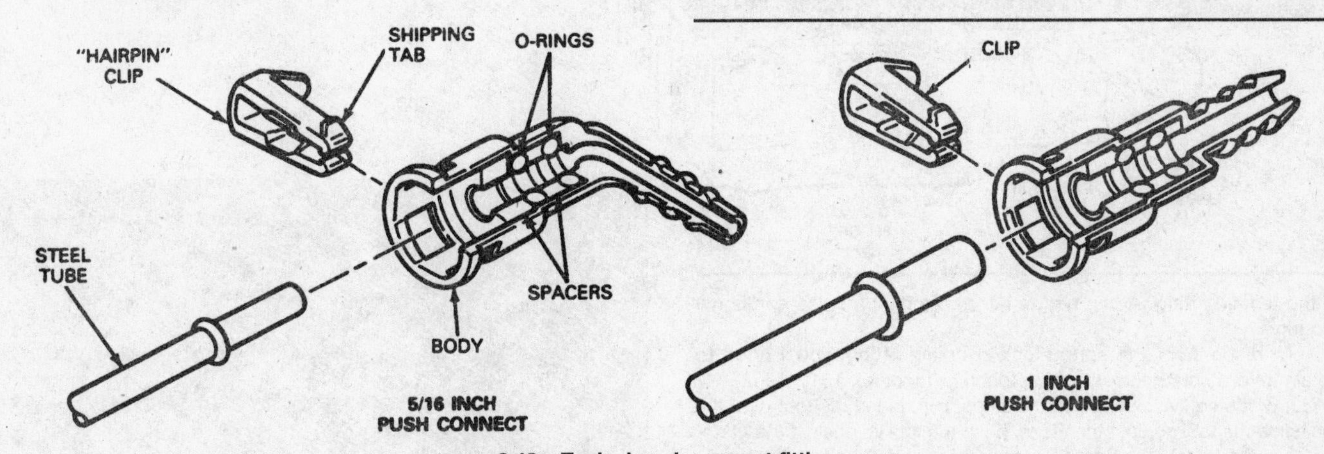

3.48a Typical push connect fittings

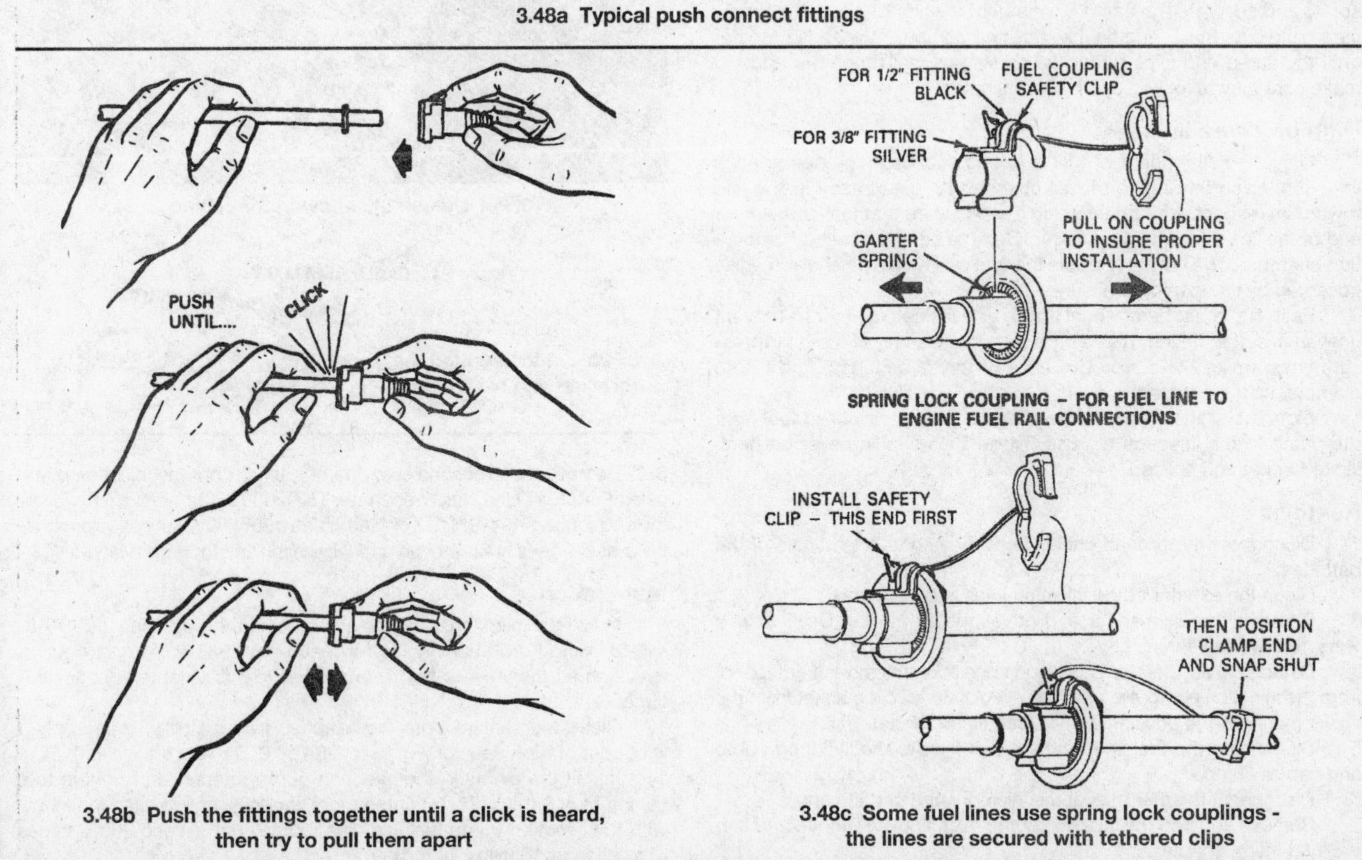

3.48b Push the fittings together until a click is heard, then try to pull them apart

3.48c Some fuel lines use spring lock couplings - the lines are secured with tethered clips

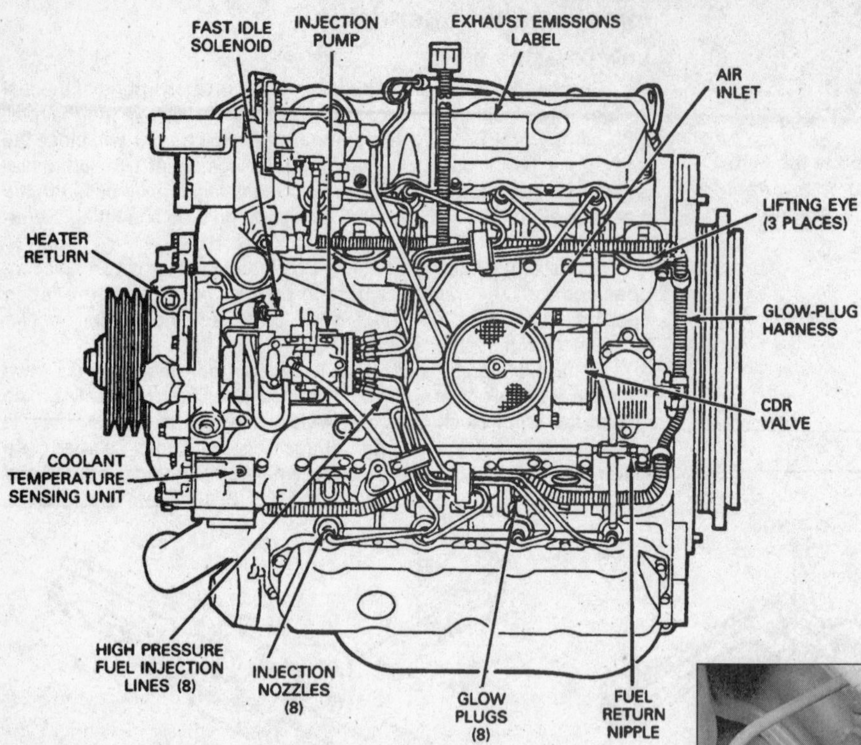

3.49 Top view of engine shows routing of fuel injection lines (non-DI Turbo)

of the female fitting. A clip and tether assembly provides additional security.

To disconnect the spring lock coupling fitting, you'll need to obtain a Ford spring lock coupling tool (Ford tool no. T81P-19623-G1 or G2) or it's equivalent. Unclip the safety clip, place the tool over the line, slide the tool against the fitting, push the tool lip against the garter spring and separate the fitting. To connect the fitting, place the female flared end over the O-ring on the male end and push the ends together until the flared end slips under the garter spring. Tug on the ends to make sure they're locked and install the safety clip.

High pressure lines

Only a small amount of fuel is forced into each line during each injection. Each tiny pulse of fuel pushes the fuel already in the line toward its respective nozzle, forcing a small amount of fuel at the other end of the line through the spring-loaded nozzle and into the combustion chamber. The nozzle restricts the fuel and acts as a shut-off valve governed by its opening pressure.

Each line must be of equal length and inside diameter to prevent uneven injection timing. If you have to replace some of the high pressure lines, make SURE you buy lines of the same length and inside diameter as the originals.

Eight metal lines **(see illustration)** carry the pressurized fuel from the pump to the injection nozzles. These lines should never be bent, kinked or cut and spliced.

Removal

1 Disconnect the ground cables from the negative terminals of the batteries.

2 Clean the exterior of the injection lines and injectors.

3 Remove the air cleaner and cover the air intake. On E-Series vans, remove the engine cover.

4 Detach the accelerator cable (and speed control cable, if equipped) from the injection pump. Remove the accelerator cable bracket from the intake manifold and position it aside **(see illustrations)**.

5 Disconnect the fuel line from the fuel filter to the injection pump and cap all fittings.

6 Disconnect the fuel lines at the injectors and cap the ends.

7 Remove and cap the injection pump inlet elbow. Remove and cap the inlet fitting adapter.

3.50a Detach the accelerator cable . . .

CABLE ASSEMBLY
BRACKET

3.50b . . . then unbolt the bracket and set it aside

8 Remove the injection lines, one at a time, from the injection pump using Ford Fuel Line Nut Wrench no. T83T-9396-A or equivalent. Follow the sequence 5-6-4-8-3-1-7-2 **(see illustration)**. Cap all of the open fittings as the lines are removed and label each line for reinstallation.

Installation

9 Install the injection lines one at a time, in the sequence 2-7-1-3-8-4-6-5. Using Ford Fuel Line Nut Wrench no. T83T-9396-A or equivalent, tighten the lines to the torque listed in this Chapter's specifications.

10 Clean old sealant from the injection pump elbow, then apply a light coat of Pipe Sealant (Ford no. D8AZ-19554-A or equivalent).

11 Install the elbow in the injection pump adapter and tighten to a minimum of 6 ft-lbs. Then tighten more, if necessary, to align the elbow with the injection pump fuel inlet line, but do not exceed 360-degrees of rotation or 10 ft-lbs. of torque.

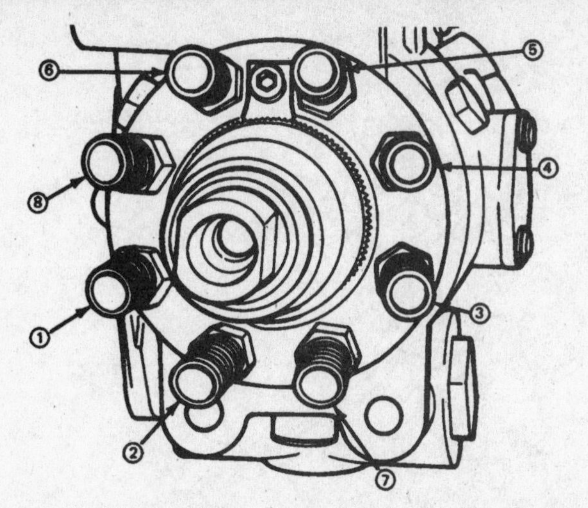

3.51 **Cylinder numbering sequence on the injection pump (non-DI Turbo)**

12 Install the remaining components in the reverse order of removal.
13 Start the engine and check for fuel leaks. If necessary, purge air from the lines by operating the starter while the fittings are loosened at the injectors. Crank the engine until bubble-free fuel comes from the lines.

Fuel injection nozzles

1994 and earlier (except Direct Injection Turbo) models

Special high pressure fuel injection nozzles (injectors) are used in all non-DI Turbo diesel engines covered in this Chapter. The injector nozzle **(see illustration)** sprays fuel when the fuel reaches a predetermined pressure. Unlike a gasoline fuel injector, which sprays fuel into the intake port, the diesel nozzle sprays fuel into a pre-combustion chamber.

As fuel is injected from the nozzle, it's swirled just before it passes around the head of the valve, producing a high-velocity, narrow-cone, atomized spray which promotes more efficient combustion. The nozzle assembly is a precisely-machined assembly, matched and preset at the factory.

Do NOT disassemble fuel injectors! They're assembled to precise tolerances and may not work properly once you've taken them apart.

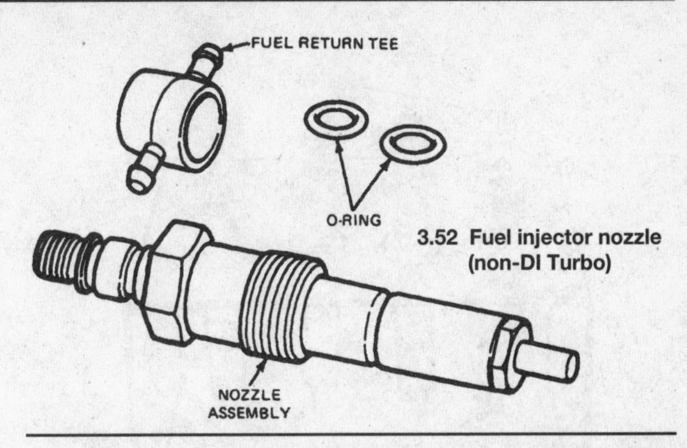

3.52 **Fuel injector nozzle (non-DI Turbo)**

New or rebuilt injectors are available through your Ford dealer or auto parts store. When you replace worn injectors, make sure you get the correct units for your engine. Though they may look identical, all are NOT interchangeable.

1994 and later Direct Injection Turbo models

The Direct Injection model fuel injection nozzles are electronically controlled, hydraulically operated fuel injectors. The Powertrain Control Module (PCM) controls the amount of time the injectors are energized. When energized, high pressure oil forces the piston down, injecting fuel into the combustion chamber. The fuel injectors are located beneath the valve covers, placing the fuel injector tip near the center of the combustion chamber. Do not try to service or disassemble these injectors. If defective, they must be replaced. **Caution:** *Do not pierce any of the wires in the harness. The red striped wires carry 115 volts DC and can cause a severe shock. Do not attempt to install an injector without first removing all fuel and oil from the cylinder. Failure to do so can cause a hydraulic lock in the engine and bend a connecting rod.*

Fuel injection nozzle(s) - removal and installation

1994 and earlier (except Direct Injection Turbo) models

1 Clean the injectors and lines.
2 Remove the injection lines as described previously.
3 Remove the line sensors **(see illustration)** and cap the open fittings. On F-Series trucks, the sensor is on the number one injector line.

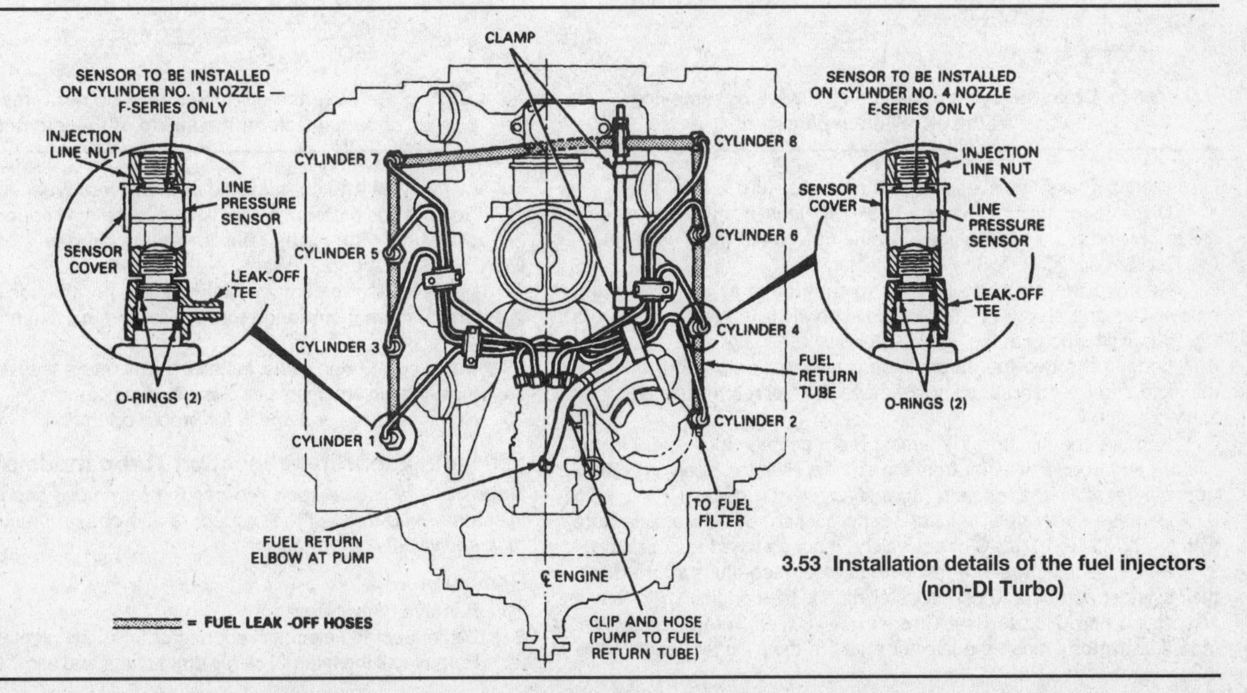

3.53 **Installation details of the fuel injectors (non-DI Turbo)**

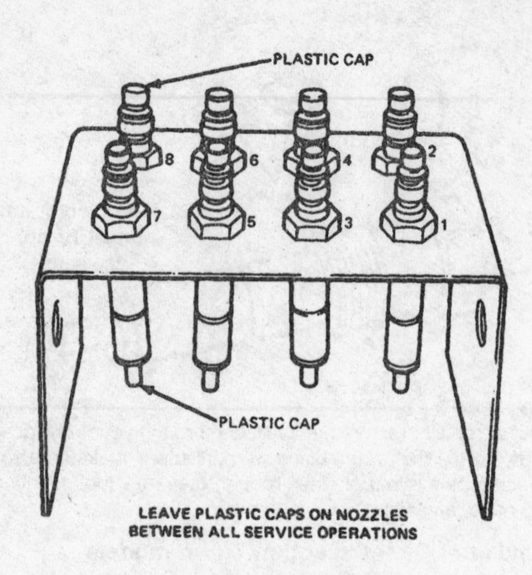

3.54a Store the injectors in order so they can be returned to their original cylinders

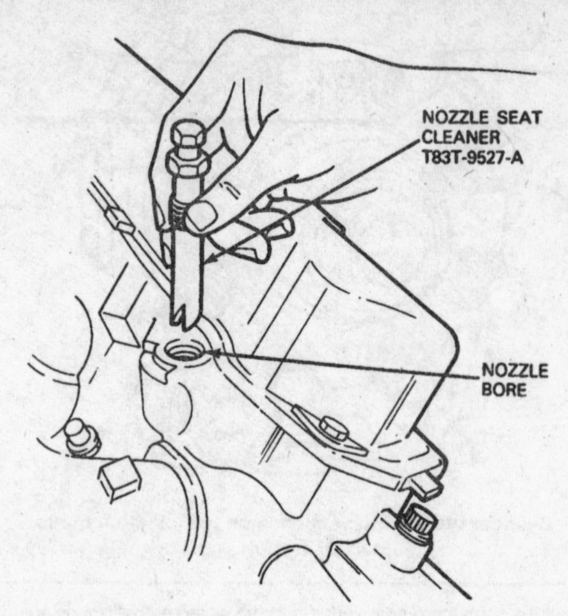

3.54b Clean the injector bores with a nozzle seat cleaner

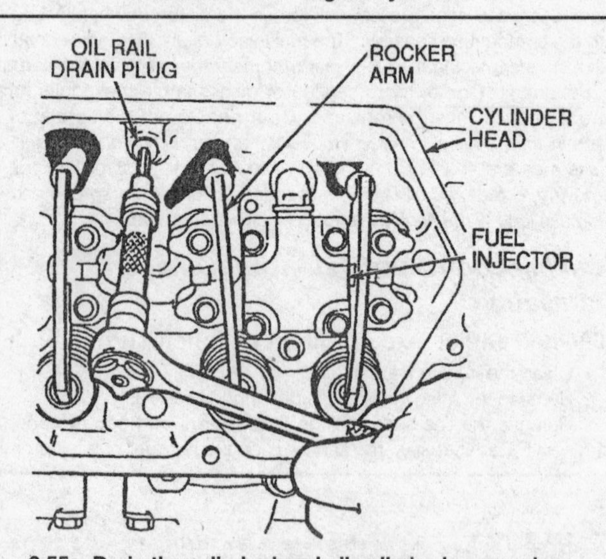

3.55a Drain the cylinder head oil galleries by removing all the oil rail drain plugs

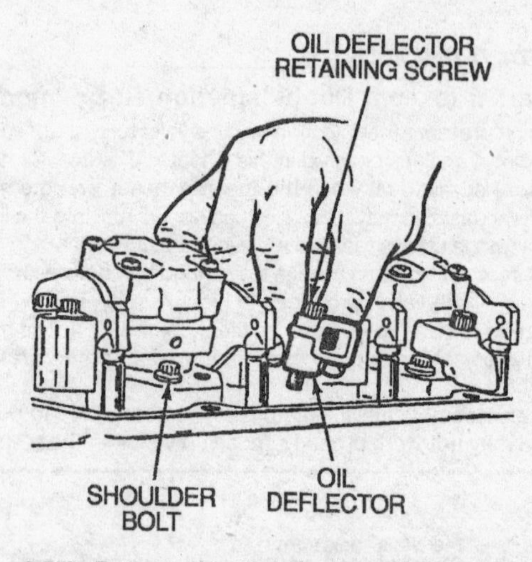

3.55b Remove the oil deflectors - do not remove the shoulder bolt on the inside of the injector

On E-Series trucks, the sensor is on the number four injector line.

4 Disconnect the fuel return (leak-off) tees from each injector assembly and place them aside. Cap the open ends with Ford Protective Cap Set no. T83T-9395-A or equivalent.

5 Remove the pump-to-fuel return tube hose at the fuel return elbow. Cap the elbow at the pump. Disconnect the return (leak-off) tee-to-fuel filter hose at the leak off tee.

6 Loosen the two fuel return tube retaining clamps (at the intake manifold and the lifting bracket). Remove the fuel return lines and tees as an assembly.

7 Remove the injectors by turning them counterclockwise. Lift the injector and copper washer from the engine. Plug the openings. **Caution:** *Do not strike the nozzle tip against any hard surface.*

8 Remove the copper injector nozzle gasket from the nozzle bore with O-ring T71P-19703-C or equivalent, if not attached to nozzle tip.

9 Place the injectors in a fabricated holder **(see illustration).** Mark their cylinder numbers to permit installation in their original cylinders.

10 Using Ford Nozzle Seat Cleaner no. T83T-9527-A or equivalent **(see illustration)**, clean the injector bores in the cylinder heads. Blow out any particles with compressed air. **Warning:** *Wear eye protection.*

11 Remove the protective caps and install a new copper gasket nozzle assembly, with a small dab of multi-purpose grease (Ford no. DOAZ-19584-AA or equivalent).

12 Apply anti-seize compound on the injector threads and install the injectors. Tighten them to the torque listed in this Chapter's specifications.

13 Using new O-ring seals, install the fuel return tees and lines.

14 Install the injector lines as described previously.

15 Run the engine and check for proper operation.

1994 and later Direct Injection Turbo models

Note: *Several special tools are needed to remove and install the fuel injectors. Read through the procedure and obtain the special tools, or their equivalents before beginning.*

Removal

1 Remove the valve cover.

2 Disconnect the electrical connector from the injector.

3 Remove all the internal oil rail drain plugs and allow the oil to drain

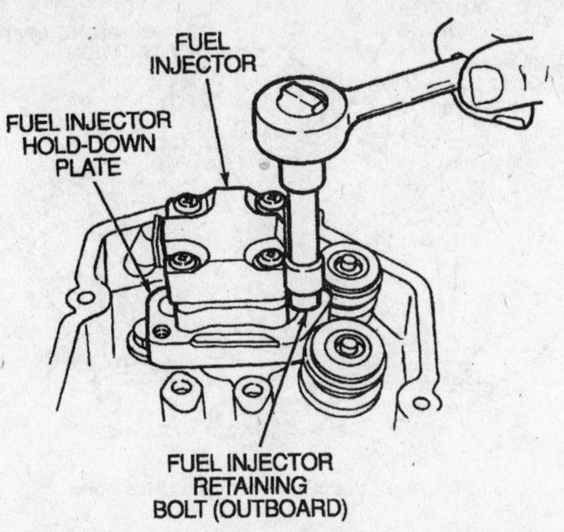

3.55c Remove only the outside injector retaining bolt

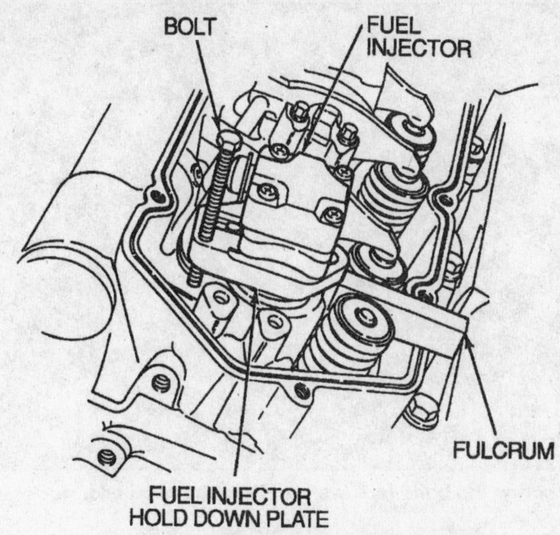

3.55d Use the special tool to pull the injector from the cylinder head

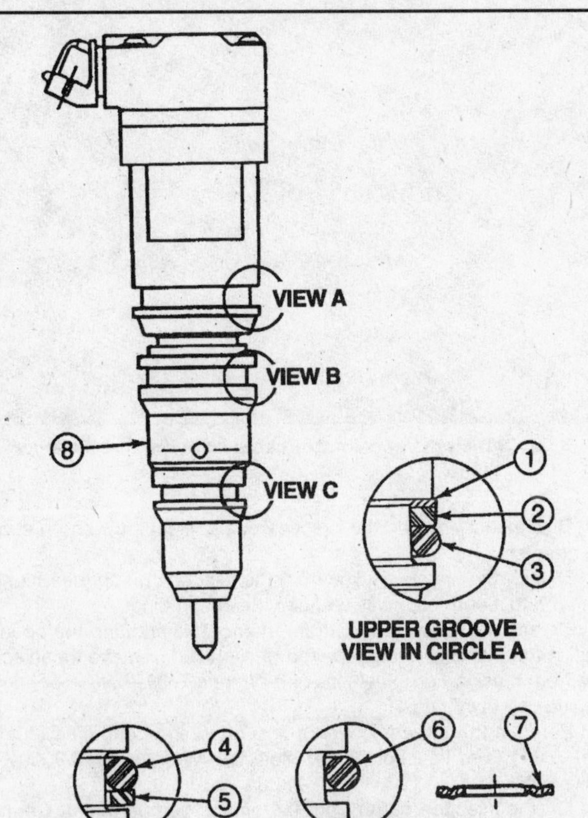

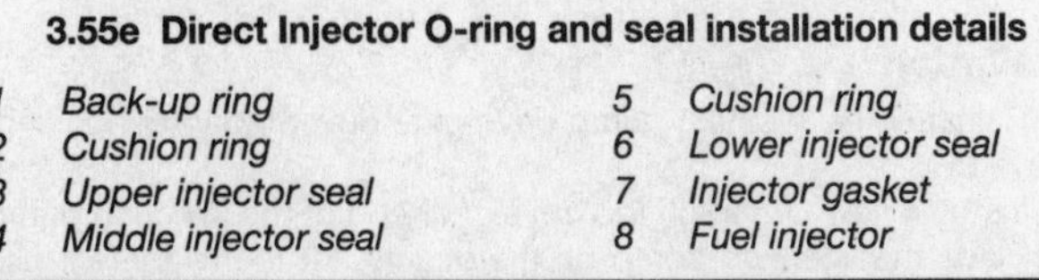

3.55e Direct Injector O-ring and seal installation details

1. *Back-up ring*
2. *Cushion ring*
3. *Upper injector seal*
4. *Middle injector seal*
5. *Cushion ring*
6. *Lower injector seal*
7. *Injector gasket*
8. *Fuel injector*

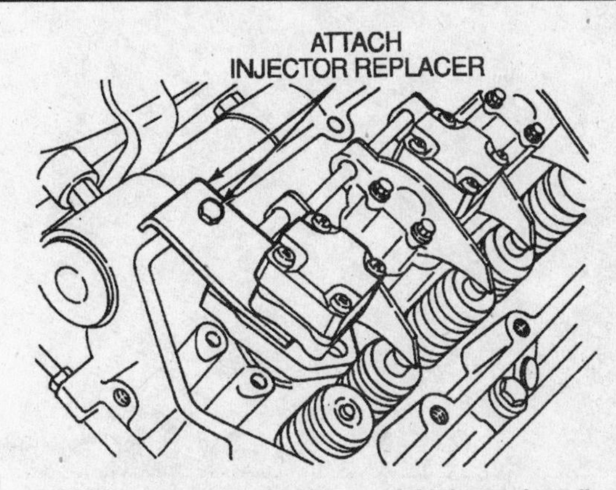

3.56f If necessary, use the special injector installer to press the injector into the bore

from the oil galleries **(see illustration)**. **Caution:** *If the plugs are not removed prior to removing the injectors, oil will enter the combustion chamber and could cause hydrostatic lock and severe engine damage.*

4 Remove the retaining screws and oil deflectors **(see illustration)**.

5 Remove the outboard injector retaining bolt **(see illustration)**.

6 Remove the injector using Ford special tool no. T94-9000-AH1 **(see illustration)**. Retrieve all three O-rings and the copper washer from the bore. **Note:** *It may be necessary to remove the outer half of the heater case for access to the cylinder no. 4 injector.*

7 Mark the injectors, cap the ends and store them in a holder so they can be kept clean and returned to their original locations.

8 Normally the injector sleeves do not require removal. There are several special tools required to extract the injector sleeves from the cylinder heads. A dealership should be able to handle this operation if it is necessary.

9 Be sure to seal the injector bores in the cylinder head to prevent debris from entering.

Installation

10 Carefully clean the injector sleeve with a bottle brush and wipe it with a clean cloth to ensure a good O-ring seal.

11 Install new O-rings, cushion rings and a new copper washer as shown **(see illustration)**. Apply engine oil to the O-rings and insert the injector into the cylinder head. Press the injector in by hand until injector hold-down plate is under the head of the shoulder bolt. Use Ford special tool no. T94T-9000-AH2 to seat the injector **(see illustration)**. Do not strike the top of the injector.

12 Install the injector retaining bolt, the oil deflector and the oil rail drain plugs. Tighten the bolts to the torque listed in this Chapter's Specifications.

13 The remainder of installation is the reverse of removal.

3.56 Unscrew the bolts (arrows) and detach the oil filler neck

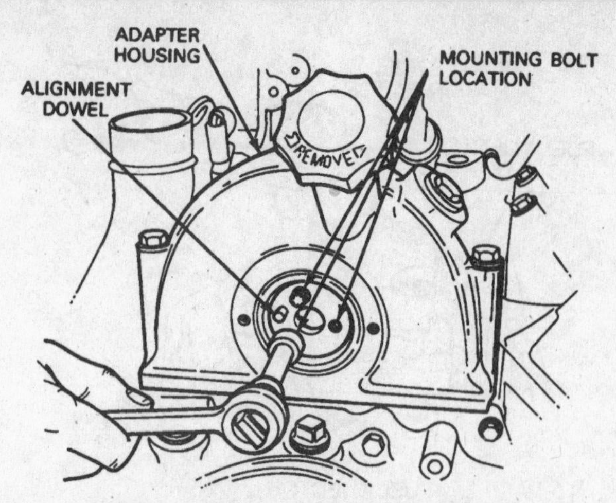

3.57 Unbolt the pump from the drive gear

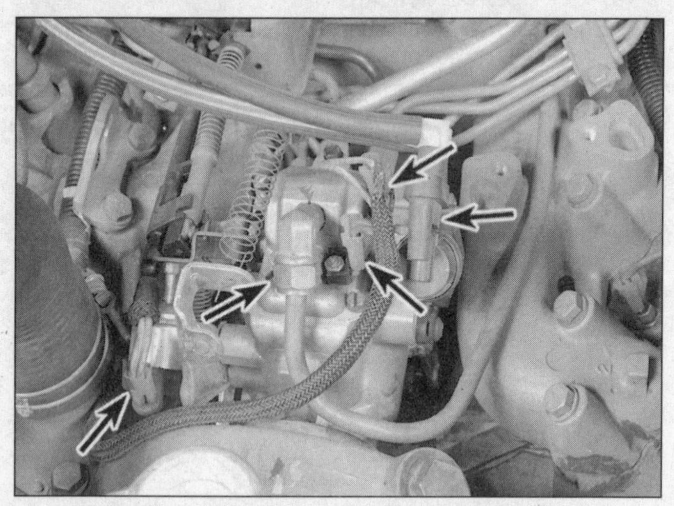

3.58 Label and detach the hoses and electrical connectors from the pump

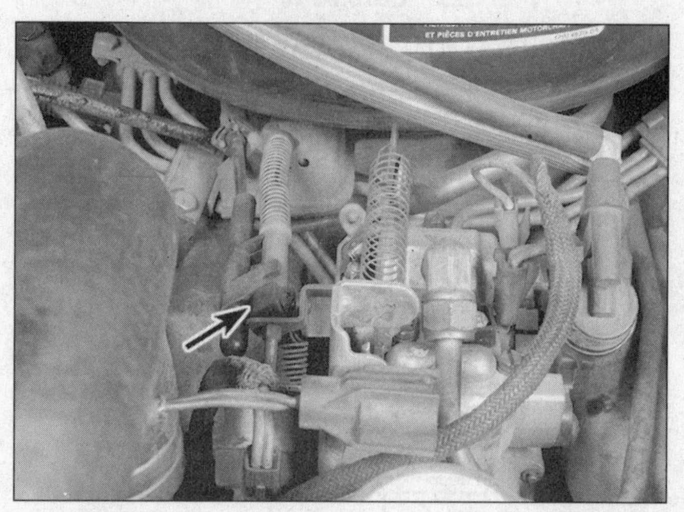

3.59 Detach the accelerator cable from the throttle lover

Injection pump, 1994 and earlier (except Direct Injection Turbo) models - removal and installation

Removal

1 Allow the engine to cool completely, then clean the exterior of the injection pump. **Caution:** *Do not clean the engine when it is still warm.*

2 Disconnect the ground cables from the negative terminals of the batteries.

3 On "E-Series" trucks, remove the engine cover.

4 Remove the engine oil filler neck **(see illustration)**.

5 Remove the injection pump-to-drive gear bolts **(see illustration)**.

6 Detach the electrical connectors from the injection pump **(see illustration)**.

7 Disconnect the accelerator cable (and speed control cable, if equipped) from the throttle lever **(see illustration)**.

8 Remove the air cleaner and cover the opening with Ford Intake Manifold Cover no. T83T-9424-A or a cloth.

9 Detach the accelerator cable bracket, with cables, from the intake manifold and set it aside **(see illustration)**.

10 On "E-Series" trucks, disconnect the fuel inlet and return lines from the fuel filter. Then unbolt the fuel filter bracket and remove the filter assembly. **Caution:** *Cap all the open lines and fittings with Ford Fuel System Protective Cap Set T83T-9395-A or equivalent.*

11 Remove the fuel filter-to-injection pump fuel line and cap the fittings.

12 Disconnect and cap the injection pump inlet elbow and pump fitting adapter.

13 Detach the fuel return line from the injection pump **(see illustration)**, rotate it out of the way and cap the open fittings.

14 Detach the injection lines from the nozzles and cap the open fittings. If the pump is being repaired or replaced, remove the injection lines from it using Ford Fuel Line Nut Wrench T83T-9396-A or equivalent and cap the fittings.

15 Remove the three injection pump attaching bolts **(see illustration)**, using Ford Injection Pump Mounting Wrench T86T-9000-C or equivalent.

16 Lift the injection pump from the engine compartment. **Caution:** *Do not lift or carry the injection pump by the lines.*

17 Unbolt the injection pump drive gear cover (adapter housing), if necessary. Refer to Timing gears -removal, inspection and installation in this Chapter.

Installation

18 Reinstall the injection pump drive gear cover (adapter housing), if removed.

19 Install a new O-ring on the drive gear end of the injection pump.

20 Place the injection pump on the engine. Fit the alignment dowel on the injection pump into the alignment hole on the drive gear **(see illustration in Step 5)**.

21 Install the three drive gear-to-injection pump bolts and tighten to the torque listed in this Chapter's Specifications.

22 Install the injection pump-to-adapter nuts. Align the scribe lines

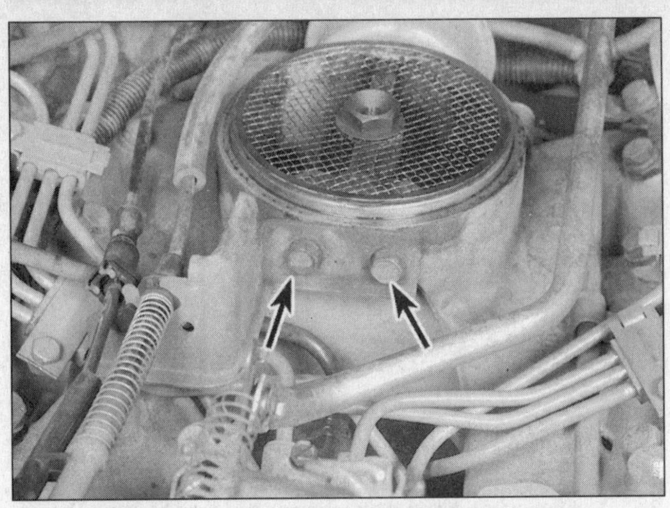

3.60 Unbolt the accelerator cable bracket (arrows)

3.61 Disconnect the fuel return line from the elbow fitting

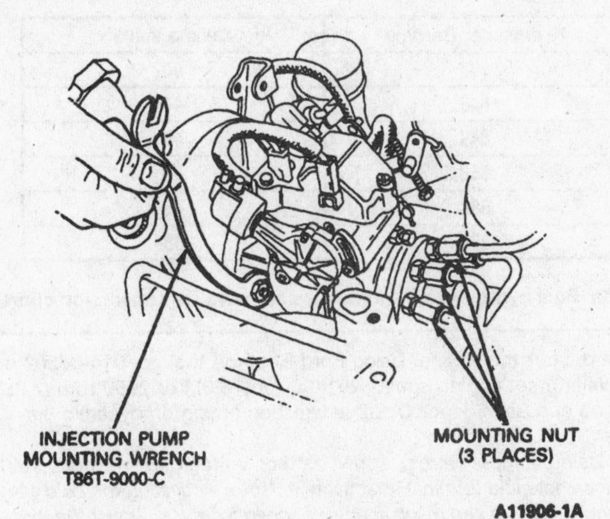

3.62 A special wrench may be required to loosen the pump mounting bolts

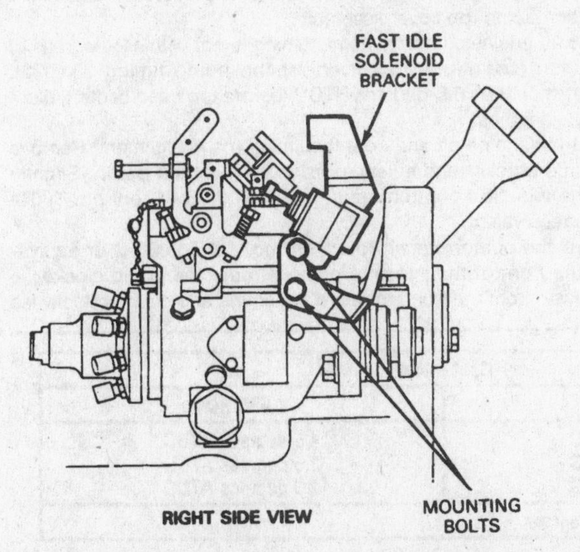

3.63 Fast idle solenoid details

on the injection pump flange and injection pump adapter and tighten to the torque listed in this Chapter's Specifications.

23 Install the fuel lines. When tightening the fuel line-to-nozzle fittings, use Ford Fuel Line Wrench T83T-9396-A or equivalent. Tighten them to the torque listed in this Chapter's Specifications.

24 Install the injection pump fitting adapter with a new O-ring.

25 Thoroughly clean the injection pump elbow threads. Start the elbow into the pump adapter, then coat the threads with Ford Pipe Sealant D8AZ-19554-A or equivalent.

26 Tighten the elbow to a minimum of 6 ft-lb., then tighten further, if necessary, until the elbow aligns with the fuel inlet line on the pump. Do not exceed one full turn.

27 Install the remaining components in the reverse order of removal. Refer to the *Timing gears - removal, inspection and installation* in this Chapter when installing the camshaft timing gear cover (pump adapter).

28 If necessary, purge the injector lines of air by loosening the connectors at the injectors one turn, then operate the starter until the bubbles stop flowing from the fitting. Tighten the fittings, then start the engine and check for oil leaks. **Warning:** *Avoid the high pressure fuel spray. Wear gloves and eye protection - fuel under pressure can penetrate skin.*

29 Check and adjust injection pump timing as described in the Section entitled *Injection pump - timing.*

Injection pump, 1994 and earlier (except Direct Injection Turbo) models - repairs

The fuel injection high pressure pump is a precision instrument with extremely small clearances and requires expensive equipment to test and calibrate it. Numerous special tools and procedures are necessary to service the pump and internal repairs should be performed by trained personnel using the proper equipment.

Check with your authorized Ford dealer or a Stanadyne Roosa-Master repair station for injection pump calibration and repairs.

Injection pump, 1994 and earlier (except Direct Injection Turbo) models - timing

Note: *The pump must be static-timed to get the engine running after assembly. Once the engine is assembled and running, the engine should be timed dynamically (with the engine running).*

Static timing

1 Remove the fast idle bracket and solenoid from the injection pump **(see illustration).**

2 Using Ford Injection Pump Mounting Wrench no. T86T-9000-C (see the injection pump removal procedure) or equivalent, slightly loosen the three nuts that attach the injection pump to the engine.

3.64 Ford tool no. T83T-9000-C injection pump rotating tool

3 Using Ford Pump Rotating Tool no. T83T-9000-C (or equivalent) **(see illustration)**, turn the injection pump until the timing mark on the injection pump mounting flange aligns with the mark on the pump gear housing **(see illustration)** within 0.030 inch.

4 Remove the tool and tighten the pump mounting nuts securely. Recheck the timing marks to verify they are still aligned.

5 Install the fast idle bracket and solenoid and tighten the bolts securely.

Dynamic timing

Note: *Diesel engines are very sensitive to slight changes in timing, and dynamic timing is more accurate than static timing. Although the dynamic timing procedure is not extremely difficult, the procedure requires several expensive tools. If you don't have ready access to these items, we recommend you take the vehicle to a dealer service department or diesel specialist for dynamic timing.*

1 Obtain a fuel sample and check the cetane value using Ford Rotunda Tool no. 078-00121 supplied with Ford Rotunda Tool no. 078-00200 or equivalent (6.9L engines only). Cetane value is determined by measuring the specific gravity (density) of the fuel with the fuel temperature between 75 and 95-degrees F. Poor quality fuel can affect diesel engine performance. Check the cetane value as follows:

a) *Fill the hydrometer container with fuel until the hydrometer floats.*
b) *Gently spin the unit to break surface tension. Read the number at the lowest point of the fuel level in the hydrometer. Record the number and refer to the cetane value conversion chart* **(see illustration).**

2 Run the engine to bring it up to normal operating temperature (192 to 212-degrees F). **Note:** *Temperature is critical to obtain accurate timing.*

3 Stop the engine and install a Dynamic Timing Meter (Ford Rotunda no. 078-00100 or equivalent for 6.9L engines, or no. 078-0020 for 7.3L engines), by placing the magnetic pickup in the hole on the timing pointer scale. Insert the pickup until it almost touches the vibration damper. **Note:** *To ensure accurate readings, clean the damper and remove any rust.*

4 On 6.9L engines, disconnect the glow plug from the number one cylinder using Ford tool no. D83T-6002-A or equivalent. Install the luminosity probe and tighten it to 12 foot-lbs. Install the photocell over the probe.

5 On 7.3L engines, connect the clamp of the Timing Meter Adapter (no. 078-00201, or equivalent) to the pressure line sensor on the injector nozzle (no. 1 on F-series, no. 4 on E-series).

6 Connect the dynamic timing meter to the battery and dial in 20-degrees of offset on the meter. Unplug the cold start advance solenoid connector from the solenoid terminal. **Caution:** *Keep the wiring away from the drivebelts.*

7 With the transmission in Neutral and the rear wheels raised off the ground, start the engine. Using Ford Rotunda tool no. 014-00302 or equivalent, set engine speed to 1400 rpm (6.9L) or 2000 rpm (7.3L) with no accessory load. Observe injection timing on dynamic timing meter.

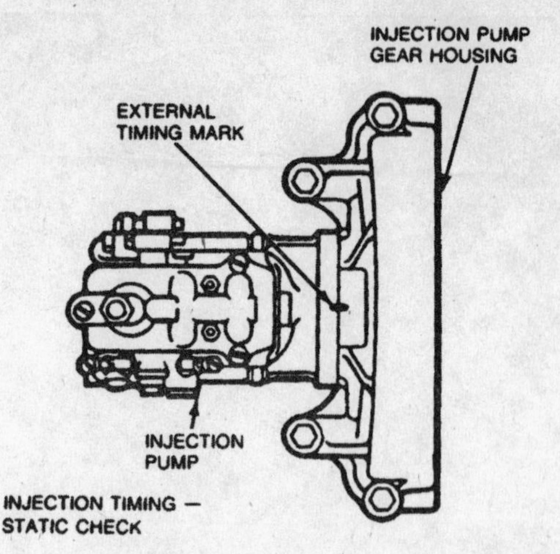

3.65 The timing marks on the pump and the gear housing should be aligned

Hydrometer Reading	Cetane Value
.837	50
.846	47
.849	46
.858	43
.862	42
.876	38

3.66a Fuel hydrometer reading-to-cetane value conversion chart

8 Using jumper wires, apply battery voltage to the cold start advance solenoid terminal to activate it. **Note:** *Activating the cold start advance solenoid can result in engine speed increase. Adjust the throttle control to maintain 1400 rpm (6.9L) or 2000 (7.3L) (if necessary).*

9 Check the timing. The timing should be advanced at least 2.5-degrees on 6.9L engines or 1-degree on 7.3L engines before the timing obtained in Step 7. If the advance is less than 2.5-degrees, replace the fuel injection pump top cover assembly.

10 On 6.9L engines, if the dynamic timing is not within 2-degrees of specification **(see illustration)**, adjust the pump timing. On 7.3L engines, it should be 8.5-degrees BTDC (before top dead center), plus-or-minus 2-degrees.

11 Shut the engine off and note the timing mark alignment. Remove the fast idle bracket and solenoid from the injection pump. Slightly loosen the injection pump mounting nuts with Ford tool no. T86T-9000-C or equivalent.

12 Install the pump rotating tool (Ford no. T83T-9000-C or equivalent) on the front of the injection pump. Rotate the pump clockwise (when viewed from the front) to retard the timing and counterclockwise

Fuel Cetane Value	Calibration	
	4-68J ROO	4-68X ROO
38-42 43-46 47 or greater	3.5 degrees ATDC 2.5 degrees ATDC 1.5 degrees ATDC	4.5 degrees ATDC 3.5 degrees ATDC 2.5 degrees ATDC
*Installation or resetting tolerance for dynamic timing is ± 1 degree. Service limit is ± 2 degrees.		

3.66b Adjust the timing to the fuel cetane value (6.9L engines only)

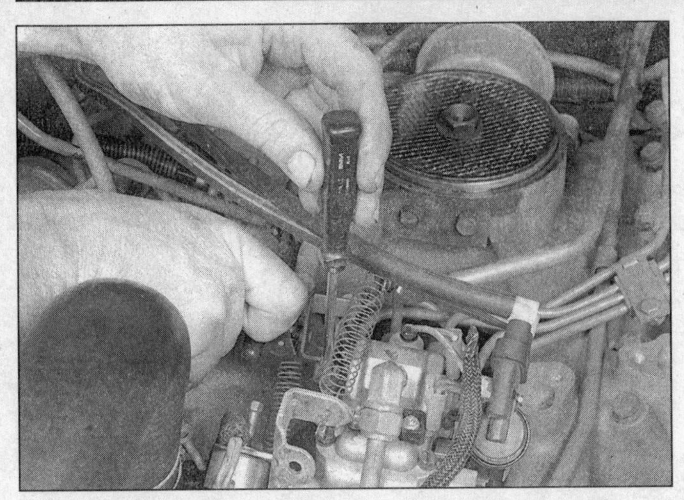

3.67 Turn the screw clockwise to increase idle speed

to advance it. If necessary, tap the tool lightly with a soft-face hammer if the pump sticks. **Note:** *Two degrees of dynamic timing is approximately 0.030 inch of timing mark movement.*

13 Remove the rotating tool and tighten the pump mounting nuts securely. Start the engine and recheck the timing. Repeat the procedure as necessary until timing is within one degree of Specification.

14 Turn the engine OFF and remove the test equipment. Install the glow plugs as described in the Electrical System portion of this Chapter.

15 Run the engine and check for power and smoothness.

Idle speed - 1994 and earlier (except Direct Injection Turbo) models

Warm engine idle speed adjustment

1 Warm the engine up to normal operating temperature and shut it off. Set the parking brake and block the wheels to prevent the vehicle from moving.

2 Remove the air cleaner. **Caution:** *Do not drop anything into the air intake.*

3 Connect a special diesel tachometer (Ford Rotunda 055-00108 or equivalent) to the engine.

4 Ensure that the warm (curb) idle speed adjusting screw is against its stop **(see illustration)**. If not, correct the throttle linkage as necessary.

5 Start the engine in Neutral and measure idle speed. Compare the reading to that listed on the VECI label under the hood. If the VECI label is missing, refer to the idle speed listed in this Chapter's Specifications.

6 Set the idle speed (if necessary) by turning the adjusting screw **(see illustration in Step 4)**. Turning the screw clockwise increases the idle speed and counterclockwise decreases it. **Note:** *Some models have a locknut on the adjusting screw.*

7 Remove the tachometer and reinstall the air cleaner.

Cold engine idle speed adjustment

1 Warm the engine up to normal operating temperature and shut it off. Set the parking brake and block the wheels to prevent the vehicle from moving.

2 Remove the air cleaner. **Caution:** *Do not drop anything into the air intake.*

3 Connect a special diesel tachometer (Ford Rotunda 055-00108 or equivalent) to the engine.

4 Ensure that the warm (curb) idle speed is adjusted properly as described above.

5 Disconnect the wire from the cold (fast) idle speed solenoid.

6 Using a jumper wire, apply battery voltage (+) to the solenoid terminal. Momentarily rev the engine to set the solenoid.

7 Check the idle speed on the tachometer. Compare the reading to that listed in this Chapter's Specifications. Adjust as necessary by turning the solenoid plunger in or out.

8 Recheck idle speed, then disconnect the tester and install the air cleaner.

Turbocharger (1994 and later Direct Injection Turbo models)

General information

The turbocharger pressurizes the air entering the combustion chamber by using an exhaust gas-driven turbine. Compressing more air into the combustion chamber increases power output, improves fuel efficiency and performance at high altitudes. The turbocharger identification number is located on an ID plate riveted to the housing. This number should be referred to if replacement is required. **Note:** *The turbocharger is normally covered by the Federally mandated emissions warranty. Check with a dealer service department concerning coverage.*

Check

Warning: *All checks must be made with the engine off and cool to the touch or personal injury could result. Operating the engine without the turbocharger ducts installed is dangerous and may result in damage to the turbine wheel.*

1 The integrity of the air intake and exhaust system must be maintained at all times for the turbocharger to perform properly. If the engine lacks power, is smoking excessively or is noisy, check for a restricted air filter, collapsed or leaking intake air ducts and a leaking exhaust system.

2 If the turbocharger is suspected of failure, remove the air intake duct and the exhaust outlet pipe so you can get a clear view of the compressor and turbine wheels. Spin the wheels by hand, the wheels should spin freely and not contact the housing. Spin the wheels while pushing in from each end, again the wheels should not contact the housing.

3 If necessary, remove the backpressure control valve, mount a dial indicator to the housing and check the shaft endplay. Endplay should be 0.0008 to 0.004-inch.

4 A small amount of oil residue in the compressor inlet is normal. Excessive oil in the compressor inlet may indicate a problem with the crankcase breather system or an internal engine failure. Excessive oil in the turbine inlet, but not the compressor outlet indicates a turbocharger seal failure or a oil drain-back problem. This condition is usually accompanied by large amounts of blue-gray smoke.

5 The turbocharger requires a pressurized supply of oil to the compressor and turbine wheel bearings to operate properly. Always check the oil supply and drain passages in the pedestal assembly if the turbocharger has experienced a bearing or seal failure.

Removal and installation

1 Disconnect the cables from both negative battery terminals.

2 Remove the air cleaner and intake air ducts from the turbocharger.

3 Remove the exhaust outlet pipe clamp from the turbocharger.

4 Raise the vehicle and support it securely on jackstands.

5 Disconnect the exhaust outlet pipe from the catalytic converter and the transmission brace.

6 Loosen (but do not remove) the left and right exhaust manifold-to-turbo inlet pipe bolts.

7 Remove the lower bolts from the left and right side inlet pipes at the compressor manifold.

8 Lower the vehicle and remove the two upper bolts form the left and right side inlet pipes at the compressor manifold.

9 Remove the compressor manifold hose clamps and disconnect the manifold from the turbocharger. Remove the right engine lift hook, if necessary.

10 Remove the bolts retaining the pedestal to the engine block, disconnect any electrical connectors and remove the turbocharger assembly. Cap the open passages in the engine block while the turbocharger is off.

11 Replace the oil supply and drain passage O-rings and install the turbocharger assembly by reversing the removal procedure.

3.68 Turbocharger and related components (Direct Injection Turbo models)

1. *Turbocharger assembly*
2. *Exhaust turbo inlet adapter gasket*
3. *Compressor duct elbow clamp*
4. *Compressor duct elbow*
5. *Turbocharger compressor inlet pipe*
6. *Compressor inlet duct tube*
7. *Compressor inlet duct tube screw*
8. *Breather hose elbow clamp*
9. *Breather hose elbow*
10. *Air inlet bracket*
11. *Turbocharger exhaust inlet adapter*
12. *Turbocharger oil supply O-ring*
13. *Turbocharger oil drain O-ring*
14. *Pedestal assembly*
15. *Turbocharger manifold hose clamp*
16. *Turbocharger manifold hose*
17. *Compressor manifold*
18. *Turbocharger compressor outlet seal*
19. *Turbocharger compressor outlet clamp*

3.69 The control switch is located near the rear of the intake manifold

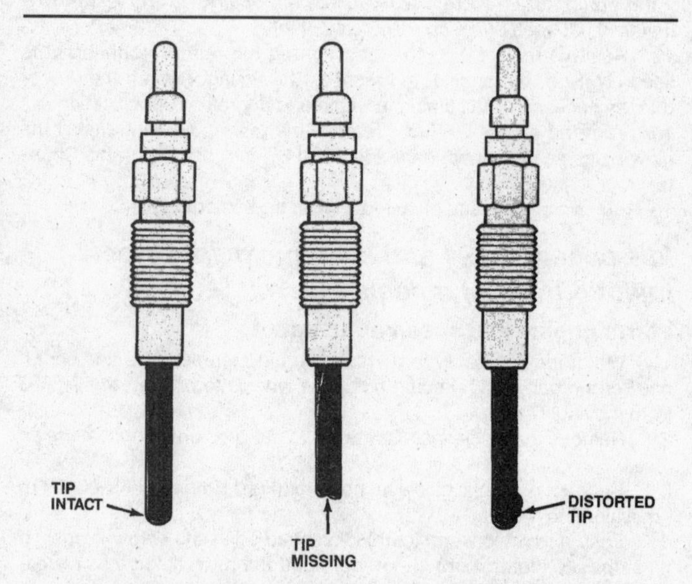

3.70 Glow plug tips tell a lot about operating conditions

Electrical system

General information and precautions

The electrical requirements of diesel engines are quite different from gasoline engines. Diesels need extra current to operate the glow plugs and a heavy duty starter. But once the engine is running, very little power is needed by the engine.

In order to provide the power for increased starting loads, dual batteries are installed on diesel powered models. The charging and starting systems are basically the same as gasoline powered models, although the components are slightly larger.

Always observe the following precautions when working on the electrical system:

a) Be extremely careful when servicing devices with solid-state components. They are easily damaged if checked, connected or handled improperly.

b) Never leave the ignition ON for long periods of time with the engine OFF.

c) Always disconnect the negative cables first and connect them last or the battery may be shorted by the tool being used to loosen the clamp bolts.

Glow plug system - diagnosis

General information

Ford diesel engines utilize an electrical glow plug system to preheat the air inside the pre-combustion chamber to aid in the starting of the engine.

On 6.9 liter models through 1986, the system consists of eight glow plugs (one for each cylinder), a control switch, power relay and after glow relay. This includes a wait lamp latching relay, wait lamp and a wiring harness, which incorporates eight fusible links (one for each glow plug) located between the wiring harness and the glow plug terminals. The system is activated when the ignition switch is turned to the ON position.

On 1987 6.9 liter models and all 7.3 liter (non-DI) models, the glow plug system consists of an engine-mounted solid state glow plug controller, a glow plug harness assembly and eight Positive Temperature Coefficient (PTC) glow plugs (one for each cylinder). The power relay is mounted above the electronic control module and the entire unit is protected by a plastic cover. The whole assembly is located on the rear of the intake manifold.

Operation of the glow plug system is completely automatic. Never bypass the timed pulse function of the glow plug system. A constant 12 volts to the glow plugs will cause them to overheat and fail, possibly resulting in costly engine repair.

If, after diagnosis procedures are followed, you determine that a component must be replaced, the glow plugs should be disconnected until the system has been re-checked with a test lamp for correct operation (as described later in this Section.

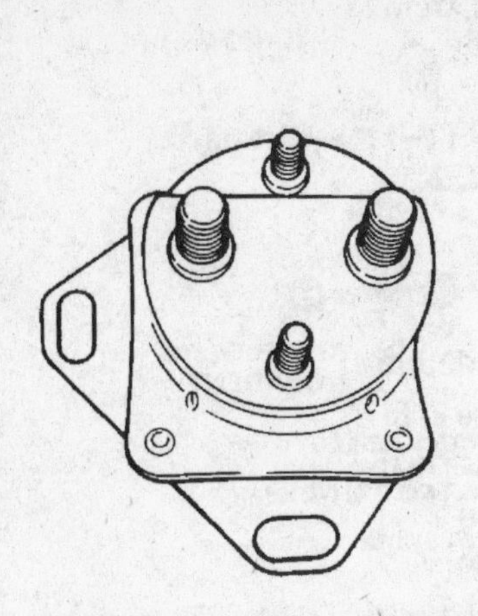

3.71 The power relay is located on the right inner fender

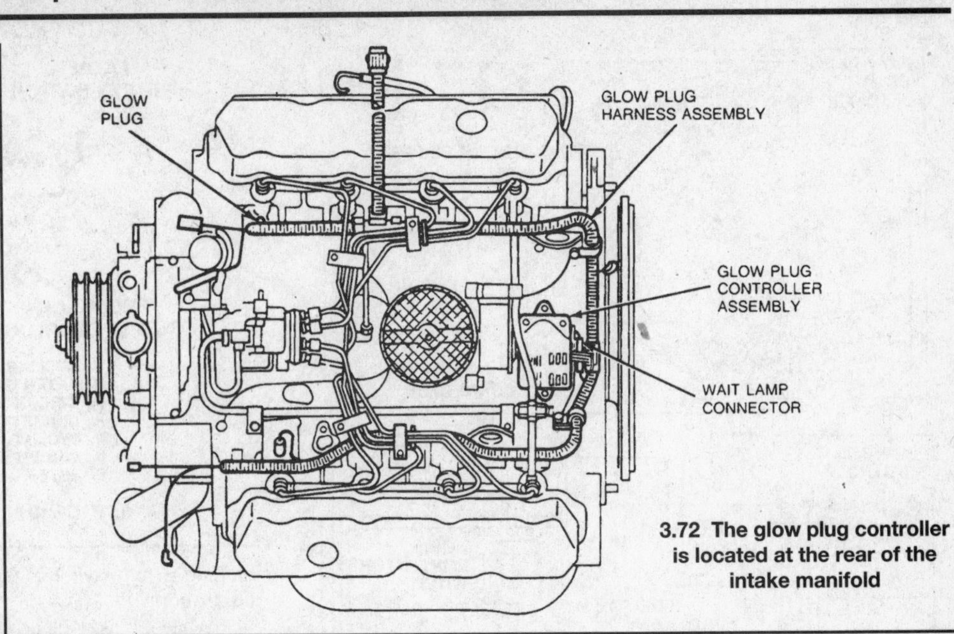

3.72 The glow plug controller is located at the rear of the intake manifold

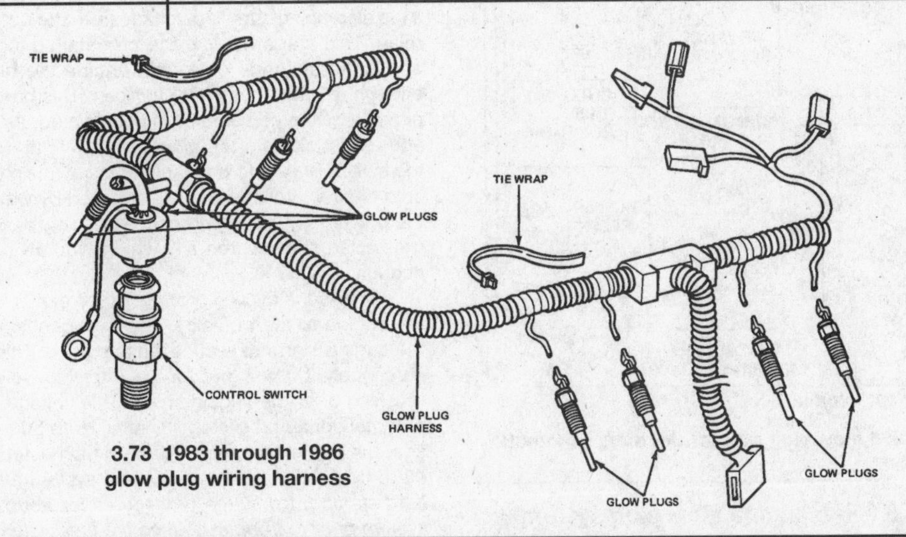

3.73 1983 through 1986 glow plug wiring harness

Component description

After-glow relay (pre-1987 6.9 liter only) - activated by the alternator after the engine has started, this relay sends power to the glow plugs for a short time to reduce start-up smoke.

Control Switch (pre-1987 6.9 liter only) - The control switch is located near the rear of the engine **(see illustration)**. It senses the temperature of the engine coolant and determines the length of time that the glow plugs will be on. The length of time varies between 4 to 10 seconds.

Glow plugs - The glow plugs are designed to heat up extremely fast. When 12-volts are applied to the glow plug, rapid heating of the glow plug tip occurs, which in turn heats up the air inside the pre-chamber. As the fuel is injected into the pre-chamber, it is ignited. The rapid burning of the fuel forces it through the small opening in the pre-chamber into the cylinder where it mixes with more hot air and a second combustion takes place. To prevent overheating of the glow plug, a cycling device is used in the circuit. **Note:** *If the tip is missing from a glow plug, the pump timing or fuel quality is probably incorrect. If the tip is distorted, the glow plugs are probably staying on too long* **(see illustration)**. *Glow plugs are not interchangeable between the early and later model systems.*

Power relay (pre-1987 6.9 liter only) - The power relay **(see illustration)** controls the flow of battery voltage to the glow plugs. On most 6.9 liter engines, it is located near the battery on the right inner fender in the engine compartment.

Solid state electronic glow plug controller (1987 through 1994 non-DI Turbo models only) - Controls glow plug on-time by monitoring their resistance to determine temperature and cycling the current to the glow plugs. Located at the rear of the intake manifold **(see illustration)**.

Powertrain Control Module (PCM) controlled glow plug system (1994 and later Direct Injection Turbo models) - The PCM controls the glow plug relay on-off time. The system consists of eight glow plugs, a glow plug relay, the PCM, an engine oil temperature sensor and a barometric pressure sensor. The PCM decreases the glow plug on-time as oil temperature and barometric pressure increase. The glow plugs are located under the valve covers.

Wait lamp - Located in the instrument cluster, this informs the driver when the glow plugs are hot enough for the engine to be started. The wait lamp will come on when the ignition switch is in the On position, only when the engine is below 165 degrees F. For a bulb check, the bulb will come on when the ignition switch is in the Start position.

Wait lamp latching relay (pre-1987 6.9 liter only) - Controls wait lamp, and turns it On and Off.

Wiring harness - Connects all components of the glow plug system. The harness **(see illustration)** incorporates eight fusible wires (one for each glow plug) as a safety device to protect against a direct short of the glow plugs. If a fusible wire should become damaged, it will be necessary to replace the complete glow plug wiring harness. **Note:** *The chassis wiring harness to the glow plug wiring harness* **(see illustrations)** *also incorporates two replaceable fusible links, one for each glow plug band.*

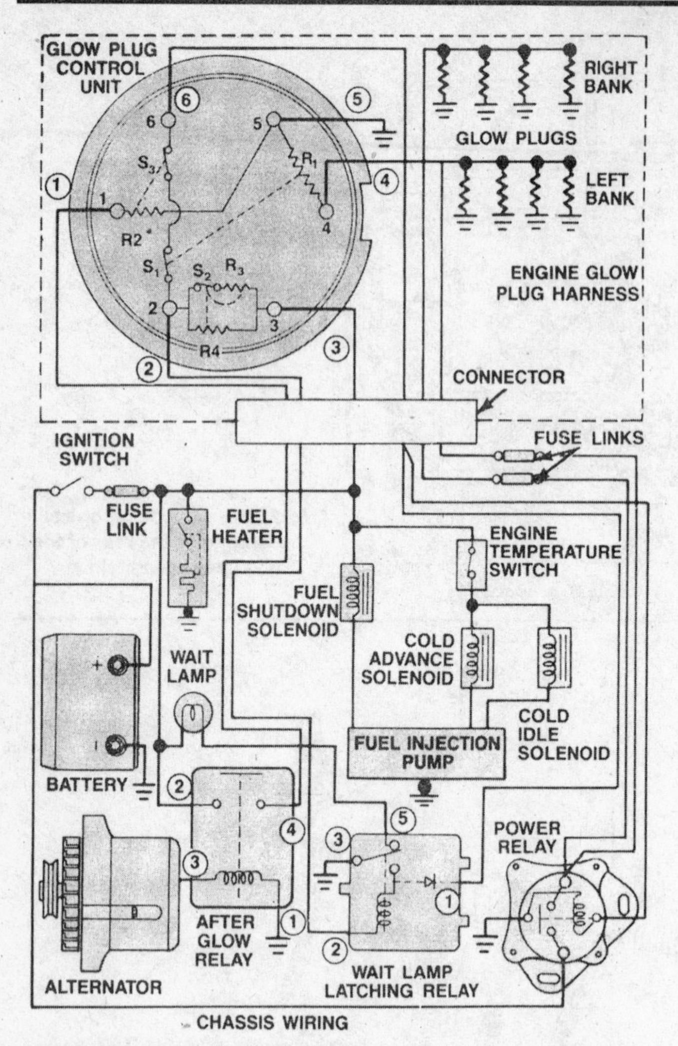

3.74 1983 through 1986 glow plug system electrical schematic

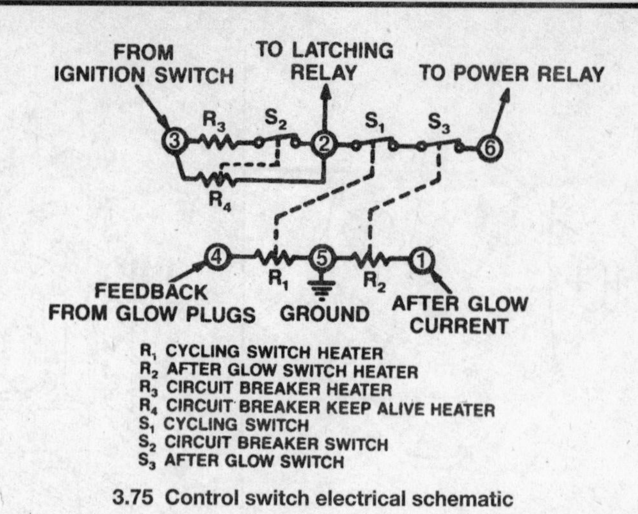

3.75 Control switch electrical schematic

1983 through 1986 6.9 liter engine glow plug system diagnosis

When the ignition switch is turned to the On position, power flows through the ignition switch to pin three of the control switch. The power then flows through resistor R3, switch S2, pin No. 2 (to the latching relay), switches S1, S3 and pin 6 (to the power relay) **(see illustration).**

While the glow plugs are heating, the resistance heater on switch #1 is also being supplied voltage from the output terminal of the power relay. This starts heating the bi-metal spring on switch #1 and after about ten seconds on a cold engine the bi-metal spring is heated enough to cause switch #1 to open. This breaks the voltage supply to pin #6 and the power relay is deactivated. As soon as there is no voltage feedback from the glow plugs to heat the bi-metal spring on #1, the switch begins to cool and it closes again for a few seconds (with a cold engine). Voltage is again supplied by way of pin #6 to the input of the power relay which will cause the glow plugs to heat again and heat the resistor for switch #1, which would open again when it is hot enough.

This is the cycling process of the glow plugs. The glow plugs help the engine to start quickly, and they continue to operate briefly after the engine starts to reduce white smoke. This cycling would continue indefinitely if it was not for the after-glow signal from the alternator, received after the engine starts. This voltage signal starts heating the bi-metal spring on switch #3. After 20 to 90 seconds of receiving voltage, the bi-metal spring will have been heated enough to cause switch #3 to open and shut the glow plug system off. **Note:** *If the problem is hard starting, follow the procedures for troubleshooting the glow plug system prior to troubleshooting the fuel system.*

The fast start glow plug system diagnostic procedure describes methods for checking the operation of the glow plug system. Use of these procedures will help in correcting problems. Results of one basic diagnostic test will direct you to one of six specific diagnostic procedures.

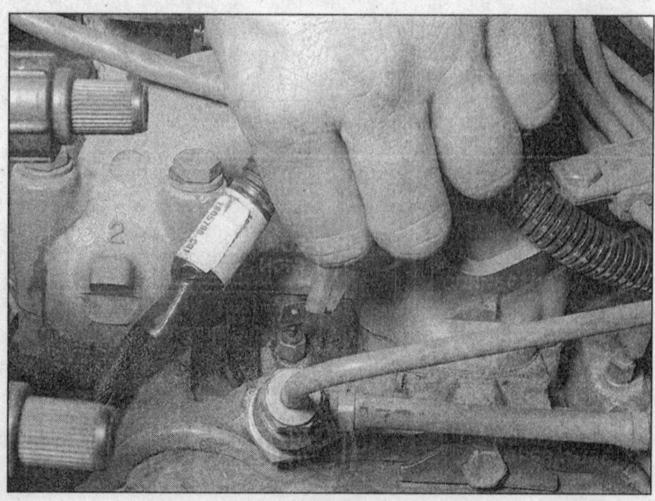

3.76a Unplug the glow plug connector

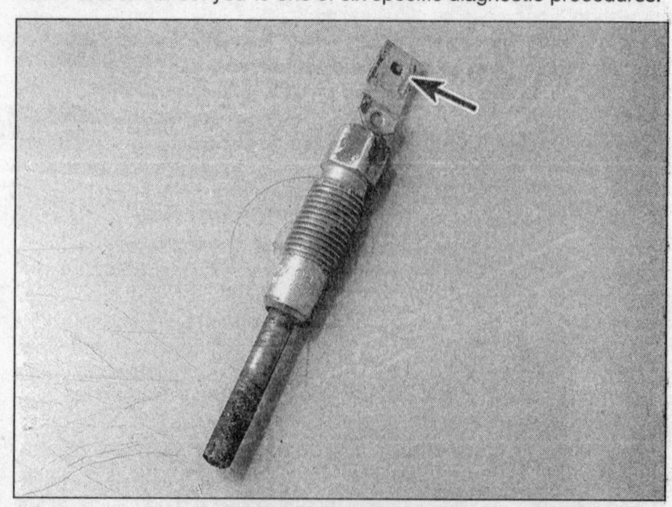

3.76b Connect a test light to the glow plug terminal (arrow) - glow plug removed for clarity

3.76c Using a deep socket, remove the glow plug

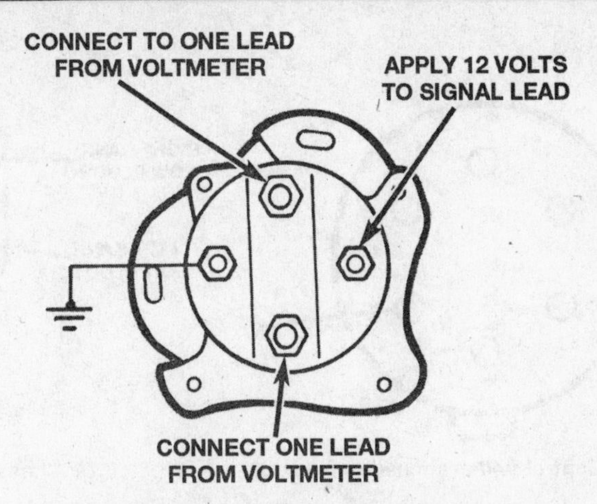

3.77 Connect a test light between the glow plug output terminal and ground

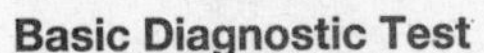

Basic Diagnostic Test

To test the glow plugs, unplug the electrical connector **(see illustration)** and hook up a test light between a battery positive terminal and the terminal on the glow plug **(see illustration)**. **Caution:** *Whenever disconnecting the glow plug wires from the glow plugs, insulate them from contacting the engine and causing a possible short circuit.* If the light comes on, the glow plug is OK. If the light fails to come on, replace the glow plug **(see illustration)**. Tighten the glow plug to the torque listed in this Chapter's Specifications.

To diagnose the glow plug system, connect a 12-volt test lamp between the power relay output (to glow plugs) and ground **(see illustration)**. Check test lamp signals as follows:

Engine cold, ignition switch ON, engine not started

1 Test lamp will light when ignition switch is turned ON.

2 Lamp will be lit for four to ten seconds (depending upon engine coolant temperature).

3 Test lamp will then repeat ON-OFF cycle.

After engine is started

1 Test lamp continues to repeat ON-OFF cycle for 20 to 90 seconds.

2 Lamp will cycle for a maximum of 90 seconds after engine is started.

Engine stopped, then restarted

1 Test lamp will light if coolant temperature at the engine temperature switch has fallen below 165-degrees F between starts. The system will operate for a shorter period of time than with a completely cold engine.

2 Test lamp will remain off if coolant at the control switch probe is at 165-degrees F or higher.

Pinpoint diagnostic test

Using the results of the basic diagnostic test, you can identify one of the six diagnostic procedures to resolve the problem. A basic test result is described at the start of each diagnostic procedure. Pick the procedure that best fits your basic test results. Then proceed with the particular tests. After completing a specific repair, repeat the basic diagnostic test to verify that the system is operating properly.

1 Test lamp signal is correct

a) *Check voltage to each glow plug - Remove leads from glow plugs. Turn ignition switch to On (with the test lamp still connected to the power relay). Check voltage between each glow plug lead and ground. There should be 11-volts minimum at each lead whenever the test lamp lights. If voltage is not detected at any or all leads, check the fusible links, repair vehicle wiring harness or replace the glow plug harness.*

b) *Check glow plug resistance - Turn the ignition switch Off. Using an ohmmeter, measure resistance between glow plug terminal and shell. Replace any glow plug with a reading of 2-ohms or more. Less than 2-ohms is OK. The glow plug system is functioning properly if both tests check out correctly.*

2 Test lamp on continuously

a) *Isolate damaged component - Remove all leads from glow plugs before proceeding with this test. With the test lamp still connected to the power relay, turn ignition switch ON. DO NOT START THE ENGINE. Remove harness plug from control switch. If test lamp stays on, remove signal lead (lead from control switch) from power relay. If test lamp goes out, replace the glow plug wiring harness. If the test lamp stays on, replace the power relay.* **Caution:** *Do not attach leads to glow plugs until system has been rechecked for correct operation.*

b) *Check for damage to glow plugs and harness - Check the continuity of the fusible links in wiring harness. With the ignition switch OFF, check the continuity between each glow plug connector and power relay output terminal. If any reading is 1-ohm or more, replace the wiring harness. Less than 1-ohm is OK.*

c) *Check glow plug resistance - With the ignition switch Off, measure the resistance between the glow plug terminal and shell. Replace any glow plug where resistance is 2-ohms or higher. Less than 2-ohms is OK.*

3 Lamp does not light

a) *This is normal and expected with a warm engine (at or above 165 degrees F. If the engine is below normal operating temperature, turn the ignition Off. Wait five minutes, then turn the ignition ON again. Observe the monitor lamp. If the lamp still does not light, proceed with this test. If the lamp now lights, go to step 5.*

b) *Check the battery state of charge - If the battery will crank the engine, it will operate the glow plug system.*

c) *Replace the fuse in the circuit from ignition switch to control switch - Retest the system. If the system still will not function, but the fuse did not blow, proceed to the next step. If the system will not function and the replacement fuse blew, remove signal lead from the power relay, replace the fuse, and turn the ignition switch to ON. If the fuse does not blow - replace the power relay. If the fuse blows again, separate glow plug harness to chassis harness connector. Replace the fuse, turn ignition ON. If the fuse blows, repair vehicle wiring. If the fuse does not blow, replace the glow plug wiring harness.*

d) *Check battery power to power relay - ignition switch Off. Connect a voltmeter between the power relay battery terminal (input) and ground. If less than 11 volts is detected, repair the wiring between the battery and power relay. If 11 volts or greater is detected, check for power to the control switch (next step).*

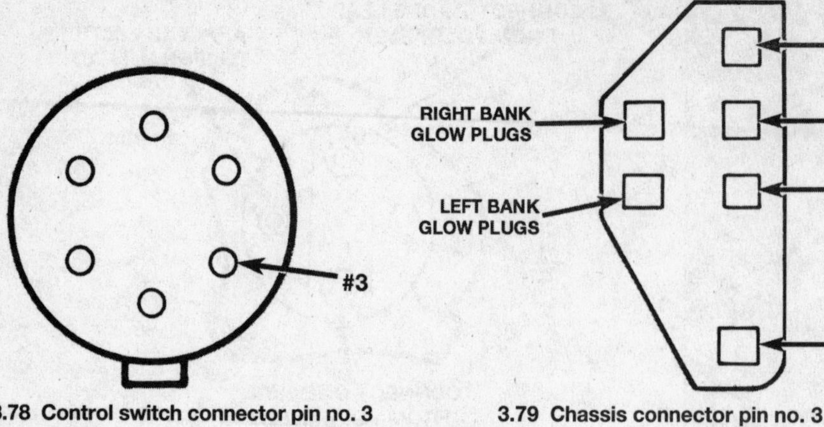

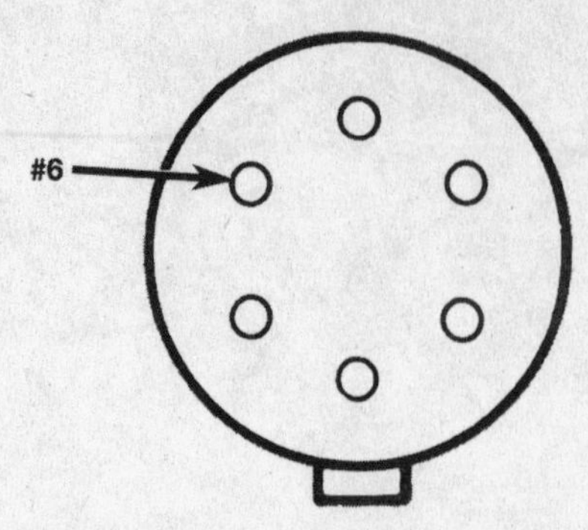

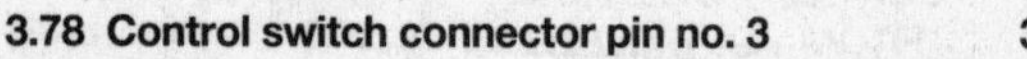

3.78 Control switch connector pin no. 3

3.79 Chassis connector pin no. 3

3.80 Control switch connector pin no. 6

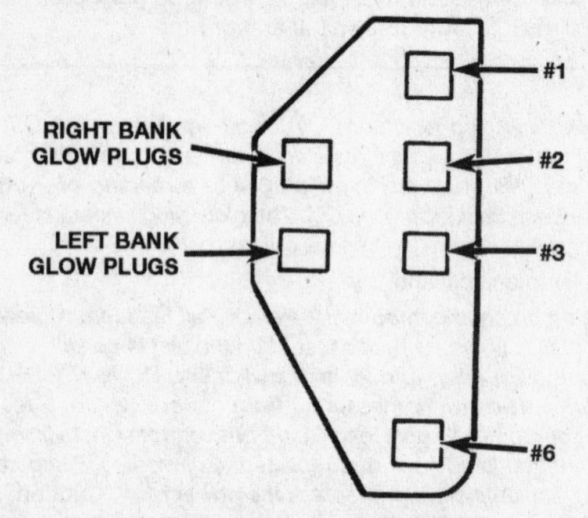

3.81 Chassis side connector pin no. 6

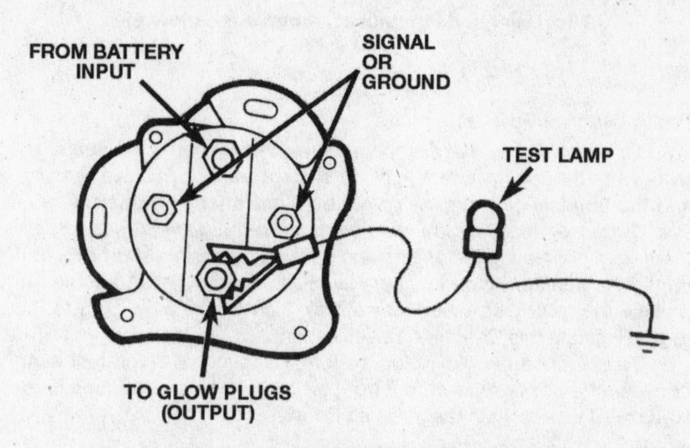

3.82 Checking power relay for short

e) *Check power to control switch - With the ignition switch ON and the harness plug removed from the control switch, check the voltage between pin-3 of the control switch connector* **(see illustration)**, *and ground. If a minimum of 11-volts is not detected, open glow plug harness-to-chassis connector and test for voltage at pin-3 on chassis side of connector* **(see illustration)**. *If a reading of less than 11-volts is still detected, repair vehicle wiring. If 11-volts or greater is detected, replace the glow plug wiring harness.*

f) *If 11-volts or greater is detected at the control switch, proceed to the next step.*

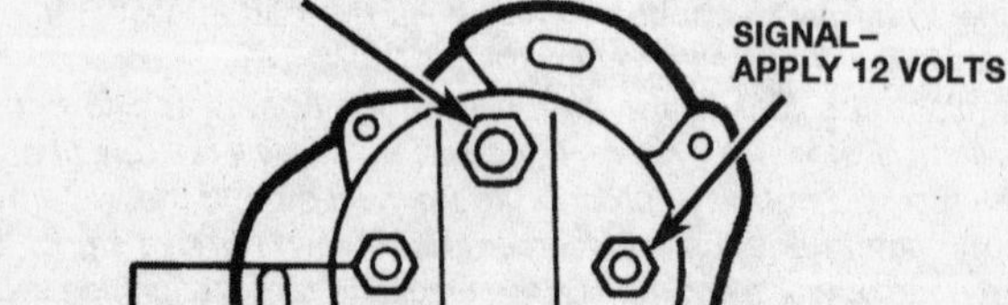

3.83 Checking power relay function

g) *Check wiring harness continuity - With the harness leads at control switch disconnected, disconnect the power relay signal lead, and check continuity between the signal lead and pin six of the control switch connector* **(see illustration)**.

h) *If less than 1-ohm resistance is observed, proceed to the next step.*

i) *If the resistance is 1-ohm or greater, separate the glow plug harness to chassis connector and check for continuity between chassis-side pin six* **(see illustration)** *and the power relay trigger lead. If reading is less than 1-ohm, replace the glow plug wiring harness. If reading is greater than 1-ohm, repair the vehicle wiring.*

j) *Check the power relay for a short - Disconnect all leads except ground to the power relay. Connect a voltmeter between the input and output terminals and apply 12-volts to the signal lead input terminal* **(see illustration)**. *If voltage is detected by voltmeter, replace the power relay. If no voltage is detected, proceed to the next step.*

k) *Check function of power relay - Disconnect all leads to the power relay except ground. Connect ohmmeter across input and output terminals of the power relay* **(see illustration)**. *Apply battery voltage to the signal lead input terminal of the power relay. If there is less than 1-ohm resistance, proceed to the next step.*

l) *If resistance is 1-ohm or greater, check ground leak or replace the power relay.*

m) *Check after glow voltage - Remove the harness plug at the control switch and connect all leads to power relay. Turn the ignition switch ON - DO NOT START ENGINE. Check for voltage at the control switch harness plug pin one* **(see illustration)**. *If no voltage is detected, replace the glow plug control switch. If voltage is detected, repair alternator or vehicle wiring and replace after glow relay, as necessary.*

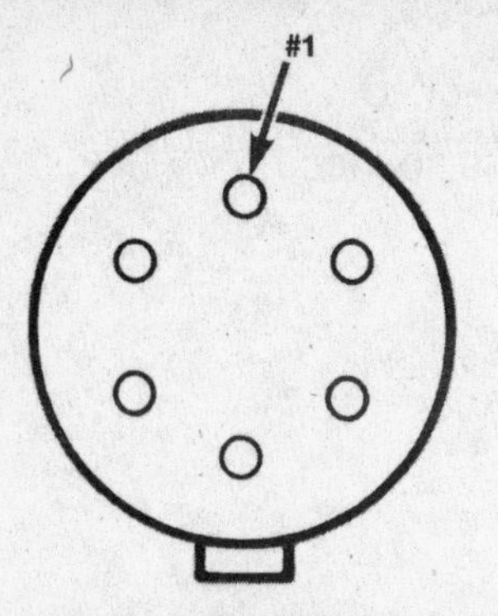

3.84 Control switch connector pin no. 1

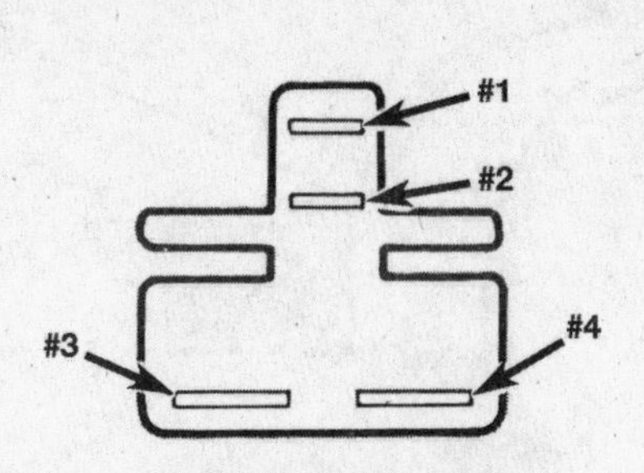

3.85 After-glow relay harness connector

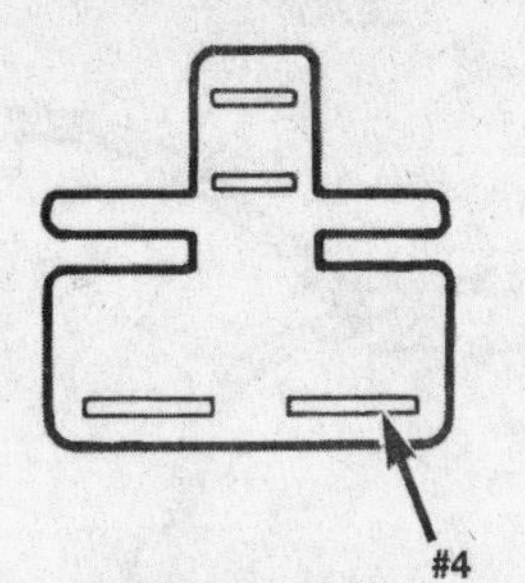

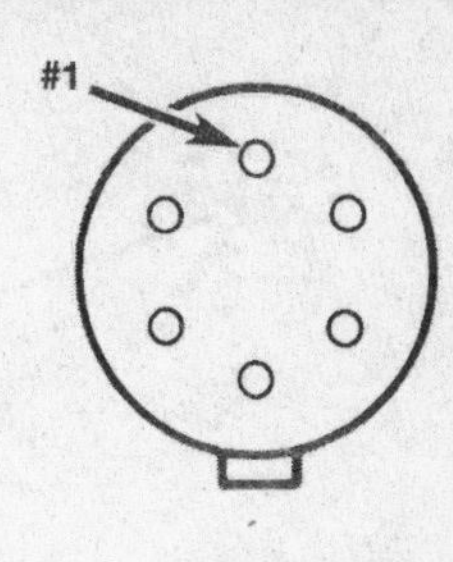

3.86 After-glow relay harness connector pin no. 4 and control switch connector pin no. 1

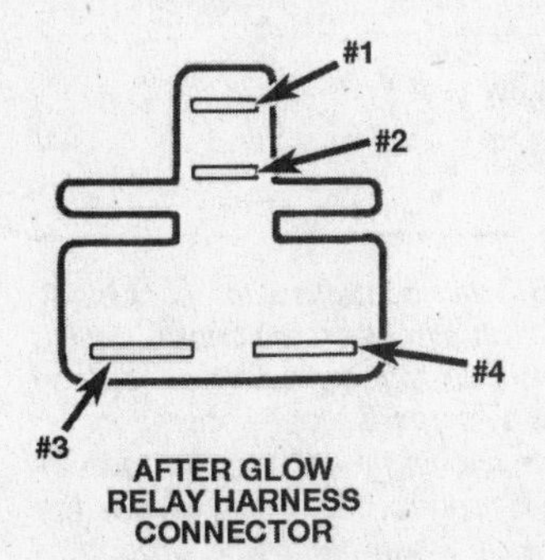

CHASSIS
WIRING HARNESS
CONNECTOR

#1

3.87 After-glow relay harness connector pin no. 4 and chassis connector pin no. 1

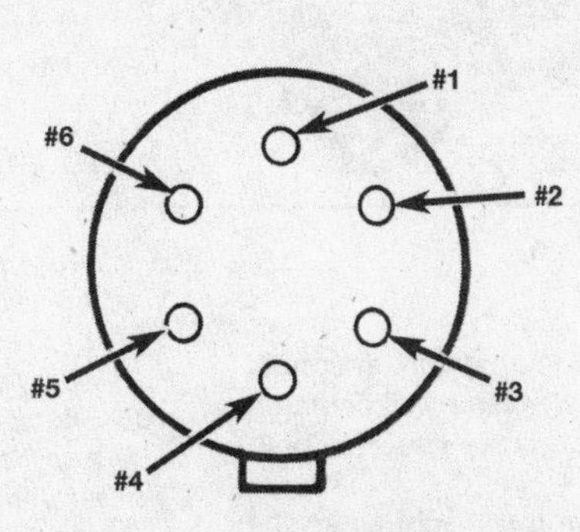

3.88 Control switch harness plug sockets no. 4 and 5

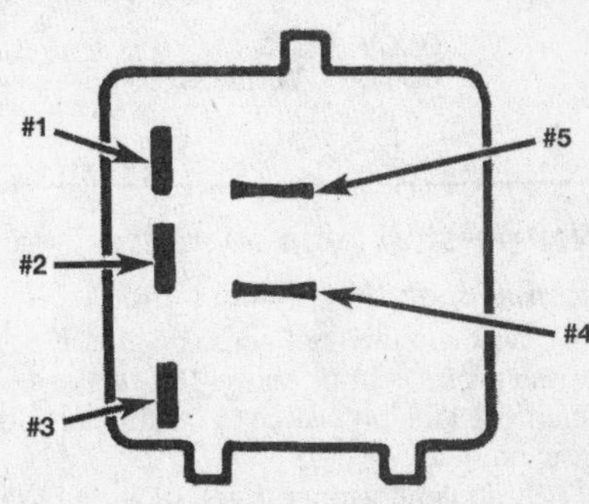

3.89 Latching relay harness connector

4 Test lamp continues to cycle beyond the maximum of 90 seconds following engine start

a) *Check after glow voltage - Remove harness plug at the control switch. Start engine and check for voltage at control switch harness*
b) *Pull harness connector from after-glow relay - Turn the ignition switch Off. Disconnect wiring at the after glow relay and check connector terminal #1 for ground* **(see illustration)**. *If the ground is not OK, repair or replace wiring as required. If the ground is OK, turn the ignition switch to On and check for voltage at terminal #2. 11 volts is the minimum acceptable reading. If the reading is less than 11 volts, repair the wiring as required. If the voltage was a minimum of 11 volts, start the engine and check for voltage at terminal #3. The reading should be 6.5 to 7.5-volts. If it is not, repair the wiring or alternator as required.*
c) *If the harness connector checks out OK, check for resistance between harness connector terminal #4 and control switch terminal #1* **(see illustration)**. *If resistance is less than 1-ohm, replace the after-glow relay. If resistance is 1-ohm or greater, disconnect the chassis-to-engine glow plug harness connector, and check for resistance between terminal #1 of the chassis connector and after-glow relay connector terminal #4* **(see illustration)**. *If resistance is less than 1-ohm, replace the glow plug harness. If it is 1-ohm or greater, repair the vehicle wiring.*

5 Test lamp lights when ignition is first turned on, then goes off and stays off for an extended period of time.

a) *This is normal and expected with a partially-warmed engine (coolant temperature of 140-degrees F or above). If the engine is below 140-degrees F turn the ignition OFF. Wait five minutes and turn ignition switch ON again. Observe the test lamp. If the lamp does not light, follow the troubleshooting procedure in Step No. 3. If the lamp does light, proceed to next step.*
b) *Check wiring harness resistance - Disconnect harness plug at the control switch, and connectors at power relay and glow plugs. Check resistance between the control switch connector pin four* **(see illustration)**, *and the power relay output terminal connector. If resistance is 1-ohm or greater, repair harness ground connection, or replace harness as necessary. If resistance is less than 1 ohm, proceed to next step.*
c) *Check wait lamp latching relay resistance - Disconnect the harness connector at latching relay. Check resistance across terminals 1 and 2 of latching relay with red probe (+) on terminal 2 and black probe (-) on terminal 1. If less than 45 ohms, replace latching relay, and recheck system. If 45 ohms or greater, go to next step.*
d) *Check power relay resistance - Turn the ignition switch Off. Disconnect the harness plug at control switch only. All other leads must be connected. Check the resistance between control switch connector pin 6 and ground. If the resistance is 2.5 ohms or more, proceed to next step. If the resistance is less than 2.5 ohms, replace power relay. Recheck the system for proper operation.*
e) *Check after glow voltage - Remove harness plug at the control switch. Turn the ignition switch to On. Do not start engine.*
f) *Check for voltage at harness plug pin 1.*
g) *If voltage reading is 0, replace the control switch.*
h) *The problem should be corrected. Recheck for proper operation.*
i) *If voltage is greater than 0 volts, separate the engine glow plug harness-to-chassis harness connector. Turn the ignition switch ON, do not start engine. Check for voltage on chassis side of connector at pin-1.*
j) *If no voltage is detected, replace glow plug wiring harness.*
k) *The problems should be corrected. Recheck for proper operation.*
l) *If voltage is detected, remove harness plug from after glow relay, and retest.*
m) *If voltage is still detected, repair or replace faulty vehicle wiring between after glow relay and glow plug wiring harness.*
n) *If voltage is not detected, replace the after glow relay.*
o) *The problems should be corrected. Recheck for proper operation.*
p) *If voltage is detected, repair wiring to alternator or repair the alternator.*
q) *If voltage is not detected, the problem is solved. Recheck the system for correct operation.*

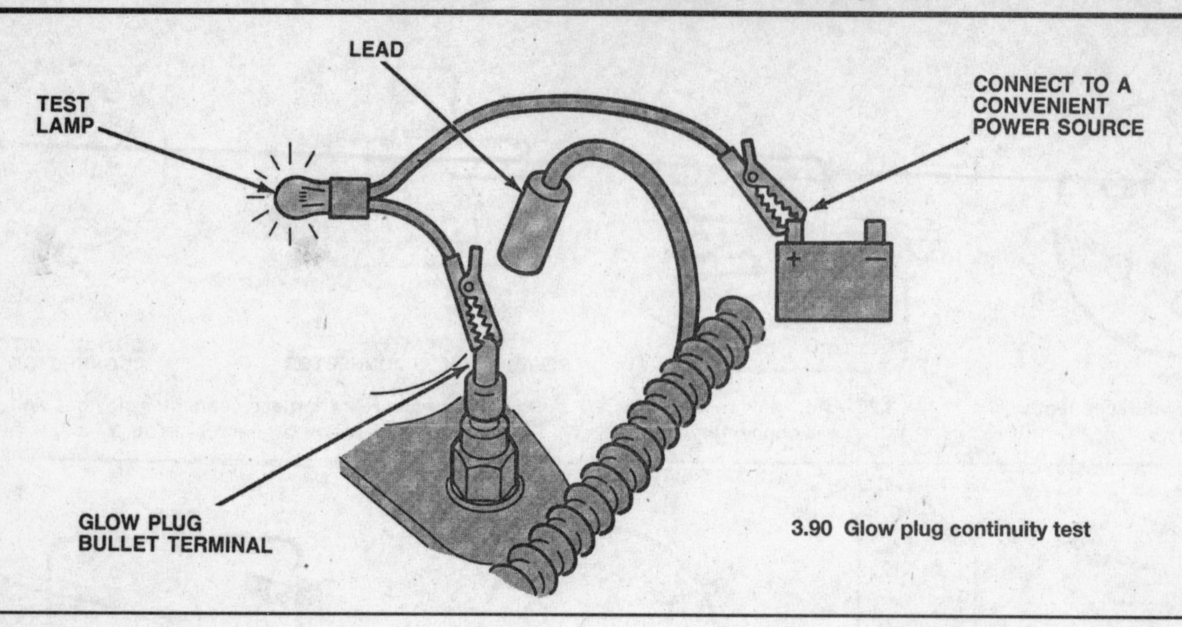

3.90 Glow plug continuity test

Wait-to-start lamp operation

Normal operation of wait lamp

Wait lamp will not come on at all if coolant temperature at control switch probe is at or above 165-degrees. The wait lamp will come on when the ignition switch is in the start position, for a check of bulb function.

If coolant temperature is low the Wait lamp will light when the ignition switch is turned ON.

Lamp will remain lit for 4 to 10 seconds depending upon engine coolant temperature.

Lamp will then remain off until ignition switch is turned OFF, then ON again.

Check wait lamp operation - refer to the above description of normal operation of wait lamp. If the wait lamp operation is incorrect, check glow plug system operation using test lamp, as described previously. If improper test lamp signal is observed, troubleshoot and repair glow plug system, as previously described. If glow plug system operation is OK, but wait lamp operation is faulty, proceed to one of the following three tests.

1 No wait lamp

a) *Check lit bulb - If bulb is damaged, replace and recheck wait lamp operation. If bulb is OK, proceed to the next step.*

b) *Jump from harness at latching relay - Turn the ignition switch On. Disconnect the harness connector from latching relay. Place the jumper wire between harness connector pin 5 and ground. If lamp is not lit, repair wiring in circuit from ignition switch to wait lamp to latching relay. If lamp is lit, proceed to next step.*

c) *Check latching relay ground circuit - Turn the ignition switch On. Disconnect the latching relay harness plug. Place the jumper wire between pins 3 and 5 in latching relay harness connector* **(see illustration)**. *If wait lamp remains off, repair latching relay ground connection and retest. If wait lamp comes on, replace the latching relay and retest.*

2 Wait lamp On with ignition switch On

a) *Check for grounded lead between wait lamp and latching relay. Turn the ignition switch ON. Disconnect the harness connector at the latching relay. If the lamp remains on, repair ground in the wire between the wait lamp and latching relay and retest. If lamp is off, proceed to next step.*

b) *Check voltage to latching relay from control switch - Turn the ignition switch On. Disconnect harness connector at latching relay. With a voltmeter, check voltage between connector pin 2 and ground. If reading is 11-volts or more, proceed to the next step. If reading is less than 11 volts, turn ignition switch Off. Disconnect harness plug at control switch. Check the continuity between latching relay connector plug pin 2 and control switch plug pin 2. If resistance reading is less than 1-ohm, replace the control switch and retest. If resistance is 1-ohm or higher, repair or replace wiring as necessary. Recheck the system for proper operation.*

c) *Check continuity to ground at the latching relay harness socket - turn the ignition switch OFF. Disconnect the connector at the latching relay. Check resistance between pin-1 and ground. If resistance is 5-ohms or less, replace the latching relay and recheck for proper operation. If resistance is greater then 5 ohms, repair or replace the wiring between the latching relay and power relay.* **CAUTION:** *When working on glow plug system, be careful not to short out any components.*

7.3 liter engine glow plug system diagnosis - 1994 and earlier (except Direct Injection Turbo) models

Glow Plug Continuity Check

1 With the engine and ignition switch OFF, disconnect the wire from a glow plug.

2 Connect a test light to the positive terminal of a battery and to a glow plug terminal **(see illustration)**.

3 If the test light goes ON, the glow plug is OK.

4 If the test light doesn't go ON, replace the glow plug.

5 Repeat this test for each glow plug.

Wiring Harness Check

1 With the engine and ignition switch OFF, disconnect the wire from the glow plugs.

2 Remove the cover from the control module by squeezing the sides together.

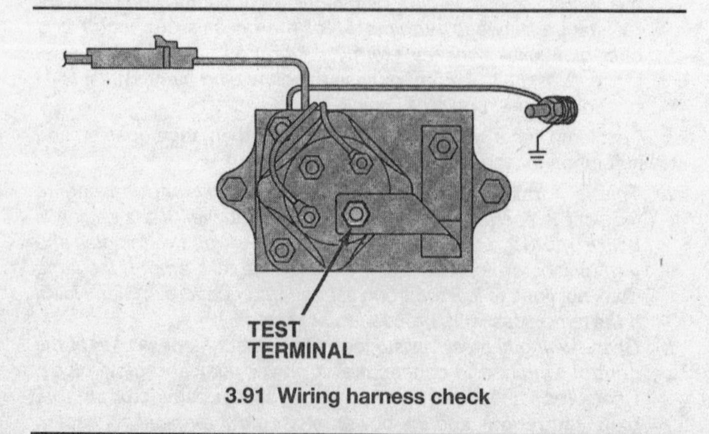

3.91 Wiring harness check

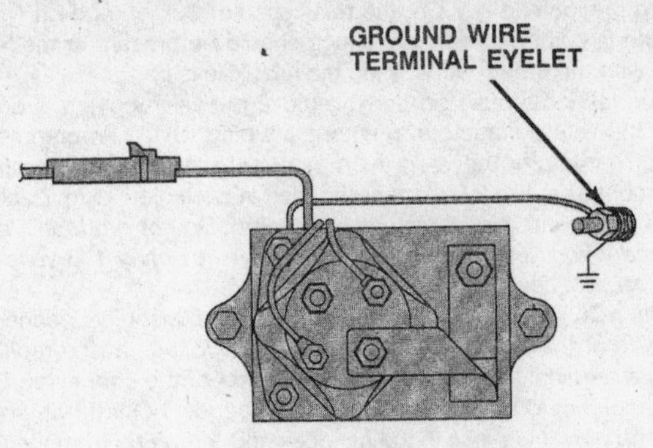

3.92 Control unit ground test

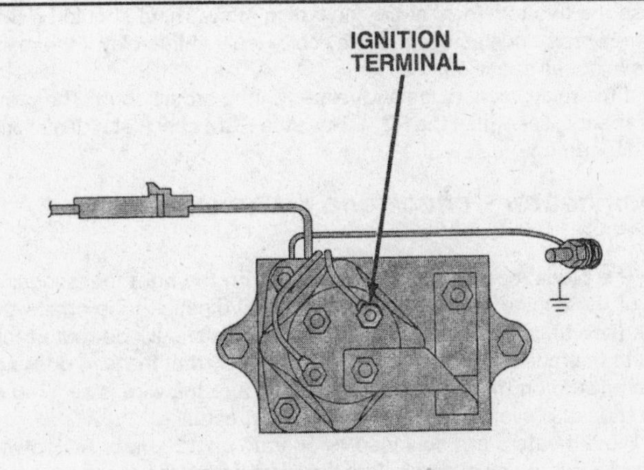

3.94 Ignition switch voltage test

3 Using an ohmmeter or continuity tester, check for continuity between each glow plug lead and the test terminal **(see illustration)** on the control unit. Repair any open circuits as necessary.
4 Reconnect the glow plug wires.

Control unit ground test

1 With the engine and ignition switch OFF, connect an ohmmeter to the ground wire terminal eyelet and to the ground terminal on a battery **(see illustration)**.
2 If the resistance is more than one ohm, clean or repair the ground connection.
3 If the resistance is more than one ohm, perform the *supply voltage test*.

Supply voltage test

1 With the engine and ignition switch OFF, connect a voltmeter to the control unit power terminal **(see illustration)**.
2 If there is less than ten volts, repair the wiring or recharge the battery.
3 If there is more than ten volts, go to *Ignition switch voltage test*.

Ignition switch voltage test

1 Connect the voltmeter to the ignition terminal **(see illustration)** on the control unit and to ground.
2 Turn the ignition switch ON, all accessories OFF.
3 If there is less than eight volts, check the fusible link, recharge the battery or repair the wiring.
4 If there is more than eight volts, go to the *Control unit functional test*.

Control unit functional test

1 With the engine and ignition switch OFF, connect a test light to the test terminal on the control unit **(see illustration)**.

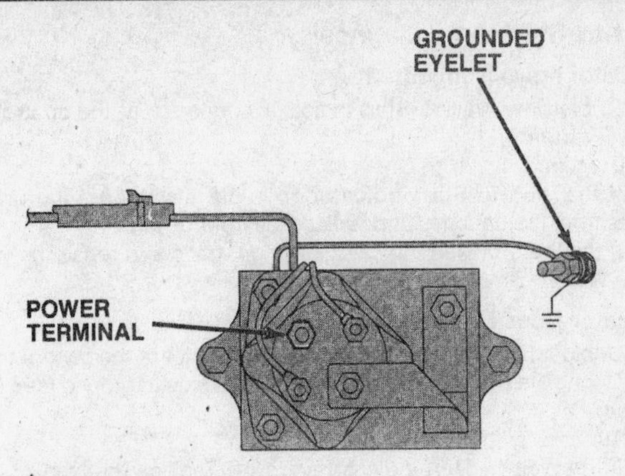

3.93 Supply voltage test

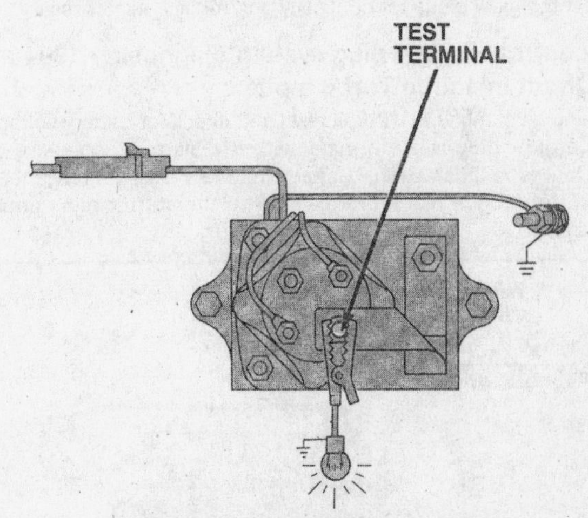

3.95 Control unit functional test

2 Have an assistant turn on the ignition switch. Compare the amount of time the test light stays on to the Test Light Chart **(see illustration)**. **Note:** *The Wait-to-start/test light may not come on if the engine temperature is at or near normal operating temperature. Total test light ON time is measured from the beginning of the ON cycle to the end of the last ON-OFF cycle.*
3 If the test light time is incorrect, disconnect the ground cables from the batteries and replace the control unit.
4 If the test light time is correct, the control unit is OK.

Control Unit* Temperature F	Wait-To-Start Lamp On Time (seconds)	Test Light Total Time (seconds)
-20 F	7-15	35-70
0 F	7-12	25-60
35 F	5-12	15-35
70 F	3-5	7-15
105 F	1-3	3-5
140 F	1 or less	1-3

*Temperature of Control Unit, NOT ambient Temperature

3.96 Test light chart

Wait-to-Start lamp diagnosis

Indicator stays illuminated

1 Unplug the Wait-to-Start indicator connector at the control unit **(see illustration)**.
2 Turn ignition switch On.
3 If the Wait-to-Start indicator goes off, disconnect the ground cables from the batteries and replace the control unit.
4 If the Wait-to-Start indicator stays on, trace and repair the wiring to the Wait-to-Start indicator.

Indicator Does Not Illuminate

1 Unplug the Wait-to-Start indicator connector at the control unit.
2 Connect a jumper wire from the harness side to ground **(see illustration)**.
3 Turn the ignition switch On.
4 If the Wait-to-Start indicator stays off, replace the bulb or trace and repair the wiring.
5 If the Wait-to-Start indicator illuminates, go to the Hard Starting checks.

Indicator flashes rapidly

1 This occurs often when two or more glow plugs are burned out.
2 Test the glow plugs for continuity and replace as necessary.

PCM controlled glow plug system diagnosis - 1994 and later Direct Injection Turbo models

1 Using a voltmeter or 12-volt test light, check for battery voltage at the large glow plug relay terminal with the black/orange wire connected to it **(see illustration)**. Battery voltage should be present at all times. If not, check black/orange wire to the starter relay and the fusible links.

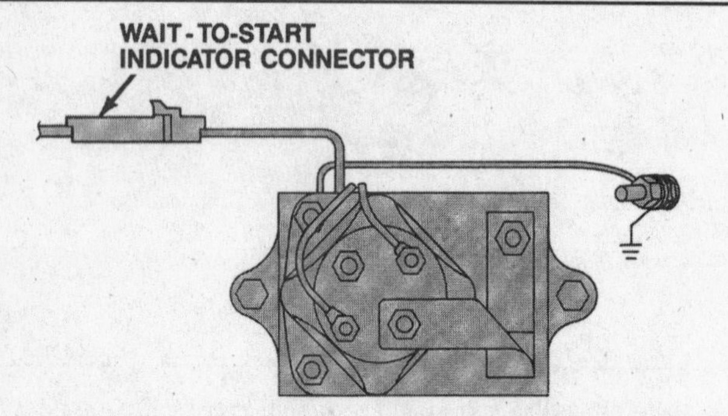

3.97 Wait-to-start indicator test (indicator illuminated)

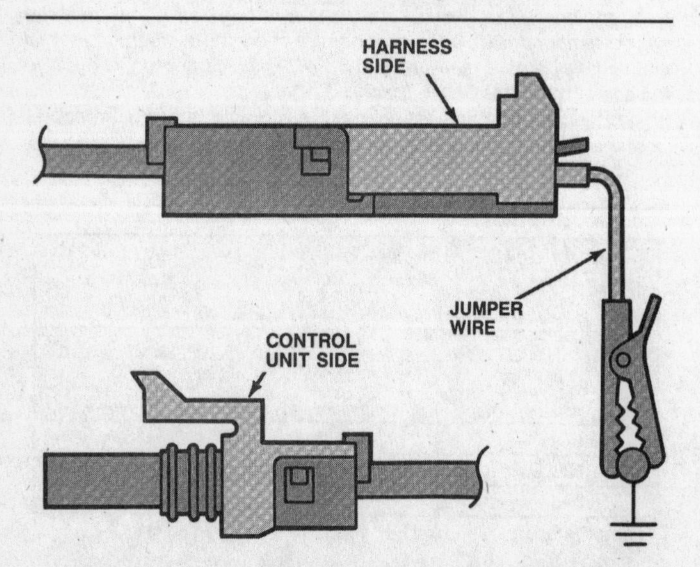

3.98 Wait-to-start indicator test (indicator not illuminated)

2 Turn the ignition key On, the relay should click on and off (when the engine is cold) and battery voltage should be present at the large terminal with the brown wires when the relay clicks on.
3 If the relay operates properly, remove the valve covers and disconnect the wire connectors at each glow plug. Using an ohmmeter, check for continuity between each glow plug terminal and ground. There should be 0.1 to 2.0 ohms resistance at each glow plug. Replace any glow plugs with high resistance. **Warning:** *Do not probe the large red-striped wires to the fuel injectors. These wires carry 115-volts DC and can cause a severe shock.*
4 If the relay does not operate, disconnect the electrical connector at the base of the relay and check for battery voltage at the red/light green wire terminal, on the wiring harness side of the connector. Battery voltage should be present with the ignition key On. If not, check the 30-amp glow plug fuse in the fuse box and the wiring from the fuse box to the connector.
6 If battery voltage was present in Step 4, measure the resistance across the two terminals at the glow plug relay. There should be 5 to 15 ohms resistance across the relay coil. Replace the relay if the resistance is not as specified.
7 If the relay, glow plugs and related wiring are all good. The problem probably lies within the PCM. Have the PCM check at a dealer service department.

Block heater - check and replacement

Check

The block (coolant) heater is located on the right (passenger's) side of the engine adjacent to where the oil dipstick tube enters the block **(see illustration)**. The block heater wire should be routed forward to the radiator grille opening in such a way that the wire does not chafe or burn on the exhaust manifold. Be sure the wire is secured at both ends to prevent it from being pulled out of place.

Block heaters are designed to operate on 120-volts AC power. Check for available voltage before troubleshooting the block heater. A block heater can be tested in or out of the engine, however, the easiest method is with the unit in place.

First, check the cord for continuity. Unplug both ends and test the wires with an ohmmeter or continuity tester. There are three separate wires - the center one is a ground and the two outer ones are "live". Look for damage such as opens and shorts, and replace the wire if necessary.

If the wires are OK, connect an ohmmeter to the two terminals of the block heater. There should be a very low ohms reading (near zero). If the ohmmeter shows the heating element to be open (infinite ohms), replace the block heater.

If the previous checks indicate no problems, choose a double electrical outlet and plug a work light into it. Then plug the heater into the other side of the outlet. The light should dim slightly. If it does, current is flowing to the block heater. Check for heating by feeling near the block heater. If heat is given off, the unit is working.

If the light does not dim and no heat is given off, replace the block heater.

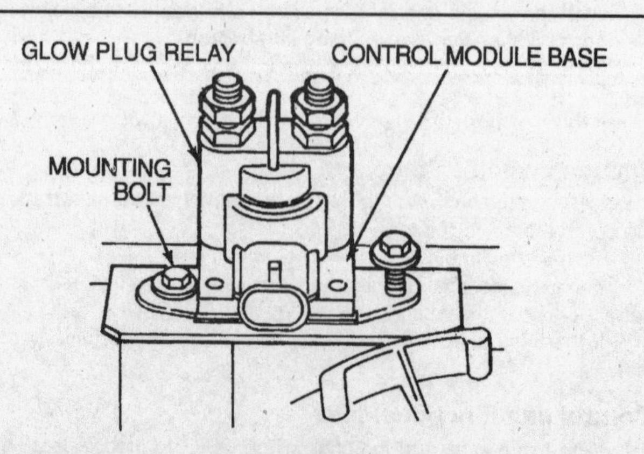

3.99 Glow plug relay - 1994 and later Direct Injection Turbo models

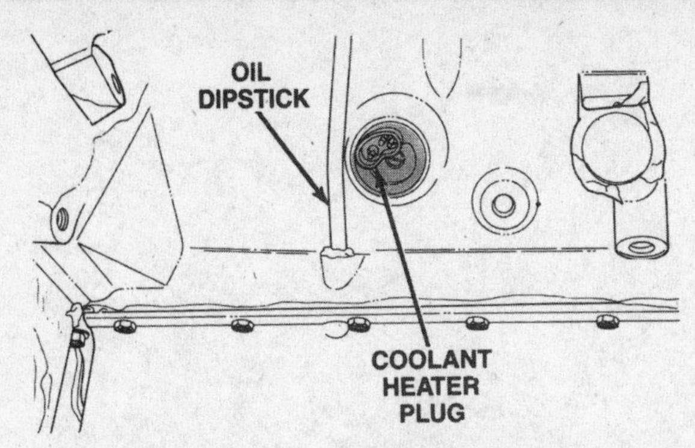

3.100a The block (coolant) heater plug is located low on the side of the engine block

Replacement

Warning: *Allow the engine to cool completely before following this procedure.*

1 Drain the coolant (see the Maintenance portion of this Chapter).
2 Unplug the power cord from the block heater.
3 Loosen the center screw **(see illustration)** and pull the block heater out of the engine. It may be necessary to push it to one side to clear the "tee" bar retainer.
4 Clean the sealing surfaces in the engine block and install the block heater. Be sure the seal is intact and tighten the center bolt securely.
5 Connect the power cord.
6 Refill and bleed the cooling system.
7 Test the block heater as described above.
8 Run the engine and check for coolant leaks.

Fuel heater

1994 and earlier (except Direct Injection Turbo) models

General information

An electric fuel heater is incorporated into the top of fuel filter

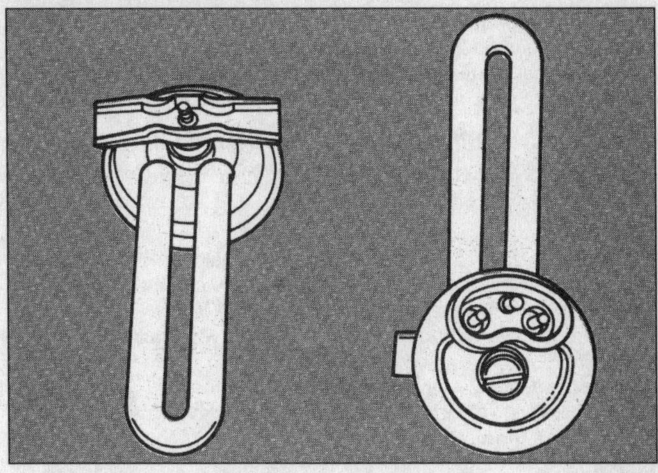

3.100b Loosen the center screw, pull the heater out then move it sideways until the 'Tee' bar clears the opening

assembly on late models. When ambient temperatures drop below 30-degrees, it is designed to warm the fuel as it enters the filter to prevent clogging due to low temperature "jelling".

Test

1 Disconnect the electrical connector at the fuel heater **(see illustration)**.
2 Hold a test light probe to the female socket of the fuel heater harness wire and connect the test light wire to ground.
3 Turn the ignition switch On and check for power.
4 If the test light comes on, the circuit to the heater is OK.
5 If the test light doesn't come on, check the fusible link.
6 If the fusible link is burned out, check the wiring and if the wiring is OK, replace the heating element.
7 If the fusible link is OK, check and repair the wiring.

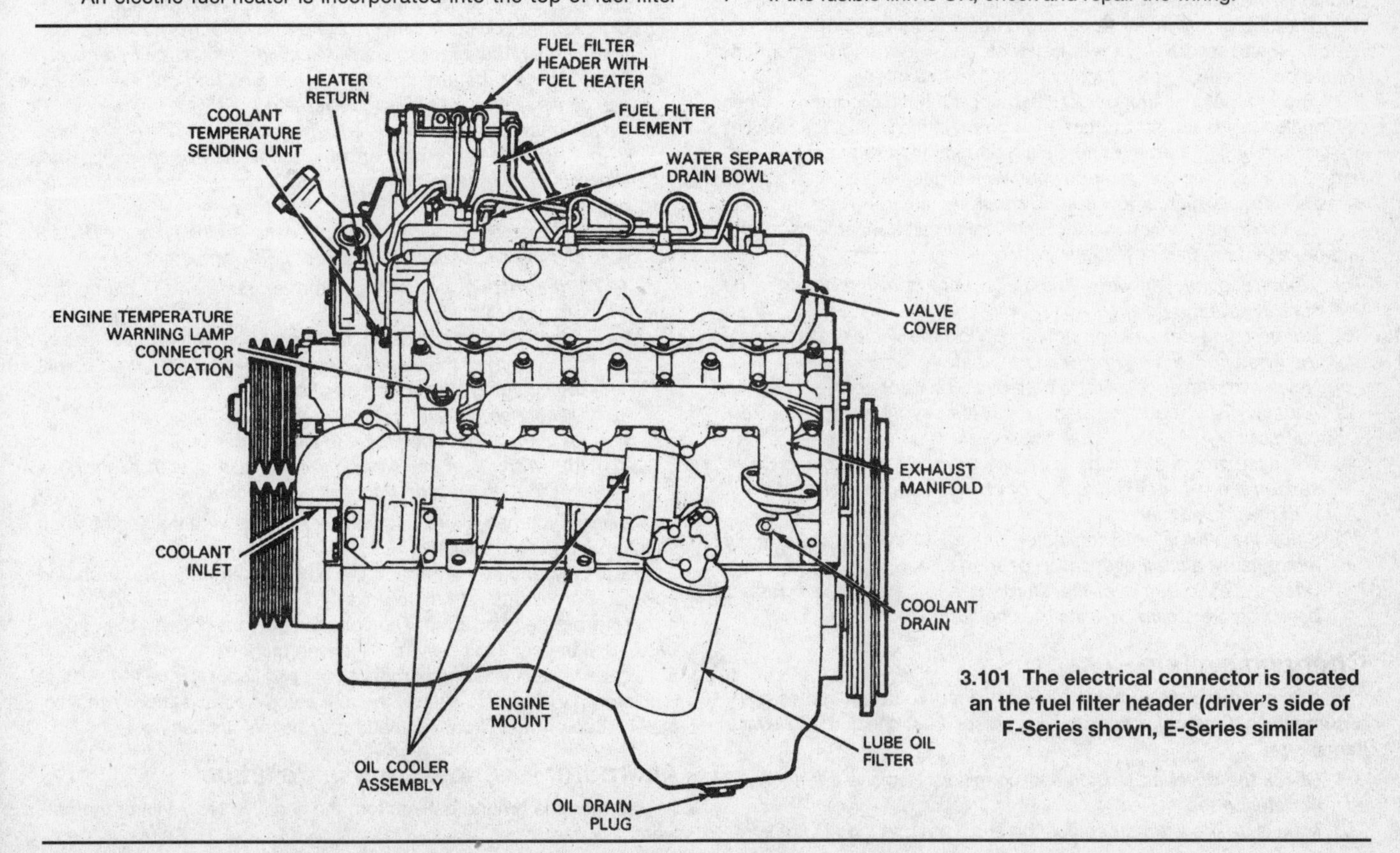

3.101 The electrical connector is located an the fuel filter header (driver's side of F-Series shown, E-Series similar

8 With the temperature below 30-degrees, start the engine and place your hand on the top of the fuel filter housing. If there is voltage at the terminal (as tested above) and the heating element doesn't get warm, replace the element **(see illustration 3.9c)**.

1994 and later direct Injection Turbo models

Note: *The fuel heater is controlled by the Powertrain Control Module (PCM).*

Removal

1 Drain the fuel from the fuel filter/water separator. Refer to the cutaway illustration of the fuel filter/water separator and the previous section on replacing the fuel filter.
2 Remove the cap from the assembly and withdraw the filter element.
3 Disconnect the wiring from the heating element.
4 Unscrew the standpipe and remove it from the housing.
5 Remove the heater element.

Installation

6 The remainder of installation is the reverse of removal. Note that the gasket used is bevel cut and must be reinstalled properly to avoid leaks.

Charging system - general information and precautions

The charging system includes the alternator, an internal voltage regulator, a charge indicator, the batteries, fusible links and the wire between all the components. The charging system supplies electrical power for the vehicle when the engine is running. The alternator is driven by a drivebelt at the front of the engine.

The purpose of the voltage regulator is to limit the alternator's voltage to a preset value. This prevents power surges, circuit overloads, etc., during peak voltage output.

A fusible link is a short length of insulated wire integral with the engine compartment wiring harness. The link is smaller in diameter than the circuit it protects. Production fusible links and their identification flags are identified by the flag color. See the owner's manual for additional information regarding fusible links.

The charging system doesn't ordinarily require periodic maintenance. However, the drivebelt, batteries and wires and connections should be inspected at the factory recommended intervals.

The alternator warning light (if equipped) should come on when the ignition key is turned to Start, then go off immediately. If it remains on with the engine running, there's a malfunction in the charging system. Some vehicles are also equipped with a gauge. If the gauge indicates abnormally high or low output, check the charging system.

Be very careful when making electrical circuit connections to protect the alternator and note the following:

a) When reconnecting wires to the alternator from the batteries, be sure to note the polarity.
b) Before using arc welding equipment on the vehicle, disconnect the wires from the alternator and the battery terminals.
c) Never start the engine with a battery charger connected.
d) Always disconnect the battery cables before using a battery charger.
e) The alternator is turned by an engine drivebelt which could cause serious injury if your hands, hair or clothes become entangled in it with the engine running.
f) Since the alternator is connected directly to the battery, it could arc or cause a fire if overloaded or shorted out.
g) Wrap a plastic bag over the alternator and secure it with rubber bands before steam cleaning the engine.

Charging system - check

1 If a charging system malfunction occurs, don't immediately assume the alternator is causing the problem. First check the following items:

a) Check the drivebelt tension and condition. Replace it if it's worn or deteriorated.
b) Make sure the alternator mounting and adjustment bolts are tight.

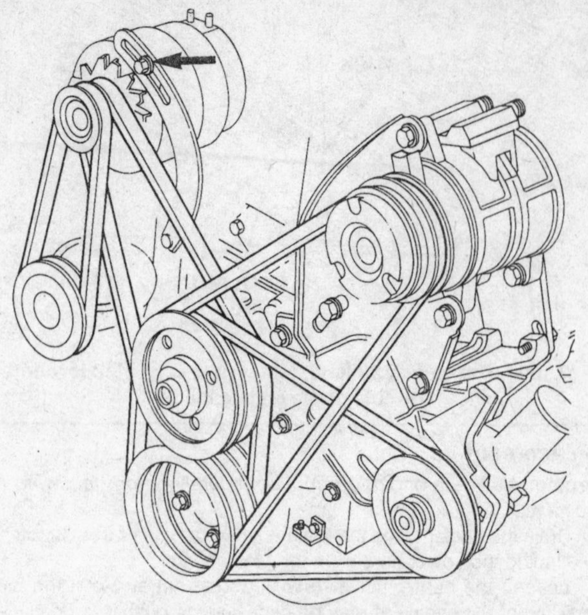

3.102 Remove the adjustment bolt (arrow)

c) Inspect the alternator wiring harness and the connectors at the alternator and voltage regulator. They must be in good condition and tight.
d) Check the fusible link (if equipped) located between the starter solenoid and the alternator. If it's burned, determine the cause, repair the circuit and replace the link (the engine won't start and/or the accessories won't work if the fusible link blows). Sometimes a fusible link may look good, but still be bad. If in doubt, remove it and check for continuity.
e) Start the engine and check the alternator for abnormal noises (a shrieking or squealing sound may indicate a bad bearing).
f) Check the specific gravity of the battery electrolyte. If it's low, charge the battery (doesn't apply to maintenance free batteries).
g) Make sure the battery is fully charged (one bad cell in a battery can cause overcharging by the alternator).
h) Disconnect the battery cables (negative first, then positive). Inspect the battery posts and the cable clamps for corrosion. Clean them thoroughly if necessary. Reconnect the cable to the negative terminal.
i) With the key off, connect a test light between the negative battery post and the disconnected negative cable clamp.
    1) If the test light does not come on, reattach the clamp and proceed to the next step.
    2) If the test light comes on, there is a short (drain) in the electrical system of the vehicle. The short must be repaired before the charging system can be checked.
    3) Disconnect the alternator wiring harness.
        (a) If the light goes out, the alternator is bad.
        (b) If the light stays on, pull each fuse until the light goes out (this will tell you which component is shorted).

2 Using a voltmeter, check the battery voltage with the engine off. It should be approximately 12-volts.
3 Start the engine and check the battery voltage again. It should now be approximately 14 to 15 Volts.
4 Turn on the headlights. The voltage should drop, and then come back up, if the charging system is working properly.
5 If the voltage reading is more than specified, replace the voltage regulator. If the voltage is less, the alternator diodes, stator or rectifier may be bad or the voltage regulator may be malfunctioning.

Alternator - removal and installation

1 Detach the ground cables from the negative terminals of the batteries.

3.103 Remove the mounting bolt (arrow)

3.104 Unplug the electrical connector from the fuel shut-off solenoid terminal (arrow)

2 Detach the wires from the alternator.
3 Remove the adjustment bolt **(see illustration)**, then remove the alternator drivebelt (see the *Maintenance* portion of this Chapter).
4 Remove the lower mounting bolt and separate the alternator from the engine **(see illustration)**.
5 If you're replacing the alternator, take the old one with you when purchasing the replacement. Make sure the new/rebuilt unit looks identical to the old one. Look at the terminals - they should be the same in number, size and location as the terminals on the old alternator. Finally, look at the identification numbers stamped into the housing or printed on a tag attached to the housing. Make sure the numbers are the same on both alternators.
6 Many new/rebuilt alternators DO NOT have a pulley installed, so you may have to switch the pulley from the old unit to the new/rebuilt one. When buying an alternator, find out the shop's policy regarding pulleys - some shops will perform this service free of charge.
7 Installation is the reverse of removal.
8 After the alternator is installed, check the drivebelt tension.
9 Check the charging voltage to verify proper operation of the alternator.

Starting system - general information and precautions

The sole function of the starting system is to turn over the engine quickly enough to allow it to start.

The starting system consists of the batteries, the starter motor, the starter solenoid and the wires connecting them.

The starter motor/solenoid assembly is installed on the lower part of the engine, next to the transmission bellhousing.

When the ignition key is turned to the Start position, the starter solenoid is actuated through the starter control circuit. The starter solenoid then connects the battery to the starter. The batteries supply the electrical energy to the starter motor, which does the actual work of cranking the engine.

The starter motor on a vehicle equipped with a manual transmission can only be operated when the clutch pedal is depressed; the starter on a vehicle equipped with an automatic transmission can only be operated when the shift lever is in Park or Neutral.

Always observe the following precautions when working on the starting system:

a) *Excessive cranking of the starter motor can overheat it and cause serious damage. Never operate the starter motor for more than 15 seconds at a time without pausing to allow it to cool for at least two minutes.*
b) *The starter is connected directly to the batteries and could arc or cause a fire if mishandled, overloaded or shorted out.*
c) *Always detach the cable from the negative terminal of the batteries before working on the starting system.*

Starter motor - check

1 If the starter motor doesn't turn at all when the switch is operated, make sure the shift lever is in Neutral or Park (automatic transmission) or the clutch pedal is depressed (manual transmission).
2 Make sure the batteries are charged and all cables, both at the battery and starter solenoid terminals, are clean and secure.
3 If the starter motor spins but the engine doesn't turn, the overrunning clutch in the starter motor is slipping and the starter motor must be rebuilt or replaced.
4 If, when the switch is actuated, the starter motor doesn't operate at all but the solenoid clicks, then the problem lies with either the batteries, the solenoid contacts or the starter motor itself (or the engine is seized).
5 If the solenoid plunger cannot be heard when the switch is actuated, either the batteries are bad, the fusible link is burned, the circuit is open or the solenoid is defective.
6 To check the solenoid, connect a jumper wire between the battery (+) and the ignition switch wire terminal (the small terminal) on the solenoid. If the starter motor now operates, the solenoid is okay and the problem is in the ignition switch, neutral start switch or the wires.
7 If the starter motor still does not operate, remove the starter for disassembly, testing and repair.
8 If the starter motor cranks the engine at an abnormally slow speed, first make sure the batteries are charged and all terminal connections are tight. If the engine is partially seized, or has the wrong viscosity oil in it, it'll crank slowly.
9 Run the engine until normal operating temperature is reached, then temporarily disconnect the wire from the fuel shut-off solenoid **(see illustration)** to prevent the engine from starting during the test.
10 Connect a voltmeter positive lead to a positive battery post and connect the negative lead to the negative post.
11 Crank the engine and take the voltmeter readings as soon as a steady figure is indicated. Do not allow the starter motor to turn for more than 15 seconds at a time. A reading of 9-volts or more, with the starter motor turning at normal cranking speed, is normal. If the reading is 9-volts or more but the cranking speed is slow, the starter is faulty. If the reading is less than 9-volts and the cranking speed is slow, the starter motor is bad or the battery is discharged or the engine is tight.

Starter motor - removal and installation

1 Detach the cables from the negative terminals of the batteries.
2 Block the rear wheels and apply the parking brake. Raise the front of the vehicle and support it securely on jackstands.

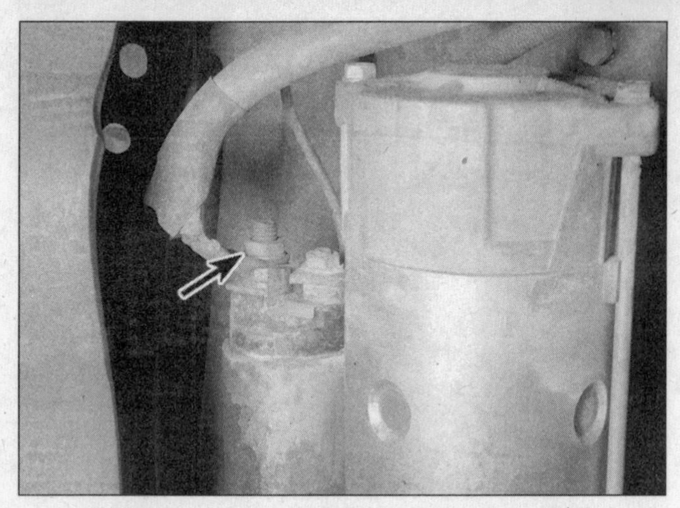

3.105 Working from below, disconnect the cable(s) from the starter solenoid (arrow)

3.106 Remove the bolts (arrows) and remove the starter (third bolt hidden from view)

3 Clearly label, then disconnect, the wires from the terminals on the starter motor solenoid **(see illustration)**.
4 Remove the mounting bolts and detach the starter **(see illustration)**.
5 Installation is the reverse of removal. Be sure to snug all the bolts while holding the starter in place and fully inserted into the pilot hole.
6 Check the operation of the starter.

Emission control systems

General information

Compared to gasoline engines, diesel engines produce very little carbon monoxide and unburned hydrocarbons. Additionally, diesel fuel is not as volatile as gasoline; at normal ambient temperatures diesel fuel does not evaporate readily. Therefore, fewer emission controls are required on diesel powered vehicles.

On the Ford diesel engines, internal modifications intended to decrease emissions are included in the design. This reduces the need for external add-on emission control devices.

Because of a federally-mandated extended warranty which covers the emission control system, check with a dealer about warranty coverage before working on any part of the system. Expensive items such as the injection pump and injectors are among the components covered. Once the warranty has expired, you may wish to do the work yourself.

A Vehicle Emissions Control Information (VECI) label is located in the engine compartment. This label contains important emissions specifications and adjustment information, as well as an emission control system schematic diagram. When servicing the engine or emissions systems, the VECI label should be checked for specific information on your vehicle.

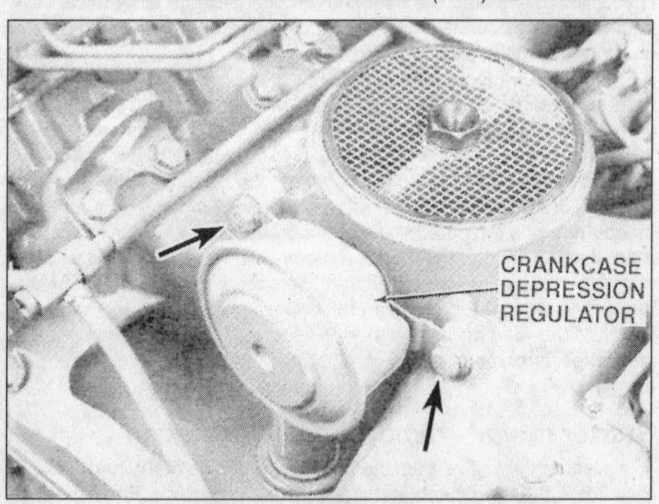

3.107 Remove the mounting bolts (arrows)

Crankcase Depression Regulator (CDR)

Internal combustion engines produce a certain volume of crankcase vapors during operation. Combustion pressure forces a small percentage of the combustion gases past the piston rings and into the crankcase.

The Crankcase Depression Regulator (CDR) routes crankcase vapors (blow-by) from the top of the valley cover into the intake manifold for reburning. It is a diaphragm-type pressure differential regulator that maintains near-atmospheric pressure in the crankcase.

Removal and installation

1 On "F-Series" trucks, open the hood. On "E-Series" trucks, remove the engine cover.
2 Remove the air cleaner and cover the air intake opening.
3 Remove the two CDR valve mounting bolts **(see illustration)** and remove the valve.
4 Install a new seal ring and valley cover grommet **(see illustration)** onto the valve and position the valve onto the engine.
5 Install the two mounting bolts and tighten them securely.
6 Remove the cover and install the air cleaner and engine cover, if applicable.

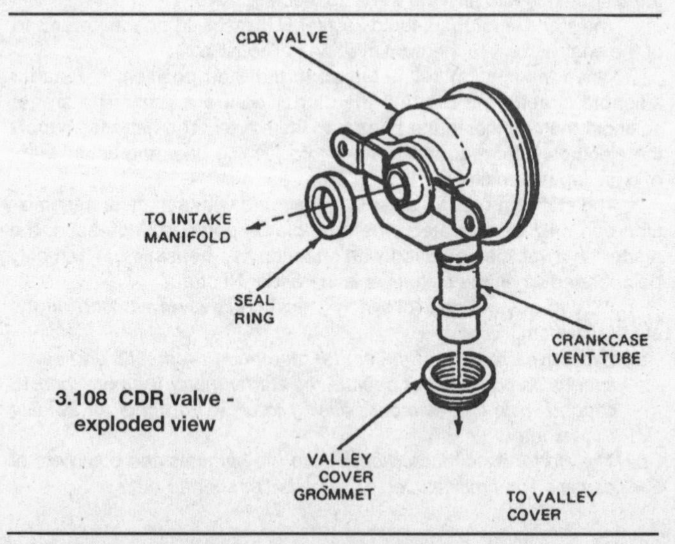

3.108 CDR valve - exploded view

In-vehicle engine repairs

General information

This part of Chapter 3 is devoted to in-vehicle repair procedures for Ford V8 diesel engines. All information concerning engine removal and installation and engine block and cylinder head overhaul can be found in Chapter 4.

Since the repair procedures included in this Chapter are based on the assumption that the engine is still installed in the vehicle, if they are being used during a complete engine overhaul (with the engine already out of the vehicle and on a stand) many of the steps included here will not apply.

The Specifications included in this Chapter apply only to the procedures found here. The specifications necessary for rebuilding the block and cylinder heads are included in Chapter 4.

Ford diesel engines are naturally-aspirated (non-turbocharged), 90-degree, overhead-valve V8s. Cylinder numbers 1, 3, 5 and 7 are on the left (passenger's side) bank; numbers 2, 4, 6 and 8 are on the right (driver's side) bank. The number 1 cylinder is at the water pump end. Firing order is 1-2-7-3-4-5-6-8.

The diesel V8 engines used in Ford light trucks come in displacements of 6.9 and 7.3 liters. Engines produced in model years 1983 through 1987 displace 6.9 liters (420 cubic inches), 1988 and newer models displace 7.3 liters (444 cubic inches). Almost all removal, inspection, installation and replacement procedures are the same for both models, with the exceptions noted within this Chapter.

Neither engine uses a carburetor or conventional gasoline-type fuel injection; they use a fuel system known as indirect injection (IDI). On a conventional gasoline engine, fuel is either drawn into the bore of a carburetor, or it's injected into a throttle body or directly into the intake port; on IDI-equipped diesels, fuel is injected indirectly into a pressurized "pre-combustion" chamber. There's no throttle valve - power and speed are controlled by the amount and timing of the injected fuel. Because there's no throttle valve, there's very little intake manifold vacuum, so both engines are equipped with a vacuum pump to operate vacuum controlled accessories.

There are no spark plugs or high voltage ignition system. Diesel ignition occurs when the heat developed in the combustion chamber during compression ignites the injected fuel. In effect, the injection nozzles, which are calibrated to open at a specified fuel pressure, take the place of the spark plugs.

Ford light truck diesel engines use two batteries, a heavy duty starter and "glow plugs" to speed up cold starts. Because diesel fuel must reach a certain temperature before it will ignite, diesel engines can take a little longer to fire up when they're cold, hence the extra battery and heavy duty starter motor to withstand the slightly longer cranking period. Glow plugs - small resistor-like heating elements which protrude into the pre-combustion chambers - heat up the diesel fuel in the pre-combustion chamber until the cylinder temperature is high enough to sustain combustion. A sensor in the cooling system signals the glow plug controller how long to leave the glow plugs on, based on coolant temperature.

Repair operations possible with the engine in the vehicle

Many major repair operations can be accomplished without removing the engine from the vehicle.

Clean the engine compartment and the exterior of the engine with some type of pressure washer before any work is done. A clean engine will make the job easier and will help keep dirt out of the internal areas of the engine. **Caution:** *Due to the exact tolerances in the injection pump, never clean or wash the engine unless it is cold. Failure to do so could result in damage to the injection pump.*

Depending on the components involved, it may be a good idea to remove the hood to improve access to the engine as repairs are performed.

If oil or coolant leaks develop, indicating a need for gasket or seal replacement, the repairs can generally be made with the engine in the vehicle. The oil pan gasket, the cylinder head gaskets, intake and exhaust manifold gaskets, timing cover gaskets and the crankshaft oil seals are accessible with the engine in place.

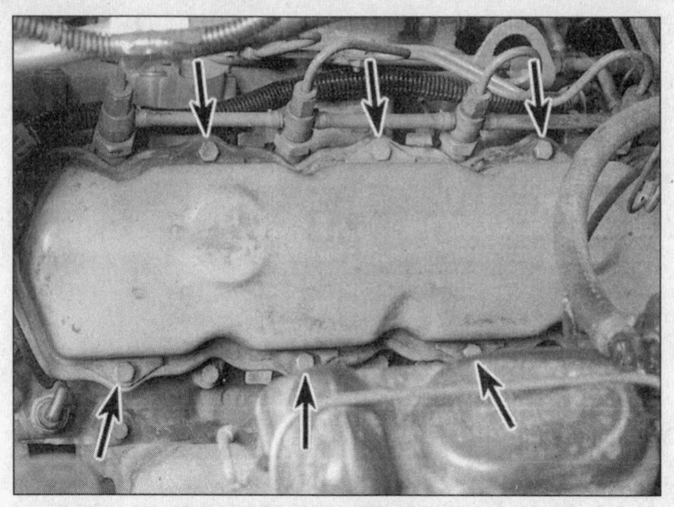

3.109 Valve cover bolt locations (arrows)

Exterior engine components, such as the water pump, the starter motor, the alternator and the fuel injection components, as well as the intake and exhaust manifolds, can be removed for repair with the engine in place.

Since the cylinder heads can be removed without pulling the engine, valve component servicing can also be accomplished with the engine in the vehicle.

Replacement of, repairs to or inspection of the timing chain and sprockets and the oil pump are all possible with the engine in place.

Valve covers - removal and installation

Removal

1 Disconnect the negative cables from the batteries.

2 Remove the air cleaner assembly.

3 Detach any accessory components that are in the way of valve cover removal. If you are working on the passenger's side cover on an "F" series truck, remove the fuel filter.

4 Remove the breather tube or CD valve.

5 Remove the valve cover mounting bolts **(see illustration)**.

6 On 1994 and later Direct Injection Turbo models, disconnect the electrical connectors from the valve cover gasket.

7 Remove the valve cover. On some models it may be necessary to raise the engine off its mounts slightly. **Note:** *If the cover is stuck to the head, bump the cover with a block of wood and a hammer to release it. If it still will not come loose, try to slip a flexible putty knife between the head and cover to break the seal. Do not pry at the cover-to-head joint or damage to the sealing surface and cover flange will result and oil leaks will develop.*

Installation

8 The mating surfaces of each cylinder head and valve cover must be perfectly clean when the covers are installed. Use a gasket scraper to remove all traces of sealant or old gasket, then wipe the mating surfaces with a cloth saturated with lacquer thinner or acetone. If there is sealant or oil on the mating surfaces when the cover is installed, oil leaks may develop.

9 Make sure the threaded holes are clean. Run a tap into them to remove corrosion and restore damaged threads.

10 Mate the new gaskets to the covers before the covers are installed. Apply a thin coat of RTV sealant to the cover flange, then position the gasket inside the cover lip and allow the sealant to set up so the gasket adheres to the cover (if the sealant is not allowed to set, the gasket may fall out of the cover as it is installed on the engine). On

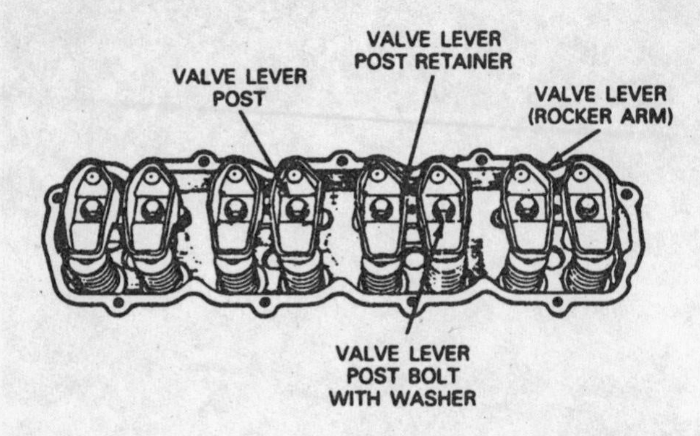

3.110 Rocker arm details (non-DI Turbo)

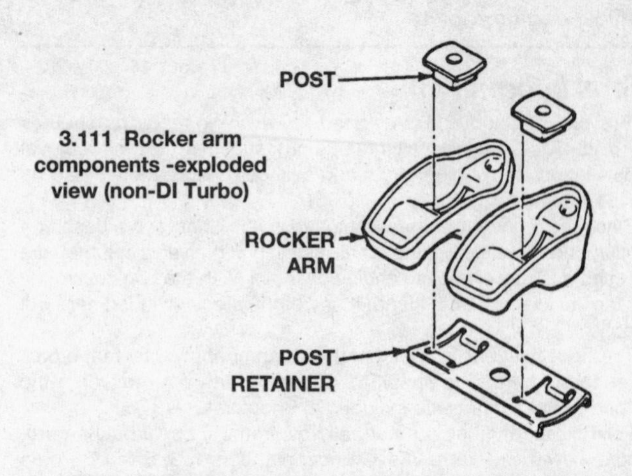

3.111 Rocker arm components -exploded view (non-DI Turbo)

1994 and later Direct Injection Turbo models, connect the wiring connectors to the valve cover gasket before placing the valve cover on the cylinder head.

11 Carefully position the cover on the head and install the bolts.

12 Tighten the nuts/bolts in three steps to the torque listed in this Chapter's Specifications.

13 The remaining installation steps are the reverse of removal.

14 Start the engine and check carefully for oil leaks as the engine warms up.

Rocker arms and pushrods

1994 and earlier (except Direct Injection Turbo) models

Removal

1 Refer to the previous Section and detach the valve covers from the cylinder heads.

2 Beginning at the front of one cylinder head, loosen and remove the rocker arm post bolts **(see illustration)**. Store the parts separately in marked containers to ensure that they will be reinstalled in their original locations.

3 Lift off the rocker arms, posts and post retainers **(see illustration)** and store them in the marked containers with the bolts (they must be reinstalled in their original locations).

4 Remove the pushrods and store them in order to make sure they will not get mixed up during installation.

Inspection

5 Check each rocker arm for wear, cracks and other damage, especially where the pushrods and valve stems contact the rocker arm faces.

6 Make sure the hole at the pushrod end of each rocker arm is open.

7 Check each rocker arm pivot area for wear, cracks and galling. If the rocker arms are worn or damaged, replace them with new ones and use new posts as well.

8 Inspect the pushrods for cracks and excessive wear at the ends. Roll each pushrod across a piece of plate glass to see if it is bent (if it wobbles, it is bent).

Installation

9 Lubricate the lower end of each pushrod with clean engine oil or moly-base grease and install them in their original locations. Install the copper-colored ends of the pushrods toward the rocker arms. Make sure each pushrod seats completely in the lifter socket.

10 Apply moly-base grease (Ford no. D0AZ-19584-AA or equivalent) to the ends of the valve stems and the upper ends of the pushrods before positioning the rocker arms.

11 Set the rocker arms in place, then install the posts and bolts. Apply moly-base grease to the contact surfaces to prevent damage to the mating surfaces before engine oil pressure builds up.

12 Using a socket and breaker bar on the vibration damper bolt, turn the crankshaft by hand until the timing mark is at the 11 o'clock position as viewed from the front of the engine.

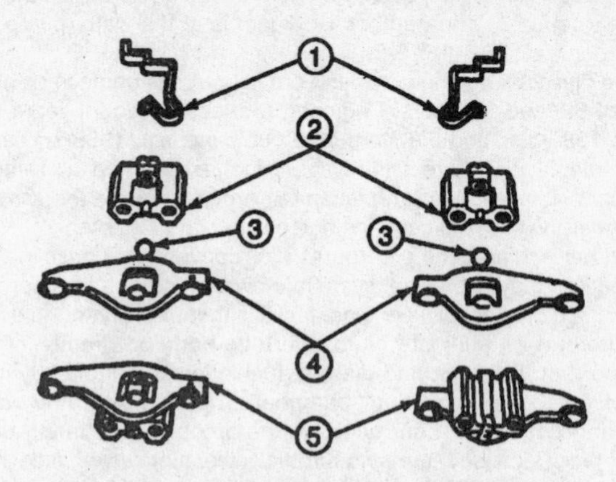

3.112a 1994 and later Direct Injection Turbo rocker arm components

1 *Retaining clip*
2 *Rocker arm pedestal*
3 *Rocker arm ball*
4 *Rocker arm*
5 *Rocker arm assembly complete*

13 Install all of the rocker arms, posts and bolts and tighten them to the torque listed in this Chapter's Specifications.

14 Install the remaining components in the reverse order of removal.

15 Refer to the appropriate Section and install the valve covers. Start the engine, listen for unusual valve train noises and check for oil leaks at the valve cover joints.

1994 and later Direct Injection Turbo models

Removal

1 Remove the valve covers and gaskets.

2 Remove the rocker arm retaining bolts.

3 Remove the rocker arms and pushrods.

4 Arrange to store the rocker arms and pushrods in an organized manner so they can be reinstalled in their original positions.

Inspection

5 Disassemble each rocker arm by removing the retaining clips from the rocker arm assembly **(see illustration)**.

6 Remove the rocker arm and steel ball from the pedestal. Be careful not to lose the steel ball.

7 Keep all components together and properly organized so they can be reassembled in their original locations.

8 Inspect the components as described above.

9 To reassemble a rocker arm, place the steel ball the rocker arm cup. Lubricate the ball with clean engine oil.

10 Place the rocker arm pedestal on the steel ball and snap the retaining clip over the pedestal groove **(see illustration)**. Assemble the remaining rocker arms.

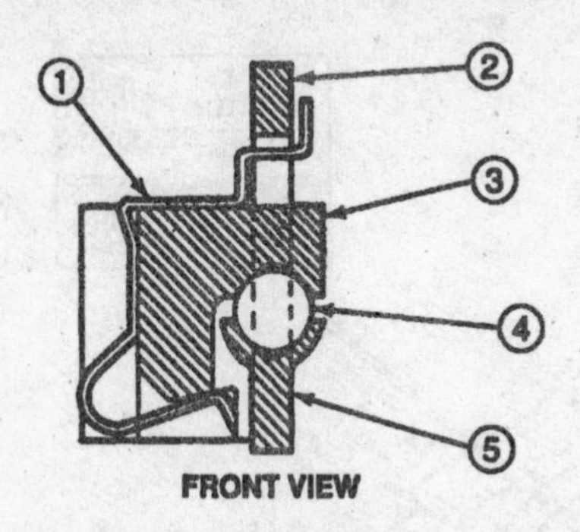

3.112b 1994 and later Direct Injection Turbo rocker assembly details

1 Retaining clip
2 Rocker arm
3 Pedestal
4 Ball
5 Rocker arm assembly

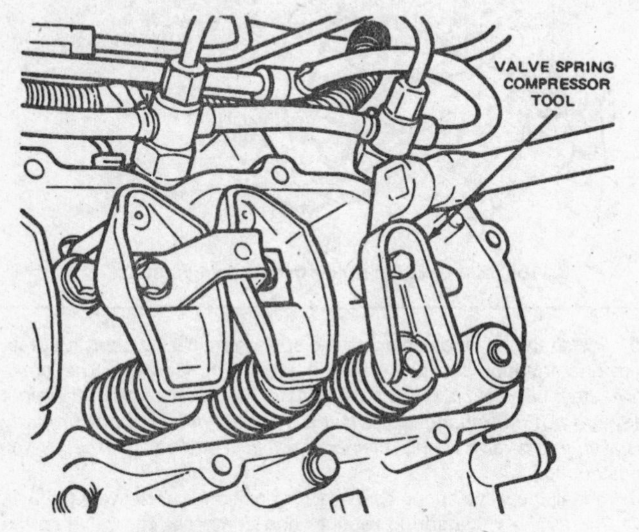

3.114a Compress the valve spring with Ford tool T83T-6513-A or equivalent

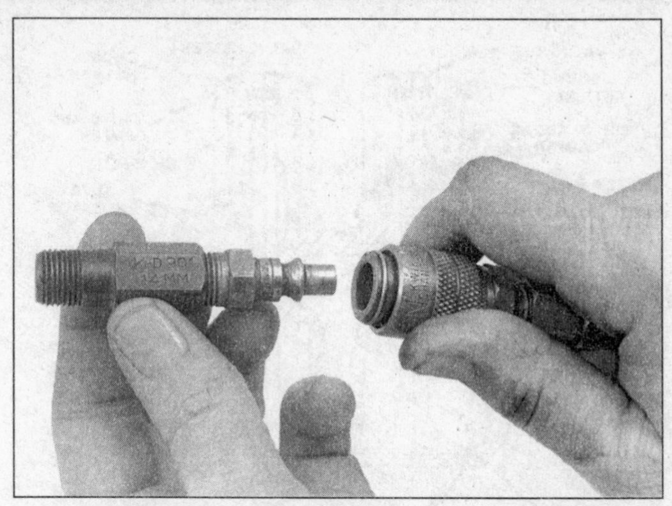

3.113 This is what the air hose adapter that threads into the spark plug hole looks like - they're commonly available from auto parts stores

3.114b Use a valve spring compressor to compress the springs, then remove the keepers from the valve stem with a magnet or small needle-nose pliers

Installation

11 Lubricate each push rod end with clean engine oil and place them in their original positions with the copper painted ends up.

12 Rotate the crankshaft until the mark on the damper is at the 11 o'clock position. **Caution:** *Failure to position the engine with the mark on the crankshaft damper at 11 o'clock before tightening the rocker arms could result in bent valves and other engine damage.*

13 Install the rocker arms and tighten the bolts to 20 ft-lb.

14 Install the valve cover gaskets, wiring connectors and valve covers.

Valve springs, retainers and seals - replacement

Note: *Broken valve springs and defective valve stem seals can be replaced without removing the cylinder head, providing damage to the valve or valve seat have not occurred. Two special tools and a compressed air source are normally required to perform this operation, so read through this Section carefully and rent or buy the tools before beginning the job. If compressed air is not available, a length of nylon rope can be used to keep the valves from falling into the cylinder during this procedure.*

1 Remove the valve cover from the affected cylinder head. If all of the valve stem seals are being replaced, remove both rocker arm covers.

2 Remove the glow plug(s) from the cylinder(s) with the defective component(s), using Ford glow plug socket no. D83T-6002-A or equivalent. If all of the valve stem seals are being replaced, all of the glow plugs should be removed.

3 Thread an adapter (from Ford Rotunda compression tester no. 014-00701, or equivalent) into the glow plug hole and connect an air hose from a compressed air source to it **(see illustration)**. **Note:** *Some diesel cylinder compression gauges utilize a screw-in fitting that may work with your air hose quick-disconnect fitting.*

4 Remove the bolt, lever post and rocker arm for the valve with the defective part and pull out the pushrod. If all of the valve stem seals are being replaced, all of the rocker arms and pushrods should be removed. Refer to the previous Section as necessary.

5 Apply compressed air to the cylinder. The valves should be held in place by the air pressure. If the valve faces or seats are in poor condition, leaks may prevent the air pressure from retaining the valves, in this case remove the cylinder heads for repair. **Caution:** *The crankshaft may turn when air pressure is applied.*

6 Stuff shop rags into the cylinder head holes above and below the valves to prevent parts and tools from falling into the engine, then use a valve spring compressor (Ford no. T83T-6513-A or equivalent) to compress the spring/damper assembly **(see illustration)**. Remove the keepers with a pair of small needle-nose pliers or a magnet **(see illustration)**. **Note:** *A couple of different types of tools are available for compressing the valve springs with the head in place. One type grips the lower spring coils and presses on the retainer as the knob is turned, while the type shown utilizes the rocker arm bolt for leverage.*

7 Remove the spring retainer and valve spring/damper assembly. **Note:** *It may be necessary to tap the valve stem end with a soft-face hammer to loosen the valve retainers. Remove the valve stem seal and valve rotator* **(see illustration)**. **Note:** *If air pressure fails to hold the*

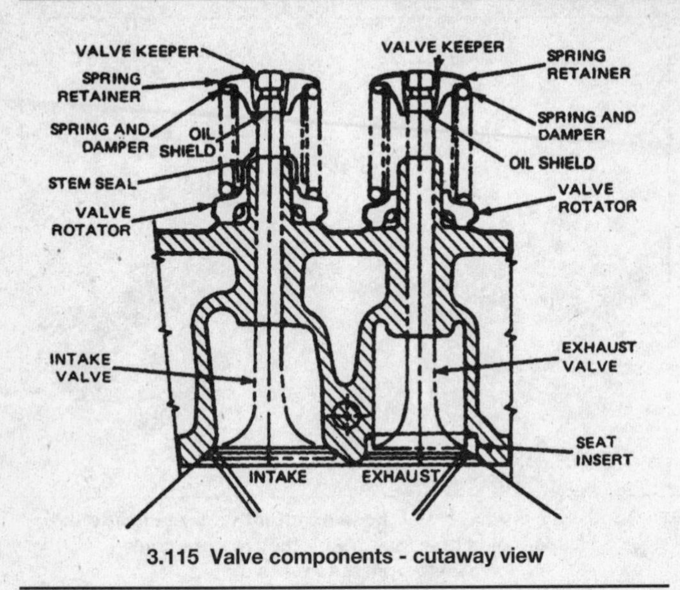

3.115 Valve components - cutaway view

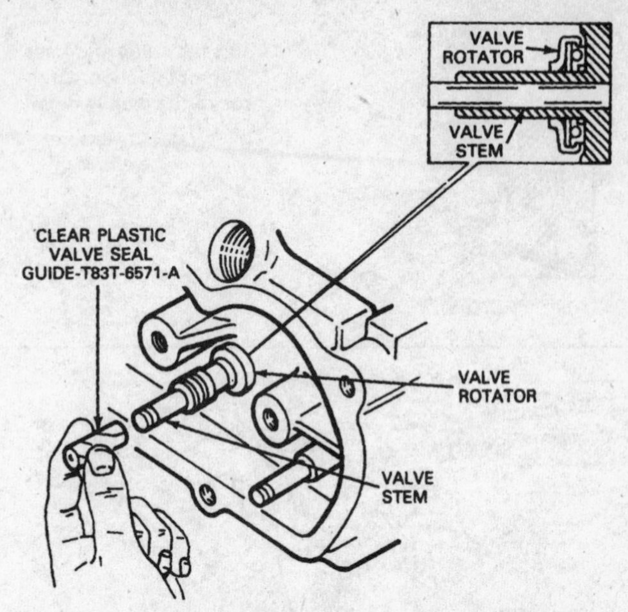

3.116a Slip the seal guide over the valve stem . . .

valve in the closed position during this operation, the valve face or seat is probably damaged. If so, the cylinder head will have to be removed for additional repair operations.

8 Wrap a rubber band or tape around the top of the valve stem so the valve will not fall into the combustion chamber, then release the air pressure.

9 Inspect the valve stem for damage. Rotate the valve in the guide and check the end for eccentric movement, which would indicate that the valve is bent.

10 Move the valve up-and-down in the guide and make sure it doesn't bind. If the valve stem binds, either the valve is bent or the guide is damaged. In either case, the head will have to be removed for repair.

11 Inspect the rocker arm components for wear. Look for cracks and wear marks and grooves and replace as necessary.

12 Reapply air pressure to the cylinder to retain the valve in the closed position, then remove the tape or rubber band from the valve stem.

13 Lubricate the valve stem with engine oil and install the valve rotator.

14 For the intake valves only, install valve seal guides (Ford no. T83T-6571-A, or equivalent) over the valve stems **(see illustrations)**. Using Ford Valve Stem Seal Tool T83T-6571-A (or equivalent), press the seal into place gently by hand until it is fully seated on the guide. Remove the tool and seal guide.

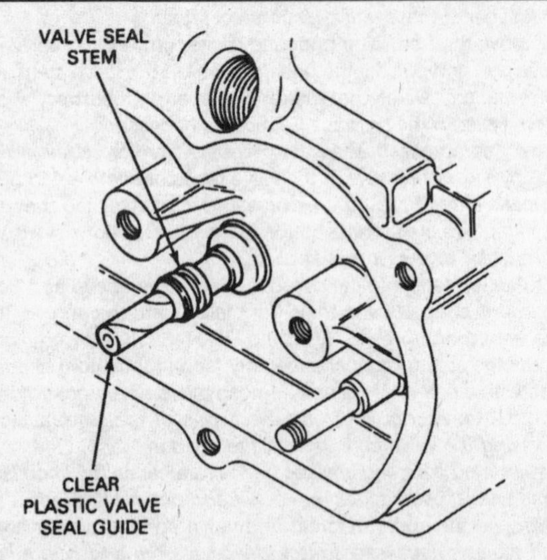

3.116b . . . then slip the valve seal into place

15 Install the nylon oil shield in the spring retainer by applying pressure until it snaps into place **(see illustration)**. **Caution:** *The intake valve stem oil shield is much smaller than the exhaust shield; therefore they are not interchangeable. If the nylon oil shield is not properly installed in the valve spring retainer, it will float and cause excessive oil consumption.*

16 Install the valve spring assembly in position over the valve.

17 Install the valve spring retainer and compress the valve spring assembly.

18 Position the keepers in the upper groove. Apply a small dab of grease to the inside of each keeper to hold it in place if necessary. Remove the pressure from the spring tool and make sure the keepers are seated.

19 Disconnect the air hose and remove the adapter from the glow plug hole. Apply multi-purpose grease (Ford no. DOAZ-19584-AA, or equivalent) to the tips of the valve stem and pushrods.

20 Repeat the above process for each valve, as needed. Refer to the appropriate Section and install the rocker arms and pushrods.

21 Install the glow plugs using Ford socket no. D83T-6002-A or equivalent and tighten them to the torque listed in this Chapter's Specifications.

22 Refer to the appropriate Section and install the valve covers.

23 Start and run the engine, then check for oil leaks and unusual sounds coming from the valve cover area.

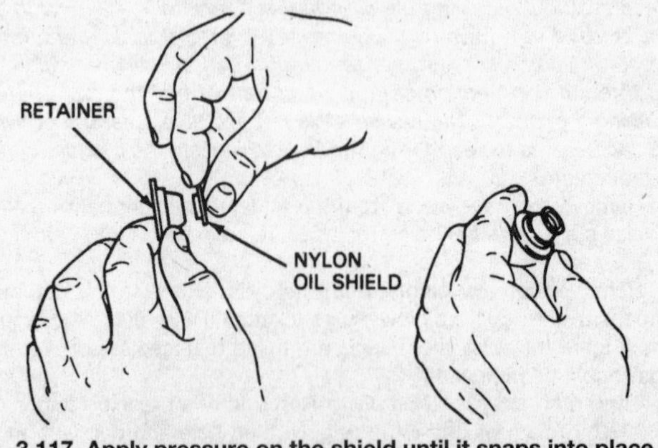

3.117 Apply pressure on the shield until it snaps into place

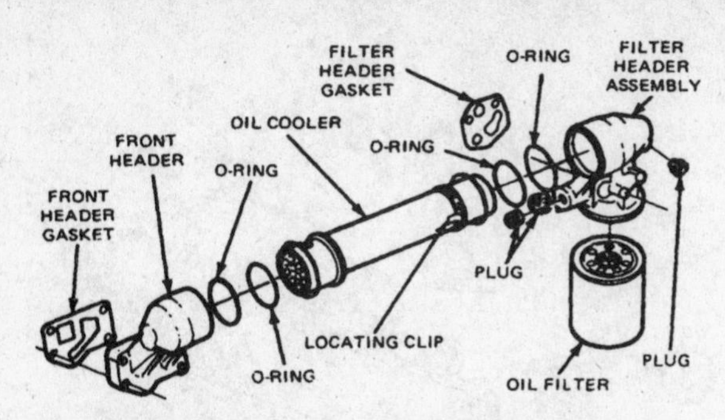

3.118 Typical oil cooler assembly - exploded view

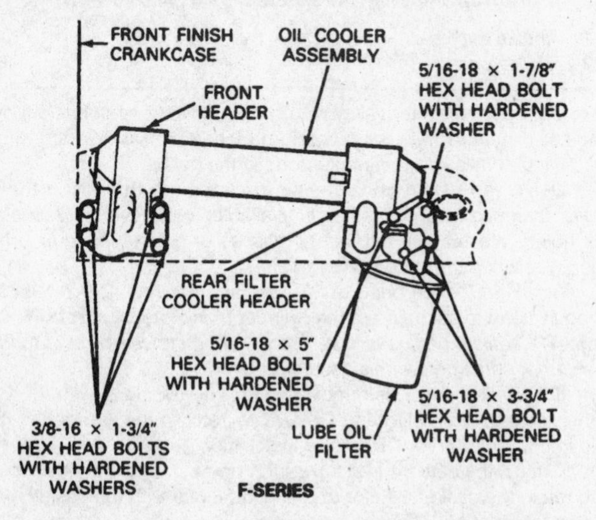

3.119b Oil cooler installation details (F-Series truck)

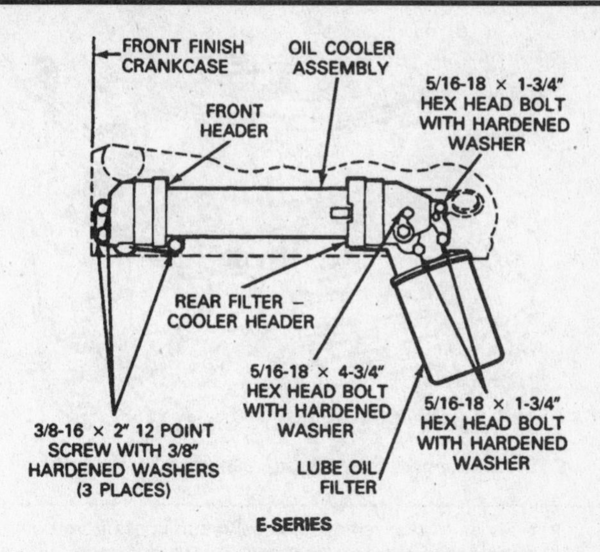

3.119a Oil cooler installation details (E-series truck)

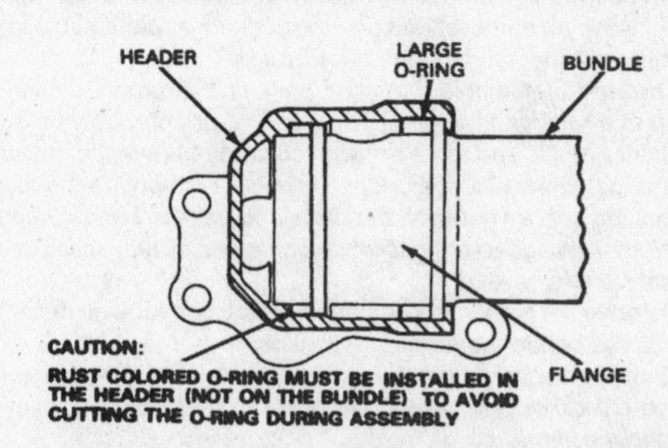

3.120 Apply oil to the O-rings and install them carefully

Oil cooler - removal and installation

Removal

1 Disconnect the ground cables from the negative terminals of the batteries.

2 Drain the cooling system.

3 Remove the radiator fan shroud and engine cooling fan/clutch assembly.

4 Block the rear wheels to prevent the vehicle from rolling. Set the parking brake and place the transmission in Park (automatic) or first gear (manual).

5 Raise the front of the vehicle and support it securely on jackstands.

6 Drain the engine oil and remove the oil filter. Temporarily leave the drain plug out.

F-Series trucks only

7 Remove the driver's side engine mount-to-frame nut.

8 Raise the driver's side of the engine slightly and insert a one-inch block of wood between the mount and frame. Use an overhead hoist or raise the engine from below using a floor jack with a large wood block to protect the oil pan.

All models

9 Remove the oil cooler mounting bolts and remove the oil cooler.

Installation

10 Assemble the oil cooler with new gaskets and O-rings **(see illustration)**. Place the oil cooler into position on the engine and install the mounting bolts **(see illustrations)**. Tighten the bolts to the torque listed in this Chapter's Specifications. **Caution:** *All oil coolers use four O-rings - be sure to use ones designed for this application. The inner O-ring must be installed on the header (end piece), not on the bundle, to avoid cutting the inner O-ring during assembly* **(see illustration).**

11 Install the oil filter and drain plug. On "F-Series" trucks, remove the wood block from the engine mount and install the mounting nut.

12 Add engine oil as described in the maintenance portion of this Chapter. Prime the lubricating system by operating the starter motor with a remote starter switch. When the oil pressure gauge or light registers pressure, the system is primed. **Caution:** *Using the dash-mounted "ignition" switch may cause the engine to start before oil has circulated.*

13 Refill and bleed the cooling system as described in the maintenance portion of this Chapter.

14 Reinstall any remaining components in the reverse order of removal.

15 Run the engine and check for oil and coolant leaks.

Intake manifold and valley cover (non-DI Turbo) - removal and installation

Removal

1 Disconnect the ground cables from the negative battery terminals. On E-Series trucks, remove the engine cover.

2 Remove the air cleaner assembly and cover the opening with a cloth or Ford cover no. T83T-9424-A.

3 Label and disconnect any fuel lines, wires and hoses from the vehicle to the intake manifold. On E-Series trucks, disconnect the fuel lines from the fuel filter and unbolt the fuel filter and bracket as an assembly.

4 Remove the fuel injection pump as described previously in this Chapter.

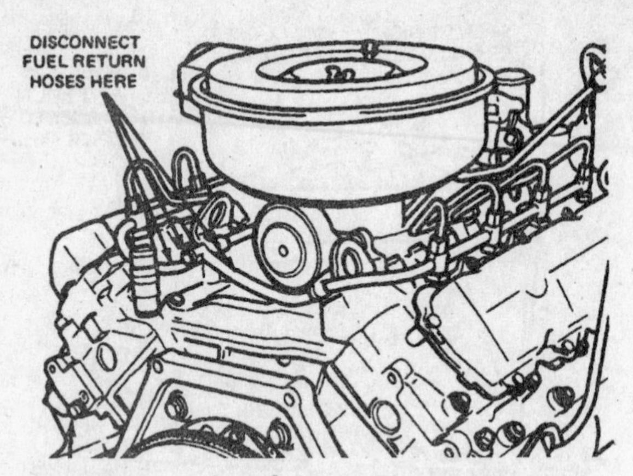

3.121 Disconnect the fuel return lines and ground strap

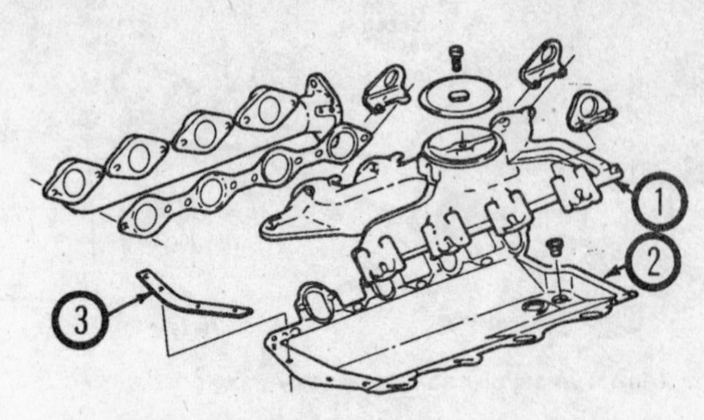

3.122 Intake manifold details - exploded view

1 *Intake manifold*
2 *Valley pan*
3 *End seal*

5 On F-Series trucks, remove the fuel return hoses from both rear injection nozzles **(see illustration)** and remove the return hose that goes to the fuel tank.

6 On all models, remove the glow plug harness and controller and the engine wiring harness. Be sure to detach the ground cable from the rear of the cylinder head on the driver's side.

7 Loosen the manifold mounting bolts in 1/2-turn increments until they can be removed by hand. The manifold will probably be stuck to the cylinder heads and force may be required to break the gasket seal. A pry bar can be positioned under a cast-in lug to pry up the manifold. **Caution:** *Do not pry between the block and manifold or the heads and manifold or damage to the gasket sealing surfaces may result and vacuum leaks could develop.*

8 Remove the crankcase depression reduction tube as described in the *Emission control* portion of this Chapter.

9 Remove the valley pan strap from the front of the engine. **Note:** *Remove the valley pan* **(see illustration)** *to replace the gasket or for access to the lifters.*

10 Remove the valley pan drain plug and remove the valley pan.

Installation

Note: *The mating surfaces of the cylinder heads, block, valley pan and manifold must be perfectly clean when the components are installed. Gasket removal solvents in aerosol cans are available at most auto parts stores and may be helpful when removing old gasket material that is stuck to the mating surfaces. Be sure to follow the directions printed on the container.*

11 Use a gasket scraper to remove all traces of sealant and old gasket material, then wipe the mating surfaces with a cloth saturated with lacquer thinner or acetone. If there is old sealant or oil on the mating surfaces when the manifold is installed, oil or vacuum leaks may develop. Cover the lifter valley with shop rags to keep debris out of the engine. If necessary, use a vacuum cleaner to remove any gasket material that falls into the intake ports in the heads.

12 Use a tap of the correct size to chase the threads in the bolt holes, then use compressed air (if available) to remove the debris from the holes. **Warning:** *Wear safety glasses or a face shield to protect your eyes when using compressed air.*

13 Apply a 1/8-inch bead of RTV sealant (Ford no. D6AZ-19562-AA or equivalent) to each end of the cylinder block **(see illustration). Caution:** *RTV sealant begins to cure in about 15 minutes - be sure to install the components before this occurs.*

14 If the valley pan was removed, install it now along with the valley pan strap and drain plug and CDR valve (refer to the appropriate Section in this Chapter for CDR valve installation details).

15 Extend the sealant bead 1/2-inch up each cylinder head to seal and retain the gaskets. Refer to the instructions with the gasket set for further information.

16 Position the intake manifold gaskets on the cylinder heads. Make sure all intake port openings, coolant passage holes and bolt holes are aligned correctly.

17 Carefully set the manifold in place. Do not disturb the gaskets and do not move the manifold fore-and-aft after it makes contact.

18 Install the manifold bolts and tighten them to the torque listed in this Chapter's Specifications following the recommended sequence **(see illustration)**. First tighten the bolts working side-to-side from the center out, then go around the perimeter as shown.

19 The remaining installation steps are the reverse of removal. Refer to the fuel system Section in this Chapter for information on purging air from the system. Start the engine and check carefully for oil, air and coolant leaks at the intake manifold joints.

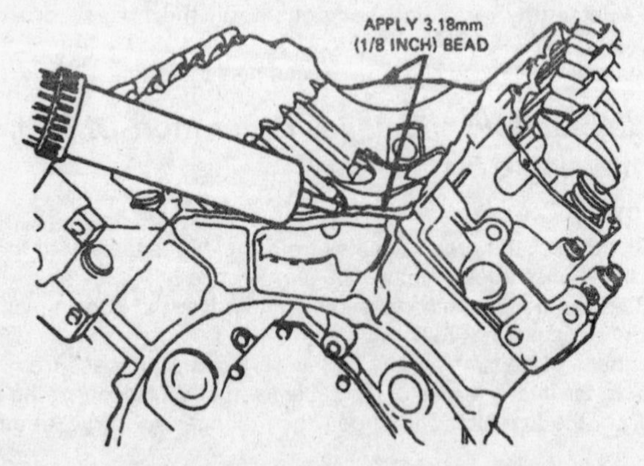

3.123 Apply a bead of RTV "sealant as shown here

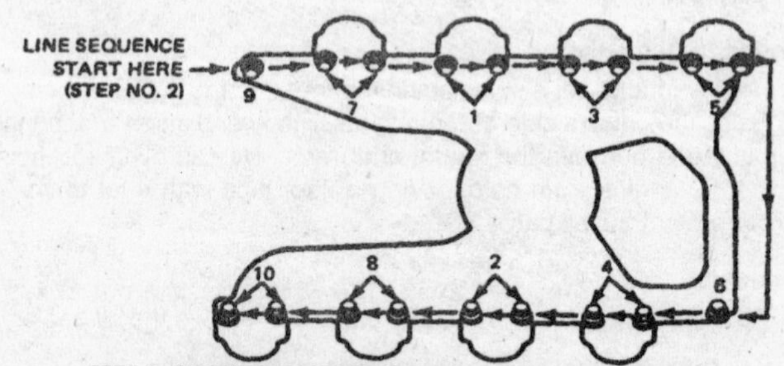

3.124 Intake manifold bolt tightening sequence

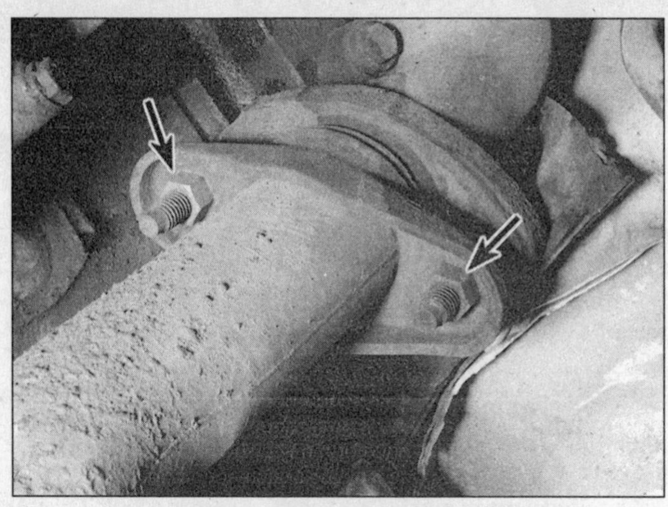

3.125 Remove the manifold-to-pipe nuts (arrows)

Exhaust manifolds - removal and installation

Warning: *Allow the engine to cool completely before following this procedure.*

Removal

1 Disconnect the ground cables from the negative terminals of the battery. On E-Series trucks, remove the engine cover.

2 Block the rear wheels and apply the parking brake. On manual transmission models, engage low gear. On automatics, engage Park.

3 Raise the front of the vehicle and support it securely on jackstands.

E-Series trucks - right (passenger's) side only

4 Remove the fan shroud from the radiator.

5 Unbolt the dipstick tube(s) for the engine oil and automatic transmission (if equipped) and remove the dipstick tube(s).

6 Remove the nuts attaching the right front engine mount to the frame.

7 Place a wood block under the oil pan and raise the engine slightly with a floor jack until the fuel filter header contacts the body sheet metal. Place a wood block between the mount and frame and remove the jack. **Warning:** *Do not place any part of your body where it could be crushed if the jack slips.*

All models

8 Disconnect the exhaust pipes from the exhaust manifolds **(see illustration). Note:** *Often a short period of soaking with penetrating oil is necessary to remove frozen exhaust pipe attaching nuts. Use caution not to apply excessive force to frozen nuts, which could shear off the exhaust manifold studs.*

9 Remove the manifold mounting bolts **(see illustration).**

Installation

10 Installation is basically the reverse of the removal procedure. Clean the manifold and head gasket surfaces of old gasket material, then install new gaskets. Do not use any gasket cement or sealer on exhaust system gaskets.

11 Apply anti-seize compound to the bolt threads. Install all the manifold bolts and tighten them to the torque listed in this Chapter's Specifications in two steps. First, work from the center to the ends, then tighten them from the rear to the front **(see illustration).**

12 Apply anti-seize compound to the exhaust manifold-to-exhaust pipe studs, and use a new exhaust "doughnut" gasket, if equipped. Install the pipes and tighten the nuts securely.

13 Reinstall any remaining components. Lower the vehicle and reconnect the batteries. Run the engine and check for exhaust leaks. On E-Series trucks, install the engine cover.

Cylinder heads - removal and installation

1994 and earlier (except Direct Injection Turbo) models

Removal

Caution: *Allow the engine to cool completely before following this procedure.*

1 Disconnect the ground cables from the negative battery terminals.

2 Remove the engine cooling fan and shroud as described in the cooling system portion of this Chapter. Note that the cooling system must be drained to prevent coolant from getting into internal areas of the engine when the manifold and heads are removed.

3 Remove the valve covers as described previously in this Chapter.

4 Remove the rocker arms and pushrods as described previously in this Chapter.

5 Label and disconnect wiring and hoses as necessary. Then remove the alternator and vacuum pump and brackets - refer to the appropriate Sections in this Chapter.

6 Disconnect and cap the fuel lines at the fuel filter, then remove the fuel injection pump, nozzles and glow plugs as described in the Fuel system and Electrical system portions of this Chapter.

7 Remove the intake manifold as described previously in this Chapter.

8 Detach the exhaust manifolds as described previously in this Chapter.

9 Loosen the head bolts in 1/4-turn increments until they can be removed by hand. Work from bolt-to-bolt in a pattern that is the reverse of the tightening sequence.

10 Attach lifting brackets (Ford no. T70P-6000 or equivalent) to the cylinder heads. Connect a lifting sling to the brackets and a hoist, or

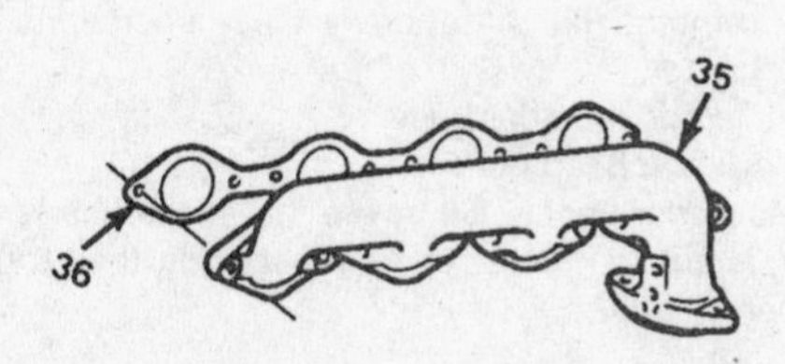

3.126 Remove the manifold bolts (arrows)

1 *Exhaust manifold* 2 *Gasket*

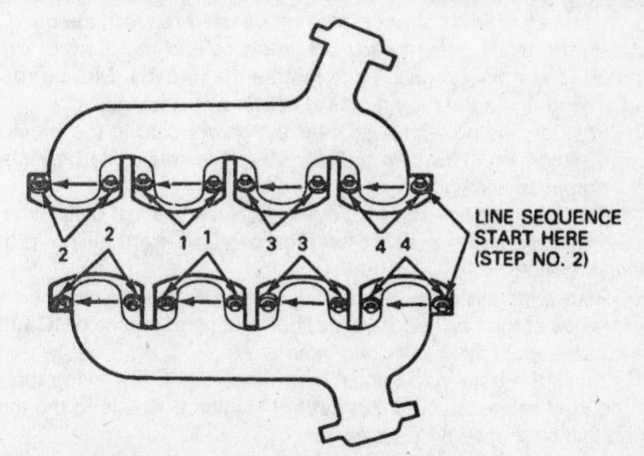

3.127 Exhaust manifold bolt tightening sequence

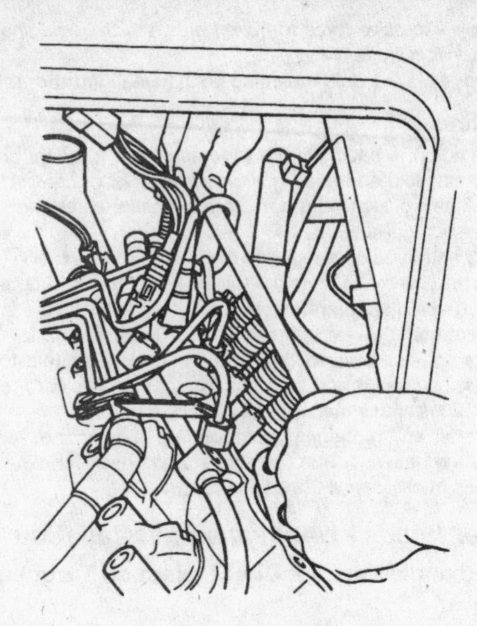

3.128 Attach lifting brackets and carefully lift the cylinder head (E-Series shown, F-Series similar)

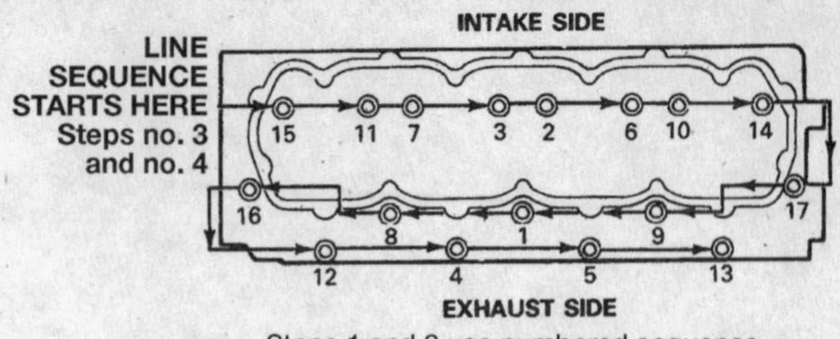

3.129 Cylinder head bolt tightening sequence

use a bar and chain with an assistant. Lift the heads off the engine **(see illustration)**. If resistance is felt, do not pry between the head and block as damage to the mating surfaces will result; also avoid damaging the cylinder head locating dowels. To dislodge the head, place a block of wood against the end of it and strike the wood block with a hammer. Store the heads on blocks of wood to prevent damage to the gasket sealing surfaces.

11 Cylinder head disassembly and inspection procedures are covered in detail in Chapter 4.

Installation

12 The mating surfaces of the cylinder heads and block must be perfectly clean when the heads are installed.

13 Use a gasket scraper to remove all traces of carbon and old gasket material, then wipe the mating surfaces with a cloth saturated with lacquer thinner or acetone. If there is oil on the mating surfaces when the heads are installed, the gaskets may not seal correctly and leaks may develop. When working on the block, cover the lifter valley with shop rags to keep debris out of the engine. Use a vacuum cleaner to remove any debris that falls into the cylinders.

14 Check the block and head mating surfaces for nicks, deep scratches and other damage. If damage is slight, it can be removed with emery cloth. If it is excessive, machining may be the only alternative.

15 Use a tap of the correct size to chase the threads in the head bolt holes in the block. Mount each bolt in a vise and run a die down the threads to remove corrosion and restore the threads. Dirt, corrosion, sealant and damaged threads will affect torque readings.

16 Position the new gaskets over the dowel pins in the block. Be sure the marking *"This side up"* is visible. Gaskets must be installed dry - don't use sealant.

17 Carefully position the heads on the block without disturbing the gaskets. Use care to prevent the pre-chambers from falling into the cylinder bores.

18 Before installing the head bolts, coat the threads, washers and undersides of the bolt heads with a light coating of engine oil. **Caution:** *Do not use anti-seize compound, grease, etc.*

19 Install the bolts and tighten them finger tight. Following the recommended sequence **(see illustration)**, tighten the bolts to the torque listed in this Chapter's Specifications.

20 The remaining installation steps are the reverse of removal.

1994 and later Direct Injection Turbo models

Removal

Warning: *The air conditioning system is under high pressure. DO NOT loosen any fittings or remove any components until after the system has been discharged. Air conditioning refrigerant should be properly discharged into an EPA-approved container at a dealership service department or an automotive air conditioning facility. Always wear eye protection when disconnecting air conditioning system fittings.*

1 If equipped with air conditioning, have the system discharged.
2 Disconnect the negative battery cable from both batteries.
3 Remove the radiator. Remove the turbocharger.
4 Remove the crankcase breather with it's two screws.
5 Disconnect the air conditioning compressor wiring.
6 Remove the air conditioning compressor bracket bolts.
7 Disconnect the vacuum hose at the vacuum pump.
8 Disconnect the refrigerant lines from the air conditioning compressor.
9 Remove the lines from the power steering pump.
10 Remove the air conditioning compressor bracket and the accessory brackets.
11 Remove the valve covers and gaskets.
12 Drain the fuel and water separator.
13 Disconnect the fuel lines from the rear of both cylinder heads.
14 Remove the banjo bolt from the fuel pump and remove the fuel line assembly.
15 Remove the fuel line nut from the intake manifold stud.
16 Disconnect the fuel return lines.
17 Remove the drivebelt, disconnect the electrical connectors from the alternator and remove the alternator with it's bracket.
18 Remove the dipstick tube.
19 Remove the MAP sensor and move it aside.
20 Remove the high pressure oil supply lines from the cylinder heads.
21 Remove the left turbocharger exhaust inlet pipe bolts from the exhaust manifold.
22 Remove the exhaust back pressure line.
23 Remove the outer half of the heater plenum
24 Remove any ground straps attached to the cylinder heads.
25 Remove the glow plugs, injectors, rocker arm assemblies and push rods. Refer to the sections that cover these operations in detail elsewhere in this chapter.
26 Remove the injector shoulder bolts.
27 Remove the cylinder head bolts.
28 Remove the cylinder heads. Be aware that these heads are very heavy. Get an assistant to help or use a hoist to lift them out of the engine compartment.

Installation

29 Follow the above installation procedures for 1994 and earlier models with the following additions:

a) *Following the recommended sequence* **(see illustration)**, *tighten the cylinder head bolts to the torque listed in this Chapter's Specifications.*

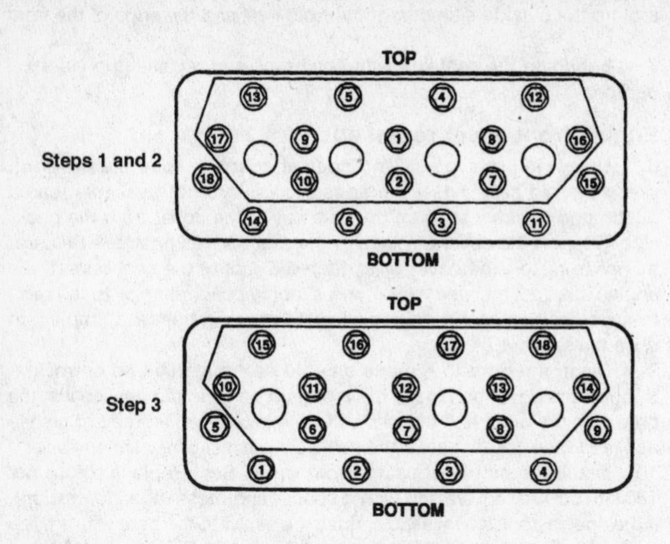

3.130 1994 and later Direct Injection Turbo cylinder head bolt torque sequence - note Steps one and two use a different torque sequence than Step three

b) *Rotate the crankshaft until the mark on the damper is at the 11 o'clock position.* **Caution:** *Failure to position the engine with the mark on the crankshaft damper at 11 o'clock before tightening the rocker arms could result in bent valves and other engine damage.*

c) *Remove the fuel and oil from the injector bores before installing the injectors.* **Caution:** *Failure to remove the fuel and oil from the injector bores could result in hydrostatic lock and severe engine damage.*

Vibration damper - removal and installation

Removal

1 Disconnect the ground cables from the negative terminals of the batteries.

2 Remove the fan shroud and engine cooling fan/clutch assembly (refer to the *Cooling system* portion of this Chapter).

3 Remove the drivebelts.

4 Block the rear wheels to prevent the vehicle from rolling. Set the parking brake and place the transmission in Park (automatic) or first gear (manual).

5 Raise the front of the vehicle and support it securely on jackstands.

6 Unbolt the crankshaft pulley and remove it from the engine compartment.

7 Loosen the vibration damper bolt **(see illustration)**. If available, use Ford Crank/Cam Gear and Damper Remover T83T-6316-A or equivalent to hold the damper stationary. If you don't have access to one of these tools, remove the flywheel/driveplate inspection cover and wedge a screwdriver into the ring gear teeth.

8 Using a special puller (Ford no. T83T-6316-A or equivalent), remove the vibration damper **(see illustration)**. A two-jaw puller can be used instead.

9 Replace the crankshaft front oil seal (see next Section).

Installation

10 Lubricate the seal contact surface on the vibration damper with clean engine oil and guide the vibration damper onto the crankshaft. Be sure the key in the crankshaft is seated properly and align the keyway in the damper with the key.

11 Using a special installation tool (Ford no. T83T-6316-B or equivalent), press the vibration damper onto the crankshaft **(see illustration)** until it is fully seated.

12 Apply RTV sealant (Ford no. D6AZ-19562-AA or equivalent) to the engine side of the washer, then install the vibration damper retaining

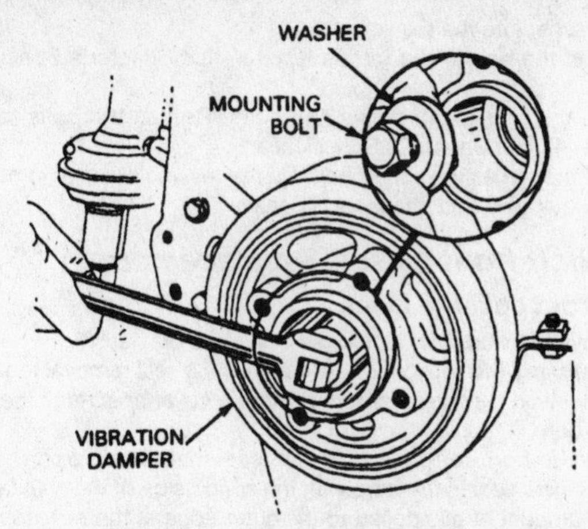

3.131 Remove the mounting bolt and washer

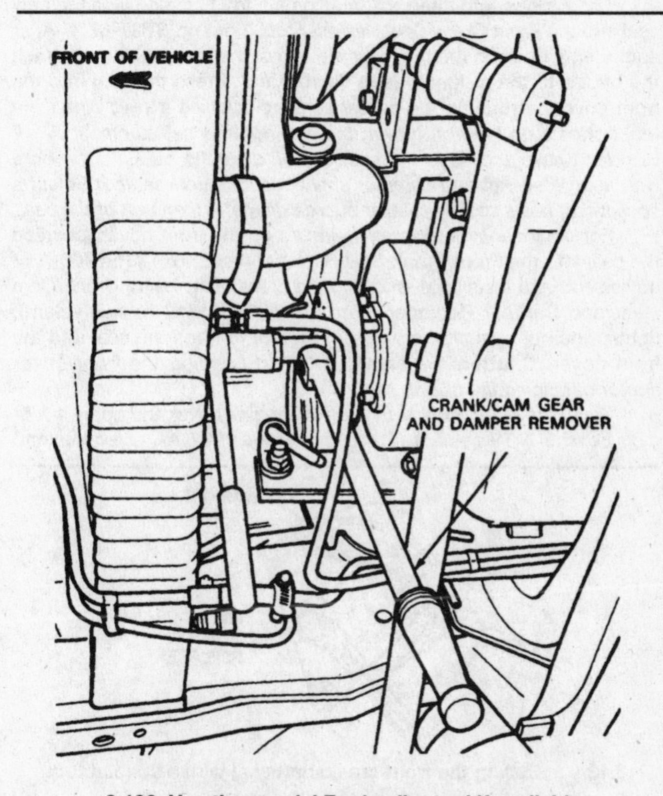

3.132 Use the special Ford puller tool if available (if not, a two-jaw puller will work)

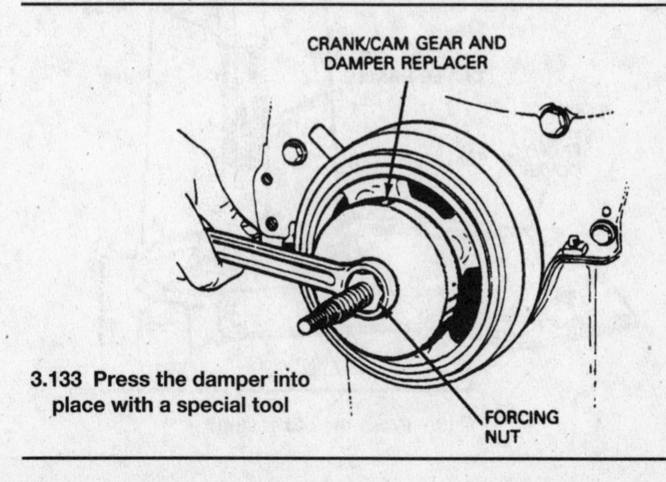

3.133 Press the damper into place with a special tool

bolt and washer into the crankshaft.

13 Tighten the bolt to the torque listed in this Chapter's Specifications.

14 Install the crankshaft pulley and bolts. Tighten the bolts to the torque listed in this Chapter's Specifications.

15 Install the remaining components in the reverse order of removal.

16 Run the engine and check for oil leaks.

Crankshaft front oil seal - replacement

Engine front cover in place

1 Remove the vibration damper as described previously.

2 Carefully pry the seal out of the cover with a seal removal tool or a large screwdriver. Be careful not to distort the cover or scratch the wall of the seal bore.

3 Clean the bore to remove any old seal material and corrosion. Position the new seal in the bore with the spring side of the seal facing in. A small amount of oil applied to the outer edge of the new seal will make installation easier.

4 For engines with three weldnuts on the front cover, place the new seal into the Front Crank Seal Replacer tool (Ford no. T83T-6700-A, or equivalent). Position the tool over the end of the crankshaft and attach the bridge to the weldnuts **(see illustration)**. Press the seal into the front cover by rotating the center screw clockwise. **Note:** *When the tool bottoms on the front cover, the seal depth is set automatically. If you don't have access to the special tool, drive the seal into the bore with a large socket and hammer until it's completely seated. Select a socket that has a slightly smaller outside diameter than that of the seal.*

5 For engines without three weldnuts on the front cover, position the seal into the Front Crank Seal Replacer (Ford no. T83T-6700-A or equivalent) and install the end onto the crankshaft. Using Crank/Cam Gear and Damper Replacer (Ford no. T83T6316-B or equivalent), tighten the nut against the washer and tool to force the seal into the front cover. **Caution:** *Be careful to avoid bending the front cover and/or damaging the oil pan seal.*

6 Clean the outer surface of the front engine cover and apply a 1/8-inch bead of RTV sealant (Ford no. D6AZ-19562-AA or equivalent) around the outside diameter of the front seal and the edge of the front cover.

7 Lubricate the seal lips with engine oil and reinstall the vibration damper.

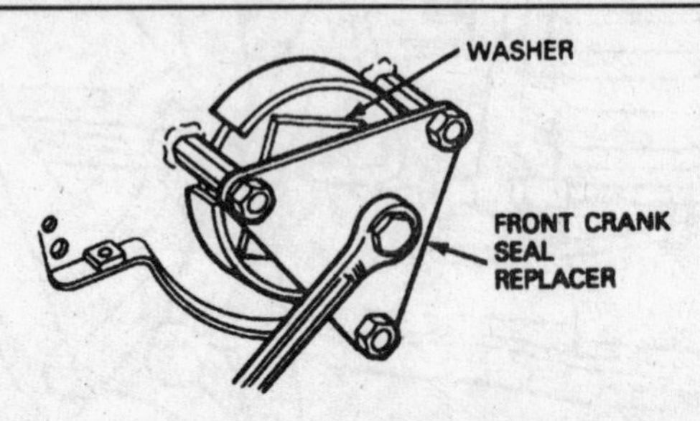

3.134 Installing the front crankshaft seal with a special tool

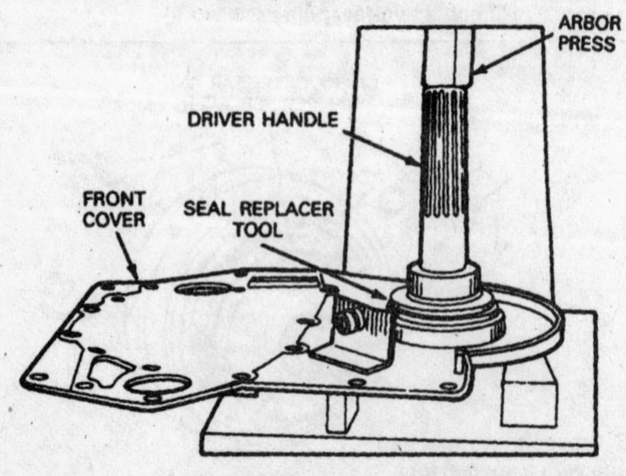

3.135 Press the old seal out

Engine front cover removed

8 Using an arbor press and special adapters **(see illustration)**, press the old seal out. If the special tools are not available, use a punch and hammer to drive the seal out of the cover from the backside. Support the cover as close to the seal bore as possible. Be careful not to distort the cover or scratch the wall of the seal bore. If the engine has accumulated a lot of miles, apply penetrating oil to the seal-to-cover joint on each side and allow it to soak in before attempting to drive the seal out.

9 Clean the bore to remove any old seal material and corrosion. Support the cover on blocks of wood and position the new seal in the bore with the open end of the seal facing in. A small amount of grease applied to the outer edge of the new seal will make installation easier.

10 Install the new seal using Front Crank Seal Replacer (Ford no. T83T-6700-A or equivalent) with a spacer and press **(see illustration)**. If the special tool is unavailable, drive the seal into the bore with a large socket and hammer until it is completely seated. Select a socket that has an outside diameter that is slightly smaller than that of the seal (a section of pipe can be used if a socket is not available).

11 Reinstall the engine front cover.

Engine front cover - removal and installation

Note: *On 1994 and later Direct Injection Turbo models, remove the engine prior to front cover removal. In addition, the oil pan, oil pump screen, cover and tube must be removed before the front cover can be removed.*

Removal

1 Disconnect the ground cables from the negative terminals of the batteries and from the front of the engine.

2 Drain the cooling system (refer to the *Maintenance* portion of this Chapter).

3 Remove the air cleaner assembly and cover the opening with a cloth.

4 Remove the radiator fan shroud, engine cooling fan and water pump as described in the *Cooling System* portion of this Chapter.

5 Remove the fuel injection pump and drive gear housing as detailed in the *Fuel system* portion of this Chapter.

6 Remove the vibration damper as described previously in this Chapter.

7 Remove the five bolts attaching the front cover to the engine block and oil pan. Lower the vehicle, if necessary, and remove the remaining bolts attaching the front cover to the engine block and remove the front cover.

8 Replace the crankshaft front oil seal as described in the previous Section.

9 Thoroughly clean all components and remove any traces of dirt, grease and old gasket material.

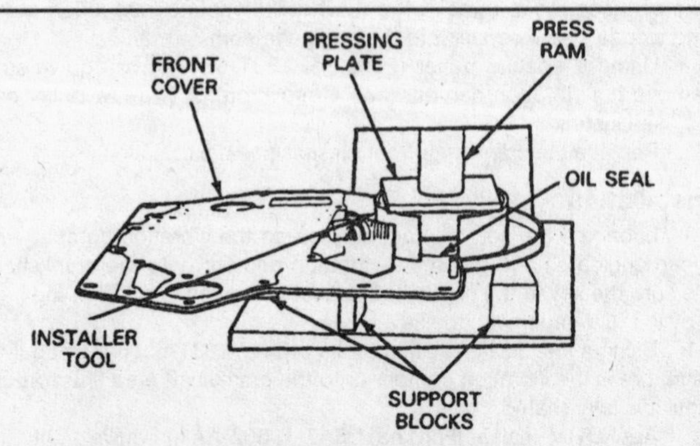

3.136 Press a new seal in

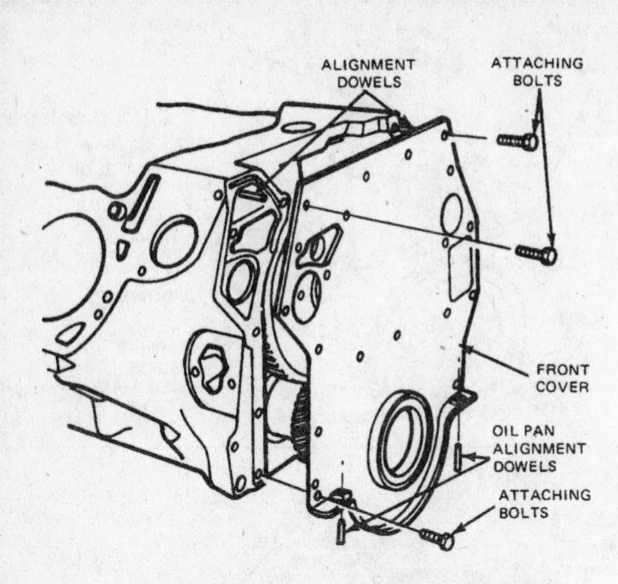

3.137 Make alignment dowels to guide the front cover into position

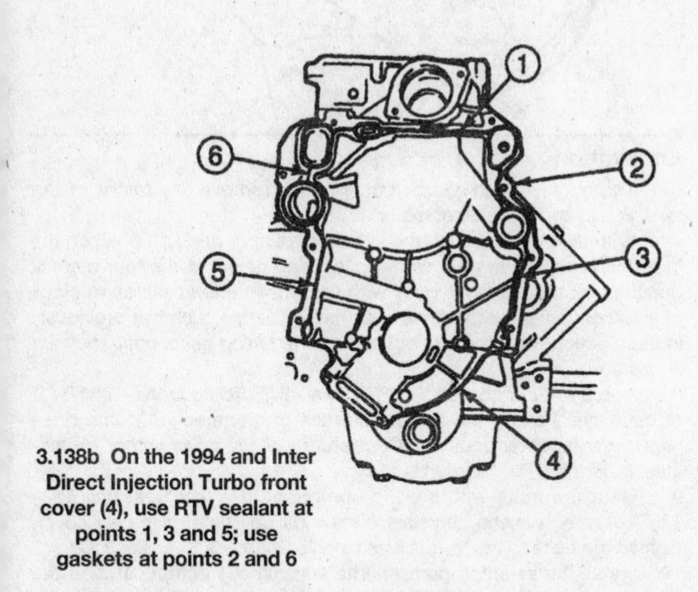

3.138b On the 1994 and Inter Direct Injection Turbo front cover (4), use RTV sealant at points 1, 3 and 5; use gaskets at points 2 and 6

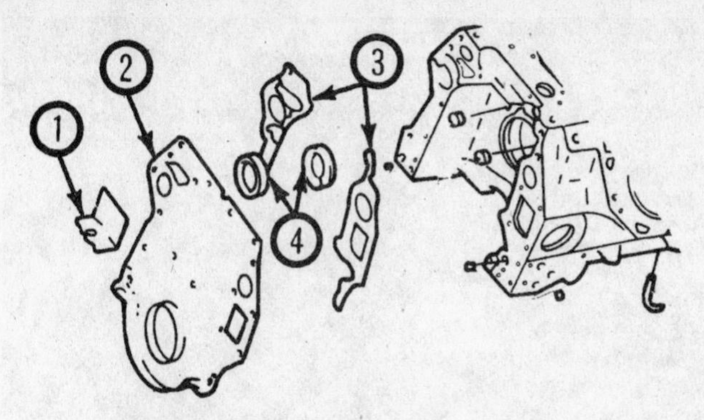

3.138a Front cover components - exploded view

1	*Bracket*	*3*	*Gaskets*
2	*Front cover*	*4*	*Seals*

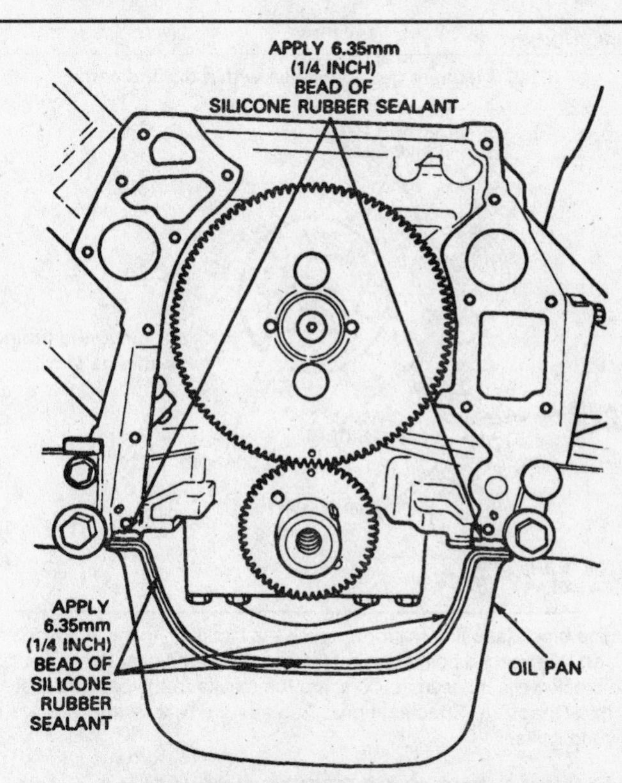

3.139 Apply sealant as shown here

Installation

10 Install fabricated alignment dowels **(see illustration)** on the oil pan and engine block to aid in gasket alignment.

11 Apply a thin film of RTV sealant and install the gaskets on the block. Apply a 1/4-inch bead of RTV to the front edge of the oil pan and at the front cover-to-oil pan joint **(see illustrations)**.

12 Install the front engine cover onto the oil pan dowels first and start the three bolts by hand.

13 Install the water pump as described in the *Cooling system* portion of this Chapter and temporarily hand tighten the bolts.

14 Remove any remaining alignment dowels, install and tighten the water pump and front cover bolts to the torque listed in this Chapter's Specifications.

15 Install the remaining components in the reverse order of removal. Refer to the appropriate Sections in this Chapter for additional installation instructions.

16 Refill and bleed the cooling system as described in the *Maintenance* portion of this Chapter. Check all of the fluid levels, then start the engine and check for oil, coolant and fuel leaks.

Timing gears - inspection, removal and installation

Inspection - all timing gears

1 Visually inspect the gears for chipped teeth and/or worn areas. Replace as necessary.

2 To check gear backlash, install a dial indicator and bracketry (Ford nos. D78P-4201-G and D78P-4201-F, or equivalents) on the

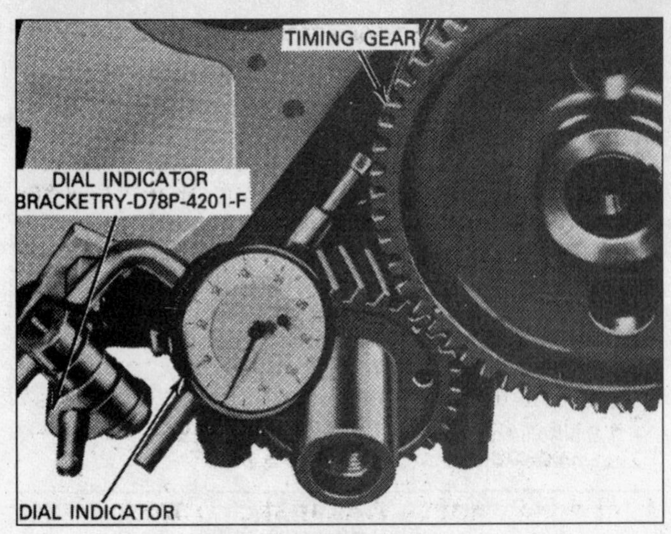

3.140 Measure gear backlash with a dial indicator

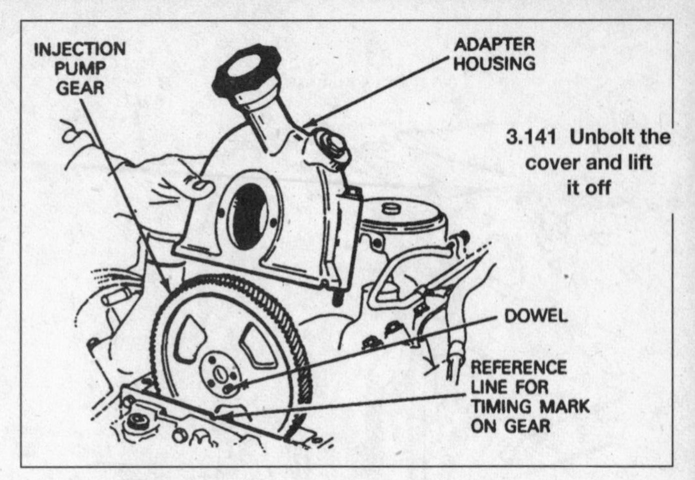

3.141 Unbolt the cover and lift it off

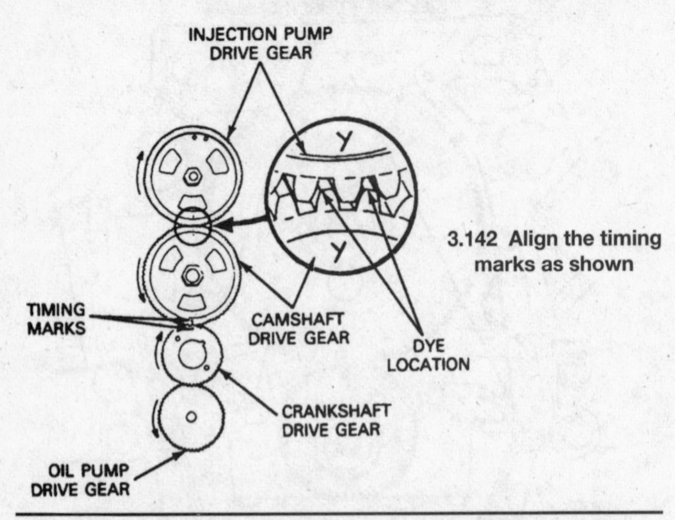

3.142 Align the timing marks as shown

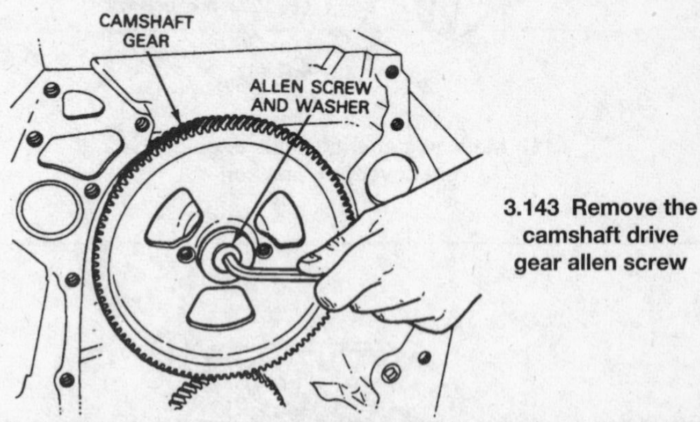

3.143 Remove the camshaft drive gear allen screw

engine block **(see illustration)**. Check the backlash (play) at six different equally spaced points around the gear. Hold the gear firmly against the block while measuring. Compare the results to the backlash listed in this Chapter's Specifications. Replace any gears that are worn beyond limits.

Injection pump drive gear and cover

Removal

1 Remove the injection pump and glow plugs (refer to the *Fuel system* portion of this Chapter).

2 Remove the injection pump drive gear cover, also known as an adapter housing **(see illustration)**.

3 Using a breaker bar and socket on the vibration damper bolt, turn the crankshaft clockwise until the number one cylinder is at top dead center on the compression stroke. When the crankshaft is positioned correctly, the injection pump drive gear dowel will be in the four o'clock position and the scribe line on the vibration damper should be at TDC. Also, the "Y" marks on the pump gear and camshaft gear should align with the "O" marks on the crankshaft and camshaft gears **(see illustration)**.

4 Slide the injection pump gear back (don't remove) to expose the top of the camshaft gear when looking down into the cover. In addition to the "Y", the gear teeth adjacent to the "Y" on the gear are permanently dyed.

5 Remove the injection pump drive gear. **Caution:** *Do not remove the gear unless the procedures described in Steps 3 and 4 are followed. Also, do not turn the crankshaft.*

Installation

6 Thoroughly clean all components and remove any traces of dirt and grease and old gasket material.

7 With the scribe line on the vibration damper on TDC, position the injection pump drive gear with the locating dowel at the four o'clock position. Install the drive gear with the drawn line at the six o'clock position and align all the drive gear timing marks as shown previously in this Section. **Caution:** *Do not disturb the timing gears once they are in position.*

8 Apply a 1/8-inch bead of RTV sealant (Ford no.D6AZ-19562-AA or equivalent) along the bottom surface of the injection pump drive gear cover (adapter housing). **Caution:** *Install the cover within 15 minutes, before the RTV hardens.*

9 Coat the bolts with sealing compound (Perfect Seal no. B5A-19554-A or equivalent), then install the injection pump drive gear cover. Tighten the bolts to the torque listed in this Chapter's Specifications.

10 Install the injection pump and the remaining components in the reverse order of removal. Refer to the appropriate Sections in this Chapter for further information.

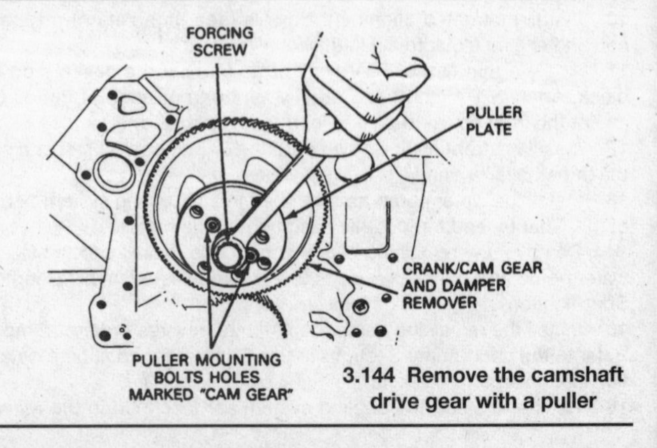

3.144 Remove the camshaft drive gear with a puller

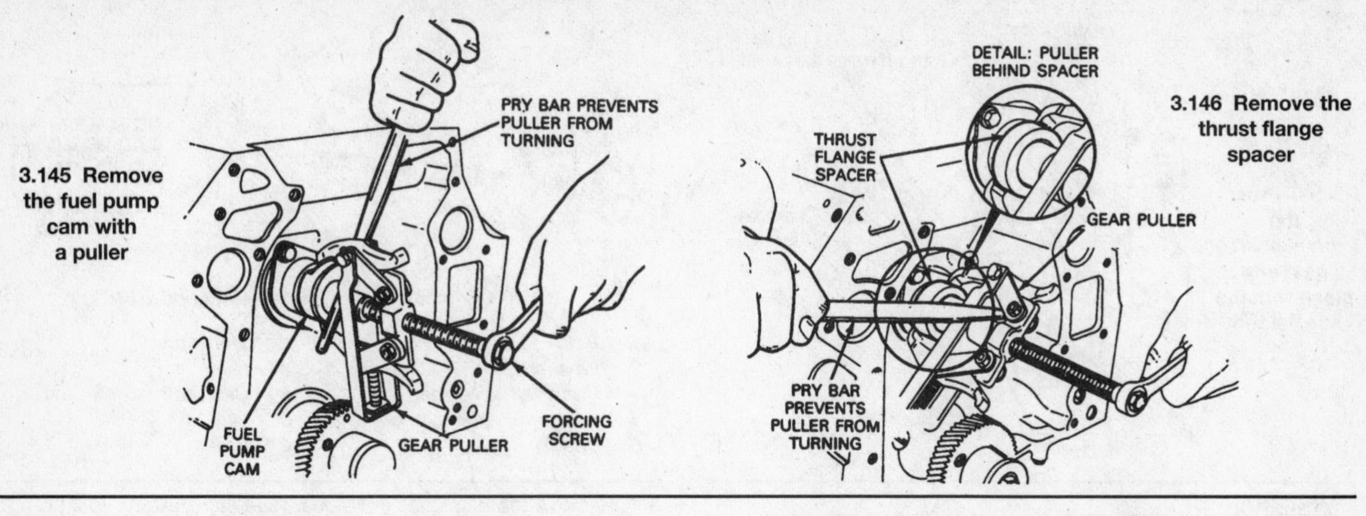

3.145 Remove the fuel pump cam with a puller

3.146 Remove the thrust flange spacer

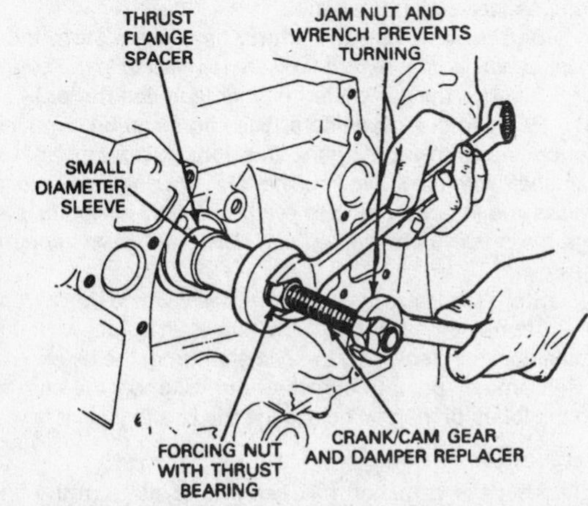

3.147 Press the thrust flange spacer into position

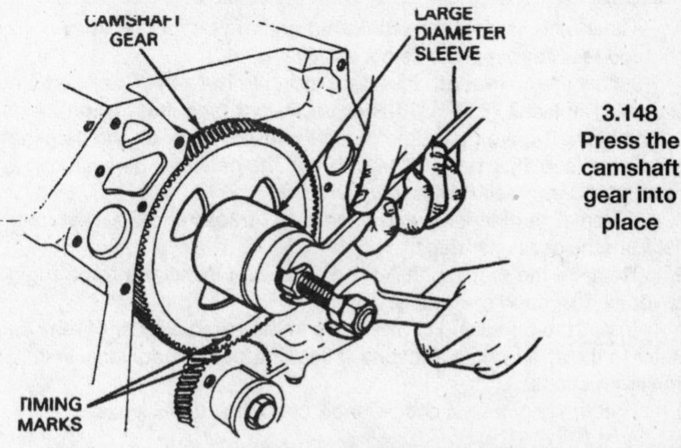

3.148 Press the camshaft gear into place

11 Run the engine and check for oil, coolant and fuel leaks.

Camshaft drive gear

Removal

1 Remove the front engine cover as described in a previous Section.
2 Remove the camshaft Allen screw and washer **(see illustration)**.
3 Remove the gear **(see illustration)**, using Ford Crank/Cam Gear and Damper Remover T83T-6316-A or equivalent.
4 Remove the mechanical fuel pump, if equipped (refer to the *Fuel system* portion of this Chapter).
5 Using Ford Gear Puller T77F-4220-B1 or equivalent **(see illustrations)**, remove the fuel pump cam and thrust flange spacer, if necessary.
6 Remove the bolts attaching the thrust plate and remove the thrust plate, if necessary.

Installation

7 Install a new thrust plate, if removed, and tighten the bolts to the torque listed in this Chapter's Specifications.
8 Inspect the fuel pump cam. If it is scored or otherwise damaged, replace the cam and the fuel pump.
9 Install the spacer and fuel pump cam onto the camshaft thrust flange **(see illustration)** using Ford Crank/Cam Gear and Damper Replacer T83T-6316-B or equivalent.
10 Clean and inspect the camshaft drive gear. Look for cracks and worn, chipped or missing teeth and replace as necessary.
11 Install the camshaft drive gear against the fuel pump cam **(see illustration)**, using Ford Crank/Cam Gear and Damper Replacer T83T-6316-B or equivalent. Align the timing mark with the mark on the crankshaft drive gear.
12 Install the camshaft Allen screw and tighten it to the torque listed in this Chapter's Specifications.

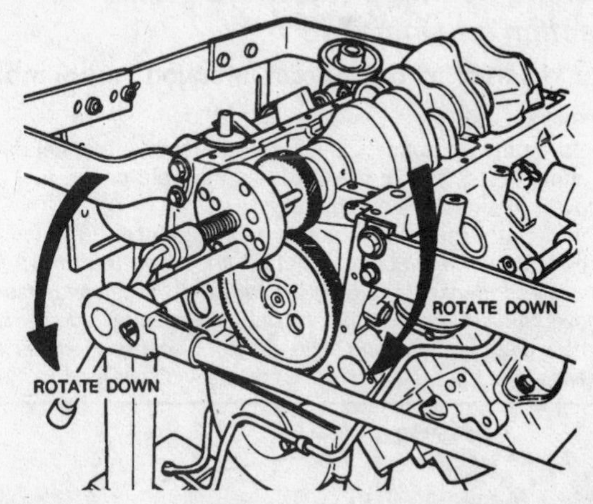

3.149 Hold the crankshaft from turning and rotate the screw in the gear puller to remove the crankshaft gear (engine shown removed from vehicle)

13 Install the remaining components in the reverse order of removal. Refer to the appropriate Sections in this Chapter for additional installation instructions.
14 Run the engine and check for oil, coolant and fuel leaks.

Crankshaft drive gear

Removal

1 Remove the engine front cover as described previously.
2 Remove the crankshaft gear **(see illustration)**, using Ford Crank/Cam Gear and Damper Remover T83T-6316-A or equivalent).

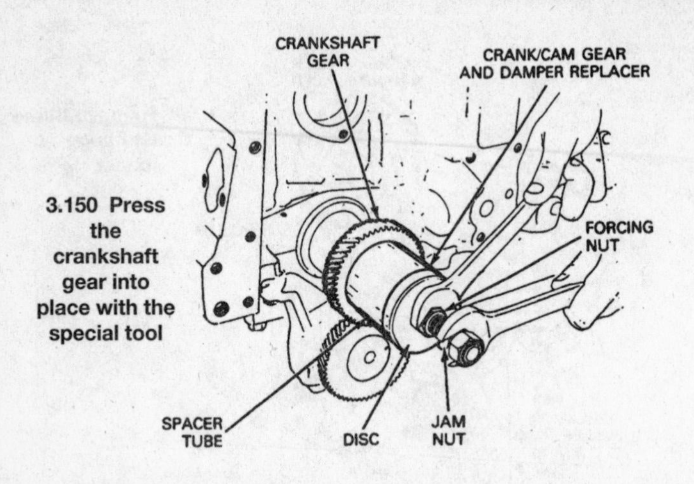

3.150 Press the crankshaft gear into place with the special tool

3.151 Camshaft components - exploded view

1. *Bolt*
2. *Washer*
3. *Camshaft drive gear*
4. *Fuel pump cam*
5. *Spacer*
6. *Thrust plate*
7. *Key*
8. *Camshaft*
9. *Lifter*
10. *Pushrod*

Installation

3 Clean and inspect the crankshaft gear. Look for cracks, worn or chipped teeth and replace as necessary.

4 Install the crankshaft drive gear, using Ford Crank/Cam Gear and Damper Replacer T83T-6316-B or equivalent **(see illustration)**. Align the crankshaft drive gear timing mark with the mark on the camshaft drive gear **(see illustration 3.148)**. **Note:** *The gear may be heated in an oven to 350-degrees F to ease installation.*

5 Clean all components and remove any traces of old gasket material from the sealing surfaces.

6 Replace the crankshaft front oil seal and install the front engine cover as described previously.

7 Install the remaining components in the reverse order of removal. Refer to the appropriate Sections in this Chapter for additional installation instructions.

8 Run the engine and check for oil, coolant and fuel leaks.

Camshaft, bearings and lifters - removal, inspection and installation

1994 and earlier (except Direct Injection Turbo) models

Removal

1 Refer to the appropriate Sections and remove the fuel injection pump and adapter, intake manifold, engine front cover, fuel pump, valve covers, rocker arms and pushrods.

2 The radiator must be removed as well. **Note:** *If the vehicle is equipped with air conditioning it will be necessary to remove the air conditioning condenser to remove the camshaft. If the condenser must be removed the system must first be depressurized by a dealer service department or air conditioning shop. Do not disconnect any air conditioning lines until the system has been properly discharged.*

3 Remove the camshaft drive gear **(see illustration)**, fuel pump cam, spacer and thrust plate.

4 Before removing the lifters, arrange to store them in a clearly labeled box to ensure that they are reinstalled in their original locations.

5 Remove the valve lifter retainer **(see illustration)**.

6 Remove the valve lifters. The lifters can be removed with a magnet or even with your fingers. **Caution:** *Do not attempt to withdraw the camshaft with the lifters in place. Do not use pliers to remove the lifters unless you intend to replace them with new ones. The pliers may damage the precision machined and hardened lifters, rendering them useless.*

7 Store the lifters in order of removal where they will not get dirty.

8 Thread a long bolt into the camshaft sprocket bolt hole to use as a handle when removing the camshaft from the block.

9 Carefully pull the camshaft out. Support the cam near the block so the lobes do not nick or gouge the bearings as it is withdrawn.

Inspection

10 After the camshaft has been removed from the engine, cleaned with solvent and dried, inspect the bearing journals for uneven wear, pitting and evidence of seizure. If the journals are damaged, the bearing inserts in the block are probably damaged as well. Both the camshaft and bearings will have to be replaced. Replacement of the camshaft bearings requires special tools and techniques which place it beyond the scope of the home mechanic. The block will have to be removed from the vehicle and taken to an automotive machine shop for this procedure.

11 Measure the camshaft bearing inside diameter with a bore gauge and record the results.

12 Measure the camshaft bearing journals with a micrometer to determine if they are excessively worn or out-of-round and record the results **(see illustration)**.

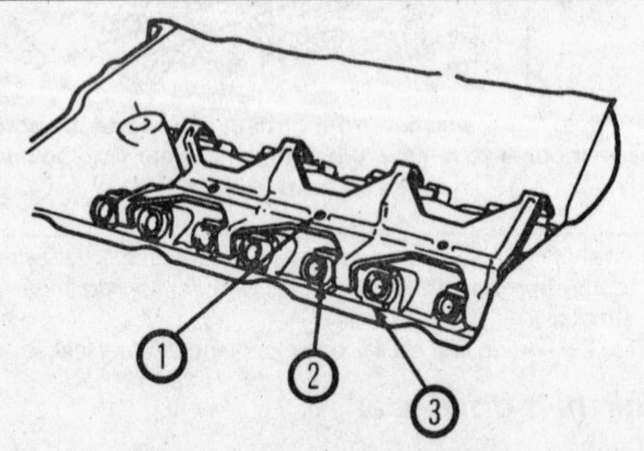

3.152 Remove the guide retainer

1. *Guide retainer*
2. *Lifter*
3. *Lifter guide*

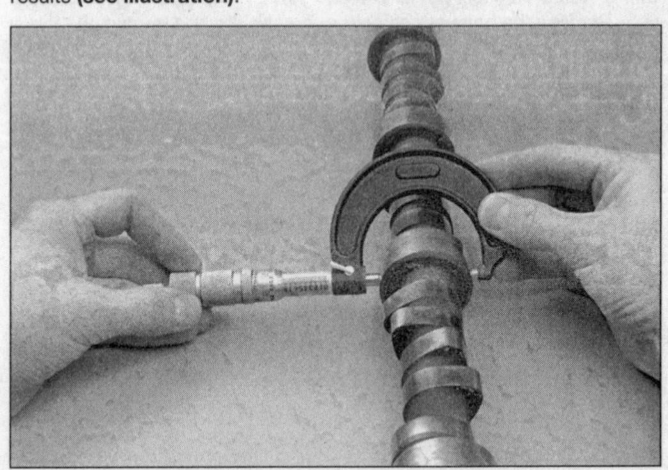

3.153 Measure the camshaft bearing journals

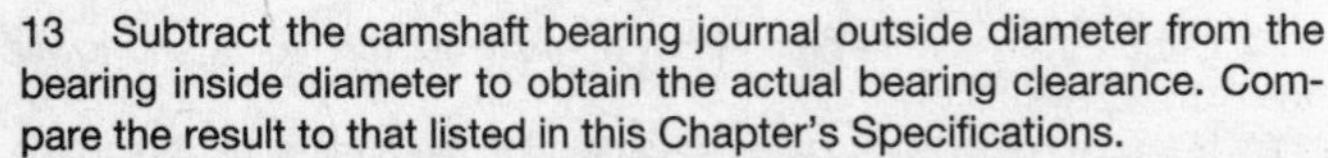

3.154 Check the lifter rollers for wear and damage (typical roller lifter shown)

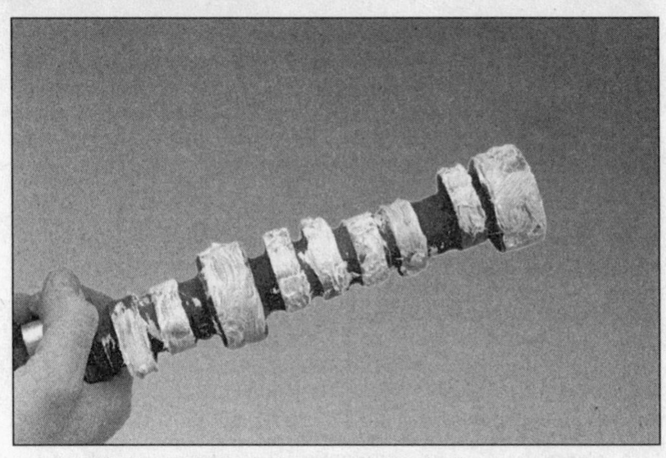

3.155a Lubricate the camshaft prior to assembly

13 Subtract the camshaft bearing journal outside diameter from the bearing inside diameter to obtain the actual bearing clearance. Compare the result to that listed in this Chapter's Specifications.

14 Check the camshaft lobes for heat discoloration, score marks, chipped areas, pitting and uneven wear. If the lobes are in good condition and the bearing journals are within the specified limits, the camshaft can be reused.

15 Clean the lifters with solvent and dry them thoroughly without mixing them up.

16 Check each lifter wall and pushrod seat for scuffing, score marks and uneven wear. If the lifter walls are damaged or worn (which is not very likely), inspect the lifter bores in the engine block as well. If the pushrod seats are worn, check the pushrod ends.

17 Check the rollers carefully for wear and damage and make sure they turn freely without excessive play **(see illustration)**.

18 Used roller lifters can be reinstalled with a new camshaft and the original camshaft can be used if new lifters are installed.

Installation

19 Lubricate the camshaft bearing journals with engine oil and the cam lobes with moly-base grease (Ford no. D0AZ-19584-AA or equivalent) or engine assembly lube **(see illustration)**.

20 Slide the camshaft into the engine. Support the cam near the block and be careful not to scrape or nick the bearings.

21 Install a new camshaft thrust plate and tighten the bolts to the torque listed in this Chapter's Specifications. Be sure to lubricate the thrust plate with engine oil before assembly.

22 To check camshaft endplay, push the camshaft toward the rear of the engine. Mount a dial indicator, using Ford Bracket no. D78P-4201-F (or equivalent), so that the indicator point is on the camshaft gear attaching bolt **(see illustration)**. Zero the dial indicator, then pull the camshaft forward with a screwdriver. Record the reading and compare it to the one listed in this Chapter's Specifications.

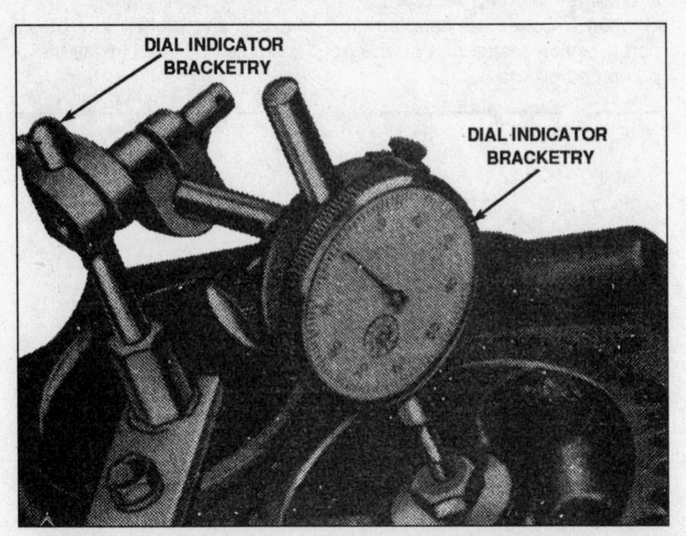

3.155b Double-check camshaft endplay with a dial indicator

23 Install the spacer and fuel pump cam against the camshaft thrust flange using Ford Crank/Cam Gear and Damper Replacer no. T83T-6316-B or equivalent.

24 Install the camshaft drive gear against the fuel pump cam, aligning the timing mark with the mark on the crankshaft drive gear as described in the previous Section.

25 Lubricate the lifters with clean engine oil and install them in the block. If the original lifters are being reinstalled, be sure to return them to their original locations.

26 The remaining installation steps are the reverse of removal.

27 Before starting and running the engine, change the oil and install a new oil filter.

1994 and later Direct Injection Turbo models

Note: *On 1994 and later Direct Injection Turbo models, the engine must be removed from the vehicle before the camshaft can be removed.*

Lifters - removal and installation

1 Remove the cylinder heads as described earlier in this Chapter.

2 Remove the two valve lifter guide retainer screws and the retainer **(see illustration)**.

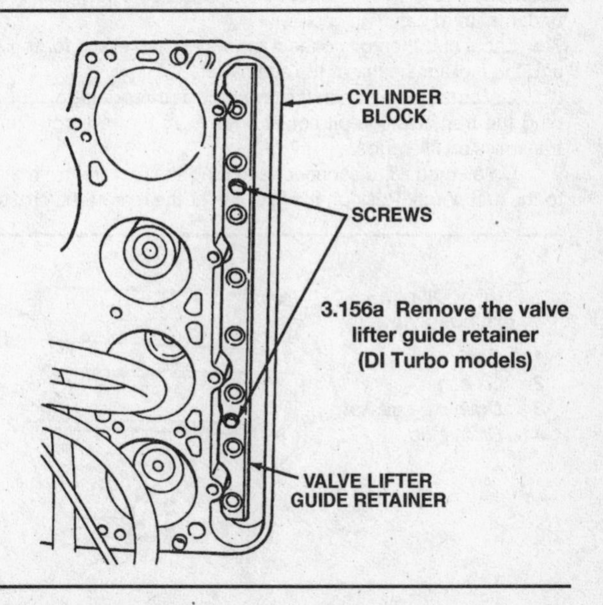

3.156a Remove the valve lifter guide retainer (DI Turbo models)

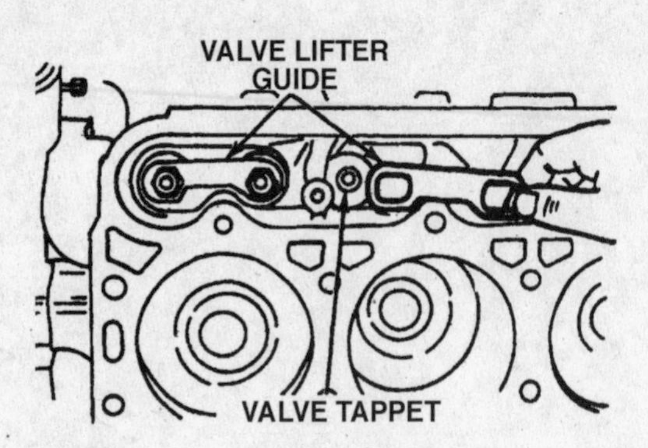

3.156b Remove the valve lifter guides (DI Turbo models)

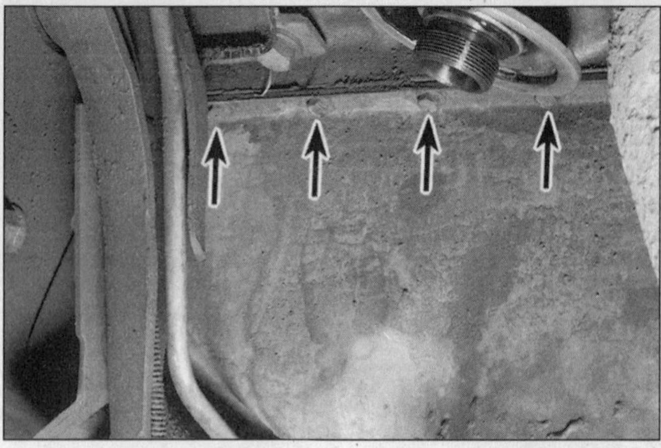

3.157a Remove the bolts around the perimeter of the oil pan (arrows)

3 Remove the valve lifter guides **(see illustration)**.
4 Using a strong magnet or a lifter removal tool, withdraw the lifters.
5 Be sure to keep all the components organized so they can be returned to their original positions in the engine.
6 Lubricate the lifters and their bores with clean engine oil and install the lifters.
7 Install the valve lifter guides and the guide retainer. Tighten the guide retainer bolts to the standard torque value listed in Chapter 1.
8 The remainder of installation is the reverse of removal.

Oil pan - removal and installation

Note: *On 1994 and later Direct Injection Turbo models, the engine must be removed from the vehicle before the oil pan can be removed.*

Removal

1 Disconnect the ground cables from the negative terminals of the batteries.
2 Block the rear wheels to prevent the vehicle from rolling. Set the parking brake and place the transmission in Park (automatic) or first gear (manual). Raise the vehicle and support it securely on jackstands.
3 Drain the coolant and engine oil and remove the oil filter (see the *Maintenance* Section in the front of this Chapter).
4 Remove the air cleaner and cover the air intake with a cloth.
5 Remove the engine cooling fan as described in the *Cooling system* Section of this Chapter. Disconnect the lower radiator hose.
6 Disconnect the power steering return hose at the pump. Plug the open ends to prevent contamination. Be sure to place a drain pan underneath to catch any spills.
7 Label and then disconnect the wiring harnesses for the alternator and the fuel line heater at the alternator.
8 On automatic transmission equipped models, disconnect and plug the transmission oil cooler lines from the radiator. Remove the transmission filler tube.
9 On all models, disconnect and plug the fuel line from the chassis to the fuel pump. Position the inlet line at the rear of the crossmember.
10 Unbolt the exhaust pipes from the exhaust manifolds and lower the exhaust system.
11 Remove the nuts and washers attaching the engine mounts to the front crossmember.
12 Raise the engine until the transmission contacts the body. Place wood blocks between the mounts and frame to hold the engine in a raised

position. Use a 2 3/4-inch block on the left (driver's) side and a 2-inch block on the right (passenger's) side. When the blocks are in place, lower the engine onto them. On F-Series trucks, use a sling and hoist to raise the engine. On E-Series trucks, use special lifting brackets (Ford Rotunda no. 014-00312 or equivalent). **Warning:** *Never place any part of your body under the engine where it could be crushed if the lifting device fails.*
13 Remove the lower bellhousing cover.
14 Remove the oil pan bolts **(see illustrations)**.
15 On F-Series trucks, unbolt the oil pump and pickup tube and drop them in the oil pan. Remove the oil pan by inserting two screwdrivers through the crankcase dowel holes on the left (driver's) side of the engine and carefully pry the oil pan off **(see illustration)**. **Note:** *You may have to turn the crankshaft to allow the oil pan to clear the counterweights.*

Installation

16 Thoroughly clean all components, including the mounting surfaces of the oil pan and engine block of old gasket material and sealer.
17 On E-Series trucks, install the oil pump and pickup, if removed, and prime the pump with oil.
18 On F-Series trucks, place the oil pump and pickup into the oil pan. Position the oil pan on the front crossmember and install the oil pump and pickup.

3.157b Oil pan - exploded view

1 *Gasket*
2 *Oil pan*
3 *Drain plug gasket*
4 *Drain plug*

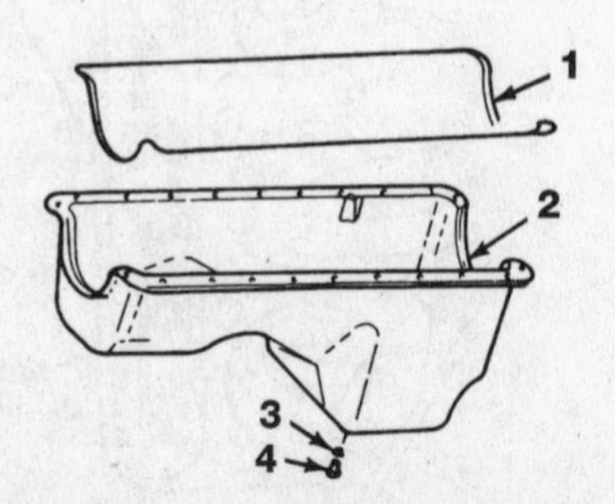

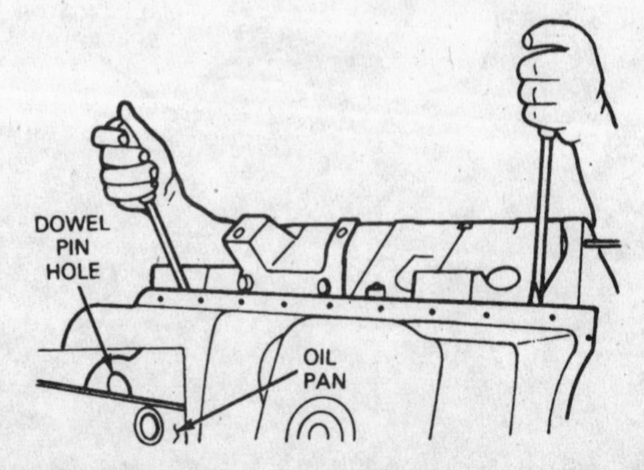

3.158 Carefully pry the oil pan off (engine removed for clarity)

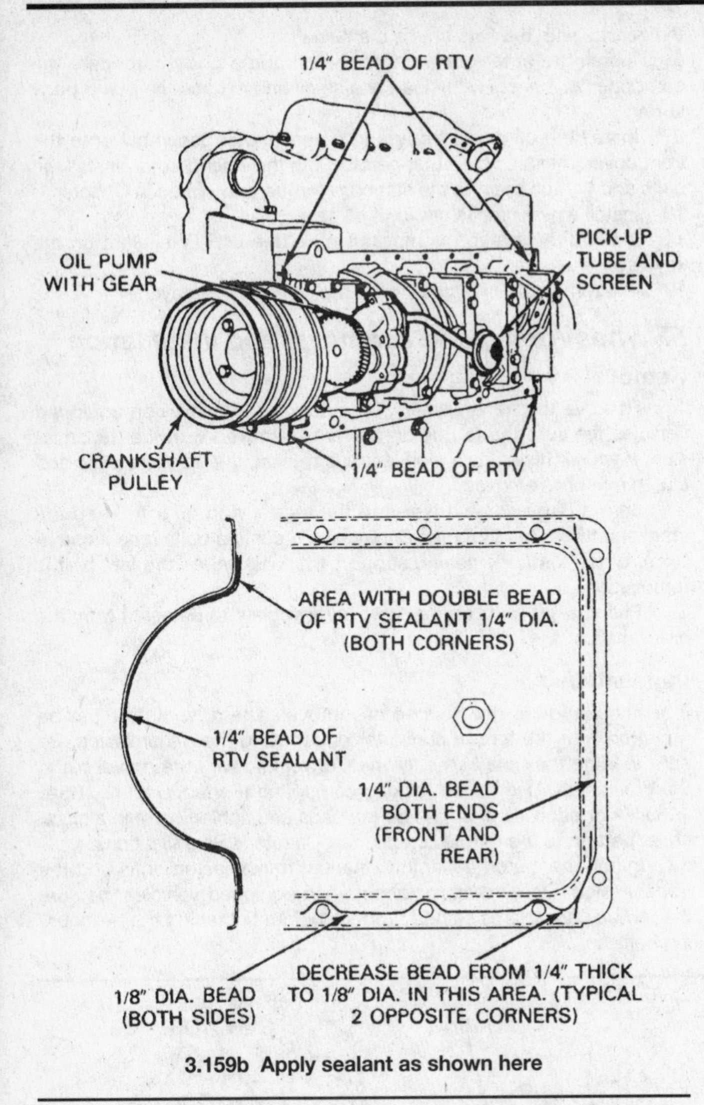

3.159b Apply sealant as shown here

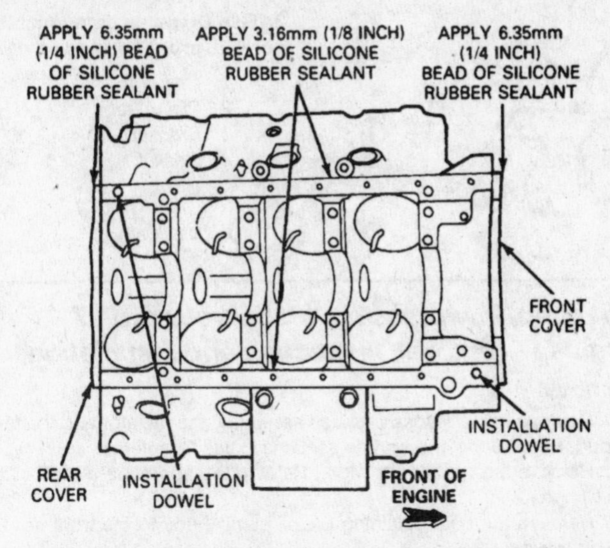

3.159a Use alignment dowels and apply sealant as shown

19 Make two locating dowels and insert them into the block on opposite corners **(see illustration)**. Apply a bead of RTV sealant to the oil pan sealing flange. Use a 1/8-inch bead on the sides and a 1/4-inch bead on the ends **(see illustration)**. Lift the pan into position, then install the bolts finger tight and remove the two locating dowels. **Caution:** *Be careful not to disturb the sealant or oil leaks may occur.*

20 Starting at the ends and alternating from side-to-side towards the center, tighten the bolts to the torque listed in this Chapter's Specifications.

21 The remainder of the installation procedure is the reverse of removal. Fill the pan with fresh oil, install a new oil filter, refill and bleed the cooling system, check the power steering fluid level and, on automatic transmission equipped models, add transmission fluid as necessary. Start the engine and check for leaks before placing the vehicle back in service.

Oil pump and pickup 1994 and earlier (except Direct Injection Turbo) models

1 Remove the oil pan as described in the previous Section. On F-Series trucks, the oil pump and pickup **(see illustration)** are removed with the oil pan. Remove the bolts and lower the pump and pickup.

2 On E-Series trucks, unbolt the oil pickup tube and oil pump **(see illustration)** and lower them from the vehicle.

3 Before installation, replace the gasket between the oil pump and pickup. Then pour oil into the pump to assist in priming it upon startup.

Oil Pick-Up Tube Installation

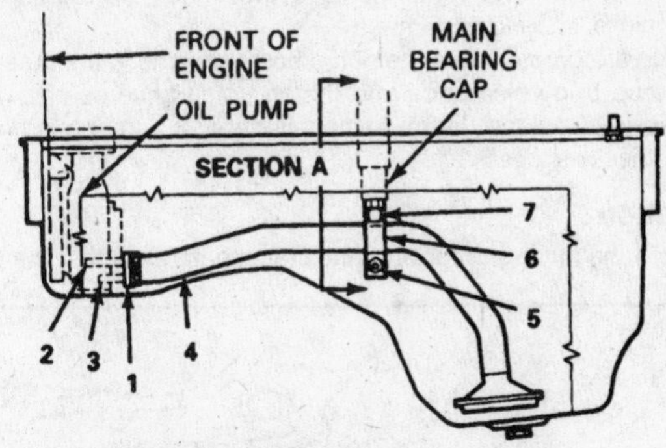

1. OIL PICK-UP TUBE MOUNTING GASKET
2. 5/16"-18 × 2" BOLT AND 5/16" HARDENED WASHER
3. 5/16"-18 × 1-1/2" BOLT AND 5/16" HARDENED WASHER
4. OIL PICK-UP TUBE ASSEMBLY
5. 5/16"-18 × 0.930 BOLT W/WASHER
6. OIL TUBE BRACKET
7. 5/16"-18 NUT AND 5/16" LOCK AND HARDENED WASHERS

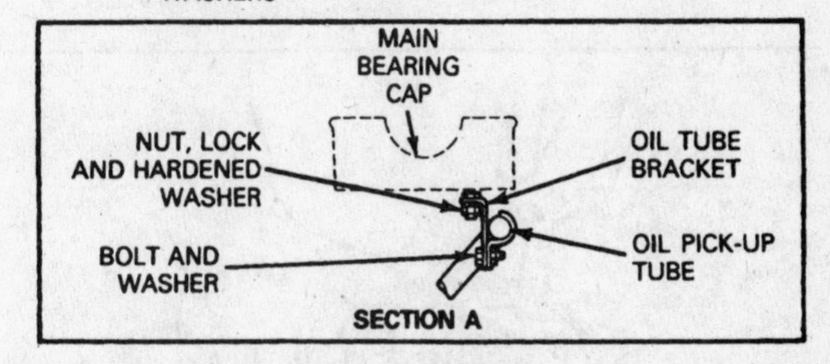

3.160a Oil pickup tube mounting details

4 Position the pump and engage the gear on the pump with the drive gear on the crankshaft. Install the mounting bolts and tighten them to the torque listed in this Chapter's Specifications.

5 Install the oil pan as described in the previous Section.

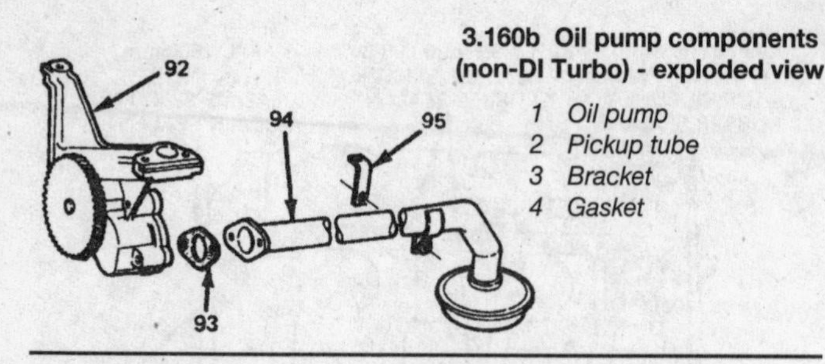

3.160b Oil pump components (non-DI Turbo) - exploded view

1 Oil pump
2 Pickup tube
3 Bracket
4 Gasket

Oil pump, low pressure 1995 through 1997 engines - removal, inspection and installation

Removal

1 Remove the radiator fan/clutch assembly, the radiator and the fan shroud. Refer to the appropriate sections in this Chapter.

2 Remove the vibration damper. Refer to the appropriate section in this Chapter.

3 Remove the bolts retaining the oil pump body to the front cover **(see illustration)**.

4 Remove the oil pump body and the inner and outer gerotors **(see illustration)**. Mark the relative position of the gerotors so they can be installed in their original positions.

Inspection

5 Clean and dry the components. Inspect the gerotors and pump body for nicks, scoring or damage.

6 Place the gerotors into the pump body and check the outer gerotor-to-pump body clearance and the end clearance as shown **(see illustrations)**. Replace the oil pump assembly if damage is noted or the clearance is excessive.

Installation

7 Install the inner gerotor onto the crankshaft, engaging the flats in the gerotor with the flats on the crankshaft.

8 Install the outer gerotor into the oil pump body. Lubricate the components liberally with clean engine oil and replace the pump body O-ring.

9 Install the oil pump body and outer gerotor assembly onto the front cover, meshing the outer gerotor with the inner gerotor. Install the bolts and tighten them to the standard torque value listed in Chapter 1.

10 Install a new crankshaft front oil seal.

11 Install the vibration damper using a dab of RTV sealant on the keyway.

12 The remainder of installation is the reverse of removal.

Flywheel/driveplate - removal and installation

Removal

1 Remove the transmission. On automatic transmission equipped vehicles, be sure the torque converter is removed with the transmission. If your vehicle has a manual transmission, the pressure plate and clutch must be removed.

2 Jam a large screwdriver into the starter ring gear to keep the crankshaft from turning, then remove the mounting bolts **(see illustration)**. Since it's fairly heavy, support the flywheel as the last bolt is removed.

3 Pull straight back on the flywheel/driveplate to detach it from the crankshaft.

Installation

4 Installation is the reverse of removal. The driveplate must be mounted with the torque converter pads facing the transmission. Be sure to align the hole in the flywheel/driveplate with the dowel pin in the crankshaft. Use thread locking compound (Perfect Seal no. B5A-19554-A or equivalent) on the bolt threads and tighten them in a criss-cross pattern to the torque listed in this Chapter's Specifications.

5 Install the clutch assembly (manual transmission only) and the transmission. On automatic transmission equipped vehicles, be sure the torque converter is seated properly before tightening the bellhousing bolts.

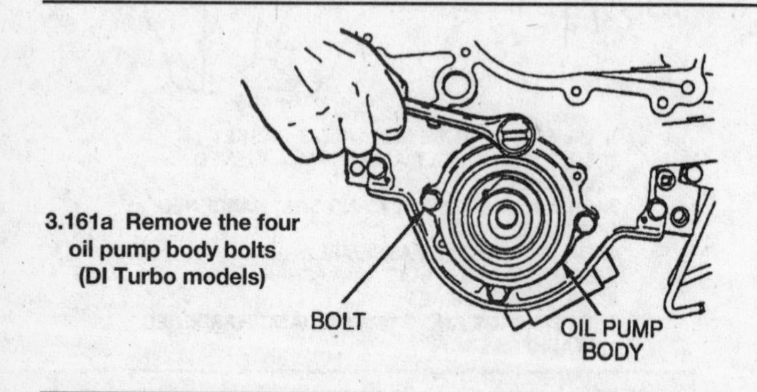

3.161a Remove the four oil pump body bolts (DI Turbo models)

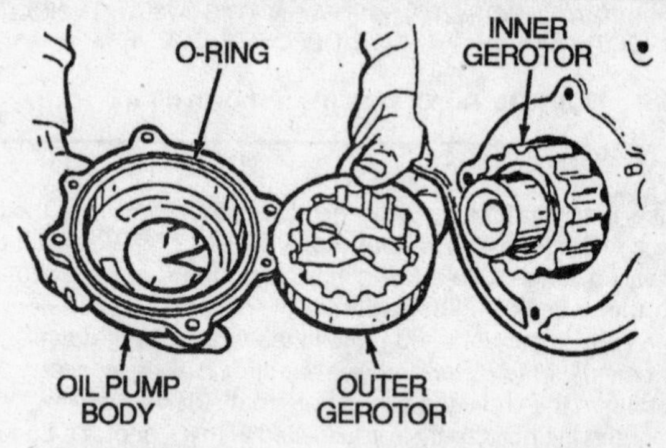

3.161b Oil pump components (DI Turbo models)

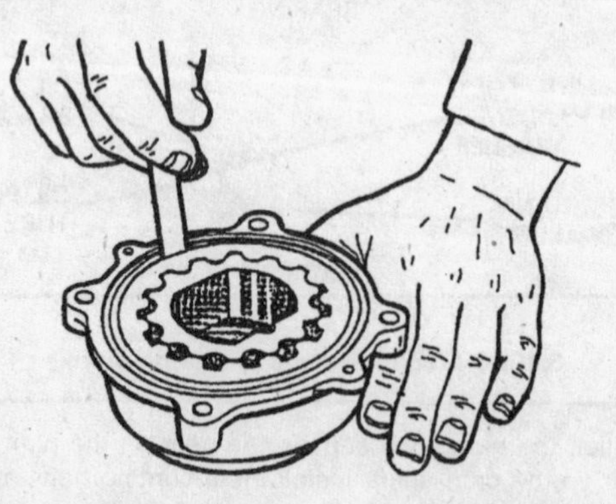

3.161c Using a feeler gauge check the clearance between the outer gerotor and the pump body - it should be 0.028 to 0.032-inch

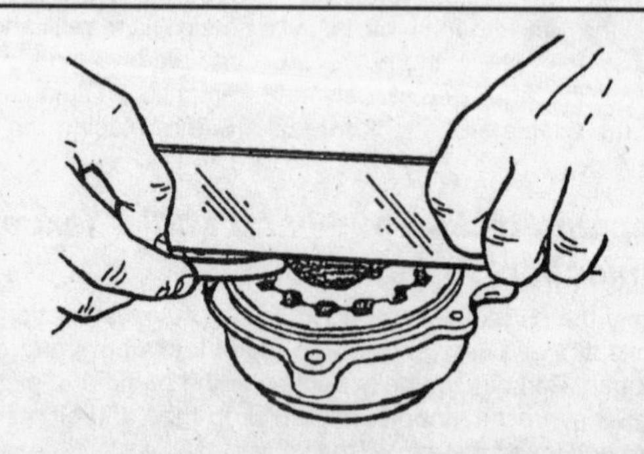

3.161d Place a straight-edge over the pump body and gerotors and measure the gerotor endplay - it should be 0.001 to 0.003-inch

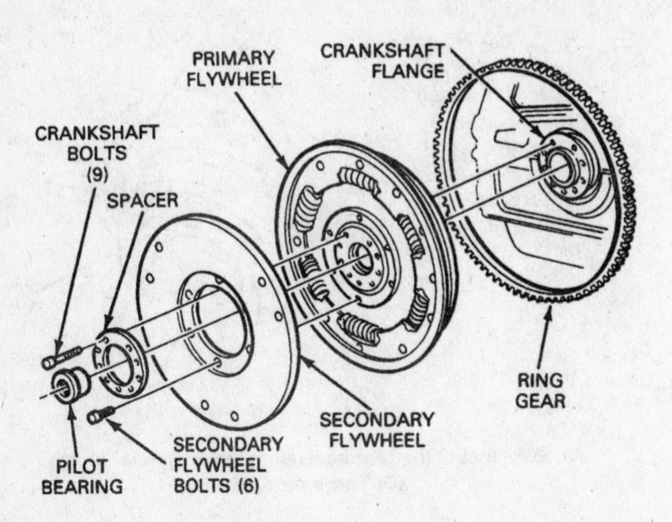

3.162 Manual transmission flywheel components - exploded view

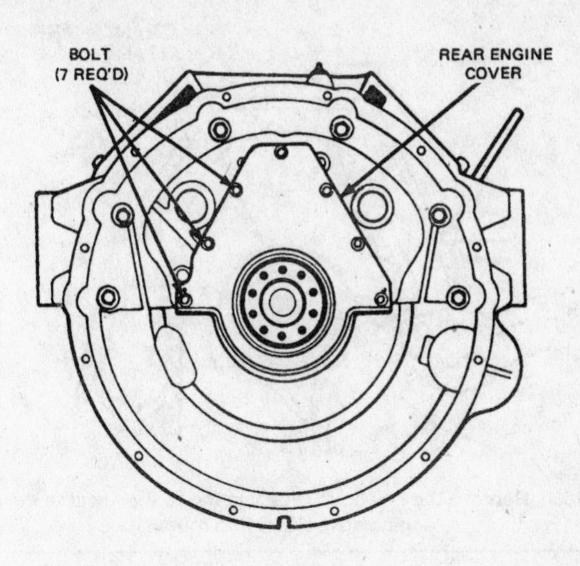

3.163 Remove the bolts from the rear engine cover

6 Add transmission fluid as needed, run the engine and check for oil leaks and proper transmission operation.

Crankshaft rear oil seal - replacement

1994 and earlier (except Direct Injection Turbo) models

1 Remove the flywheel/driveplate as described in the previous Section.

2 Remove the rear engine cover **(see illustration)**.

3 Carefully pry the seal out of the cover with a seal removal tool or a large screwdriver. Be careful not to distort the cover or scratch the wall of the seal bore.

3 Thoroughly clean the rear engine cover and all gasket surfaces. Be sure to clean the seal bore to remove any old seal material and corrosion.

4 Coat the rear engine cover seal bore with Gasket and Trim Adhesive (Ford no. D7AZ-19B508-AA or equivalent).

5 Position the new seal in the bore with the spring side of the seal facing in. Using an arbor press and Ford Rear Crankshaft Seal Replacer no. T83T-6701-A or equivalent, install the new seal **(see illustration)**. If the special equipment is not available, take the rear engine cover to a Ford dealer or automotive machine shop for seal installation. **Note:** *The seal must be installed from the engine block side, flush with the seal bore inner surface.*

6 Apply a 1/8-inch bead of RTV sealant around the outside diameter of the rear seal and edge of the rear cover.

7 Position a Rear Crankshaft Seal Pilot (Ford no. T83T-6701-B or equivalent) onto the crankshaft.

8 Apply Gasket and Trim Adhesive (Ford no. D7AZ-19B508-AA or equivalent) to the engine block and rear cover gasket surfaces. Position a new rear cover gasket to the engine block.

9 Apply a 1/4-inch bead of RTV sealant (Ford no. D6AZ-19562-AA or equivalent) at the corners of the oil pan and on the oil pan sealing surface **(see illustration)**.

10 Position the rear cover on the engine and install the mounting bolts. Remove the Seal Pilot Tool and tighten the bolts to the torque listed in this Chapter's Specifications.

11 Install the flywheel/driveplate, clutch (if equipped) and transmission.

12 Run the engine and check for oil leaks.

1994 and later Direct Injection Turbo models

1 Remove the transmission.

2 Remove the flywheel. Use **caution** - it's very heavy.

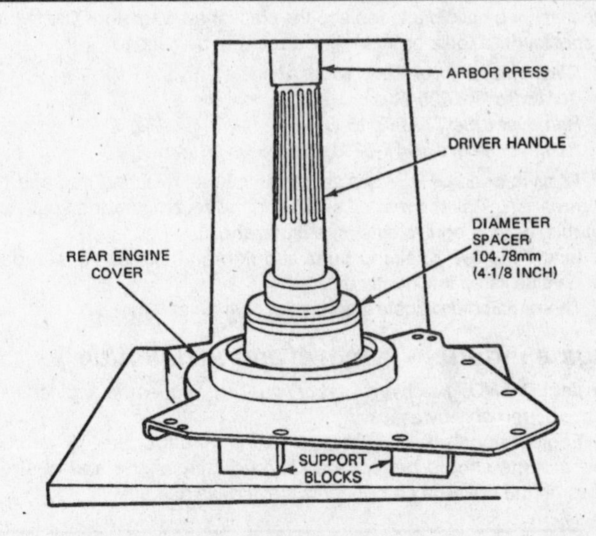

3.164a Press out the old seal and press in a new one

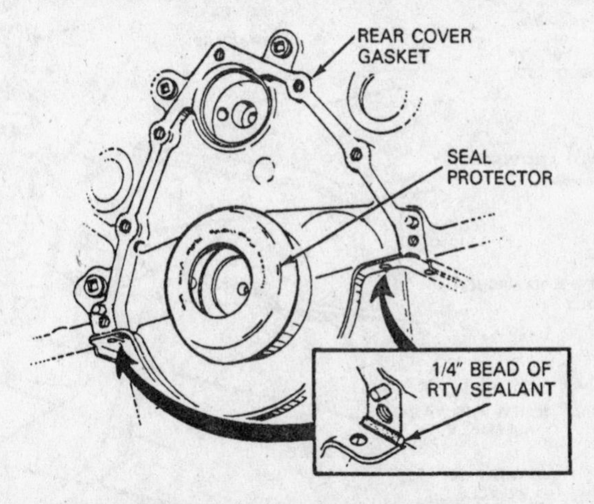

3.164b Apply sealant as shown here

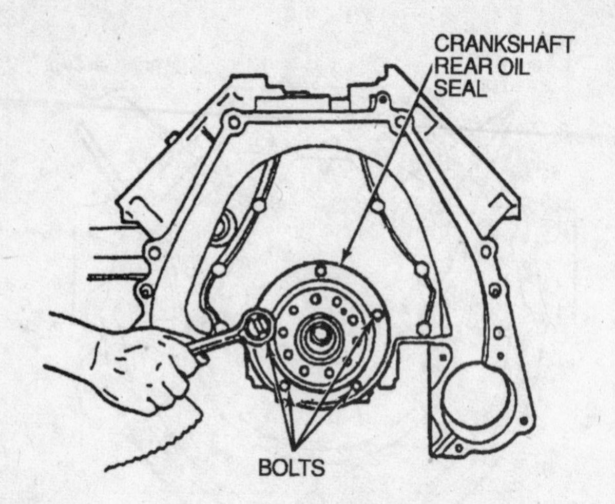

3.165a Remove the five bolts retaining the rear oil seal to the rear engine cover (DI Turbo models)

1 *Guide pins (T94P-7000-P)*
2 *Seal replacer (T94P-6701-AH4)*
3 *Seal replacer (T94P-6701-AH3)*
4 *Driver sleeve (T79T-6316-AH4)*
5 *Crankshaft rear oil seal*

3.165b Install the rear seal using these special tools (DI Turbo models)

3 Remove the five rear oil seal retaining bolts and pull rear oil seal from the crankshaft **(see illustration)**.

4 Inspect the wear ring on the crankshaft for wear or damage. If necessary, it's possible to replace the crankshaft wear ring. The following special Ford tools or their equivalents will be needed:

a) *Wear ring remover T94T-6701-AH1*
b) *Screw T84T-7025-B*
c) *Remover tube T77J-7025-B*
d) *Remover sleeve T94T-6701-AH2*

5 Apply a bead of RTV sealant to the oil real retaining ring and the bolt threads. Install the rear oil seal using the recommended Ford special tools, or their equivalents **(see illustration)**.

6 Install the five retaining bolts and tighten them to the standard torque value listed in Chapter 1.

7 The remainder of installation is the reverse of removal.

Engine mounts - removal and installation

Warning: *DO NOT place any part of your body under the engine when it is supported only by a jack.*

1 Engine mounts seldom require attention, but broken or deteriorated mounts should be replaced immediately or the added strain placed on the driveline components may cause damage.

Check

2 During the check, the engine must be raised slightly to remove the weight from the mounts.

3 Raise the vehicle and support it securely on jackstands, then position the jack under the engine oil pan. Place a large block of wood between the jack head and the oil pan, then carefully raise the engine just enough to take the weight off the mounts.

4 Check the mounts to see if the rubber is cracked, hardened or separated from the metal plates. Sometimes the rubber will split right down the center. Rubber preservative may be applied to the mounts to slow deterioration.

5 Check for relative movement between the mount plates and the engine or frame (use a large screwdriver or pry bar to attempt to move the mounts). If movement is noted, lower the engine and tighten the mount fasteners.

Replacement - F-Series trucks

6 Disconnect the ground cables from the negative terminals of the batteries.

7 Unbolt the fan shroud halves and set them aside.

8 Block the rear wheels and set the parking brake. Raise the front of the vehicle and support it securely on jackstands.

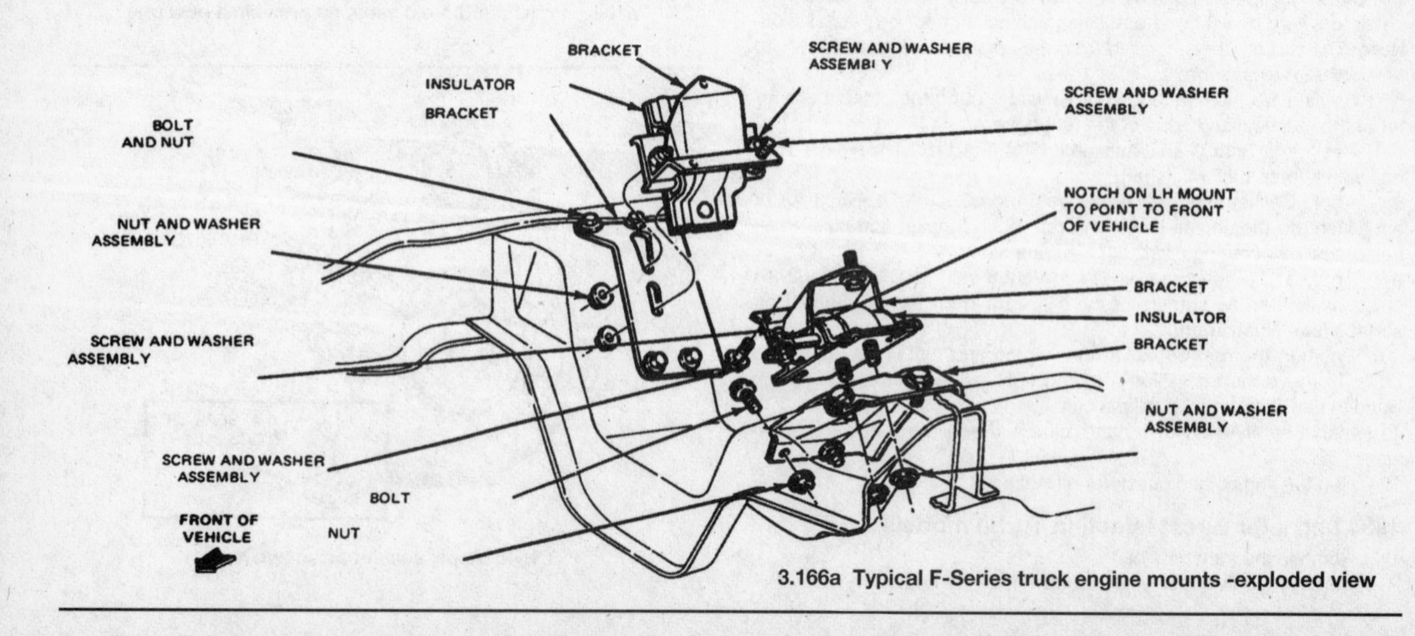

3.166a Typical F-Series truck engine mounts -exploded view

9 Remove the nuts that attach the mounts to the crossmember **(see illustrations)**.
10 Disconnect the exhaust pipes at the exhaust manifolds.
11 Remove the bolts attaching the mounts to the engine block.
12 Lower the vehicle.
13 Attach a lifting sling (Ford Rotunda no. 014-00312 or equivalent) to the engine and a hoist. Raise the engine slightly, then remove the mounts.
14 Installation is the reverse of removal. Use thread locking compound on the mount bolts/nuts and be sure to tighten them securely.

Replacement - E-Series trucks

Warning: *The refrigerant lines are under high pressure - have the system discharged by an automotive air conditioning specialist before following this procedure. Look for a shop that recycles the refrigerant gases to reduce environmental damage.*
15 Disconnect the ground cables from the negative terminals of the batteries.
16 Unbolt the fan shroud halves and set them aside.
17 Remove the vacuum pump drivebelt (see the Maintenance portion of this Chapter).
18 Remove the alternator and adjusting bracket (see the Electrical System portion of this Chapter.
19 On air conditioned models, remove the air conditioning compressor. Cap all open fittings. **Warning:** *Be sure the refrigerant gas in the system has been discharged. Always wear eye protection as a precaution against residual pressure.*
20 Remove the engine cover.
21 Remove the air cleaner and cover the intake with a cloth.
22 Detach the line from the fuel pump and fuel filter. Remove the fuel filter-to-injection pump line and the fuel filter return line. Remove the fuel filter assembly and bracket as a unit. Cap the open fittings. Refer to the Fuel system portion of this Chapter as necessary.
23 Remove the kickdown rod from the injection pump.
24 Block the rear wheels and set the parking brake. Raise the front of the vehicle and support it securely on jackstands.

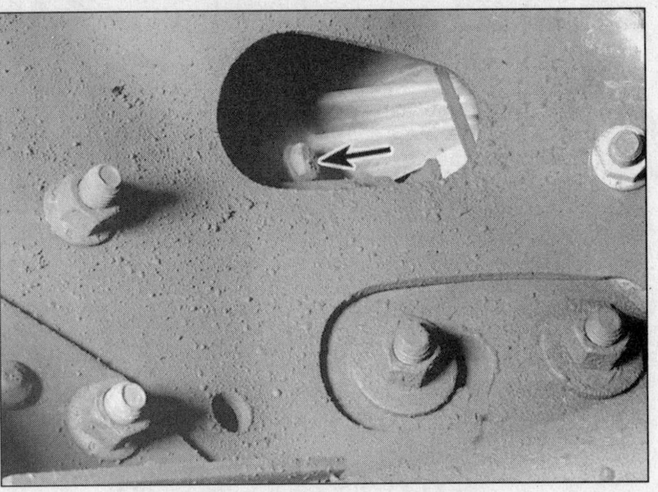

3.166b Don't overlook the nut (arrow) hidden in this pocket under the crossmember

25 Disconnect the ground cables from the lower front of the engine.
26 Remove the nuts from the engine mounts at the front crossmember **(see illustration)**.
27 Disconnect the transmission kickdown rod from the transmission.
28 Lower the vehicle.
29 Attach lifting brackets (Ford Rotunda no. 014-00312 or equivalent) to the engine. Using a crane-type engine hoist, lift the engine until it touches the body.
30 Remove the engine mount and bracket assemblies.
31 Installation is the reverse of removal. Use thread locking compound on the mount bolts/nuts and be sure to tighten them securely.
32 Have the air conditioning system (if equipped) evacuated, charged and leak tested by the shop that discharged it.

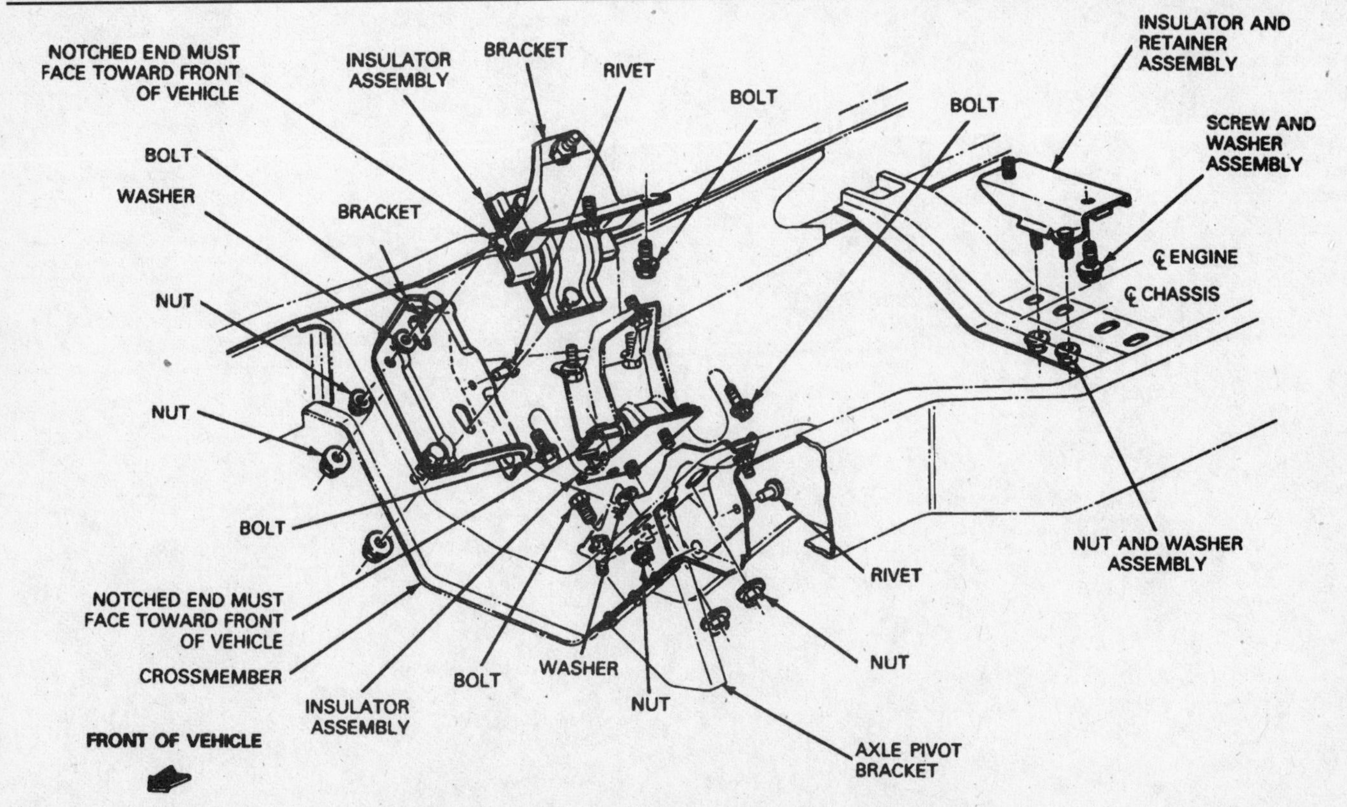

3.167 Typical E-Series truck engine mounts - exploded view

Vacuum pump - removal and installation

1 Remove the drivebelt as described in the Maintenance portion of this Chapter.

2 Unbolt the vacuum pump **(see illustration)**.

3 Detach the hose clamp and vacuum hose, then lift the pump from the engine compartment.

4 Installation is the reverse of removal. Be sure to tighten the bolts securely and adjust the drivebelt as described in the *Maintenance* portion of this Chapter.

3.168 Remove the vacuum pump mounting bolts (arrows)

Chapter 4
Engine overhaul procedures

Specifications

GM 5.7L V8 engines

General

Bore and stroke	4.057 x 3.385 in
Compression ratio	22.5:1
Compression	380 to 390 psi (warmed up)
Firing order	1-8-4-3-6-5-7-2
Static timing	Top Dead Center (TDC)
Type of fuel	
Above 20-degrees F	No. 2 Diesel
Below 20-degrees F	No. 1 Diesel
Oil pressure (engine at operating pressure)	
Idle	7 psi
1500 rpm	30 psi

Engine block

Piston displacement	5.7L (350 CID)
Cylinder bore	4.056 to 4.058 in
Taper limit	Not available
Out-of-round limit	Not available
Deck warpage limit	Not available

Pistons and rings

Piston diameter	4.051 in
Piston-to-bore clearance - selective fit	
1978 through 1981	0.005 to 0.006 in
1982 and later	0.0035 to 0.0045 in
Piston groove width	
Top compression ring	0.0838 to 0.0840 in
Bottom compression ring	
1978 through 1983	0.0798 to 0.0808 in
1984 and later	0.0810 to 0.0820 in
Oil control ring	0.1886 to 0.1896 in
Piston ring-to-groove clearance	
Top compression ring	0.005 to 0.007 in
Bottom compression ring	0.003 to 0.005 in
Piston ring end gap	
Top compression ring	
1978 through 1981	0.015 to 0.025 in
1982 and later	0.019 to 0.027 in
Bottom compression ring	
1978 through 1981	0.015 to 0.025 in
1982 and later	0.013 to 0.021 in

Crankshaft and connecting rods

Main journal	
Diameter	3.0003 to 2.9993 in
Taper limit - maximum per inch	Not available
Out-of-round limit	0.0005 in
Runout limit	Not available
Main bearing oil clearance	
Standard	
No. 1, 2, 3 and 4	0.0005 to 0.0021 in
No. 5	0.0020 to 0.0034 in
Service limit	Not available
Connecting rod journal	
Diameter	2.1248 to 2.1238 in
Taper limit - maximum per inch	Not available
Out-of-round limit	0.0005 in
Connecting rod bearing oil clearance	
Standard	0.0005 to 0.0026 in
Service limit	Not available
Connecting rod side clearance (big end)	
Standard	0.006 to 0.020 in
Service limit	Not available
Crankshaft endplay	
Standard	0.0035 to 0.0135 in
Service limit	Not available
Flywheel clutch face runout limit	Not available

Camshaft

Bearing journal diameter	
No. 1	2.0365 to 2.0357 in
No. 2	2.0165 to 2.0157 in
No. 3	1.9965 to 1.9957 in
No. 4	1.9765 to 1.9757 in
No. 5	1.9565 to 1.9557 in
Bearing oil clearance	
Standard	0.0020 to 0.0058 in
Service limit	Not available
Lobe lift	Not available
Endplay	0.011 to 0.017 in

Cylinder heads and valve train

Head warpage limit	0.003
Valve seat angle	
Intake	45-degrees
Exhaust	31-degrees
Valve seat width	
Intake	0.075 to 0.098 in
Exhaust	0.037 to 0.075 in
Valve seat runout limit	0.002 in
Valve face angle	
Intake	44-degrees
Exhaust	30-degrees
Valve face runout limit	0.002 in
Minimum valve margin width	0.0312 in
Valve stem diameter	
Intake	0.3425 to 0.3432 in
Exhaust	0.3420 to 0.3427 in
Valve guide diameter	Not available
Valve stem-to-guide clearance	
Intake	
Standard	0.0010 to 0.0027 in
Service limit	Not available
Exhaust	
Standard	0.0015 to 0.0032 in
Service limit	Not available
Valve spring free length	2.09 in
Valve spring installed height	Not available
Valve spring under load	
85 to 95 lbs	1.670 in
203 to 217 lbs	1.220 in
Valve spring inside diameter	1.065 to 1.041 in

Valve adjustment	Hydraulic
Pushrod length	7.718 in

Torque specifications* | **Ft-lbs**

Connecting rod nuts	42
Main bearing cap bolts	120
Glow plug	12
Flywheel-to-torque converter	40
Flywheel-to-crankshaft	60
Vacuum pump clamp bolt	17
Starter bolts	35
Engine mount-to-frame	50
Engine mount-to-block	75
Vibration damper bolt	200 to 310
Oil pan drain bolt	30
Oil pan bolts	10
Oil pump bolts	35
Oil pump cover bolts/screws	10

* *Additional torque specifications can be found in Chapter 2*

GM 6.2L and 6.5L V8 engines

General

Compression	380 to 400 psi (warmed up)
Firing order	1-8-7-2-6-5-4-3
Static timing	Top Dead Center (TDC)
Type of fuel	
Above 20-degrees F	No. 2 Diesel
Below 20-degrees F	No. 1 Diesel
Oil pressure (engine at operating temperature)	
Idle	10 psi
2000 rpm	45 psi

Engine block

Cylinder bore diameter	
6.2L	3.9789 to 3.9824 in
6.5L	4.057 to 4.058 in
Taper limit	0.0008 in
Out-of-round limit	0.0008 in
Deck warpage limit	
Across deck length	.006 in
Across deck width	.003 in

** **Note:** *Cylinder bores numbers 7 and 8 are marked one size class smaller than actual size i.e. a bore measuring "B" class is stamped "A" class*

Pistons and rings

Piston diameter	selective fit
Piston-to-bore clearance - selective fit	
Bohn pistons (1982 through 1992)	
Cylinders 1 through 6	0.0035 to 0.0045 in
Cylinders 7 and 8	0.0040 to 0.0050 in
Zollner pistons (1982 through 1986)	
Cylinders 1 through 6	0.0044 to 0.0054 in
Cylinders 7 and 8	0.0049 to 0.0059 in
Zollner pistons (1987 through 1992)	
Cylinders 1 through 6	0.0035 to 0.0045 in
Cylinders 7 and 8	0.0040 to 0.0050 in
Zollner pistons (1993 and later)	
6.2L	
Cylinders 1 through 6	0.0025 to 0.0035 in
Cylinders 7 and 8	0.0030 to 0.0040 in
6.5L	
Cylinders 1 through 6	0.0037 to 0.0047 in
Cylinders 7 and 8	0.0042 to 0.0052 in
Piston groove width	Not available
Service limit	Not available
Piston ring-to-groove clearance	
Top compression ring	0.0030 to 0.0070 in
Bottom compression ring	0.0015 to 0.0031 in
Oil control ring	0.0016 to 0.0038 in
Service limit	0.0010 in

Pistons and rings (continued)

Specification	Value
Piston ring end gap	
Top compression ring	
1982 through 1992	0.012 to 0.022 in
1993 and later	0.010 to 0.020 in
Bottom compression ring	0.030 to 0.040 in
Oil control ring	0.010 to 0.020 in

Crankshaft and connecting rod

Specification	Value
Main journal	
Diameter	
1982 through 1993	
No. 1 through 4	2.9495 to 2.9504 in
No. 5	2.9492 to 2.9502 in
1994 and later	
No. 1 through 4	2.9517 to 2.9520 in
No. 5	2.9515 to 2.9518 in
Taper limit - maximum per inch	0.0002 in
Out-of-round limit	0.0002 in
Runout limit	Not available
Main bearing oil clearance	
Standard	
No. 1 through 4	0.0018 to 0.0033 in
No. 5	0.0022 to 0.0037 in
Service limit	Not available
Connecting rod journal	
Diameter	
1982 through 1993	2.3982 to 2.3991 in
1994 and later	2.399 to 2.401 in
Taper limit - maximum per inch	0.0002 in
Out-of-round limit	0.0002 in
Connecting rod bearing oil clearance	
Standard	0.0018 to 0.0039 in
Service limit	Not available
Connecting rod side clearance (big end)	
Standard	0.006 to 0.025 in
Service limit	Not available
Crankshaft endplay	
Standard	0.10 to 0.25 in
Service limit	Not available
Flywheel clutch face runout limit	Not available

Camshaft

Specification	Value
Bearing journal diameter	
No. 1 through 4	2.1641 to 2.1663 in
No. 5	2.0067 to 2.0088 in
Bearing oil clearance	
Standard	0.0010 to 0.0046 in
Service limit	Not available
Lobe lift	
Intake	0.2808 in
Exhaust	0.2808 in
Endplay	0.002 to 0.012 in
Timing chain freeplay	
New chain	0.500 in
Used chain	0.800 in

Cylinder heads and valve train

Specification	Value
Head warpage limit	0.003
Valve seat angle	
Intake	46-degrees
Exhaust	46-degrees
Valve seat width	
Intake	0.035 to 0.060 in
Exhaust	0.062 to 0.093 in
Valve seat runout limit	0.002 in
Valve face angle	
Intake	45-degrees
Exhaust	45-degrees
Valve face runout limit	0.002 in
Minimum valve margin width	0.0312 in

Valve stem diameter	
Intake	0.3410 to 0.3417 in
Exhaust	0.3715 to 0.3722 in
Valve guide diameter	
Intake	0.3427 to 0.3437 in
Exhaust	0.3732 to 0.3742 in
Valve stem-to-guide clearance	
Intake	
Standard	0.0010 to 0.0027 in
Service limit	Not available
Exhaust	
Standard	0.0010 to 0.0027 in
Service limit	Not available
Valve spring free length	2.09 in
Valve spring installed height	1.81 in
Valve spring under load	
80 lbs.	1.81 in
230 lbs.	1.39 in
Valve adjustment	Hydraulic
Pre-chamber protrusion	0.004 maximum

Torque specifications*	**Ft-lbs**
Bellhousing bolts	25 to 35
Connecting rod nuts	44 to 52
Main bearing cap bolts	
Inner	110
Outer	100
Glow plug	12
Flywheel-to-torque converter	40
Flywheel-to-crankshaft	66
Vacuum pump clamp bolt	17
Starter bolts	35
Engine mount-to-frame	50
Engine mount-to-block	75
Vibration damper bolt	200
Oil pan drain bolt	20

** Additional torque specifications can be found in Chapter 2*

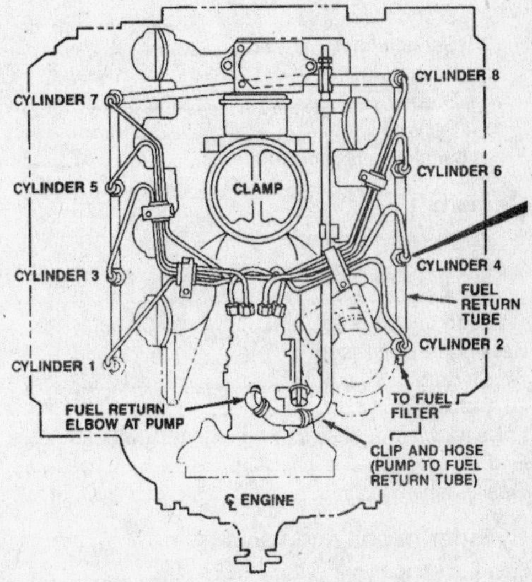

Ford 6.9L V8 engine

General

Bore and stroke	4.00 x 4.18 in.
Compression ratio	20.7:1
Compression	260 to 440 psi
Firing order	1-2-7-3-4-5-6-8
Type of fuel	
Above 20-degrees F	No. 2 Diesel
Below-20 degrees F	No. 1 Diesel
Oil pressure (engine at operating pressure)	
1983 through 1985	
2000 rpm	40 to 60 psi
1986 and later	
2000 rpm	40 to 70 psi

Engine block

Piston displacement	6.9L (420 CID)
Cylinder bore	3.9995 to 4.0015 in
Out-of-round limit	0.0003 in
Service limit	0.012 in
Deck warpage limit	0.006 in

Pistons and rings

Piston diameter	3.9935 to 3.9955 in
Piston-to-bore clearance	
(selective fit)	0.0055 to 0.0075 in
Piston ring-to-groove clearance	
Top compression ring	0.002 to 0.004 in
Bottom compression ring	0.002 to 0.004 in
Oil control ring	0.001 to 0.003 in

Ford diesel cylinder arrangement

Pistons and rings (continued)

Piston ring end gap	
Top compression ring	0.014 to 0.024 in
Bottom compression ring	0.060 to 0.070 in
Oil control ring	0.010 to 0.024 in

Crankshaft and connecting rods

Main journal	
Diameter	3.1228 to 3.1236 in
Taper limit - maximum per inch	0.0005 in
Out-of-round limit	0.0002 in
Runout limit	0.002 in
Main bearing oil clearance	
Standard	0.0018 to 0.0036 in
Service limit	0.0010 in
Connecting rod journal	
Diameter	2.4980 to 2.4990 in
Taper limit - maximum per inch	0.0005 in
Out-of-round limit	0.0003 in
Connecting rod bearing oil clearance	
Standard	0.0011 to 0.0026 in
Service limit	0.0010 in
Connecting rod side clearance (big end)	
1983 through 1985	0.008 to 0.020 in
1986 and later	0.012 to 0.024 in
Crankshaft endplay	
1983 through 1985	
Standard	0.001 to 0.009 in
Service limit	0.012 in
1986 and later	
Standard	0.002 to 0.009 in
Service limit	0.012 in
Flywheel clutch face runout limit	0.008 in

Camshaft

Bearing journal diameter	
1983	2.0990 to 2.1000 in
1984	2.1010 to 2.1045 in
1985 and later	2.1020 to 2.1055 in
Bearing oil clearance	
Standard	0.0010 to 0.0055 in
Service limit	Not available
Lobe lift	Not available
Endplay	0.001 to 0.009 in
Gear backlash	0.0055 to 0.0100 in

Cylinder heads and valve train

Head warpage limit	0.006 in
Valve seat angle	
Intake	30-degrees
Exhaust	37.5-degrees
Valve seat width	
Intake	0.080 to 0.095 in
Exhaust	0.080 to 0.095 in
Valve seat runout limit	0.002 in
Valve face angle	
Intake	30-degrees
Exhaust	37.5-degrees
Valve face runout limit	0.002 in
Minimum valve margin width	0.0312 in
Valve stem diameter	
Intake	0.37165 to 0.37235 in
Exhaust	0.37165 to 0.37235 in
Valve guide diameter	
Intake	Not available
Exhaust	Not available
Valve stem-to-guide clearance	
Intake	
Standard	0.0012 to 0.0029 in
Service limit	0.0055 in

Exhaust	
Standard	0.0012 to 0.0029 in
Service limit	0.0055 in
Valve spring free length	2.04 in
Valve spring installed height	Not available
Valve spring under load @ 60 lbs.	1.798 in
Valve adjustment	Hydraulic

Torque specifications*	**Ft-lbs**
Bellhousing bolts	25 to 35
Crankcase front cover (standard sizes)**	
1/4 x 20 UNC	7
5/16 x 18 UNC	14
3/8 x 16 UNC	24
7/16 x 14 UNC	38
1/2 x 13 UNC	60
Connecting rod nuts	
First step	38
Second step	51
Main bearing cap bolts	
First step	75
Second step	95
Glow plug	12
Flywheel-to-torque converter	40
Flywheel-to-crankshaft	47
Starter bolts	35
Engine mount-to-frame	50
Engine mount-to-block	75
Vibration damper bolt	90
Oil pan drain bolt	28
Oil pan bolts	7
Oil pan bolts (two in rear)	14
Oil cooler plug	15
Valve cover bolts	6

** Additional torque specifications can be found in Chapter 2*

*** The torque values listed for these size bolts are standard values that can be applied to a bolt(s) of the exact same dimensions used on other components*

Ford 7.3L V8 engine

General

Bore and stroke	4.11 x 4.11 in
Compression ratio	
1994 and earlier (except Direct Injection Turbo)	21.5:1
1994 and later (Direct Injection Turbo)	17.5:1
Compression	260 to 440 psi
Firing order	1-2-7-3-4-5-6-8
Type of fuel	
Above 20 degrees F	No. 2 Diesel
Below 20 degrees F	No. 1 Diesel
Oil pressure (engine at operating temperature)	
2000 rpm	40 to 70 psi

Engine block

Piston displacement	7.3L (444 CID)
Cylinder bore	
1994 and earlier (except Direct Injection Turbo)	
Cylinders 1 through 6	4.1095 to 4.1115 in
Cylinders 7 and 8	4.1100 to 4.1120 in
1994 and later (Direct Injection Turbo)	4.1096 to 4.1103
Out-of-round limit	0.0003 in
Service limit	0.012 in
Deck warpage limit	0.006 in
Pistons and rings	
Piston diameter	
1994 and earlier (except Direct Injection Turbo)	
Cylinders 1 through 6	4.10350 to 4.10400 in
1994 and later (Direct Injection Turbo)	4.10450 to 4.10500 in

Engine block (continued)

Specification	Value
Piston-to-bore clearance (selective fit)	
1994 and earlier (except Direct Injection Turbo)	0.0055 to 0.0085 in
1994 and later (Direct Injection Turbo)	0.0044 to 0.0057 in
Piston ring-to-groove clearance	
1994 and earlier (except Direct Injection Turbo)	
Top compression ring	0.002 to 0.004 in
Bottom compression ring	0.002 to 0.004 in
Oil control ring	0.001 to 0.003 in
1994 and later (Direct Injection Turbo)	
Second ring only	0.002 to 0.004 in
Piston ring end gap	
1994 and earlier (except Direct Injection Turbo)	
Top compression ring	0.013 to 0.045 in
Bottom compression ring	0.060 to 0.085 in
1994 and later (Direct Injection Turbo)	
Top compression ring	0.014 to 0.024 in
Bottom compression ring	0.062 to 0.072 in
Oil control ring	0.012 to 0.024 in

Crankshaft and flywheel

Specification	Value
Main journal	
Diameter	3.1228 to 3.1236 in
Taper limit - maximum per inch	0.0005 in
Out-of-round limit	0.0002 in
Runout limit	0.002 in
Main bearing oil clearance	
Standard	0.0018 to 0.0036 in
Service limit	0.0010 in
Connecting rod journal	
Diameter	2.4980 to 2.4990 in
Taper limit - maximum per inch	0.0005 in
Out-of-round limit	0.0003 in
Connecting rod bearing oil clearance	
Standard	0.0011 to 0.0026 in
Service limit	Not available
Connecting rod side clearance (big end)	
Standard	0.012 to 0.024 in
Service limit	Not available
Crankshaft endplay	
Standard	0.0025 to 0.0085 in
Service limit	0.012 in
Flywheel clutch face runout limit	0.008 in

Camshaft

Specification	Value
Bearing journal diameter	2.1015 to 2.1025 in
Bearing oil clearance	
1994 and earlier (except Direct Injection Turbo)	0.0015 to 0.0035 in
1994 and later (Direct Injection Turbo)	0.002 to 0.006 in
Lobe lift	
1994 and earlier (except Direct Injection Turbo)	Not available
1994 and later (Direct Injection Turbo)	
Intake	0.2535 in
Exhaust	0.2531 in
Gear backlash	
1994 and earlier (except Direct Injection Turbo)	0.0015 to 0.013 in
1994 and later (Direct Injection Turbo)	0.0055 to 0.010 in
End play	
1994 and earlier (except Direct Injection Turbo)	0.002 to 0.009 in
1994 and later (Direct Injection Turbo)	0.002 to 0.008 in
Head warpage limit	
1994 and earlier (except Direct Injection Turbo)	0.006 in
1994 and later (Direct Injection Turbo)	0.004 in

Cylinder heads and valve train

Specification	Value
Head warpage limit	0.006 in
Valve seat angle	
Intake	30-degrees
Exhaust	37.5-degrees

Valve seat width	
Intake	0.065 to 0.095 in
Exhaust	0.065 to 0.095 in
Valve seat runout limit	0.002 in
Valve face angle	
Intake	30-degrees
Exhaust	37.5-degrees
Valve face runout limit	0.0015 in
Minimum valve margin width	
1994 and earlier (except Direct Injection Turbo)	
Intake	0.112 in
Exhaust	0.053 in
1994 and later (Direct Injection Turbo)	
Intake	0.066 in
Exhaust	0.054 in
Valve stem diameter	
1994 and earlier (except Direct Injection Turbo)	0.37165 to 0.37235 in
1994 and later (Direct Injection Turbo)	0.31185 to 0.31255 in
Valve guide bore diameter	
1994 and earlier (except Direct Injection Turbo)	0.3736 to 0.3746 in
1994 and later (Direct Injection Turbo)	0.3141 to 0.3151 in
Valve stem to guide clearance	0.0055 in
Valve spring free length	2.075 in.
Valve spring installed height	
Intake	1.767 in
Exhaust	1.833 in
Valve adjustment	Hydraulic
Pre-chamber insert protrusion	-0.0025 to +0.0025 in

Torque specifications*	**Ft-lbs**
Bellhousing-to-engine bolts	
1983 through 1991	25 to 35
1995	
Automatic	41 to 56
Manual	35 to 50
1996 and later	39 to 53
Connecting rod nuts	
1983 through 1994 (except Direct Injection Turbo)	
First step	38
Second step	51
1994 (Direct Injection Turbo) and 1995	
First step	52
Second step	80
1996 and later	70
Engine mount-to-frame bolts	
1983 through 1994 (except Direct Injection Turbo)	35
1994 and later (Direct Injection Turbo)	71 to 94
Engine mount-to-block bolts	75
Flywheel-to-crankshaft bolts	
1983 through 1994 (except Direct Injection Turbo)	47
1994 (Direct Injection Turbo) and 1995	
Automatic	23 to 39
Manual	89
1996 and later	89
Glow plugs	
1994 and earlier (except Direct Injection Turbo)	12
1994 and later (Direct Injection Turbo)	14
Main bearing cap bolts	
First step	75
Second step	90
Oil pan bolts - 1994 and earlier (except Direct Injection Turbo)	
All except rear two	7
Rear two	14
Vibration damper bolt	
1994 and earlier (except Direct Injection Turbo)	90
1994 and later (Direct Injection Turbo)	212
Valve cover bolts	6

**Additional torque specifications can be found in Chapter 2. If not listed, standard torque specifications can be found in Chapter 1.*

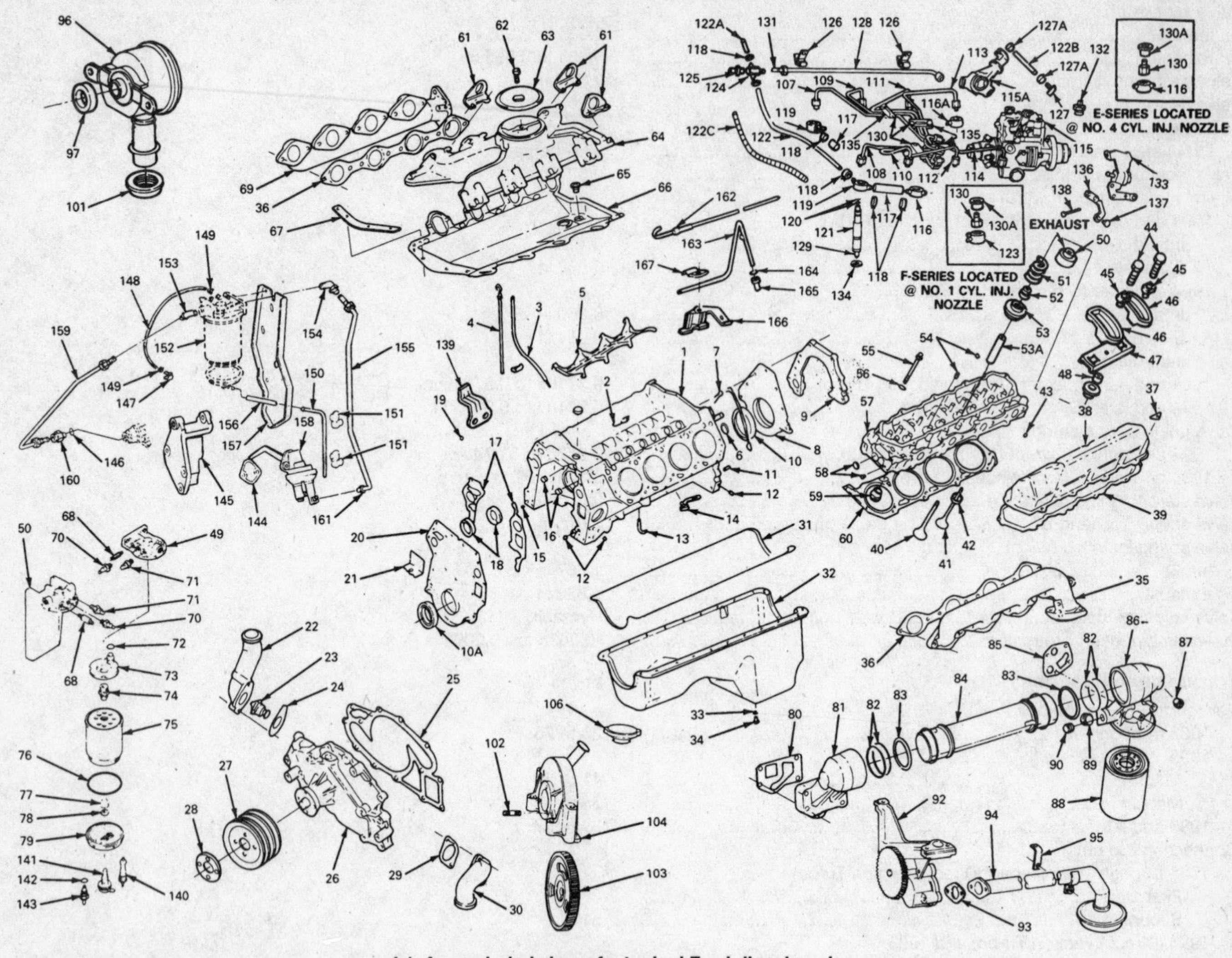

4.1 An exploded view of a typical Ford diesel engine

1 Cylinder block assembly
2 Tappet guide
3 Oil level gauge tube assembly (F-series)
4 Oil level gauge (F-series)
5 Tappet guide retainer
6 Engine plug (1-1/2 inch outside diameter)
7 Rear cover gasket
8 Rear engine cover assembly
9 Flywheel-to-transmission adapter
10 Crankshaft rear oil seal
10a Front oil seal
11 Flywheel adapter dowel pin
12 Pipe plug, 1/8 NPTF
13 Piston cooling jet
14 Block heater assembly
15 Front cover plate dowel pin
16 Cup plug
17 Front cover plate gasket
18 Camshaft bearing kit
19 Oil indicator hole ball 11/32 inch
20 Front cover plate
21 Timing indicator (part of front cover)
22 Water outlet connection
23 Thermostat
24 Water outlet gasket
25 Water pump gasket
26 Water pump
27 Water pump pulley
28 Fan spacer
29 Water inlet gasket
30 Water inlet connection
31 RTV sealant
32 Oil pan
33 Oil pan drain gasket
34 Oil pan drain plug
35 Exhaust manifold, left
36 Exhaust manifold gasket
37 Valve cover washer
38 Valve cover
39 Valve cover gasket
40 Intake valve (8)
41 Exhaust valve
42 Exhaust valve seat insert
43 Ball type plug, 13/32v in (8)
44 Valve lever bolt and washer
45 Valve lever post
46 Valve lever
47 Valve lever post retainer
48 Valve spring retainer lock
49 Valve spring retainer (36)
50 Oil shield (Exhaust)
51 Valve spring with damper (16)
52 Valve stem – intake seal (8)
53 Assembly valve rotator (16)
54 Plug 1/2 inch NPTF (4)
55 Cylinder head bolt (34)
56 Cylinder head bolt washer (34)
57 Cylinder head assembly (2)
58 Pug 1/4 inch
59 Combustion chamber insert (8)
60 Cylinder head gasket (2)
61 Lifting eye (3)
62 Bolt thread – air cleaner stud insert
63 Intake manifold screen
64 Intake manifold
65 Valley pan drain plug
66 Gasket and valley pan
67 Valley pan strap
68 Fuel priming valve and cap
69 Exhaust manifold -right
70 Continuous vent with check valve
71 Vacuum switch (fuel filter element replacement indicator)
72 Fuel heater O-ring
73 Fuel heater
74 Threaded insert
75 Fuel filter element
76 Drain bowl O-ring
77 Drain valve stem cap

78 Drain valve seal
79 Water separator drain bolt
80 Oil cooler, front header gasket
81 Oil cooler, front header
82 Oil cooler O-ring (2)
83 Oil cooler O-ring (2)
84 Oil cooler
85 Oil cooler rear header gasket
86 Oil cooler rear header
87 Plug – 1/4 inch
88 Oil filter
89 Plug – 1/2 inch
90 Plug – 1/2 inch
91 Oil pump assembly
92 Oil pump assembly
93 Oil pick-up gasket
94 Pick-up tube
95 Oil pick-up bracket
96 CDR valve
97 CDR valve seal ring
100 Crankcase vent tube
101 Valley cover grommet
102 Injection pump mounting stud
103 Injection pump drive gear
104 Injection pump adapter housing
106 Oil filler cap
107 Pipe with nuts pump to cylinder 8
108 Pipe with nuts pump to cylinder 7
109 Pipe with nuts pump to cylinder 6
110 Pipe with nuts pump to cylinder 5
111 Pipe with nuts pump to cylinder 4
112 Pipe with nuts pump to cylinder 3
113 Pipe with nuts pump to cylinder 2
114 Pipe with nuts pump to cylinder 1
115 Injection pump
115a Vacuum modulator valve (automatic transmission)
116 Fuel return tee
116a Fuel return elbow
117 Hose
118 Clip
119 Fuel return tee
120 O-rings
121 Injection nozzle holder
122 Fuel return hose
122a Fuel return hose
122b Pump-to-fuel return tube hose
122c Rear fuel return hose guard
123 Fuel return tee
124 Fuel return junction fitting
125 Fuel return nipple
126 Clamp
127 Elbow
127a Clip
128 Tube
129 Nozzle tip
130 Fuel line pressure sensor
130a Cover
131 Fuel return sleeve seal (2)
132 Temperature switch
133 Bracket and solenoid-fast idle
134 Gasket nozzle (8)
135 Clamp
136 Kickdown lever (automatic transmission)
137 Kickdown lever screw
138 Adjusting screw (kickdown lever)
139 Oil level tube support bracket (F-series)
140 Vent/valve assembly
141 Manual drain valve
142 Water sensor O-ring
143 Water sensor probe
144 Fuel pump supply gasket
145 Alternator bracket
146 Sealing O-ring
147 Fuel return tee (at nozzle)
148 Hose, 3/16 inch inside diameter x 10 inch long
149 Hose clip
150 Water drain tube
151 Drain tube clamp (Z)
152 Fuel filter/water separator element
153 Elbow
154 Fuel supply pump-to-filter header elbow
155 Fuel pump-to-fuel header tube (with two nuts and two sleeves)
156 Hose, 3/16 inch x 2 5/16 inch long
157 Fuel filter header mounting bracket
158 Fuel supply pump
159 Filter-to-injection pump tube (with two nuts and two sleeves)
160 Connector fitting
161 Inverted flare tube nut
162 Oil level gauge
163 Tube
164 O-ring
165 Oil level gauge tube, lower
166 Oil lever gauge tube bracket
167 Oil level gauge tube retainer

General information and diagnosis

General information

Included in this portion of Chapter 2 are the general overhaul procedures for the cylinder head(s) and internal engine components.

The information ranges from advice concerning preparation for an overhaul and the purchase of replacement parts to detailed, step-by-step procedures covering removal and installation of internal engine components and the inspection of parts.

The following Sections have been written based on the assumption that the engine has been removed from the vehicle. For information concerning in-vehicle engine repair, as well as removal and installation of the external components necessary for the overhaul, see Chapter 2 (GM) or Chapter 3 (Ford).

The Specifications included in this Part are only those necessary for the inspection and overhaul procedures which follow. Refer to Chapter 1, 2 and 3 for additional Specifications.

This manual contains information on the General Motors 5.7L, 6.2L and 6.5L diesel engines and the Ford 6.9L and 7.3L diesel engines.

The GM 5.7L diesel, produced mainly for passenger cars, but found in some light duty trucks, was discontinued after the 1985 model year. The GM 6.2L was introduced in 1982 and designed for light duty truck use only. In 1992 GM introduced the 6.5L Turbo diesel, the engine block, cylinder heads and fuel injection system is similar to the 6.2L, with the addition of a turbocharger for increased power and efficiency.

Ford introduced the 6.9L diesel in 1983 and upgraded it to 7.3L in 1988. In 1994 Ford introduced the 7.3L Direct Injection (DI) turbocharged diesel, coined as the "Power-stroke" diesel. The naturally-aspirated 7.3L diesel was also produced in 1994 and discontinued for 1995. The two 7.3L diesel engines share little in common, if you have a 1994 model make sure you know exactly which version your dealing with.

Diesel engines are similar to the gasoline engines in many ways, but the cylinder heads, valve sizes, crankshaft, pistons connecting rods and piston pins are different. Because of the high compression ratio required in the diesel engine to ignite the fuel mixture, the engine components are constructed of a much more heavy-duty design than the gasoline engine.

The special alloy steel pre-chamber inserts in the cylinder head combustion chambers are serviced separately from the cylinder head. With the cylinder head removed, the pre-chamber inserts can be pushed out after removing the glow plugs and the injection nozzles.

Diagnosis

Correct diagnosis is an essential part of every repair; without it you can only cure the problem by accident.

Sometimes a simple service item will cause symptoms similar to a worn out or defective engine. Be sure the engine is serviced to manufacturer's specifications (see Chapter 2 or 3) before you begin with the following diagnosis procedures.

If you are concerned about the condition of your engine because of decreased performance or fuel economy, perform a compression test. If the engine is making unusual noises, perform an oil pressure test and noise diagnosis. Computer controlled engines may be checked with a scan tool to eliminate that system as the source of the problem (this is normally done by a dealer service department).

One of the more common reasons people rebuild their engines is because of oil consumption. Before you decide that the engine needs an overhaul based on oil consumption, make sure that oil leaks aren't responsible. If you park the vehicle in the same place everyday on pavement, look for oil stains or puddles of oil. To check more accurately, place a large piece of cardboard under the engine overnight. Compare the color and feel of the oil on the dipstick to the fluids found under the vehicle to verify that it's oil (transmission fluid is slightly red).

If any drips are evident, put the vehicle on a hoist and carefully inspect the underside. Sometimes leaks will only occur when the engine is hot, under load or on an incline, so look for signs of leakage as well as active drips. If a significant oil leak is found, correct it before you take oil consumption measurements.

Excessive oil consumption is an indication that cylinders, pistons, rings, valve stems, seals and/or valve guides may be worn. A clogged crankcase ventilation system can also cause this problem.

Every engine uses oil at a different rate. However, if an engine uses a quart of oil in 700 miles or less, or has visible blue exhaust smoke, it definitely needs repair.

To measure oil consumption accurately, park the vehicle on a level surface and shut off the engine. Wait about 15 minutes to allow the oil to drain down into the sump. Wipe the dipstick and insert it into the dipstick tube until it hits the stop. Then withdraw it carefully and read the level before the oil has a chance to flow. Fill the sump exactly to the full mark with the correct grade and viscosity of oil and write down the mileage shown on the odometer. Then monitor the oil level, using this same checking procedure until one quart of oil has been consumed and note the mileage.

Every moving part in the vehicle can make noise. Owners frequently blame the engine for a noise that is actually in the transmission or driveline.

Set the parking brake and place the transmission in neutral (Park on automatics). Start the engine with the hood open and determine if the noise is actually coming from the engine. Rev the engine slightly; does the noise increase directly with engine speed? If the noise sounds like it's coming from the engine and it varies with engine speed, it probably is an engine or driveplate/torque converter noise (automatic transmissions only).

What kind of noise is it? If it's a squealing sound, check drivebelt tension. Spray some belt dressing (available at auto parts stores) on the belt(s); if the noise goes away, adjust or replace the belt(s) as necessary. Sometimes it's necessary to remove the drivebelt briefly to eliminate the engine accessories as a source of noise. With the belt removed and the engine off, turn each accessory by hand and listen for sounds. Then run the engine briefly and listen for the noise.

Knocking or ticking noises are the most common types of internal engine noises. Noises that occur at crankshaft speed are usually caused by crankshaft, connecting rod and bearing problems, so begin your search on the lower part of the engine. Noises that occur at half of crankshaft speed usually involve the camshaft, lifters, rocker arms, valves, springs and mechanical fuel pump pushrod. Listen for these sounds near the top of the engine.

To pinpoint the source of the noise, a mechanic's stethoscope is best. If you don't have one, improvise with a four-foot length of hose held to your ear. You can also hold the handle of a large screwdriver to your ear and touch the tip to the suspected areas.

Move the listening device around until the sound is loudest. Think about which components are in the area of the noise and which part could produce this sound.

If the noise is in the top of the engine, remove the valve cover from the affected side and briefly start the engine, allowing it to idle slowly so it doesn't fling too much oil. See if all the valves appear to be opening the same amount and if the pushrods are rotating slowly, as they should. Press down on each rocker arm with your thumb right above the valve. If the sound changes or goes away, you've found which cylinder has the problem. Remove the rocker arm and pushrod. Inspect them carefully for wear and cracks. If no other problems are found, the lifter is probably defective.

If the noise is in the lower portion of the engine, check for piston slap, piston pin, connecting rod, main bearing and piston ring noises.

Piston slap is loudest on a cold engine and quiets down as the engine warms up. Listen for a dull, hollow sound in the cylinder wall just below the head. If you suspect that a piston has a hole in it, remove the dipstick and listen in the opening. Hold a piece of rubber hose between your ear and the tube. You should be able to hear and feel combustion gasses escaping if the piston has a hole or is cracked.

Main bearing knocks have a low-pitched knock deep within the engine that will sound loudest when the engine is first started. It will also be quite noticeable under heavy load.

Connecting rod bearings knock loudest when the accelerator is pressed briefly and then quickly released. The noise is usually caused by excessive bearing wear or insufficient oil pressure.

Piston rings that are loose in their grooves or broken make a chattering noise that is loudest during acceleration. This should be confirmed by a dealership service department. A compression check will tell you what mechanical condition the upper end, pistons, rings, valves, head gaskets) of your engine is in. Specifically, it can tell you if the compression is down due to leakage caused by worn piston rings, defective valves and seats or a blown head gasket. **Note:** *The engine must be at normal operating temperature and the battery must be fully charged for this check. Refer to the Compression check procedure in this Chapter for more information.*

Engine oil pressure provides a fairly good indication of bearing condition in an engine. As bearing surfaces wear, oil clearances increase. This increased clearance allows the oil to flow through the bearings more readily, which results in lower oil pressure. Oil pumps also wear, which causes an additional loss of pressure.

Pull out the dipstick and check the oil level. If the oil is dirty, contaminated with gasoline from short trips, or too thin (low viscosity) for the season, change it.

Check the oil pressure with a gauge installed in place of the oil pressure sending unit. Allow the engine to reach normal operating temperature before performing the test. If the pressure is extremely low, the bearings and/or oil pump are probably worn out.

Worn camshaft lobes are fairly common on high-mileage engines. If the engine is low on power, runs rough and/or backfires constantly through the intake or exhaust, suspect crossed ignition wires, a cracked distributor cap or a worn camshaft. These problems won't show up on a compression test (unless a valve doesn't open at all), but will show up on a power balance test.

To check for worn camshaft lobes, remove the valve covers. Then remove the rocker arms or rotate them aside from the pushrods.

Mount a dial indicator so it bears on the end of the pushrod. Using a socket and ratchet on the vibration damper bolt, slowly turn the crankshaft clockwise through two full rotations (720-degrees) while observing the dial indicator. Note the high and low readings and subtract the low from the high to obtain the lift. Record the measurements for each cylinder and note whether it was for exhaust or intake.

Compare the measurements of all the intake lobes; they should be within 0.005 inch of each other. Repeat this check for the exhaust lobes as well.

When the surface hardening wears off of a camshaft lobe, the metal below it scuffs away rapidly. Usually, worn lobes will be several tenths of an inch below the good ones.

If one or more lobes are worn, replace the camshaft and lifters as an assembly. Never use old lifters on a new camshaft.

A cracked block or cylinder head and/or a blown head gasket will cause a loss of power, overheating and a host of other problems. These problems are usually brought on by severe overheating or freezing due to insufficient antifreeze.

If you suspect internal engine leakage, check the oil on the dipstick for contamination of the oil by coolant. The oil level may increase and the oil will appear milky in color. Sometimes, oil will also get into the radiator; it will usually float on the top of the coolant. Occasionally steam and coolant will come out of the exhaust pipe, even when the engine is warmed up, because of a leaking head gasket. **Note:** *Don't confuse this with the condensation vapor normally present when an engine is warming up in cool weather.*

The cooling system and engine may be checked for leaks with a pressure tester. Follow the instructions provided by the tool manufacturer. Correct any external leaks in the hoses, water pump and radiator, etc. If no external leaks are found, look and listen for signs of leakage on the engine. **Note:** *A leaking heater core will cause a hidden pressure loss. Clamp off the hoses going to it at the firewall to eliminate this source of leakage.*

Another device that is useful for determining if there is a crack in the engine or a blown head gasket is a combustion leak block tester. Combustion leak block testers use a blue colored fluid to test for combus-

4.2a On 5.7L GM diesel engines, insert an oil pressure gauge into the hole where the oil pressure sending unit (arrow) is located; on 6.2L and 6.5L engines, the oil pressure sending unit is located at the rear of the engine

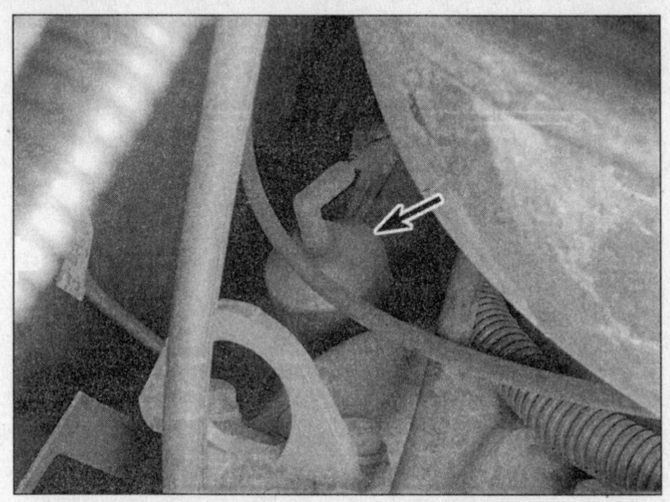

4.2b On Ford diesel engines, the oil pressure sending unit is located at the rear of the engine (arrow)

tion gases in the cooling system, which indicates a compression leak from a cylinder into the coolant. Be sure to follow the instructions included with the tester. A sample of gasses present in the top of the radiator is drawn into the tester. If any combustion gases are present in the sample taken, the test fluid will change color to yellow. Block testers and extra test fluid are readily available from most auto parts stores.

If the engine is overheating but testing indicates no cracks, blown gaskets or other internal problems, carefully inspect the cooling system for problems and correct as necessary. Frequently, partially clogged radiators, stuck thermostats and defective water pumps cause overheating.

When an internal coolant leak is found, the cylinder heads should be removed for a thorough inspection. If a gasket has blown, have an automotive machine shop check for warpage on both heads and resurface as necessary. If no warpage is found, have both cylinder heads checked for cracks. If tests indicate internal leakage, but the heads check out OK, have the block checked.

Correct the cause of the failure, such as a clogged radiator, before the vehicle is put back in service. Otherwise, the problem will likely reoccur.

Engine overhaul - general information

It's not always easy to determine when, or if, an engine should be completely overhauled, as a number of factors must be considered.

High mileage is not necessarily an indication that an overhaul is needed, while low mileage doesn't preclude the need for an overhaul. Frequency of servicing is probably the most important consideration. An engine that's had regular and frequent oil and filter changes, as well as other required maintenance, will most likely give many thousands of miles of reliable service. Conversely, a neglected engine may require an overhaul very early in its life.

Before beginning any work, perform the diagnostic checks in the previous Section. Excessive oil consumption is an indication that piston rings, valve seals and/or valve guides are in need of attention. Make sure that oil leaks aren't responsible before deciding that the rings and/or guides are bad. Perform a cylinder compression check to determine the extent of the work required (see the next Section in this Chapter).

Check the oil pressure with a gauge installed in place of the oil pressure sending unit **(see illustrations)** and compare it to this Chapter's Specifications. If it's extremely low, the bearings and/or oil pump are probably worn out.

Loss of power, rough running, knocking or metallic engine noises, excessive valve train noise and high fuel consumption rates may also point to the need for an overhaul, especially if they're all present at the same time. If a complete service doesn't remedy the situation, major mechanical work is the only solution.

An engine overhaul involves restoring the internal parts to the specifications of a new engine. During an overhaul, the piston rings are replaced and the cylinder walls are reconditioned (rebored and/or honed). If a rebore is done by an automotive machine shop, new oversize pistons will also be installed. The main bearings, connecting rod bearings and camshaft bearings are generally replaced with new ones and, if necessary, the crankshaft may be reground to restore the journals. Generally, the valves are serviced as well, since they're usually in less-than-perfect condition at this point. While the engine is being overhauled, other components, such as the starter and alternator can be rebuilt as well. The end result should be a like-new engine that will give many trouble-free miles. **Note:** *Critical cooling system components such as the hoses, drivebelts, thermostat and water pump MUST be replaced with new parts when an engine is overhauled. The radiator should be checked carefully to ensure that it isn't clogged or leaking. Also, we don't recommend overhauling the oil pump - always install a new one when an engine is rebuilt.*

Before beginning the engine overhaul, read through the entire procedure to familiarize yourself with the scope and requirements of the job. Overhauling an engine isn't difficult if you follow all of the instructions carefully, have the necessary tools and equipment and pay close attention to all specifications; however, it is time consuming. Plan on the vehicle being tied up for a minimum of two weeks, especially if parts must be taken to an automotive machine shop for repair or reconditioning. Check on availability of parts and make sure that any necessary special tools and equipment are obtained in advance. Most work can be done with typical hand tools, although a number of precision measuring tools are required for inspecting parts to determine if they must be replaced. Often an automotive machine shop will handle the inspection of parts and offer advice concerning reconditioning and replacement. **Note:** *Always wait until the engine has been completely disassembled and all components, especially the engine block, have been inspected before deciding what service and repair operations must be performed by an automotive machine shop. Since the block's condition will be the major factor to consider when determining whether to overhaul the original engine or buy a rebuilt one, never purchase parts or have machine work done on other components until the block has been thoroughly inspected. As a general rule, time is the primary cost of an overhaul, so it doesn't pay to install worn or substandard parts.*

As a final note, to ensure maximum life and minimum trouble from a rebuilt engine, everything must be assembled with care in a spotlessly clean environment.

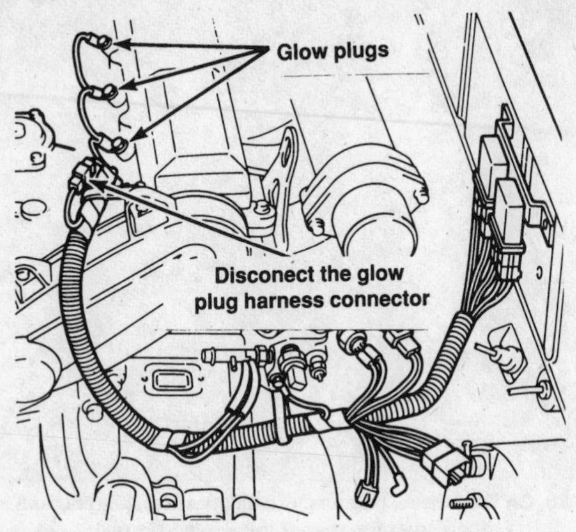

4.3 Disconnect the glow plug harness on Ford diesel engines

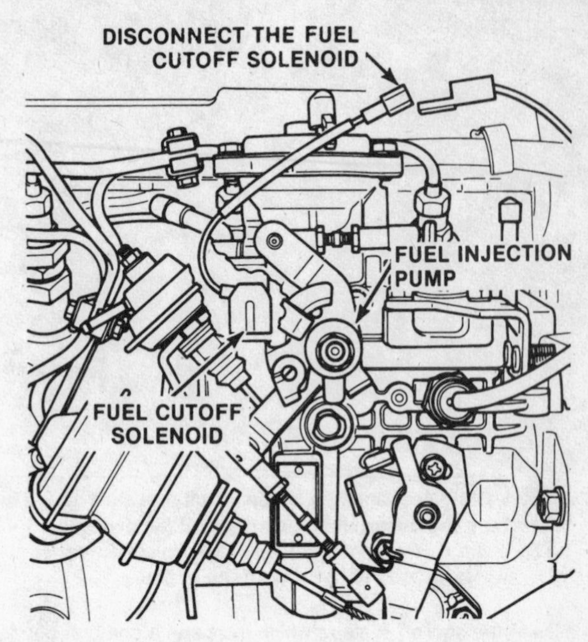

4.4 Disconnect the fuel cut-off solenoid on Ford diesel engines

Compression check

1 A compression check will tell you what mechanical condition the upper end (pistons, rings, valves, head gaskets) of your engine is in. Specifically, it can tell you if the compression is down due to leakage caused by worn piston rings, defective valves and seats or a blown head gasket. **Note:** *The methods for testing the compression will vary among the different sizes and models of diesel engines. On GM diesel engines, make sure the engine is fully warmed up. On Ford diesel engines, some will have to be warmed up while other engines will be tested cool. Consult the owner's manual or factory service manual to determine the exact temperature conditions for testing.* **Note:** *In order to receive an accurate compression reading, the battery must have an adequate charge and the engine must crank at approximately 200 rpm.*
2 Begin by cleaning the area around the glow plugs before you remove them (compressed air should be used, if available, otherwise a small brush or even a bicycle tire pump will work). The idea is to prevent dirt from getting into the cylinders as the compression check is being done.
3 Remove all of the glow plugs from the engine.
4 On Ford diesel engines, disconnect the glow plug harness from the engine **(see illustration).**
5 On GM diesel engines, remove the air cleaner and install an air crossover cover onto the engine.
6 Disconnect the fuel cutoff solenoid from the injection pump (GM models) or the fuel cutoff solenoid wire harness on Ford pumps **(see illustration).**
7 Install the compression gauge in the number one glow plug hole **(see illustrations).**
8 Crank the engine over at least six compression strokes (or "puffs") and watch the gauge. The compression should build up quickly in a healthy engine. Low compression on the first stroke, followed by gradually increasing pressure on successive strokes, indicates worn piston rings. A low compression reading on the first stroke, which doesn't build up during successive strokes, indicates leaking valves or a blown head gasket (a cracked head could also be the cause). Deposits on the undersides of the valve heads can also cause low compression. Record the highest gauge reading obtained.
9 Repeat the procedure for the remaining cylinders and compare the results to this Chapter's Specifications. **Note:** *Never add engine oil to a cylinder during a compression test on a GM or Ford diesel engine as extensive engine damage can result.*
10 If two adjacent cylinders have equally low compression, there's a strong possibility that the head gasket between them is blown. The appearance of coolant in the combustion chambers or the crankcase would verify this condition.

4.5 On Ford diesel engines, install a special adapter into the glow plug port (arrow) . . .

4.6 . . . and couple the compression gauge onto the adapter

4.7 On GM diesel engines, install a special adapter into the glow plug port (arrow) . . .

4.8 . . . and couple the compression gauge onto the adapter

11 If one cylinder is 20 percent lower than the others, and the engine has a slightly rough idle, a worn exhaust lobe on the camshaft could be the cause.
12 If compression is way down or varies greatly between cylinders, it would be a good idea to have a crankcase pressure test performed by an automotive repair shop. This test will pinpoint exactly where the leakage is occurring and how severe it is.

Engine removal - methods and precautions

If you've decided that an engine must be removed for overhaul or major repair work, several preliminary steps should be taken.

Locating a suitable place to work is extremely important. Adequate work space, along with storage space for the vehicle, will be needed. If a shop or garage isn't available, at the very least a flat, level, clean work surface made of concrete or asphalt is required.

Cleaning the engine compartment and engine before beginning the removal procedure will help keep tools clean and organized.

An engine hoist or A-frame will also be necessary. Make sure the equipment is rated in excess of the combined weight of the engine and accessories. Safety is of primary importance, considering the potential hazards involved in lifting the engine out of the vehicle.

If the engine is being removed by a novice, a helper should be available. Advice and aid from someone more experienced would also be helpful. There are many instances when one person cannot simultaneously perform all of the operations required when lifting the engine out of the vehicle.

Plan the operation ahead of time. Arrange for or obtain all of the tools and equipment you'll need prior to beginning the job. Some of the equipment necessary to perform engine removal and installation safely and with relative ease are (in addition to an engine hoist) a heavy duty floor jack, complete sets of wrenches and sockets as described in the front of this manual, wooden blocks and plenty of rags and cleaning solvent for mopping up spilled oil, coolant and gasoline. If the hoist must be rented, make sure that you arrange for it in advance and perform all of the operations possible without it beforehand. This will save you money and time.

Plan for the vehicle to be out of use for quite a while. A machine shop will be required to perform some of the work which the do-it-yourselfer can't accomplish without special equipment. These shops often have a busy schedule, so it would be a good idea to consult them before removing the engine in order to accurately estimate the amount of time required to rebuild or repair components that may need work.

Always be extremely careful when removing and installing the engine. Serious injury can result from careless actions. Plan ahead, take your time and a job of this nature, although major, can be accomplished successfully.

Engine - removal and installation

Warning: *The air conditioning system is under high pressure! Have a dealer service department or service station discharge the system before disconnecting any air conditioning system hoses or fittings.*

Removal

1 Disconnect the negative cable from the battery.
2 Cover the fenders and cowl and remove the hood. Special pads are available to protect the fenders, but an old bedspread or blanket will also work.
3 Remove the air cleaner assembly.
4 Drain the cooling system (see Chapter 2 for GM or Chapter 3 for Ford).
5 Label the vacuum lines, emissions system hoses, wiring connectors, ground straps and fuel lines, to ensure correct reinstallation, then detach them. Pieces of masking tape with numbers or letters written on them work well **(see illustration)**. If there's any possibility of confusion, make a sketch of the engine compartment and clearly label the lines, hoses and wires.

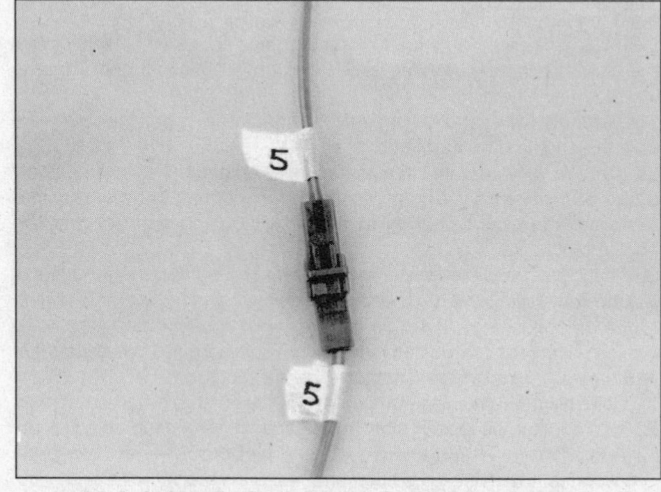

4.9 Label each wire before unplugging the connector

6 Label and detach all coolant hoses from the engine.
7 Remove the cooling fan, shroud and radiator (see Chapter 2 for GM or Chapter 3 for Ford).
8 Remove the drivebelts.
9 Disconnect the fuel lines running from the engine to the chassis. Plug or cap all open fittings/lines. **Warning:** *Gasoline is extremely flammable, so extra precautions must be taken when working on any part of the fuel system. DO NOT smoke or allow open flames or bare light bulbs near the vehicle. Also, don't work in a garage if a natural gas appliance with a pilot light is present.*
10 Disconnect the throttle linkage.
11 On power steering equipped vehicles, unbolt the power steering pump. Leave the lines/hoses attached and make sure the pump is kept in an upright position in the engine compartment (use wire or rope to restrain it out of the way).
12 On air conditioned models, unbolt the compressor and set it aside. Do not disconnect the hoses.
13 Drain the engine oil (Chapter 2 for GM or Chapter 3 for Ford) and remove the filter.
14 Remove the starter motor.
15 Remove the alternator.
16 Unbolt the exhaust system from the engine.
17 If you're working on a vehicle with an automatic transmission, remove the torque converter-to-driveplate fasteners.
18 Support the transmission with a jack. Position a block of wood between them to prevent damage to the transmission. Special transmission jacks with safety chains are available - use one if possible.
19 Attach an engine sling or a length of chain to the lifting brackets on the engine.
20 Roll the hoist into position and connect the sling to it. Take up the slack in the sling or chain, but don't lift the engine. **Warning:** *DO NOT place any part of your body under the engine when it's supported only by a hoist or other lifting device.*
21 Remove the transmission-to-engine block bolts.
22 Remove the engine mount-to-frame bolts.
23 Recheck to be sure nothing is still connecting the engine to the transmission or vehicle. Disconnect anything still remaining.
24 Raise the engine slightly. Carefully work it forward to separate it from the transmission. If you're working on a vehicle with an automatic transmission, be sure the torque converter stays in the transmission (clamp a pair of vise-grips to the housing to keep the converter from sliding out). If you're working on a vehicle with a manual transmission, the input shaft must be completely disengaged from the clutch. Slowly raise the engine out of the engine compartment. Check carefully to make sure nothing is hanging up.
25 Remove the flywheel/driveplate and mount the engine on an engine stand.

Installation

26 Check the engine and transmission mounts. If they're worn or damaged, replace them.
27 If you're working on a manual transmission equipped vehicle, install the clutch and pressure plate (see Chapter 7). Now is a good time to install a new clutch.
28 Carefully lower the engine into the engine compartment - make sure the engine mounts line up.
29 If you're working on a manual transmission equipped vehicle, apply a dab of high-temperature grease to the input shaft and guide it into the crankshaft pilot bearing until the bellhousing is flush with the engine block.
30 Install the transmission-to-engine bolts and tighten them securely. **Caution:** *DO NOT use the bolts to force the transmission and engine together! If you're working on an automatic transmission-equipped vehicle, install the torque converter-to-driveplate bolts, tightening them to the torque listed in this Chapters Specifications.*
31 Reinstall the remaining components in the reverse order of removal.
32 Add coolant, oil, power steering and transmission fluid as needed.
33 Run the engine and check for leaks and proper operation of all accessories, then install the hood and test drive the vehicle.
34 Have the air conditioning system recharged and leak tested.

Engine rebuilding alternatives

The do-it-yourselfer is faced with a number of options when performing an engine overhaul. The decision to replace the engine block, piston/connecting rod assemblies and crankshaft depends on a number of factors, with the number one consideration being the condition of the block. Other considerations are cost, access to machine shop facilities, parts availability, time required to complete the project and the extent of prior mechanical experience on the part of the do-it-yourselfer.

Some of the rebuilding alternatives include:

Individual parts - If the inspection procedures reveal that the engine block and most engine components are in reusable condition, purchasing individual parts may be the most economical alternative. The block, crankshaft and piston/connecting rod assemblies should all be inspected carefully. Even if the block shows little wear, the cylinder bores should be surface honed.

Short block - A short block consists of an engine block with a crankshaft and piston/connecting rod assemblies already installed. All new bearings are incorporated and all clearances will be correct. The existing camshaft, valve train components, cylinder head(s) and external parts can be bolted to the short block with little or no machine shop work necessary.

Long block - A long block consists of a short block plus an oil pump, oil pan, cylinder head(s), rocker arm cover(s), camshaft and valve train components, timing sprockets and chain or gears and timing cover. All components are installed with new bearings, seals and gaskets incorporated throughout. The installation of manifolds and external parts is all that's necessary.

Give careful thought to which alternative is best for you and discuss the situation with local automotive machine shops, auto parts dealers and experienced rebuilders before ordering or purchasing replacement parts.

Engine overhaul - disassembly sequence

1 It's much easier to disassemble and work on the engine if it's mounted on a portable engine stand. A stand can often be rented quite cheaply from an equipment rental yard. Before the engine is mounted on a stand, the flywheel/driveplate should be removed from the engine.
2 If a stand isn't available, it's possible to disassemble the engine with it blocked up on the floor. Be extra careful not to tip or drop the engine when working without a stand.
3 If you're going to obtain a rebuilt engine, all external components must come off first, to be transferred to the replacement engine, just as they will if you're doing a complete engine overhaul yourself. These include:

Alternator and brackets
Emissions control components
Injection pump or fuel distributor
Thermostat and housing cover
Water pump
Fuel lines and clamps
Intake/exhaust manifolds
Oil filter
Engine mounts
Clutch and flywheel/driveplate
Engine rear plate

Note: *When removing the external components from the engine, pay close attention to details that may be helpful or important during installation. Note the installed position of gaskets, seals, spacers, pins, brackets, washers, bolts and other small items.*
4 If you're obtaining a short block, which consists of the engine block, crankshaft, pistons and connecting rods all assembled, then the cylinder head(s), oil pan and oil pump will have to be removed as well. See *Engine rebuilding alternatives* for additional information regarding the different possibilities to be considered.

5 If you're planning a complete overhaul, the engine must be disassembled and the internal components removed in the following order:

Rocker arm cover(s)
Intake and exhaust manifolds
Rocker arms and pushrods
Valve lifters
Cylinder head(s)
Timing cover
Timing gears/chain and sprockets
Camshaft
Oil pan
Oil pump
Piston/connecting rod assemblies
Crankshaft and main bearings

6 Before beginning the disassembly and overhaul procedures, make sure the following items are available. Also, refer to *Engine overhaul - reassembly sequence* for a list of tools and materials needed for engine reassembly.

Common hand tools
Small cardboard boxes or plastic bags for storing parts
Gasket scraper
Ridge reamer
Vibration damper puller
Micrometers
Telescoping gauges
Dial indicator set
Valve spring compressor
Cylinder surfacing hone
Piston ring groove cleaning tool
Electric drill motor
Tap and die set
Wire brushes
Oil gallery brushes
Cleaning solvent

4.10 A small plastic bag, with an appropriate label, can be used to store the valve train components so they can be kept together and reinstalled in the original location

Cylinder head - disassembly

Warning: *Use extreme care when removing the cylinder head bolts from the engine. Failure to remove the cylinder head bolts in the correct order can cause the head to warp, or in extreme cases the head will crack. Refer to Chapter 2 for GM cylinder head removal and Chapter 3 for Ford cylinder head removal.*

Note: *New and rebuilt cylinder heads are commonly available for most engines at dealerships and auto parts stores. Due to the fact that some specialized tools are necessary for the disassembly and inspection procedures, and replacement parts may not be readily available, it may be more practical and economical for the home mechanic to purchase replacement heads rather than taking the time to disassemble, inspect and recondition the originals.*

1 Cylinder head disassembly involves removal of the intake and exhaust valves and related components. If they're still in place, remove the rocker arm nuts, pivot balls and rocker arms from the cylinder head studs. Label the parts or store them separately so they can be reinstalled in their original locations.

2 Before the valves are removed, arrange to label and store them, along with their related components, so they can be kept separate and reinstalled in the same valve guides they are removed from **(see illustration)**.

3 Compress the springs on the first valve with a spring compressor and remove the keepers **(see illustration)**. Carefully release the valve spring compressor and remove the retainer, the spring and the spring seat (if used).

4 Pull the valve out of the head, then remove the oil seal from the guide. If the valve binds in the guide (won't pull through), push it back into the head and deburr the area around the keeper groove with a fine file or whetstone **(see illustration)**.

5 Repeat the procedure for the remaining valves. Remember to keep all the parts for each valve together so they can be reinstalled in the same locations.

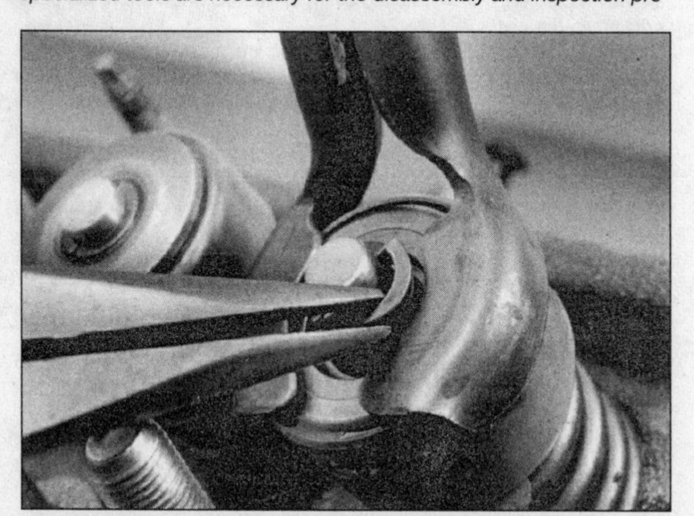

4.11 Use a valve spring compressor to compress the spring, then remove the keepers from the valve stem

4.12 If the valve won't pull through the guide, deburr the edge of the stem end and the area around the top of the keeper groove with a file or whetstone

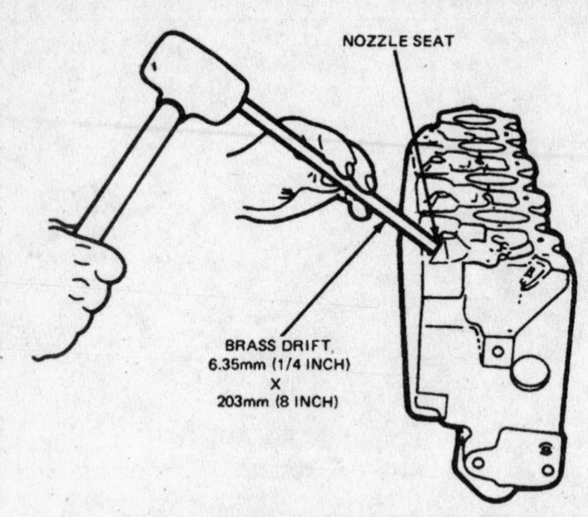

4.13 On Ford diesel engine cylinder heads, remove the pre-combustion chamber by tapping on it with a brass drift punch and hammer

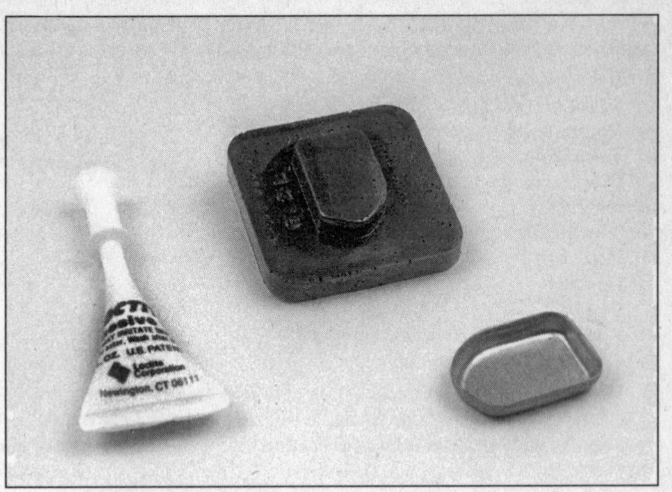

4.14 The special plug insert kit is available from a GM parts department; it consists of a brass plug, sealer and a plastic plug driver

6 Once the valves and related components have been removed and stored in an organized manner, the head should be thoroughly cleaned and inspected. If a complete engine overhaul is being done, finish the engine disassembly procedures before beginning the cylinder head cleaning and inspection process.

Cylinder head - cleaning and inspection

The GM and Ford diesel engines covered by this manual incorporate pre-combustion chambers in the cylinder heads to induce the air and fuel to mix by swirling. This divides the combustion chamber into the pre-combustion chamber (larger area) and the space between the piston and cylinder head (smaller area). Close piston clearances produces high turbulence in the ante chamber (larger area), promoting rapid combustion. The charge is forced out onto the piston (smaller area), agitating the entire mixture and resulting in more complete combustion. This design has an advantage over the more typical constant load system because it allows a broader operating range and also provides less noise and effective emission control. On GM diesel engines, pre-combustion chambers must not be recessed into the cylinder head or protrude out of the cylinder head by more than 0.004 inch. On Ford diesel engines, remove the pre-combustion chamber **(see illustration)** with a hammer and a brass punch (1/4 x 8 inches) to avoid any erroneous readings of the cylinder head surface. Ford pre-chambers, must not protrude out of the cylinder head or be recessed into the cylinder head more than 0.0025 inch.

1 Thorough cleaning of the cylinder head(s) and related valve train components, followed by a detailed inspection, will enable you to decide how much valve service work must be done during the engine overhaul. **Note:** *If the engine was severely overheated, the cylinder head is probably warped (see Step 12).*

Cleaning

2 Scrape all traces of old gasket material and sealing compound off the head gasket, intake manifold and exhaust manifold sealing surfaces. Be very careful not to gouge the cylinder head. Special gasket removal solvents that soften gaskets and make removal much easier are available at auto parts stores.

3 Remove all built-up scale from the coolant passages.

4 Run a stiff wire brush through the various holes to remove deposits that may have formed in them.

5 Run an appropriate size tap into each of the threaded holes to remove corrosion and thread sealant that may be present. If compressed air is available, use it to clear the holes of debris produced by this operation. **Warning:** *Wear eye protection when using compressed air!*

6 Clean the rocker arm pivot stud threads with a wire brush.

7 Clean the cylinder head with solvent and dry it thoroughly. Compressed air will speed the drying process and ensure that all holes and recessed areas are clean. **Note:** *Decarbonizing chemicals are available and may prove very useful when cleaning cylinder heads and valve train components. They are very caustic and should be used with caution. Be sure to follow the instructions on the container.*

8 Clean the rocker arms, pivot balls, nuts and pushrods with solvent and dry them thoroughly (don't mix them up during the cleaning process). Compressed air will speed the drying process and can be used to clean out the oil passages.

9 Clean all the valve springs, spring seats, keepers and retainers (or rotators) with solvent and dry them thoroughly. Do the components from one valve at a time to avoid mixing up the parts.

10 Scrape off any heavy deposits that may have formed on the valves, then use a motorized wire brush to remove deposits from the valve heads and stems. Again, make sure the valves don't get mixed up.

Inspection

Note 1: *Be sure to perform all of the following inspection procedures before concluding that machine shop work is required. Make a list of the items that need attention.*

Note 2: *On diesel engines, a warped cylinder head is often the result of an overheated engine. A cracked cylinder head is often the result of air pockets in the water jacket.*

Note 3: *On the 6.2L GM diesel engine cylinder heads (1982 and 1983 only), a special core plug has been developed to prevent leakage from the head gasket. One of the plugs is located in the rear lower corner on the left head and the other plug is located on the front lower corner on the right head. A special brass plug kit (Part number 14079353) has been developed to seal this problem area* **(see illustration)**. *It is available at a GM dealership parts department. 1985 and later years of the 6.2L have this hole eliminated. Follow these steps:*

a) *Make sure the cylinder head is clean and the area around the hole is free of rust and head gasket material.*
b) *Use a wire brush and thoroughly clean the sides of the hole to prepare a suitable surface for the glue to bond* **(see illustration)**.
c) *Apply a bead of sealing compound (GM #1052624 - contained in the kit) around the perimeter of the core plug and position it in the cylinder head* **(see illustration)**.
d) *Use the plastic plug driver from the kit and drive the core plug into the cylinder head, making sure it is perfectly flush with the surface of the cylinder head* **(see illustration)**.
e) *Clean any excess sealer from the area and check to make sure the plug is level and does not protrude above the level of the cylinder head surface* **(see illustration)**.

4.15 Thoroughly clean the area around the hole with a wire brush

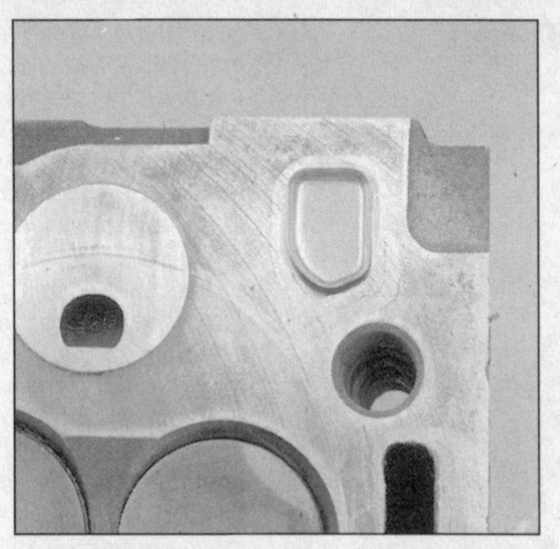

4.16 Coat the perimeter of the plug with sealer and insert it in the cylinder head

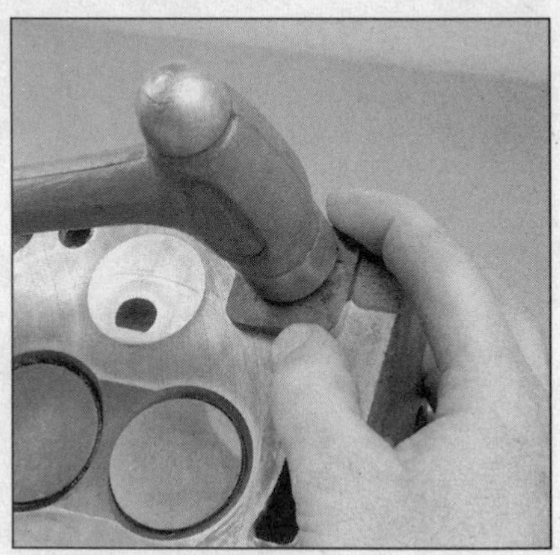

4.17 Use a plastic plug driver and force the plug into the cylinder head; drive the plug level with the surface of the cylinder head

11 Inspect the head very carefully for cracks, evidence of coolant leakage and other damage. If cracks are found, check with an automotive machine shop concerning repair. If repair isn't possible, a new cylinder head should be obtained. **Note:** *Check carefully for hairline cracks in the prechamber area on all engines. The cracks on the face of the pre-chamber start at the edge of the fire slot* **(see illustration)** *and proceed toward the circular impression of the head gasket bead. These cracks are a form of stress relief and are completely harmless up to a length of 3/16 of an inch. Cracks longer than this approach the head gasket sealing bead. Replace any pre-chambers that display the excess cracks with new parts.*

4.18 Inspect the brass plug to make sure it is not protruding above the cylinder head

Year and engine type	Standard	0.010 oversize
5.7L GM diesel engine		
1982 and 1983	22515655	22517846
1980 and 1981	22505979	Not available
1978 and 1979	558677	Not available
6.2L GM diesel engine		
1982 LH-6 (Vin Code C)	14067526	14069540
1982 LL-4 (Vin Code J)	214067527	14069541
1983 and 1984 (all 6.2L)	14067526	14069540
1985 and 1986 LH6 (Vin Code C)	14067526	14069540
1985 and 1986 LH6 (VIN code J)	23500131	23500271

12 Using a precision machinist's straightedge and feeler gauge, check the cylinder head gasket sealing surface for warpage **(see illustration)**. If the warpage exceeds the allowable limit listed in this Chapter's Specifications, replace the cylinder head. The manufacturer (neither Ford nor GM) recommend resurfacing cylinder heads on their diesel engines because there is danger of the valves colliding with the pistons if too much material is removed. Consult with a very experienced machine shop if considering cylinder head machining. **Note:** *On GM diesel engines, the pre-chambers must not be recessed into the head*

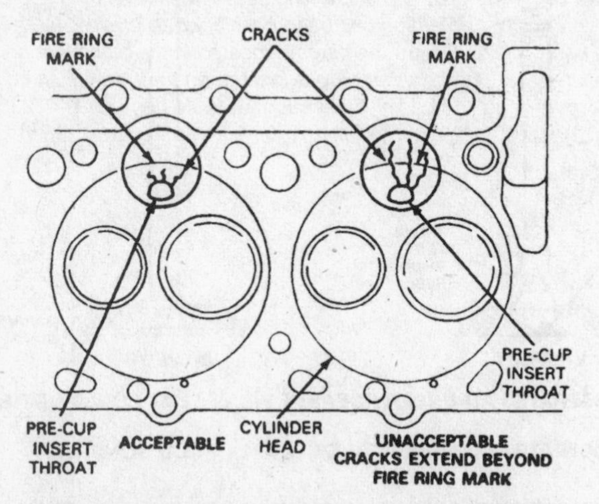

4.19 Check the pre-chamber area very carefully for any cracks that exceed the acceptable limit

4.20 Check the cylinder head gasket surface for warpage, by trying to slip a feeler gauge under the straightedge (see this Chapter's Specifications for the maximum warpage allowed and use a feeler gauge of that thickness)

4.21 A dial indicator can be used to determine the valve stem-to-guide clearance (move the valve stem as indicated by the arrows)

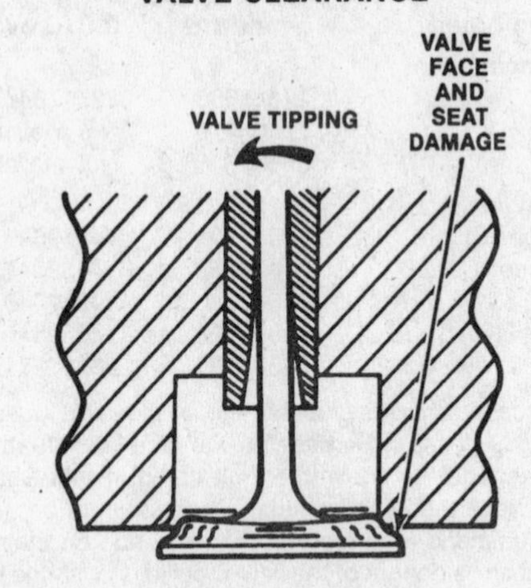

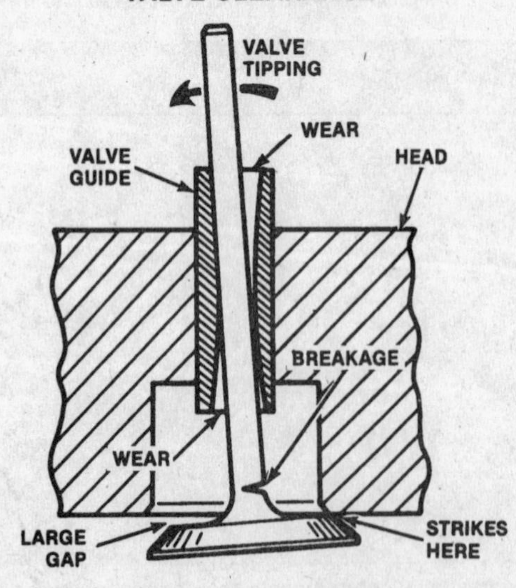

4.22 The effects of excessive valve guide clearance on the wear of the valve(s)

or protrude out of the head more than 0.004 inch or a head gasket leak may result. This measurement must be taken at two or more points on the pre-chamber where the head gasket sealing ring mates with the pre-chamber. Use a straightedge or a dial caliper to measure the difference. If it exceeds 0.004 inch either way (protruded or recessed), have the pre-chamber serviced by a professional machine shop. On Ford diesel engines, remove the pre-chambers using a 1/4 inch x 8 inch brass punch or drift and a hammer **(see illustration 4.13)**. *To avoid erroneous readings, remove the pre-chamber inserts prior to checking the cylinder head surface for warpage.*

13 Examine the valve seats in each of the combustion chambers. If they're pitted, cracked or burned, the head will require valve service that's beyond the scope of the home mechanic. Clean and inspect pre-chambers and ports for cracks and deformations. Replace any damaged pre-chambers with new units.

14 Check the valve stem-to-guide clearance by measuring the lateral movement of the valve stem with a dial indicator attached securely to the head **(see illustration)**. The valve must be in the guide and approximately 1/16-inch off the seat. The total valve stem movement indicated by the gauge needle must be divided by two to obtain the actual clearance. After this is done, if there's still some doubt regarding the condition of the valve guides they should be checked by an automotive machine shop (the cost should be minimal). **Note:** *Excessive guide clearance prevents adequate cooling of the valve through the guide and allows the valve to tilt or tip. This can cause valve breakage at high engine speed. These conditions prevent the valve from seating properly and combustion leakage will occur* **(see illustration)**. *Diesel engines rely on extremely high pressures in the combustion chamber to heat the air that ignites the fuel. Any leakage of the pressure will severely affect the ignition of the fuel. If the leakage past the valve face is excessive, the fuel will not ignite. This is much more critical than a gasoline engine because the air/fuel ratio would normally ignite even with leaky valves because it has a spark to set the fuel off.*

Valves

15 Carefully inspect each valve face for uneven wear, deformation, cracks, pits and burned areas **(see illustration)**. Check the valve stem for scuffing and galling and the neck for cracks. Rotate the valve and check for any obvious indication that it's bent. Look for pits and excessive wear on the end of the stem. The presence of any of these conditions indicates the need for valve service by an automotive machine shop. **Note:** *When a diesel is rebuilt because of high mileage, look carefully for problems such as peened valve faces and worn valve stems.*

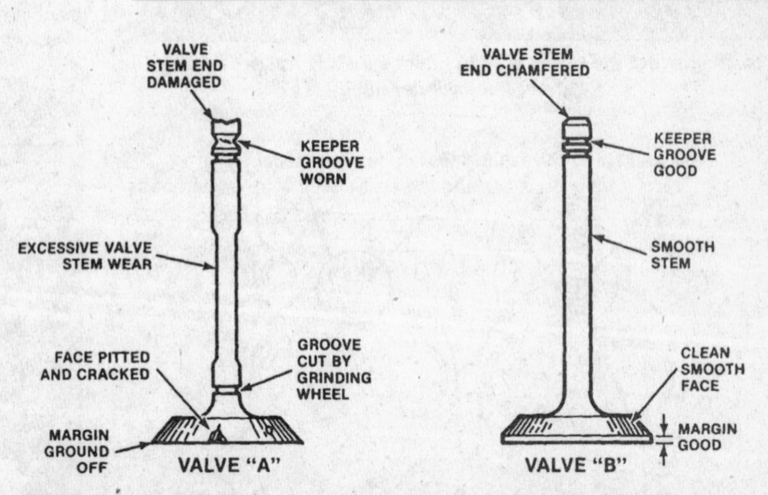

4.23 The condition of a severely worn valve versus a good valve (wear exaggerated for clarity)

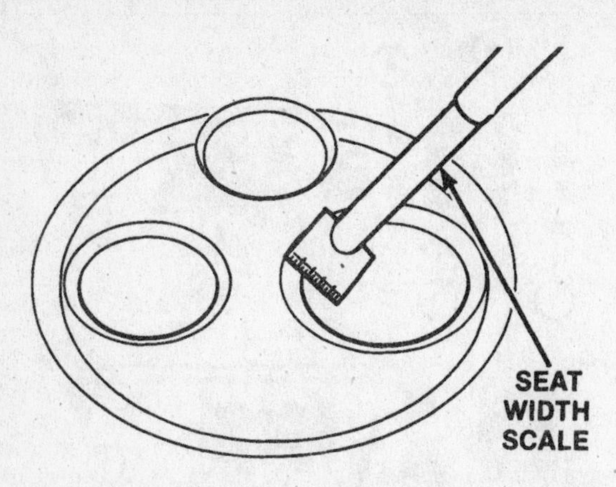

4.24 Check the valve seat width using a special tool or a finely graduated ruler

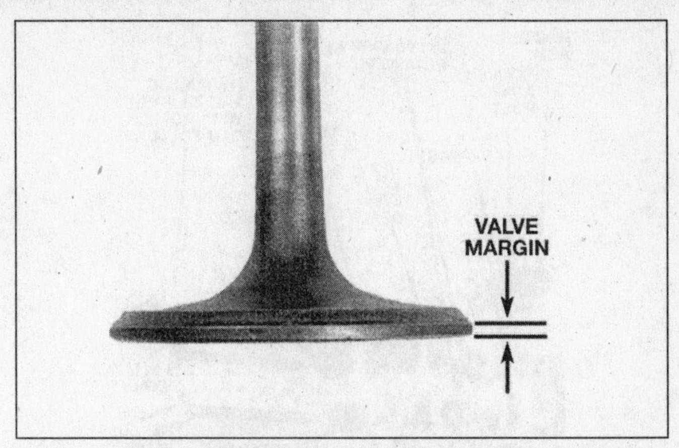

4.25 The valve margin width must not be less than specified - if it is, do not reuse the valve

4.26 Measure the free length of each valve spring with a dial or vernier caliper

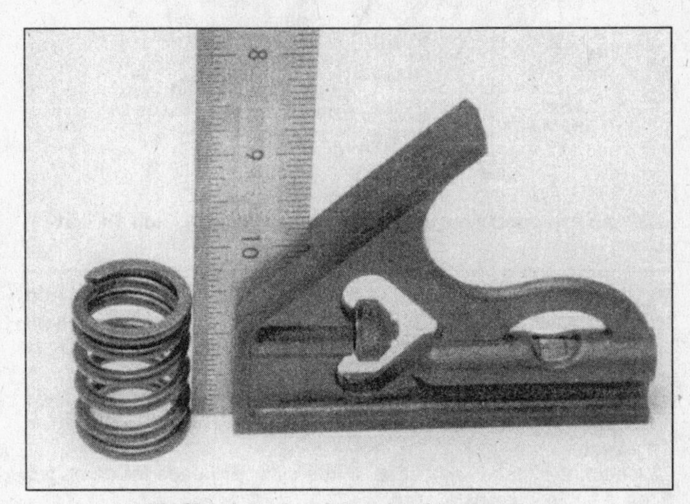

4.27 Check each valve spring for squareness

16 Measure the seat width and margin width on each valve **(see illustrations)**. Any seat or valve out of specification will have to be replaced with a new one.

17 Check each valve spring for wear (on the ends) and pits. Measure the free length and compare it to the Specifications **(see illustration)**. Any springs that are shorter than specified have sagged and should not be reused. The tension of all springs should be checked with a special fixture before deciding that they're suitable for use in a rebuilt engine (take the springs to an automotive machine shop for this check).

18 Stand each spring on a flat surface and check it for squareness **(see illustration)**. If any of the springs are distorted or sagged, replace all of them with new parts.

19 Check the spring retainers (or rotators) and keepers for obvious wear and cracks. Any questionable parts should be replaced with new ones, as extensive damage will occur if they fail during engine operation. Make sure the rotators operate smoothly with no binding or excessive play **(see illustrations)**.

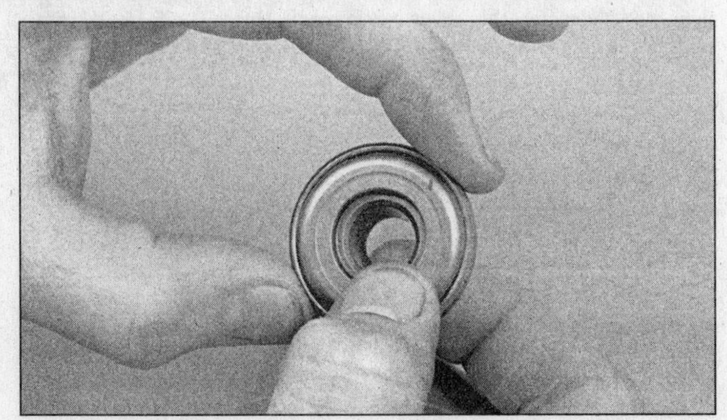

4.28 The valve rotators can be checked by turning the inner and outer sections in opposite directions - feel for smooth movement and excessive play

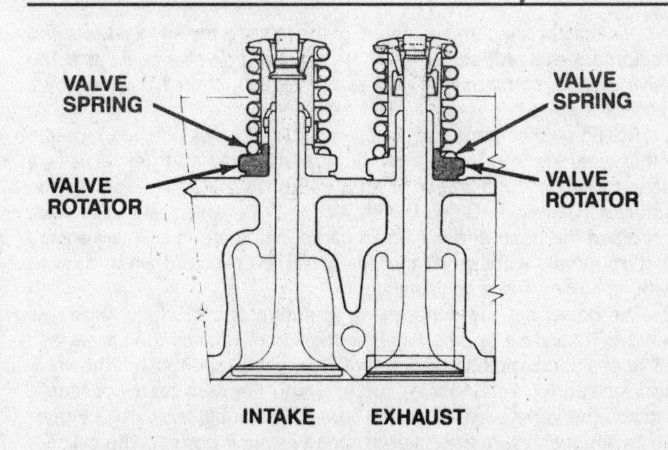

4.29 On Ford 7.3L (non-DI Turbo) diesel engines, the valve rotators are located on the bottom of the valve spring assembly

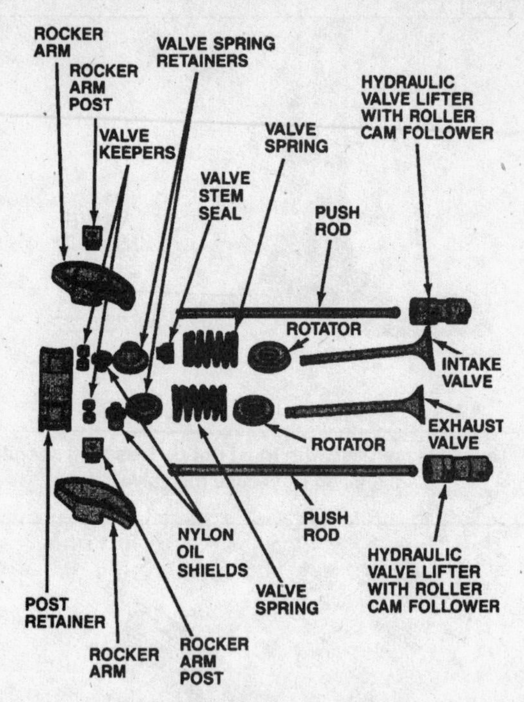

4.30 An exploded view of the valve train on a 7.3L (non-DI Turbo) Ford engine

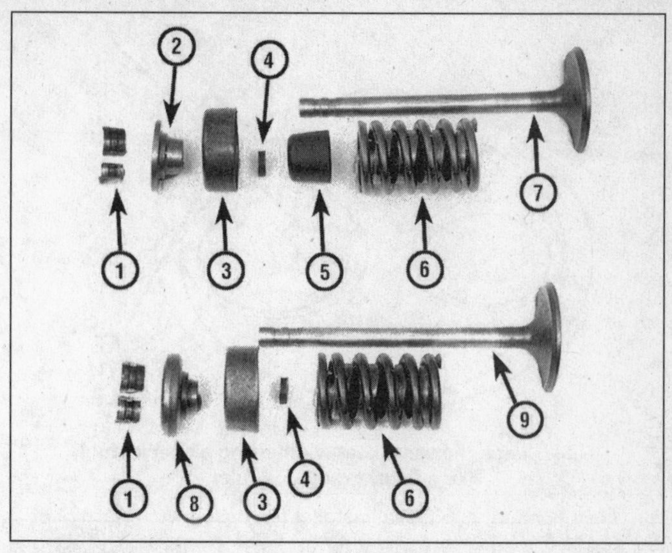

4.31 An exploded view of a typical valve pair on a GM engine

1 *Keepers*
2 *Retainer*
3 *Oil shield*
4 *O-ring oil seal*
5 *Umbrella seal*
6 *Spring and damper*
7 *Intake valve*
8 *Retainer/rotator*
9 *Exhaust valve*

20 Check the rocker arm faces (the areas that contact the pushrod ends and valve stems) for pits, wear, galling, score marks and rough spots. Check the rocker arm pivot contact areas and pivot balls as well. Look for cracks in each rocker arm and nut.
21 Inspect the pushrod ends for scuffing and excessive wear. Roll each pushrod on a flat surface, like a piece of plate glass, to determine if it's bent.
22 Check the rocker arm studs in the cylinder heads for damaged threads and secure installation.
23 Any damaged or excessively worn parts must be replaced with new ones.
24 If the inspection process indicates that the valve components are in generally poor condition and worn beyond the limits specified, which is usually the case in an engine that's being overhauled, reassemble the valves in the cylinder head and refer to the next Section for valve servicing recommendations.

Valves - servicing

1 Because of the complex nature of the job and the special tools and equipment needed, servicing of the valves, the valve seats and the valve guides, commonly known as a valve job, should be done by a professional.
2 The home mechanic can remove and disassemble the head, do the initial cleaning and inspection, then reassemble and deliver it to a dealer service department or an automotive machine shop for the actual service work. Doing the inspection will enable you to see what condition the head and valvetrain components are in and will ensure that you know what work and new parts are required when dealing with an automotive machine shop.
3 The dealer service department, or automotive machine shop, will remove the valves and springs, recondition or replace the valves and valve seats, recondition the valve guides, check and replace the valve springs, spring retainers or rotators and keepers (as necessary), replace the valve seals with new ones, reassemble the valve components and make sure the installed spring height is correct. The cylinder head gasket surface will also be resurfaced if it's warped.
4 After the valve job has been performed by a professional, the head will be in like-new condition. When the head is returned, be sure to clean it again before installation on the engine to remove any metal particles and abrasive grit that may still be present from the valve service or head resurfacing operations. Use compressed air, if available, to blow out all the oil holes and passages.

Cylinder head - reassembly

1 Regardless of whether or not the head was sent to an automotive repair shop for valve servicing, make sure it's clean before beginning reassembly.
2 If the head was sent out for valve servicing, the valves and related components will already be in place. Begin the reassembly procedure with Step 8. **Note:** *If the pre-chambers were removed for servicing, install them into their respective places in the cylinder head. Apply a light coat of steering linkage lube or equivalent to the mounting edge of the pre-chamber. Tap the unit with a plastic tipped hammer. Be sure the notches are correctly aligned.*
3 Install new seals on each of the intake valve guides. Using a hammer and a deep socket or seal installation tool, gently tap each seal into place until it's completely seated on the guide **(see illustration)**. Don't twist or cock the seals during installation or they won't seal properly on the valve stems. The umbrella-type seals (if used) are installed over the valves after the valves are in place. **Note:** *The 5.7L GM diesel engine uses either a standard or an oversize valve stem seal:*

Intake

Standard to 0.005 in (oversize)	Gray
0.10 to 0.013 in. (oversize)	Orange

Exhaust

Standard to 0.005 in. (oversize)	Ivory
0.10 to 0.013 in. (oversize)	Blue

4 Beginning at one end of the head, lubricate and install the first valve. Apply moly-base grease or clean engine oil to the valve stem.
5 Drop the spring seat or shim(s) over the valve guide and set the valve spring and retainer (or rotator) in place. Make sure that the retainer/rotator is facing in the correct direction.
6 Compress the springs with a valve spring compressor and carefully install the keepers in the upper groove, then slowly release the compressor and make sure the keepers seat properly. Apply a small dab of grease to each keeper to hold it in place, if necessary.

4.32 Use a hammer and a seal installer (or, as shown here, a deep socket) to drive the new seal onto the valve guide; tap the seal into place until it's completely seated on the guide

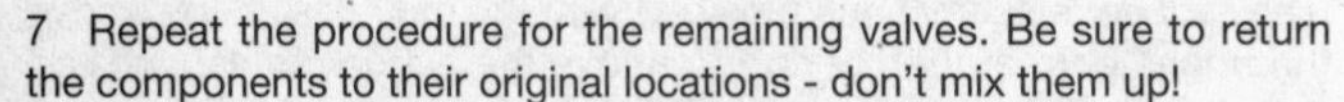

7 Repeat the procedure for the remaining valves. Be sure to return the components to their original locations - don't mix them up!

8 Check the installed valve spring height with a ruler graduated in 1/32-inch increments or a dial caliper. If the head was sent out for service work, the installed height should be correct (but don't automatically assume that it is). The measurement is taken from the top of each spring seat or shim(s) to the bottom of the retainer **(see illustration)**. If the height is greater than the figure listed in this Chapter's Specifications, shims can be added under the springs to correct it. **Caution:** *Don't, under any circumstances, shim the springs to the point where the installed height is less than specified.*

9 On GM 5.7L diesel engines, measure the valve stem height. (This procedure requires a special tool, BT-6428; if you don't have this tool, have this and the next step done at an automotive machine shop with the right tool.) Install the special tool and make sure the surface of the cylinder head is clean and free of any gasket material or burrs. There should be a 0.015-inch clearance on each valve between the gauge surface and the end of the valve stem. If the valve stem clearance is less than 0.015 in, remove the tip of the valve stem on a valve refacing machine to insure a smooth, 90-degree edge on the end. Also break the sharp edge from the valve stem tip after removing the excess metal.

4.33 Be sure to check the valve spring installed height (the distance from the top of the seat/shims to the top of the spring or spring shield)

10 On GM 5.7L diesel engines, measure the rotator height using the same special tool (or have this step done by an automotive machine shop). The clearance between the valve rotator and the gauge should be at least 0.030 inch. If any valve stem end is less than 0.005 inch above the rotator, the valve is too short and a new valve must be installed.

11 Apply moly-base grease to the rocker arm faces and the pivots, then install the rocker arms and pivots.

Pistons/connecting rods - removal

Note 1: *Prior to removing the piston/connecting rod assemblies, remove the cylinder head(s), the oil pan and the oil pump by referring to the appropriate Sections in either Chapter 2 for GM or Chapter 3 for Ford.*

Note 2: *On 6.9L and 7.3L Ford diesel engines, if the pistons are to be removed, it is necessary to remove the piston oil cooling nozzles first. Their purpose is to provide a spray of oil into the bottom of the piston* **(see illustration)** *to help absorb some of the heat that occurs from the high temperature combustion. Remember, a diesel engine is exposed to much more heat and pressure than a gasoline engine!*

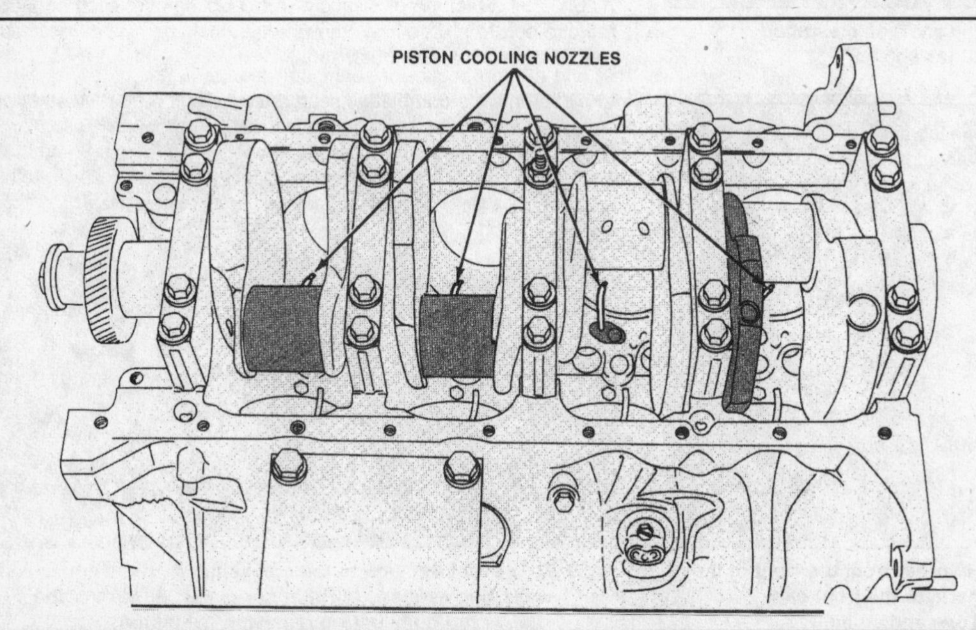

4.34 Ford diesel engines are equipped with piston cooling nozzles to help lower the temperature of the piston and cylinders

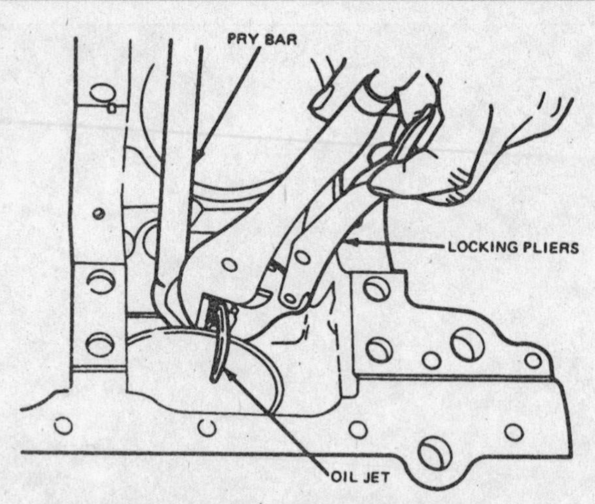

4.35 Use a pair of locking pliers and a prybar to lift the oil cooling jet from the engine block

4.36 A ridge reamer is required to remove the ridge from the top of each cylinder - do this before removing the pistons

1 On Ford diesel engines, the piston oil cooling jets (nozzles) can be removed by clamping a pair of locking pliers onto the jet and placing a prybar under the pliers and lifting up **(see illustration)**.

2 Use your fingernail to feel if a ridge has formed at the upper limit of ring travel (about 1/4-inch down from the top of each cylinder). If carbon deposits or cylinder wear have produced ridges, they must be completely re- moved with a special tool **(see illustration)**. Follow the manufacturer's instructions provided with the tool. Failure to remove the ridges before attempting to remove the piston/connecting rod assemblies may result in piston breakage.

3 After the cylinder ridges have been removed, turn the engine upside-down so the crankshaft is facing up.

4 Before the connecting rods are removed, check the endplay with feeler gauges. Slide them between the first connecting rod and the crankshaft throw until the play is removed **(see illustration)**. The endplay is equal to the thickness of the feeler gauge(s). If the endplay exceeds the service limit, new connecting rods will be required. If new rods (or a new crankshaft) are installed, the endplay may fall under the specified minimum (if it does, the rods will have to be machined to restore it - consult an automotive machine shop for advice if necessary). Repeat the procedure for the remaining connecting rods.

5 Check the connecting rods and caps for identification marks. If they aren't plainly marked, use a small center-punch **(see illustration)** to make the appropriate number of indentations on each rod and cap (1, 2, 3, etc., depending on the engine type and cylinder they're associated with).

6 Loosen each of the connecting rod cap nuts 1/2-turn at a time until they can be removed by hand. Remove the number one connecting rod cap and bearing insert. Don't drop the bearing insert out of the cap.

7 Slip a short length of plastic or rubber hose over each connecting rod cap bolt to protect the crankshaft journal and cylinder wall as the piston is removed **(see illustration)**.

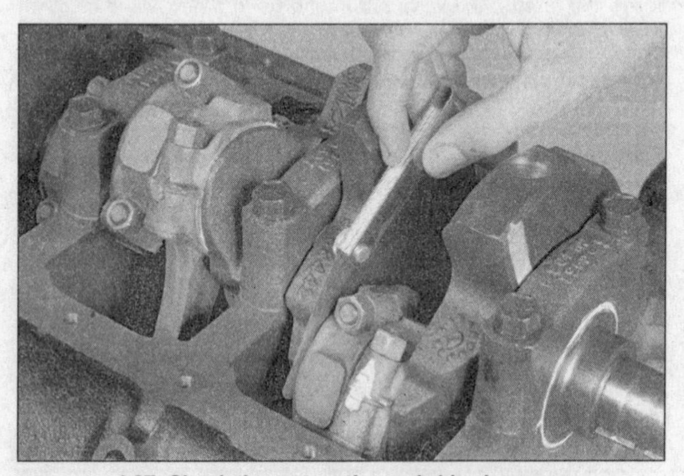

4.37 Check the connecting rod side clearance with a feeler gauges shown

4.38 Mark the rod bearing caps in order from the front of the engine to the rear (one mark for the front cap, two for the second one and so on)

4.39 To prevent damage to the crankshaft journals and cylinder walls, slip sections of rubber or plastic hose over the rod bolts before removing the pistons

4.40 Checking crankshaft endplay with a dial indicator

8 Remove the bearing insert and push the connecting rod/piston assembly out through the top of the engine. Use a wooden hammer handle to push on the upper bearing surface in the connecting rod. If resistance is felt, double-check to make sure that all of the ridge was removed from the cylinder.
9 Repeat the procedure for the remaining cylinders.
10 After removal, reassemble the connecting rod caps and bearing inserts in their respective connecting rods and install the cap nuts finger tight. Leaving the old bearing inserts in place until reassembly will help prevent the connecting rod bearing surfaces from being accidentally nicked or gouged.

Crankshaft - removal

Note: *The crankshaft can be removed only after the engine has been removed from the vehicle. It's assumed that the flywheel or driveplate, vibration damper, timing chain, oil pan, oil pump and piston/connecting rod assemblies have already been removed. If your engine is equipped with a one-piece rear main oil seal, the seal housing must be unbolted and separated from the block before proceeding with crankshaft removal.*

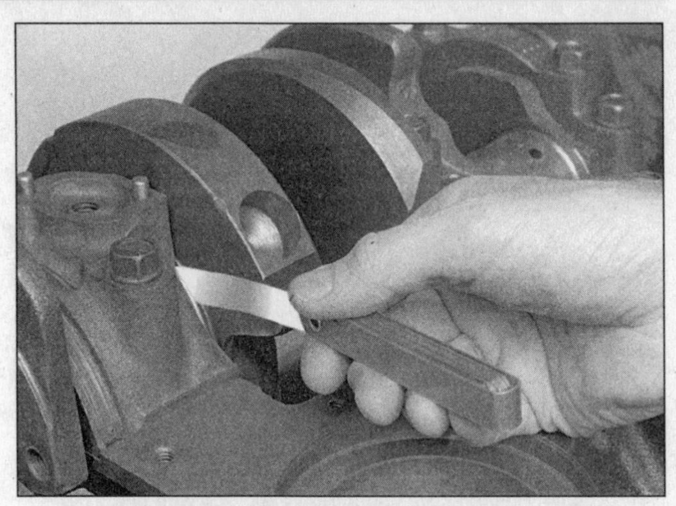

4.41 Checking crankshaft endplay with a feeler gauge

1 Using a large drift punch tap the outer edge of the core plug sideways in the bore. Then using a pair of pliers, pull the core plug from the engine block **(see illustration)**.
2 Push the crankshaft all the way to the rear and zero the dial indicator. Next, pry the crankshaft to the front as far as possible and check the reading on the dial indicator. The distance that it moves is the endplay. If it's greater than limit listed in this Chapter's Specifications, check the crankshaft thrust surfaces for wear. If no wear is evident, new main bearings should correct the endplay.
3 If a dial indicator isn't available, feeler gauges can be used. Gently pry or push the crankshaft all the way to the front of the engine. Slip feeler gauges between the crankshaft and the front face of the thrust main bearing (usually the center [number 3] main bearing) to determine the clearance **(see illustration)**.
4 Check the main bearing caps to see if they're marked to indicate their locations. They should be numbered consecutively from the front of the engine to the rear. If they aren't, mark them with number stamping dies or a center-punch **(see illustration)**. Main bearing caps generally have a cast-in arrow, which points to the front of the engine **(see illustration)**. Loosen the main bearing cap bolts 1/4-turn at a time each, until they can be removed by hand. Note if any stud bolts are used and make sure they're returned to their original locations when the crankshaft is reinstalled.
5 Gently tap the caps with a soft-face hammer, then separate them from the engine block. If necessary, use the bolts as levers to remove the caps. Try not to drop the bearing inserts if they come out with the caps.

4.42 Use a center-punch or number stamping dies to mark the main bearing caps to ensure installation in their original locations on the block (make the punch marks near one of the bolt heads)

4.43 The arrow on the main bearing cap indicates the front of the engine

4.44 Pull the core plugs from the block with pliers

6 Carefully lift the crankshaft out of the engine. It may be a good idea to have an assistant available, since the crankshaft is quite heavy. With the bearing inserts in place in the engine block and main bearing caps, return the caps to their respective locations on the engine block and tighten the bolts finger tight.

Engine block - cleaning

Caution: *The core plugs (also known as freeze plugs or soft plugs) may be difficult or impossible to retrieve if they're driven into the block coolant passages.*

1 Using a large drift punch, tap the outer edge of the core plug sideways in the bore. Then using a pair of pliers, pull the core plug from the engine block **(see illustration).**

2 Using a gasket scraper, remove all traces of gasket material from the engine block. Be very careful not to nick or gouge the gasket sealing surfaces.

3 Remove the main bearing caps and separate the bearing inserts from the caps and the engine block. Tag the bearings, indicating which cylinder they were removed from and whether they were in the cap or the block, then set them aside.

4 Remove all of the threaded oil gallery plugs from the block. The plugs are usually very tight - they may have to be drilled out and the holes retapped. Use new plugs when the engine is reassembled.

5 If the engine is extremely dirty it should be taken to an automotive machine shop to be steam cleaned or hot tanked.

6 After the block is returned, clean all oil holes and oil galleries one more time. Brushes specifically designed for this purpose are available at most auto parts stores. Flush the passages with warm water until the water runs clear, dry the block thoroughly and wipe all machined surfaces with a light, rust preventive oil. If you have access to compressed air, use it to speed the drying process and to blow out all the oil holes and galleries. **Warning:** *Wear eye protection when using compressed air!*

7 If the block isn't extremely dirty or sludged up, you can do an adequate cleaning job with hot soapy water and a stiff brush. Take plenty of time and do a thorough job. Regardless of the cleaning method used, be sure to clean all oil holes and galleries very thoroughly, dry the block completely and coat all machined surfaces with light oil.

8 The threaded holes in the block must be clean to ensure accurate torque readings during reassembly. Run the proper size tap into each of the holes to remove rust, corrosion, thread sealant or sludge and restore damaged threads **(see illustration)**. If possible, use compressed air to clear the holes of debris produced by this operation. Now is a good time to clean the threads on the head bolts and the main bearing cap bolts as well.

9 Reinstall the main bearing caps and tighten the bolts finger tight.

10 After coating the sealing surfaces of the new core plugs with Permatex no. 2 sealant, install them in the engine block **(see illustration)**. Make sure they're driven in straight and seated properly or leakage could result. Special tools are available for this purpose, but a large socket, with an outside diameter that will just slip into the core plug, a 1/2-inch drive extension and a hammer will work just as well.

11 Apply non-hardening sealant (such as Permatex no. 2 or Teflon pipe sealant) to the new oil gallery plugs and thread them into the holes in the block. Make sure they're tightened securely.

12 If the engine isn't going to be reassembled right away, cover it with a large plastic trash bag to keep it clean.

Engine block - inspection

1 Before the block is inspected, it should be cleaned as described in the previous Section.

2 Visually check the block for cracks, rust and corrosion. Look for stripped threads in the threaded holes. It's also a good idea to have the block checked for hidden cracks by an automotive machine shop that has the special equipment to do this type of work. If defects are found, have the block repaired, if possible, or replaced.

3 Check the cylinder bores for scuffing and scoring.

4 Measure the diameter of each cylinder at the top (just under the ridge area), center and bottom of the cylinder bore, parallel to the crankshaft axis **(see illustrations)**.

5 Next, measure each cylinder's diameter at the same three locations

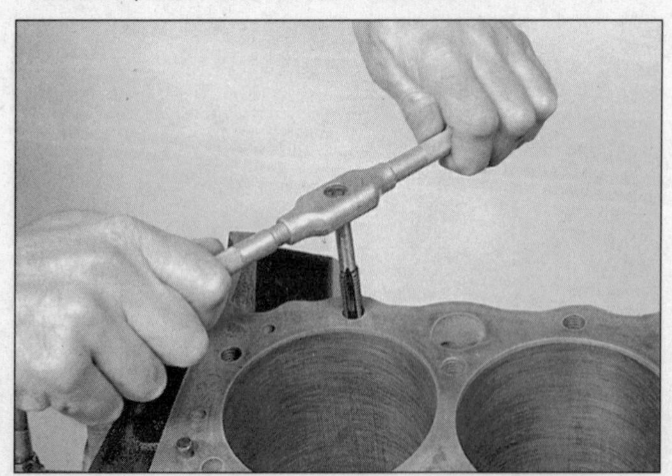

4.45 All bolt holes in the block - particularly the main bearing cap and head bolt holes - should be cleaned and restored with a tap (be sure to remove debris from, the holes after this is done)

4.46 A large socket on an extension can be used to drive the new core plugs into the bores

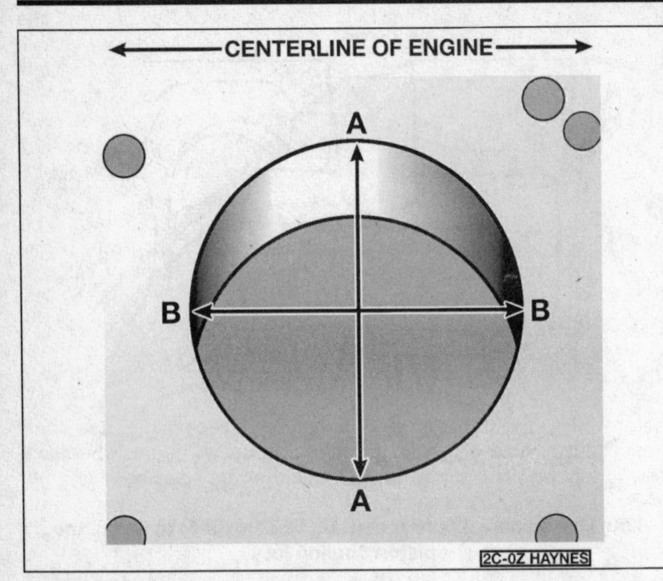

4.47a Measure the diameter of each cylinder at a right angle to the engine centerline (A), and parallel to the engine centerline (B) - out-of-round is the difference between A and B; taper is the difference between A and B at the top of the cylinder and A and B at the bottom of the cylinder

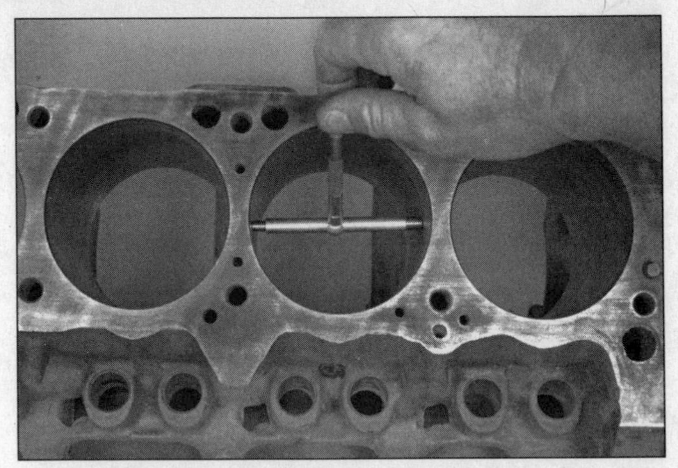

4.47b The ability to "feel" when the telescoping gauge is at the correct point will be developed over time, so work slowly and repeat the check until you're satisfied the bore measurement is accurate

across the crankshaft axis. Compare the results to this Chapter's Specifications.

6 If the required precision measuring tools aren't available, the piston-to-cylinder clearances can be obtained, though not quite as accurately, using feeler gauge stock. Feeler gauge stock comes in 12-inch lengths and various thicknesses and is generally available at auto parts stores.

7 To check the clearance, select a feeler gauge and slip it into the cylinder along with the matching piston. The piston must be positioned exactly as it normally would be. The feeler gauge must be between the piston and cylinder on one of the thrust faces (90-degrees to the piston pin bore).

8 The piston should slip through the cylinder (with the feeler gauge in place) with moderate pressure.

9 If it falls through or slides through easily, the clearance is excessive and a new piston will be required. If the piston binds at the lower end of the cylinder and is loose toward the top, the cylinder is tapered. If tight spots are encountered as the piston/feeler gauge is rotated in the cylinder, the cylinder is out-of-round.

10 Repeat the procedure for the remaining pistons and cylinders.

11 If the cylinder walls are badly scuffed or scored, or if they're out-of-round or tapered beyond the limits given in this Chapter's Specifications, have the engine block rebored and honed at an automotive machine shop. If a rebore is done, oversize pistons and rings will be required.

12 If the cylinders are in reasonably good condition and not worn to the outside of the limits, and if the piston-to-cylinder clearances can be maintained properly, then they don't have to be rebored. Honing is all that's necessary.

Cylinder honing

1 Prior to engine reassembly, the cylinder bores must be honed so the new piston rings will seat correctly and provide the best possible combustion chamber seal. **Note:** *If you don't have the tools or don't want to tackle the honing operation, most automotive machine shops will do it for a reasonable fee.*

2 Before honing the cylinders, install the main bearing caps and tighten the bolts to the torque listed in this Chapter's Specifications.

3 Two types of cylinder hones are commonly available - the flex hone or "bottle brush" type and the more traditional surfacing hone with spring-loaded stones. Both will do the job, but for the less experienced mechanic the "bottle brush" hone will probably be easier to use. You'll also need some kerosene or honing oil, rags and an electric drill motor. Proceed as follows:

a) *Mount the hone in the drill motor, compress the stones and slip it into the first cylinder* **(see illustration)**. *Be sure to wear safety goggles or a face shield!*

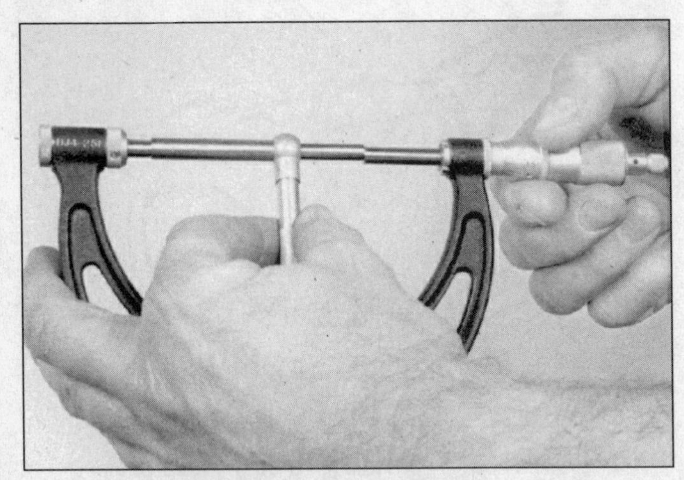

4.47c The gauge is than measured with a micrometer to determine the bore size

4.48 A "bottle brush" hone will produce better results if you've never honed cylinders before

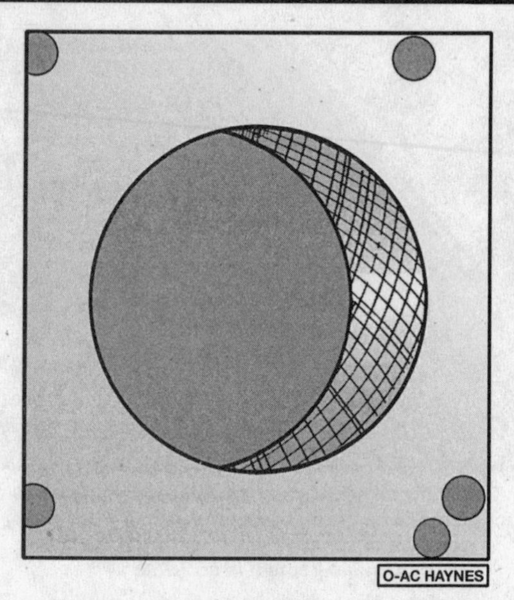

4.49 The cylinder hone should leave a smooth, crosshatch pattern with the lines intersecting at approximately a 60-degree angle for the Ford engines and a 45 to 65-degree angle for the GM engines

b) *Lubricate the cylinder with plenty of honing oil, turn on the drill and move the hone up-and-down in the cylinder at a pace that will produce a fine crosshatch pattern on the cylinder walls. Ideally, the crosshatch lines should intersect at approximately a 60-degree angle on all Ford diesel engines and 45 to 65-degrees on all GM diesel engines* **(see illustration)**. *Be sure to use plenty of lubricant and don't take off any more material than is absolutely necessary to produce the desired finish.* **Note:** *Piston ring manufacturers may specify a smaller or larger crosshatch angle - read and follow any instructions included with the new rings.*

c) *Don't withdraw the hone from the cylinder while it's running. Instead, shut off the drill and continue moving the hone up-and-down in the cylinder until it comes to a complete stop, then compress the stones and withdraw the hone. If you're using a "bottle brush" type hone, stop the drill motor, then turn the chuck in the normal direction of rotation while withdrawing the hone from the cylinder.*

d) *Wipe the oil out of the cylinder and repeat the procedure for the remaining cylinders.*

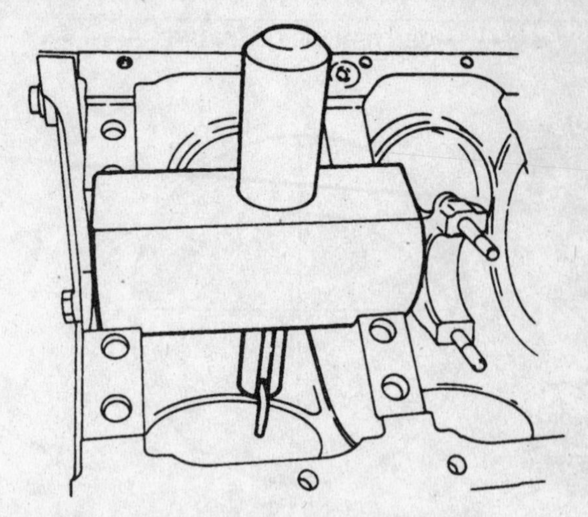

4.50 Use a special Ford tool (OTC D83T-6134A) to install the piston cooling jet

4 After the honing job is complete, chamfer the top edges of the cylinder bores with a small file so the rings won't catch when the pistons are installed. Be very careful not to nick the cylinder walls with the end of the file.

5 The entire engine block must be washed again very thoroughly with warm, soapy water to remove all traces of the abrasive grit produced during the honing operation. **Note:** *The bores can be considered clean when a lint-free white cloth - dampened with clean engine oil- used to wipe them out doesn't pick up any more honing residue, which will show up as gray areas on the cloth. Be sure to run a brush through all oil holes and galleries and flush them with running water.*

6 After rinsing, dry the block and apply a coat of light rust preventive oil to all machined surfaces.

7 On Ford 6.9L and 7.3L diesel engines, install the piston oil cooling jets into the engine block. Use a special tool positioned over the saddles of the main bearing **(see illustration)**. Place the new oil cooling jet into the tool and align it directly over the hole.

8 Start the jet by lightly tapping the tool with a hammer **(see illustration)**. Continue hitting the tool with a hammer until the drive bottoms onto the body of the tool.

9 Verify that the cooling jet is properly aligned by using a special tool. The pointer must locate in the target hole **(see illustration)**. If not, slightly bend the cooling jet tube until the pointer is aligned correctly.

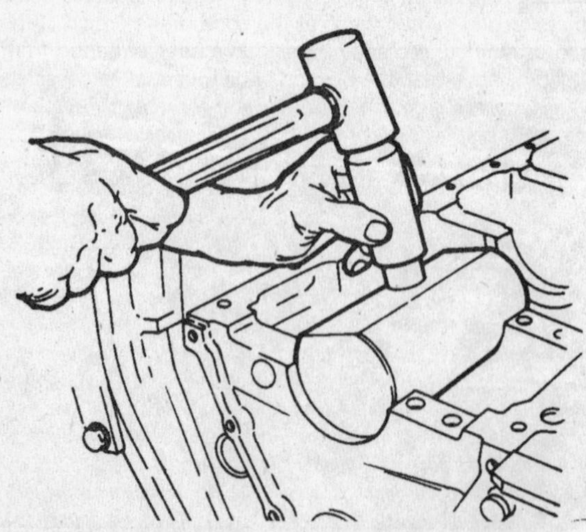

4.51 Start the jet by tapping the tool lightly

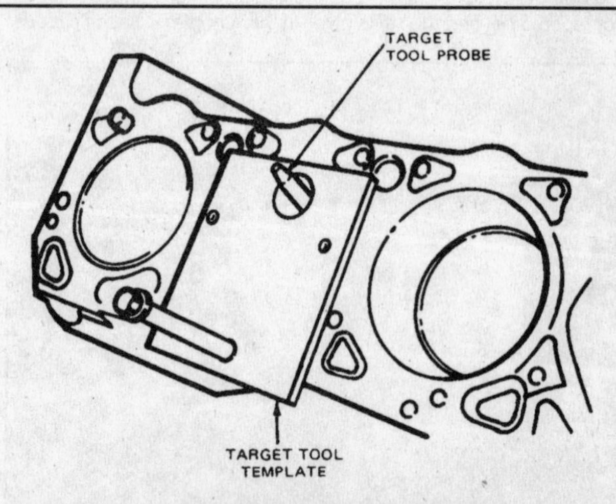

4.52 Use a special Ford tool (OTC D83T-6134B) to align the jet in the engine block

4.53 The piston ring grooves can be cleaned with a special tool, as shown here, . . .

4.54 . . . or a section of a broken ring

10 Wrap the block in a plastic trash bag to keep it clean and set it aside until reassembly.

Pistons/connecting rods - inspection

Note: *On 6.2L and 6.5L GM diesel engines, the pistons are match fitted to each cylinder bore of the engine. The size code is stamped on the piston face. Size codes A, B, C, D, E and G are used to match the piston and cylinder bore. "A" size pistons are matched with "A" size cylinder bores and "B" size pistons with "B" size pistons, etc. The size codes are stamped on the cylinder case pan rail and beside the proper cylinder. Bohn pistons are identified by the word BOHNNA LITE near the pin boss while Zollner pistons are identified by the letter Z with a circle around it also near the pin boss.*

Note: *1992 and 1993 6.2L VIN C engines may contain both standard and 0.003 inch oversized connecting rod bearings in the same block. Check the individual identification marks on each bearing insert.*

1 Before the inspection process can be carried out, the piston/connecting rod assemblies must be cleaned and the original piston rings removed from the pistons. **Note:** *Always use new piston rings when the engine is reassembled.*

2 Using a piston ring installation tool, carefully remove the rings from the pistons. Be careful not to nick or gouge the pistons in the process.

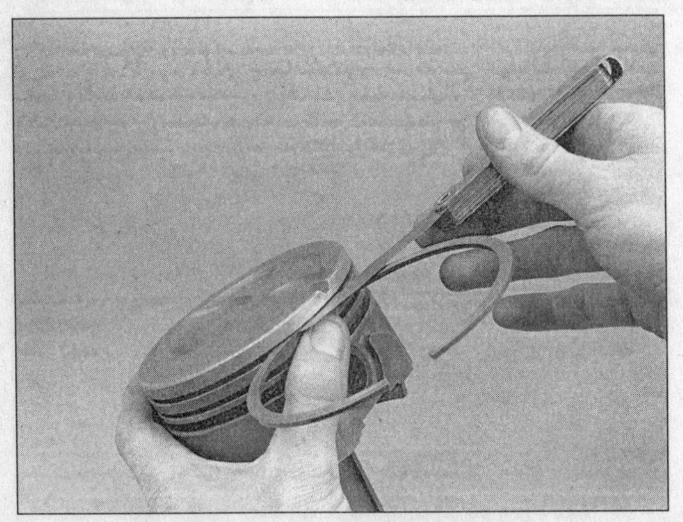

4.55 Check the ring side clearance with a feeler gauge at several points around the groove

3 Scrape all traces of carbon from the top of the piston. A handheld wire brush or a piece of fine emery cloth can be used once the majority of the deposits have been scraped away. Do not, under any circumstances, use a wire brush mounted in a drill motor to remove deposits from the pistons. The piston material is soft and may be eroded away by the wire brush.

4 Use a piston ring groove cleaning tool to remove carbon deposits from the ring grooves **(see illustration)**. If a tool isn't available, a piece broken off the old ring will do the job. Be very careful to remove only the carbon deposits - don't remove any metal and do not nick or scratch the sides of the ring grooves **(see illustration)**.

5 Once the deposits have been removed, clean the piston/rod assemblies with solvent and dry them with compressed air (if available). Make sure the oil return holes in the back sides of the ring grooves are clear.

6 If the pistons and cylinder walls aren't damaged or worn excessively, and if the engine block is not rebored, new pistons won't be necessary. Normal piston wear appears as even vertical wear on the piston thrust surfaces and slight looseness of the top ring in its groove. New piston rings, however, should always be used when an engine is rebuilt.

7 Carefully inspect each piston for cracks around the skirt, at the pin bosses and at the ring lands.

8 Look for scoring and scuffing on the thrust faces of the skirt, holes in the piston crown and burned areas at the edge of the crown. If the skirt is scored or scuffed, the engine may have been suffering from overheating and/or abnormal combustion, which caused excessively high operating temperatures. The cooling and lubrication systems should be checked thoroughly. A hole in the piston crown is an indication that abnormal combustion (preignition) was occurring. If only one or two pistons show heavy scuffing, look for an improper piston-to-bore clearance. If any of the above problems exist, the causes must be corrected or the damage will occur again. The causes may include intake air leaks, head gasket leaks, improper clearances and EGR system malfunctions.

9 Corrosion of the piston, in the form of small pits, indicates that coolant is leaking into the combustion chamber and/or the crankcase. Again, the cause must be corrected or the problem may persist in the rebuilt engine.

10 Measure the piston ring side clearance by laying a new piston ring in each ring groove and slipping a feeler gauge in beside it **(see illustration)**. Check the clearance at three or four locations around each groove. Be sure to use the correct ring for each groove - they are different. If the side clearance is greater than the figure listed in this Chapter's Specifications, new pistons will have to be used.

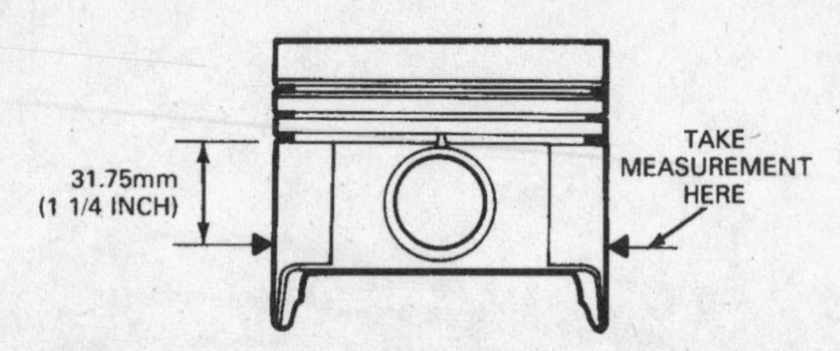

4.56a On Ford 6.9L and 7.3L engines, measure the piston diameter 1-1/4-inches from the lower ring land

11 Check the piston-to-bore clearance by measuring the bore (earlier in this Chapter) and the piston diameter. Make sure the pistons and bores are correctly matched. Measure the piston across the skirt, at a 90-degree angle with the piston pin **(see illustrations)**. Subtract the piston diameter from the bore diameter to obtain the clearance. If it's greater than specified, the block will have to be rebored and new pistons and rings installed.

12 Check the piston-to-rod clearance by twisting the piston and rod in opposite directions. Any noticeable play indicates excessive wear, which must be corrected. The piston/connecting rod assemblies should be taken to an automotive machine shop to have the pistons and rods resized and new pins installed.

13 If the pistons must be removed from the connecting rods for any reason, they should be taken to an automotive machine shop. While they are there have the connecting rods checked for bend and twist, since automotive machine shops have special equipment for this purpose. **Note:** *Unless new pistons and/or connecting rods must be installed, do not disassemble the pistons and connecting rods.*

14 Check the connecting rods for cracks and other damage. Temporarily remove the rod caps, lift out the old bearing inserts, wipe the rod and cap bearing surfaces clean and inspect them for nicks, gouges and scratches. After checking the rods, replace the old bearings, slip the caps into place and tighten the nuts finger tight. **Note:** *If the engine is being rebuilt because of a connecting rod knock, be sure to install new rods.*

Crankshaft - inspection

1 Remove all burrs from the crankshaft oil holes with a stone, file or scraper **(see illustration)**.

2 Clean the crankshaft with solvent and dry it with compressed air (if available). Be sure to clean the oil holes with a stiff brush **(see illustration)** and flush them with solvent.

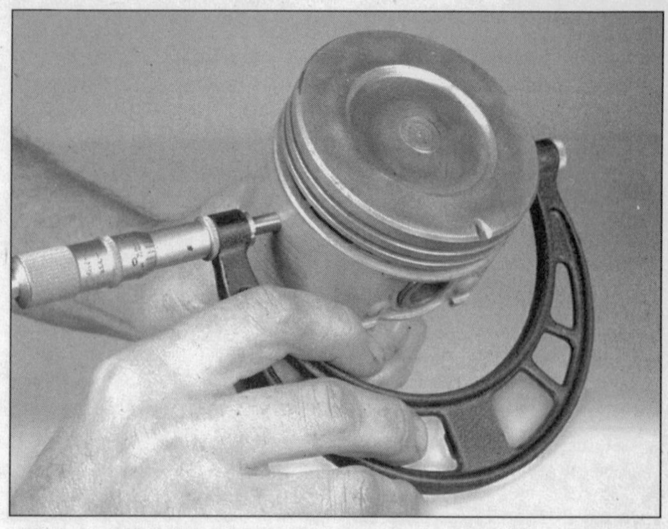

4.56b On GM 5.7L and 6.2L engines, measure the piston diameter 3/4 inch below the center of the piston pin hole

3 Check the main and connecting rod bearing journals for uneven wear, scoring, pits and cracks.

4 Rub a penny across each journal several times **(see illustration)**. If a journal picks up copper from the penny, it's too rough and must be reground.

5 Check the rest of the crankshaft for cracks and other damage. It should be magnafluxed to reveal hidden cracks - an automotive machine shop will handle the procedure.

6 Using a micrometer, measure the diameter of the main and connecting rod journals and compare the results to this Chapter's Specifications **(see illustration)**. By measuring the diameter at a number of points around each journal's circumference, you'll be able to determine whether or not the journal is out-of-round. Take the measurement at each end of the journal, near the crank throws, to determine if the journal is tapered.

7 If the crankshaft journals are damaged, tapered, out-of-round or worn beyond the limits given in the Specifications, have the crankshaft reground by an automotive machine shop. Be sure to use the correct size bearing inserts if the crankshaft is reconditioned.

8 Check the oil seal journals at each end of the crankshaft for wear and damage. If the seal has worn a groove in the journal, or if it's nicked or scratched, the new seal may leak when the engine is reassembled. In some cases, an automotive machine shop may be able to repair the journal by pressing on a thin sleeve. If repair isn't fea-

4.57 The oil holes should be chamfered so sharp edges don't gouge or scratch the new bearings

4.58 Rubbing a penny lengthwise on each journal will reveal its condition - if copper rubs off and is embedded in the crankshaft, the journals should be reground

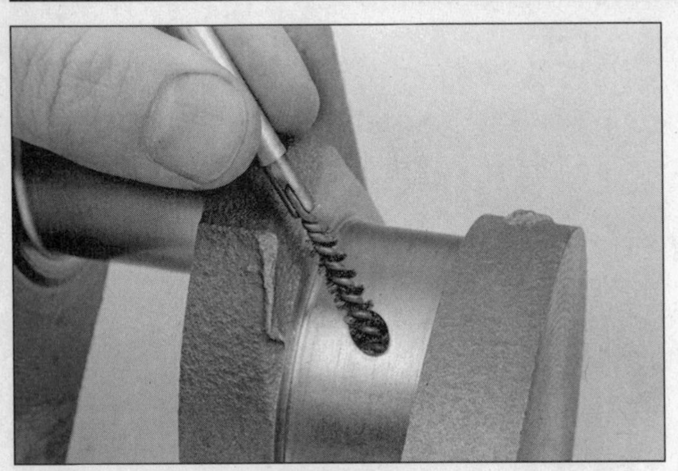

4.59 Use a wire or stiff plastic bristle brush to clean the oil passages in the crankshaft

4.60 Measure the diameter of each crankshaft journal at several points to detect taper and out-of-round conditions

sible, a new or different crankshaft should be installed.
9 Refer to the next Section and examine the main and rod bearing inserts.

Main and connecting rod bearings - inspection

1 Even though the main and connecting rod bearings should be replaced with new ones during the engine overhaul, the old bearings should be retained for close examination, as they may reveal valuable information about the condition of the engine.
2 Bearing failure occurs because of lack of lubrication, the presence of dirt or other foreign particles, overloading the engine and corrosion. Regardless of the cause of bearing failure, it must be corrected before the engine is reassembled to prevent it from happening again.
3 When examining the bearings, remove them from the engine block, the main bearing caps, the connecting rods and the rod caps and lay them out on a clean surface in the same general position as their location in the engine. This will enable you to match any bearing problems with the corresponding crankshaft journal.
4 Dirt and other foreign particles get into the engine in a variety of ways. It may be left in the engine during assembly, or it may pass through filters or the crankcase ventilation system. It may get into the oil, and from there into the bearings. Metal chips from machining operations and normal engine wear are often present. Abrasives are sometimes left in engine components after reconditioning, especially when parts are not thoroughly cleaned using the proper cleaning methods. Whatever the source, these foreign objects often end up embedded in the soft bearing material and are easily recognized. Large particles will not embed in the bearing and will score or gouge the bearing and journal. The best prevention for this cause of bearing failure is to clean all parts thoroughly and keep everything spotlessly clean during engine assembly. Frequent and regular engine oil and filter changes are also recommended.
5 Lack of lubrication (or lubrication breakdown) has a number of interrelated causes. Excessive heat (which thins the oil), overloading (which squeezes the oil from the bearing face) and oil leakage or throw off (from excessive bearing clearances, worn oil pump or high engine speeds) all contribute to lubrication breakdown. Blocked oil passages, which usually are the result of misaligned oil holes in a bearing shell, will also oil starve a bearing and destroy it. When lack of lubrication is the cause of bearing failure, the bearing material is wiped or extruded from the steel backing of the bearing. Temperatures may increase to the point where the steel backing turns blue from overheating.
6 Driving habits can have a definite effect on bearing life. Full throttle, low speed operation (lugging the engine) puts very high loads on bearings, which tends to squeeze out the oil film. These loads cause the bearings to flex, which produces fine cracks in the bearing face (fatigue failure). Eventually the bearing material will loosen in pieces and tear away from the steel backing. Short trip driving leads to corrosion of bearings because insufficient engine heat is produced to drive off the condensed water and corrosive gases. These products collect in the engine oil, forming acid and sludge. As the oil is carried to the engine bearings, the acid attacks and corrodes the bearing material.
7 Incorrect bearing installation during engine assembly will lead to bearing failure as well. Tight fitting bearings leave insufficient bearing oil clearance and will result in oil starvation. Dirt or foreign particles trapped behind a bearing insert result in high spots on the bearing which lead to failure.

Engine overhaul - reassembly sequence

1 Before beginning engine reassembly, make sure you have all the necessary new parts, gaskets and seals as well as the following items on hand:

Common hand tools
A 1/2-inch drive torque wrench
Piston ring installation tool
Piston ring compressor
Vibration damper installation tool
Short lengths of rubber or plastic hose to fit over connecting rod bolts
Plastigage
Feeler gauges
A fine-tooth file
New engine oil
Engine assembly lube or moly-base grease
Gasket sealant
Thread locking compound

2 In order to save time and avoid problems, engine reassembly must be done in the following general order:

New camshaft bearings (must be done by automotive machine shop)
Piston rings
Crankshaft and main bearings
Piston/connecting rod assemblies
Oil pump
Camshaft and lifters
Oil pan
Timing chain and sprockets
Cylinder head(s), pushrods and rocker arms
Timing cover
Intake and exhaust manifolds
Rocker arm cover(s)
Engine rear plate
Flywheel/driveplate

4.61 When checking piston ring end gap, the ring must be square in the cylinder bore (this is done by pushing the ring down with the top of a piston as shown)

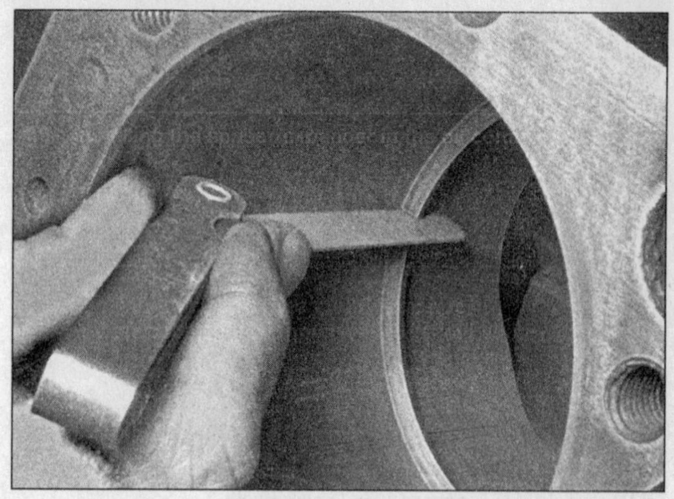

4.62 With the ring square in the cylinder, measure the and gap with a feeler gauge

Piston rings - installation

1 Before installing the new piston rings, the ring end gaps must be checked. It's assumed that the piston ring side clearance has been checked and verified correct (earlier in this Chapter).

2 Lay out the piston/connecting rod assemblies and the new ring sets so the ring sets will be matched with the same piston and cylinder during the end gap measurement and engine assembly.

3 Insert the top (number one) ring into the first cylinder and square it up with the cylinder walls by pushing it in with the top of the piston **(see illustration)**. The ring should be near the bottom of the cylinder, at the lower limit of ring travel.

4 To measure the end gap, slip feeler gauges between the ends of the ring until a gauge equal to the gap width is found **(see illustration)**. The feeler gauge should slide between the ring ends with a slight amount of drag. Compare the measurement to this Chapter's Specifications. If the gap is larger or smaller than specified, double-check to make sure you have the correct rings before proceeding.

5 If the gap is too small, it must be enlarged or the ring ends may come in contact with each other during engine operation, which can cause serious damage to the engine. The end gap can be increased by filing the ring ends very carefully with a fine file. Mount the file in a vise equipped with soft jaws, slip the ring over the file with the ends contacting the file face and slowly move the ring to remove material from the ends. When performing this operation, file only from the outside in **(see illustration)**.

6 Excess end gap isn't critical unless it's greater than 0.040-inch. Again, double-check to make sure you have the correct rings for your engine.

7 Repeat the procedure for each ring that will be installed in the first cylinder and for each ring in the remaining cylinders. Remember to keep rings, pistons and cylinders matched up.

8 Once the ring end gaps have been checked/corrected, the rings can be installed on the pistons.

9 The oil control ring (lowest one on the piston) is usually installed first. It's normally composed of three separate components, although some 5.7L engines have only two components. Slip the spacer/expander into the groove **(see illustration)**. If an anti-rotation tang is used, make sure it's inserted into the drilled hole in the ring groove. Next, install the lower side rail. Don't use a piston ring installation tool on the oil ring side rails, as they may be damaged. Instead, place one end of the side rail into the groove between the spacer/expander and the ring land, hold it firmly in place and slide a finger around the piston while pushing the rail into the groove **(see illustration)**. Next, install the upper side rail in the same manner.

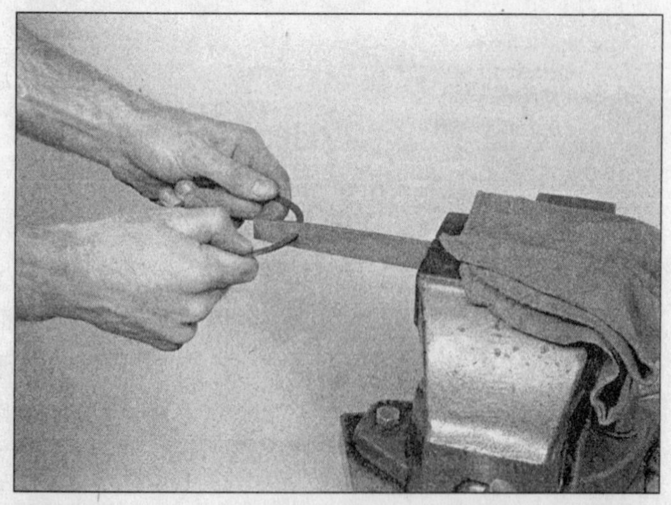

4.63 If the end gap is too small, clamp a file in a vise and file the ring ends (from the outside in only) to enlarge the gap slightly

4.64 Installing the spacer/expander in the oil control ring groove

Engine bearing distress analysis

DEBRIS

Babbitt bearing embedded with debris from machinings

Microscopic detail of debris

Microscopic detail of gouges

Overplated copper alloy bearing gouged by cast iron debris

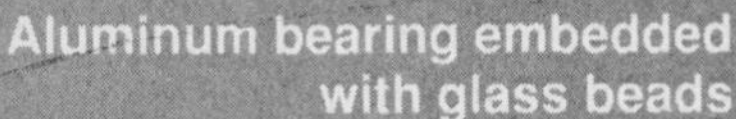

Aluminum bearing embedded with glass beads

Microscopic detail of glass beads

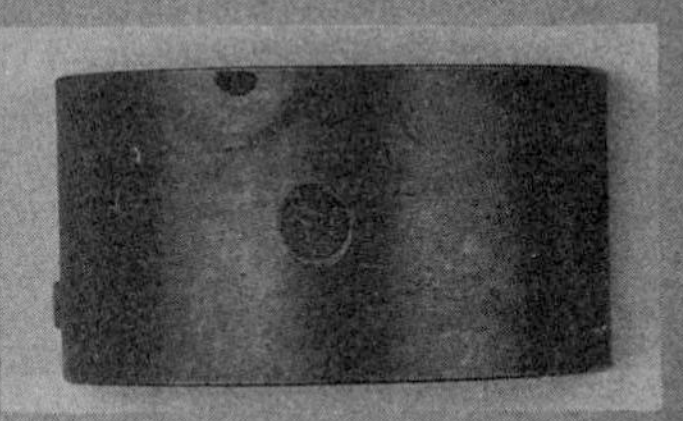

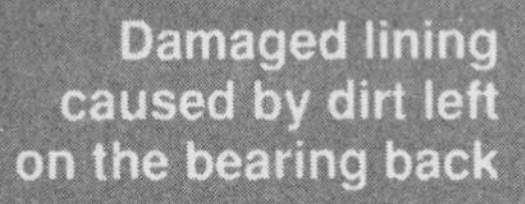

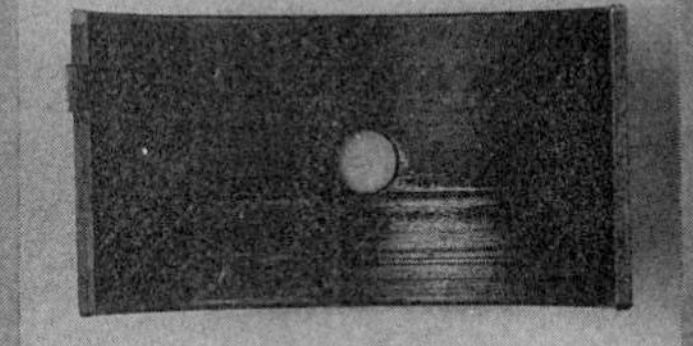

Damaged lining caused by dirt left on the bearing back

MISASSEMBLY

Result of a lower half assembled as an upper – blocking the oil flow

Excessive oil clearance is indicated by a short contact arc

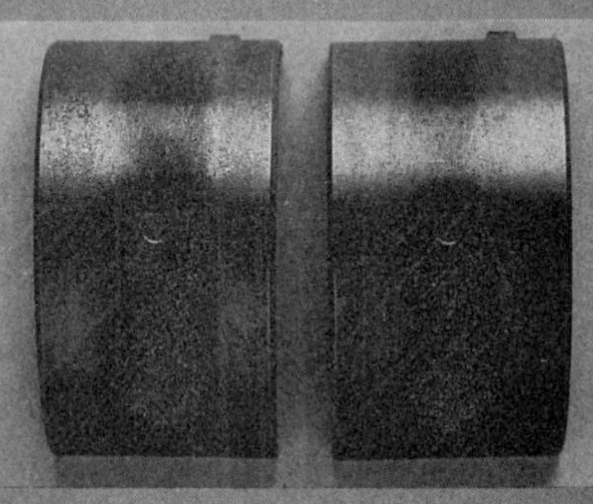

Polished and oil-stained backs are a result of a poor fit in the housing bore

Result of a wrong, reversed or shifted cap

Damage from excessive idling which resulted in an oil film unable to support the load imposed

Damaged upper connecting rod bearings caused by engine lugging; the lower main bearings (not shown) were similarly affected

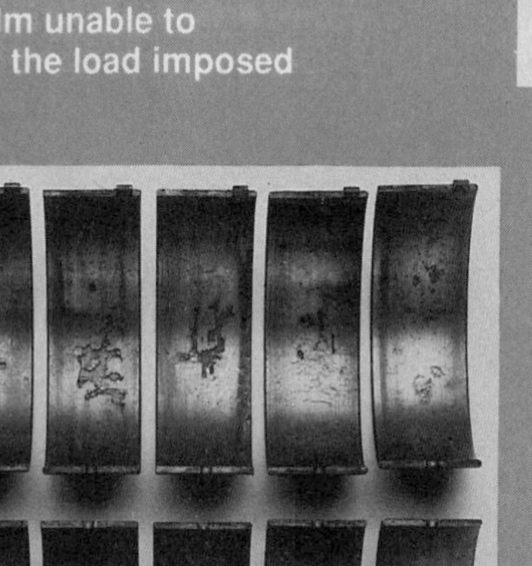

OVERLOADING

The damage shown in these upper and lower connecting rod bearings was caused by engine operation at higher-than-rated speed under load

CORROSION

Microscopic detail of corrosion

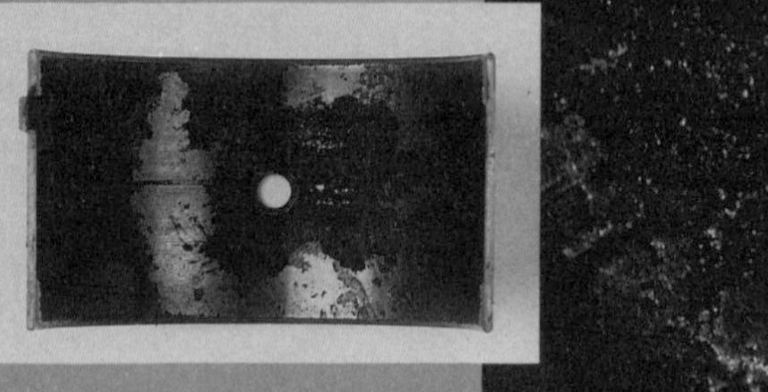

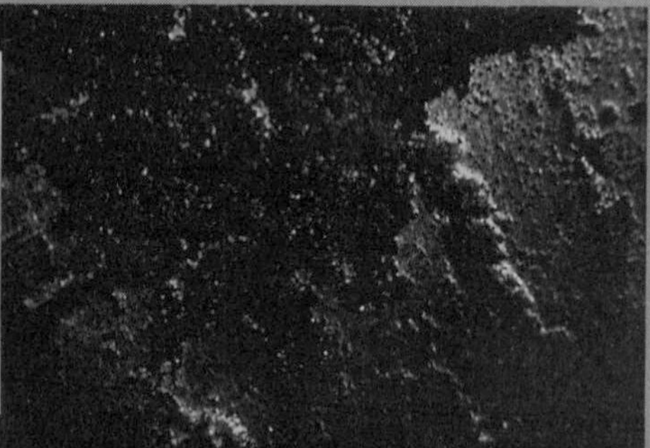

Corrosion is an acid attack on the bearing lining generally caused by inadequate maintenance, extremely hot or cold operation, or inferior oils or fuels

Result of dry start: The bearings on the left, farthest from the oil pump, show more damage

LUBRICATION

Severe wear as a result of inadequate oil clearance

Result of a low oil supply or oil starvation

Microscopic detail of cavitation

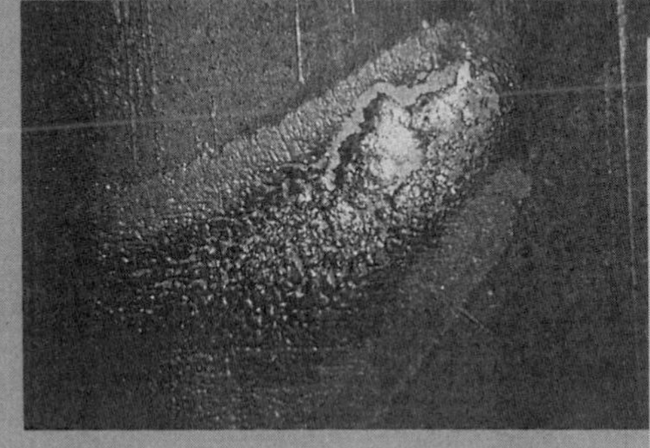

Example of cavitation – a surface erosion caused by pressure changes in the oil film

Damage from excessive thrust or insufficient axial clearance

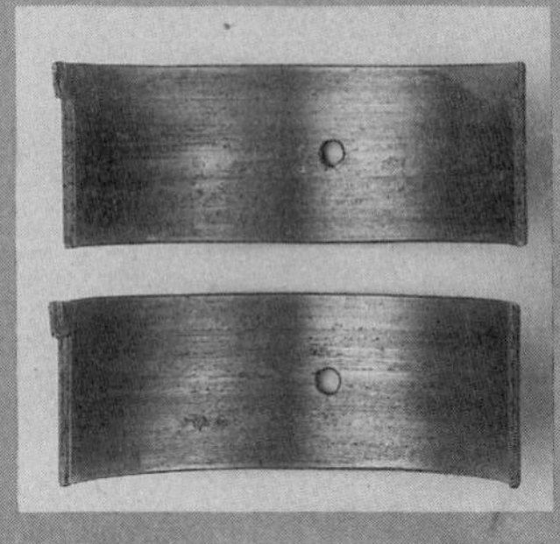

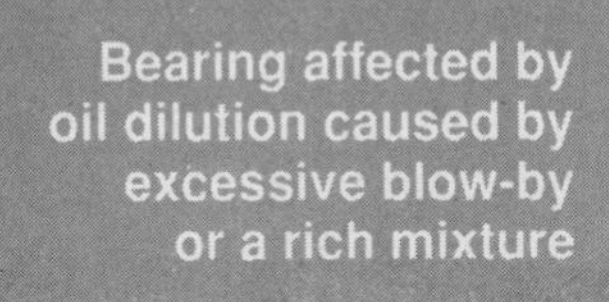

Bearing affected by oil dilution caused by excessive blow-by or a rich mixture

MISALIGNMENT

A warped crankshaft caused this pattern of severe wear in the center, diminishing toward the ends

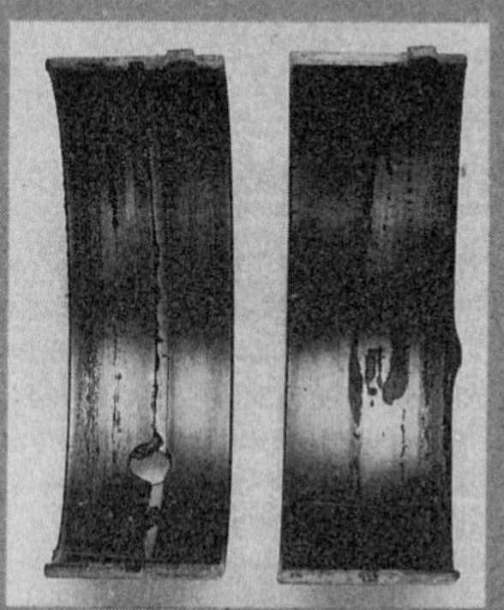

A tapered housing bore caused the damage along one edge of this pair

A poorly finished crankshaft caused the equally spaced scoring shown

A bent connecting rod led to the damage in the "V" pattern

For more detailed information, please refer to the Federal-Mogul Engine Bearing Service Manual

Copy and photographs courtesy of Federal-Mogul Corporation

10 After the three oil ring components have been installed, check to make sure that both the upper and lower side rails can be turned smoothly in the ring groove.
11 The number two (middle) ring is installed next. It's usually stamped with a mark which must face up, toward the top of the piston. **Note:** *Always follow the instructions printed on the ring package or box - different manufacturers may require different approaches. Do not mix up the top and middle rings, as they have different cross-sections.*
12 Use a piston ring installation tool and make sure the identification mark is facing the top of the piston, then slip the ring into the middle groove on the piston **(see illustration)**. Don't expand the ring any more than necessary to slide it over the piston.
13 Install the number one (top) ring in the same manner. Make sure the mark is facing up. Be careful not to confuse the number one and number two rings.
14 Repeat the procedure for the remaining pistons and rings.

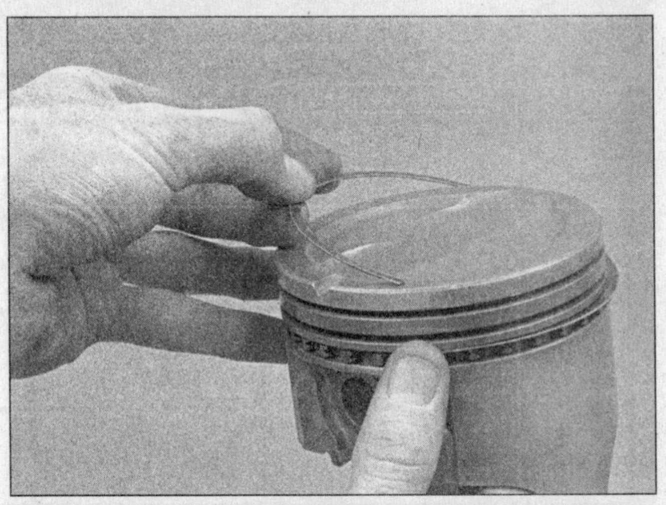

4.65 DO NOT use a piston ring installation tool when installing the oil ring side rails

Crankshaft - installation and main bearing oil clearance check

1 Crankshaft installation is the first step in engine reassembly. It's assumed at this point that the engine block and crankshaft have been cleaned, inspected and repaired or reconditioned.
2 Position the engine with the bottom facing up.
3 Remove the main bearing cap bolts and lift out the caps. Lay them out in the proper order to ensure correct installation. **Note:** *The 6.2L and 6.5L GM diesel engine block is designed to match fit the cylinder bore with the piston. This is accomplished by dividing the total tolerance size range into six different categories. Each one is designated with a letter* **(see illustration)** *that is stamped on the pan rail adjacent to the cylinder i.e. "A" size pistons for "A" size cylinder bores. The center or number 3 bearing is the thrust bearing. The main bearings are select fitted to each of the 5 main bearing bores. The proper size code is stamped on the pan rail at the corresponding main bearing bulkhead* **(see previous illustration)**. *It will be stamped either 1, 2 or 3 on the side rail. Each of the sizes is matched to the corresponding size in the lower half (case) only. The upper main bearings are match fitted from the specifications from the crankshaft main journal (cap). Measure the crankshaft main journal diameter and consult the chart* **(see illustration)** *to obtain the correct bearing for the upper (cap) insert. Use the number designation to obtain the correct bearing for the lower half (case).*

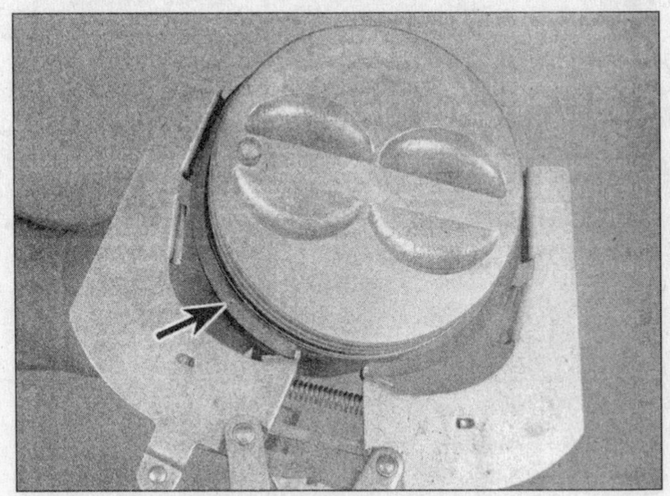

4.66 Installing the compression rings with a ring expander -the mark (arrow) must face up

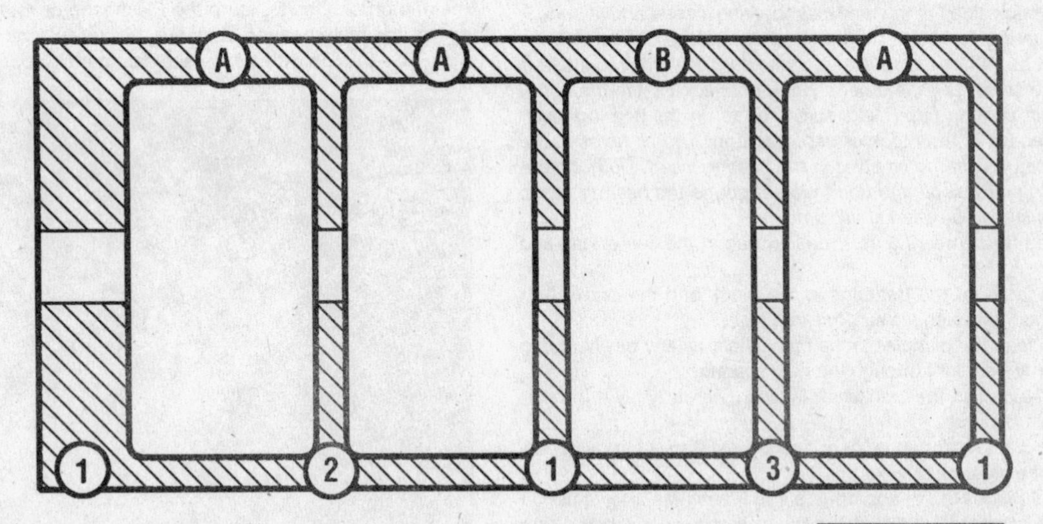

4.67 On GM 6.2L and 6.5L engines, check the block for the correct letter and number designations in order to match the correct bearings

Crankshaft Main Journal Diameter		Cylinder & Case Main Bearing Bore Diameter		
		(3) 3.1434 to 3.1437	(2) 3.1431 to 3.1434	(1) 3.1328 to 3.1431
Front, Front Intermediate Center & Rear Intermediate Main Bearing	Blue 2.9495 to 2.9498	0.001 U.S. in case 0.001 U.S. in cap	0.0005 U.S. in case 0.001 U.S. in cap	std in case 0.001 U.S. in cap
	Orange 2.9498 to 2.9501	0.001 U.S. in case 0.0005 U.S. in cap	0.0005 U.S. in case 0.0005 U.S. in cap	std in case 0.0005 U.S. in cap
	White 2.9501 to 2.9504	0.001 U.S. in case std in cap	0.0005 U.S. in case std in cap	std in case std in cap
Rear Main Bearing	Blue 2.9493 to 2.9496	0.001 U.S. in case 0.001 in cap	0.0005 U.S. in case 0.001 U.S. in cape	std in case 0.001 U.S. in cap
	Orange 2.9496 to 2.9499	0.001 U.S. in case 0.0005 U.S. in cap	0.0005 U.S. in case 0.0005 U.S. in cap	std in case 0.0005 U.S. in cap
	White 2.9499 to 2.9502	0.001 U.S. in case std in cap	0.0005 U.S. in case std in cap	std in case std in cap

10330-4-4.68 HAYNES

4.68 GM 6.2L and 6.5L crankshaft bearing chart

4 If they're still in place, remove the original bearing inserts from the block and the main bearing caps. Wipe the bearing surfaces of the block and caps with a clean, lint-free cloth. They must be kept spotlessly clean.

Main bearing oil clearance check

5 Clean the back sides of the new main bearing inserts and lay one in each main bearing saddle in the block. If one of the bearing inserts from each set has a large groove in it, make sure the grooved insert is installed in the block. Lay the other bearing from each set in the corresponding main bearing cap. Make sure the tab on the bearing insert fits into the recess in the block or cap. **Caution:** *The oil holes in the block must line up with the oil holes in the bearing insert. Do not hammer the bearing into place and don't nick or gouge the bearing faces. No lubrication should be used at this time.*
6 The flanged thrust bearing must be installed in the center cap and saddle.
7 Clean the faces of the bearings in the block and the crankshaft main bearing journals with a clean, lint-free cloth.
8 Check or clean the oil holes in the crankshaft, as any dirt here can go only one way - straight through the new bearings.
9 Once you're certain the crankshaft is clean, carefully lay it in position in the main bearings.
10 Before the crankshaft can be permanently installed, the main bearing oil clearance must be checked.
11 Cut several pieces of the appropriate size Plastigage (they must be slightly shorter than the width of the main bearings) and place one piece on each crankshaft main bearing journal, parallel with the journal axis **(see illustration)**.
12 Clean the faces of the bearings in the caps and install the caps in their respective positions (don't mix them up) with the arrows pointing toward the front of the engine. Don't disturb the Plastigage.
13 Starting with the center main and working out toward the ends, tighten the main bearing cap bolts, in three steps, to the torque listed in this Chapter's Specifications. Don't rotate the crankshaft at any time during this operation.
14 Remove the bolts and carefully lift off the main bearing caps. Keep them in order. Don't disturb the Plastigage or rotate the crankshaft. If any of the main bearing caps are difficult to remove, tap them gently from side-to-side with a soft-face hammer to loosen them.

4.69 Lay the Plastigage strips (arrow) on the main bearing journals, parallel to the crankshaft centerline

4.70 Compare the width of the crushed Plastigage to the scale on the envelope to determine the main bearing oil clearance (always take the measurement at the widest point of the Plastigage); be sure to use the correct scale - standard and metric ones are included

15 Compare the width of the crushed Plastigage on each journal to the scale printed on the Plastigage envelope to obtain the main bearing oil clearance **(see illustration)**. Check the Specifications to make sure it's correct.

16 If the clearance is not as specified, the bearing inserts may be the wrong size (which means different ones will be required). Before deciding that different inserts are needed, make sure that no dirt or oil was between the bearing inserts and the caps or block when the clearance was measured. If the Plastigage was wider at one end than the other, the journal may be tapered (refer to the Crankshaft - inspection procedure in this Chapter).

17 Carefully scrape all traces of the Plastigage material off the main bearing journals and/or the bearing faces. Use your fingernail or the edge of a credit card - don't nick or scratch the bearing faces.

Final crankshaft installation

18 Carefully lift the crankshaft out of the engine.

19 Clean the bearing faces in the block, then apply a thin, uniform layer of moly-base grease or engine assembly lube to each of the bearing surfaces. Be sure to coat the thrust faces as well as the journal face of the thrust bearing. If the engine is equipped with a two-piece rear main oil seal, refer to the next Section in this Chapter and install the seal halves in the cap and block.

20 Make sure the crankshaft journals are clean, then lay the crankshaft back in place in the block.

21 Clean the faces of the bearings in the caps, then apply lubricant to them.

22 Install the caps in their respective positions with the arrows pointing toward the front of the engine.

23 Install the bolts.

24 Tighten all except the thrust bearing cap bolts to the specified torque (work from the center out and approach the final torque in three steps).

25 Tighten the thrust bearing cap bolts to 10-to-12 ft-lbs.

26 Tap the ends of the crankshaft forward and backward with a lead or brass hammer to line up the main bearing and crankshaft thrust surfaces.

27 Retighten all main bearing cap bolts to the specified torque, starting with the center main and working out toward the ends.

28 On manual transmission equipped models, install a new pilot bearing in the end of the crankshaft.

29 Rotate the crankshaft a number of times by hand to check for any obvious binding.

30 The final step is to check the crankshaft endplay with a feeler gauge or a dial indicator as described earlier in this Chapter. The endplay should be correct if the crankshaft thrust faces aren't worn or damaged and new bearings have been installed.

31 If you are working on an engine with a one-piece rear main oil seal, refer to the next Section and install the new seal, then bolt the housing to the block.

4.71 When correctly installed, the ends of the seal should extend out of the block slightly

4.72 Set the seal in the groove, but don't depress it below the bearing surface (the seal must contact the crankshaft journal)

Rear main oil seal installation

Split-type (two-piece) seal

1 Inspect the rear main bearing cap and engine block mating surfaces, as well as the seal grooves, for nicks, burrs and scratches. Remove any defects with a fine file or deburring tool.

2 Install one seal section in the block **(see illustrations)**. If the seal is made of rope, pack the seal into the area using a special tool (GM tool J33153) until it is packed to a depth of 1/4 to 3/8-inch. Trim the excess, but leave one end of the seal protruding from the block approximately 1/4 to 3/8-inch and make sure it's completely seated **(see illustrations)**.

3 Repeat the procedure to install the remaining seal half in the rear main bearing cap. In this case, leave the opposite end of the seal protruding from the cap the same distance the block seal is protruding from the block.

4.73a Trim the seal ends flush with the block . . .

4.73b . . . but leave the inner edge (arrow) protruding slightly

4.74 Lubricate the seal with assembly lube or moly based grease

4.75 When applying sealant, be sure it gets into each corner and onto the vertical cap-to-block mating surfaces or oil leaks will result

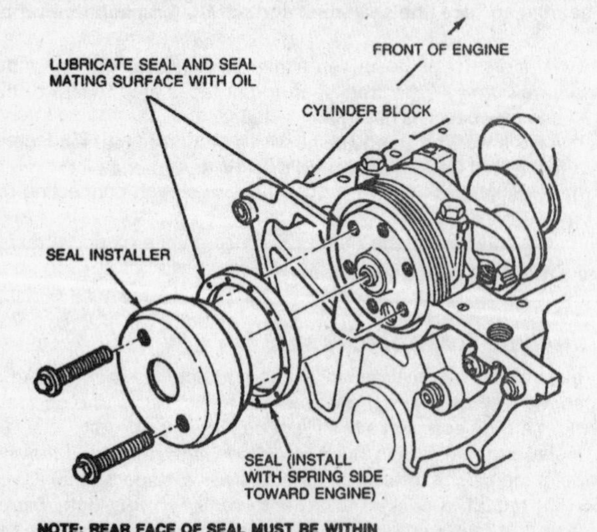

4.76 One-piece rear seals are best installed using a special tool such as the one shown here, but a standard seal driver of the correct size will also work if used carefully (Ford shown, GM similar)

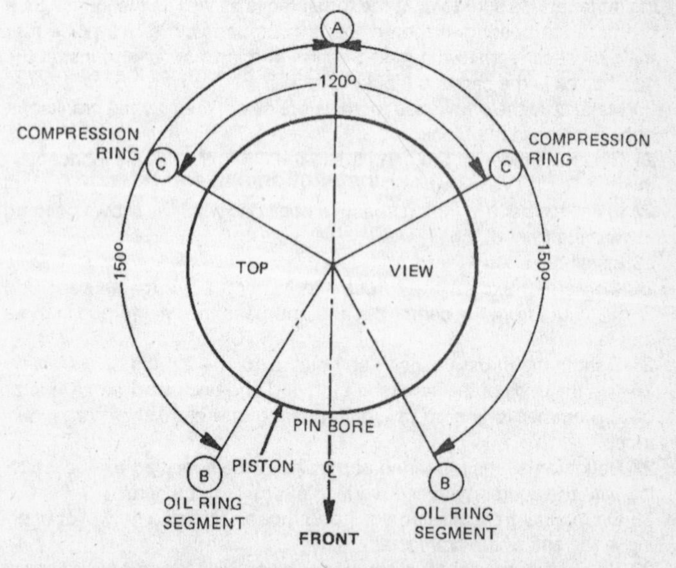

4.77 Piston ring spacing on Ford diesel engines

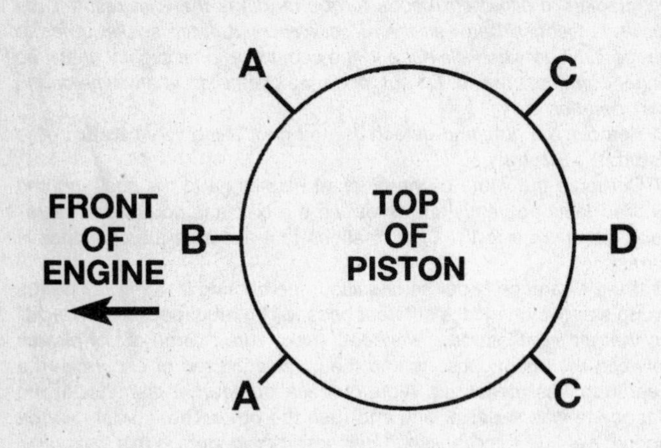

4.78 Piston ring spacing on GM diesel engines

A Oil ring rail gaps
B Second compression ring gap
C Oil ring spacer gap (position in-between marks)
D Top compression ring gap

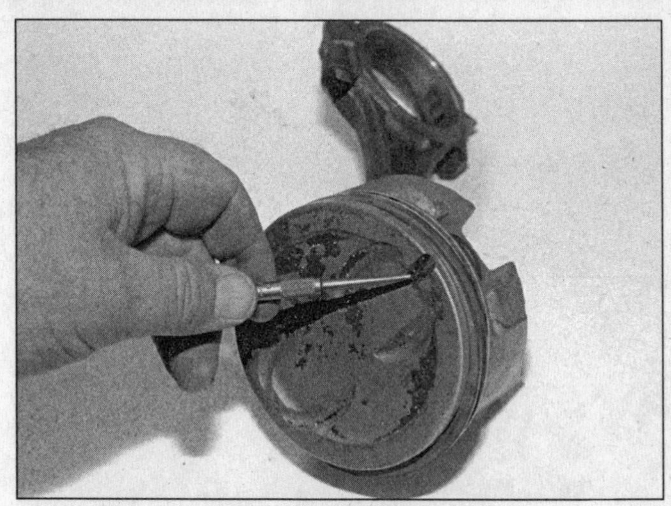

4.79a On GM engines, the notch in each piston must face toward the FRONT of the engine as the pistons are installed

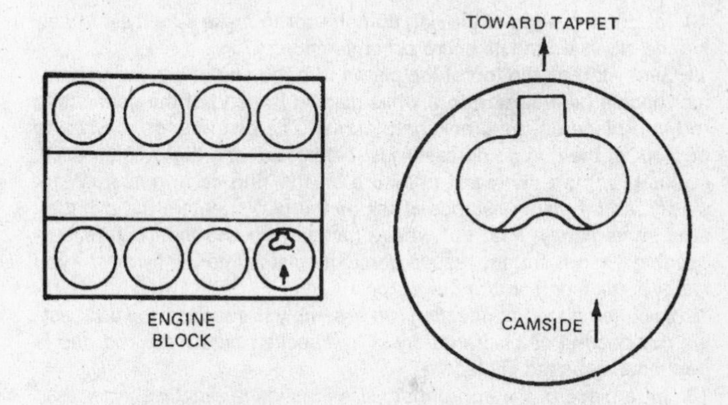

4.79b Piston orientation on Ford diesel engines

4 During final installation of the crankshaft (after the main bearing oil clearances have been checked with Plastigage) as described in the previous Section, apply a thin, even coat of anaerobic-type gasket sealant to the mating surfaces of the cap or block **(see illustrations)**. Don't get any sealant on the bearing face, crankshaft journal, seal ends or seal lips. Also, lubricate the seal lips with moly-base grease or engine assembly lube.

Note: *To prevent the possibility of cylinder block or main bearing cap damage, the main bearing caps must be tapped onto the engine block with a brass hammer or a leather mallet before the main bearing bolts are tightened down. DO NOT use the bolts to pull the main bearing cap and block together.*

One-piece seal

5 Some models are equipped with a one-piece seal that fits into a housing attached to the block. The crankshaft must be installed first and the main bearing caps bolted in place, then the new seal should be installed in the housing and the housing bolted to the block **(see illustration)**. Refer to Chapter 2 (GM) or Chapter 3 (Ford) for the rear main seal installation procedure.

Pistons/connecting rods - installation and rod bearing oil clearance check

1 Before installing the piston/connecting rod assemblies, the cylinder walls must be perfectly clean, the top edge of each cylinder must be chamfered, and the crankshaft must be in place.

2 Remove the cap from the end of the number one connecting rod (refer to the marks made during removal). Remove the original bearing inserts and wipe the bearing surfaces of the connecting rod and cap with a clean, lint-free cloth. They must be kept spotlessly clean.

Connecting rod bearing oil clearance check

3 Clean the back side of the new upper bearing insert, then lay it in place in the connecting rod. Make sure the tab on the bearing fits into the recess in the rod. Don't hammer the bearing insert into place and be very careful not to nick or gouge the bearing face. Don't lubricate the bearing at this time.

4 Clean the back side of the other bearing insert and install it in the rod cap. Again, make sure the tab on the bearing fits into the recess in the cap, and don't apply any lubricant. It's critically important that the mating surfaces of the bearing and connecting rod are perfectly clean and oil free when they're assembled.

5 Position the piston compression ring gaps at a 120-degree interval around the piston on Ford engines **(see illustration)** and 180-degrees around the piston on GM engines **(see illustration)**.

6 Slip a section of plastic or rubber hose over each connecting rod cap bolt.

7 Lubricate the piston and rings with clean engine oil and attach a piston ring compressor to the piston. Leave the skirt protruding about 1/4-inch to guide the piston into the cylinder. The rings must be compressed until they're flush with the piston.

8 Rotate the crankshaft until the number one connecting rod journal is at BDC (bottom dead center) and apply a coat of engine oil to the cylinder walls.

9 With the mark or notch on top of the piston **(see illustrations)** facing the front of the engine, gently insert the piston/connecting rod assembly into the number one cylinder bore and rest the bottom edge of the ring compressor on the engine block. **Note:** *On 1978 through 1981 GM 5.7L engines, the larger valve depression must be positioned toward the rear on the rear half of the engine (cylinder numbers 5, 6 7 and 8), and toward the front on the front half of the engine (cylinder numbers 1, 2, 3 and 4). On 1982 through 1984 GM 5.7L engines, the piston pin is positioned .045-inch off center to allow smoother running. If this offset is on the wrong side the engine will develop a rapping noise once it is warmed up. Be sure to install the piston with the notch forward.*

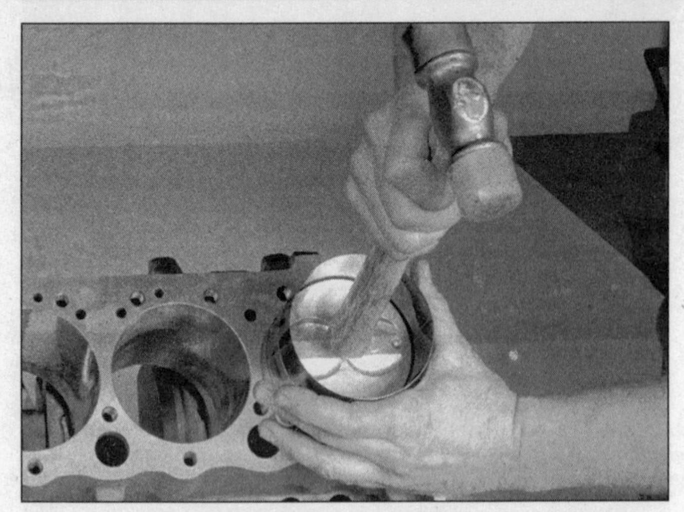

4.80 Drive the piston gently into the cylinder bore with the end of a wooden or plastic hammer handle

10 Tap the top edge of the ring compressor to make sure it's contacting the block around its entire circumference.

11 Gently tap on the top of the piston with the end of a wooden hammer handle **(see illustration)** while guiding the end of the connecting rod into place on the crankshaft journal. The piston rings may try to pop out of the ring compressor just before entering the cylinder bore, so keep some downward pressure on the ring compressor. Work slowly, and if any resistance is felt as the piston enters the cylinder, stop immediately. Find out what's hanging up and fix it before proceeding. Do not, for any reason, force the piston into the cylinder - you might break a ring and/or the piston.

12 Once the piston/connecting rod assembly is installed, the connecting rod bearing oil clearance must be checked before the rod cap is permanently bolted in place.

13 Cut a piece of the appropriate size Plastigage slightly shorter than the width of the connecting rod bearing and lay it in place on the number one connecting rod journal, parallel with the journal axis **(see illustration)**.

14 Clean the connecting rod cap bearing face, remove the protective hoses from the connecting rod bolts and install the rod cap. Make sure the mating mark on the cap is on the same side as the mark on the connecting rod.

15 Install the nuts and tighten them to the torque listed in this Chapter's Specifications, working up to it in three steps. **Note:** *Use a thin-wall socket to avoid erroneous torque readings that can result if the socket is wedged between the rod cap and nut. If the socket tends to wedge itself between the nut and the cap, lift up on it slightly until it no longer contacts the cap. Do not rotate the crankshaft at any time during this operation.*

16 Remove the nuts and detach the rod cap, being very careful not to disturb the Plastigage.

17 Compare the width of the crushed Plastigage to the scale printed on the Plastigage envelope to obtain the oil clearance **(see illustration)**. Compare it to the Specifications to make sure the clearance is correct.

18 If the clearance is not as specified, the bearing inserts may be the wrong size (which means different ones will be required). Before deciding that different inserts are needed, make sure that no dirt or oil was between the bearing inserts and the connecting rod or cap when the clearance was measured. Also, recheck the journal diameter. If the Plastigage was wider at one end than the other, the journal may be tapered (refer to the *Crankshaft-inspection* procedure in this Chapter).

Final connecting rod installation

19 Carefully scrape all traces of the Plastigage material off the rod journal and/or bearing face. Be very careful not to scratch the bearing - use your fingernail or the edge of a credit card.

20 Make sure the bearing faces are perfectly clean, then apply a uniform layer of clean moly-base grease or engine assembly lube to both of them. You'll have to push the piston into the cylinder to expose the face of the bearing insert in the connecting rod - be sure to slip the protective hoses over the rod bolts first.

21 Slide the connecting rod back into place on the journal, remove the protective hoses from the rod cap bolts, install the rod cap and tighten the nuts to the specified torque. Again, work up to the torque in three steps.

22 Repeat the entire procedure for the remaining pistons/connecting rods.

23 The important points to remember are:

- a) *Keep the back sides of the bearing inserts and the insides of the connecting rods and caps perfectly clean when assembling them.*
- b) *Make sure you have the correct piston/rod assembly for each cylinder.*
- c) *The notch or mark on the piston must face the proper direction.*
- d) *Lubricate the cylinder walls with clean oil.*
- e) *Lubricate the bearing faces when installing the rod caps after the oil clearance has been checked.*

24 After all the piston/connecting rod assemblies have been properly installed, rotate the crankshaft a number of times by hand to check for any obvious binding.

4.81 Lay the Plastigage strips on each rod bearing journal, parallel to the crankshaft centerline

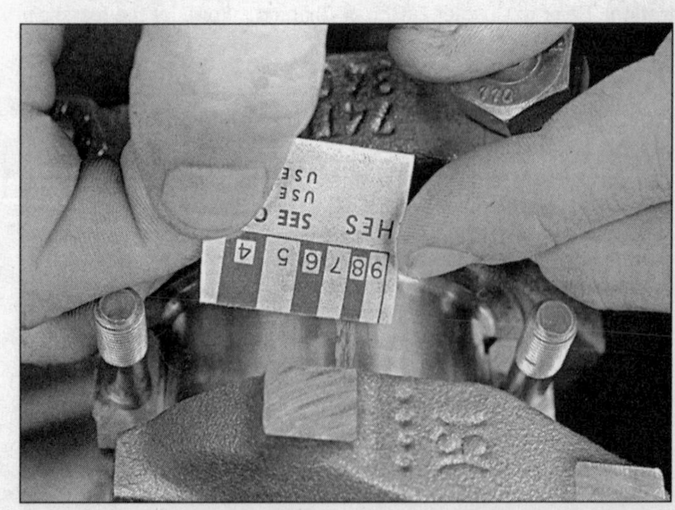

4.82 Measuring the width of the crushed Plastigage to determine the rod bearing oil clearance (be sure to use the correct scale - standard and metric ones are included)

25 As a final step, the connecting rod endplay must be checked as described under *Piston/connecting rods - removal* earlier in this Chapter.
26 Compare the measured endplay to the Specifications to make sure it's correct. If it was correct before disassembly and the original crankshaft and rods were reinstalled, it should still be right. If new rods or a new crankshaft were installed, the endplay may be inadequate. If so, the rods will have to be removed and taken to an automotive machine shop for resizing.

Initial start-up and break-in after overhaul

Warning: *Have a fire extinguisher handy when starting the engine for the first time.*
1 Once the engine has been installed in the vehicle, double-check the engine oil and coolant levels.
2 With the glow plugs out of the engine (see the *Compression check* procedure earlier in this Chapter), crank the engine until oil pressure registers on the gauge or the light goes out.
3 Install the glow plugs.
4 With the injector pump statically timed, start the engine. It may take a few moments for the fuel system to build up pressure, but the engine should start without a great deal of effort.
5 After the engine starts, it should be allowed to warm up to normal operating temperature. While the engine is warming up, make a thorough check for fuel, oil and coolant leaks.
6 Shut the engine off and recheck the engine oil and coolant levels.
7 Drive the vehicle to an area with minimum traffic, accelerate at full throttle from 30 to 50 mph, then allow the vehicle to slow to 30 mph with the throttle closed. Repeat the procedure 10 or 12 times. This will load the piston rings and cause them to seat properly against the cylinder walls. Check again for oil and coolant leaks.
8 Drive the vehicle gently for the first 500 miles (no sustained high speeds) and keep a constant check on the oil level. It is not unusual for an engine to use oil during the break-in period.
9 At approximately 500 to 600 miles, change the oil and filter.
10 For the next few hundred miles, drive the vehicle normally. Do not pamper it or abuse it.
11 After 2000 miles, change the oil and filter again and consider the engine broken in.

Notes

Index

D

E

F

G

H

I

L

M

O

P

R

S

T

U

V

Haynes Automotive Manuals

NOTE: New manuals are added to this list on a periodic basis. If you do not see a listing for your vehicle, consult your local Haynes dealer for the latest product information.

ACURA
*12020 Integra '86 thru '89 & Legend '86 thru '90

AMC
Jeep CJ - *see JEEP (50020)*
14020 Concord/Hornet/Gremlin/Spirit '70 thru '83
14025 (Renault) Alliance & Encore '83 thru '87

AUDI
15020 4000 all models '80 thru '87
15025 5000 all models '77 thru '83
15026 5000 all models '84 thru '88

AUSTIN
Healey Sprite - *see MG Midget (66015)*

BMW
*18020 3/5 Series '82 thru '92
*18021 3 Series except 325iX models '92 thru '97
18025 320i all 4 cyl models '75 thru '83
18035 528i & 530i all models '75 thru '80
18050 1500 thru 2002 except Turbo '59 thru '77

BUICK
Century (FWD) - *see GM (38005)*
*19020 Buick, Oldsmobile & Pontiac Full-size (Front wheel drive) '85 thru '98
Buick Electra, LeSabre and Park Avenue; Oldsmobile Delta 88 Royale, Ninety Eight and Regency; Pontiac Bonneville
19025 Buick Oldsmobile & Pontiac Full-size (Rear wheel drive)
Buick Estate '70 thru '90, Electra'70 thru '84, LeSabre '70 thru '85, Limited '74 thru '79
Oldsmobile Custom Cruiser '70 thru '90, Delta 88 '70 thru '85,Ninety-eight '70 thru '84
Pontiac Bonneville '70 thru '81, Catalina '70 thru '81, Grandville '70 thru '75, Parisienne '83 thru '86
19030 Mid-size Regal & Century '74 thru '87
Regal - *see GENERAL MOTORS (38010)*
Skyhawk - *see GM (38030)*
Skylark - *see GM (38020, 38025)*
Somerset - *see GENERAL MOTORS (38025)*

CADILLAC
*21030 Cadillac Rear Wheel Drive '70 thru '93
Cimarron, Eldorado & Seville - *see GM (38015, 38030)*

CHEVROLET
10305 Chevrolet Engine Overhaul Manual
*24010 Astro & GMC Safari Mini-vans '85 thru '93
24015 Camaro V8 all models '70 thru '81
24016 Camaro all models '82 thru '92
Cavalier - *see GM (38015)*
Celebrity - *see GM (38005)*
24017 Camaro & Firebird '93 thru '97
24020 Chevelle, Malibu, El Camino '69 thru '87
24024 Chevette & Pontiac T1000 '76 thru '87
Citation - *see GENERAL MOTORS (38020)*
*24032 Corsica/Beretta all models '87 thru '96
24040 Corvette all V8 models '68 thru '82
*24041 Corvette all models '84 thru '96
24045 Full-size Sedans Caprice, Impala, Biscayne, Bel Air & Wagons '69 thru '90
24046 Impala SS & Caprice and Buick Roadmaster '91 thru '96
Lumina '90 thru '94 - *see GM (38010)*
24048 Lumina & Monte Carlo '95 thru '98
Lumina APV - *see GM (38038)*
24050 Luv Pick-up all 2WD & 4WD '72 thru '82
24055 Monte Carlo all models '70 thru '88
Monte Carlo '95 thru '98 - *see LUMINA*
24059 Nova all V8 models '69 thru '79
*24060 Nova/Geo Prizm '85 thru '92
24064 Pick-ups '67 thru '87 - Chevrolet & GMC, all V8 & in-line 6 cyl, 2WD & 4WD '67 thru '87; Suburbans, Blazers & Jimmys '67 thru '91
*24065 Pick-ups '88 thru '98 - Chevrolet & GMC, all full-size models '88 thru '98; Blazer & Jimmy '92 thru '94; Suburban '92 thru '98; Tahoe & Yukon '95 thru '98
*24070 S-10 & GMC S-15 Pick-ups '82 thru '93
24071 S-10, Gmc S-15 & Jimmy '94 thru '96
*24075 Sprint & Geo Metro '85 thru '94
*24080 Vans - Chevrolet & GMC '68 thru '96

CHRYSLER
10310 Chrysler Engine Overhaul Manual
*25015 Chrysler Cirrus, Dodge Stratus, Plymouth Breeze, '95 thru '98
*25020 Full-size Front-Wheel Drive '88 thru '93
K-Cars - *see DODGE Aries (30008)*
Laser - *see DODGE Daytona (30030)*
25025 Chrysler LHS, Concorde & New Yorker, Dodge Intrepid, Eagle Vision, '93 thru '97
*25030 Chrysler/Plym. Mid-size '82 thru '95
Rear-wheel Drive - *see DODGE (30050)*

DATSUN
28005 200SX all models '80 thru '83
28007 B-210 all models '73 thru '78
28009 210 all models '78 thru '82
28012 240Z, 260Z & 280Z Coupe '70 thru '78
28014 280ZX Coupe & 2+2 '79 thru '83
300ZX - *see NISSAN (72010)*
28016 310 all models '78 thru '82
28018 510 & PL521 Pick-up '68 thru '73
28020 510 all models '78 thru '81
28022 620 Series Pick-up all models '73 thru '79
720 Series Pick-up - *NISSAN (72030)*
28025 810/Maxima all gas models, '77 thru '84

DODGE
400 & 600 - *see CHRYSLER (25030)*
*30008 Aries & Plymouth Reliant '81 thru '89
30010 Caravan & Ply. Voyager '84 thru '95
*30011 Caravan & Ply. Voyager '96 thru '98
30012 Challenger/Plymouth Saporro '78 thru '83
Challenger '67-'76 - *see DART (30025)*
30016 Colt/Plymouth Champ '78 thru '87
*30020 Dakota Pick-ups all models '87 thru '96
30025 Dart, Challenger/Plymouth Barracuda & Valiant 6 cyl models '67 thru '76
*30030 Daytona & Chrysler Laser '84 thru '89
Intrepid - *see Chrysler (25025)*
*30034 Dodge & Plymouth Neon '95 thru '97
*30035 Omni & Plymouth Horizon '78 thru '90
30040 Pick-ups all full-size models '74 thru '93
*30041 Pick-ups all full-size models '94 thru '96
*30045 Ram 50/D50 Pick-ups & Raider and Plymouth Arrow Pick-ups '79 thru '93
30050 Dodge/Ply./Chrysler RWD '71 thru '89
*30055 Shadow/Plymouth Sundance '87 thru '94
*30060 Spirit & Plymouth Acclaim '89 thru '95
*30065 Vans - Dodge & Plymouth '71 thru '96

EAGLE
Talon - *see MITSUBISHI Eclipse (68030)*
Vision - *see CHRYSLER (25025)*

FIAT
34010 124 Sport Coupe & Spider '68 thru '78
34025 X1/9 all models '74 thru '80

FORD
10355 Ford Automatic Transmission Overhaul
10320 Ford Engine Overhaul Manual
*36004 Aerostar Mini-vans '86 thru '96
Aspire - *see FORD Festiva (36030)*
*36006 Contour/Mercury Mystique '95 thru '98
36008 Courier Pick-up all models '72 thru '82
36012 Crown Victoria & Mercury Grand Marquis '88 thru '96
36016 Escort/Mercury Lynx '81 thru '90
*36020 Escort/Mercury Tracer '91 thru '96
Expedition - *see FORD Pick-up (36059)*
*36024 Explorer & Mazda Navajo '91 thru '95
36028 Fairmont & Mercury Zephyr '78 thru '83
36030 Festiva & Aspire '88 thru '97
36032 Fiesta all models '77 thru '80
36036 Ford & Mercury Full-size, Ford LTD & Mercury Marquis ('75 thru '82); Ford Custom 500,Country Squire, Crown Victoria & Mercury Colony Park ('75 thru '87); Ford LTD Crown Victoria & Mercury Gran Marquis ('83 thru '87)
36040 Granada & Mercury Monarch '75 thru '80
36044 Ford & Mercury Mid-size, Ford Thunderbird & Mercury Cougar ('75 thru '82); Ford LTD & Mercury Marquis ('83 thru '86); Ford Torino,Gran Torino, Elite, Ranchero pick-up, LTD II, Mercury Montego, Comet, XR-7 & Lincoln Versailles ('75 thru '86)
36048 Mustang V8 all models '64-1/2 thru '73
36049 Mustang II 4 cyl, V6 & V8 '74 thru '78
36050 Mustang & Mercury Capri incl. Turbo Mustang, '79 thru '93; Capri, '79 thru '86
*36051 Mustang all models '94 thru '97
36054 Pick-ups and Bronco '73 thru '79
*36058 Pick-ups and Bronco '80 thru '96
*36059 Pick-ups, Expedition & Lincoln Navigator '97 thru '98
36062 Pinto & Mercury Bobcat '75 thru '80
36066 Probe all models '89 thru '92
*36070 Ranger/Bronco II gas models '83 thru '92
*36071 Ford Ranger '93 thru '97 & Mazda Pick-ups '94 thru '97
*36074 Taurus & Mercury Sable '86 thru '95
*36075 Taurus & Mercury Sable '96 thru '98
*36078 Tempo & Mercury Topaz '84 thru '94
36082 Thunderbird/Mercury Cougar '83 thru '88
*36086 Thunderbird/Mercury Cougar '89 and '97
36090 Vans all V8 Econoline models '69 thru '91
*36094 Vans full size '92 thru '95
*36097 Windstar Mini-van '95 thru '98

GENERAL MOTORS
*10360 GM Automatic Transmission Overhaul
*38005 Buick Century, Chevrolet Celebrity, Olds Cutlass Ciera & Pontiac 6000 all models '82 thru '96
*38010 Buick Regal, Chevrolet Lumina, Oldsmobile Cutlass Supreme & Pontiac Grand Prix front wheel drive '88 thru '95
*38015 Buick Skyhawk, Cadillac Cimarron, Chevrolet Cavalier, Oldsmobile Firenza Pontiac J-2000 & Sunbird '82 thru '94
*38016 Chevrolet Cavalier & Pontiac Sunfire '95 thru '98
38020 Buick Skylark, Chevrolet Citation, Olds Omega, Pontiac Phoenix '80 thru '85
38025 Buick Skylark & Somerset, Olds Achieva, Calais & Pontiac Grand Am '85 thru '95
38030 Cadillac Eldorado & Oldsmobile Toronado '71 thru '85, Seville '80 thru '85, Buick Riviera '79 thru '85
*38035 Chevrolet Lumina APV, Oldsmobile Silhouette & Pontiac Trans Sport '90 thru '95
General Motors Full-size Rear-wheel Drive - *see BUICK (19025)*

GEO
Metro - *see CHEVROLET Sprint (24075)*
Prizm - *see CHEVROLET (24060) or TOYOTA (92036)*
*40030 Storm all models '90 thru '93
Tracker - *see SUZUKI Samurai (90010)*

GMC
Safari - *see CHEVROLET ASTRO (24010)*
Vans & Pick-ups - *see CHEVROLET*

HONDA
42010 Accord CVCC all models '76 thru '83
42011 Accord all models '84 thru '89
42012 Accord all models '90 thru '93
*42013 Accord all models '94 thru '95
42020 Civic 1200 all models '73 thru '79
42021 Civic 1300 & 1500 CVCC '80 thru '83
42022 Civic 1500 CVCC all models '75 thru '79
42023 Civic all models '84 thru '91
42024 Civic & del Sol '92 thru '95
Passport - *see ISUZU Rodeo (47017)*
*42040 Prelude CVCC all models '79 thru '89

HYUNDAI
*43015 Excel all models '86 thru '94

ISUZU
Hombre - *see CHEVROLET S-10 (24071)*
*47017 Rodeo '91 thru '97, Amigo '89 thru '94, Honda Passport '95 thru '97
*47020 Trooper '84 thru '91, Pick-up '81 thru '93

JAGUAR
*49010 XJ6 all 6 cyl models '68 thru '86
*49011 XJ6 all models '88 thru '94
*49015 XJ12 & XJS all 12 cyl models '72 thru '85

JEEP
*50010 Cherokee, Comanche & Wagoneer Limited all models '84 thru '96
50020 CJ all models '49 thru '86
*50025 Grand Cherokee all models '93 thru '98
*50029 Grand Wagoneer & Pick-up '72 thru '91
*50030 Wrangler all models '87 thru '95

LINCOLN
Navigator - *see FORD Pick-up (36059)*
59010 Rear Wheel Drive all models '70 thru '96

MAZDA
61010 GLC (rear wheel drive) '77 thru '83
61011 GLC (front wheel drive) '81 thru '85
*61015 323 & Protegé '90 thru '97
*61016 MX-5 Miata '90 thru '97
*61020 MPV all models '89 thru '94
Navajo - *see FORD Explorer (36024)*
61030 Pick-ups '72 thru '93
Pick-ups '94 on - *see Ford (36071)*
61035 RX-7 all models '79 thru '85
*61036 RX-7 all models '86 thru '91
61040 626 (rear wheel drive) '79 thru '82
*61041 626 & MX-6 (front wheel drive) '83 thru '91

MERCEDES-BENZ
63012 123 Series Diesel '76 thru '85
*63015 190 Series 4-cyl gas models, '84 thru '88
63020 230, 250 & 280 6 cyl sohc '68 thru '72
63025 280 123 Series gas models '77 thru '81
63030 350 & 450 all models '71 thru '80

MERCURY
See FORD Listing

MG
66010 MGB Roadster & GT Coupe '62 thru '80
66015 MG Midget & Austin Healey Sprite Roadster '58 thru '80

MITSUBISHI
*68020 Cordia, Tredia, Galant, Precis & Mirage '83 thru '93
*68030 Eclipse, Eagle Talon & Plymouth Laser '90 thru '94
*68040 Pick-up '83 thru '96, Montero '83 thru '93

NISSAN
72010 300ZX all models incl. Turbo '84 thru '89
*72015 Altima all models '93 thru '97
*72020 Maxima all models '85 thru '91
*72030 Pick-ups '80 thru '96, Pathfinder '87 thru '95
72040 Pulsar all models '83 thru '86
72050 Sentra all models '82 thru '94
*72051 Sentra & 200SX all models '95 thru '98
*72060 Stanza all models '82 thru '90

OLDSMOBILE
*73015 Cutlass '74 thru '88
For other OLDSMOBILE titles, see BUICK, CHEVROLET or GENERAL MOTORS listing.

PLYMOUTH
For PLYMOUTH titles, see DODGE.

PONTIAC
79008 Fiero all models '84 thru '88
79018 Firebird V8 models except Turbo '70 thru '81
79019 Firebird all models '82 thru '92
For other PONTIAC titles, see BUICK, CHEVROLET or GENERAL MOTORS listing.

PORSCHE
*80020 911 Coupe & Targa models '65 thru '89
80025 914 all 4 cyl models '69 thru '76
80030 924 all models incl. Turbo '76 thru '82
*80035 944 all models incl. Turbo '83 thru '89

RENAULT
Alliance, Encore - *see AMC (14020)*

SAAB
*84010 900 including Turbo '79 thru '88

SATURN
*87010 Saturn all models '91 thru '96

SUBARU
89002 1100, 1300, 1400 & 1600 '71 thru '79
*89003 1600 & 1800 2WD & 4WD '80 thru '94

SUZUKI
*90010 Samurai/Sidekick/Geo Tracker '86 thru '96

TOYOTA
92005 Camry all models '83 thru '91
*92006 Camry all models '92 thru '96
92015 Celica Rear Wheel Drive '71 thru '85
*92020 Celica Front Wheel Drive '86 thru '93
92025 Celica Supra all models '79 thru '92
92030 Corolla all models '75 thru '79
92032 Corolla rear wheel drive models '80 thru '87
*92035 Corolla front wheel drive models '84 thru '92
*92036 Corolla & Geo Prizm '93 thru '97
92040 Corolla Tercel all models '80 thru '82
92045 Corona all models '74 thru '82
92050 Cressida all models '78 thru '82
92055 Land Cruiser Series FJ40, 43, 45 & 55 '68 thru '82
*92056 Land Cruiser Series FJ60, 62, 80 & FZJ80 '68 thru '82
*92065 MR2 all models '85 thru '87
92070 Pick-up all models '69 thru '78
*92075 Pick-up all models '79 thru '95
*92076 Tacoma '95 thru '98, 4Runner '96 thru '98, T100 '93 thru '98
*92080 Previa all models '91 thru '95
92085 Tercel all models '87 thru '94

TRIUMPH
94007 Spitfire all models '62 thru '81
94010 TR7 all models '75 thru '81

VW
96008 Beetle & Karmann Ghia '54 thru '79
96012 Dasher all gasoline models '74 thru '81
*96016 Rabbit, Jetta, Scirocco, & Pick-up gas models '74 thru '91 & Convertible '80 thru '92
*96017 Golf & Jetta '93 thru '97
96020 Rabbit, Jetta, Pick-up diesel '77 thru '84
96030 Transporter 1600 all models '68 thru '79
96035 Transporter 1700, 1800, 2000 '72 thru '79
96040 Type 3 1500 & 1600 '63 thru '73
96045 Vanagon air-cooled models '80 thru '83

VOLVO
97010 120, 130 Series & 1800 Sports '61 thru '73
97015 140 Series all models '66 thru '74
*97020 240 Series all models '76 thru '93
97025 260 Series all models '75 thru '82
*97040 740 & 760 Series all models '82 thru '88

TECHBOOK MANUALS
10205 Automotive Computer Codes
10210 Automotive Emissions Control Manual
10215 Fuel Injection Manual, 1978 thru 1985
10220 Fuel Injection Manual, 1986 thru 1996
10225 Holley Carburetor Manual
10230 Rochester Carburetor Manual
10240 Weber/Zenith/Stromberg/SU Carburetor
10305 Chevrolet Engine Overhaul Manual
10310 Chrysler Engine Overhaul Manual
10320 Ford Engine Overhaul Manual
10330 GM and Ford Diesel Engine Repair
10340 Small Engine Repair Manual
10345 Suspension, Steering & Driveline
10355 Ford Automatic Transmission Overhaul
10360 GM Automatic Transmission Overhaul
10405 Automotive Body Repair & Painting
10410 Automotive Brake Manual
10415 Automotive Detaling Manual
10420 Automotive Eelectrical Manual
10425 Automotive Heating & Air Conditioning
10430 Automotive Reference Dictionary
10435 Automotive Tools Manual
10440 Used Car Buying Guide
10445 Welding Manual
10450 ATV Basics

SPANISH MANUALS
98903 Reparación de Carrocería & Pintura
98905 Códigos Automotrices de la Computadora
98910 Frenos Automotriz
98915 Inyección de Combustible 1986 al 1994
99040 Chevrolet & GMC Camionetas '67 al '87
99041 Chevrolet & GMC Camionetas '88 al '95
99042 Chevrolet Camionetas Cerradas '68 al '95
99055 Dodge Caravan/Ply. Voyager '84 al '95
99075 Ford Camionetas y Bronco '80 al '94
99077 Ford Camionetas Cerradas '69 al '91
99083 Ford Modelos de Tamaño Grande '75 al '87
99088 Ford Modelos de Tamaño Mediano '75 al '86
99091 Ford Taurus & Mercury Sable '75 al '86
99095 GM Modelos de Tamaño Grande '70 al '90
99100 GM Modelos de Tamaño Mediano '70 al '88
99110 Nissan Camionetas '80 al '96, Pathfinder '87 al '95
99118 Nissan Sentra '82 al '94
99125 Toyota Camionetas y 4-Runner '79 al '95

** Listings shown with an asterisk (*) indicate model coverage as of this printing. These titles will be periodically updated to include later model years - consult your Haynes dealer for more information.*

Nearly 100 Haynes motorcycle manuals also available

5-98

Haynes North America, Inc., 861 Lawrence Drive, Newbury Park, CA 91320 • (805) 498-6703